STATICS AND STRENGTH OF MATERIALS

KARL K. STEVENS

Florida Atlantic University

Prentice-Hall, Inc. Englewood Cliffs, New Jersey 07632

Library of Congress Cataloging in Publication Data

STEVENS, KARL K.
Statics and strength of materials.

Bibliography: p.
Includes index.
1.–Statics. 2.–Strength of materials. I.–Title.
TA351.S73 620.1'12 77-25910
ISBN 0-13-844688-1

10 9 8 7 6 5 4 3 2 1

Printed in the United States of America

PRENTICE-HALL INTERNATIONAL, INC., *London*
PRENTICE-HALL OF AUSTRALIA PTY. LIMITED, *Sydney*
PRENTICE-HALL OF CANADA, LTD., *Toronto*
PRENTICE-HALL OF INDIA PRIVATE LIMITED, *New Delhi*
PRENTICE-HALL OF JAPAN, INC., *Tokyo*
PRENTICE-HALL OF SOUTHEAST ASIA PTE. LTD., *Singapore*
WHITEHALL BOOKS LIMITED, *Wellington, New Zealand*

CONTENTS

PREFACE

This book is designed for use in a first undergraduate course covering the subject matter traditionally referred to as statics and strength of materials. A combined statics and strength of materials course has been offered at The Ohio State University for a number of years. This text is an outgrowth of my experiences, and to some degree those of my former colleagues, in teaching this course.

The subjects of statics and strength of materials have much in common in that they are both concerned primarily with bodies in equilibrium. Consequently, equilibrium has been made the central theme of this book. This has made it possible to present a highly unified treatment of the statics of rigid and deformable bodies.

Chapters 1 through 4 are devoted to subject matter covered in traditional statics courses. This material provides a complete treatment of vector statics, and serves as a background for later studies in dynamics. Although the material is presented from a vector viewpoint, I have not hesitated to drop formal vector methods wherever their use is not warranted. Consequently, scalar-geometric methods are used for most two-dimensional and for some simple three-dimensional problems.

A thorough treatment of vectors and vector algebra is presented in Chapter 2, with emphasis upon applications to statics. The concepts of moments and couples are also introduced in this chapter. Equilibrium is treated in Chapter 3, followed by a study of statically equivalent force systems in Chapter 4. Chapter 4 also includes a discussion of distributed forces and the related concepts of center of gravity, centroids, and area moments of inertia.

The consideration of equilibrium before the study of statically equivalent force systems and resultants is a major departure from tradition, but offers several advantages. Most importantly, it exposes the student to meaningful problems early in the course without first building up a backlog of abstract work regarding the properties of force systems. Early work with forces and moments as they relate to meaningful physical problems gives the student a better feeling for these quantities and their physical effects. Once the principles of equilibrium have been presented, the study of statically equivalent force systems follows as a natural consequence of the need to know how to handle distributed forces. I believe that this places the importance of static equivalence in better perspective than when it is considered before equilibrium. Furthermore, statically equivalent force systems can then be defined in terms of their effect upon equilibrium. This very clearly brings out the limited nature of static equivalence, which is important for the work in later chapters involving deformable bodies.

The work on deformable bodies is contained in Chapters 5 through 11. With equilibrium being the common thread throughout, the difference between problems involving rigid and deformable bodies lies fundamentally in the degree of sophistication required in the mathematical modeling of the problem. Accordingly, the determination of stresses and deformations is approached from the viewpoint that these are indeterminate problems for which the material properties and geometry of the deformations must be taken into account in order to obtain a solution.

The concepts of stress and strain are introduced in Chapter 5, along with a discussion of material properties. Chapters 6 through 10 deal with axially-loaded members, torsion, bending, combined loadings, and buckling, respectively. Both elastic and inelastic responses are considered. A brief, but complete, discussion of the common methods of experimental strain and stress analysis is given in Chapter 11.

Every attempt has been made to present the material in a consistent and orderly fashion. The fundamentals are presented as succinctly as possible, consistent with good understanding. On the other hand, considerable discussion has been devoted to the meaning and interpretation of the fundamentals and the procedures for applying them to engineering problems. As a further aid to the student, there are numerous illustrative problems with complete explanations designed to reinforce the concepts presented. There are approximately 800 exercise problems, with answers to the even-numbered problems provided. While my concerns lay primarily with making the material easily understood by beginning students, individual instructors will find that they have considerable latitude to expand upon subjects and to add their own flavoring to the presentation.

At the time of writing, the transition between the traditional US-British system of units and the Système International d'Unités (SI) is in progress. Accordingly, both systems are used to an approximately equal extent. This has made necessary

several compromises to avoid confusion. For example, the practice recommended in SI of using a space instead of a comma to set apart groups of three digits on either side of the decimal point is also used for quantities expressed in US-British units. In problems using SI, bodies are sometimes described in terms of their weight or specific weight, although, strictly speaking, they should be described in terms of their mass or mass density, since mass, and not force, is the fundamental quantity in SI. A complete discussion of units of measure is included in Chapter 1.

This book contains sufficient material for a five-semester-hour course. For courses of fewer credits, it will be necessary to delete certain material. The content of combined courses in statics and strength of materials apparently varies widely from school to school. Consequently, the material has been organized in such a way that topics can be easily deleted to fit course content. In particular, discussions of the inelastic response of members is placed in separate sections, and the exercise problems are placed at the end of the appropriate section instead of at the end of the chapters. Aside from the emphasis upon equilibrium, the work on rigid and deformable bodies has not been intermixed. Thus, this book should serve equally well as a text for separate courses in statics and strength of materials.

A project of this magnitude cannot be completed alone. I am deeply indebted to Ms. Josephine French for typing the several drafts. My dear friend and former graduate student, Dr. Leonard Sung, checked the solutions to all the example problems and assisted with the solutions of the exercise problems. Assistance with the exercise problems was also provided by Mr. Shah Malik, Mr. Jim Lester, and Dr. Kuang-shi Ju. Professor Terry Richard offered numerous helpful comments and suggestions on the chapter on Experimental Strain and Stress Analysis. The photographs, except where acknowledged otherwise, were taken by Mr. Scott Vibberts. Special thanks are due my anonymous reviewers, whose comments and suggestions were most helpful. The patience and guidance of all those I worked with at Prentice-Hall are greatly appreciated.

K. K. STEVENS

chapter one INTRODUCTION

1.1 SCOPE

Let us begin with a brief overview of where we are going and what we hope to accomplish in the following chapters.

In this text we shall be concerned with the determination of the forces acting upon material bodies and their mechanical response to these forces. By mechanical response, we shall generally mean the deformation of a body and the stresses produced within it. Our interest in such problems stems from the fact that they pervade science and technology and are of direct concern to engineers, designers, and scientists.

For instance, structural engineers must be able to determine the loads acting upon a building and to design the building so that it responds to these loads in an acceptable manner. The same is true for other structures such as nuclear reactors, bridges, highways, machine parts, airplanes, pipelines, storage tanks, and electrical transmission lines. Designers of electrical machinery must account for forces produced by the magnetic and electrical fields present, and orthopedic surgeons require knowledge of the forces to which the human body is subjected in order to better repair bone fractures and to design artificial limbs and replacement parts for bones and joints. Mechanical engineers must determine the forces acting upon and within machines in order to properly design gears, bearings, and other machine elements. Agricultural engineers are faced with the problem of determining the effects of the forces to which fruits, vegetables, and grains are subjected during harvesting, handling, and storage.

These are but a few examples of problems involving the determination of forces acting upon physical systems and their response to these forces. How many other examples can you think of? Obviously, problems like these are innumerable.

Many of the problems we have mentioned are complex and cannot be treated in all generality in an introductory text such as this. Consequently, it will be necessary to restrict the types of problems to be considered.

First, we shall consider only static problems involving bodies at rest, or more precisely, bodies in equilibrium. A large number of the problems encountered in practice fall into this category. The behavior of bodies in motion is treated in texts on dynamics and vibrations.

Second, we shall consider only solid bodies. Gasses and liquids will not be considered per se, although we shall consider the loadings they impart to a solid body, such as a container or dam.

Third, although actual structures and machines are usually complex, they often consist of simple component parts such as rods, shafts, beams, and columns. We shall consider only the response of these basic structural elements and simple combinations thereof. This will enable us to solve many problems that are important in their own right and will provide the fundamentals necessary for consideration of more complex problems.

We have now identified in general terms the class of problems to be considered in this text. But what information will be needed to solve these problems? This question can probably best be answered by considering the steps involved in the solution of an engineering problem.

The first step in any engineering analysis is to clearly identify and focus attention upon the particular system or subsystem of interest. For example, suppose that we wish to find the forces that a driver exerts upon the seat of an automobile. Should we consider the driver? The seat? The automobile? The combination of driver, seat, and automobile? Or just what? A clear definition and statement of the problem usually help resolve this question.

After identifying the system of interest, the next step is to formulate an idealized model of the problem. The idea is to simplify the problem as much as possible by considering only those factors that have an important bearing upon the quantities to be determined, while neglecting those which do not. For instance, the weight of a floor beam might be only a very small fraction of the total load it supports and could, therefore, be neglected when determining the deflection (sag) of the beam.

Once a model has been decided upon, it is analyzed mathematically by using whatever physical laws are applicable. The results obtained are then examined to see if they make good sense and correspond to physical reality. A further check of the validity of the model may be obtained by performing critical experiments, or by comparing the results with those obtained from analyses based on more sophisticated models. If the model is found to be deficient, it is modified and the process repeated until satisfactory results are obtained.

What does the preceding discussion tell us about the things we must know in order to solve statics problems? Obviously, we must know how to construct appropriate models of problems, we must know the physical laws which apply, and we must know how to describe the quantities of interest in a form amenable to mathematical analysis.

The remaining sections of this chapter are devoted to a brief discussion of the basic concepts and physical laws that apply to statics problems. This will be followed in Chapter 2 by a study of how to describe certain quantities of interest, forces in particular, in a convenient mathematical form.

Once this background material has been mastered, we shall be in a position to solve some engineering problems. There will be two phases to our work. The first involves the determination of all the forces acting upon the system of interest; the second involves the determination of the response to these forces. Chapters 3 and 4 are devoted to the first phase. Here we shall develop and learn to apply the mathematical conditions for equilibrium and, at the same time, begin to develop some insight into problem modeling. Equilibrium is the single most important topic to be considered in this text, and the results presented in these chapters will be used throughout the remainder of our work.

Chapters 5–11 are devoted to the determination of the response of load-carrying members to applied loadings. In particular, we shall consider procedures for analyzing a member to see if it can fulfill its intended purpose and for designing a member so that it can safely support a given loading. The mechanical properties of common engineering materials will also be discussed, insofar as they relate to the problems of interest.

1.2 BASIC CONCEPTS AND PRINCIPLES

The study of the forces acting upon physical systems and their response to these forces forms the branch of science known as *mechanics*. Mechanics is one of the oldest branches of science, and some of its principles, such as the principle of the lever, probably came into play during man's earliest attempts to cope with his environment.

Although simple principles of mechanics were undoubtedly used in prehistoric times, it is unlikely that they were understood. The study of mechanics dates to the time of Aristotle (384–322 BC) and Archimedes (287–212 BC). Despite this early start, a satisfactory formulation of the principles of mechanics was not available until Sir Isaac Newton postulated the three laws of motion which are now the basis of most engineering applications of mechanics. These laws were presented in his *Philosophiae Naturalis Principia Mathematica* (The Mathematical Principles of Natural Science) published in 1686. The modern theories of relativity and quantum mechanics show Newton's laws to be inexact. However, the innumerable problems that have been solved successfully using these laws testify to the fact that their precision is extremely high, except in problems involving subatomic particles or velocities approaching the speed of light.

It will not be one of our objectives to trace in detail the history of the developments in mechanics. Those who are interested will find a number of good books available on this important and fascinating subject.

There are three main divisions of mechanics: *statics*, *kinematics*, and *dynamics*. Statics is concerned with systems in equilibrium and with the force interactions that operate to establish equilibrium. Kinematics is the study of the geometry of motion. Dynamics is concerned with the relationships between force interactions and the motions they produce. As we have already mentioned, we shall be concerned only with the statics of solid bodies.

Basic concepts. Mechanics is based upon the concepts of force, mass, length, and time. Although each of us has a degree of familiarity with these concepts as a result of our physical intuition and experiences, these concepts cannot truly be defined. Rather, we give a rough indication of their physical significance and leave their actual description to be determined from the laws and postulates that describe the relationships between them. Thus, we may speak of the mass of an object as being a measure of the quantity of matter in it, but this statement can hardly be construed as being a precise definition of mass.

Our intuitive concepts of mass, length, and time will be adequate for problems to be considered in this text, particularly since time does not enter into statics problems and mass enters only indirectly. Force, however, plays a dominant role in statics and merits further comment.

The concept of force. The concept of force in mechanics provides a simple and convenient way to describe the very complex physical interactions between systems that alter or tend to alter the motion or state of rest of the systems. The key word here is interaction, which implies that two systems always participate in the creation of a force.

For example, when we stretch a rubber band, there is a force interaction between the band and our hands. This interaction causes the band to stretch, and it also produces the effect that we experience as a resistance to this stretching. The situation is illustrated schematically in Figure 1.1, where the arrows $\mathbf{F}_1$ represent the effect of the force interaction on our hands and the arrows $\mathbf{F}_2$ represent its effect on the band. According to Newton's third law, force interactions have an equal and opposite effect on the interacting systems. Thus, the effects $\mathbf{F}_1$ and $\mathbf{F}_2$ shown in Figure 1.1 are of equal magnitude and opposite direction. In other words, whatever tendency there is for our hands to stretch the band, the band has the same tendency to resist the stretching and hold our hands together.

Although the word force is properly used to denote an interaction between systems, it is more commonly used to denote the individual effects of the interaction. Thus, the individual effects $\mathbf{F}_1$ and $\mathbf{F}_2$ shown in Figure 1.1 would each be called a force. We shall also use the word force in this context, but in so doing it is

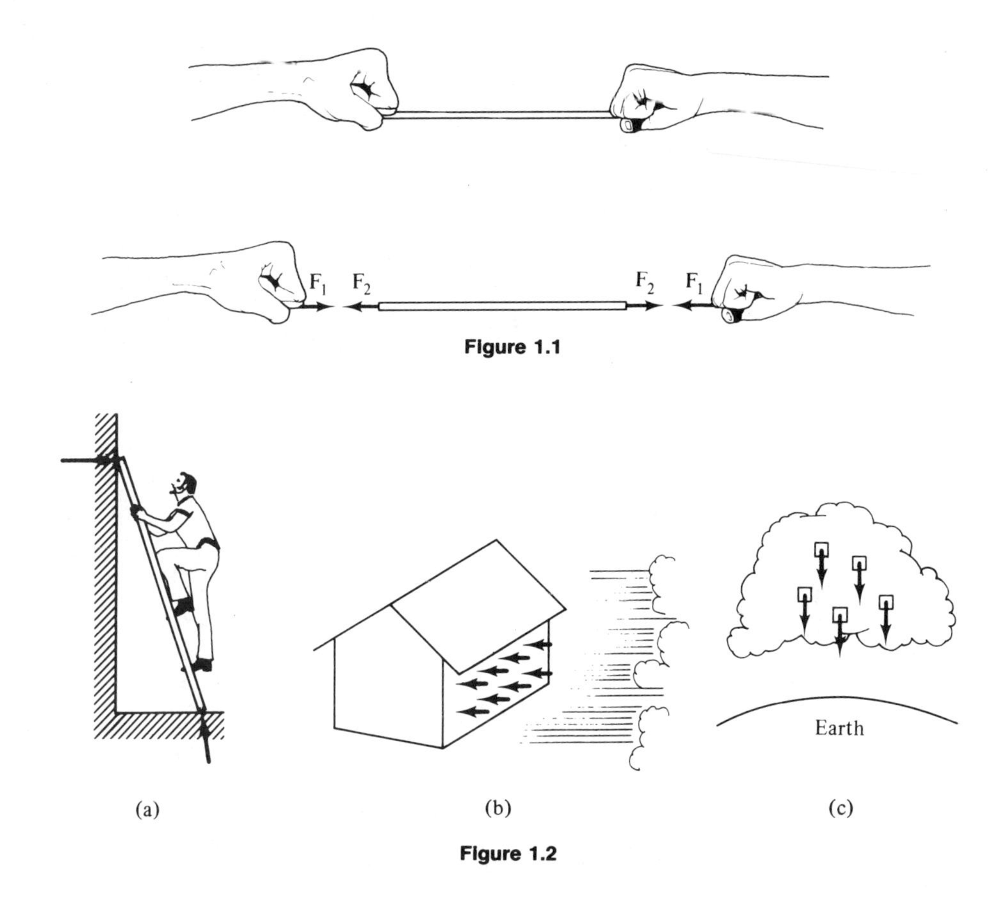

Figure 1.1

Figure 1.2

essential to remember that we are talking about only one aspect of an interaction between two systems.

Force interactions arise when bodies come into direct physical contact, but they can also occur between systems that are separated from one another. Gravitational, electrical, and magnetic force interactions are of this latter type. Forces may act essentially at a point, as in the case of a ladder resting against a wall [Figure 1.2(a)], or they may be distributed over an area, as in the case of the wind loading on the side of a building [Figure 1.2(b)]. Some forces, such as the gravitational attraction between an object and the earth [Figure 1.2(c)], are distributed throughout a volume.

The physical effect of a force depends upon its *magnitude*, *orientation*, *sense*, and *point of application*, and these four factors must be specified in order to completely describe the force. For example, the behavior of the block shown in Figure 1.3(a) clearly depends upon the magnitude of the force exerted upon it by the man via the rope and pulley. If the force is too small, the block will remain at rest, but if the man pulls hard enough, the block can be lifted. The block can also be lifted if the rope is attached to a different point, as in Figure 1.3(b), but the block will rotate as it lifts. If the sense of the force is reversed [Figure 1.3(c)], the block will only be pressed more firmly against the surface upon which it rests.

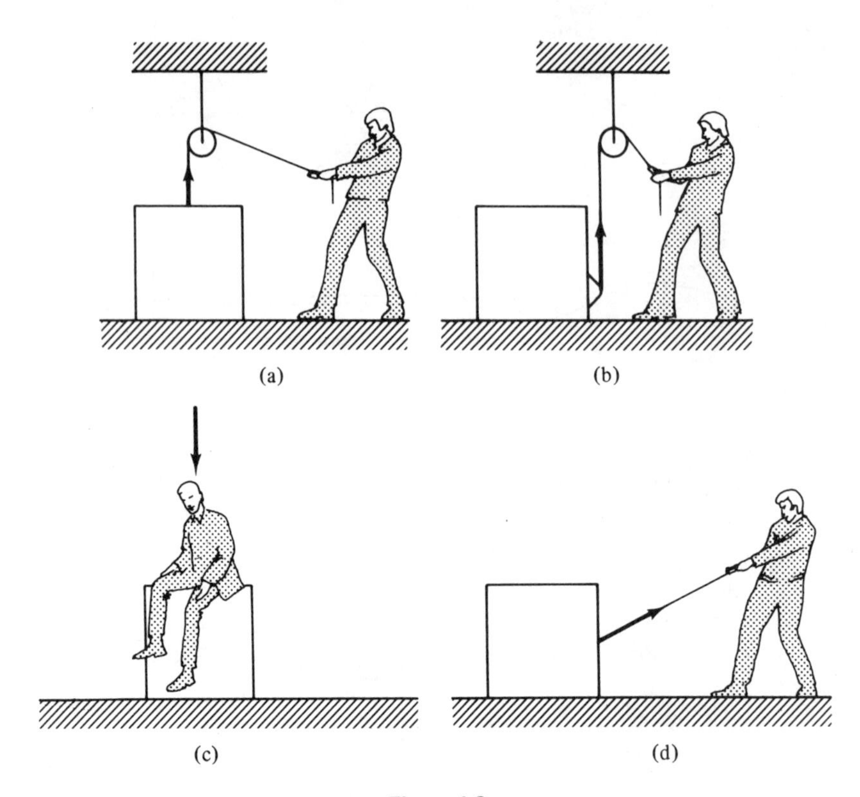

Figure 1.3

Finally, if the orientation of the force is changed, as in Figure 1.3(d), the block can be slid, but not lifted. In each of these cases, the force produces a different physical effect.

Forces have one other very important property: They combine according to the rules for vector addition. This is one of the fundamental postulates of mechanics, and it is based upon experimental evidence.

Quantities that have magnitude, orientation, and sense, and which combine according to the rules of vector addition, are known in mathematics as *vectors*. Thus, force interactions are vector quantities and can be represented by directed line segments (arrows), as we have been doing. The significance of this result is twofold. First, the vector character of forces will be important to us in developing a physical feeling for them and the actions they produce. Second, all the rules of vector algebra can be used in mathematical operations involving forces. This will be helpful in the derivation of basic equations and in the solution of problems. Vectors and vector algebra will be discussed in Chapter 2.

Weight and mass. The force exerted on a body because of its gravitational interaction with the earth is called the *weight* of the body. As will be shown in Chapter 4, the magnitude W of the weight of a body with mass m is given by the relation

$$W = mg \tag{1.1}$$

where g is the so-called *acceleration of gravity*. The value of g at or near the surface of the earth is 32.2 feet/second^2 (9.81 metres/second^2).

The distinction between weight and mass is widely misunderstood, mostly because of the common usage versus the standard technical usage of these terms. For example, grocery store scales are calibrated in units of force, usually pounds and ounces, and we speak of 5 pounds of potatoes as being a certain quantity of the vegetable. What we really mean, however, is the mass of potatoes which experiences a force of magnitude 5 pounds due to the gravitational attraction of the earth.

The mass of an object can be determined by balancing it against the standard unit of mass, the kilogram, on a balance scale (Figure 1.4). (The kilogram will be discussed in Section 1.3.) From Eq. (1.1), we find that the weight of the standard kilogram is $W = (1\text{ kg})\,g$ and the weight of the mass m is $W = mg$. When the scale is balanced, the tendency of each of these forces to rotate the scale arm about the fulcrum will be equal, and we have $(1\text{ kg})\,ga = mgb$, or $m = a/b$ kilograms. Since g does not appear in this result, the scale really measures mass instead of weight. This is why balance scales are used in commerce.

Even though a balance scale measures mass, it can be calibrated to read in force units. The grocery scale is an example. Herein lies the source of much of the confusion between mass and weight. From the balance condition for the scale, we find that the weight W of the object is equal to $g(a/b)$ units of force. Since g varies from place to place, the calibration of the scale in force units would be valid only at a particular location. Any change in the weight of the object resulting from a change in g would not be detected by the scale because it actually measures mass.

Force can be measured with a spring scale (Figure 1.5). Experiments show that the elongation of a properly constructed spring is proportional to the force applied to it. The spring scale is based on this principle. If a standard kilogram is suspended from the scale, the elongation of the spring will correspond to g units of force.

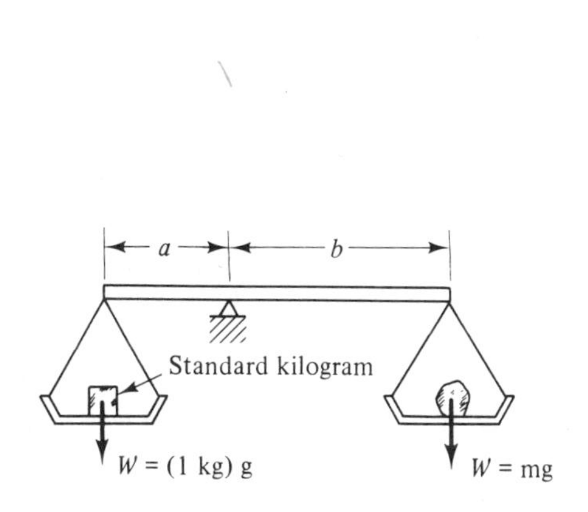

Figure 1.4

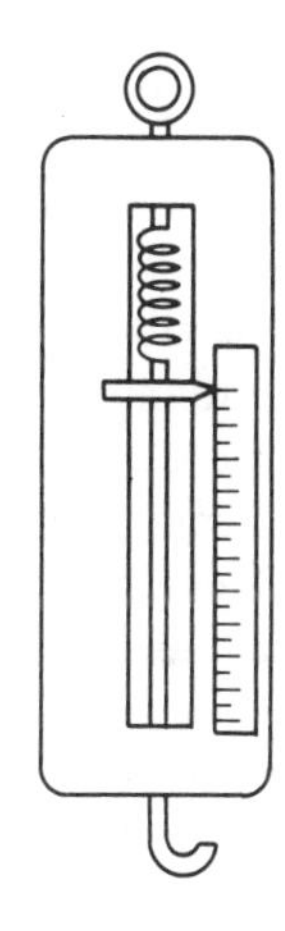

Figure 1.5

Basic principles. The relationship between the concepts of force, mass, space, and time are defined by several fundamental postulates based upon experimental evidence. These are the law of vector combination of forces mentioned previously, Newton's laws of motion, and laws, such as Newton's law of gravitation and Coulomb's law, which describe the nature of certain types of force interactions. These laws are sufficient to determine the response of a body to the forces acting upon it, if supplemented with information concerning the geometry of the deformations of the body and the resistive behavior of the material of which it is comprised. Further discussion of these basic laws and principles will be deferred until the chapters in which they are first used.

1.3 UNITS

In order to carry out calculations and measurements, one must first have a system of units and measures. Measurement systems are devised by choosing arbitrary standard units for a certain set of fundamental physical quantities such as mass, length, and time. For instance, the ancient Egyptians divided the period from sunrise to sunset into 12 hours of day and used the hour as their standard unit of time. The standard unit of length used by the Romans was the foot, which was derived originally from the length of a man's foot. When first introduced by the French Republican Government in 1793, the standard unit of length called the metre was defined as a specific fraction of the earth's circumference. The metre was later redefined as the distance between two marks on a certain platinum-iridium bar when at the temperature of melting ice.

Standards. Obviously, "standard" units of length and time based upon the length of a man's foot and the period from sunrise to sunset are subject to variation and are not really standard. In order to eliminate ambiguities, standard units are now defined in terms of natural physical constants whenever possible.

The *metre*, which is the standard unit of length, is currently defined in terms of the wavelength of the orange–red line of the krypton–86 atom. The frequency of the radiation emitted by the cesium–133 atom in a transition between two of its fundamental hyperfine energy levels is used to define the standard unit of time, the *second*. The *kilogram* is the standard unit of mass. It is a platinum–iridium cylinder, 39 mm high and 39 mm in diameter, preserved at the International Bureau of Weights and Measures near Paris. The *degree kelvin* is the standard unit of temperature and is defined in terms of the triple point of water.

Once an acceptable standard unit for a quantity has been established, measurements can be made by comparison with the standard. Other units, which are multiples or submultiples of the standard unit, may be introduced as a matter of convenience. The Roman system of units, for example, included the inch, which was defined as one–twelfth of a foot, and the mile, which was equal to 5000 feet.

US–British system of units and SI. Numerous systems of units have been devised and used, but over the years the majority of the countries have come to use a metric system of measurement. A modernized version of this system, called the *Système International d'Unités* (abbreviated SI), was established by international agreement in 1960. The United States is moving toward the use of SI, but much of the engineering work in this country is still based upon what we shall call the US–British system of units.

In the US–British system of units, force, length, time, temperature, electrical current, and luminous intensity are taken to be fundamental quantities, and the corresponding basic units are the pound, foot, second, degree fahrenheit, ampere, and candela. The units of all other physical quantities are derived in terms of these six basic units. For example, the unit of mass, as derived from Newton's second law, is the $\text{lb}\cdot\text{sec}^2/\text{ft}$, sometimes called the *slug*.

The fundamental quantities in SI are mass, length, time, temperature, electrical current, and luminous intensity. The corresponding basic units are the kilogram, metre, second, degree kelvin (usually converted into degree celsius in common use), ampere, and candela. As in the US–British system, the units of all other physical quantities are expressed in terms of these basic units. Thus, the unit of force, as derived from Newton's second law, is the $\text{kg}\cdot\text{m/s}^2$. This unit is called the *newton*.

Multiples and submultiples of the basic units are used in both the US–British system and SI. For example, in the US–British system we find the inch, yard, and mile as units of length and the kip (1 kip = 1000 pounds) as a unit of force. In SI we find the millimetre (10^{-3} metres) and kilometre (10^3 metres) as units of length and the tonne (1 tonne = 1000 kilograms) as a unit of mass.

One of the major advantages of SI over the US–British system is that multiples and submultiples of the basic units are related by powers of 10 (the time units are an exception). This greatly simplifies the conversion between units because our number system is also based upon powers of 10. For example, since there are 100 centimetres in 1 metre, metres can be converted to centimetres simply by shifting the decimal point. In contrast, since there are 36 inches in a yard, conversion from yards to inches requires a multiplication by 36.

In view of the increasing use of SI in the United States and the need for engineers to be conversant with both SI and the US–British system of units, the

Table 1.1 Basic Units

Physical Quantity	*Name of SI Unit*	*SI Symbol*	*Name of US–British Unit*	*US–British Symbol*
length	metre	m	foot	ft
force	newton	N	pound	lb
mass	kilogram	kg	slug	$\text{lb}\cdot\text{s}^2/\text{ft}$
time	second	s	second	s
ordinary temperature	degree celsius	°C	degree fahrenheit	°F

Table 1.2 Equivalents of Measure

length	1 m = 3.281 ft	1 ft = 3.048×10^{-1} m
force	1 N = 2.248×10^{-1} lb	1 lb = 4.448 N
mass	1 kg = 6.854×10^{-2} slugs	1 slug = 1.459×10 kg
ordinary temperature	t_C(°C) and t_F (°F) are related by $t_C = 5/9\,(t_F - 32)$	

following scheme is used in this text. Physical constants, material properties, and similar data are stated in US–British units, with the equivalent in SI given in parentheses. Approximately one-half of the example and exercise problems involve US–British units, with SI used in the remainder. Table 1.1 is a list of some of the basic units in both systems. Various equivalents of measure are given in Table 1.2.

Rules for use of SI. Because rules and recommended practices for the use of SI differ in some respects from those for the US–British system, they will be discussed briefly.

In SI, it is recommended that numerical values be kept between 0.1 and 1000 by the use of appropriate multipliers involving powers of 10. These multipliers are denoted by certain prefixes. The most common prefixes are listed in Table 1.3. Prefixes that shift the decimal point fewer than three places, such as centi (10^{-2}), are generally to be avoided.

A space, instead of a comma, is used to separate numbers representing multiples of 1000. This is done on both sides of the decimal point. For example, we write 36 420.076 8, not 36,420.0768. Four-digit numbers need not involve a space. Thus, we may write 3642 instead of 3 642. To avoid confusion, we shall also use these conventions for quantities expressed in US–British units.

In SI, the kilogram is used exclusively as a unit of mass. This is in contrast to common metric usage in which it is also used as a unit of force. Since mass, and not force, is the fundamental quantity in SI, the mass of a body should be specified instead of its weight. The weight can then be calculated by multiplying the mass by the acceleration of gravity, in accordance with the relationship $W = mg$ given in Eq. (1.1).

Table 1.3 SI Prefixes

Factor	*Prefix*	*SI Symbol*
10^{9}	giga	G
10^{6}	mega	M
10^{3}	kilo	k
10^{-3}	milli	m
10^{-6}	micro	μ
10^{-9}	nano	n

1.4 COMPUTATIONS

In this section we shall consider several factors that frequently arise in computations.

Conversion of units. It is often necessary to convert from one set of units to another. This may be accomplished as follows. Suppose that the magnitude of the moment of a force is given as 60 lb·ft and we wish to express it in the units N·m. We first write

$$M = 60 \text{ lb·ft}$$

Since we want to replace feet by metres, we need to multiply by an expression containing metres in the numerator and feet in the denominator, and similarly for the conversion from pounds to newtons. From Table 1.2 we find that

$$1 \text{ m} = 3.281 \text{ ft} \qquad \text{or} \qquad \frac{1 \text{ m}}{3.281 \text{ ft}} = 1$$

and

$$1 \text{ N} = 0.2248 \text{ lb} \qquad \text{or} \qquad \frac{1 \text{ N}}{0.2248 \text{ lb}} = 1$$

Multiplying the value of M by these ratios, we have

$$M = (60 \text{ lb·ft})\left(\frac{1 \text{ m}}{3.281 \text{ ft}}\right)\left(\frac{1 \text{ N}}{0.2248 \text{ lb}}\right)$$

Canceling out the units that appear in both the numerator and denominator and performing the numerical computations, we obtain

$$M = 60(1.356) = 81.35 \text{ N·m}$$

Note that the value of M has not been changed because each of the ratios by which it was multiplied is equal to unity.

The factor 1.356 by which the value of moment expressed in lb·ft is multiplied to obtain the value expressed in N·m is called a *conversion factor*. Conversion factors for a number of quantities are given in Table A.14 of the Appendix.

Dimensional homogeneity. The principle of dimensional homogeneity states that basic equations representing physical phenomena, and all results derived from them, must be valid for all systems of units. This principle is of considerable theoretical and practical importance. It implies that all quantities must be expressed in the same units before their numerical values can be added or subtracted. Also, the units of quantities on the left side of an equality sign must be the same as those on the right side. If these conditions cannot be satisfied after performing any necessary conversion of units, the results in question are in error. Note, however, that satisfaction of the requirements of dimensional homogeneity does not guarantee that the results are correct. For example, the results could be off by a numerical factor, which would not be revealed in a check of the units.

Accuracy of computations. Now that electronic calculators are commonplace, it is possible to perform computations to a degree of precision unheard of in the days of the slide rule. Much of this precision is an illusion, however. Generally speaking, the computed value of a quantity can be no more accurate than the factors involved in it. This is a very important observation, particularly since the data in engineering problems are often not known to a high degree of precision. It is not uncommon, for example, for the values of certain material properties to be known only to within $\pm 10\%$, and the degree of uncertainty is often higher. Clearly, it would be nonsense to list to five or six decimal places a result involving a factor with a possible error of this magnitude. Furthermore, any precision that exceeds that which can be reasonably obtained when measuring the quantity in question has little or no practical significance.

We shall not be concerned in this text with a detailed analysis of errors in computations. However, all numerical results should at least be checked to see if the implied degree of precision is within reason.

1.5 CLOSURE

In this chapter we have discussed in general terms the basic concepts and principles upon which statics is based. At this point these ideas undoubtedly still seem a bit vague, and they must remain so until we see them used over and over again in the following chapters. It is only through repeated application that the basic principles can be truly understood and appreciated.

PROBLEMS

1.1 The acceleration of gravity is approximately 0.18% less in San Francisco than in Juneau, Alaska. If gold were valued by weight instead of mass, what would be the loss in value of a million-dollar shipment from Juneau to San Francisco?

1.2 At points on or above the surface of the earth, the acceleration of gravity is inversely proportional to the square of the distance from the earth's center. At what altitude will the weight of an object be one-half that on the earth's surface? The earth's radius is approximately 3960 miles.

1.3 Show that if the mass of an object is measured in kilograms and the acceleration of gravity is measured in metres per second squared, its weight will be in newtons.

1.4 The spring in a spring scale elongates 2 in. when a block with a mass of 0.5 slug is suspended from it. Calibrate the scale by determining the elongation of the spring per pound of force applied to it.

1.5 What is the range (in metres) of a golfer who can drive a ball 250 yards?

1.6 A motorist is traveling at a speed of 80 km/hr on a highway whose posted speed limit is 55 mi/hr. Is he speeding?

1.7 Express your weight in newtons, your height in metres, and your waist line in millimetres. What is your mass in slugs? In kilograms? In tonnes?

1.8 If 1 ft^3 of steel has a weight of 490 lb on the surface of the earth, what is its density (mass per unit volume) in kg/m^3?

1.9 The advertised fuel consumption rate of one make of automobile is 10 kilometres per litre, while that for another make is 32 miles per gallon. Which automobile is the most economical from the standpoint of fuel consumption? (1 US gallon = 3.785 litres.)

1.10 A weather reporter states that the temperature is 3°C. What is the temperature in °F?

chapter two

VECTORS AND VECTOR QUANTITIES

2.1 INTRODUCTION

Before attempting to apply the basic ideas and principles of Chapter 1 to the solution of problems, we must first learn how to describe forces and related vector quantities in an appropriate and convenient mathematical form. We must also learn how to perform certain basic operations with these quantities, such as addition and subtraction. These topics will be considered in this chapter.

The basic properties of vectors will be discussed, along with vector algebra and its application to statics problems. Only those aspects of vector analysis for which we shall actually have use will be considered, and primary emphasis will be placed upon applications. For those who have no prior training in vector analysis, the material in this chapter will serve as an introduction to the subject. Those who have some background in vectors will find it a convenient review and source of reference. A more detailed and rigorous treatment of vector analysis may be found in the many available texts on this subject.

2.2 VECTORS

Definition and basic properties of vectors. Quantities such as mass, length, time, and temperature have only a magnitude and are called *scalars*. Scalars are real numbers and obey the rules of elementary algebra. In contrast, vector quantities have both a *magnitude* and *direction* and combine according to the parallelogram law (this law will be discussed in Section 2.3). The direction of a vector includes both its *orientation* in space and its *sense*. We have already shown in Chapter 1 that forces are vector quantities. Other examples of vector quantities that occur in physical problems are velocity, acceleration, electric field intensity, and magnetic induction. Vectors can also be used to define the position in space of one point relative to another.

A vector can be represented graphically by a directed line segment (arrow). For example, Figure 2.1(a) shows the representation of a 50-lb push exerted on the crate by the man. The length of the arrow represents the magnitude of the vector, which is defined by a positive number indicating the number of units of measure involved. The force vector shown in Figure 2.1(a) has a magnitude of 50 lb. The orientation of the vector is defined by the angle between the line along which it lies, or *line of action*, and some convenient reference line. In Figure 2.1(a) the vector is oriented 30° above the horizontal. The sense of the vector is indicated by the arrowhead. Figure 2.1(b) shows the vector representation of the velocity of an aircraft in level flight traveling northeast at 800 km/hr.

Since vectors and scalars combine according to different rules, it is essential to distinguish between them. In this text, vector quantities will be denoted by letters in boldface type. The magnitude of a vector, which is a scalar, will be represented by a letter in lightface italic type or by bars enclosing the boldface vector symbol. Thus, A or $|\mathbf{A}|$ denotes the magnitude of the vector **A**.

If two vectors have the same line of action, they are said to be *collinear*. Vectors that lie in the same plane are called *coplanar*, while those whose lines of action intersect at a common point are referred to as *concurrent* vectors. Two vectors **A** and **B** are *equal* ($\mathbf{A} = \mathbf{B}$) if they have the same magnitude, orientation, and sense; they need not be collinear.

It is important to note that equal vectors are not necessarily *equivalent* in the sense that they correspond to the same physical effect. For example, in Section 1.2 we found that the effect of a force depends not only upon its magnitude and direction, but also upon its point of application. Thus, two equal forces that act at different points on a body may produce entirely different responses. For example,

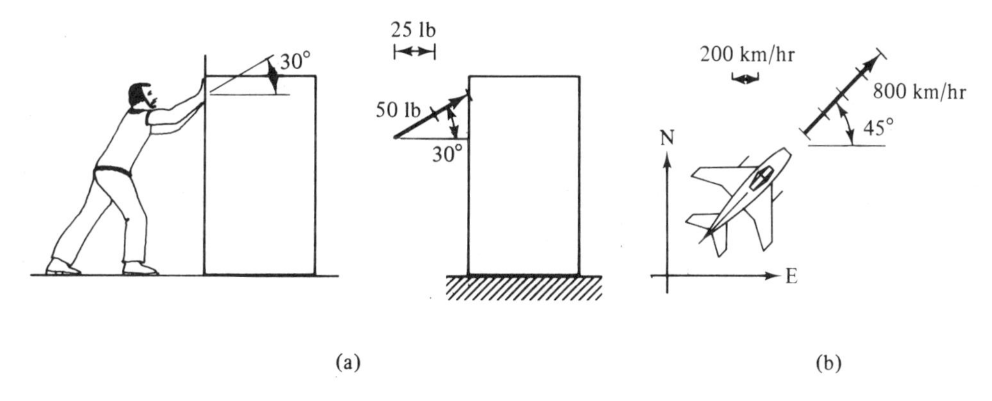

Figure 2.1

a diving board will deflect considerably more if one stands at the tip instead of at the base, even though the force exerted upon the board has the same magnitude and direction in each instance. This distinction between the equality and equivalence of two forces is very important, and will be discussed in detail in Chapter 4.

Multiplication of a vector by a scalar. If a vector **A** is multiplied by a scalar (number) h, the product $h\mathbf{A}$ is defined as a vector whose magnitude is $|h|$ times the magnitude of **A** and whose sense is equal or opposite to that of **A**, depending upon whether h is positive or negative (Figure 2.2). Note that multiplication of a vector by (-1) merely reverses its sense. The rule for multiplication of a vector and a scalar also covers division of a vector by a scalar h, since this operation is equivalent to multiplication by the factor $1/h$, provided $h \neq 0$. Multiplication of a vector by a scalar obeys the same laws as the multiplicatication of two scalars.

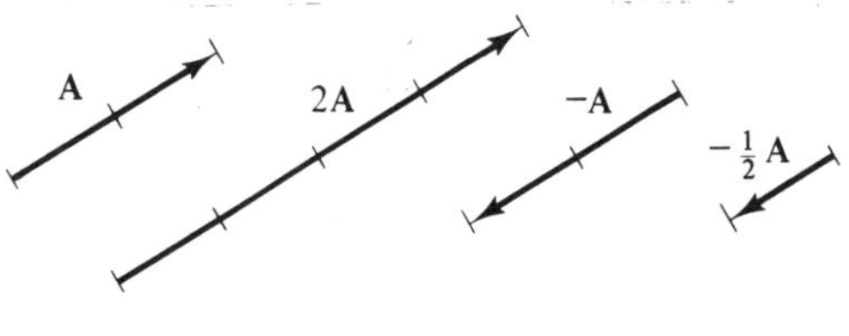

Figure 2.2

Unit vectors. Vectors with a magnitude of unity are called *unit vectors* and will be denoted by the boldface letter **e**. Any vector can be turned into a unit vector simply by dividing by its magnitude. Thus,

$$\mathbf{e}_A = \frac{\mathbf{A}}{A} \tag{2.1}$$

where the subscript denotes the vector with which the unit vector is associated. The set of units used to describe a vector **A** is the same as that used for its magnitude; therefore, the unit vector will be without units, or *dimensionless*. A special set of unit vectors, directed along coordinate axes and denoted by **i**, **j**, and **k**, will be introduced in Section 2.5.

Equation (2.1) can be rewritten as

$$\mathbf{A} = A\mathbf{e}_A \tag{2.2}$$

This relation indicates that the magnitude and direction of a vector can be expressed separately. The scalar A defines the magnitude of the vector **A**, and the dimensionless unit vector $\mathbf{e}_A$ defines its direction in space.

Zero vector. The *zero*, or *null*, *vector* **0** has zero magnitude by definition; its direction is undefined. It will be convenient to think of **0** as having all possible directions, so that it is parallel or perpendicular to any other vector.

PROBLEMS

2.1 The man is pulling on a sled with a force of 50 lb as shown. Represent this force by a vector. Use whatever scale is convenient.

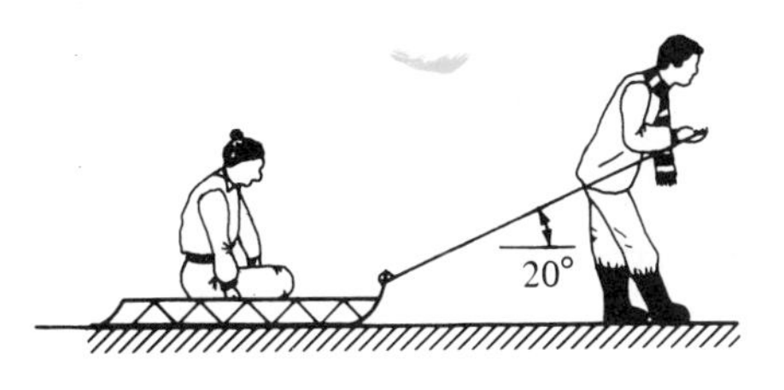

Problem 2.1

2.2 The outboard motor shown develops a thrust of 100 N and is rotated 30° with respect to the boat's centerline. Using a vector, represent the force exerted upon the boat. Also represent the force that the 90-kg man exerts upon the seat. Use whatever scale is convenient.

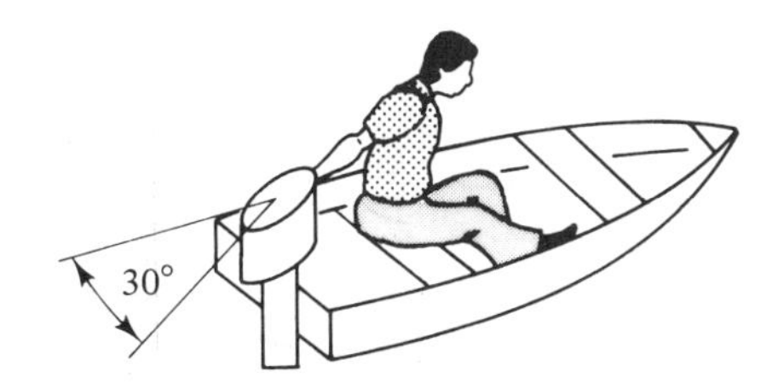

Problem 2.2

2.3 Two ships are spotted on a radar screen. Ship A is 40 km due east of the radar location and ship B is 30 km due west and 40 km due north of it. Represent the positions of the ships by vectors.

2.4 Automobile A is traveling northeast at 55 mph and automobile B is traveling due east at 30 mph. Represent the velocities of each auto by vectors. How does the velocity vector of auto B change if its speed is doubled? How does it change if B reverses directions?

2.3 ADDITION, SUBTRACTION, AND RESOLUTION OF VECTORS

Vector addition and subtraction. The sum of two vectors **A** and **B** is defined by the *parallelogram law*, which is illustrated in Figure 2.3(a). According to this law, the vectors are placed tail to tail and a parallelogram is constructed by using them as two of the sides. The sum, $\mathbf{S} = \mathbf{A} + \mathbf{B}$, corresponds to the diagonal of the parallelogram. The magnitude and orientation of **S** can be obtained graphically or by using trigonometry; its sense is determined by the construction.

It is evident from Figure 2.3(a) that the sum can also be obtained by laying out the two vectors tip to tail and constructing the vector with the tail at the starting point necessary to form a closed triangle. This construction, which is illustrated in Figures 2.3(b) and 2.3(c), is known as the *triangle rule* for vector addition. Note that the order in which the vectors are added is immaterial.

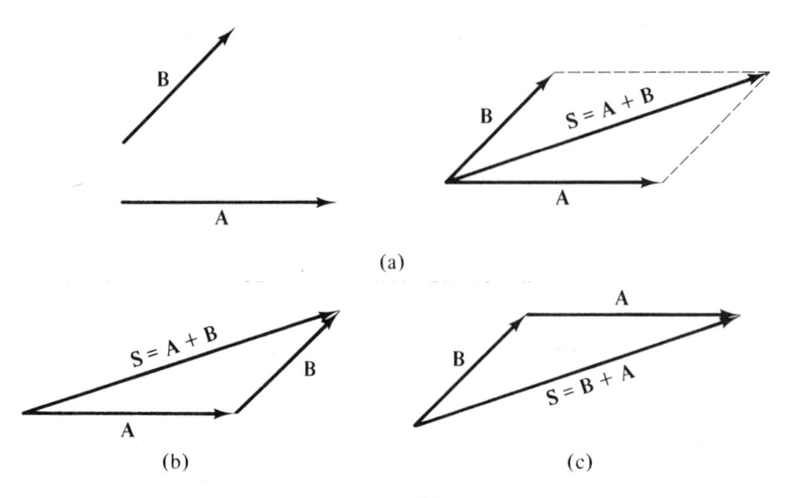

Figure 2.3

The subtraction of a vector can be defined as the addition of the corresponding negative vector. Thus, the difference of two vectors, $\mathbf{A} - \mathbf{B}$, can be rewritten as $\mathbf{A} + (-\mathbf{B})$ and the rules of addition applied.

If several vectors are to be added or subtracted, the triangle rule or parallelogram law can be applied to them two at a time. The procedure for addition using the triangle rule is illustrated in Figure 2.4. The resulting diagram reveals that the sum can also be obtained by laying out all the vectors tip to tail and constructing the vector with the tail at the starting point necessary to close the loop. This is called the *polygon rule* for vector addition because a polygon is formed when the vectors all lie in the same plane. Although this rule applies for any vectors, it is obviously practical to use it only when the vectors are coplanar.

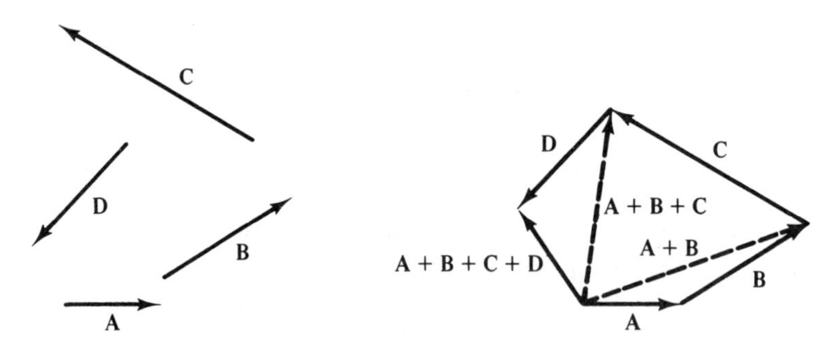

Figure 2.4

Addition and subtraction of vectors obey the same laws as addition and subtraction of scalars:

$$
\begin{aligned}
\mathbf{A} + \mathbf{B} &= \mathbf{B} + \mathbf{A} \qquad \text{(commutative law)} \\
\mathbf{A} + (\mathbf{B} + \mathbf{C}) &= (\mathbf{A} + \mathbf{B}) + \mathbf{C} \qquad \text{(associative law)} \\
\mathbf{A} - \mathbf{B} &= \mathbf{A} + (-\mathbf{B}) \\
\mathbf{A} + \mathbf{0} &= \mathbf{A}, \qquad \mathbf{A} - \mathbf{A} = \mathbf{0}
\end{aligned}
\tag{2.3}
$$

These laws follow directly from the stated rules for vector addition and subtraction and are easily proved by constructing the corresponding vector diagrams.

Resolution of a vector. Just as two or more vectors can be added to obtain a single vector, a single vector can be broken down into two or more constituent parts called *components*, or *component vectors*. This operation, which is known as *resolution* of a vector, is just the reverse of vector addition.

There are two reasons for resolving vectors into components. First, it is done for physical reasons. For example, consider the force **F** applied to the wrench shown in Figure 2.5. Only the component of **F** that is perpendicular to the handle of the wrench is effective in turning it; therefore, we would need to know this component when determining the tendency of **F** to rotate the bolt. Second, the computations involved in vector algebra can often be simplified by first resolving the vectors into components that are mutually perpendicular, as in Figure 2.5. Such components are called *rectangular*, or *orthogonal, components*. The advantages of using them will be demonstrated in Section 2.5.

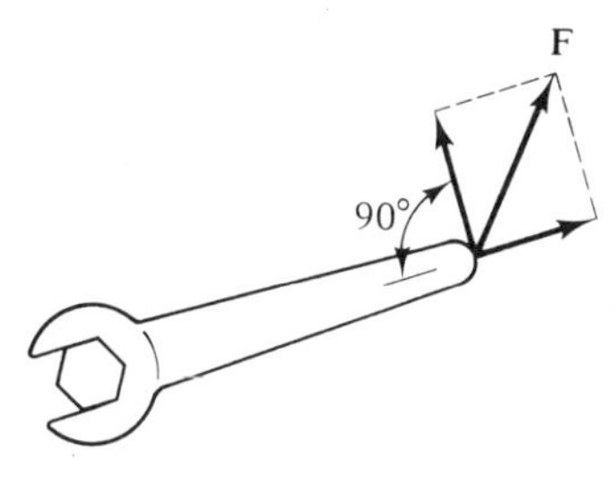

Figure 2.5

Resolution of a vector into components is accomplished by using the rules of vector addition and the fact that the sum of the components must equal the original vector. Naturally, the exact procedure depends upon what is known about the components. Two typical cases are considered in Examples 2.3 and 2.4.

Example 2.1. Addition of Vectors. Find the sum **S** of the two forces **P** and **Q** acting upon the bracket in Figure 2.6(a).

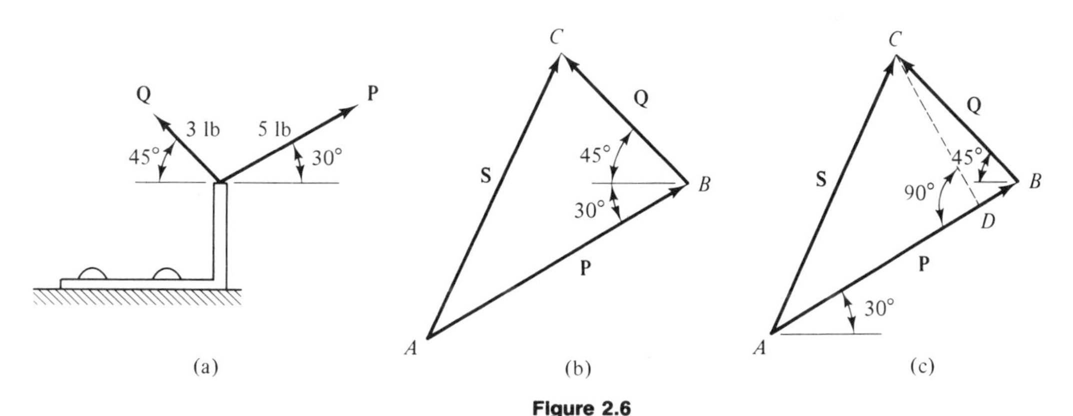

Figure 2.6

Trigonometric Solution. We first sketch the vector triangle (or parallelogram) as shown in Figure 2.6(b). It need not be accurately constructed, but it should be sketched approximately to scale so that the figure isn't misleading. The sense of **S** is determined by the sketch. It remains to find its magnitude and orientation.

Two sides of the triangle and the included angle are known. Using the law of cosines, we have

$$S^2 = P^2 + Q^2 - 2PQ \cos B$$

$$S^2 = (5 \text{ lb})^2 + (3 \text{ lb})^2 - 2(5 \text{ lb})(3 \text{ lb})\cos 75°$$

$$S = 5.12 \text{ lb}$$

Applying the law of sines, we write

$$\frac{\sin A}{Q} = \frac{\sin 75°}{S}$$

from which

$$\sin A = \left(\frac{3 \text{ lb}}{5.12 \text{ lb}}\right)(\sin 75°)$$

$$\sin A = 0.566$$

$$A = 34.5°$$

Thus,

S = 5.12 lb ∠ 64.5° ***Answer***

Alternate Trigonometric Solution. Instead of using the law of sines and cosines, we can proceed as follows. We form right triangles by dropping a perpendicular from C to AB, as shown in Figure 2.6(c). For triangle BCD, we have

$$DB = Q \cos 75° = (3 \text{ lb})\cos 75° = 0.78 \text{ lb}$$

$$CD = Q \sin 75° = (3 \text{ lb})\sin 75° = 2.90 \text{ lb}$$

so

$$AD = AB - DB = 5.00 \text{ lb} - 0.78 \text{ lb} = 4.22 \text{ lb}$$

From triangle ACD, we get

$$S^2 = (CD)^2 + (AD)^2 = (2.90 \text{ lb})^2 + (4.22 \text{ lb})^2$$

$$S = 5.12 \text{ lb}$$

$$\sin A = \frac{CD}{S} = \frac{2.90 \text{ lb}}{5.12 \text{ lb}} = 0.566$$

$$A = 34.5°$$

This is the same result obtained previously. Note that the angle A could also have been determined from expressions for $\cos A$ or $\tan A$.

Graphical Solution. We lay out the vector triangle to a convenient scale, say 2 in. = 1 lb. The scale is arbitrary, but the larger the triangle, the greater the accuracy attainable. The triangle will not be reconstructed here. Referring to our previous triangle, Figure 2.6(b), we would measure the length AC: $AC = 10.24$ in. Thus, $S = 10.24$ in. (1 lb/2 in.) = 5.12 lb. The angle that **S** makes with the horizontal, as measured with a protractor, is found to be 64.5°.

Example 2.2. Addition and Subtraction of Vectors. Given the three vectors of Figure 2.7(a), find the vector $\mathbf{T} = \mathbf{P} + \mathbf{Q} - \mathbf{R}$.

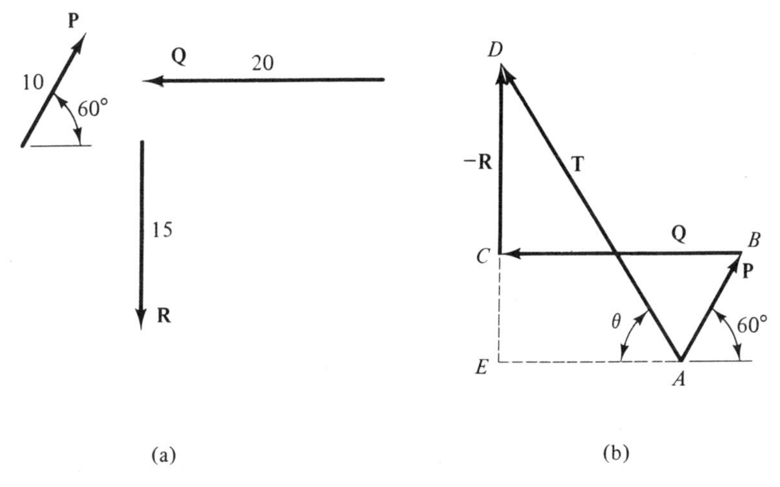

Figure 2.7

Solution. Starting with the vector **P**, we sketch the vector polygon as shown in Figure 2.7(b). Note that **R** is subtracted by adding −**R**. From the geometry of the polygon, we find

$$CB = AB \cos 60° + AD \cos \theta$$
$$ED = AD \sin \theta = AB \sin 60° + CD$$

where θ defines the orientation of **T**. Rearranging these equations and dividing the second equation by the first, we obtain

$$\frac{AD \sin \theta}{AD \cos \theta} = \tan \theta = \frac{AB \sin 60° + CD}{CB - AB \cos 60°}$$
$$\tan\theta = \frac{10 \sin 60° + 15}{20 - 10 \cos 60°} = 1.577$$
$$\theta = 57.6°$$

The magnitude of **T** can now be determined from either of our first two equations. Using the first equation, we get

$$T = AD = \frac{CB - AB\cos 60°}{\cos\theta} = \frac{20 - 10\cos 60°}{\cos 57.6°} = 28.0$$

Thus,

$$\mathbf{T} = 28.0 \quad 57.6° \qquad \textit{Answer}$$

Of course, the solution could also be obtained graphically.

Example 2.3. Resolution of a Vector. Resolve the force **F** in Figure 2.8(a) into rectangular components that are horizontal and vertical.

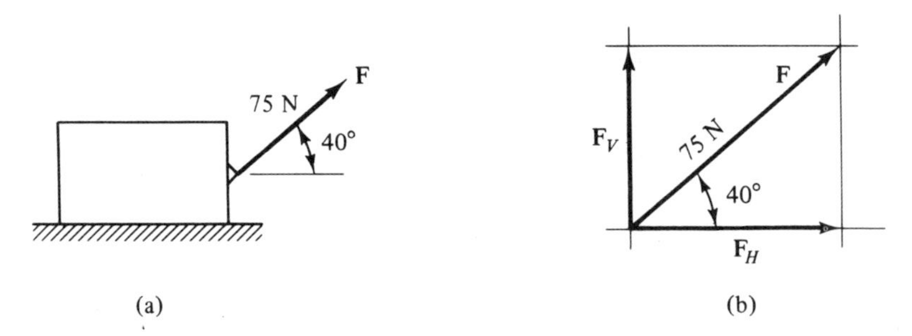

Figure 2.8

Solution. Let $\mathbf{F}_H$ and $\mathbf{F}_V$ denote the horizontal and vertical components of **F**, respectively. Recognizing that $\mathbf{F}_H + \mathbf{F}_V = \mathbf{F}$, we lay out lines at the tip and tail of **F** which are parallel to the lines of action of the component vectors [Figure 2.8(b)]. This forms the vector parallelogram (a rectangle in this case) from which the components can be determined. Using simple trigonometry, we have

$$F_H = (75 \text{ N})\cos 40° = 57.5 \text{ N}$$

$$F_V = (75 \text{ N})\sin 40° = 48.2 \text{ N} \qquad \textit{Answer}$$

The sense of the components is defined by the vector diagram.

Example 2.4. Resolution of a Vector. Two tug boats [Figure 2.9(a)] are assisting a freighter out of a harbor. (a) How hard should each tug push so that their combined effort is a 100 000-lb force in the direction the ship is heading? (b) If tug A remains oriented as shown, how should tug B be oriented so that the force it must exert is a minimum? Assume that the forces exerted by the tugs are parallel to the direction they are heading.

Solution. (a) The problem requires that we find the magnitudes of the component vectors of the 100 000-lb force with lines of action oriented in the

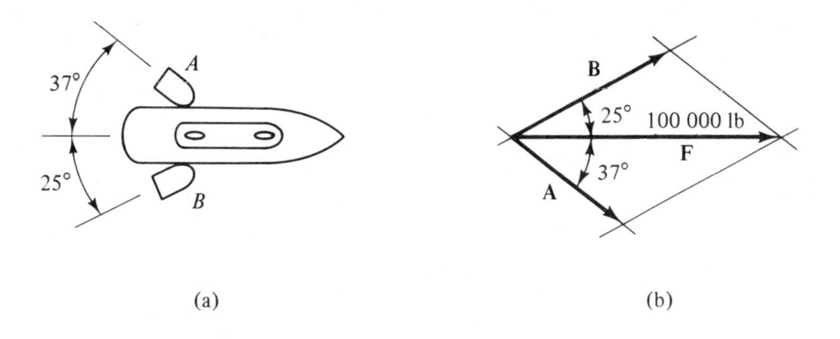

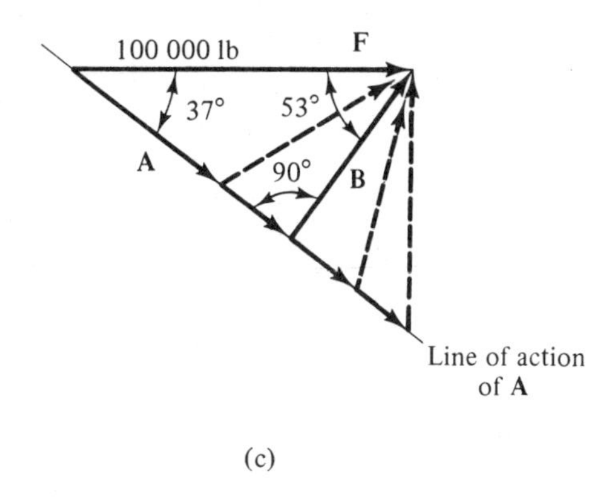

Figure 2.9

directions the tugs push. Let **A** and **B** be the forces exerted by tugs A and B, respectively, and let **F** be their sum ($F = 100\,000$ lb). We lay out lines at the tip and tail of **F** with the orientation of **A** and **B** to form a vector parallelogram [Figure 2.9(b)]. Considering one of the triangles formed, we have from the law of sines

$$\frac{\sin 37^\circ}{B} = \frac{\sin 118^\circ}{F} = \frac{\sin 25^\circ}{A}$$

Solving for A and B, we get

$$A = 47\ 900 \text{ lb} \qquad \textbf{\textit{Answer}}$$

$$B = 68\ 200 \text{ lb} \qquad \textbf{\textit{Answer}}$$

(b) Here we use the triangle rule. We know that **A** must lie along a line that makes an angle of 37° with **F** [Figure 2.9(c)]. We also know that the tail of **B** coincides with the tip of **A** and the tip of **B** coincides with the tip of **F**, since **A** + **B** = **F**. However, we do not know where **A** ends and **B** begins. Several possibilities are shown. Clearly, B is smallest when **B** is perpendicular to **A**. Thus, tug B should be oriented at an angle of 53° with respect to the longitudinal axis of the ship. The magnitudes of the forces that the tugs must exert in this case are easily determined from the vector triangle: $A = 79\ 900$ lb and $B = 60\ 200$ lb.

PROBLEMS

2.5 to 2.7 Find the sum of the vectors shown (a) graphically and (b) analytically.

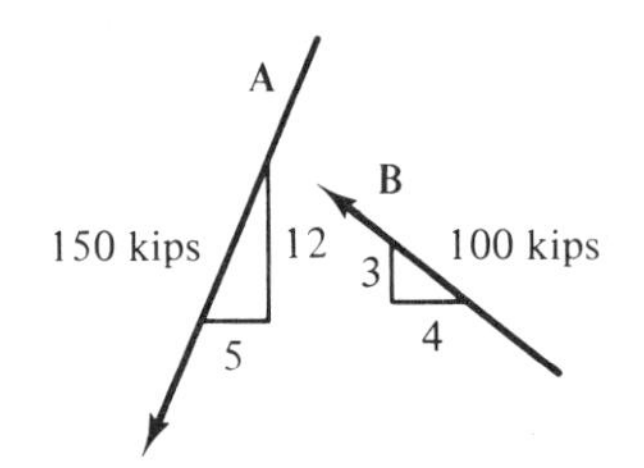

Problem 2.5

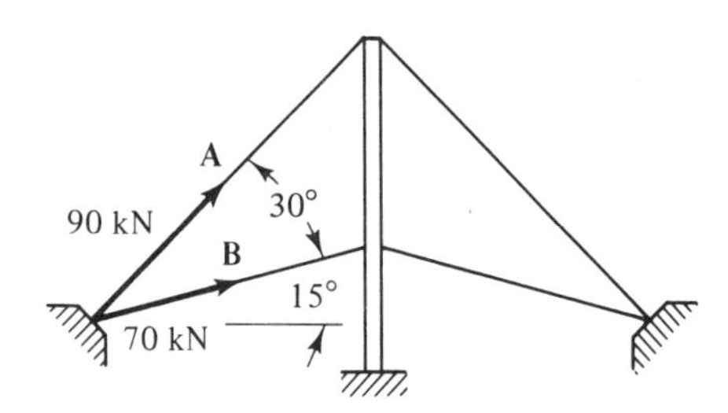

Problem 2.6

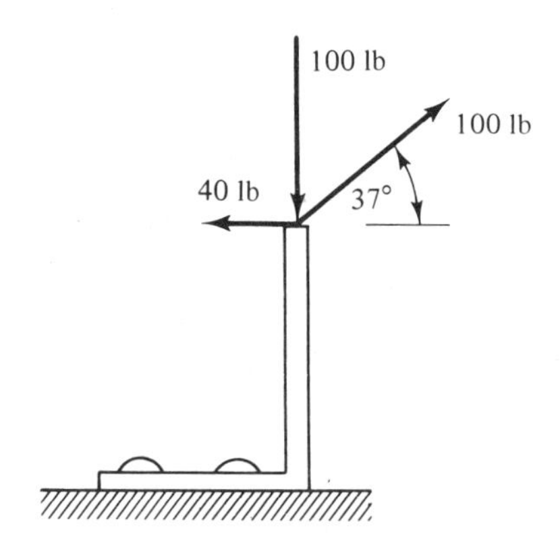

Problem 2.7

2.8 Find **A** − **B** and **B** − **A** for the vectors of Problem 2.5

2.9 For the vectors in Problem 2.6, find **A** − **B** and **B** − **A**.

2.10 In Problem 2.3, determine the vector that represents the position of ship B relative to ship A.

2.11 A string exerts a 12-lb force on a kite as shown. Determine the horizontal and vertical components of the force.

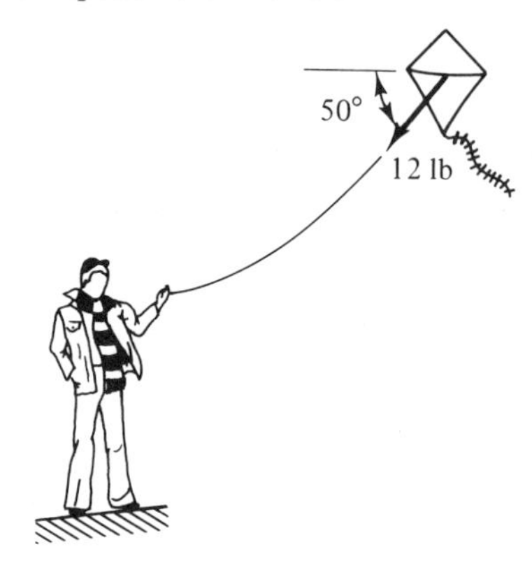

Problem 2.11

2.12 Determine the horizontal and vertical components of the forces **A** and **B** in Problem 2.6 and resolve **A** into a vertical component and a component collinear with **B**.

2.13 The total force **F** acting upon the airplane shown is known to be horizontal. Determine (a) the magnitude T of the thrust and (b) the total force **F**.

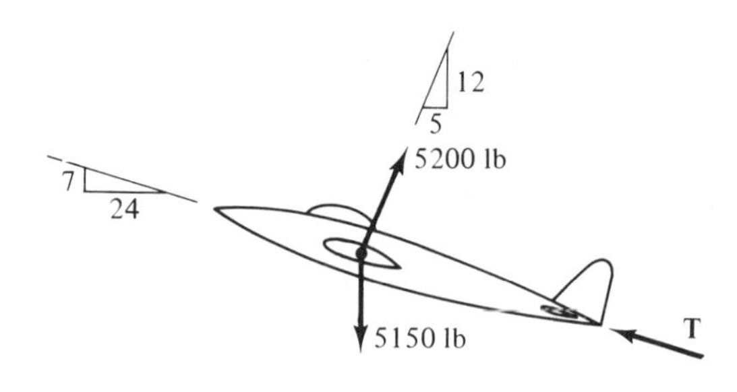

Problem 2.13

2.14 If the sum of the four forces shown is zero, determine the magnitudes of forces **A** and **B**.

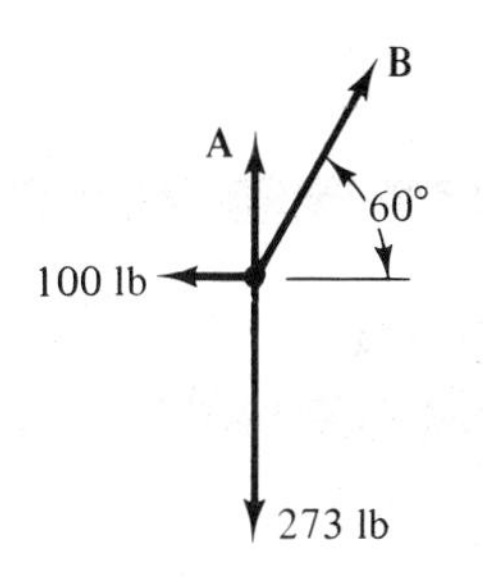

Problem 2.14

2.15 Two tow trucks pull on a disabled truck as shown. Determine the magnitude of the force **P** so that the total force **F** acting lies along the longitudinal axis of the truck. What is the magnitude of **F**?

2.16 Knowing that the sum of the forces acting at point A on the lamp shown must be zero, determine the magnitudes of the tensions $\mathbf{T}_1$ and $\mathbf{T}_2$ on the cords for $\theta = 45°$. For what value of θ will the tension $\mathbf{T}_2$ be minimum? What is the corresponding magnitude of $\mathbf{T}_2$?

2.17 and 2.18 Resolve the vectors shown into components along the coordinate axes.

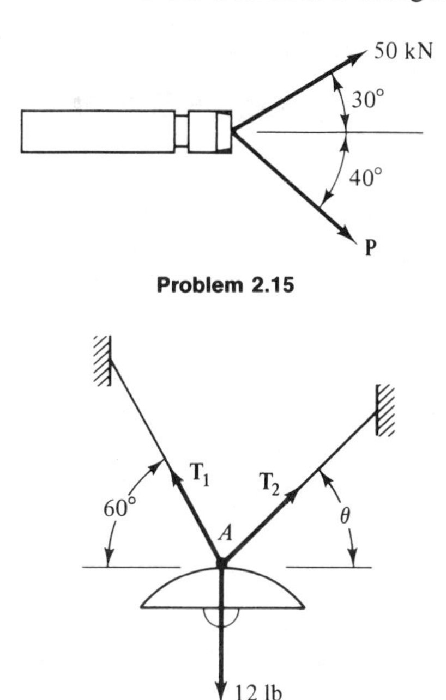

Problem 2.15

Problem 2.16

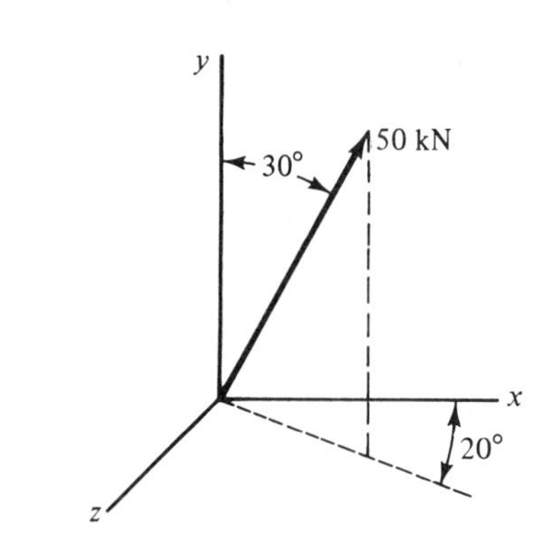

Problem 2.17

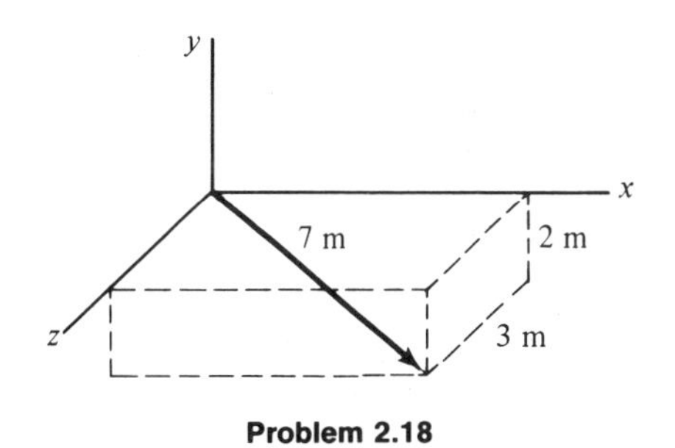

Problem 2.18

2.4 SCALAR PRODUCT OF TWO VECTORS

There are two forms of vector multiplication. One is the scalar product, which we shall consider in this section, and the other is the vector product, which we shall consider in Section 2.7. The scalar product is closley associated with the concept of the projection of a vector onto a line or axis, which enters into some of the most important and useful results in vector analysis.

The *scalar product* of two vectors **A** and **B** is a number denoted by $\mathbf{A} \cdot \mathbf{B}$ and defined by the relation

$$\mathbf{A} \cdot \mathbf{B} = AB \cos \theta \tag{2.4}$$

where A and B are the respective magnitudes of the vectors and θ is the angle between them. The angle between two vectors is defined as the smaller of the two angles between their positive directions, so $0 \leqslant \theta \leqslant 180°$. The scalar product is also called the *dot* or *inner product*.

Interpreted physically, the quantity $B \cos \theta$ in Eq. (2.4) represents the *projection* of **B** onto **A** (Figure 2.10), which we write as

$$\text{Proj}_{\mathbf{A}}\, \mathbf{B} = B \cos \theta$$

Similarly, the quantity $A \cos \theta$ can be interpreted as the projection of **A** onto **B**:

$$\text{Proj}_{\mathbf{B}}\, \mathbf{A} = A \cos \theta$$

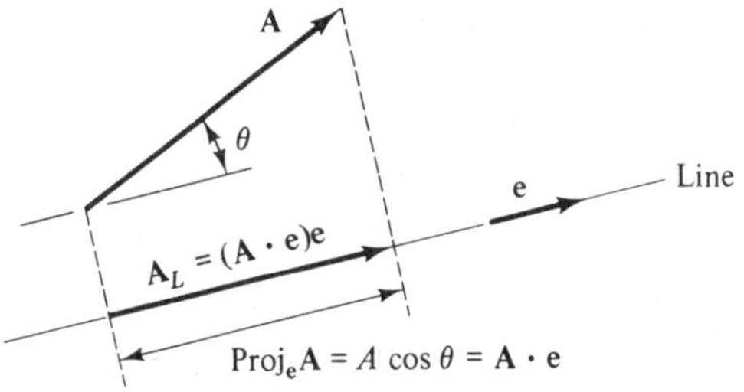

The projections will be either positive or negative, depending upon whether **A** and **B** form an acute angle ($\theta < 90°$) or obtuse angle ($\theta > 90°$). If $\theta = 90°$, the projections are zero and so is the dot product $\mathbf{A} \cdot \mathbf{B}$.

Figure 2.10

The preceding results indicate that the dot product of two vectors can also be expressed as

$$\mathbf{A} \cdot \mathbf{B} = A\ \text{Proj}_{\mathbf{A}}\, \mathbf{B} = B\ \text{Proj}_{\mathbf{B}}\, \mathbf{A} \tag{2.5}$$

If one of the vectors is a unit vector **e** along a given line or axis, the dot product gives the projection of the second vector onto that line or axis (Figure 2.11), since

$$\mathbf{A} \cdot \mathbf{e} = (1)\ \text{Proj}_{\mathbf{e}}\, \mathbf{A} = A \cos \theta$$

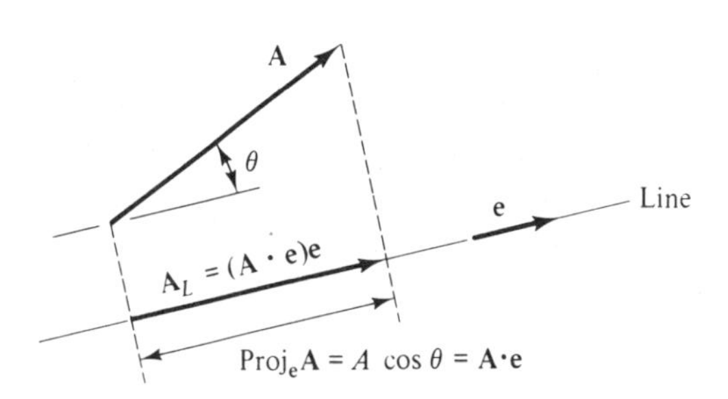

Figure 2.11

The component $\mathbf{A}_L$ of the vector along the line can then be written as

$$\mathbf{A}_L = (\mathbf{A} \cdot \mathbf{e})\mathbf{e} \tag{2.6}$$

where the projection $\mathbf{A} \cdot \mathbf{e}$ defines its magnitude and sense and the unit vector **e** defines its orientation in space.

The scalar product obeys the same laws as the product of real numbers:

$$\mathbf{A} \cdot \mathbf{B} = \mathbf{B} \cdot \mathbf{A} \qquad \text{(commutative law)}$$

$$\mathbf{A} \cdot (\mathbf{B} + \mathbf{C}) = \mathbf{A} \cdot \mathbf{B} + \mathbf{A} \cdot \mathbf{C} \qquad \text{(distributive law)}$$

$$(h\mathbf{A}) \cdot \mathbf{B} = \mathbf{A} \cdot (h\mathbf{B}) = h(\mathbf{A} \cdot \mathbf{B});\ h \text{ is any scalar} \tag{2.7}$$

$$\mathbf{A} \cdot \mathbf{A} = A^2$$

$$\mathbf{A} \cdot \mathbf{B} = 0 \text{ when } \mathbf{A} \text{ is perpendicular to } \mathbf{B}$$

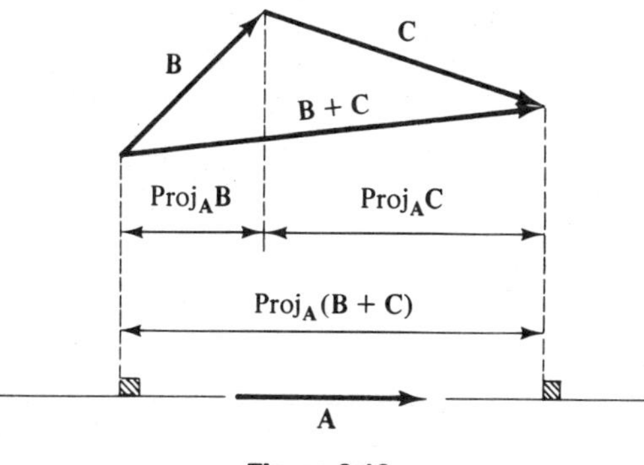

Figure 2.12

All but the second of these laws follow directly from the definition of the scalar product, Eq. (2.4). To prove the second law, we note from Eq. (2.5) that

$$\mathbf{A} \cdot (\mathbf{B} + \mathbf{C}) = A \text{ Proj}_{\mathbf{A}} (\mathbf{B} + \mathbf{C})$$

From Figure 2.12, we see that

$$\text{Proj}_{\mathbf{A}} (\mathbf{B} + \mathbf{C}) = \text{Proj}_{\mathbf{A}} \mathbf{B} + \text{Proj}_{\mathbf{A}} \mathbf{C}$$

so

$$\mathbf{A} \cdot (\mathbf{B} + \mathbf{C}) = A \text{ Proj}_{\mathbf{A}} \mathbf{B} + A \text{ Proj}_{\mathbf{A}} \mathbf{C} = \mathbf{A} \cdot \mathbf{B} + \mathbf{A} \cdot \mathbf{C}$$

Example 2.5. Projections of a Vector. The cables shown in Figure 2.13(a) exert forces **P** and **Q** on the block, as indicated. Determine (a) the projection of **Q** onto **P**, (b) the projection of **P** onto **Q**, (c) the dot product of **P** and **Q**, and (d) the projection of **Q** onto the horizontal.

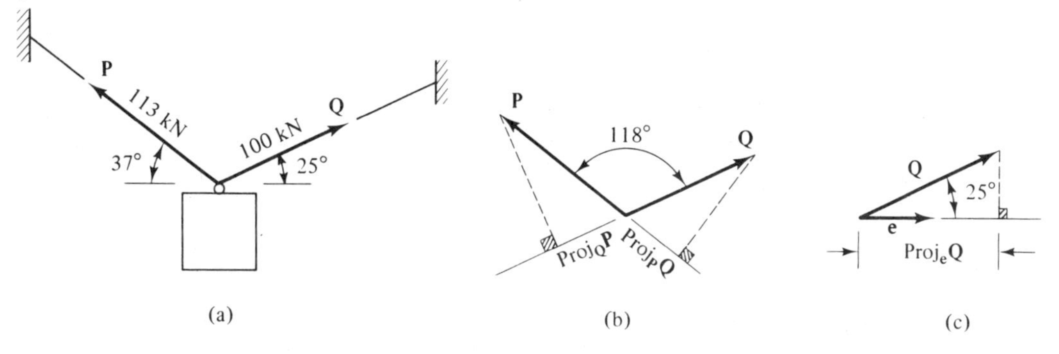

Figure 2.13

Solution. (a) The angle between the two vectors is $\theta = 118°$; so

$$\text{Proj}_{\mathbf{P}} \mathbf{Q} = Q \cos \theta = (100 \text{ kN})\cos 118° = -47.0 \text{ kN} \qquad \textit{Answer}$$

(b)

$$\text{Proj}_{\mathbf{Q}} \mathbf{P} = P \cos \theta = (113 \text{ kN})\cos 118° = -53.1 \text{ kN} \qquad \textit{Answer}$$

The two projections are shown in Figure 2.13(b).

(c) From Eq. (2.5), we have

$$\mathbf{P} \cdot \mathbf{Q} = Q \text{ Proj}_{\mathbf{Q}} \mathbf{P} = (100 \times 10^3 \text{ N})(-53.1 \times 10^3 \text{ N})$$
$$= -5.31 \times 10^9 \text{ N}^2 \text{ or } -5.31 \text{ GN}^2 \qquad \textit{Answer}$$

The dot product can also be computed by using Eq. (2.4).

(d) Let the positive direction along the horizontal be to the right, as indicated by the unit vector **e** in Figure 2.13(c). The angle between **Q** and **e** is $\theta = 25°$; so

$$\text{Proj}_{\mathbf{e}}\,\mathbf{Q} = Q\cos\theta = (100\text{ kN})\cos 25° = 90.6\text{ kN} \qquad \textit{Answer}$$

If the positive direction along the horizontal is taken to be to the left, the sign of the projection will be reversed.

PROBLEMS

2.19 and 2.20 For the vectors **A** and **B** shown find (a) the projection of **A** onto **B**, (b) the projection of **B** onto **A**, (c) the dot product $\mathbf{A}\cdot\mathbf{B}$, and (d) the projections of **A** and **B** onto the vertical.

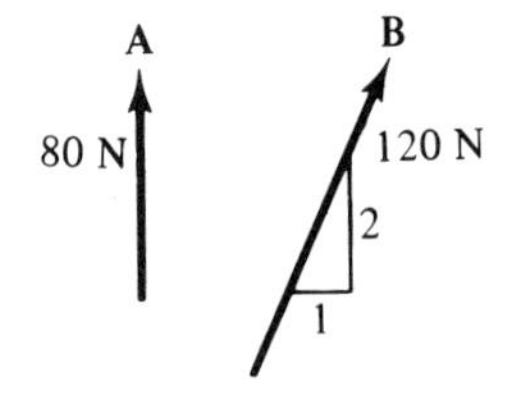

Problem 2.19

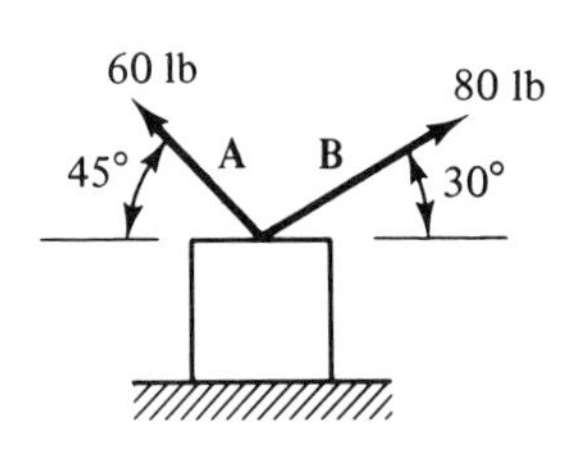

Problem 2.20

2.21 The dot product of two vectors **A** and **B** is $\mathbf{A}\cdot\mathbf{B} = 43.3$, where **A** is as shown and **B** lies along line *b-b*. Find **B** if (a) $\theta = 30°$ and (b) $B = 12$.

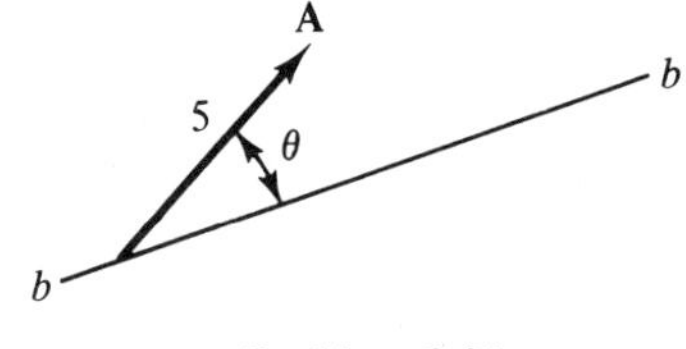

Problem 2.21

2.22 The vector **S** shown is the sum of two vectors **A** and **B**, where **A** lies along line *a-a*. If $\mathbf{A}\cdot\mathbf{S} = 173.6\text{ ft}^2$, find **A** and **B**.

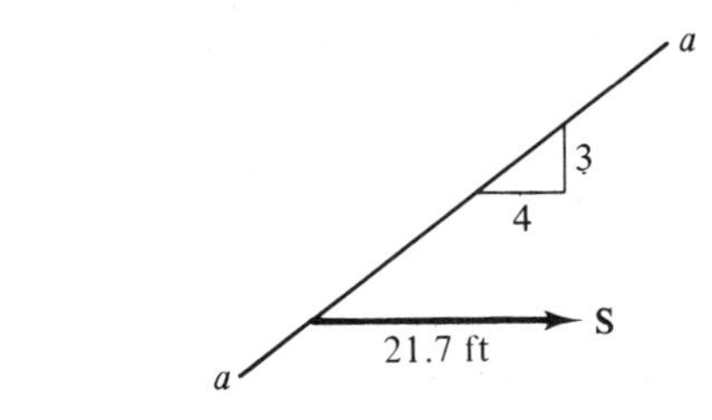

Problem 2.22

2.23 If **S** is the sum of the vectors **A** and **B** shown and $\mathbf{A}\cdot\mathbf{S} = 4\text{ N}^2$ and $\mathbf{B}\cdot\mathbf{S} = 5\text{ N}^2$, find **S** and **B**.

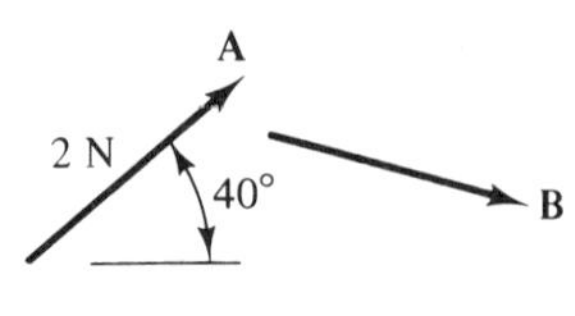

Problem 2.23

2.5 VECTORS REFERRED TO RECTANGULAR COORDINATE SYSTEMS

Up to this point, we have represented vectors by directed line segments in space. In this section we shall consider an alternate representation that is usually more convenient to use in computations.

Vectors in component form. Consider a rectangular Cartesian coordinate system with axes x, y, and z (Figure 2.14). Now let us introduce a set of unit vectors $\mathbf{i}$, $\mathbf{j}$, and $\mathbf{k}$ which define the positive directions along the coordinate axes. These vectors will be called the *base vectors*. The special notation is used to distinguish them from other unit vectors, $\mathbf{e}$. In this text we shall use only *right-handed coordinate systems* for which x, y, and z lie along the thumb, index, and middle fingers of the right hand, respectively.

Any vector $\mathbf{A}$ located within a given coordinate system can be written as the sum of its three rectangular components $\mathbf{A}_x$, $\mathbf{A}_y$, and $\mathbf{A}_z$ parallel to the coordinate axes (Figure 2.14):

$$\mathbf{A} = \mathbf{A}_x + \mathbf{A}_y + \mathbf{A}_z$$

This is easily verified by using the polygon rule for vector addition, as indicated in the figure. It is also evident from Figure 2.14 that the magnitude of the component $\mathbf{A}_x$ is the projection of $\mathbf{A}$ onto the x axis. Similarly, the magnitudes of $\mathbf{A}_y$ and $\mathbf{A}_z$ are the respective projections onto axes y and z. These projections, which we shall call the *rectangular projections* of the vector, will be denoted by A_x, A_y, and A_z:

$$\begin{aligned} A_x &= \text{Proj}_{\mathbf{i}}\,\mathbf{A} = \mathbf{A}\cdot\mathbf{i} \\ A_y &= \text{Proj}_{\mathbf{j}}\,\mathbf{A} = \mathbf{A}\cdot\mathbf{j} \\ A_z &= \text{Proj}_{\mathbf{k}}\,\mathbf{A} = \mathbf{A}\cdot\mathbf{k} \end{aligned} \tag{2.8}$$

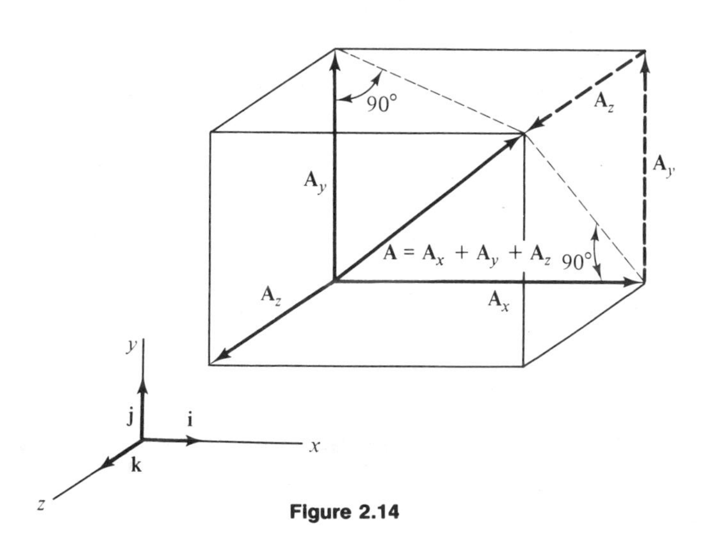

Figure 2.14

Since the component $\mathbf{A}_x$ is parallel to the base vector $\mathbf{i}$, it can be written as

$$\mathbf{A}_x = A_x\mathbf{i}$$

Similarly,

$$\mathbf{A}_y = A_y\mathbf{j} \qquad \mathbf{A}_z = A_z\mathbf{k}$$

Thus, any vector $\mathbf{A}$ can be expressed in the *component form*

$$\mathbf{A} = A_x\mathbf{i} + A_y\mathbf{j} + A_z\mathbf{k} \tag{2.9}$$

where A_x, A_y, and A_z are the projections of $\mathbf{A}$ onto the coordinate axes. The projections are shown as being positive in Figure 2.14, but, of course, they can be negative.

We now have two ways to represent a vector. It can be represented graphically by a directed line segment in space, or, according to Eq. (2.9), it can be represented by a set of three numbers (scalars) which are its projections onto the axes of a rectangular coordinate system. As we shall see in the following, the latter representation is usually the most convenient to use in computations.

For brevity, we shall write the units for vectors given in component form only at the end of the complete vector expression. For example, we write a vector $\mathbf{F} = -(3 \text{ lb})\mathbf{i} + (4 \text{ lb})\mathbf{j}$ as $\mathbf{F} = -3\mathbf{i} + 4\mathbf{j}$ lb, where it is understood that the units apply to each component of the vector.

Scalar products. The scalar product of two vectors $\mathbf{A}$ and $\mathbf{B}$ expressed in component form can be computed by using the distributive property stated in Eq. (2.7) of Section 2.4. We have

$$\begin{aligned}\mathbf{A}\cdot\mathbf{B} &= (A_x\mathbf{i} + A_y\mathbf{j} + A_z\mathbf{k})\cdot(B_x\mathbf{i} + B_y\mathbf{j} + B_z\mathbf{k})\\ &= A_xB_x\mathbf{i}\cdot\mathbf{i} + A_xB_y\mathbf{i}\cdot\mathbf{j} + A_xB_z\mathbf{i}\cdot\mathbf{k}\\ &\quad + A_yB_x\mathbf{j}\cdot\mathbf{i} + A_yB_y\mathbf{j}\cdot\mathbf{j} + A_yB_z\mathbf{j}\cdot\mathbf{k}\\ &\quad + A_zB_x\mathbf{k}\cdot\mathbf{i} + A_zB_y\mathbf{k}\cdot\mathbf{j} + A_zB_z\mathbf{k}\cdot\mathbf{k}\end{aligned}$$

Now, the base vectors $\mathbf{i}$, $\mathbf{j}$, and $\mathbf{k}$ are mutually perpendicular; so

$$\mathbf{i}\cdot\mathbf{i} = \mathbf{j}\cdot\mathbf{j} = \mathbf{k}\cdot\mathbf{k} = 1$$
$$\mathbf{i}\cdot\mathbf{j} = \mathbf{j}\cdot\mathbf{i} = \mathbf{i}\cdot\mathbf{k} = \mathbf{k}\cdot\mathbf{i} = \mathbf{j}\cdot\mathbf{k} = \mathbf{k}\cdot\mathbf{j} = 0$$

Thus,

$$\mathbf{A}\cdot\mathbf{B} = A_xB_x + A_yB_y + A_zB_z \tag{2.10}$$

This form of the dot product is particularly convenient when the angle between the two vectors is unknown, as is often the case. In fact, this angle can be determined by computing $\mathbf{A}\cdot\mathbf{B}$ from Eq. (2.10) and setting it equal to the alternate expression for the dot product, $AB\cos\theta$. These steps yield the result

$$\cos\theta = \frac{\mathbf{A}\cdot\mathbf{B}}{AB} = \frac{A_xB_x + A_yB_y + A_zB_z}{AB} \tag{2.11}$$

where θ is the angle between $\mathbf{A}$ and $\mathbf{B}$.

Magnitude and orientation of a vector in space. The magnitude of a vector **A** expressed in component form can be computed by using the scalar product. We have

$$\mathbf{A} \cdot \mathbf{A} = A^2 = A_x^2 + A_y^2 + A_z^2$$

or

$$A = \sqrt{A_x^2 + A_y^2 + A_z^2} \tag{2.12}$$

This result is just a statement of the Pythagorean theorem in three dimensions.

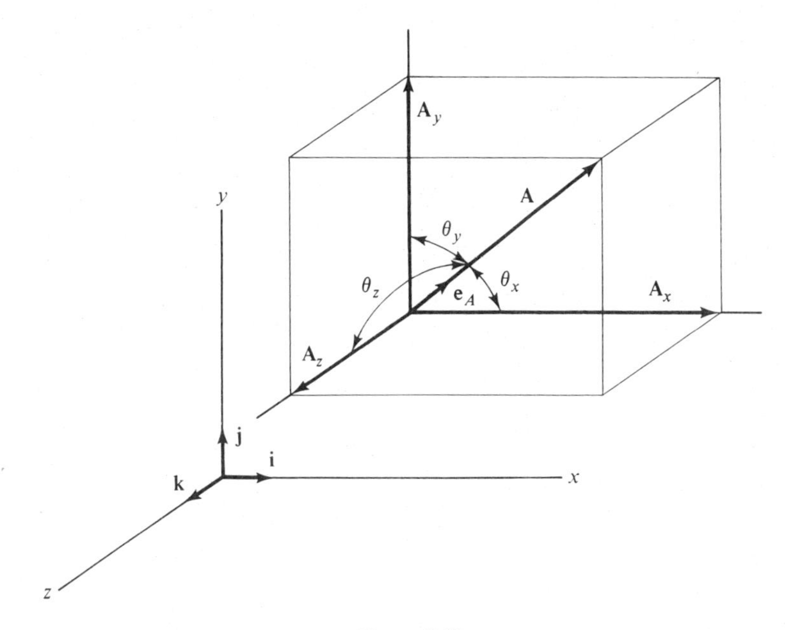

Figure 2.15

The orientation of a vector **A** in space is defined by the *direction angles* θ_x, θ_y, and θ_z between it and the positive coordinate axes (Figure 2.15). By definition, the values of these angles are restricted to the interval 0° to 180°, inclusive. The direction angles of a vector are directly related to its rectangular projections, as can be shown from Eq. (2.8) and the definition of the scalar product. We have

$$\begin{aligned} A_x &= \mathbf{A} \cdot \mathbf{i} = A \cos \theta_x \\ A_y &= \mathbf{A} \cdot \mathbf{j} = A \cos \theta_y \\ A_z &= \mathbf{A} \cdot \mathbf{k} = A \cos \theta_z \end{aligned} \tag{2.13}$$

The quantities $\cos \theta_x$, $\cos \theta_y$, and $\cos \theta_z$ are called the *direction cosines* of the vector.

According to Eq. (2.13), we can write

$$\mathbf{A} = A(\cos \theta_x \mathbf{i} + \cos \theta_y \mathbf{j} + \cos \theta_z \mathbf{k}) = A\mathbf{e}_A \tag{2.14}$$

where $\mathbf{e}_A$ is a unit vector along **A** (Figure 2.15). Now

$$\mathbf{e}_A \cdot \mathbf{e}_A = \cos^2 \theta_x + \cos^2 \theta_y + \cos^2 \theta_z = 1 \tag{2.15}$$

which indicates that the direction angles are not independent. If any two of them are given, the third must be such that Eq. (2.15) is satisfied.

Vector sums. Vector addition and subtraction is greatly simplified when the vectors are expressed in component form. We have already shown that the projection of the sum of two vectors is equal to the sum of their projections (see Figure 2.12 in Section 2.4). Thus, to add vectors, we need only add their corresponding rectangular projections:

$$\mathbf{A} + \mathbf{B} = (A_x + B_x)\mathbf{i} + (A_y + B_y)\mathbf{j} + (A_z + B_z)\mathbf{k} \tag{2.16}$$

Similarly,

$$\mathbf{A} - \mathbf{B} = (A_x - B_x)\mathbf{i} + (A_y - B_y)\mathbf{j} + (A_z - B_z)\mathbf{k} \tag{2.17}$$

These methods of addition and subtraction hold for any number of vectors, and they are particularly convenient for three-dimensional problems. Notice that the results are obtained in component form, which is usually preferable. However, the magnitude and orientation of the resulting vectors can always be obtained by using Eqs. (2.12) and (2.13), if desired.

Example 2.6. Addition of Vectors. Find the sum **S** of the two forces **P** and **Q** acting upon the bracket in Figure 2.16(a) and determine its magnitude and orientation.

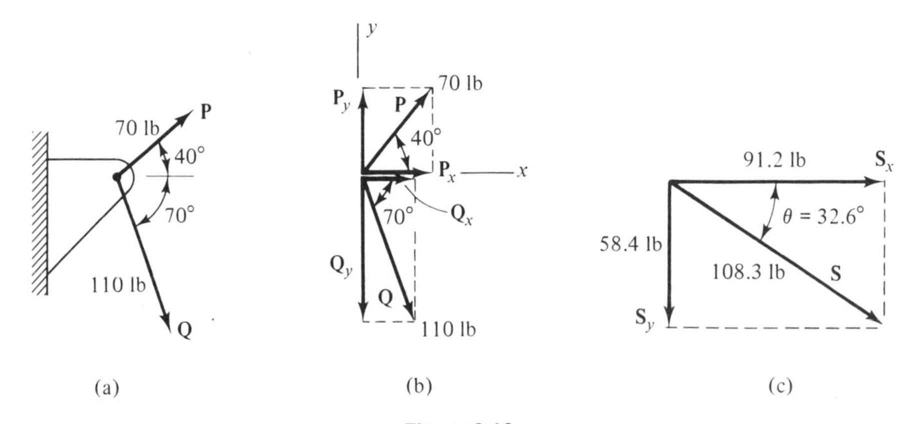

Figure 2.16

Solution. We first establish a coordinate system and sketch the component vectors [Figure 2.16(b)]. The rectangular projections of the vectors can be computed by using Eq. (2.13), but it is simpler to determine them directly from the geometry of the figure:

$$\begin{aligned} P_x &= (70 \text{ lb})\cos 40^\circ = 53.6 \text{ lb} \\ P_y &= (70 \text{ lb})\sin 40^\circ = 45.0 \text{ lb} \\ Q_x &= (110 \text{ lb})\cos 70^\circ = 37.6 \text{ lb} \\ Q_y &= -(110 \text{ lb})\sin 70^\circ = -103.4 \text{ lb} \end{aligned}$$

Note that Q_y is negative because the component vector points in the negative coordinate direction. Now,

$$S_x = P_x + Q_x = 53.6 \text{ lb} + 37.6 \text{ lb} = 91.2 \text{ lb}$$

$$S_y = P_y + Q_y = 45.0 \text{ lb} + (-103.4 \text{ lb}) = -58.4 \text{ lb}$$

so

$$\mathbf{S} = S_x\mathbf{i} + S_y\mathbf{j} = 91.2\mathbf{i} - 58.4\mathbf{j} \text{ lb} \qquad \textit{Answer}$$

The components of **S** are shown in Figure 2.16(c). Referring to the figure, or using Eq. (2.12), we have

$$S = \sqrt{S_x^2 + S_y^2} = \sqrt{(91.2)^2 + (-58.4)^2} \text{ lb} = 108.3 \text{ lb} \qquad \textit{Answer}$$

and

$$\tan\theta = \frac{58.4 \text{ lb}}{91.2 \text{ lb}} = 0.64$$

$$\theta = 32.6^\circ \qquad \textit{Answer}$$

Example 2.7. Vector Addition and Dot Product. Given the three vectors $\mathbf{A} = 3\mathbf{i} + 2\mathbf{j} - 6\mathbf{k}$, $\mathbf{B} = 2\mathbf{i} + 4\mathbf{j} + 7\mathbf{k}$, and $\mathbf{C} = -2\mathbf{i} + 9\mathbf{j} + 3\mathbf{k}$, find (a) the vector $\mathbf{D} = \mathbf{A} + \mathbf{B} - \mathbf{C}$ and its magnitude and direction angles and (b) the dot product of the vectors **A** and **B** and the angle between them.

Solution. (a)

$$\begin{aligned} D_x &= A_x + B_x - C_x = 3 + 2 - (-2) = 7 \\ D_y &= A_y + B_y - C_y = 2 + 4 - 9 = -3 \\ D_z &= A_z + B_z - C_z = -6 + 7 - 3 = -2 \end{aligned}$$

$$\mathbf{D} = 7\mathbf{i} - 3\mathbf{j} - 2\mathbf{k} \qquad \textit{Answer}$$

From Eq. (2.12), we have

$$D = \sqrt{(7)^2 + (-3)^2 + (-2)^2} = 7.87 \qquad \textit{Answer}$$

The direction angles of **D** are determined from Eq. (2.13):

$$\cos\theta_x = \frac{D_x}{D} = \frac{7}{7.87} = 0.89 \qquad \theta_x = 27.2^\circ$$

$$\cos\theta_y = \frac{D_y}{D} = \frac{-3}{7.87} = -0.38 \qquad \theta_y = 112.4° \qquad \textit{Answer}$$

$$\cos\theta_z = \frac{D_z}{D} = \frac{-2}{7.87} = -0.25 \qquad \theta_z = 104.7°$$

(b) Using Eq. (2.10), we find

$$\mathbf{A}\cdot\mathbf{B} = A_xB_x + A_yB_y + A_zB_z = 3(2) + 2(4) + (-6)(7) = -28 \qquad \textit{Answer}$$

Also,

$$A = \sqrt{(3)^2 + (2)^2 + (-6)^2} = 7$$

$$B = \sqrt{(2)^2 + (4)^2 + (7)^2} = 8.31$$

The angle between the two vectors is given by Eq. (2.11):

$$\cos\theta = \frac{\mathbf{A}\cdot\mathbf{B}}{AB} = \frac{-28}{7(8.31)} = -0.48 \qquad \theta = 118.8° \qquad \textit{Answer}$$

PROBLEMS

2.24 to 2.26 Express the vectors shown in component form.

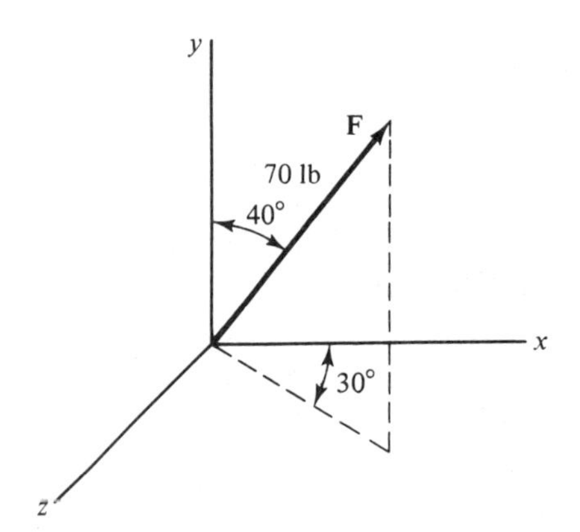

Problem 2.24

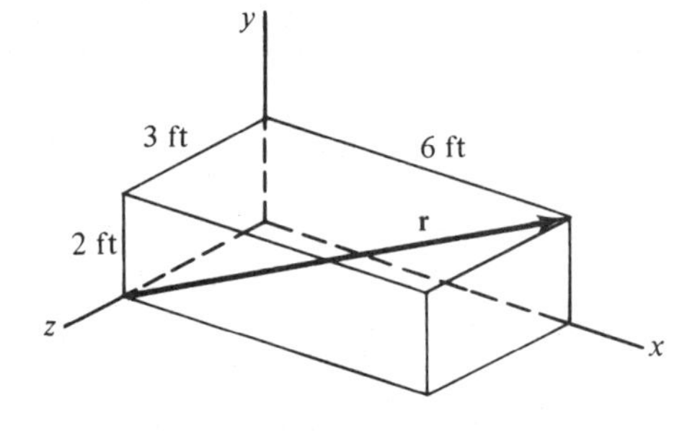

Problem 2.25

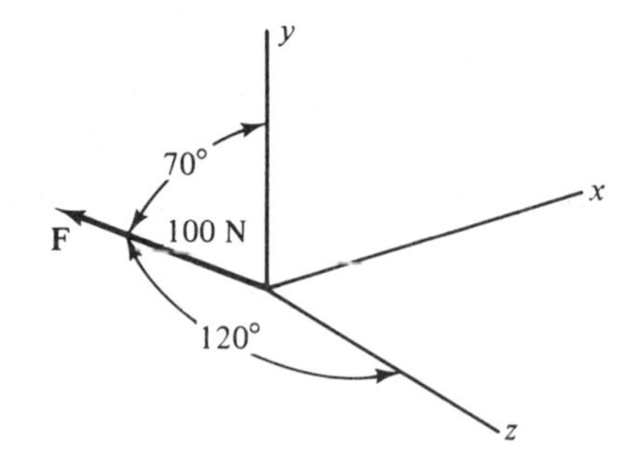

Problem 2.26

2.27 Determine the direction angles of the vector in Problem 2.24.

2.28 Determine the direction angles of the vector in Problem 2.25.

2.29 to 2.31 For the following sets of vectors **A** and **B**, find **A** + **B**, **A** − **B**, 3**A** + 2**B** and **A** · **B** by using orthogonal components. Also determine the angle between **A** and **B**.

2.29 $\mathbf{A} = 3\mathbf{i} + 4\mathbf{j}$ $\quad\mathbf{B} = 2\mathbf{i} - \mathbf{j} + 3\mathbf{k}$
2.30 $\mathbf{A} = 2\mathbf{i} + 3\mathbf{j} + 6\mathbf{k}$ $\quad\mathbf{B} = 12\mathbf{i} + 4\mathbf{j} + 3\mathbf{k}$
2.31 $\mathbf{A} = 6\mathbf{i} - 2\mathbf{j} + 3\mathbf{k}$ $\quad\mathbf{B} = -2\mathbf{i} + 4\mathbf{j} + 16\mathbf{k}$

2.32. Rework Problem 2.7 by using orthogonal components.

2.33 and 2.34 For the vectors shown, find (a) their sum **S**, (b) their dot product, and (c) the angle between them. Also determine the magnitude and direction angles of **S**. Use rectangular components.

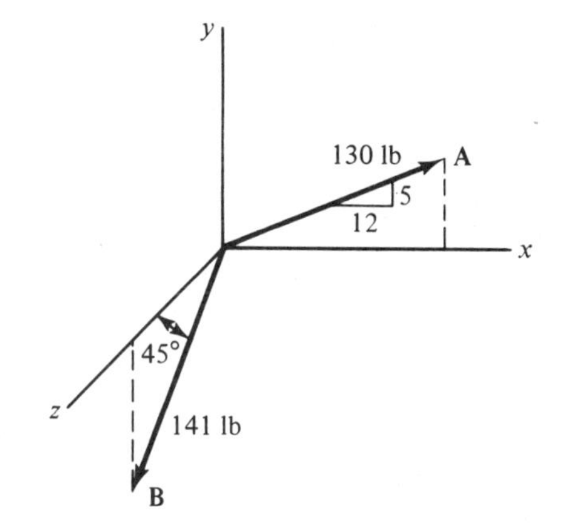

Problem 2.33

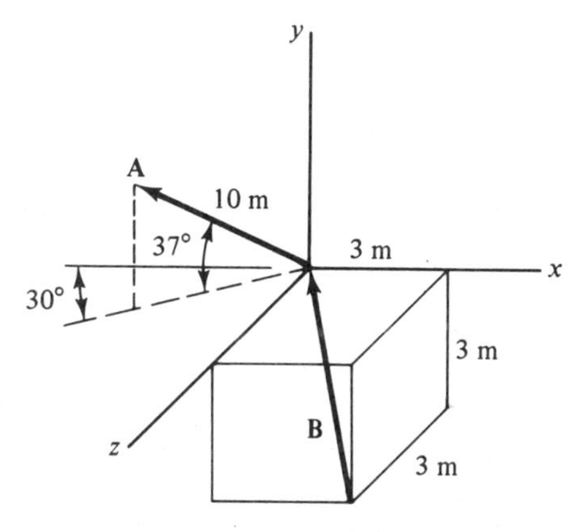

Problem 2.34

2.35 A 20-kg block is held away from a wall by two ropes as shown. It is known that the sum of the forces acting at point *B* is zero. Find the tensions $\mathbf{T}_1$ and $\mathbf{T}_2$ on the ropes by using rectangular components.

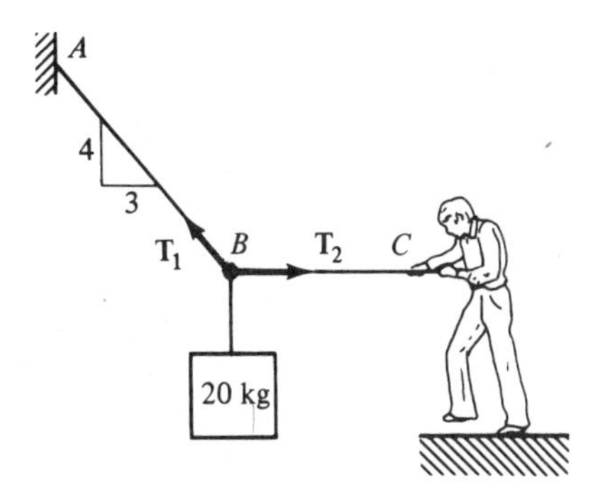

Problem 2.35

2.36 Determine the required tensions $\mathbf{T}_1$, $\mathbf{T}_2$, and $\mathbf{T}_3$ in the support wires for the pole shown if their combined effect is to be a 5-kip force acting downward along the axis of the pole. The wires are equally spaced, and each is inclined at 30° from the vertical.

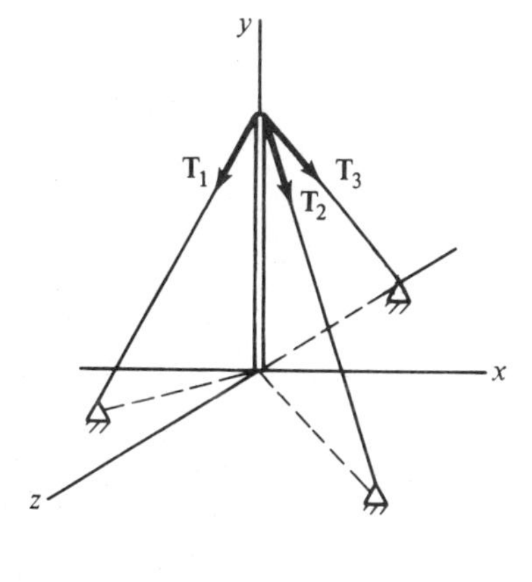

Problem 2.36

2.6 POSITION VECTORS

In this section we shall consider the use of vectors to describe the position of a point in space or the position of one point relative to another. This representation will be useful in a number of applications, and it is used extensively in the study of kinematics (the geometry of motion).

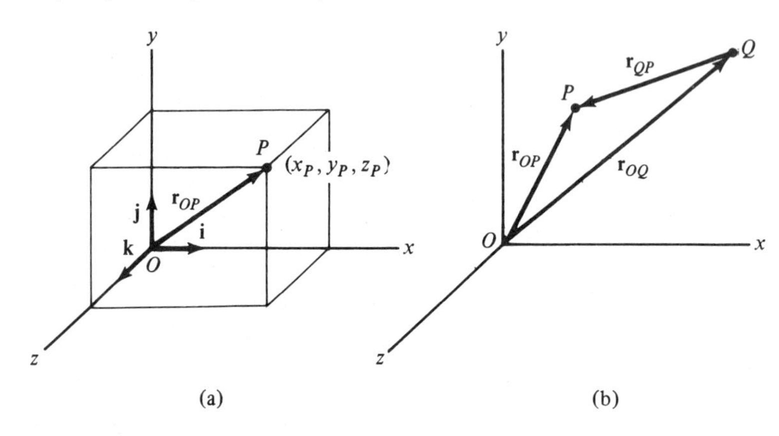

Figure 2.17

The position of a point P relative to the origin O of a given coordinate system can be defined by a *position vector* $\mathbf{r}_{OP}$, which is directed from O to P and whose rectangular projections are equal to the coordinates of the point [Figure 2.17(a)].

$$\mathbf{r}_{OP} = x_P\mathbf{i} + y_P\mathbf{j} + z_P\mathbf{k} \tag{2.18}$$

Similarly, the position of one point relative to another can be defined by a position vector between them. For example, the vector $\mathbf{r}_{QP}$ shown in Figure 2.17(b) goes from point Q to point P and defines the position of P relative to Q. Similarly, the vector $\mathbf{r}_{PQ}$ would go from point P to point Q and would define the position of Q with respect to P. Note that the first and second letters in the subscript on $\mathbf{r}$ denote the points at its tail and tip, respectively.

Referring to Figure 2.17(b), we see that

$$\mathbf{r}_{OQ} + \mathbf{r}_{QP} = \mathbf{r}_{OP}$$

or

$$\mathbf{r}_{QP} = \mathbf{r}_{OP} - \mathbf{r}_{OQ} = (x_P - x_Q)\mathbf{i} + (y_P - y_Q)\mathbf{j} + (z_P - z_Q)\mathbf{k} \tag{2.19}$$

According to this equation, the rectangular projections of a vector between two points in space are equal to the corresponding coordinates of its tip minus those of its tail. In many cases, the projections can also be obtained by inspection by determining how far, and in which direction, one must move parallel to the coordinate axes to get from the tail of the vector to the tip (see Example 2.8).

Position vectors are particularly useful for finding unit vectors along a line that passes through two known points. All that is required is to form the position vector between the points and then divide by its magnitude. Once a unit vector along a line has been established, any other vector **A** along the line can be expressed in component form by using the relationship

$$\mathbf{A} = \pm A\mathbf{e}_A$$

The plus sign is used if **A** and $\mathbf{e}_A$ have the same sense and the minus sign is used if they are of opposite sense.

Example 2.8. Determination of a Position Vector. Determine the position vector $\mathbf{r}_{AB}$ between points A and B on the wire shape shown in Figure 2.18.

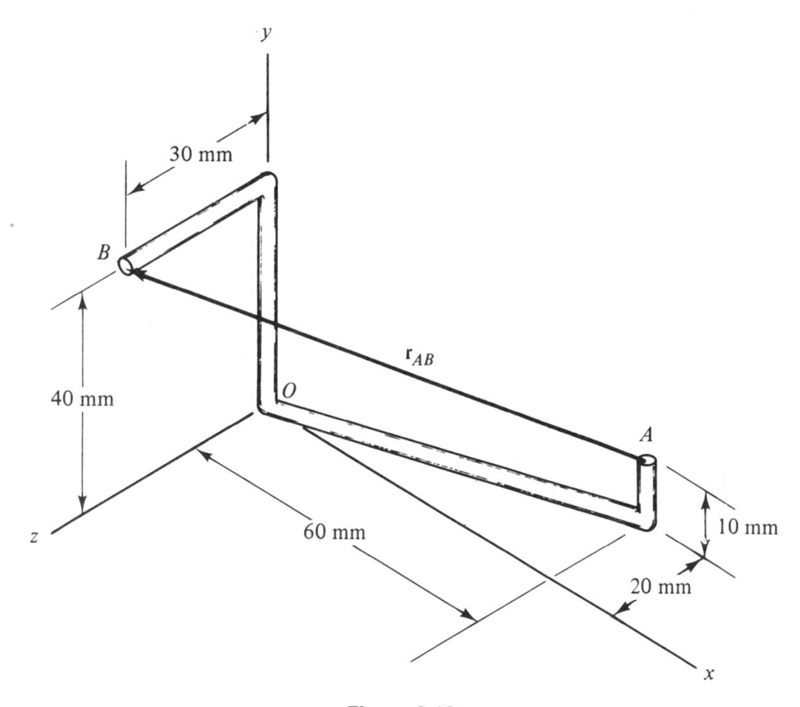

Figure 2.18

Solution. It is helpful to first sketch the vector. The rectangular projections of $\mathbf{r}_{AB}$ are the coordinates of its tip minus those of its tail; so

$$\mathbf{r}_{AB} = (0 - 60)\mathbf{i} + (40 - 10)\mathbf{j} + [30 - (-20)]\mathbf{k} \text{ mm}$$

$$= -60\mathbf{i} + 30\mathbf{j} + 50\mathbf{k} \text{ mm} \qquad \textit{Answer}$$

The position vector can also be obtained by inspection. It is clear from Figure 2.18 that we can get from A to B by going 60 mm in the negative x direction, a total of 30 mm in the positive y direction, and a total of 50 mm in the positive z direction. Thus,

$$\mathbf{r}_{AB} = -60\mathbf{i} + 30\mathbf{j} + 50\mathbf{k} \text{ mm}$$

Example 2.9. Representation of a Vector in Space. The stone slab of Figure 2.19 is partially supported by a cable AB that exerts a 10 000-lb force $\mathbf{T}$ on it as shown. (a) Express $\mathbf{T}$ in component form and (b) determine its component $\mathbf{T}_{OB}$ along diagonal OB of the slab.

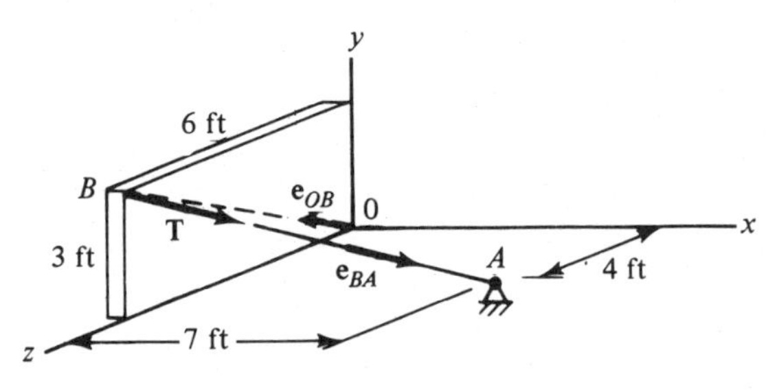

Figure 2.19

Solution. (a) We first determine the unit vector $\mathbf{e}_{BA}$ along the line of action of $\mathbf{T}$. Points B and A have coordinates (0, 3, 6) ft and (7, 0, 4) ft, respectively; so

$$\mathbf{r}_{BA} = 7\mathbf{i} - 3\mathbf{j} - 2\mathbf{k} \text{ ft}$$

$$r_{BA} = \sqrt{(7)^2 + (-3)^2 + (-2)^2} = \sqrt{62} \text{ ft}$$

$$\mathbf{e}_{BA} = \frac{\mathbf{r}_{BA}}{r_{BA}} = \frac{1}{\sqrt{62}}(7\mathbf{i} - 3\mathbf{j} - 2\mathbf{k})$$

Now $\mathbf{T}$ has the same sense as $\mathbf{e}_{BA}$; so

$$\mathbf{T} = T\mathbf{e}_{BA} = \frac{10\ 000 \text{ lb}}{\sqrt{62}}(7\mathbf{i} - 3\mathbf{j} - 2\mathbf{k})$$

$$= 8890\mathbf{i} - 3810\mathbf{j} - 2540\mathbf{k} \text{ lb} \qquad \textit{Answer}$$

(b) The component $\mathbf{T}_{OB}$ of $\mathbf{T}$ along OB is equal to the projection of $\mathbf{T}$ onto the line multiplied by a unit vector along it. Either of the unit vectors $\mathbf{e}_{OB}$ or $\mathbf{e}_{BO}$ can be used. Suppose we use $\mathbf{e}_{OB}$. Referring to Figure 2.19, we have

$$\mathbf{r}_{OB} = 3\mathbf{j} + 6\mathbf{k} \text{ ft}$$

$$r_{OB} = \sqrt{(3)^2 + (6)^2} = \sqrt{45} \text{ ft}$$

$$\mathbf{e}_{OB} = \frac{1}{\sqrt{45}}(3\mathbf{j} + 6\mathbf{k})$$

The projection of $\mathbf{T}$ onto line OB is

$$\mathbf{T} \cdot \mathbf{e}_{OB} = \frac{1}{\sqrt{45}}\left[-3810(3) - 2540(6)\right] = -3980 \text{ lb}$$

The vector component $\mathbf{T}_{OB}$ can now be written as

$$\mathbf{T}_{OB} = (\mathbf{T} \cdot \mathbf{e}_{OB})\mathbf{e}_{OB} = -3980\mathbf{e}_{OB} \text{ lb}$$

or

$$\mathbf{T}_{OB} = -1780\mathbf{j} - 3560\mathbf{k} \text{ lb} \qquad \textit{Answer}$$

PROBLEMS

2.37 and 2.38 For the given sets of points P and Q, determine (a) the position vector of each point relative to the origin, (b) the position vector of point Q relative to point P, and (c) the unit vector directed from P toward Q. Show each vector on a sketch.

2.37 P: (2 m, −1 m, 3 m) Q: (5 m, 1 m, −3 m)
2.38 P: (−2 ft, 1 ft, −1 ft) Q: (4 ft, − 3 ft, 2 ft)

2.39 According to Coulomb's law, two electric point charges exert on each other equal and opposite forces which act along the line joining them. The magnitude of these forces is proportional to the product of the charges and is inversely proportional to the square of the distance between them. The constant of proportionality is $C = 8.987 \times 10^9$ N·m²/coul². Knowing that like charges attract and unlike charges repel, write an expression for the force exerted on a 2-coulomb positive charge located at (4 m, 4 m, −2 m) by a 3-coulomb positive charge located at the origin.

2.40 Same as Problem 2.39, except that the 3-coulomb charge is negative and is located at (0 m, −2 m, 3 m).

2.41 Determine the projection of the vector **F** shown onto line OA and find its vector component along this line.

2.42. Same as Problem 2.41, except line OA is replaced by line OB.

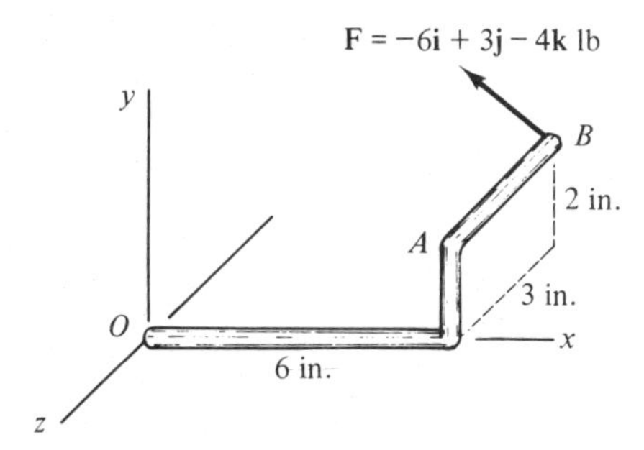

Problems 2.41 and 2.42

2.43 Determine the angle between the support cables AB and AC of the boom shown.

2.44 The tension on support cable AC of the boom shown is known to be 7.5 kN. Express the force **T** exerted on the anchor at C in component form and determine its direction angles.

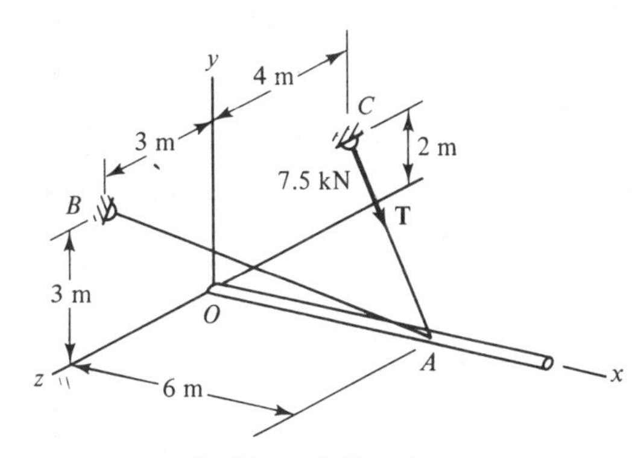

Problems 2.43 and 2.44

2.7 VECTOR PRODUCT OF TWO VECTORS

The *vector product* of two vectors **A** and **B** is a vector denoted by **A × B** and whose magnitude is given by the relation

$$|\mathbf{A} \times \mathbf{B}| = AB \sin \theta \tag{2.20}$$

In this equation, A and B are the respective magnitudes of the vectors and θ in the angle between them (Figure 2.20). The product **A × B**, which is also called the

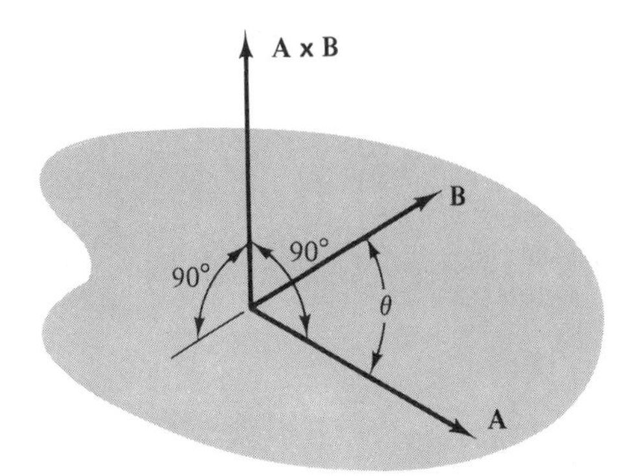

Figure 2.20

cross, or *outer*, *product*, is perpendicular to the plane formed by **A** and **B**. Its sense is defined by the so-called *right-hand screw rule*. According to this rule, if the fingers of the right hand are placed along the first vector, **A**, and then rotated toward the second vector, **B**, the thumb points in the direction of their cross product $\mathbf{A} \times \mathbf{B}$.

The vector product obeys the following laws:

$$
\begin{aligned}
\mathbf{A} \times \mathbf{B} &= -(\mathbf{B} \times \mathbf{A}) \qquad \text{(anticommutative law)} \\
\mathbf{A} \times (\mathbf{B} + \mathbf{C}) &= \mathbf{A} \times \mathbf{B} + \mathbf{A} \times \mathbf{C} \qquad \text{(distributive law)} \\
(h\mathbf{A}) \times \mathbf{B} &= \mathbf{A} \times (h\mathbf{B}) = h(\mathbf{A} \times \mathbf{B});\ h \text{ is any scalar} \\
\mathbf{A} \times \mathbf{A} &= \mathbf{0} \\
\mathbf{A} \times \mathbf{B} &= \mathbf{0} \text{ when } \mathbf{A} \text{ is parallel to } \mathbf{B}
\end{aligned}
\tag{2.21}
$$

The first law follows directly from the right-hand screw rule and indicates that the order of the factors in the vector product cannot be changed without altering its sign. This is the only deviation of the rules of vector algebra from those of ordinary algebra that we shall encounter. All the other laws follow from the definition of the vector product. Details of the proof of the distributive law are left as an exercise in Problem 2.48.

Using the distributive property of the cross product, we have for vectors expressed in component form

$$
\begin{aligned}
\mathbf{A} \times \mathbf{B} &= (A_x\mathbf{i} + A_y\mathbf{j} + A_z\mathbf{k}) \times (B_x\mathbf{i} + B_y\mathbf{j} + B_z\mathbf{k}) \\
&= A_xB_x(\mathbf{i} \times \mathbf{i}) + A_xB_y(\mathbf{i} \times \mathbf{j}) + A_xB_z(\mathbf{i} \times \mathbf{k}) \\
&\quad + A_yB_x(\mathbf{j} \times \mathbf{i}) + A_yB_y(\mathbf{j} \times \mathbf{j}) + A_yB_z(\mathbf{j} \times \mathbf{k}) \\
&\quad + A_zB_x(\mathbf{k} \times \mathbf{i}) + A_zB_y(\mathbf{k} \times \mathbf{j}) + A_zB_z(\mathbf{k} \times \mathbf{k})
\end{aligned}
$$

Since the base vectors are mutually perpendicular and have magnitude one, the cross product of any two of the base vectors is just the positive or negative of the third base vector, with the sense readily determined from the right-hand screw rule. Thus, $\mathbf{i} \times \mathbf{j} = \mathbf{k}$, $\mathbf{k} \times \mathbf{j} = -\mathbf{i}$, $\mathbf{i} \times \mathbf{k} = -\mathbf{j}$, etc. Substituting these results into the preceding expression and noting that the cross product of a vector with itself is zero, we obtain

$$\mathbf{A} \times \mathbf{B} = (A_y B_z - A_z B_y)\mathbf{i} - (A_x B_z - A_z B_x)\mathbf{j} + (A_x B_y - A_y B_x)\mathbf{k} \quad (2.22)$$

Equation (2.22) can also be expressed as a determinant with the base vectors as the elements of the first row and the rectangular projections of the first and second vectors in the cross product as the elements of the second and third row, respectively:

$$\mathbf{A} \times \mathbf{B} = \begin{vmatrix} \mathbf{i} & \mathbf{j} & \mathbf{k} \\ A_x & A_y & A_z \\ B_x & B_y & B_z \end{vmatrix} \quad (2.23)$$

Expansion of this determinant in terms of elements of the first row confirms that Eqs. (2.22) and (2.23) are the same:

$$\mathbf{A} \times \mathbf{B} = \mathbf{i}\begin{vmatrix} A_y & A_z \\ B_y & B_z \end{vmatrix} - \mathbf{j}\begin{vmatrix} A_x & A_z \\ B_x & B_z \end{vmatrix} + \mathbf{k}\begin{vmatrix} A_x & A_y \\ B_x & B_y \end{vmatrix}$$

or

$$\mathbf{A} \times \mathbf{B} = (A_y B_z - A_z B_y)\mathbf{i} - (A_x B_z - A_z B_x)\,\mathbf{j} + (A_x B_y - A_y B_x)\,\mathbf{k}$$

If one or both of the vectors have several zero components, it is usually simpler to compute the cross product term by term by using the distributive property (see Example 2.10).

Example 2.10. Vector Product of Two Vectors. Determine the vector product $\mathbf{A} \times \mathbf{B}$ of the two vectors $\mathbf{A} = 2\mathbf{i} + 4\mathbf{j}$ and $\mathbf{B} = -3\mathbf{i} + \mathbf{k}$.

Solution. Using the determinant form of the vector product, we have

$$\mathbf{A} \times \mathbf{B} = \begin{vmatrix} \mathbf{i} & \mathbf{j} & \mathbf{k} \\ 2 & 4 & 0 \\ -3 & 0 & 1 \end{vmatrix}$$

$$\begin{aligned} \mathbf{A} \times \mathbf{B} &= \mathbf{i}[(4)(1) - (0)(0)] - \mathbf{j}[(2)(1) - (0)(-3)] \\ &\quad + \mathbf{k}[(2)(0) - (4)(-3)] \\ &= 4\mathbf{i} - 2\mathbf{j} + 12\mathbf{k} \qquad \textit{Answer} \end{aligned}$$

The same result is obtained by computing the vector product term by term:

$$\begin{aligned} \mathbf{A} \times \mathbf{B} &= (2\mathbf{i} + 4\mathbf{j}) \times (-3\mathbf{i} + \mathbf{k}) \\ &= -6(\mathbf{i} \times \mathbf{i}) + 2(\mathbf{i} \times \mathbf{k}) - 12(\mathbf{j} \times \mathbf{i}) + 4(\mathbf{j} \times \mathbf{k}) \\ &= 4\mathbf{i} - 2\mathbf{j} + 12\mathbf{k} \end{aligned}$$

PROBLEMS

2.45 and 2.46 For the following sets of vectors **A** and **B**, determine the cross product **A** × **B**.

2.45 $\mathbf{A} = -2\mathbf{i} + 3\mathbf{j}$ $\mathbf{B} = -3\mathbf{i} + 4\mathbf{j} + \mathbf{k}$
2.46 $\mathbf{A} = 2\mathbf{i} + 3\mathbf{j} - \mathbf{k}$ $\mathbf{B} = 5\mathbf{i} - 9\mathbf{j} + 8\mathbf{k}$

2.47 Show that the magnitude of the cross product **A** × **B** can be interpreted geometrically as the area of a parallelogram with **A** and **B** as two of its sides.

2.48 Use the definition of the vector product given in Eq. (2.20) to construct a geometric proof of the distributive law $\mathbf{A} \times (\mathbf{B} + \mathbf{C}) = (\mathbf{A} \times \mathbf{B}) + (\mathbf{A} \times \mathbf{C})$.

2.49 Show that the three vectors $\mathbf{A} = 3\mathbf{i} + 2\mathbf{j} - \mathbf{k}$, $\mathbf{B} = \mathbf{i} + \mathbf{j}$, and $\mathbf{C} = 4\mathbf{i} - 2\mathbf{j} - 6\mathbf{k}$ all lie in the same plane. Write an expression for a unit vector $\mathbf{e}_n$ normal to this plane.

2.50 Same as Problem 2.49, except that $\mathbf{A} = 2\mathbf{i} - \mathbf{j} - 3\mathbf{k}$.

2.51 and 2.52 Determine the cross product of vectors **A** and **B** in Problems 2.33 and 2.34.

2.8 MOMENT OF A FORCE ABOUT A POINT

When a force is applied to an object, the object may tend to rotate as well as translate. For example, the friction force developed between the front wheel of an automobile and the road causes the wheel to rotate instead of simply sliding along the surface. A measure of the tendency of a force to rotate the body upon which it acts is given by the *moment* of the force. As we shall see in Chapter 3, moments play a fundamental role in the conditions for equilibrium of a body. They are also encountered in dynamics problems involving rotational motions. Moments about a point will be considered in this section, and moments about a line, or axis, will be discussed in Section 2.9.

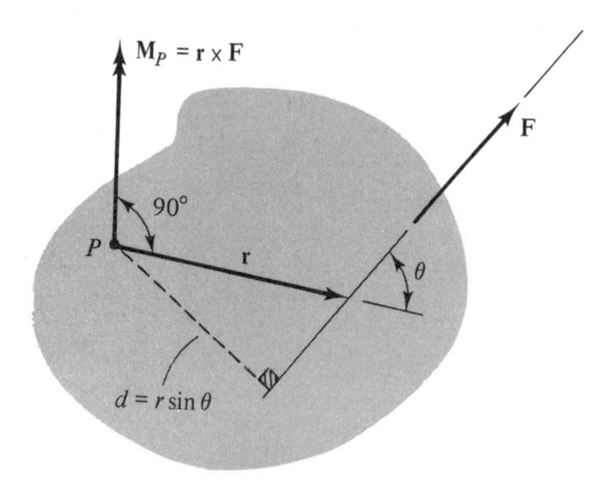

Figure 2.21

The moment of a force **F** about a point *P* is

$$\mathbf{M}_P = \mathbf{r} \times \mathbf{F} \tag{2.24}$$

where **r** is a position vector from *P* to any point on the line of action of **F** (Figure 2.21). Note that the moment is a vector quantity. It is perpendicular to the plane formed by **r** and **F**, and its sense is determined by the right-hand screw rule for vector products. We shall represent moments by two-headed arrows, as in Figure 2.21, to distinguish them from forces.

The magnitude of the moment is given by the definition of the cross product, Eq. (2.20):

$$M_P = |\mathbf{r} \times \mathbf{F}| = rF \sin \theta \tag{2.25}$$

From Figure 2.21, the quantity $r \sin \theta$ is recognized as the perpendicular distance d from the point P to the line of action of the force. Thus, the magnitude of the moment can also be expressed as

$$M_P = Fd \tag{2.26}$$

The distance d is called the *moment arm* of the force. Since $M_P = 0$ when $d = 0$, we conclude that any force whose line of action passes through a point P will have zero moment about that point. Equation (2.26) indicates that the moment has units of force times distance. Common units for moments are lb·ft, lb·in., and N·m (Newton-metres).

If a force F is expressed in terms of its component vectors $\mathbf{F}_1, \mathbf{F}_2, \mathbf{F}_3, \ldots,$ its moment is

$$\mathbf{M}_P = \mathbf{r} \times \mathbf{F} = \mathbf{r} \times (\mathbf{F}_1 + \mathbf{F}_2 + \mathbf{F}_3 + \ldots)$$

or

$$\mathbf{M}_P = (\mathbf{r} \times \mathbf{F}_1) + (\mathbf{r} \times \mathbf{F}_2) + (\mathbf{r} \times \mathbf{F}_3) + \ldots = \mathbf{M}_{P1} + \mathbf{M}_{P2} + \mathbf{M}_{P3} + \ldots \tag{2.27}$$

Thus, the moment of a force is equal to the sum of the moments of each of its components. This result, which follows directly from the distributive property of the vector product, is very useful in applications. It was derived originally (long before vector algebra was invented) by P. Varignon (1654–1722), and is known as *Varignon's theorem.*

In two-dimensional problems the moment vector is perpendicular to the page, which makes it difficult to show on sketches. An alternative is to represent the moment vector by a curved arrow in the plane of the forces, as in Figure 2.22. This arrow can be thought of as indicating the direction of rotation of the fingers of the right hand when applying the right-hand screw rule. The sign convention for moments represented by curved arrows is arbitrary; in this text it will be denoted by a curved arrow with a plus sign over the moment symbol. Thus, $\overset{+}{\curvearrowleft} M$ signifies that counterclockwise moments are considered positive.

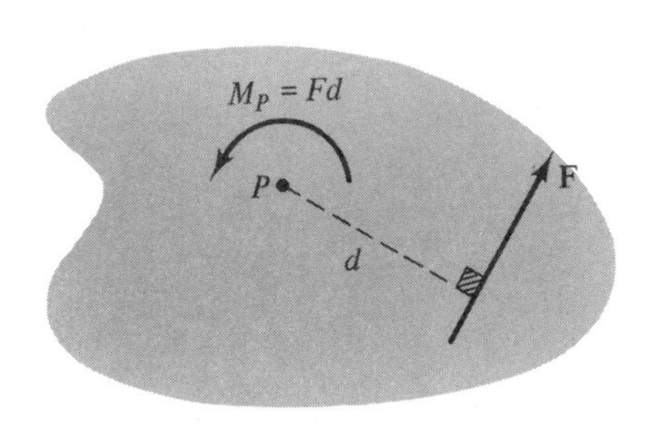

Figure 2.22

Although the moment of a force can always be determined by using cross products, it is often simpler to compute its magnitude from the relation $M_P = Fd$ and determine its orientation and sense by inspection. This is particularly true for two-dimensional problems. The curved arrow representing the moment will lie in the plane of the forces in this case, and the sense of the moment can be easily determined by noting the direction in which the force appears to be moving around the point about which the moment is computed.

Example 2.11. Moment About a Point. The door of Figure 2.23(a) is subjected to a 2-kN force as shown. Determine the moment of the force about point A.

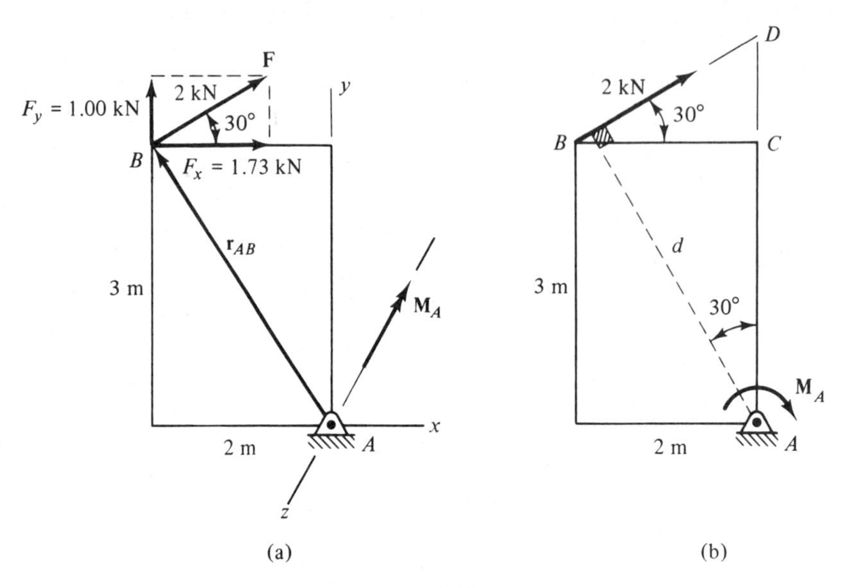

Figure 2.23

Solution. Since the problem is two-dimensional, there are several ways to approach it.

(a) *Formal Vector Approach.* We first set up a coordinate system, express the force in component form, and determine the position vector $\mathbf{r}$; $\mathbf{r}$ can go from A to any point on the line of action of $\mathbf{F}$. Point B is chosen because its coordinates are known. Using the coordinate system shown in Fig. 2.23(a), we have

$$\mathbf{F} = 1.73\mathbf{i} + 1.00\mathbf{j} \text{ kN}$$

$$\mathbf{r}_{AB} = -2\mathbf{i} + 3\mathbf{j} \text{ m}$$

$$\mathbf{M}_A = \mathbf{r}_{AB} \times \mathbf{F} = (-2\mathbf{i} + 3\mathbf{j}) \times (1.73\mathbf{i} + 1.00\mathbf{j}) \text{ kN·m}$$

$$= -2(1.00)\mathbf{k} - 3(1.73)\mathbf{k} = -7.2\mathbf{k} \text{ kN·m} \qquad \textit{Answer}$$

The determinant form of the cross product can also be used. $\mathbf{M}_A$ lies along the negative z axis, as shown.

(b) *Informal Approach.* The magnitude of the moment is $M_A = Fd$, where d is the perpendicular distance from A to the line of action of $\mathbf{F}$ [Eq. (2.26)]. From Figure 2.23(b), we find

$$CD = 2 \tan 30° \text{ m} = 1.15 \text{ m}$$

$$AD = AC + CD = 4.15 \text{ m}$$

$$d = AD \cos 30° = 3.59 \text{ m}$$

so

$$M_A = Fd = (2 \text{ kN})(3.59 \text{ m}) = 7.2 \text{ kN·m} \qquad \textit{Answer}$$

The sense of the moment is determined by inspection. Since the force appears to be moving clockwise about point A, the moment is clockwise [Figure 2.23(b)].

(c) *Informal Approach Using Components*. Instead of working with the perpendicular distance d as in (b), we can make use of the fact that the moment of the force is equal to the sum of the moments of its components (Varignon's theorem). Referring to Figure 2.23(a) and taking counterclockwise moments to be positive, we have

$$\overset{+}{\curvearrowleft} M_A = -(1.73 \text{ kN})(3 \text{ m}) - (1.00 \text{ kN})(2 \text{ m})$$
$$= -7.2 \text{ kN·m} \qquad \textit{Answer}$$

The negative answer means that $\mathbf{M}_A$ is actually clockwise, as in Figure 2.23(b).

All of the procedures used here lead to the same result. The formal approach is a bit longer than the others; the informal approach using components is the simplest. The choice of procedure in problems like this is primarily a matter of convenience and personal preference.

Example 2.12. Moment About a Point. A T-bar is subjected to a force **F** of magnitude 1414 lb, as shown in Figure 2.24. Determine the moment about the origin O.

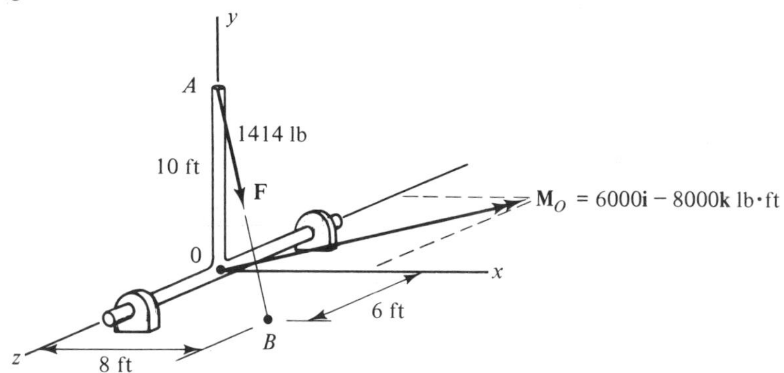

Figure 2.24

Solution. The formal vector approach is preferable because of the complicated geometry. The position vector **r** can be taken from O to either A or B. It is taken to A because this results in the simplest expression. In component form (details not given here), we have

$$\mathbf{F} = 800\mathbf{i} - 1000\mathbf{j} + 600\mathbf{k} \text{ lb}$$
$$\mathbf{r}_{OA} = 10\mathbf{j} \text{ ft}$$

so

$$\mathbf{M}_O = \mathbf{r}_{OA} \times \mathbf{F} = 10\mathbf{j} \times (800\mathbf{i} - 1000\mathbf{j} + 600\mathbf{k}) \text{ lb·ft}$$
$$\mathbf{M}_O = 6000\mathbf{i} - 8000\mathbf{k} \text{ lb·ft} \qquad \textit{Answer}$$

The moment vector is as shown.

PROBLEMS

2.53 to 2.55 For the situations shown, determine the moment of the force about point P.

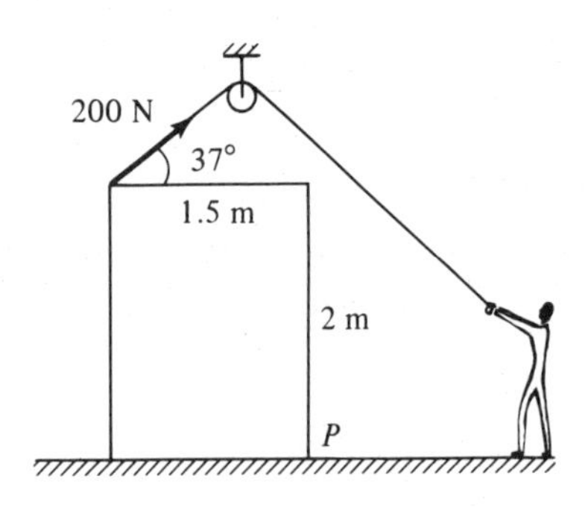

Problem 2.53

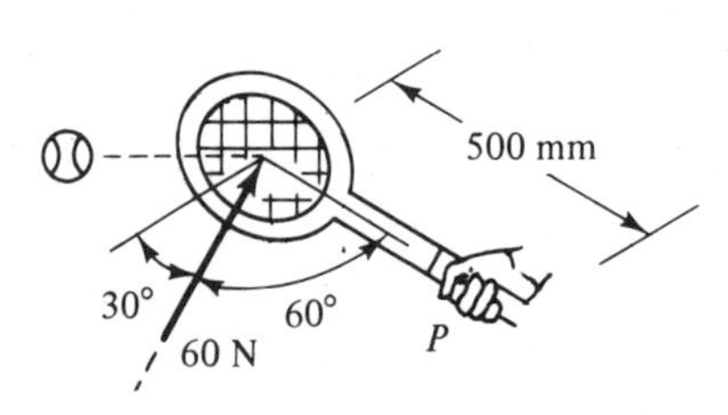

Problem 2.54

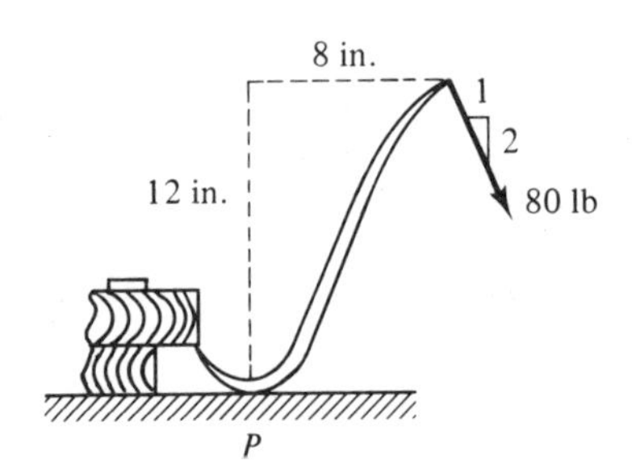

Problem 2.55

2.56 A 200-N force is applied to a bracket as shown. For $\theta = 37°$, determine (a) the moment of the 200-N force about point C, (b) the horizontal force applied at E that would produce the same moment about C, and (c) the smallest force applied at E that would produce the same moment about C.

2.57 Same as Problem 2.56, except that the moments are to be computed about point D.

2.58 For what orientation θ of the 200-N force in Problem 2.56 will the moment of this force about B be maximum? For what orientation will it be minimum? What are these maximum and minimum values of moment?

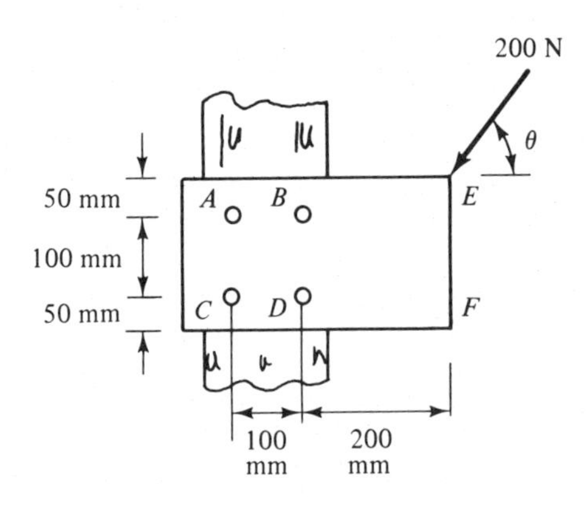

Problem 2.56

2.59 If the tensions in a belt are as shown, determine the net moment that tends to rotate the pulley about its center O.

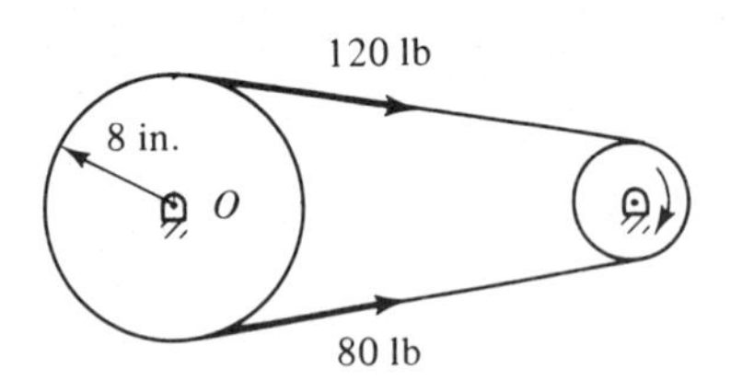

Problem 2.59

2.60 The water exerts a net horizontal force on a dam and the foundation exerts a net vertical force on it as shown. Determine the total moment about point C that tends to overturn the dam.

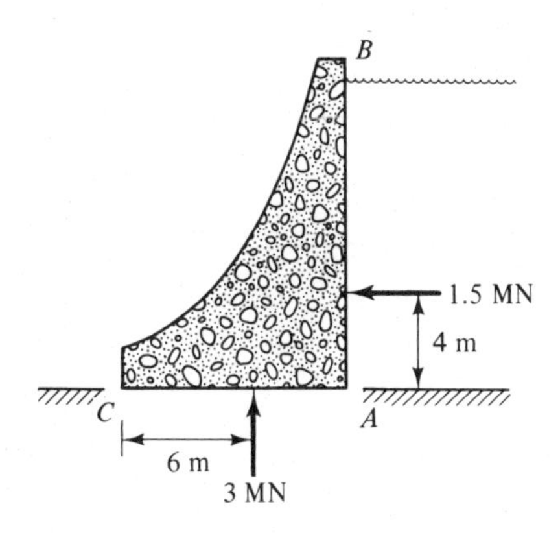

Problem 2.60

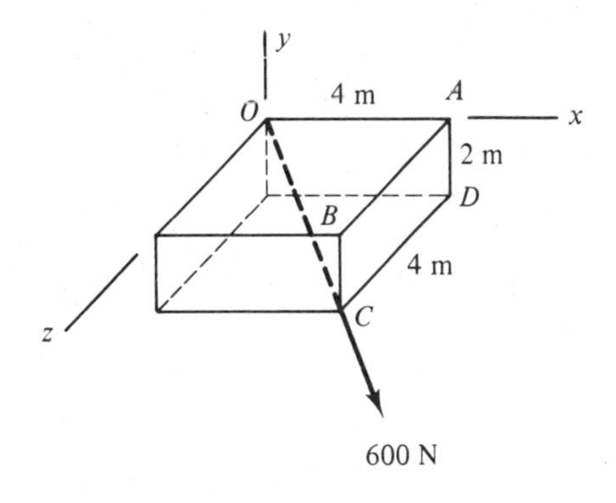

Problem 2.64

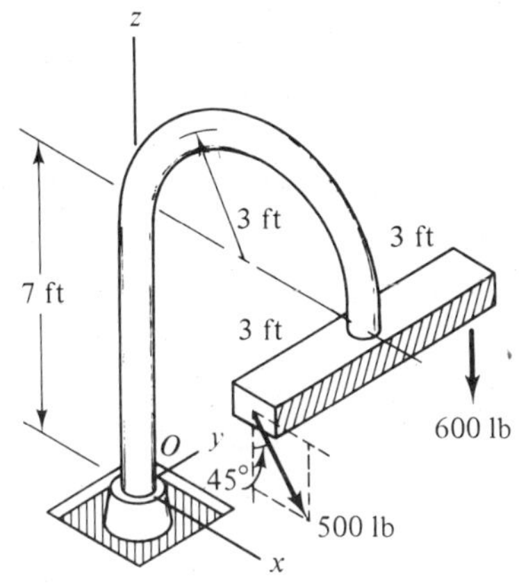

Problem 2.65

2.61 and 2.62 For the following forces, determine (a) the moment of the force about the origin O and (b) the perpendicular distance from O to the line action of the force.

2.61 $\mathbf{F} = -200\mathbf{i} - 300\mathbf{j} + 500\mathbf{k}$ lb acting at (6, 2, 3) ft

2.62 $\mathbf{F} = 4\mathbf{i} + 7\mathbf{j} - 3\mathbf{k}$ kN acting at (1, − 3, − 2) m

2.63 Determine the moment of the tension **T** about the base O of the utility pole shown.

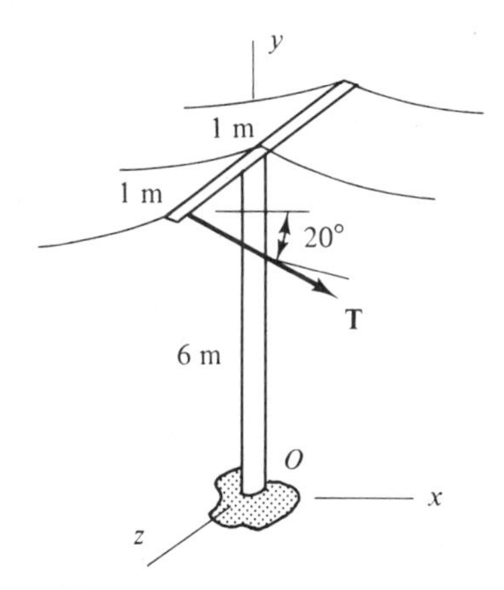

Problem 2.63

2.64 Find the moment of the force acting upon the block shown about points O, A, B, and C.

2.65 Find the total moment of the two forces shown about the base O of the lamppost.

2.66 Find the moment of the 300-lb force shown about the origin O and determine the perpendicular distance from O to the line of action of the force. The force is oriented 45° above the horizontal and 30° from a plane parallel to the $x - y$ plane.

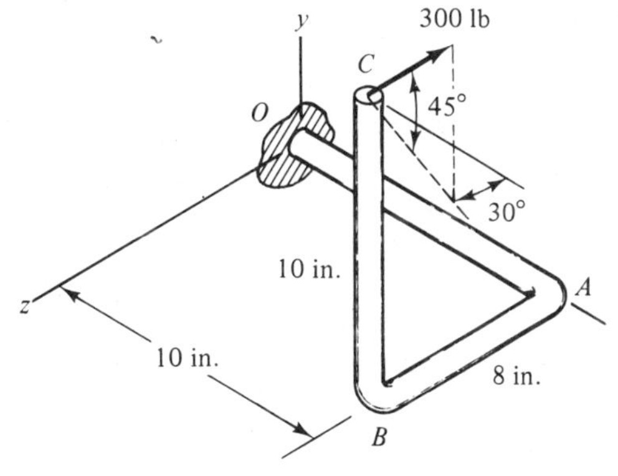

Problem 2.66

2.9 MOMENT OF A FORCE ABOUT A LINE

In many applications the body of interest is constrained to rotate about a particular line or axis. Bodies supported by bearings or hinges fall into this category; typical examples are the door of a house, the crankshaft in an automobile engine, and the bit in a drill. The tendency of a force to produce rotation about a specific line is given by the moment about the line.

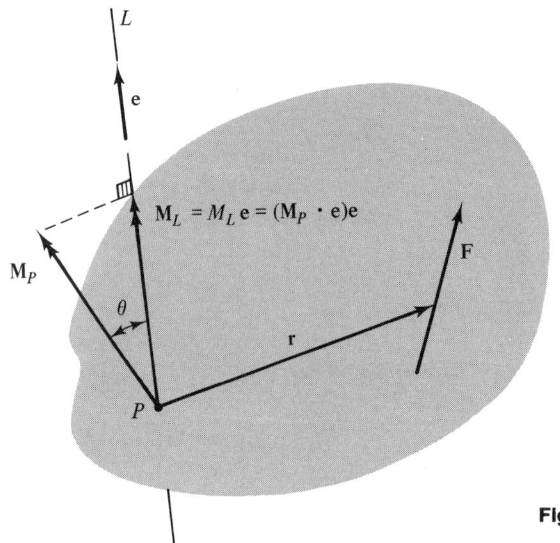

Figure 2.25

Consider a line L, with $\mathbf{M}_P$ the moment about some point P along it (Figure 2.25). The moment about the line, $\mathbf{M}_L$, is simply the component of $\mathbf{M}_P$ along L. If the angle θ between $\mathbf{M}_P$ and L is known, $\mathbf{M}_L$ can be determined directly ($M_L = M_P \cos\theta$). Otherwise, it is convenient to use the scalar product:

$$\mathbf{M}_L = M_L\mathbf{e} = (\mathbf{M}_P \cdot \mathbf{e})\mathbf{e} = \left[(\mathbf{r} \times \mathbf{F}) \cdot \mathbf{e}\right]\mathbf{e} \tag{2.28}$$

where $\mathbf{e}$ is a unit vector along the line and $M_L = \text{Proj}_{\mathbf{e}}\, \mathbf{M}_P$. The projection M_L will be positive or negative, depending upon whether $\mathbf{M}_L$ and $\mathbf{e}$ have the same or opposite sense. In the way of physical interpretation, the direction of the rotation about the line is given by the fingers of the right hand when the thumb is aligned with $\mathbf{M}_L$.

The expression for the projection M_L in Eq. (2.28) is of the form $\mathbf{A} \cdot (\mathbf{B} \times \mathbf{C})$, which is called the *scalar triple product*. This product can be determined in two steps by first performing the cross product and then the dot product, or it can be computed in one step by using the relationship

$$\mathbf{A} \cdot (\mathbf{B} \times \mathbf{C}) = \begin{vmatrix} A_x & A_y & A_z \\ B_x & B_y & B_z \\ C_x & C_y & C_z \end{vmatrix} \tag{2.29}$$

Equation (2.29) can be easily verified by performing the indicated vector operations and comparing the result with the expanded form of the determinant.

From Eq. (2.28), we see that $\mathbf{M}_L$ will be zero if $\mathbf{M}_P = \mathbf{0}$ or is perpendicular to $\mathbf{e}$. Now $\mathbf{M}_P$ is perpendicular to the plane formed by the vectors $\mathbf{r}$ and $\mathbf{F}$; so the latter condition implies that $\mathbf{r}$, $\mathbf{F}$, and the line L all lie in the same plane. Consequently, L and the line of action of the force either intersect or are parallel. The condition $\mathbf{M}_P = \mathbf{0}$ also implies an intersection of these two lines, since $\mathbf{F}$ passes through P in this case. Thus, we conclude that $\mathbf{M}_L = \mathbf{0}$ for any force whose line of action intersects, or is parallel to, the line L. Physically, this implies that only the component of $\mathbf{F}$ perpendicular to the line contributes to $\mathbf{M}_L$.

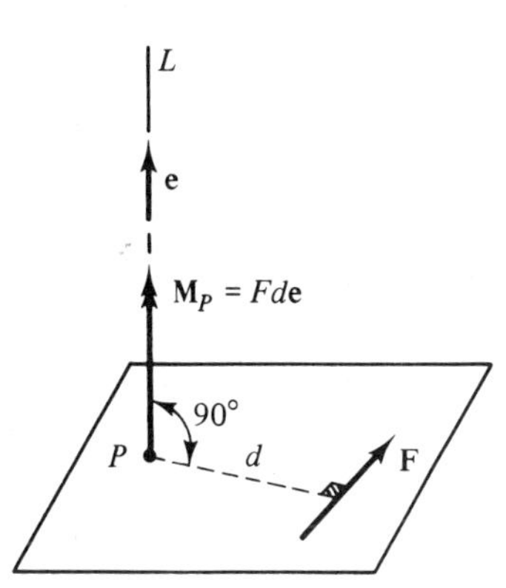

Figure 2.26

If the force is normal to the line and a perpendicular distance d away from it (Figure 2.26), the magnitude of the moment about the line is simply the magnitude of the force times the distance d:

$$|\mathbf{M}_L| = Fd \qquad \mathbf{F} \perp L \tag{2.30}$$

To show this, we choose the point P, which is arbitrary, such that it lies in the same plane as $\mathbf{F}$. Then $\mathbf{M}_P = Fd\mathbf{e}$ and $|\mathbf{M}_L| = \mathbf{M}_P \cdot \mathbf{e} = Fd$, which confirms Eq. (2.30).

Moments about a line can often be used to advantage when determining moments about a point. In this regard, we note that the components of the moment about a point P, $\mathbf{M}_P = M_{Px}\mathbf{i} + M_{Py}\mathbf{j} + M_{Pz}\mathbf{k}$, are the moments about lines through P parallel to the coordinate axes. This is easy to show. The unit vector along a line through P parallel to the x axis is $\mathbf{i}$; so the moment about this line is $(\mathbf{M}_P \cdot \mathbf{i})\mathbf{i} = M_{Px}\mathbf{i}$. Similarly, $M_{PY}\mathbf{j}$ and $M_{Pz}\mathbf{k}$ can be shown to be moments about lines through P parallel to the y and z axes, respectively. Thus, the moment about a point can be determined term by term by finding the moments about lines through the point parallel to the coordinate axes. This procedure is illustrated in Example 2.15.

Example 2.13. Moment About a Line. The stone slab of Figure 2.27 is partially supported by a cable AB that exerts a 10 000-lb force $\mathbf{T}$ on it as shown.

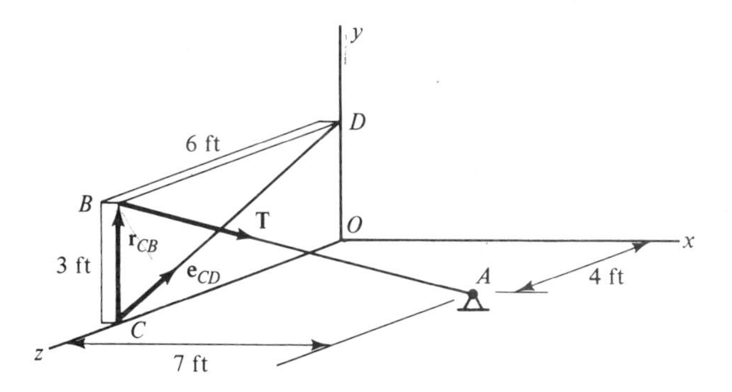

Figure 2.27

Determine the moment about (a) the z axis, (b) the diagonal OB, and (c) the diagonal CD. Note that the dimensions and loading are the same as in Example 2.9.

Solution. (a) We first determine the moment about some point on the z axis. There are two logical choices: point C and the origin. Suppose we pick point C and take the position vector from C to B:

$$\mathbf{r}_{CB} = 3\mathbf{j} \text{ ft}$$

In Example 2.9 we found that

$$\mathbf{T} = 8890\mathbf{i} - 3810\mathbf{j} - 2540\mathbf{k} \text{ lb}$$

so

$$\begin{aligned}\mathbf{M}_C &= \mathbf{r}_{CB} \times \mathbf{T} = 3\mathbf{j} \times (8890\mathbf{i} - 3810\mathbf{j} - 2540\mathbf{k}) \text{ lb·ft} \\ &= -7620\mathbf{i} - 26\ 670\mathbf{k} \text{ lb·ft}\end{aligned}$$

The moment about the z axis is just the z component of M_C:

$$\mathbf{M}_{Oz} = (\mathbf{M}_C \cdot \mathbf{k})\mathbf{k} = -26\ 670\mathbf{k} \text{ lb·ft} \qquad \textit{Answer}$$

(b) The line of action of **T** intersects OB; therefore, the moment about this line is zero.

(c) Because we already know the moment about point C on line CD, it only remains to find a unit vector along this line. Since the sense of the unit vector is arbitrary, we take it from C to D:

$$\mathbf{r}_{CD} = 3\mathbf{j} - 6\mathbf{k} \text{ ft}$$

$$r_{CD} = \sqrt{(3)^2 + (-6)^2} = \sqrt{45} \text{ ft}$$

$$\mathbf{e}_{CD} = \frac{\mathbf{r}_{CD}}{r_{CD}} = \frac{1}{\sqrt{45}}(3\mathbf{j} - 6\mathbf{k})$$

so

$$\mathbf{M}_{CD} = (\mathbf{M}_C \cdot \mathbf{e}_{CD})\mathbf{e}_{CD} = \left[(-7620\mathbf{i} - 26\ 670\mathbf{k}) \cdot \frac{1}{\sqrt{45}}(3\mathbf{j} - 6\mathbf{k})\right]\mathbf{e}_{CD}$$

$$= \frac{6}{\sqrt{45}}(26\ 670)\mathbf{e}_{CD} = 23\ 850\mathbf{e}_{CD} \text{ lb·ft}$$

or

$$\mathbf{M}_{CD} = 10\ 670\mathbf{j} - 21\ 340\mathbf{k} \text{ lb·ft} \qquad \textit{Answer}$$

Alternate Solution: The solution could also have been obtained by using the determinant form of the scalar triple product, Eq. (2.29). For part (c) of the problem, for example,

$$M_{CD} = \mathbf{e}_{CD} \cdot (\mathbf{r}_{CB} \times \mathbf{T}) = \begin{vmatrix} 0 & \frac{3}{\sqrt{45}} & \frac{-6}{\sqrt{45}} \\ 0 & 3 & 0 \\ 8890 & -3810 & -2540 \end{vmatrix}$$

Expanding the determinant, we get

$$M_{CD} = 8890(3)\frac{6}{\sqrt{45}} = 23\ 850 \text{ lb·ft}$$

so

$$\mathbf{M}_{CD} = 23\ 850\mathbf{e}_{CD} \text{ lb·ft}$$

This is the same result obtained previously.

Example 2.14. Moment About a Line. A man exerts a 180-N force on a wrench as shown in Figure 2.28. What is the moment tending to unscrew the bolt?

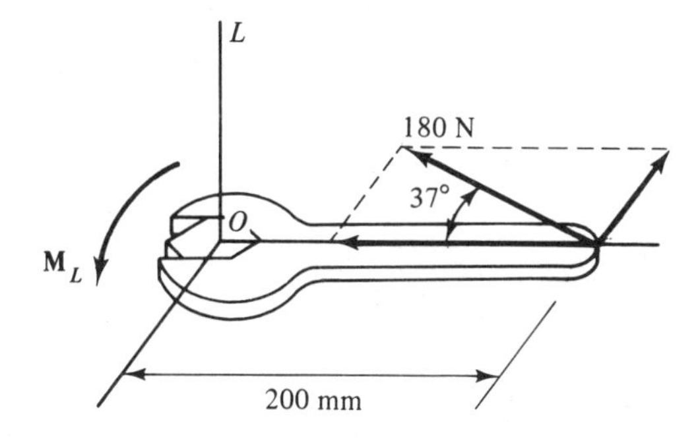

Figure 2.28

Solution. The quantity desired is the moment about the axis L of the bolt. Resolving the force into components, we see that the component along the handle will not contribute to the moment because its line of action intersects the bolt axis. The moment due to the other component is readily determined by inspection. From Eq. (2.30), we have

$$M_L = Fd = (180 \sin 37° \text{ N})(0.2 \text{ m}) = 21.7 \text{ N·m} \qquad \textit{Answer}$$

Since the force appears to be moving counterclockwise about the bolt axis when viewed from above, $\mathbf{M}_L$ can be represented either by a counterclockwise curved arrow in the plane of the wrench, as shown, or by a vector pointing upward along the bolt axis.

Example 2.15. Moment About a Point. The force **F** in Figure 2.29 is parallel to the y axis. Find its moment about the origin.

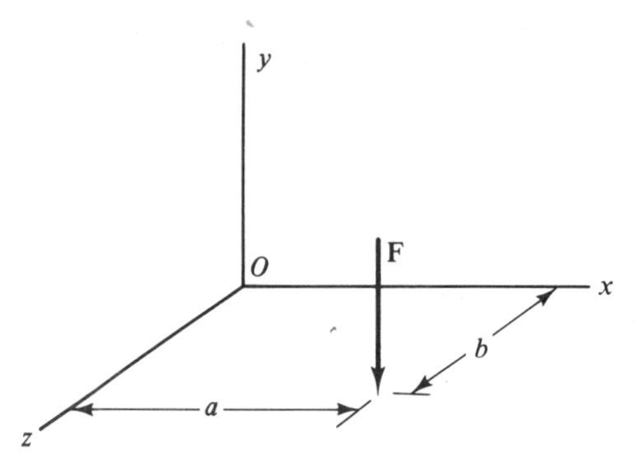

Figure 2.29

Solution. The problem is three-dimensional, but the moment can be determined by inspection by computing the moments about the coordinate axes. Since the force is perpendicular to the x and z axes and is parallel to the y axis, we have

$$|\mathbf{M}_{Ox}| = Fb \qquad |\mathbf{M}_{Oz}| = Fa \qquad |\mathbf{M}_{Oy}| = 0$$

The sense of the moments is obtained by the right-hand screw rule. If we curve the fingers of the right hand in the direction of rotation about the x axis, we see that our thumb points in the positive x direction. Thus,

$$\mathbf{M}_{Ox} = Fb\mathbf{i}$$

Similarly,

$$\mathbf{M}_{Oz} = -Fa\mathbf{k}$$

The moments about the coordinate axes are the vector components of the moment about the origin; so

$$\mathbf{M}_O = Fb\mathbf{i} - Fa\mathbf{k}$$

It is left as an exercise to show that the same result is obtained by using the cross product.

PROBLEMS

2.67 and 2.68 For the following sets of vectors, determine the scalar triple product $\mathbf{A} \cdot (\mathbf{B} \times \mathbf{C})$.

2.67 $\mathbf{A} = 10\mathbf{i} - 8\mathbf{j} + 14\mathbf{k}$, $\mathbf{B} = 3\mathbf{k}$, $\mathbf{C} = 2\mathbf{i} + 5\mathbf{j} - 8\mathbf{k}$
2.68 $\mathbf{A} = 2\mathbf{i} + 5\mathbf{j}$, $\mathbf{B} = -\mathbf{i} - 2\mathbf{j} - 3\mathbf{k}$, $\mathbf{C} = 3\mathbf{i} - 2\mathbf{j} + 8\mathbf{k}$

2.69 Determine the moment tending to unscrew the pipe shown from the fitting at O.

2.70 A 30-lb force is applied to a door as shown. What is the moment tending to

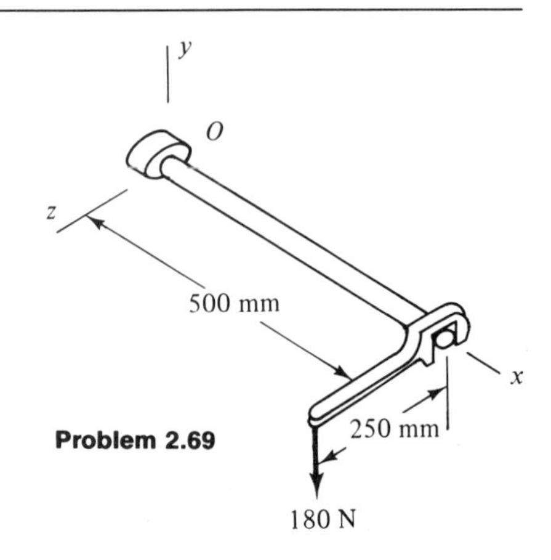

Problem 2.69

rotate the door about its hinges? The force is oriented 45° above the horizontal and lies in a plane perpendicular to the door.

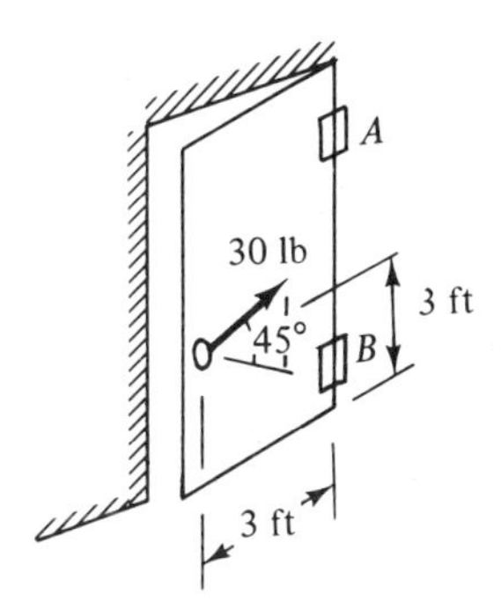

Problem 2.70

2.71 In Problem 2.63 (Section 2.8), determine the moment tending to twist the pole about its longitudinal axis.

2.72 In Problem 2.64 (Section 2.8), determine the moment of the force about (a) the x axis, (b) side AB of the block, and (c) diagonal BD.

2.73 A force $\mathbf{F} = 4\mathbf{i} - 4\mathbf{j} + 2\mathbf{k}$ kN acts at point A with coordinates $(4, -3, 0)$ m. Determine the moment of the force about a line passing through points B and C with coordinates $(0, 0, 1)$ m and $(2, 2, 0)$ m, respectively.

2.74 The tension **T** on the thread of a sewing machine is directed as shown and has magnitude 10 N. Determine the moment tending to rotate the 25-mm diameter spool about the spindle.

2.75 For the force shown, determine the moment about the origin by considering moments about the coordinate axes.

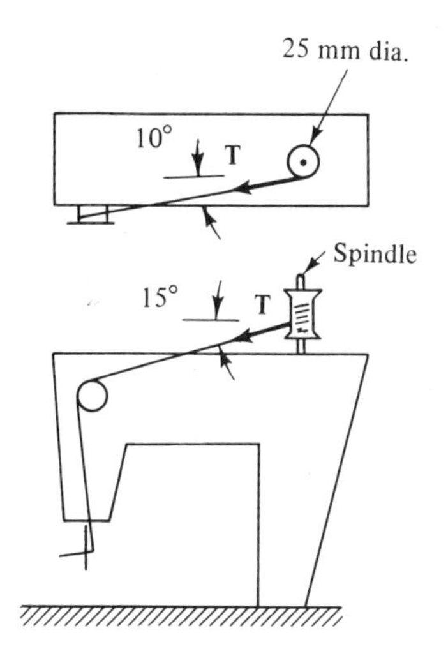

Problem 2.74

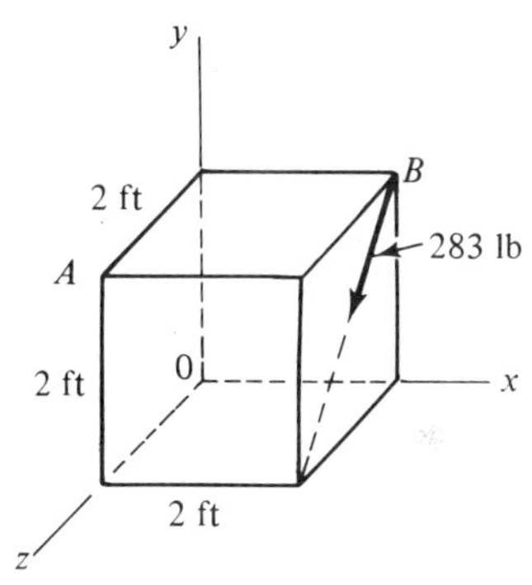

Problems 2.75 and 2.76

2.76 Compute the moment of the force shown about point A by considering moments about lines through A parallel to the coordinate axes. Check the result by using the vector product.

2.77 and 2.78 Rework Problems 2.63 and 2.64 by considering moments about the coordinate axes or lines parallel to them.

2.10 COUPLES

Figure 2.30 illustrates a situation that is not uncommon in physical problems. As shown in the figure, the fluid jets of the lawn sprinkler produce equal and opposite forces acting some distance apart. The sum of these forces is zero, but the sum of

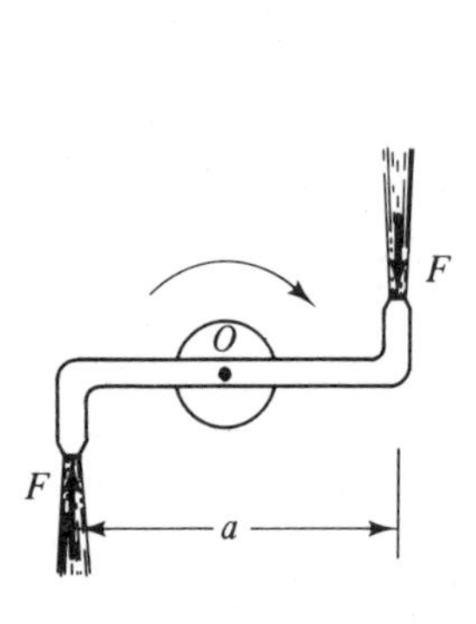

Figure 2.30

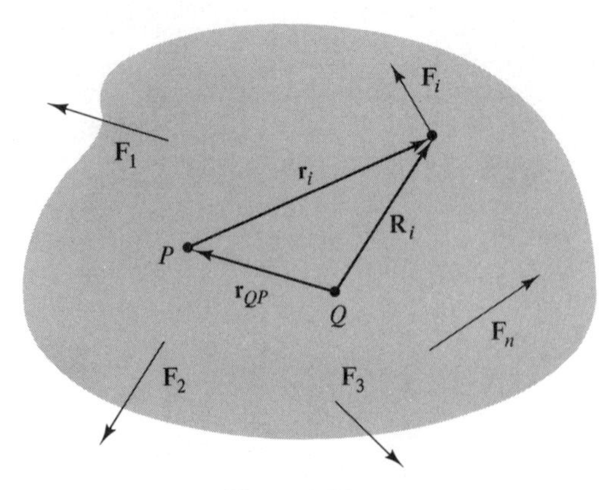

Figure 2.31

their moments about point O is nonzero. Thus, the sole physical effect of the forces is to tend to rotate the arm of the sprinkler. This type of force system, with zero total force and nonzero toal moment, is called a *couple*. Couples are extremely important in the study of mechanics because they represent a pure rotational effect.

A couple, which we shall denote by **C**, can consist of any number of forces and is represented by the total moment of the force system of which it is comprised. However, unlike the moment of a force, the moment of a couple is the same about every point. To prove this, we consider the total moment of a general force system, such as shown in Figure 2.31, about two different points P and Q. Referring to the figure, we have

$$\Sigma \mathbf{M}_P = \Sigma(\mathbf{r}_i \times \mathbf{F}_i)$$
$$\Sigma \mathbf{M}_Q = \Sigma(\mathbf{R}_i \times \mathbf{F}_i)$$

Now

$$\mathbf{R}_i = \mathbf{r}_i + \mathbf{r}_{QP}$$

so

$$\Sigma \mathbf{M}_Q = \Sigma\left[(\mathbf{r}_i + \mathbf{r}_{QP}) \times \mathbf{F}_i\right] = \Sigma \mathbf{M}_P + \mathbf{r}_{QP} \times (\Sigma \mathbf{F}_i) \tag{2.31}$$

But $\Sigma \mathbf{F}_i = \mathbf{0}$ for a couple, in which case $\Sigma \mathbf{M}_Q = \Sigma \mathbf{M}_P$. Since points P and Q are arbitrary, it follows that the total moment of a couple is the same about every point.

In the case of a simple couple consisting of two equal and opposite forces spaced a distance apart (Figure 2.32), we have for the couple vector

$$\mathbf{C} = \Sigma \mathbf{M}_P = \mathbf{r}_1 \times \mathbf{F} + \mathbf{r}_2 \times (-\mathbf{F}) = (\mathbf{r}_1 - \mathbf{r}_2) \times \mathbf{F} \tag{2.32}$$

where $\mathbf{r}_1$ and $\mathbf{r}_2$ are the position vectors from P to the lines of action of $\mathbf{F}$ and $-\mathbf{F}$, respectively. The magnitude of the couple is given by the relation

$$C = |\mathbf{r}_1 - \mathbf{r}_2| F \sin\theta = Fd \tag{2.33}$$

where d is the perpendicular distance between the two forces. Since the vector $(\mathbf{r}_1 - \mathbf{r}_2)$ lies in the plane of the forces, the couple is perpendicular to this plane. Its sense can be determined by inspection or by the right-hand screw rule.

Since the moment of a couple is the same about every point, the couple vector $\mathbf{C}$ has no definite point of application. It can be located anywhere on the body upon which it acts. In Figure 2.32, for example, it was arbitrarily placed at point P. For two-dimensional problems, couples can also be represented by curved arrows in the plane of the forces or by two equal and opposite parallel forces spaced an appropriate distance apart (Figure 2.33). Finally, we note that a couple, being a moment vector, can be combined directly with the moment of a force.

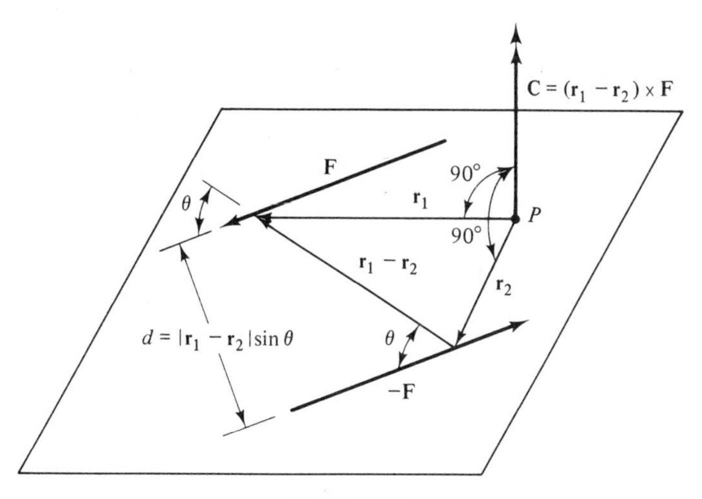

Figure 2.32

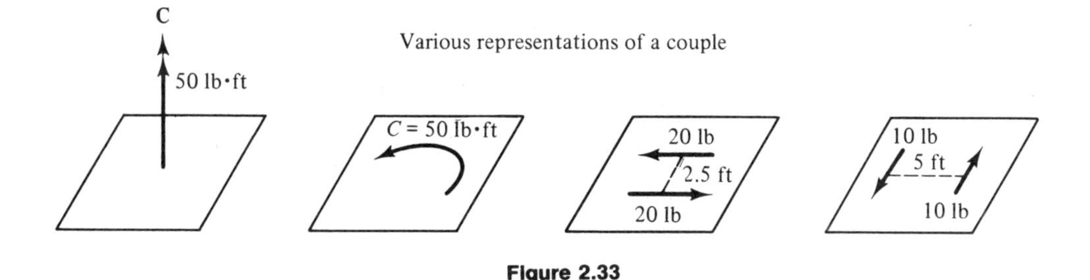

Figure 2.33

Example 2.16. Moment of a Couple. If it takes a 40-N·m couple to turn the steering wheel of a bus, what forces acting as shown in Figure 2.34 must be applied to the wheel?

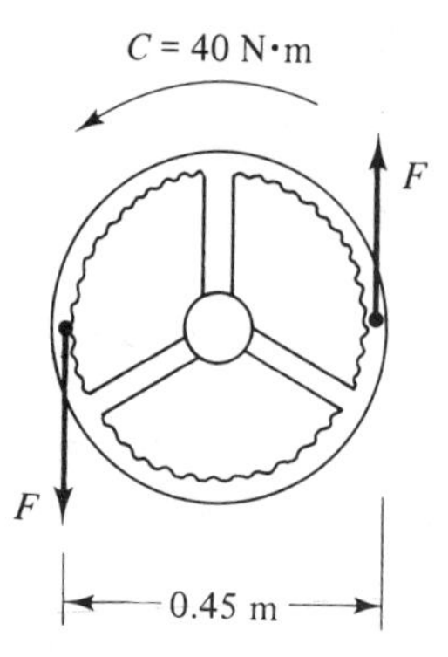

Figure 2.34

Solution. The moment of the couple is equal to the magnitude of the forces times the perpendicular distance between them. Thus,

$$F = \frac{C}{d} = \frac{40 \text{ N·m}}{0.45 \text{ m}} = 88.9 \text{ N} \qquad \textbf{\textit{Answer}}$$

Example 2.17. Addition of Couples. Express the couples acting upon the block in Figure 2.35(a) in vector form and determine their sum.

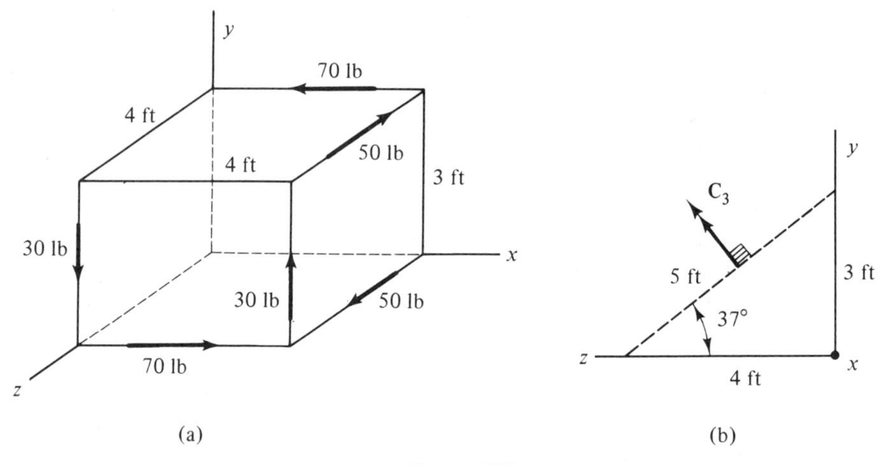

Figure 2.35

Solution. Each pair of forces forms a couple perpendicular to the plane in which the forces lie and with magnitude Fd [Eq. (2.33)].

The 50-lb forces form a couple of magnitude

$$C_1 = Fd = 50 \text{ lb } (3 \text{ ft}) = 150 \text{ lb.ft}$$

parallel to the x axis and pointing in the negative x direction. Thus,

$$\mathbf{C}_1 = -150\mathbf{i} \text{ lb.ft} \qquad \textbf{\textit{Answer}}$$

Similarly, the couple formed by the 30-lb forces is

$$\mathbf{C}_2 = 120\mathbf{k} \text{ lb·ft} \qquad \textit{Answer}$$

The couple formed by the 70-lb forces is directed as shown in Figure 2.35(b) and has magnitude

$$C_3 = Fd = 70 \text{ lb } (5 \text{ ft}) = 350 \text{ lb·ft}$$

Resolving $\mathbf{C}_3$ into components, we have

$$\mathbf{C}_3 = 350 \cos 37°\mathbf{j} + 350 \sin 37°\mathbf{k} \text{ lb·ft}$$
$$= 280\mathbf{j} + 210\mathbf{k} \text{ lb·ft} \qquad \textit{Answer}$$

The sum of the couples is

$$\mathbf{C} = \mathbf{C}_1 + \mathbf{C}_2 + \mathbf{C}_3 = -150\mathbf{i} + 280\mathbf{j} + 330\mathbf{k} \text{ lb·ft} \qquad \textit{Answer}$$

and can be shown as acting anyplace upon the block.

PROBLEMS

2.79 to 2.81 Determine whether or not the force systems shown form a couple. If they do, determine the couple and show it on a sketch. If they don't, what additional force acting at point D is necessary for the system to be a couple, and what is this couple?

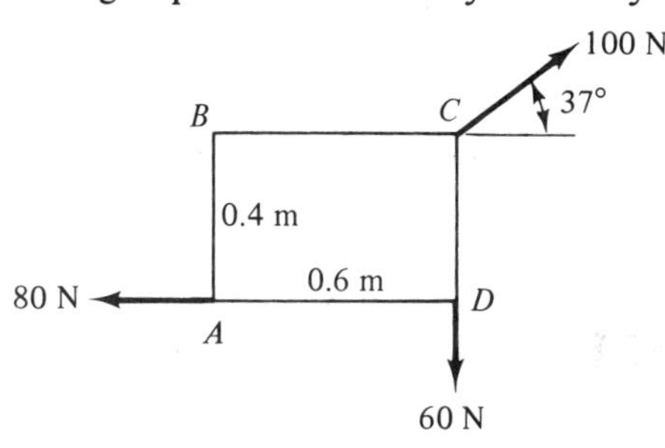

Problem 2.79

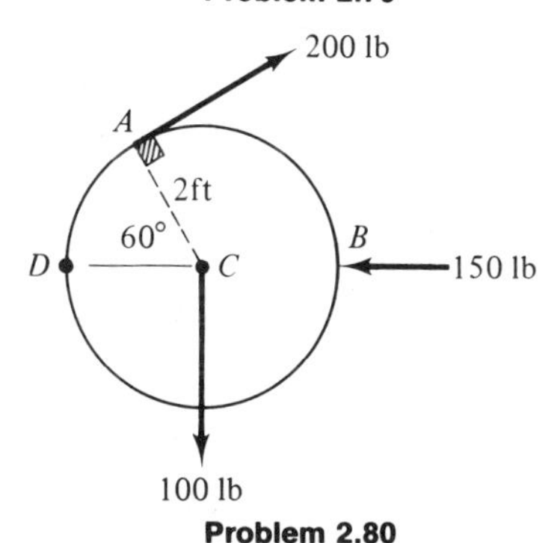

Problem 2.80

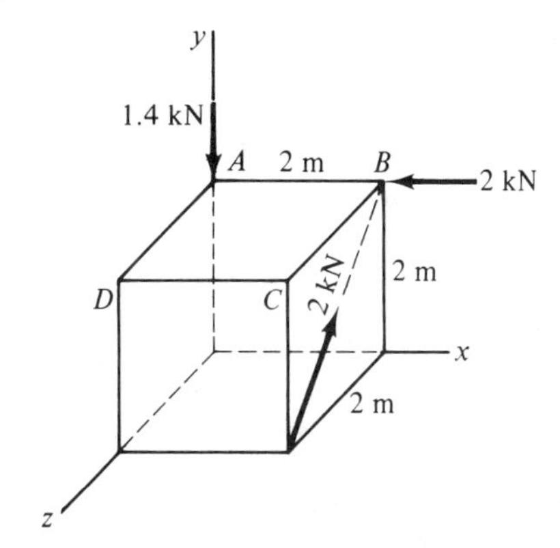

Problem 2.81

2.82 Equal and opposite tangential forces with a magnitude of 4 lb are exerted on the sides of the light bulb shown when it is turned. If the forces lie in a horizontal plane, what is the couple acting upon the bulb?

2.83 It takes a 50-lb·ft couple to close a certain pipeline valve. Determine the forces acting upon the handle as shown necessary to generate this couple.

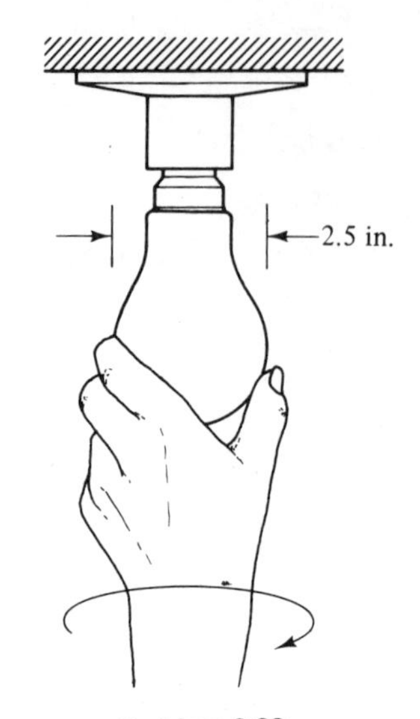

Problem 2.82

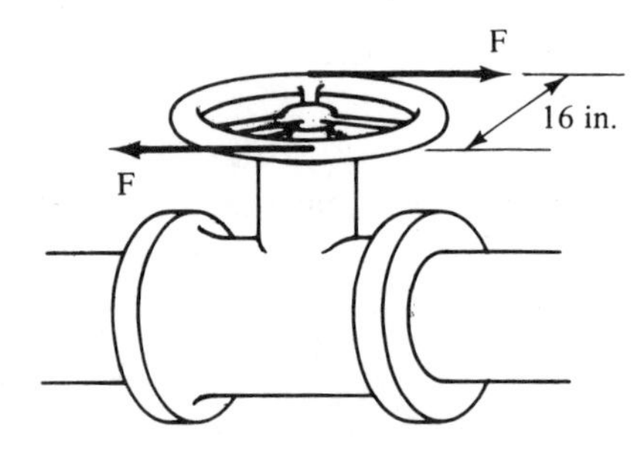

Problem 2.83

2.84 The three forces in Problem 2.79 are to be replaced by a pair of forces that produces the same couple. What are the smallest forces that can be used and where on the rectangular plate should they be applied? Show the results on a sketch.

2.85 If $C = 1500$ lb·in., determine the total couple acting upon the triangular plate shown.

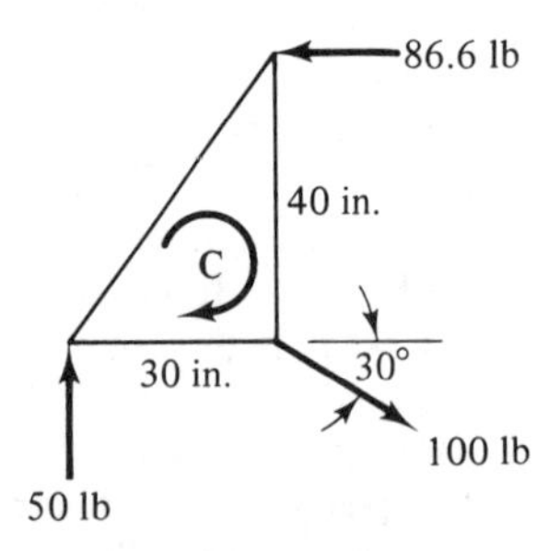

Problems 2.85 and 2.86

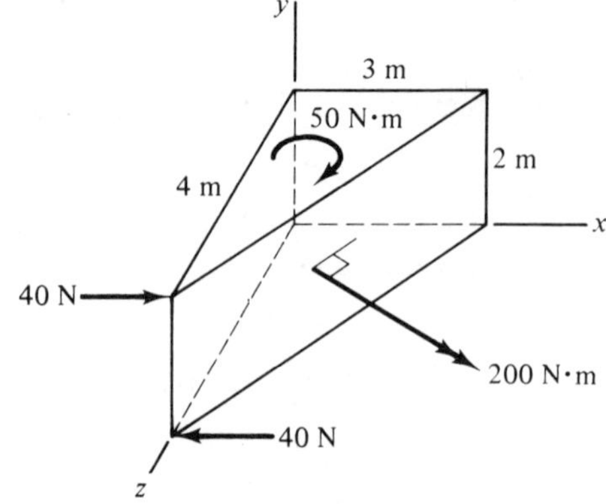

Problem 2.87

2.86 Determine the couple **C** for which the total couple acting upon the triangular plate shown is zero.

2.87 Find the sum of the couples shown.

2.88 A triangular slab is subjected to two couples and a force as shown. Determine the total moment about points *A*, *B*, and *C*.

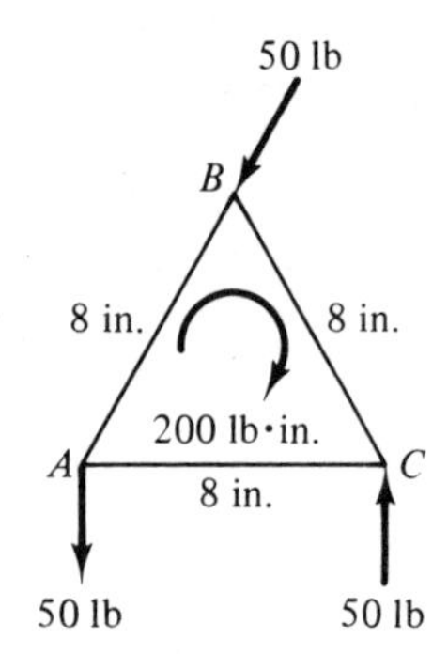

Problem 2.88

2.89 Determine the total moment due to the force and couple about points *O*, *A*, and *B* on the member shown.

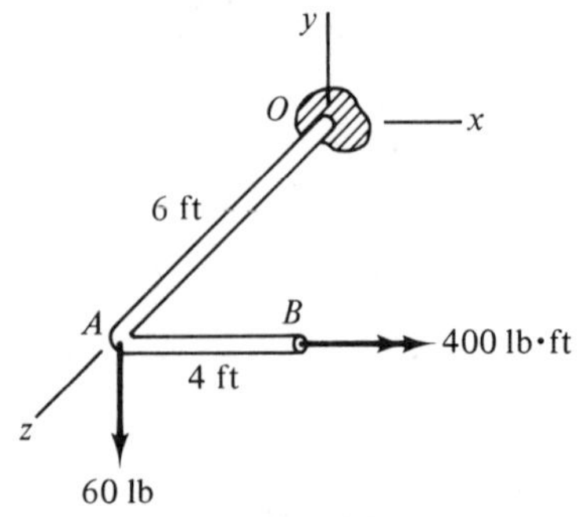

Problem 2.89

2.11 CLOSURE

In this chapter we have considered various vector operations and their application in determining certain quantities of interest. The formal vector approach to the solution of problems is very elegant and powerful, but as illustrated in a number of examples, it is often simpler to proceed from the basic definitions and simple geometric concepts. Hence, it is important that one learn to choose mathematical tools to fit the problem. In our work to follow we shall use the formal vector approach whenever it is warranted, but we shall not hesitate to drop it when it isn't.

As should be evident by now, vector algebra is primarily a convenient way of handling the geometry of a problem. The more complicated the geometry, the more useful vector algebra becomes. Also, the use of vectors and vector algebra makes it possible to state basic concepts, definitions, and equations in a general, but compact, form.

chapter three EQUILIBRIUM

3.1 INTRODUCTION

A body which is initially at rest and which remains at rest when acted upon by forces and couples is said to be in a state of *equilibrium*. In order for such a state to exist, the forces and couples acting must satisfy certain conditions. When these conditions are expressed in mathematical form, they can be used to determine information about these forces and couples which may not be known beforehand. The determination of such information, which we shall refer to as a *force analysis*, is a prerequisite for determining the response of a body to applied forces, or in designing it to withstand some prescribed loading.

Our goal in this chapter is to develop the equilibrium conditions and to learn how to use them in a force analysis. We begin with a discussion of Newton's laws of motion, upon which the equilibrium conditions are based.

3.2 NEWTON'S LAWS OF MOTION

Newton's third law. Newton's third law concerns the nature of force interactions. It states that if one body exerts a force on another, the second body exerts a force on the first which is equal in magnitude, opposite in sense, and which has the same line of action. Stated another way, we may say that for every force effect there is an equal and opposite countereffect.

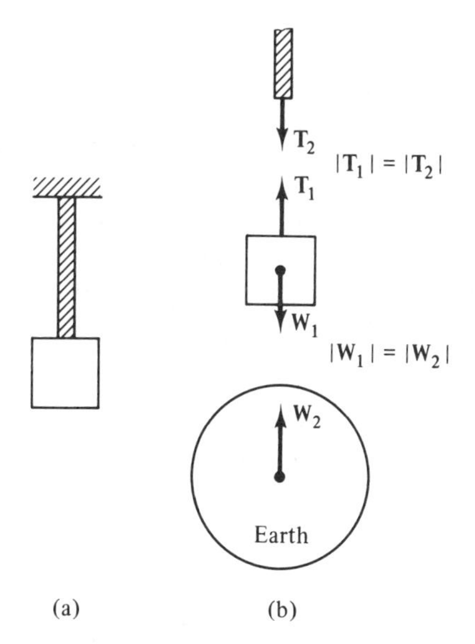

Figure 3.1

To illustrate the third law, let us consider the simple example of a block suspended from a cord [Figure 3.1(a)]. The cord prevents the block from falling down and, therefore, exerts an upward force on it, represented by $\mathbf{T}_1$ in Figure 3.1(b). The countereffect to this force is a downward force $\mathbf{T}_2$ of equal magnitude exerted on the cord by the block. There is also a gravitational interaction between the block and the earth. The force exerted on the block (its weight) is denoted by $\mathbf{W}_1$ in Figure 3.1(b). The countereffect to $\mathbf{W}_1$ is an upward force $\mathbf{W}_2$ of equal magnitude exerted on the earth by the block.

Newton's third law often seems confusing at first exposure, but it is actually very simple to apply. As indicated in Chapter 1 (Section 1.2), forces originate from an interaction between two bodies. Thus, to find the countereffect to any force, it is only necessary to determine the other body involved in that particular force interaction. In other words, we simply consider the physical origin of the force.

Most force interactions obey Newton's third law, but there are some that do not. For instance, the forces between two moving charges and between a moving charge and a magnetic field are equal and opposite, but they are not collinear. Hence, they violate the third law, at least in the form in which we have stated it.

Newton's first law. Newton's first law states that if a body is at rest or is moving in a straight line with constant speed, the sum of the forces acting upon the body must be zero.

As an illustration of the first law, let us reconsider the block shown in Figure 3.1(b). Since the block is at rest, the sum of the forces acting upon it must be zero. Thus, $T_1 - W_1 = 0$, or $T_1 = W_1$. In other words, the cord pulls up on the block with a force which is equal in magnitude to the force with which the earth pulls down on it. It is important to note that $\mathbf{T}_1$ is not the countereffect of $\mathbf{W}_1$ referred to in Newton's third law, even though the two forces are equal and opposite. As mentioned previously, the countereffect to $\mathbf{T}_1$ is the force $\mathbf{T}_2$ shown in Figure 3.1(b), and the countereffect to $\mathbf{W}_1$ is the force $\mathbf{W}_2$.

Newton's first law is a special case of his second law, which states that the acceleration of a body is proportional to the sum of the forces acting upon it. A body at rest or moving in a straight line with constant speed has no acceleration; thus, according to the second law, the sum of the forces acting upon the body must be zero. This is the same result stated in the first law. Since we shall not be dealing with accelerating bodies, we shall have no direct need for Newton's second law.

PROBLEMS

3.1. A package is sitting on a table as shown. Identify the forces acting upon the package and the countereffect to each. If the package and contents weigh 10 lb, what is the magnitude of the force exerted on the package by the table?

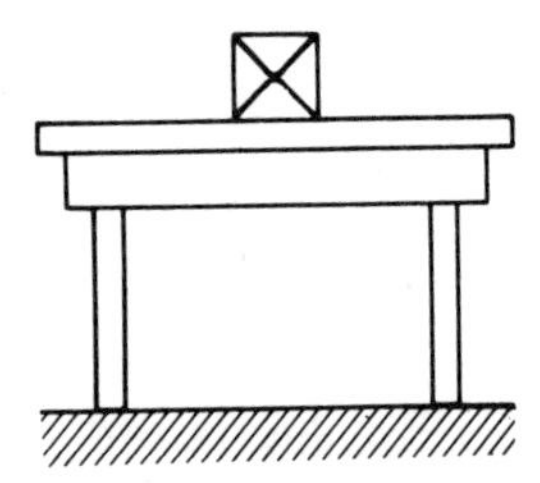

Problem 3.1

3.2. Identify the forces acting upon the hanging plant shown and the countereffect to each. If the plant and pot have a combined mass of 2 kg, what is the magnitude of the tension on the supporting cord?

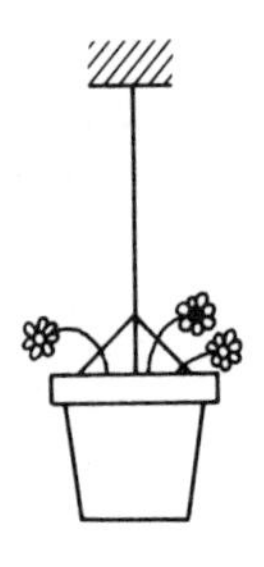

Problem 3.2

3.3. In Problem 3.1, identify the forces acting on the table and the countereffect to each. Assume the package weighs 10 lb and the table weighs 50 lb.

3.4. The pole shown is used as a lever to move a large rock. Identify the forces acting upon the pole and upon the rock. What are the countereffects to these forces? Assume the weight of the pole is negligible.

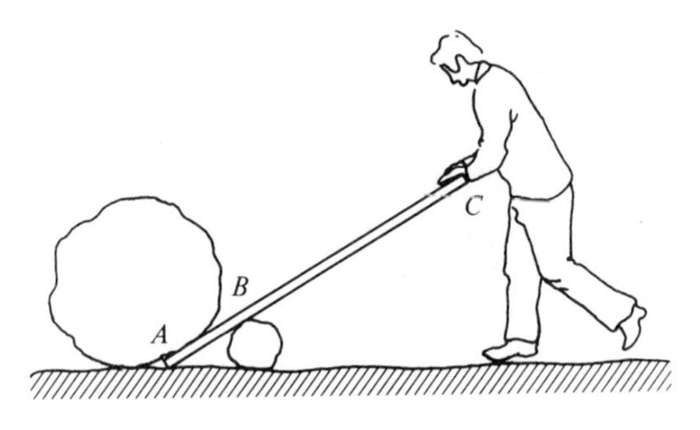

Problem 3.4

3.3 GENERAL EQUILIBRIUM CONDITIONS

In this section, we shall consider the conditions that the forces and couples acting upon a body must satisfy in order for it to be in equilibrium.

According to Newton's first law, the sum of the forces exerted on a body at rest must be zero. Notice, however, that this law says nothing about the moments, or rotational effects, of the forces. Clearly, the total moment must also be zero, else the body would rotate.

The fundamental problem here is that Newton's first law (and second law), as originally stated, applies only for very small bodies, or *particles*, with negligible dimensions and nonzero mass. However, it can be extended to bodies of finite size as follows.

Consider a system consisting of two particles, and let $\mathbf{f}_1$ and $\mathbf{f}_2$ be the forces due to the interaction between them (Figure 3.2). These forces are called *internal forces*, since they are due to interactions between bodies *within* the system.

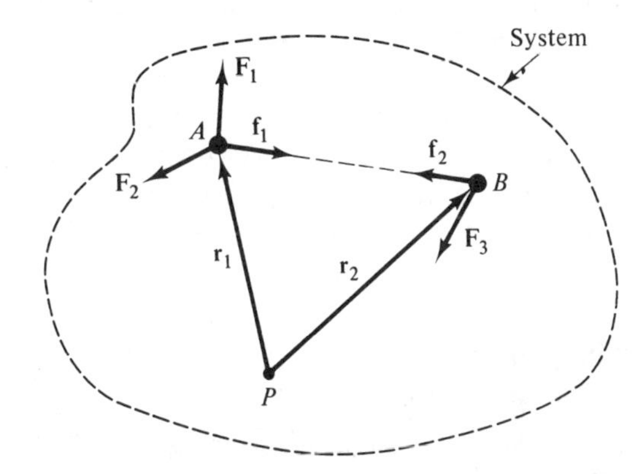

Figure 3.2

According to Newton's third law, $\mathbf{f}_1 = -\mathbf{f}_2$. Suppose that there are also forces, such as $\mathbf{F}_1$, $\mathbf{F}_2$, and $\mathbf{F}_3$, exerted on the particles due to interactions with bodies outside the system. Such forces are called *external forces*. Clearly, all the forces acting upon a particular particle must have the same point of application because a particle has negligible dimensions.

We shall say that the system is in equilibrium if each particle within it is in equilibrium. In this case, by Newton's first law, the sum of the forces acting upon each particle must be zero. For particle A we have

$$\Sigma\,\mathbf{F}_A = \mathbf{F}_1 + \mathbf{F}_2 + \mathbf{f}_1 = \mathbf{0}$$

and for particle B

$$\Sigma\,\mathbf{F}_B = \mathbf{f}_2 + \mathbf{F}_3 = \mathbf{0}$$

The total force acting upon the system is

$$\Sigma\,\mathbf{F} = \Sigma\,\mathbf{F}_A + \Sigma\,\mathbf{F}_B = \mathbf{F}_1 + \mathbf{F}_2 + \mathbf{F}_3 + \mathbf{f}_1 + \mathbf{f}_2 = \mathbf{0}$$

Now let us consider the total moment of these forces about some point P.

Referring to Figure 3.2, we have

$$\Sigma \mathbf{M}_P = \mathbf{r}_1 \times (\Sigma \mathbf{F}_A) + \mathbf{r}_2 \times (\Sigma \mathbf{F}_B)$$

But $\Sigma \mathbf{F}_A = \Sigma \mathbf{F}_B = \mathbf{0}$; so the total moment must also be zero, as stated previously.

Since the forces $\mathbf{f}_1$ and $\mathbf{f}_2$ have the same line of action, the moment condition can be rewritten as

$$\Sigma \mathbf{M}_P = \mathbf{r}_1 \times (\mathbf{F}_1 + \mathbf{F}_2 + \mathbf{f}_1 + \mathbf{f}_2) + \mathbf{r}_2 \times \mathbf{F}_3 = \mathbf{0}$$

Since $\mathbf{f}_1 = -\mathbf{f}_2$, the conditions on the forces and moments reduce to

$$\Sigma \mathbf{F} = \mathbf{F}_1 + \mathbf{F}_2 + \mathbf{F}_3 = \mathbf{0}$$

and

$$\Sigma \mathbf{M}_P = (\mathbf{r}_1 \times \mathbf{F}_1) + (\mathbf{r}_1 \times \mathbf{F}_2) + (\mathbf{r}_2 \times \mathbf{F}_3) = \mathbf{0}$$

In other words, if the system is in equilibrium, the sum of the *external* forces acting upon it is zero and so is the sum of the moments of these forces about an arbitrary point. The internal forces need not be considered because their effects cancel out.

Although we shall not go through the details, it should not be too difficult to see that the preceding results hold for a system consisting of any number of particles acted upon by any number of external forces, since the internal forces always occur as pairs of equal, opposite, and collinear forces whose effects cancel. In particular, these results apply to bodies of finite extent, since such bodies can be thought of as consisting of a large number of very small pieces, or particles. Thus, we have the following general equilibrium conditions:

If a system is in equilibrium, then

$$\Sigma \mathbf{F} = \mathbf{0} \quad \text{and} \quad \Sigma \mathbf{M}_P = \mathbf{0} \tag{3.1}$$

where $\Sigma \mathbf{F}$ is the sum of the external forces acting upon the system and $\Sigma \mathbf{M}_P$ is the total moment of these forces about an arbitrary point, including the moments of any couples which may be acting.

Equations (3.1) are *necessary* conditions for equilibrium; i.e., if the system is in equilibrium, these equations must be satisfied. They are not, in general, *sufficient* conditions for equilibrium; satisfaction of these equations does not necessarily guarantee that the system will be in equilibrium. This presents no difficulties, however, for we shall be dealing only with systems known to be in equilibrium. Equations (3.1) are both necessary and sufficient conditions for equilibrium of a rigid body. Proof that they are sufficient requires use of Newton's second law and other knowledge beyond the level of this text.

It is important to note that Eqs. (3.1) hold for any system in equilibrium, regardless of the material of which it is comprised. For example, they hold for a mass of fluid at rest, as well as for solid bodies. They also apply to moving systems under certain conditions, since Newton's first law, upon which they are based, applies to particles moving with constant velocity as well as to particles at rest. For instance, Eqs. (3.1) hold for bodies that move in a straight line at constant speed without rotation and for bodies that rotate at a constant rate about a fixed axis through their mass center. Typical examples are an airplane in straight, level flight at constant speed and the pulley on an electric motor rotating at constant speed. However, problems involving motion of any kind are usually relegated to texts on dynamics.

When expressed in component form, Eqs. (3.1) yield the six scalar equations:

$$\begin{aligned} \Sigma F_x &= 0 \qquad \Sigma F_y = 0 \qquad \Sigma F_z = 0 \\ \Sigma M_{Px} &= 0 \qquad \Sigma M_{Py} = 0 \qquad \Sigma M_{Pz} = 0 \end{aligned} \tag{3.2}$$

These equations can be used in a force analysis of a system to solve for unknown information concerning the external forces and couples acting. Since there are six equations, we can generally solve for six unknowns. If all of the unknowns concerning the external forces and couples can be determined from the equilibrium equations, the problem is said to be *statically determinate*. If not, it is said to be *statically indeterminate*. For the present, we shall consider only statically determinate problems. Statically indeterminate problems will be considered in later chapters.

When there are more unknowns than equations of equilibrium in a problem, it is tempting to try to obtain additional equations by considering moments about more than one point. Unfortunately, this procedure does not work. According to Eq. (2.31), the relationship between the total moment about two points P and Q is

$$\Sigma \mathbf{M}_Q = \Sigma \mathbf{M}_P + \mathbf{r}_{QP} \times \Sigma \mathbf{F}_i$$

But $\Sigma \mathbf{F}_i = \mathbf{0}$ and $\Sigma \mathbf{M}_P = \mathbf{0}$ for a system in equilibrium. Thus, the condition $\Sigma \mathbf{M}_Q = \mathbf{0}$ will be satisfied automatically, and no new information can be obtained from it. Consequently, we conclude that there are no more than six independent equations of equilibrium for a given system.

3.4 MODELING OF PROBLEMS—FREE-BODY DIAGRAMS

The equilibrium equations are relatively simple to apply when the system of interest is given and the nature of the forces and couples acting upon it are known. In fact, there is little more to it than the addition of forces and calculation of moments considered in Chapter 2.

Unfortunately, the situation is not so simple in practice. We are usually confronted with a physical system, such as a machine, structure, or tool, along with some applied loadings that are either known or can be estimated. It is left up to us to (1) select the specific system of interest, (2) determine the nature of all unknown external forces and couples exerted on the system by its supports, connections, gravitational attraction, etc., and (3) make whatever assumptions are reasonable to facilitate the solution of the problem. In other words, we must select the system and model the problem before we apply the equations of equilibrium.

The specific system of interest can usually be determined from a clearly stated problem definition; some of the aspects of problem modeling are discussed in the following. Throughout our discussion keep in mind that the basic idea is to simplify the problem as much as possible by considering only factors that have an important bearing upon the quantities to be determined and neglecting those that do not.

Free-body concept. Once the system to be analyzed has been selected, we imagine that it is completely isolated, or freed, from its surroundings, including any other bodies with which it may be in contact. We then draw a separate sketch showing only the system of interest and the external forces and couples acting upon it. These forces and couples include the applied loadings, the weight, and any other loadings exerted upon the system by its surroundings. All the forces and couples acting are represented by vectors (symbols are used to denote unknown quantities). The resulting sketch is called a *free-body diagram*, or FBD. This diagram focuses attention upon the particular system to be investigated and clearly indicates the forces and couples involved. In other words, the free-body diagram is the basis for the equilibrium equations.

A simple example of a free-body diagram is given in Figure 3.3. Figure 3.3(a) shows a man crossing a stream on a board, and Figure 3.3(b) shows the corresponding free-body diagram for the member. In this diagram the vectors $\mathbf{F}_A$ and $\mathbf{F}_B$ represent the upward forces that the banks of the stream exert on the board and $\mathbf{W}$ denotes its weight. The force $\mathbf{N}$ is exerted by the man's feet and is equal in magnitude to his weight.

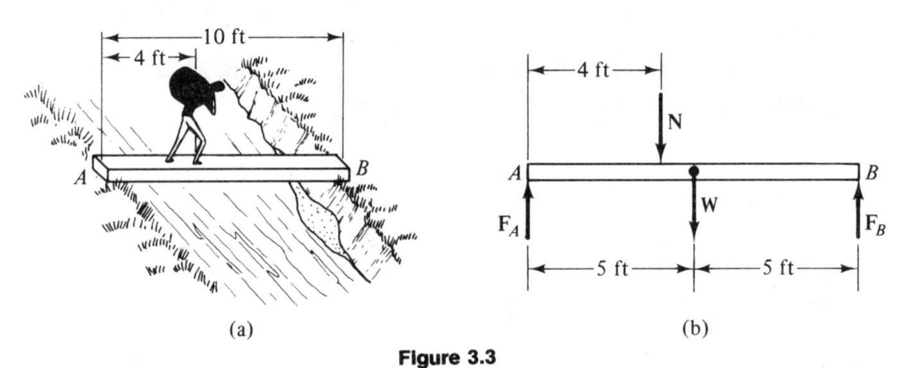

Figure 3.3

Since the loads applied to a body are usually known, it is no problem to include them in the free-body diagram. The major difficulty lies in deciding what other forces and couples of significance may be acting. This aspect of construction of the free-body diagram will be discussed next.

Weight–force interactions at a distance. As indicated in Chapter 1, the weight of a body is the force exerted on it due to the gravitational attraction of the earth. If the mass of the body is the quantity specified, its weight can be computed from the relation

$$W = mg$$

given in Eq. (1.1). Note that the weight will be in pounds if the mass is measured in slugs ($\text{lb}\cdot\text{s}^2/\text{ft}$) and the acceleration of gravity in feet per second squared. If the mass is given in kilograms and the acceleration of gravity in metres per second squared, the weight will be in newtons.

Insofar as the equilibrium of a body is concerned, its weight can be considered to be a single force acting downward toward the center of the earth with point of application at the *center of gravity G* of the body. The location of the center of gravity is often obvious from the body's geometry. If the body is symmetrical and uniform throughout, its center of gravity will lie at the intersection of the planes of symmetry. For example, the center of gravity of a uniform cylinder corresponds to its geometric center. The procedure for locating the center of gravity of irregular shaped bodies will be discussed in Chapter 4. In many instances, the weight of a body is small in comparison to the applied loads and can be neglected in the force analysis.

A body is actually subjected to forces due to gravitational interactions with every other body in the universe, not just with the earth. However, as long as we are considering systems located at or near the surface of the earth, these other bodies are usually so small or so far away that their gravitational effects are negligible. There may also be force interactions at a distance due to the presence of electrical or magnetic fields, but they will not be considered here.

Contact forces–reactions. Contact forces arise whenever a body is connected to or in contact with another body. Such forces are prevalent in engineering problems. For instance, structures and machines are held in place by various kinds of supports, and they usually consist of a number of parts that are in contact or are joined together by various types of connections.

The forces exerted on a system by contacts, supports, and connections are often distributed over an area, as in the case of a book lying on a table. These force distributions prevent translation and rotation of the system and hold it in equilibrium under the influence of the applied loads. As we shall show in Chapter 4, these effects can be represented by an equivalent force system consisting of a single force and/or a single couple. These equivalent forces and couples are called the *reactions*, and are the quantities determined in a force analysis. Their relationship to the actual distribution of force that may be acting will be examined in detail in Chapter 4.

The nature of the reactions exerted on a body by most types of contacts, connections, and supports is either intuitively obvious or can be determined with a little physical reasoning. We shall attempt to illustrate this fact in the following examples.

Suppose we wish to determine the nature of the forces that the floor exerts on the block shown in Figure 3.4(a). Because the floor prevents the block from falling through space, it obviously must exert an upward force on it. If we attempt to slide

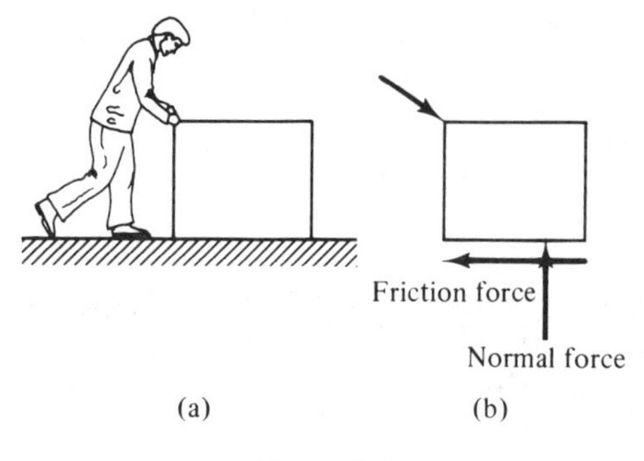

Figure 3.4

the block, there will normally be some resistance because of the friction between it and the floor. Thus, the floor can exert forces on the block which are perpendicular and parallel to the plane of contact [Figure 3.4(b)]. These forces are usually referred to as the *normal* and *friction force*, respectively. The orientation and sense of these forces are known, but their magnitudes and the location of the normal force are unknown. If the friction force is small enough to be considered negligible, we say that the bodies or surfaces in contact are *smooth*, or *frictionless*. This is an idealization; there is no such thing as a perfectly frictionless surface.

As a second example, let us consider the bar shown in Figure 3.5(a). The left end of the bar is connected to the wall by means of a *smooth pin*, or *ideal hinge*. Again, the concept of a frictionless hinge is an idealization. The right end of the bar is supported by a wire BC. We wish to determine the nature of the reactions on the bar at the supports.

Flexible members, such as wires, ropes, strings, and cables, can only "pull" on the bodies to which they are attached. The force exerted lies along the axis of the member and is equal in magnitude to the tension on it. Thus, the wire exerts a force **T** on the bar of unknown magnitude and directed from B toward C [Figure 3.5(b)].

Since the pin at A will resist translation of the bar in any direction within the plane of the figure, it can exert a force **A** on the bar of unknown magnitude and direction [Figure 3.5(b)]. Alternatively, we can say that the pin exerts two unknown components of force on the bar, as indicated by the dashed arrows in the figure. If we attempt to rotate the bar about the pin, the pin will offer no resistance because

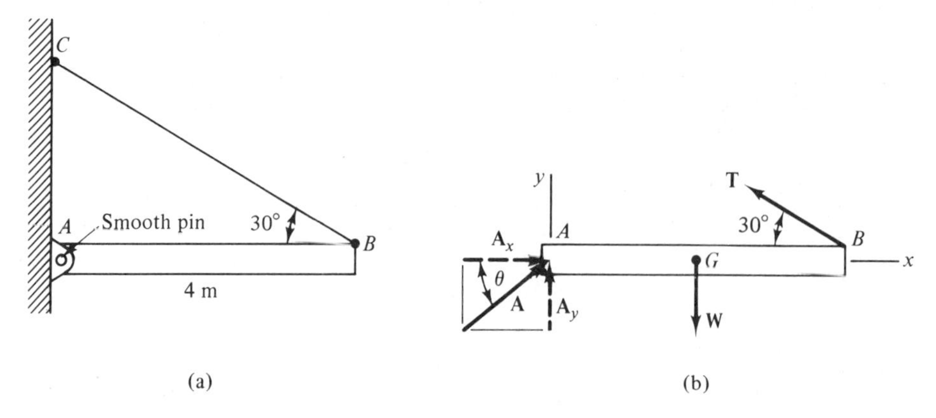

Figure 3.5

it is frictionless. Thus, there is no couple exerted on the bar. The free-body diagram is completed by adding the weight **W** of the member. It should be noted, however, that additional reactions will be present if there are loads acting that tend to move the bar out of the plane of the page.

Our final example concerns the reactions on a post embedded in a concrete slab [Figure 3.6(a)]. Obviously, the slab will resist translation and rotation of the post in any direction. Thus, it can exert a force and couple of unknown magnitude and direction on the post. Alternatively, we can say that the slab can exert three components of force and three components of couple [Figure 3.6(b)]. In either case, there are six unknowns at the supports. A support that prevents both translation and rotation of a body is called a *fixed*, or *clamped, support*.

Generalizing the results obtained for the preceding examples, we can say that if a support or connection prevents or restricts the translation of a body in a given direction, it can exert a force on the body in that direction. If it prevents or restricts the rotation of the body about a particular axis, it can exert a couple on the body about that axis.

Since a force is completely defined by its three rectangular components, the nature of the force exerted on a body can be determined by pretending to attempt to move the body in three mutually perpendicular directions and observing whether or not the support or connection would resist such motions. For two-dimensional problems, all forces lie in one plane, and only two perpendicular directions within this plane need be considered.

Similarly, the nature of the couple exerted on the body can be determined by pretending to rotate the body about three mutually perpendicular axes and observing whether or not the support or connection would resist such rotations. In two-dimensional problems, all couple vectors are perpendicular to the plane of the problem, and only rotation about an axis perpendicular to this plane need be considered.

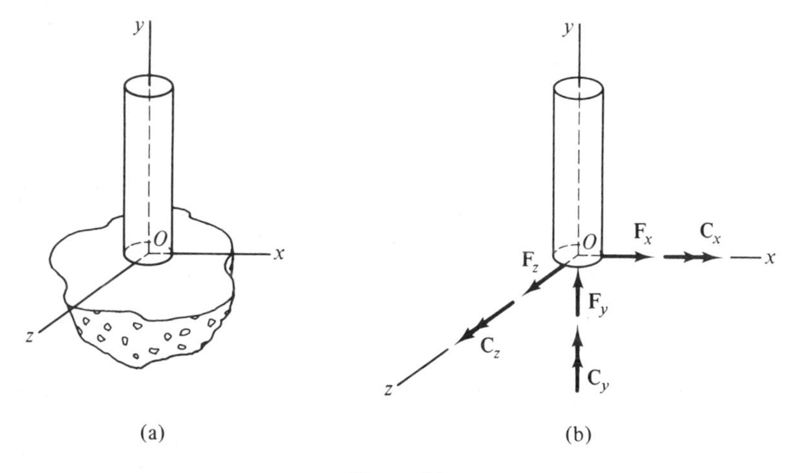

Figure 3.6

Support or connection	Reactions	
	Two-dimensional loadings	Three-dimensional loadings
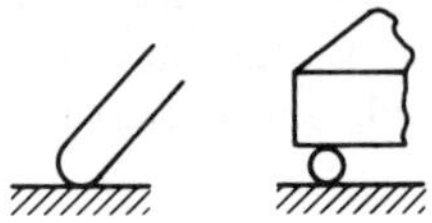 Smooth surface or ball	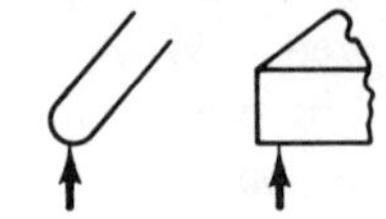 Single force normal to surface (1 unknown)	Same as two-dimensional case
 Rough surface	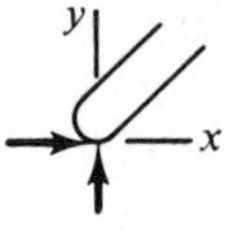 Normal force and friction force (2 unknowns)	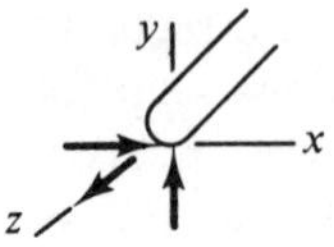 Normal force and two comp. friction (3 unknowns)
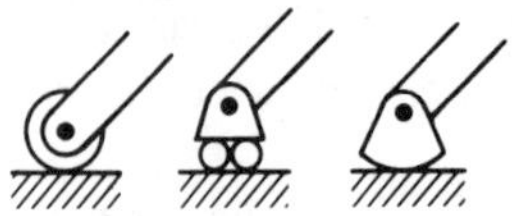 Roller or rocker	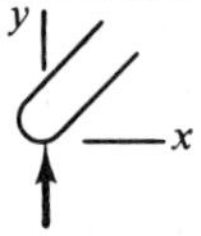 Single force normal to surface (1 unknown)	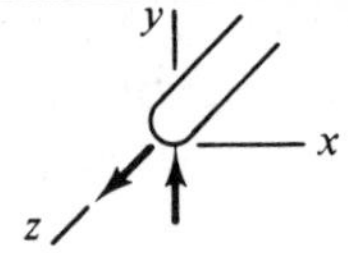 Normal force and friction force (2 unknowns)
 Cable, string, wire, thin rod, etc.	 Single force along cable (1 unknown)	Same as two-dimensional case
 Smooth pin or hinge	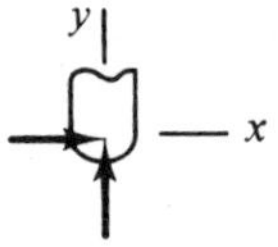 Two components of force (2 unknowns)	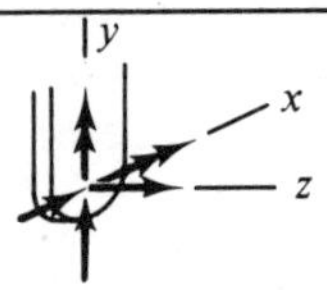 3 components of force and 2 of couple (5 unknowns)
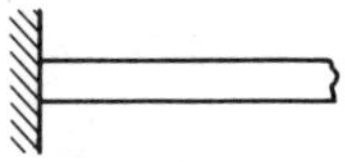 Fixed support	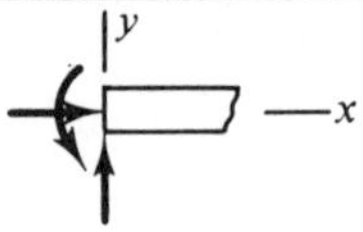 2 components of force and 1 of couple (3 unknowns)	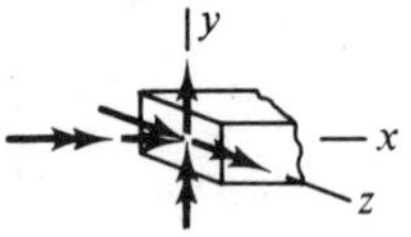 3 components of force and 3 of couple (6 unknowns)
 Ball and socket	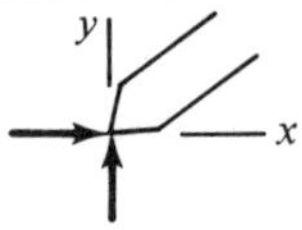 2 components of force (2 unknowns)	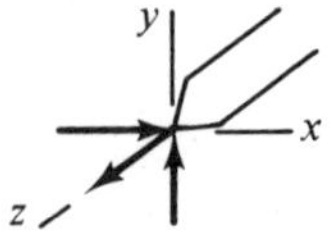 3 components of force (3 unknowns)

Figure 3.7

The reactions exerted by some common types of supports and connections are shown in Figure 3.7. Study these carefully and see if you can verify the results presented by considering the motions that each support will or will not resist. These results should not be memorized. The point we have been trying to make is that the nature of the reactions can be deduced from simple physical reasoning.

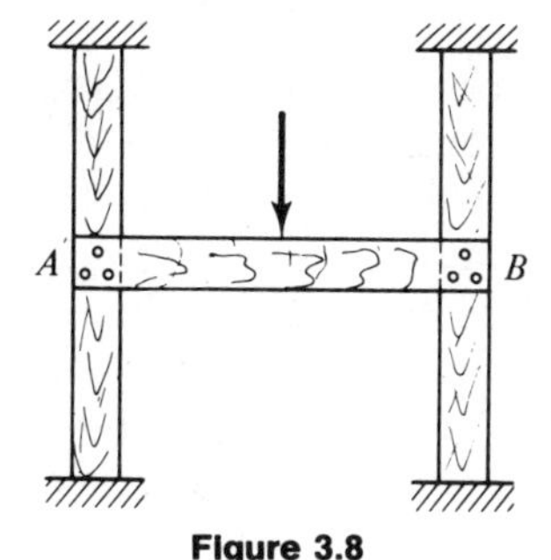

Figure 3.8

Many of the supports and connections encountered in practice do not correspond completely to any one of the idealizations shown in Figure 3.7. For example, Figure 3.8 shows a structure comprised of three boards nailed together. The connections at A and B will resist, but not completely prevent, rotation. Thus, they are not exactly equivalent to fixed supports, and they are not equivalent to pinned joints. It is a matter of engineering judgment as to how these connections should be modeled. Assuming that only a few nails are used at each connection, the resistance to rotation will likely be small. In this case, it would be reasonable to treat the connections as pinned joints.

Rigid and deformable bodies. When loads are applied to a body, the body will deform. Strictly speaking, the equilibrium equations apply only after all deformation has ceased and every part of the body has come to rest. Hence, the force analysis should be based upon the dimensions and geometry of the deformed body. In most problems, however, the deformations are very small compared to the original dimensions of the body, and little error is introduced if the force analysis is based upon the undeformed geometry.

It is no coincidence that the deformations of most engineering structures are relatively small. These structures are purposely designed this way. For example, a highway bridge that sags several feet when vehicles cross it would be of little value, even though it may be in no danger of breaking.

If the deformations of a body are neglected when performing a force analysis, this is the same as assuming the body is rigid. The concept of a *rigid body* is widely used in mechanics, but it is an idealization. All bodies deform to some extent when loaded.

Most of the problems considered in this text are such that the rigid-body assumption can be used in the force analyses. There are, however, problems for which it is essential to base the equilibrium equations upon the deformed geometry of the body. These problems will be considered in Chapter 10. Because of its importance, the rigid-body assumption will be considered in more detail in Section 3.11, after we have had a chance to make use of it in the solution of some problems.

Free-body diagrams. When constructing the free-body diagram of a system, there are several points to keep in mind.

It is important that only the external forces and couples exerted on the system be shown on the free-body diagram. Forces and couples that are internal to the system, that the system exerts on other bodies, or that can be considered negligible do not enter into the equilibrium equations for the system and should not be shown. However, the diagram should include dimensions and any other information needed to write down the equilibrium equations.

The sense of any unknown forces and couples can be chosen arbitrarily on the free-body diagram. If a positive value is obtained for these quantities, their sense is as assumed. If their values turn out to be negative, their sense is opposite to that assumed. In contrast, all known forces and couples must be shown on the free-body diagram with their proper sense. Otherwise, a sign error will result.

There are two ways of representing forces and couples on the free-body diagram. They can be shown as vectors, which are then defined by appropriate mathematical expressions, or they can be represented by their projections, with their orientation and sense defined by the figure of the free-body diagram (as in Example 3.1). The representation in terms of projections is usually the most convenient, except in some three-dimensional problems. We shall use it in most instances.

The ideas presented in this section regarding problem modeling and construction of the free-body diagram are further illustrated in the following examples.

Example 3.1. Free-Body Diagram of a Ladder. The ladder shown in Figure 3.9(a) has a mass of 25 kg and is held in place by a rope attached to the wall. The wall and floor are smooth. Draw an FBD of the ladder.

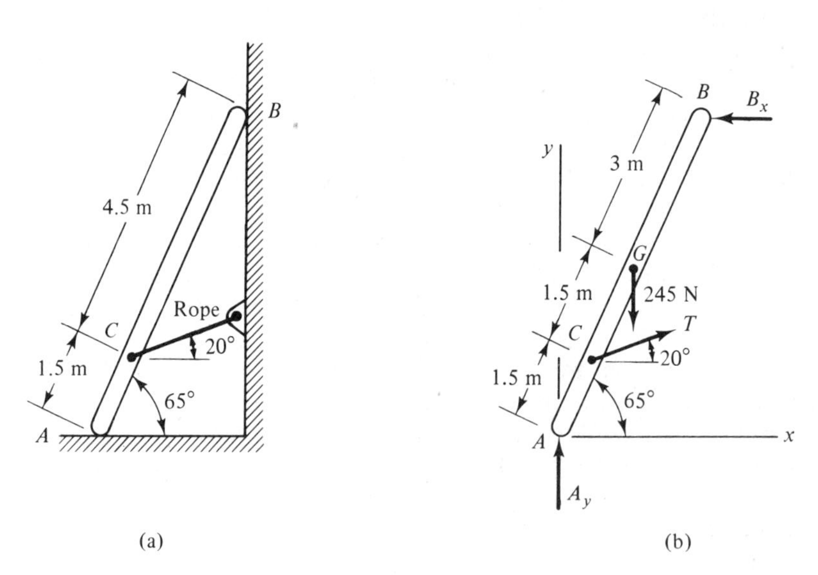

Figure 3.9

Solution. We first isolate the ladder [Figure 3.9(b)]. The forces acting upon the ladder are its weight, the reactions at the points of contact with the wall and floor, and the force due to the rope. The smooth wall and floor exert forces normal to the plane of contact, which we denote by B_x and A_y, respectively. The rope exerts a force along its axis, which we denote by T. From Eq. (1.1), the weight of the ladder is

$$W = mg = (25\text{ kg})(9.81\text{ m/s}^2) = 245\text{ N}$$

We assume the ladder is uniform; therefore, its center of gravity is located at its midpoint. The complete FBD is as shown.

Example 3.2. Free-Body Diagrams of Bars. Draw an FBD of each of the bars in the structure shown in Figure 3.10(a). All connections are smooth pins, and the weights of the bars are negligible.

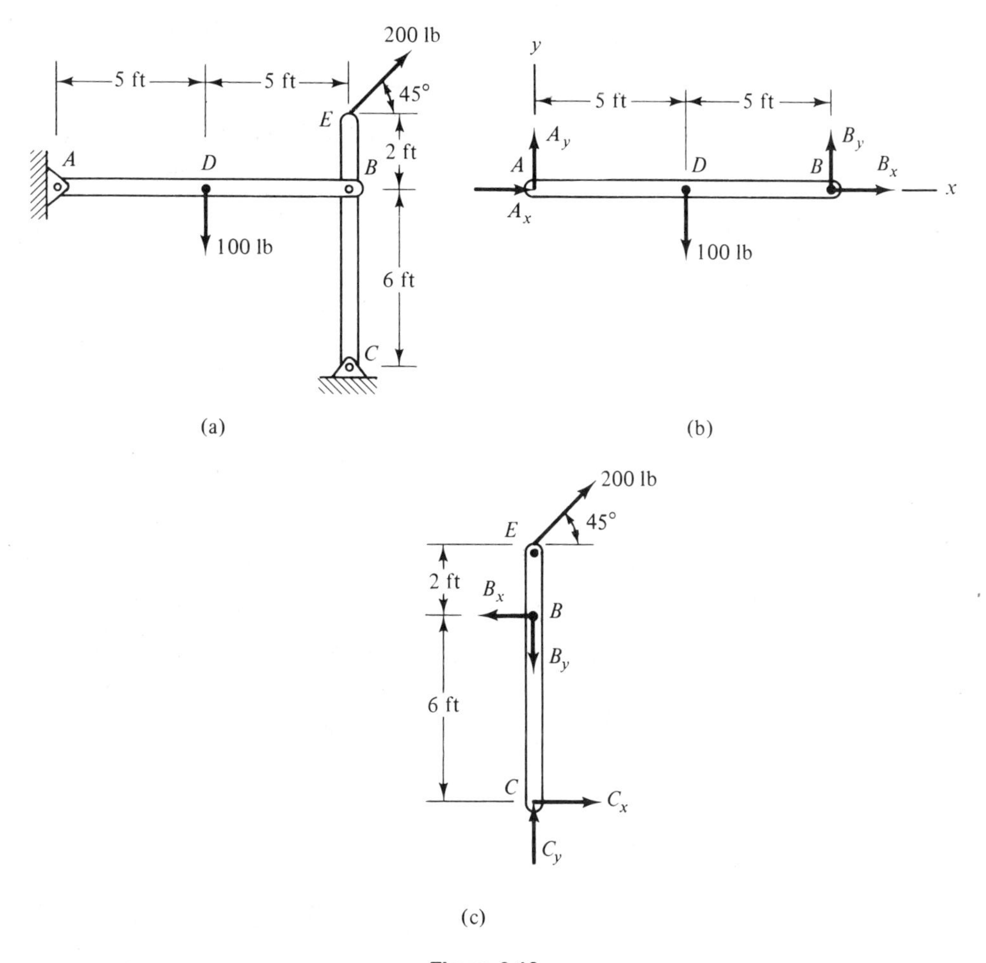

Figure 3.10

Solution. We first isolate the horizontal bar [Figure 3.10(b)]. The pin at A prevents horizontal and vertical motion of the bar and, therefore, can exert a horizontal and vertical component of force on it. The same is true of the pin at B. The resulting FBD of bar AB is as shown, where the sense of the forces has been chosen arbitrarily.

We now isolate the vertical bar [Figure 3.10(c)]. Horizontal and vertical components of force are exerted on the bar by the pin at C. The forces at B are obtained from the FBD of the horizontal bar and Newton's third law. Since the forces B_x and B_y acting upon bar AB are due to the vertical bar, the horizontal bar will exert equal and opposite forces on the vertical bar at B. Consequently, the FBD of bar CE is as shown. Note carefully the use of Newton's third law at point B. The importance of this law in constructing free-body diagrams cannot be overstated.

PROBLEMS

3.5 to 3.9 Draw a complete free-body diagram for the bodies, members, or structures shown. If no value is given for the mass or weight, it may be assumed to be negligible.

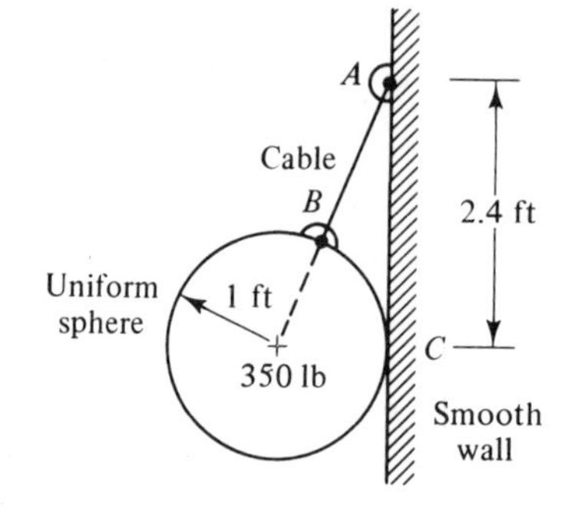

Problem 3.5

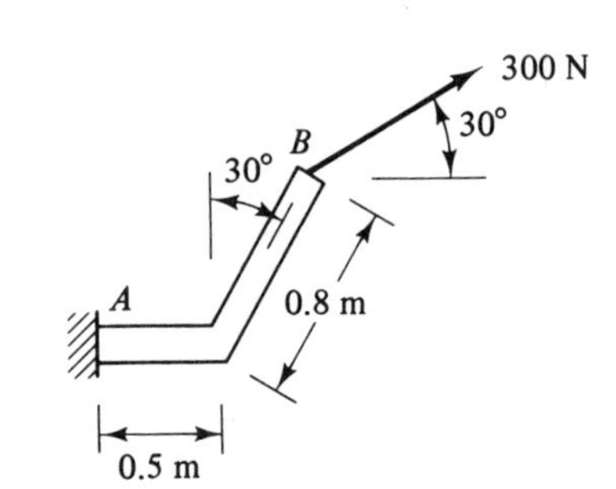

Problem 3.8

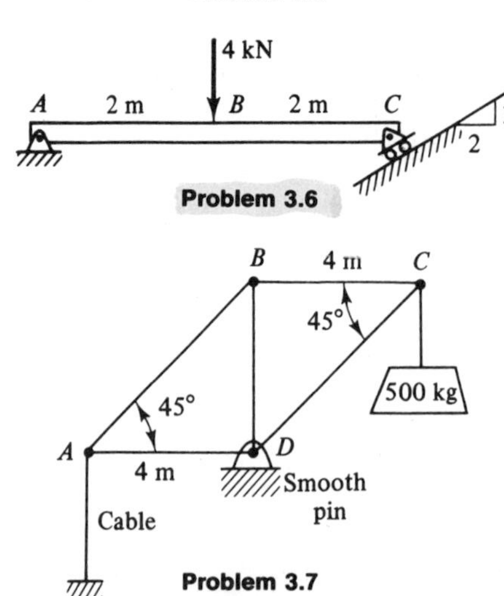

Problem 3.6

Problem 3.7

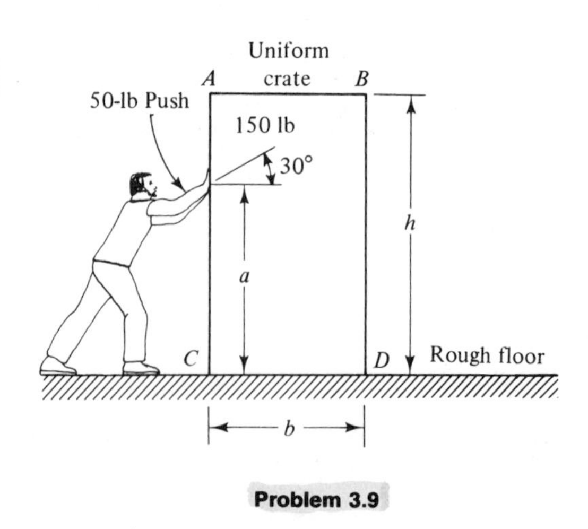

Problem 3.9

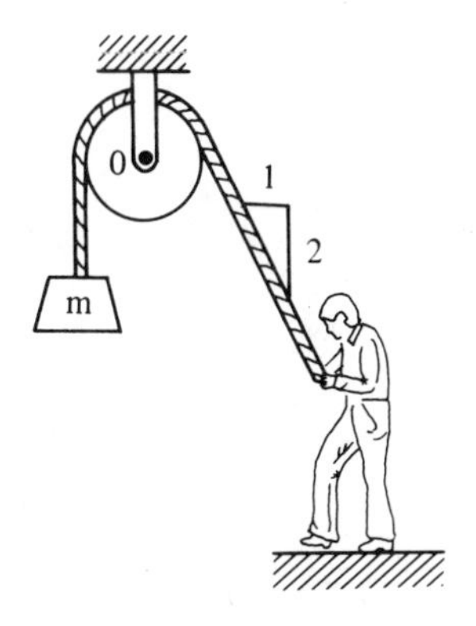

Problem 3.10

3.10 For the situation shown, draw an FBD for the system consisting of the pulley and rope; assume the bearing at O is frictionless. Show that, when the system is in equilibrium, the tension on the rope has the same magnitude on both sides of the pulley.

3.11 A slender pole with negligible mass is supported by two smooth eye-bolts and is loaded as shown. Draw an FBD for the pole.

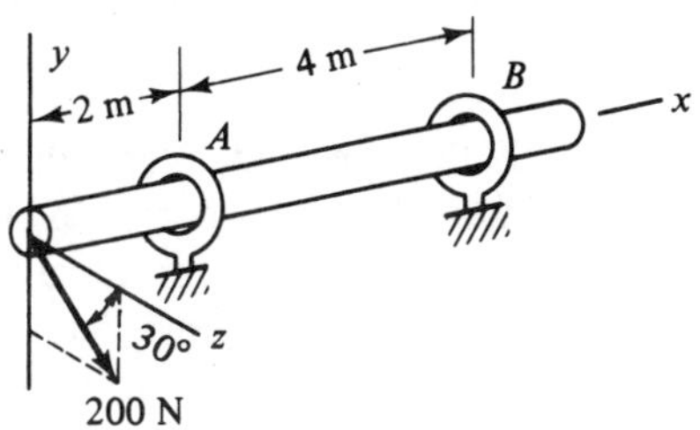

Problem 3.11

3.12 The 20-kg uniform bar shown is held in a horizontal position by a smooth hinge at one end and a vertical wire at the other. Draw an FBD for the bar.

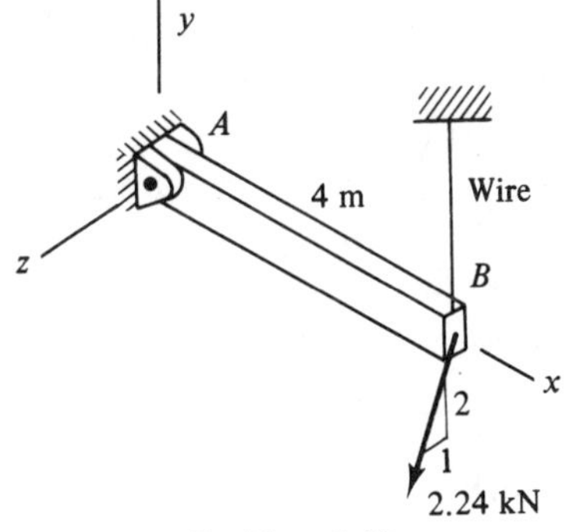

Problem 3.12

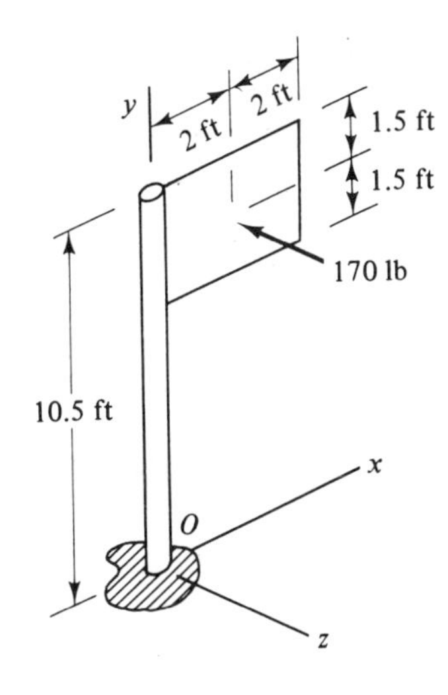

Problem 3.13

3.13 A uniform sign, which weighs 40 lb, is supported by a uniform pole with a weight of 120 lb. The wind exerts a net force of 170 lb normal to the sign and acting at its center, as shown. Draw an FBD for the combination of sign and pole.

3.14 to 3.16 For the situations shown, draw complete free-body diagrams for body ① and body ②. If no value is given for the mass or weight, it may be assumed to be negligible.

Identical 100-kg cylinders
All surfaces smooth

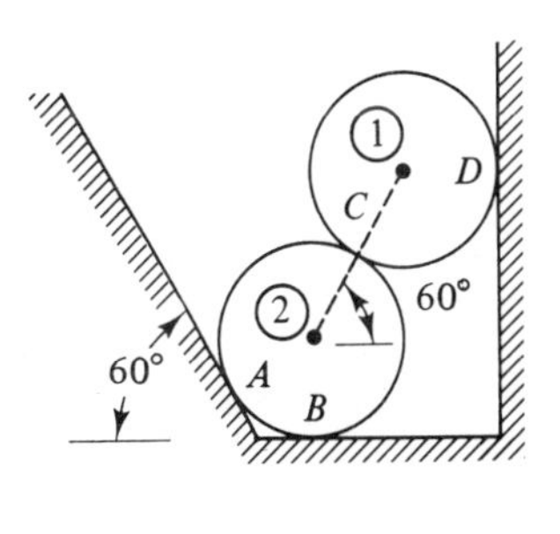

Problem 3.14

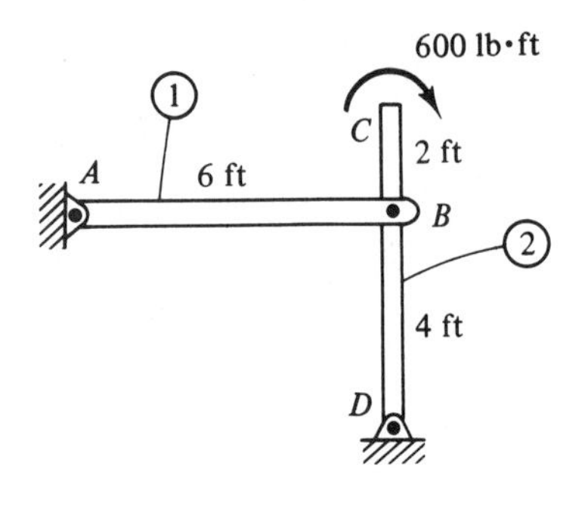

Problem 3.15

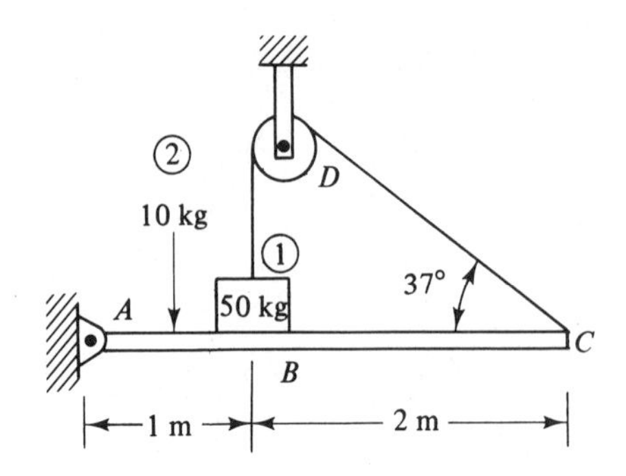

Problem 3.16

3.5 FORCE ANALYSIS OF SOME SPECIAL SYSTEMS

As indicated in Section 3.3, there are no more than six independent equations of equilibrium for a given system. There are, however, cases in which there are fewer than six independent equations because some of them are satisfied automatically. The equilibrium equations and force analysis for several of the more commonly encountered cases are considered in this section.

Concurrent force systems. Recall from Section 2.2 that the lines of action of concurrent forces all intersect at a common point. In order for a system of such forces to be in equilibrium, it is only necessary that $\Sigma\, \mathbf{F} = \mathbf{0}$. The moment equation of equilibrium, $\Sigma\, \mathbf{M}_p = \mathbf{0}$, is satisfied automatically. This can be readily seen by considering moments about the point of concurrency. Thus, there are at most three independent equations of equilibrium for concurrent force systems:

$$\Sigma\, F_x = 0 \qquad \Sigma\, F_y = 0 \qquad \Sigma\, F_z = 0 \tag{3.3}$$

Coplanar force systems. Coplanar forces all lie in the same plane, which is the same as saying that the problem is two dimensional. Let us arbitrarily choose coordinates such that the forces lie in the x-y plane. Since the forces have no z component, the equilibrium equation $\Sigma\, F_z = 0$ is satisfied automatically. Since the moments about a point P in the x-y plane will have only a z component, the total moment is $\Sigma\, \mathbf{M}_P = \Sigma\, M_{Pz}\mathbf{k}$. Thus, the equations $\Sigma\, M_{Px} = 0$ and $\Sigma\, M_{Py} = 0$ are also satisfied, and the equation $\Sigma\, M_{Pz} = 0$ can be written as $\Sigma\, M_P = 0$. This leaves the three independent equations of equilibrium

$$\Sigma\, F_x = 0 \qquad \Sigma\, F_y = 0 \qquad \Sigma\, M_P = 0 \tag{3.4}$$

These equations correspond to the sum of forces in two perpendicular directions within the plane equal to zero and the sum of moments about some point in the plane equal to zero.

It is sometimes convenient to replace one or both of the force equations of equilibrium in Eqs. (3.4) with additional moment equations. This can be done, provided we still end up with three independent equations of equilibrium.

For instance, suppose that in Eqs. (3.4) we wish to replace the condition $\Sigma F_y = 0$ with $\Sigma M_Q = 0$, where Q is a second point in the plane of the forces. This new set of equations will be valid if it can be shown to be equivalent to the original set.

From Eq. (2.31), we have

$$\Sigma \mathbf{M}_Q = \Sigma \mathbf{M}_P + \mathbf{r}_{QP} \times \left[(\Sigma F_x)\mathbf{i} + (\Sigma F_y)\mathbf{j}\right]$$

Since $\Sigma \mathbf{M}_P = \mathbf{0}$, $\Sigma \mathbf{M}_Q = \mathbf{0}$, and $\Sigma F_x = 0$, this expression reduces to

$$\mathbf{r}_{QP} \times (\Sigma F_y)\mathbf{j} = \mathbf{0}$$

which is equivalent to $\Sigma F_y = 0$ if $\mathbf{r}_{QP}$ and $\mathbf{j}$ are not parallel. Interpreted geometrically, this means that points P and Q cannot lie on a line that is perpendicular to the x direction. If this condition is satisfied, Eqs. (3.4) can be replaced with the equations

$$\Sigma F_x = 0 \qquad \Sigma M_P = 0 \qquad \Sigma M_Q = 0 \tag{3.5}$$

where P and Q are two points in the plane of the forces. In applying these results, it is important to remember that the x direction can be chosen arbitrarily.

Equations (3.4) can also be replaced by three moment equations. Let P, Q, and S be points in the plane of the forces. From Eq. (2.31), we have

$$\Sigma \mathbf{M}_Q = \Sigma \mathbf{M}_P + \mathbf{r}_{QP} \times \Sigma \mathbf{F} = \mathbf{0}$$

$$\Sigma \mathbf{M}_S = \Sigma \mathbf{M}_P + \mathbf{r}_{SP} \times \Sigma \mathbf{F} = \mathbf{0}$$

If we take $\Sigma \mathbf{M}_P = \mathbf{0}$, the first of these equations implies that either $\mathbf{r}_{QP}$ is parallel to $\Sigma \mathbf{F}$ or $\Sigma \mathbf{F} = \mathbf{0}$. Similarly, the second equation implies that either $\mathbf{r}_{SP}$ is parallel to $\Sigma \mathbf{F}$ or $\Sigma \mathbf{F} = \mathbf{0}$. These two equations are, therefore, equivalent to $\Sigma F_x = 0$ and $\Sigma F_y = 0$ if $\mathbf{r}_{QP}$ and $\mathbf{r}_{SP}$ are not both parallel to $\Sigma \mathbf{F}$. Interpreted geometrically, this means that the points P, Q, and S cannot all lie on the same line. Under this condition, Eqs. (3.4) can be replaced by the equations

$$\Sigma M_P = 0 \qquad \Sigma M_Q = 0 \qquad \Sigma M_S = 0 \tag{3.6}$$

where P, Q, and S are three points in the plane of the forces.

Equations (3.4), (3.5), and (3.6) are equivalent under the conditions stated, and the particular set of equations used in any given problem is strictly a matter of convenience and personal preference.

Two-force systems. A *two-force system*, or *member*, is one that is subjected to external forces at only two points. Cables, ropes, wires, and rods that are loaded at their ends and that are of negligible weight are examples of two-force members commonly encountered in practice.

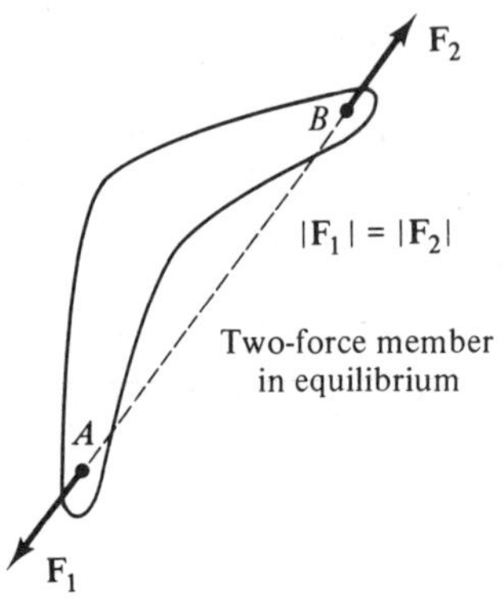

Figure 3.11

In order for a two-force system to be in equilibrium, the forces acting upon it must be equal and opposite and lie along a line connecting their points of application (Figure 3.11). Obviously, the forces must be equal and opposite in order to have $\Sigma\ \mathbf{F} = \mathbf{0}$. Furthermore, they must have the same line of action in order to satisfy the condition $\Sigma\ \mathbf{M}_P = \mathbf{0}$. This can be readily seen from Figure 3.11 by considering moments about the point of application of one of the forces. If there are several forces acting at each of the points of application, they can be added to obtain a single force at each point. The preceding statements concerning the equilibrium of the system then apply.

If the forces acting upon a straight two-force member tend to stretch it, we say that the member is in *tension*. If the forces tend to compress it, we say that the member is in *compression*.

Three-force systems. A *three-force system*, or *member*, is one that is subjected to external forces at three different points. In order for such a system to be in equilibrium, the forces must be coplanar and either concurrent or parallel.

Consider a system subjected to three forces, $\mathbf{F}_1$, $\mathbf{F}_2$, and $\mathbf{F}_3$. For equilibrium, $\mathbf{F}_1 + \mathbf{F}_2 + \mathbf{F}_3 = \mathbf{0}$, which implies that the vector polygon for the forces is a closed triangle. Interpreted geometrically, this means that the three forces must lie in the same plane. If we assume that the forces are not all parallel, the lines of action of two of them will intersect at some point P [Figure 3.12(a)]. If we take moments

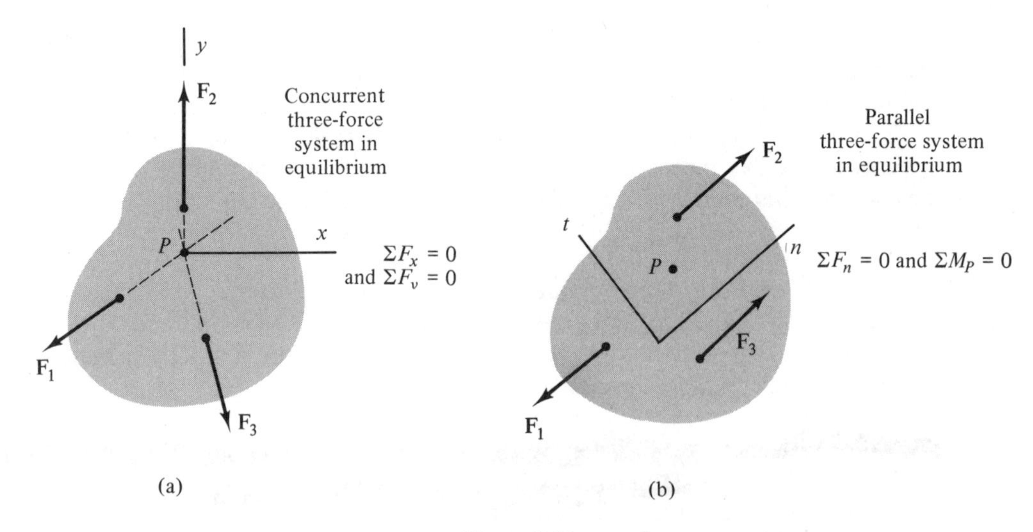

Figure 3.12

about this point, it is clear that the line of action of the third force must also pass through this point in order to satisfy the equilibrium condition $\Sigma \mathbf{M}_P = \mathbf{0}$. Thus, the forces must be concurrent. The only exception is when all three of the forces are parallel, as in Figure 3.12(b). If there are several forces acting at each of the points of application, they can be added to obtain a single force at each point. The preceding statements concerning the equilibrium of the system then apply.

Once it is recognized that the forces acting upon a three-force system must be coplanar and either concurrent or parallel, we are left with two independent equations of equilibrium. If the forces are concurrent, these equations correspond to summation of forces in two perpendicular directions within the plane equal to zero. [Figure 3.12(a)]. If the forces are parallel, we have summation of forces in one direction within the plane and summation of moments about some point in the plane equal to zero [Figure 3.12(b)]. In most problems, however, it is usually just as convenient to make use only of the fact that the forces must be coplanar. In this case, there are three independent equations of equilibrium, and Eqs. (3.4), (3.5), or (3.6) apply.

The force analysis of the various types of systems considered in this section is illustrated in the following examples. However, a word of caution is in order before we begin.

It is often tempting to try to take shortcuts in a force analysis by omitting the free-body diagram and writing down the equations of equilibrium directly from the figure in the problem statement. Do not get into this habit! Although this procedure may work for simple problems, it invariably results in confusion and errors when applied to more complex problems. A well-organized and systematic plan of attack is the key to success in a force analysis. Since the organization is provided by the free-body diagram, its importance cannot be overstated.

Example 3.3. Force Analysis of Ropes. A block with weight W is supported by two ropes as shown in Figure 3.13(a). Determine the tension on the ropes.

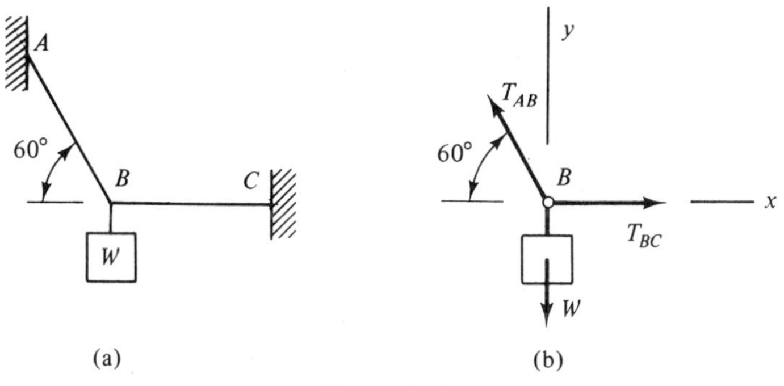

Figure 3.13

Solution. We take the block as our system, but there are other possibilities. For example, we could choose a system consisting of the block and both ropes. The

FBD of the block is as shown in Figure 3.13(b). The forces T_{AB} and T_{BC} shown are those which the ropes exert on the block; by Newton's third law, they are equal and opposite to the forces the block exerts on the ropes.

Since the forces on the block are coplanar and concurrent, there are two independent equations of equilibrium. Taking coordinates as shown, we have

$$\Sigma F_x = T_{BC} - T_{AB} \cos 60° = 0$$

$$\Sigma F_y = T_{AB} \sin 60° - W = 0$$

Solving these equations, we get

$$T_{AB} = 1.15W \qquad T_{BC} = 0.58W \qquad \textbf{\textit{Answer}}$$

The orientation and sense of the tensions are defined by the FBD. This problem can also be solved by constructing the vector polygon and determining the magnitudes of the tensions graphically or by using trigonometry.

Example 3.4. Force Analysis of a Lawn Roller. The lawn roller shown in Figure 3.14(a) has a weight of 200 lb with center of gravity at point O. Determine the pull P that the man must exert parallel to the handle to prevent the roller from rolling down the incline.

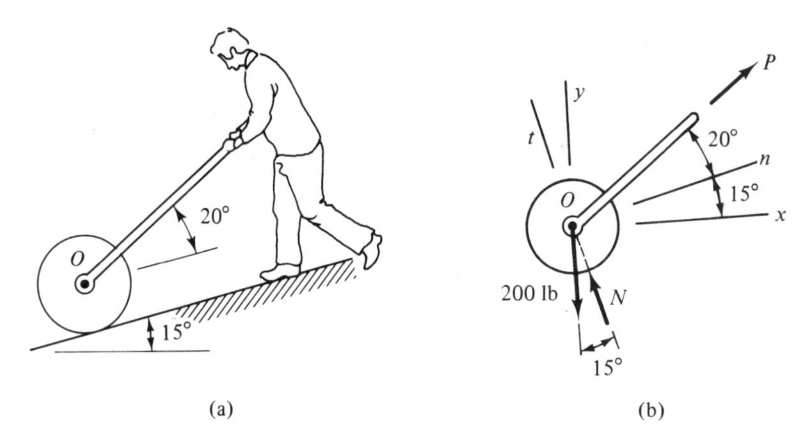

Figure 3.14

Solution. The FBD of the roller is as shown in Figure 3.14(b). Note that the roller is a three-force member; therefore, the force N exerted by the incline must be oriented such that its line of action passes through point O. Since the forces acting are concurrent and coplanar, there are two equilibrium equations, $\Sigma F_x = 0$ and $\Sigma F_y = 0$, from which to determine the two unknowns P and N.

The coordinate system used when applying the equilibrium equations is arbitrary, but a judicious choice of axes can often simplify the computations. To illustrate this, let us consider one set of axes, n and t, which are parallel and perpendicular to the plane, and another set, x and y, which are horizontal and vertical.

For the first set of axes, we have

$$\Sigma F_n = P \cos 20° - (200 \text{ lb}) \sin 15° = 0$$

from which we obtain

$$P = 55 \text{ lb} \qquad \textbf{\textit{Answer}}$$

The second equilibrium equation, $\Sigma F_t = 0$, is not needed in this problem.

For the second set of axes, we have

$$\Sigma F_x = P \cos 35° - N \sin 15° = 0$$

$$\Sigma F_y = P \sin 35° + N \cos 15° - 200 \text{ lb} = 0$$

These two equations yield the same result as our first equation, but their solution requires more work.

Example 3.5. Force Analysis of a Door. The uniform door shown in Figure 3.15(a) has a weight of 60 lb and is supported by hinges at points A and B. The construction of the hinges is such that only the bottom one can exert a force in the vertical direction. Determine the reactions on the door at the hinges.

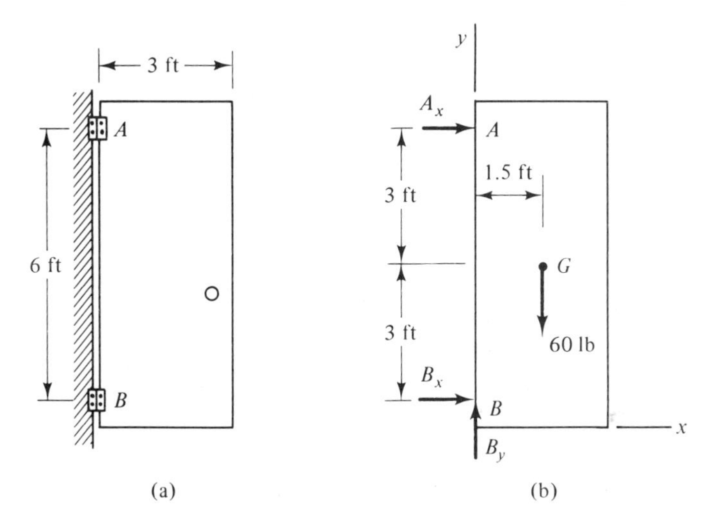

Figure 3.15

Solution. A FBD of the door is shown in Figure 3.15(b). Although the hinges can exert couples about axes parallel to the x and z axes, there is very little tendency for the door to rotate about such axes under the given loading. Thus, any couples exerted by the hinges will be negligible, and they have been omitted from the FBD. Similarly, there is no tendency for the door to move normal to the plane of the page under the given loading; therefore, the z components of the hinge forces have also been taken to be zero.

Since the forces acting upon the door are coplanar, there are three independent equations of equilibrium from which to determine the three unknowns A_x, B_x, and B_y. We have

$$\Sigma F_x = B_x + A_x = 0$$
$$\Sigma F_y = B_y - 60\ \text{lb} = 0$$
$$\overset{+}{\curvearrowleft}\ \Sigma M_A = B_x(6\text{ft}) - (60\ \text{lb})(1.5\ \text{ft}) = 0$$

Solving these equations, we get

$$B_x = 15\ \text{lb} \qquad B_y = 60\ \text{lb} \qquad A_x = -15\ \text{lb} \qquad \textit{Answer}$$

The negative value of A_x means that the horizontal force on the door at A has a sense opposite to that shown in the figure. The point about which moments are computed is, of course, arbitrary. The idea is to try to select the point that most simplifies the resulting calculations.

This problem can also be solved by summing forces in one direction, say the y direction, and summing moments about two points, such as A and B, which do not lie along a line perpendicular to the direction in which the forces are summed [see Eqs. (3.5)]. A third approach is to sum moments about three points, such as A, B, and G, which do not all lie on the same line [see Eqs. (3.6)]. It is left as an exercise to show that these approaches lead to the same results obtained previously.

Example 3.6. Force Analysis of a Tripod. The tripod shown in Figure 3.16(a) supports a 20-kg surveying instrument. The legs of the tripod are equally spaced and hinged at the top. If the mass of the tripod is negligible compared to that of the instrument, determine the reactions on the legs at the ground.

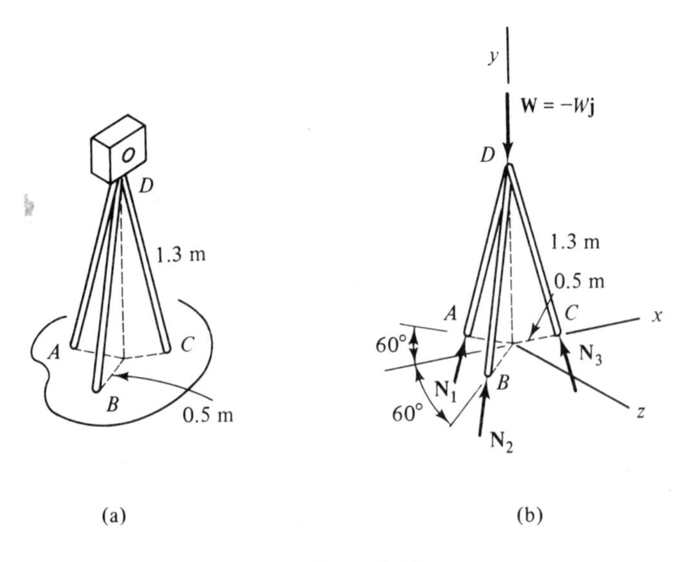

Figure 3.16

Solution. The legs of the tripod are two-force members because they have negligible weight and are loaded only at the top and bottom. Thus, the forces exerted on the legs by the ground must lie along the axes of the legs, as indicated in the FBD of Figure 3.16(b).

Taking coordinates as shown, we find from the geometry of the FBD

$$\mathbf{r}_{AD} = 0.25\mathbf{i} + 1.20\mathbf{j} + 0.43\mathbf{k}$$

and

$$\mathbf{N}_1 = N_1 \frac{\mathbf{r}_{AD}}{r_{AD}} = N_1 \frac{(0.25\mathbf{i} + 1.20\mathbf{j} + 0.43\mathbf{k})}{1.30}$$

Similarly,

$$\mathbf{N}_2 = N_2 \frac{(0.25\mathbf{i} + 1.20\mathbf{j} - 0.43\mathbf{k})}{1.30}$$

$$\mathbf{N}_3 = N_3 \frac{(-0.50\mathbf{i} + 1.20\mathbf{j})}{1.30}$$

Since the forces on the tripod are concurrent, the equilibrium equation is

$$\Sigma\, \mathbf{F} = \mathbf{N}_1 + \mathbf{N}_2 + \mathbf{N}_3 + \mathbf{W} = \mathbf{0}$$

or, in component form,

$$\Sigma\, F_x = 0.25N_1 + 0.25N_2 - 0.50N_3 = 0$$

$$\Sigma\, F_y = 1.20N_1 + 1.20N_2 + 1.20N_3 - W = 0$$

$$\Sigma\, F_z = 0.43N_1 - 0.43N_2 = 0$$

The weight of the instrument is

$$W = mg = (20 \text{ kg})(9.81 \text{ m/s}^2) = 196.2 \text{ N}$$

Substituting this value into the preceding equations and solving, we obtain

$$N_1 = N_2 = N_3 = 70.9 \text{ N} \qquad \textit{Answer}$$

Even though this problem is three-dimensional, the geometry is sufficiently simple that a solution can be obtained without the formal use of vector algebra. This can be done by first resolving the forces into horizontal and vertical components and then considering side and top views of the tripod. Application of this procedure is left as an exercise.

PROBLEMS

3.17 Determine the tension on strings *AC* and *BC* supporting the 2-kg lamp shown. (See Figure on p. 84.)

3.18 A speedboat tows a water skier and kite at constant speed and elevation as shown. If the tension on the towline at the kite has a magnitude of 900 N, what is the vertical lift and horizontal drag of the kite? The skier and skies have a combined mass of 81.6 kg. (See Figure on p.84.)

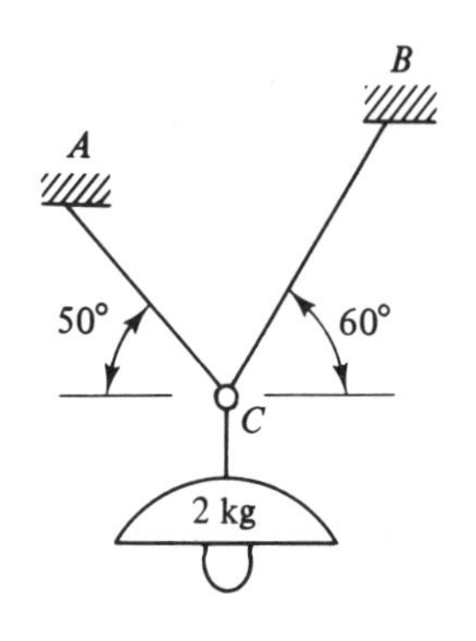

Problem 3.17

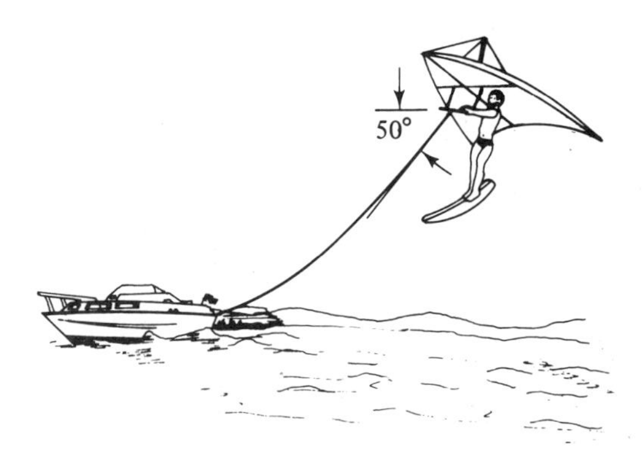

Problem 3.18

3.19 Determine the forces exerted on the uniform 100-lb cylinder shown by the walls of the smooth trough if $\theta = 40°$.

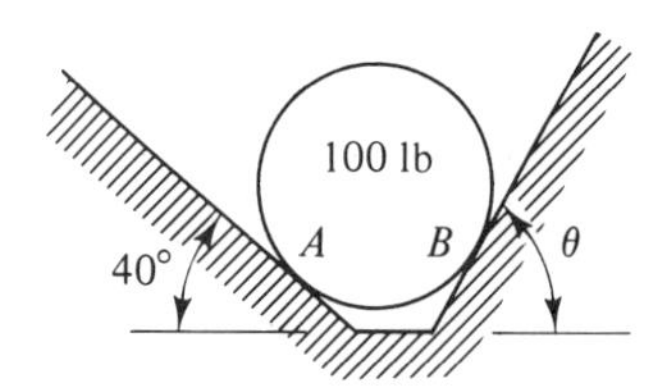

Problem 3.19

3.20 Same as Problem 3.19, except that $\theta = 65°$.

3.21 A boom with negligible weight supports a 1200-lb load as shown. Determine the compressive force on the boom and the tension on the supporting cable AB.

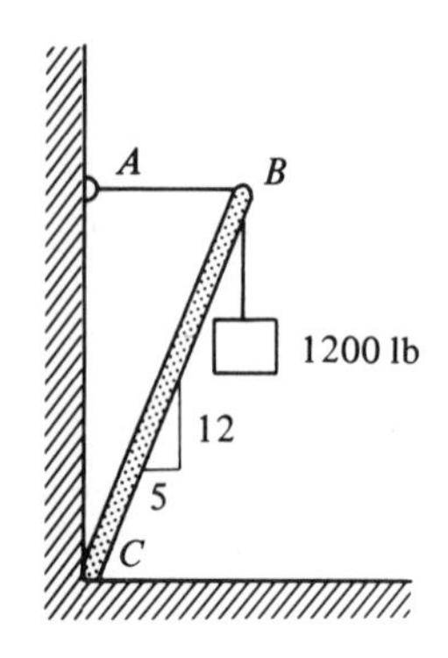

Problem 3.21

3.22 The beam shown has a mass per unit length of 25.5 kg/m and is supported by smooth pegs at A and B. Find (a) the reactions at the supports for $a = 0.5$ m and (b) the maximum value of a for which the beam can be held in equilibrium.

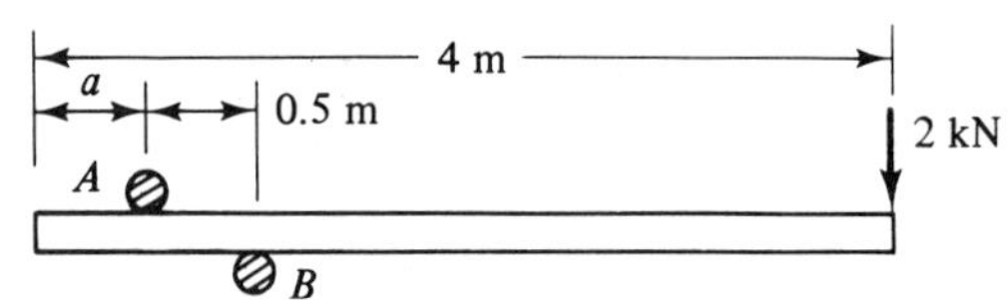

Problem 3.22

3.23 The uniform bent bar shown has a total weight of 75 lb. If $C = 50$ lb·ft, determine (a) the force F necessary for equilibrium and (b) the reactions at the pin O.

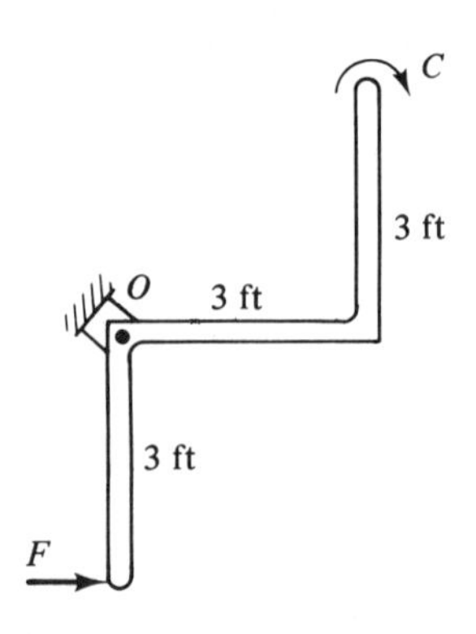

Problem 3.23

3.24 The uniform bent bar of Problem 3.23 has a total weight of 75 lb. If $F = 100$ lb, determine (a) the couple C necessary for equilibrium and (b) the reactions at the pin O.

3.25 The ladder shown weighs 50 lb, and the wall and floor are smooth. If the rope holding the ladder in place can withstand a tension of 60 lb, what distance d can a 200-lb man safely climb up the ladder?

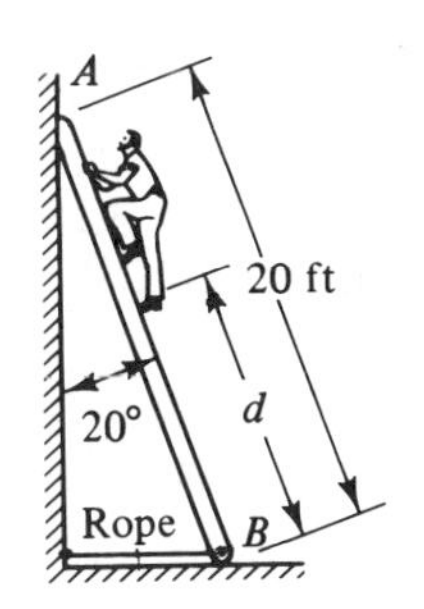

Problem 3.25

3.26 A 60-lb box is held on a smooth incline by a rope passing over a smooth peg as shown. Find (a) the tension on the rope and (b) the magnitude and location of the normal force exerted on the box by the plane.

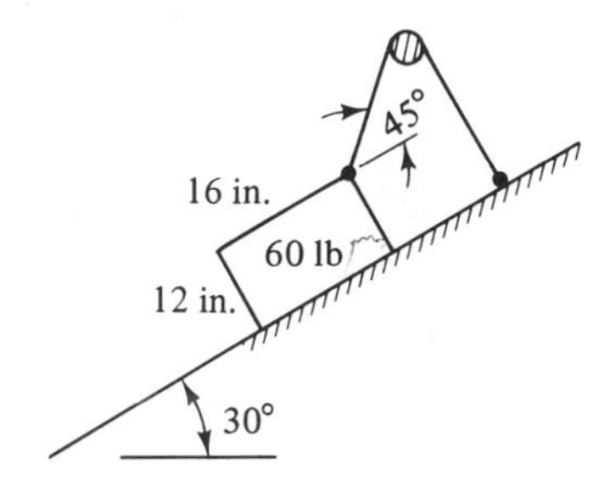

Problem 3.26

3.27 For the frictionless pulley shown, determine (a) the magnitude of the tension T necessary for equilibrium and (b) the corresponding reactions at the bearing O.

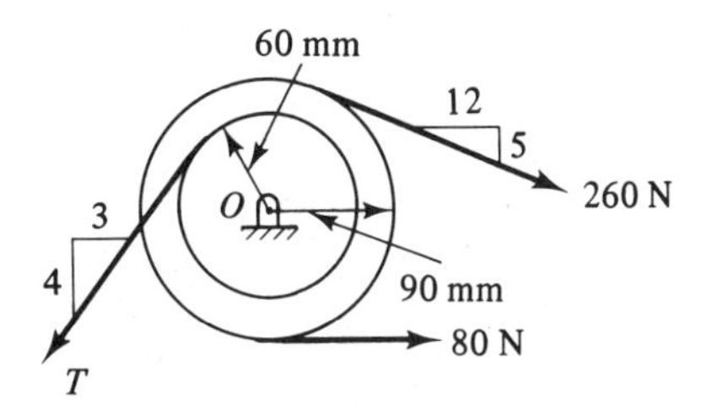

Problem 3.27

3.28 Determine the reactions at the base of the utility pole shown if the tensions on the wires supported are as indicated. The pole weighs 300 lbs.

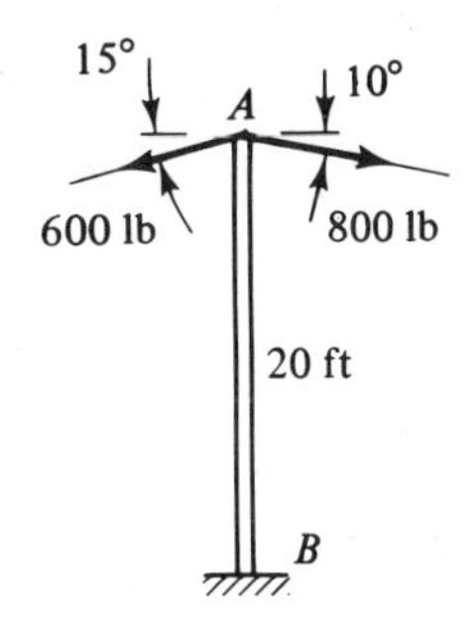

Problem 3.28

3.29 Find the reactions at the base of the diving board shown as a function of the position x of the 110-lb diver. The board is uniform and weighs 60 lb.

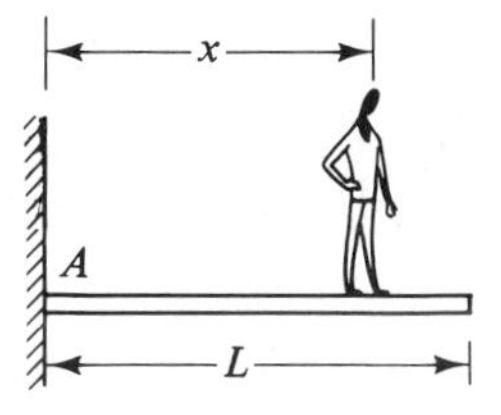

Problem 3.29

3.30 The 3-tonne truck shown hauls a 2-tonne load and travels up the incline with constant velocity. What is the steepest incline the truck can negotiate without tipping over?

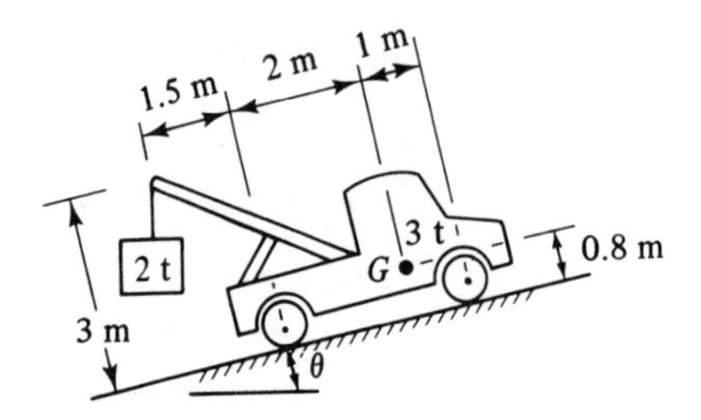

Problem 3.30

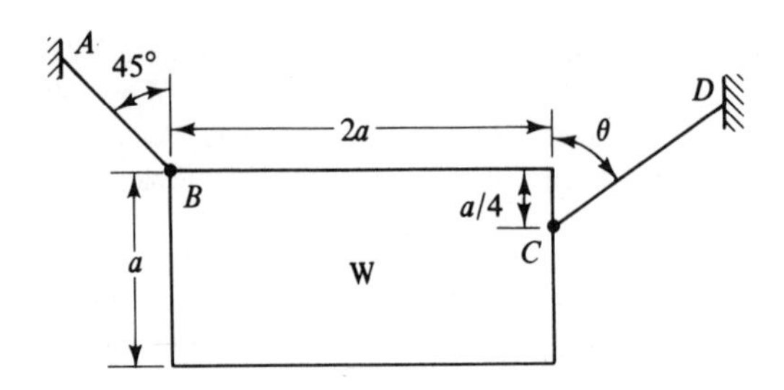

Problem 3.33

3.31 Calculate the angle θ for equilibrium of the uniform bar AB shown. The bar weighs 20 lb and all surfaces are smooth.

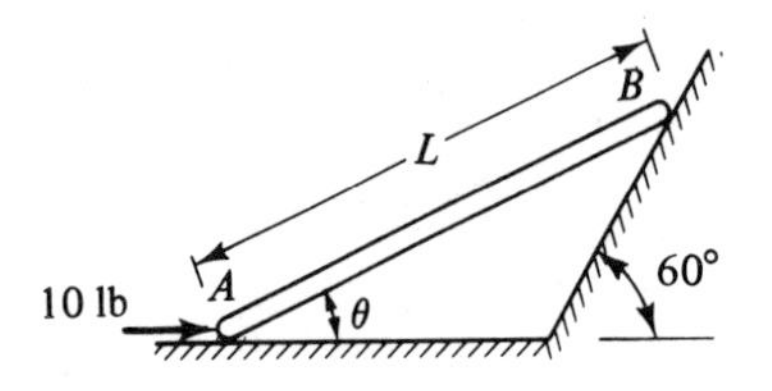

Problem 3.31

3.32 The occupant of the hammock shown has a mass of 50 kg. Determine the tension on the two support ropes.

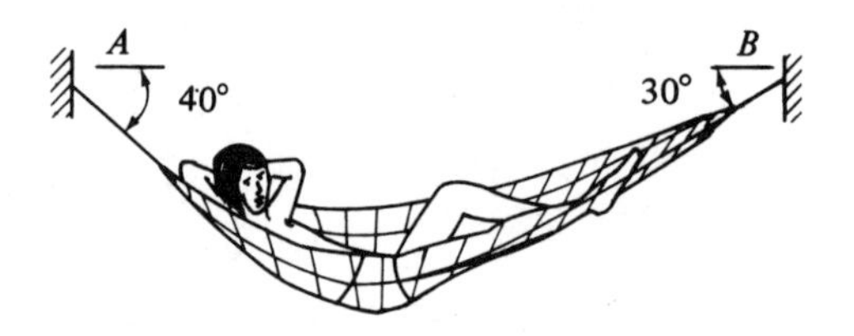

Problem 3.32

3.33 A homogeneous rectangular plate with weight W is in equilibrium when supported by two wires attached at points B and C as shown. Determine (a) the angle θ and (b) the tension on the wires.

3.34 A frame with negligible mass is supported by a smooth pin at A and a rope that passes over a smooth pulley at C as shown. Determine the tension on the cable and the reactions at A.

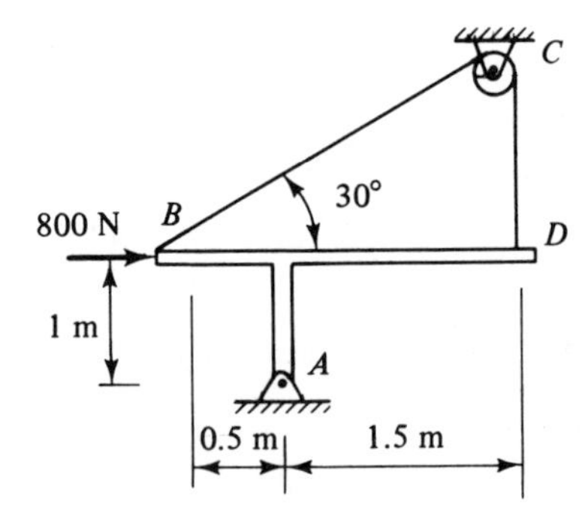

Problem 3.34

3.35 A pole is braced by three equally spaced wires as shown. Each wire is inclined at 30° from the vertical and supports a tension of 10 kips. What is the compressive force exerted on the tip of the pole?

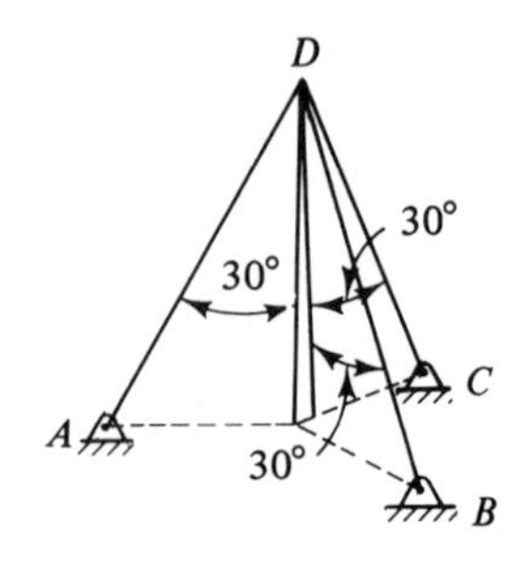

Problem 3.35

3.36 Determine the forces acting upon members AD, BD, and CD of the structure shown for $F = 0$. The members have negligible mass and are capable of supporting either tension or compression. Assume that the supports behave like ball and socket connections.

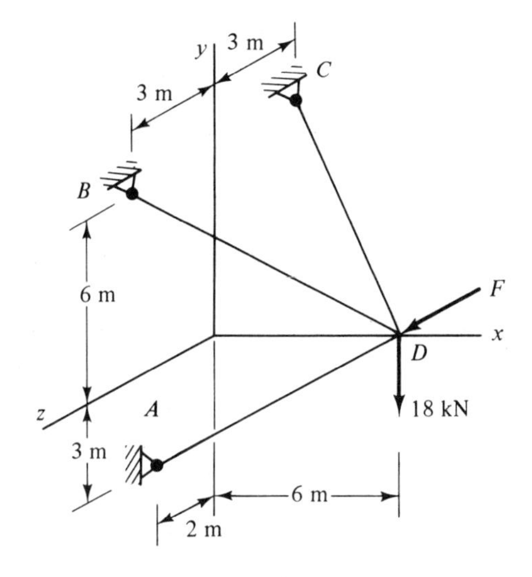

Problems 3.36 and 3.37

3.37 For the structure shown, determine the force F that must be applied parallel to the z axis so that the forces on members BD and CD are the same. What are the corresponding forces acting upon members AD, BD, and CD? The members have negligible mass and can support either tension or compression. Assume ball and socket connections.

3.38 The contents of the shopping bag shown have a combined weight of 30 lb. Determine the tension on each of the two cord handles.

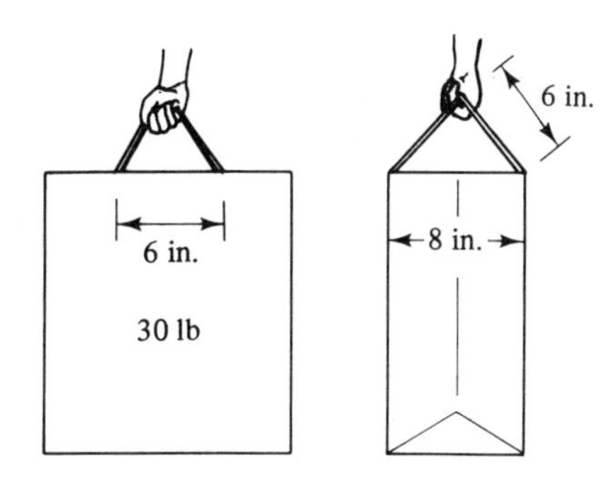

Problem 3.38

3.39 In the assembly shown, block A is constrained to slide along a smooth track that lies along the x axis. The block is held in place by two cords passing over smooth pulleys at B and C and connected to counterweights with mass m_1 and m_2. If $x = 3m$, find (a) the ratio m_1/m_2 for equilibrium and (b) the corresponding force exerted on block A by the track. Neglect the weight of the block.

3.40 In Problem 3.39, suppose that $m_1 = 1.5$ m_2. Determine (a) the position x of block A for which the system will be in equilibrium and (b) the corresponding force exerted on A by the track.

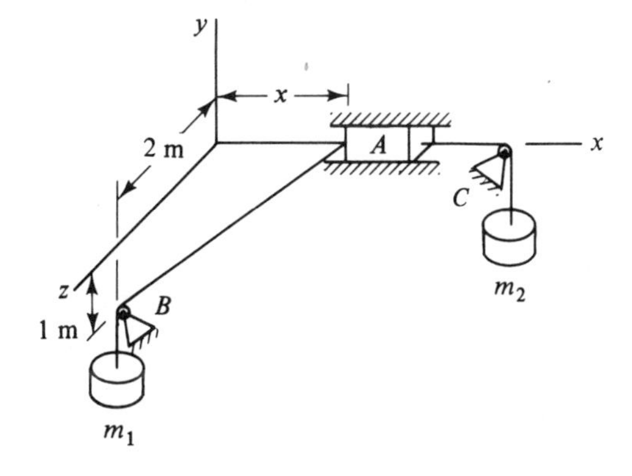

Problem 3.39

3.6 FORCE ANALYSIS OF GENERAL THREE-DIMENSIONAL SYSTEMS

Force analyses of general three-dimensional systems involve no principles not already encountered in our study of two-dimensional and other special force systems. The only difference is that all six equilibrium equations must usually be considered in the general three-dimensional case. Some of the equilibrium equations may be satisfied automatically, but these cannot be identified in general. The following examples illustrate the steps involved in the force analysis. Note that it is convenient to use a formal vector approach if the geometry of the problem is at all complicated.

Example 3.7. Force Analysis of a Boom. A safe with a weight of 10 000 lb is suspended from a boom of negligible weight in preparation for loading onto a truck [Figure 3.17(a)]. The boom is supported by a smooth pin at O and a cable at A. What are the reactions on the boom and the tension on the cable? Could the pin connection at O be replaced by a ball and socket connection?

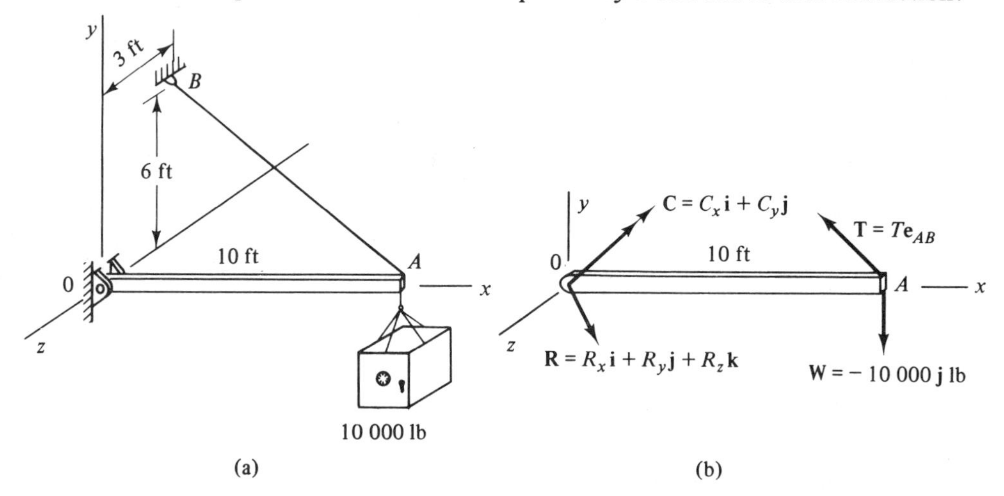

Figure 3.17

Solution. Since the pin at O can resist translation in all three coordinate directions, the force **R** exerted on the boom at O will have three unknown components. Similarly, the pin can resist rotation about the x and y axes; therefore, the couple **C** exerted on the boom will have two unknown components. Thus, the FBD of the boom is as shown in Figure 3.17(b), where **T** is the force exerted by the cable. From the geometry of the problem, we find

$$\mathbf{r}_{AB} = -10\mathbf{i} + \mathbf{j} - 3\mathbf{k} \text{ ft}$$

$$r_{AB} = \sqrt{(-10)^2 + (6)^2 + (-3)^2} = 12.04 \text{ ft}$$

$$\mathbf{T} = T\mathbf{e}_{AB} = T\frac{\mathbf{r}_{AB}}{r_{AB}} = \frac{T}{12.04}(-10\mathbf{i} + 6\mathbf{j} - 3\mathbf{k})$$

For equilibrium

$$\Sigma\,\mathbf{F} = \mathbf{R} + \mathbf{T} + \mathbf{W} = \mathbf{0}$$

or

$$\Sigma\,F_x = R_x - \frac{10}{12.04}T = 0$$

$$\Sigma\,F_y = R_y + \frac{6}{12.04}T - 10\ 000\ \text{lb} = 0$$

$$\Sigma\,F_z = R_z - \frac{3}{12.04}T = 0$$

Taking moments about point A, we have

$$\Sigma\,\mathbf{M}_A = \mathbf{C} + \mathbf{r}_{AO} \times \mathbf{R} = \mathbf{0}$$

where

$$\mathbf{r}_{AO} = -10\mathbf{i}\ \text{ft}$$

Thus,

$$\Sigma\,\mathbf{M}_A = C_x\mathbf{i} + C_y\mathbf{j} - 10\mathbf{i} \times (R_x\mathbf{i} + R_y\mathbf{j} + R_z\mathbf{k}) = \mathbf{0}$$

Performing the cross products and collecting like terms, we obtain the three scalar equations

$$\Sigma\,M_{Ax} = C_x = 0$$

$$\Sigma\,M_{Ay} = C_y + 10R_z = 0$$

$$\Sigma\,M_{Az} = -10R_y = 0$$

Solving the six equations of equilibrium for the six unknowns, we have

$$R_x = 16\ 670\ \text{lb} \qquad R_y = 0 \qquad R_z = 5000\ \text{lb}$$

$$T = 20\ 070\ \text{lb} \qquad C_x = 0 \qquad C_y = -50\ 000\ \text{lb}\cdot\text{ft}$$

Answer

If the pin at O is replaced with a ball and socket connection, $\mathbf{C} = \mathbf{0}$, because a ball and socket can't resist rotation. It is impossible to satisfy all the equilibrium equations in this case, which means that the boom cannot be held in equilibrium in the position shown. In order to maintain equilibrium, it would be necessary to add an additional constraint. Another cable attached to the boom would do, provided it was anchored to the wall at an appropriate point.

Example 3.8. Force Analysis of a Platform. Figure 3.18(a) shows a platform with weight W_P which is held in a horizontal position by three vertical wires. A man with weight W_M stands on the platform. Where should he stand so that the tension on each wire is the same?

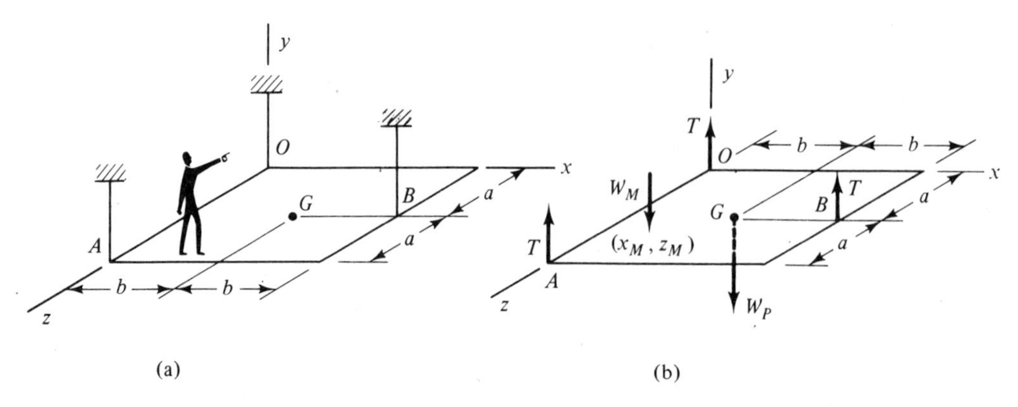

Figure 3.18

Solution. An FBD of the platform is shown in Figure 3.18(b). The position of the man is denoted by the coordinates x_M and z_M. The problem is three-dimensional, but a formal vector approach isn't necessary. The orientation of all the forces is readily defined on the FBD, and moments can be determined by inspection.

Setting the sum of forces in the vertical direction equal to zero and the sum of moments about the coordinate axes equal to zero (which is the same as summing moments about the origin and setting each component of the sum equal to zero), we obtain the equilibrium equations

$$\Sigma F_y = 3T - W_P - W_M = 0$$

$$\overset{+}{\curvearrowleft} \Sigma M_{0x} = W_M z_M + W_P a - Ta - T(2a) = 0$$

$$\overset{+}{\curvearrowleft} \Sigma M_{0z} = -W_P b - W_M x_M + T(2b) = 0$$

All other equilibrium equations are satisfied automatically. The sign convention used for moments is that of the right-hand screw rule; a moment is positive if it appears to be counterclockwise when viewed from the positive coordinate direction.

Solving for the unknowns, we have

$$T = \frac{W_P + W_M}{3} \qquad z_M = a \qquad x_M = \frac{b}{3}\left(2 - \frac{W_P}{W_M}\right) \qquad \textit{Answer}$$

Note that $x_M < 0$ if $W_P > 2W_M$. In this case there is no place the man can stand on the platform and have the tensions on the wires be the same.

Example 3.9. Force Analysis of a T-Bar. The T-bar shown in Figure 3.19(a) has a mass of 32.6 kg and is supported by smooth bearings at A and B and a smooth wall at C. Determine the force that the wall exerts on the bar.

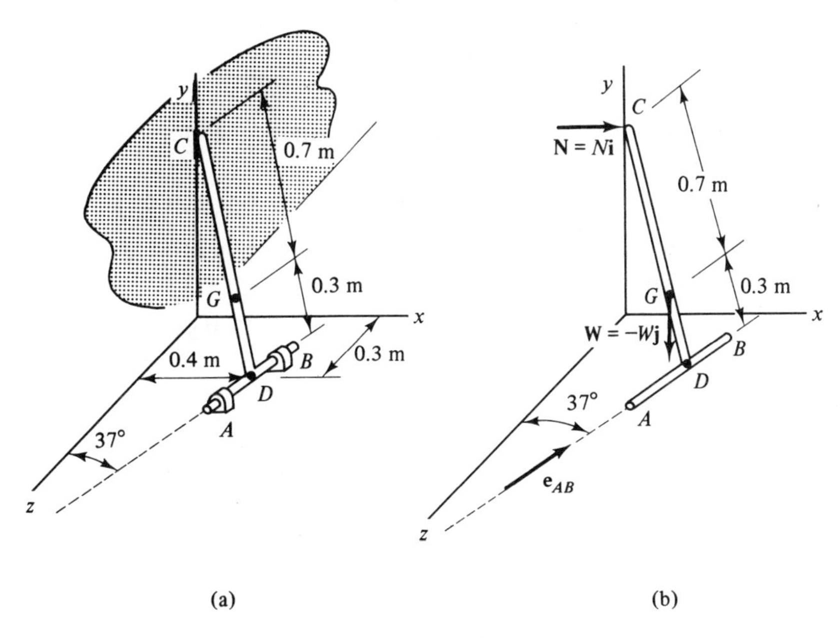

Figure 3.19

Solution. Instead of summing forces and considering moments about some point, we can obtain the solution more directly if we recognize that the sum of the moments about the bearing axis (line AB) must be zero. The only forces that have a moment about this axis are the normal force due to the wall and the weight. These forces are shown in Figure 3.19(b). Note that this figure is not an FBD because we have not shown all of the reactions on the bar.

The magnitude of the moment about line AB can be expressed as

$$M_{AB} = \mathbf{M}_D \cdot \mathbf{e}_{AB} = (\mathbf{r}_{DG} \times \mathbf{W} + \mathbf{r}_{DC} \times \mathbf{N}) \cdot \mathbf{e}_{AB}$$

From the given dimensions we see that

$$\mathbf{r}_{DG} = 0.3\mathbf{r}_{DC}$$

so

$$M_{AB} = \left[\mathbf{r}_{DC} \times (0.3\mathbf{W} + \mathbf{N})\right] \cdot \mathbf{e}_{AB}$$

The normal force is perpendicular to the wall ($y - z$ plane) and the weight is downward; so

$$\mathbf{N} = N\mathbf{i} \qquad \mathbf{W} = -(32.6\text{ kg})(9.81\text{ m/s}^2)\mathbf{j} = -320\mathbf{j}\text{ N}$$

The vertical coordinate of point C is not given, but it can be determined from the known distance DC. We have

$$\mathbf{r}_{DC} = -0.4\mathbf{i} + y_C\mathbf{j} - 0.3\mathbf{k}\text{ m}$$

$$r_{DC}^2 = (-0.4)^2 + y_C^2 + (-0.3)^2 = 1(\text{m})^2$$

from which we find

$$y_C = \sqrt{0.75} = 0.87 \text{ m}$$

The unit vector $\mathbf{e}_{AB}$ can be expressed in component form simply by resolving it into components along the coordinate axes:

$$\mathbf{e}_{AB} = \sin 37°\mathbf{i} - \cos 37°\mathbf{k} = 0.6\mathbf{i} - 0.8\mathbf{k}$$

Using the determinant form of the scalar triple product [Eq. (2.29)] for M_{AB}, we obtain the equilibrium equation

$$M_{AB} = \begin{vmatrix} 0.6 & 0 & -0.8 \\ -0.4 & y_C & -0.3 \\ N & -0.3W & 0 \end{vmatrix} = 0$$

or

$$-0.6(0.3)(0.3W) - 0.8\left[0.4(0.3W) - Ny_C\right] = 0$$

Solving for N, we get

$$N = \frac{0.15W}{0.8y_C} = \frac{0.15(320) \text{ N·m}}{0.8(0.87 \text{ m})} = 69 \text{ N} \qquad \textbf{\textit{Answer}}$$

PROBLEMS

3.41 Two beams with a mass per unit length of 50 kg/m are welded together to form the ell shape shown. Find the tension on each of the vertical supporting cables. The beams lie in a horizontal plane.

3.42 A round table with negligible mass is supported by three equally spaced legs as shown. Determine the force on each leg due to the loading indicated.

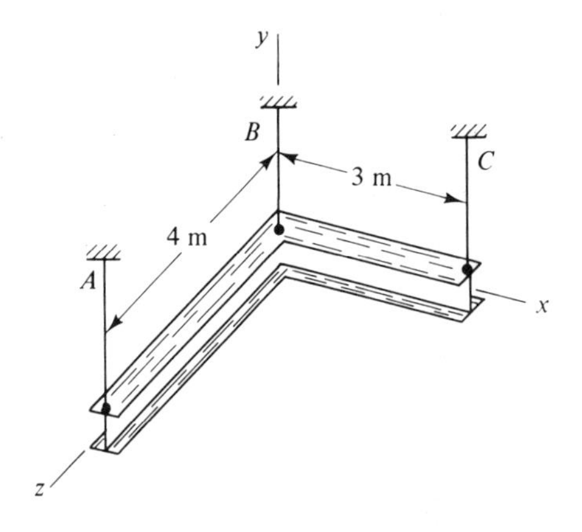

Problem 3.41

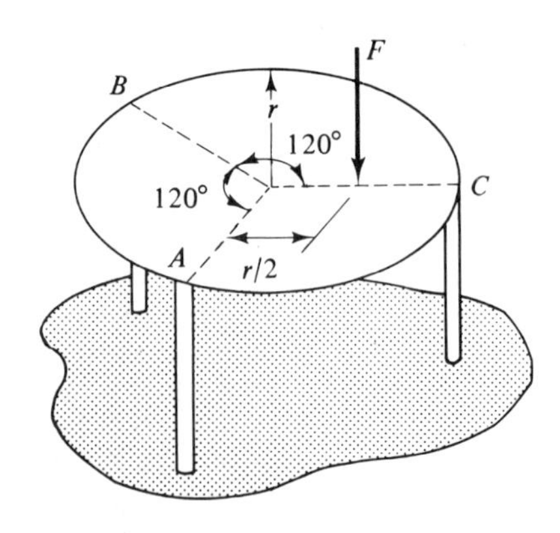

Problem 3.42

3.43 In Problem 3.42, within what region of the center of the table can the force F be applied without the force on any one of the legs exceeding the value $F/2$?

3.44 A counterweight W connected to the door shown by a rope attached at point C and passing over a smooth pulley at D serves as a door-closer. If $\theta = 20°$, what force F applied normal to the door is required to hold it open?

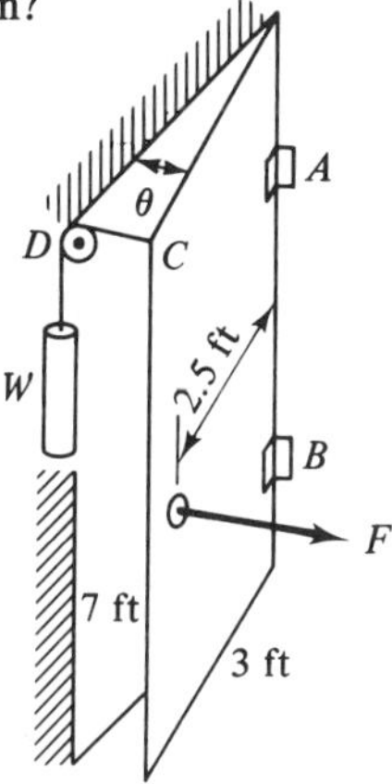

Problem 3.44

3.45 In Problem 3.44, determine the variation with the angle θ of the force F required to hold the door open. Show this variation by plotting F versus θ for $0° \leqslant \theta \leqslant 90°$.

3.46 The line shaft shown is supported by smooth bearings at A and B. Determine (a) the couple C necessary for equilibrium and (b) the reactions on the shaft at the bearings.

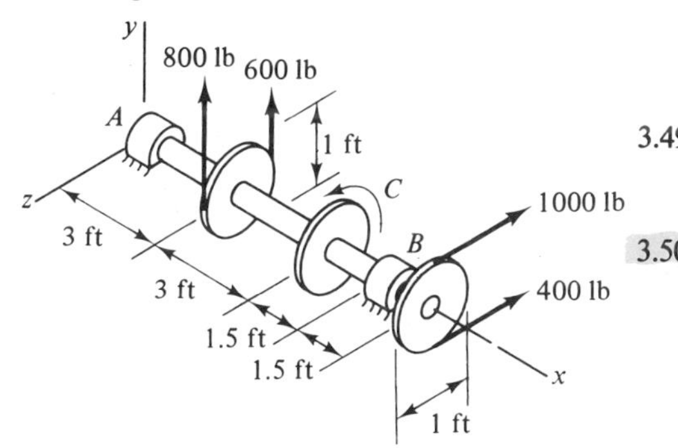

Problem 3.46

3.47 The uniform rod AB shown has a mass of 50 kg. It is supported by a ball and socket joint at B and a string DC attached to its midpoint C. The wall at A is smooth. Determine all the forces acting upon the rod.

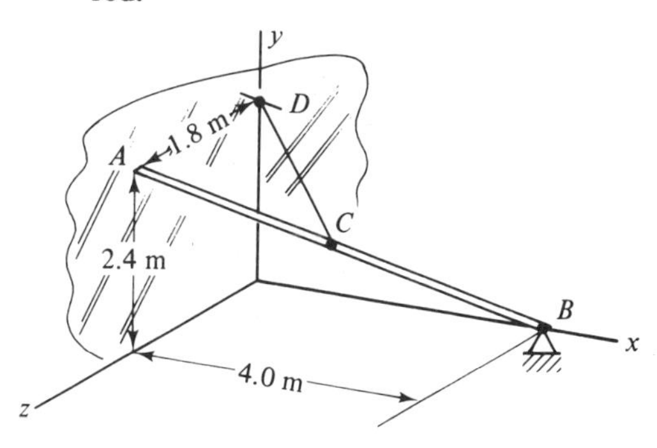

Problem 3.47

3.48 A Z-shaped member is subjected to the forces and couple shown. Determine the reactions at the wall.

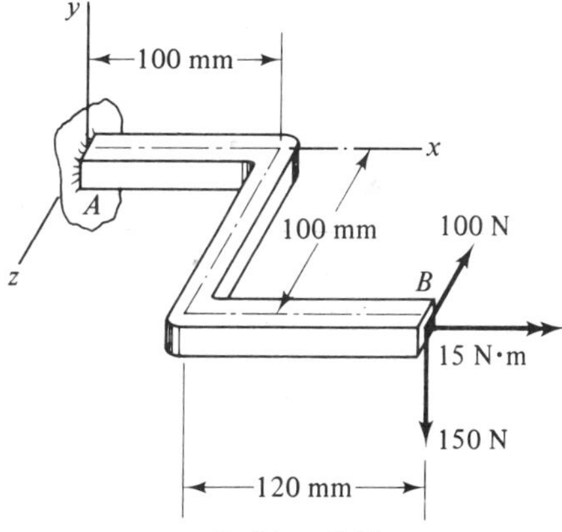

Problem 3.48

3.49 Same as Problem 3.48, except that the applied couple at B is directed vertically downward.

3.50 The 16-ft boom shown is held by a ball and socket joint at A and two cables ECF and FBG which pass around frictionless pulleys at B and C. A 900-lb load acting parallel to the z axis is applied at the top of the boom as shown. Determine the tension on each of the cables.

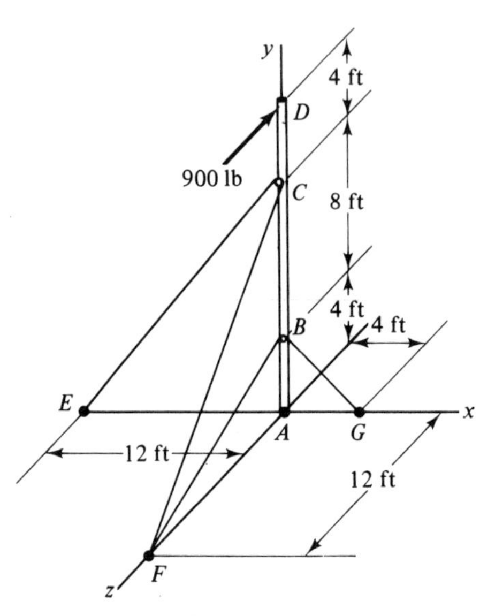

Problem 3.50

3.51 The uniform sign shown weighs 270 lb and is supported by a ball and socket joint at A and two slender rods BG and EF with negligible weight. Find the tensions on the rods and the reactions on the sign at A.

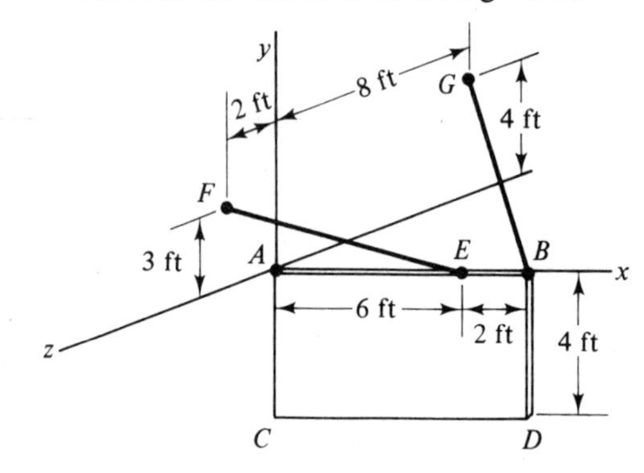

Problem 3.51

3.52 Same as Problem 3.51, except that rod EF is removed and the ball and socket joint at A is replaced by a hinge that allows rotation only about the z axis.

3.53 The gearbox shown weighs 10 000 lb and is subjected to couples C_1 and C_2. It is proposed that the box be supported by spherical rollers placed at corners A and B and at the midpoint of side CD. If $C_2 =$ 4000 lb·ft, are these supports adequate to hold the gearbox in position? If so, what are the reactions at the supports? If not, what changes could be made in the supports to correct the situation? The center of gravity of the box is at its geometric center.

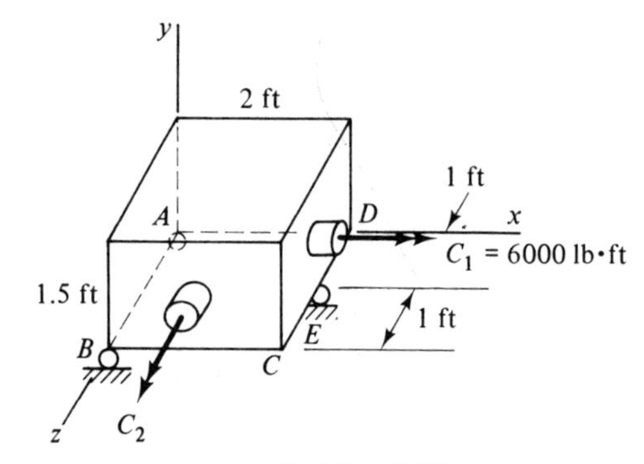

Problem 3.53

3.54 Same as Problem 3.53, except that C_2 = 12 000 lb·ft.

3.55 A 20.4-kg trap door in a horizontal floor is held partially open by a stick placed under one corner as shown. Determine the reactions on the door at the hinges A and B. Neglect any forces and couples believed to be insignificant.

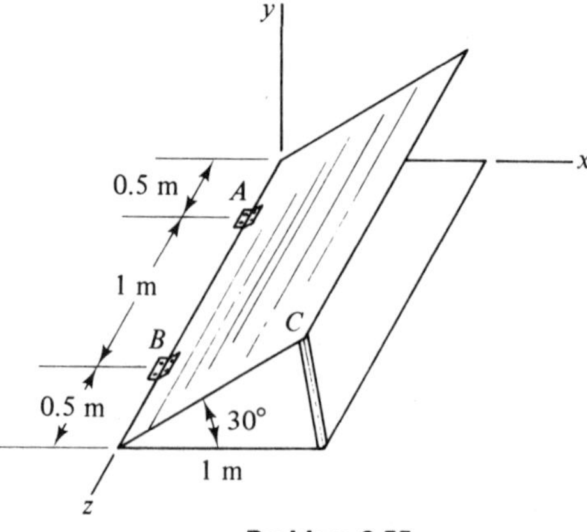

Problem 3.55

3.7. FORCE ANALYSIS OF COMPLEX SYSTEMS

Up to this point, all the force analyses encountered have been relatively simple in that only one free-body diagram had to be considered. This was because the systems consisted of only one major element. However, engineering structures and machines often consist of a number of members. The force analysis of such systems and their various elements involves no new principles, but the large number of free-body diagrams and equilibrium equations that are often involved does make it essential to have a systematic plan of attack.

Although there is no one procedure that is "best" for all problems, the following steps represent a reasonable approach in many instances:

1. Determine the reactions by considering the equilibrium of the entire structure.

2. Dismember the structure and draw FBDs of its various elements. Be sure to apply Newton's third law when considering the forces that one member exerts on another, and take advantage of any two-force members that may be present. FBDs of all the members may not be needed in order to determine the quantities of interest, but it usually does not require much extra effort to go ahead and draw them. There is less likelihood of making an error if all FBDs are drawn at one time.

3. Consider the equilibrium of as many individual members as necessary in order to determine all the desired unknowns. Try to avoid solving for unknowns that are not of interest, but this will not always be possible.

With experience, you will be able to modify these steps to best fit each different situation as it arises. These ideas are illustrated in the following examples.

Example 3.10. Force Analysis of a Bolt Cutter. What force P must be applied to the handles of the bolt cutter shown in Figure 3.20(a) if it takes a force of 75 lb to cut the bolt?

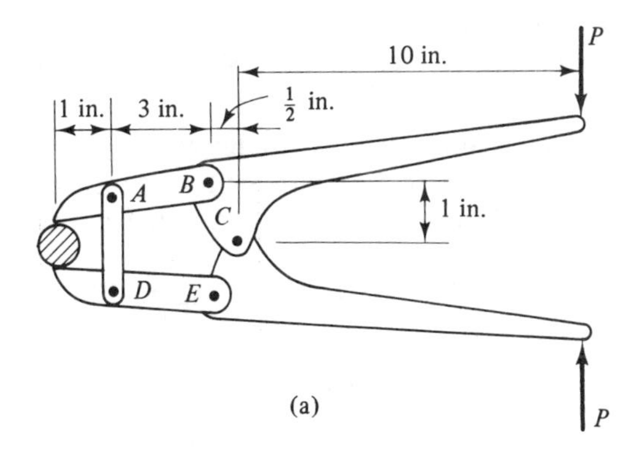

Figure 3.20 a

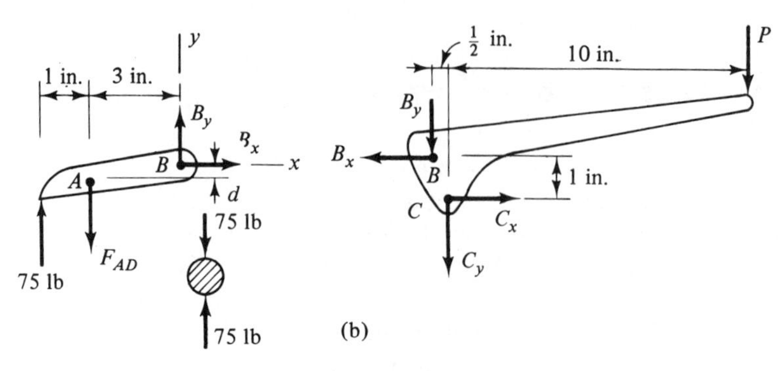

Figure 3.20b

Solution. Consideration of the equilibrium of the entire cutter gives no information other than the fact that the force P on each handle must be the same. Thus, we consider the individual parts. The FBDs of one of the handles and jaws are shown in Figure 3.20(b). Note that AD is a two-force member; therefore, the orientation of force F_{AD} is known.

For the jaw

$$\overset{+}{\curvearrowleft}\Sigma M_A = B_y(3 \text{ in.}) - 75 \text{ lb } (1 \text{ in.}) - B_x d = 0$$
$$\Sigma F_x = B_x = 0$$

from which we get

$$B_y = 25 \text{ lb} \qquad B_x = 0$$

For the handle

$$\overset{+}{\curvearrowleft}\Sigma M_C = B_y\left(\tfrac{1}{2} \text{ in.}\right) + B_x(1 \text{ in.}) - P(10 \text{ in.}) = 0$$

so

$$P = 1.25 \text{ lb} \qquad \textit{Answer}$$

Example 3.11. Force Analysis of a Frame. Determine the forces acting upon member AC of the pin-connected frame shown in Figure 3.21(a).

Solution. We shall start by considering the equilibrium of the whole frame, even though we are interested only in the forces on member AC. It is convenient to first remove the pulley so that attention can be focused on the basic structure [Figure 3.21(b)].

Referring to the FBD of the pulley [Figure 3.21(b)], we have

$$\overset{+}{\curvearrowleft}\Sigma M_C = Wr - Tr = 0$$
$$\Sigma F_x = C_x + T = 0$$
$$\Sigma F_y = C_y - W = 0$$

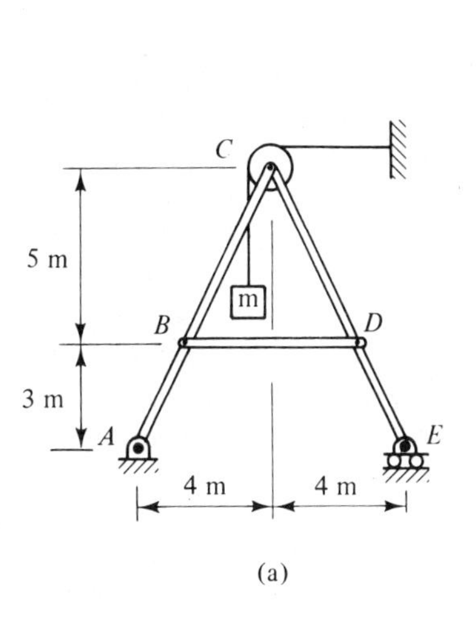

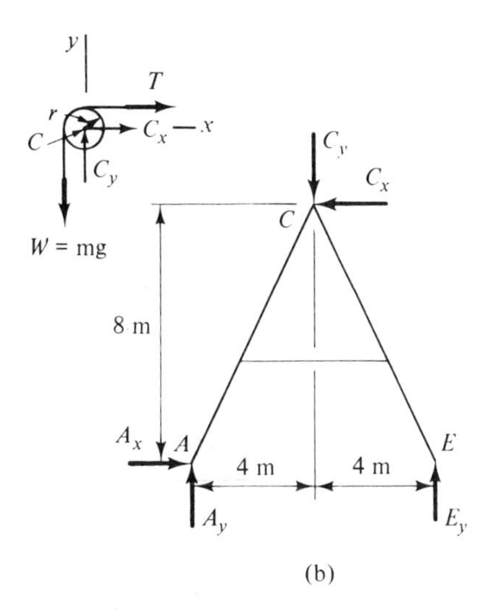

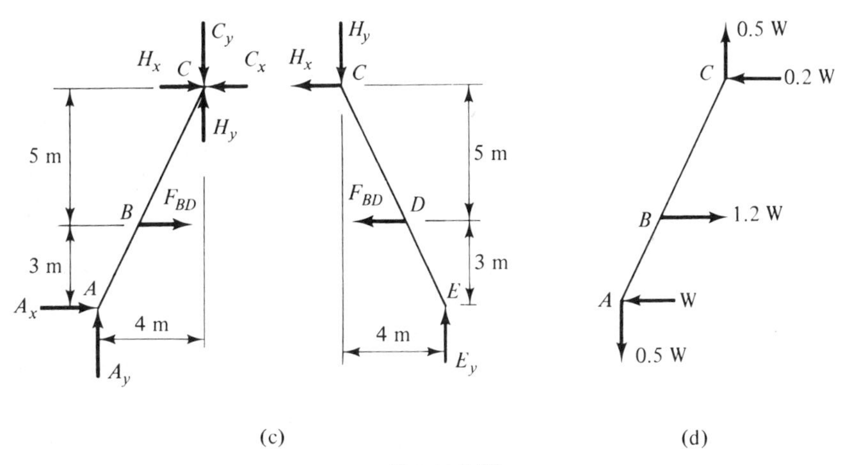

Figure 3.21

from which we obtain

$$T = W \qquad C_x = -W \qquad C_y = W$$

For equilibrium of the frame

$$\overset{+}{\curvearrowleft}\Sigma M_A = C_x(8\text{ m}) - C_y(4\text{ m}) + E_y(8\text{ m}) = 0$$
$$\Sigma F_y = A_y + E_y - C_y = 0$$
$$\Sigma F_x = A_x - C_x = 0$$

Solving these equations, we obtain for the reactions

$$A_x = -W \qquad A_y = -0.5W \qquad E_y = 1.5W \qquad \textit{Answer}$$

We now draw the FBDs of the individual members, but in doing so we find ourselves in a bit of a predicament at point C. Do we show the forces C_x and C_y as acting upon member AC or upon member EC? Actually, the members share these loads. However, the fraction of the loading that each member carries cannot be determined beforehand, and it is not necessary to do so. We need only show the forces as acting on one member or the other. The members also exert forces on each other, and the values of these forces are automatically adjusted in the analysis so that the correct total force on each member is obtained. This same situation will be encountered whenever there are three or more members connected at a common point or there are two or more members connected at a point where forces are also applied.

The FBDs of members AC and EC are shown in Figure 3.21(c), where C_x and C_y have been shown as acting upon AC. Note that we have made use of the fact that BD is a two-force member. The remaining unknowns can be obtained by considering the equilibrium of either AC or EC. We choose EC because it involves fewer forces:

$$\overset{+}{\curvearrowleft}\Sigma M_C = E_y(4\text{ m}) - F_{BD}(5\text{ m}) = 0$$
$$\Sigma F_x = -H_x - F_{BD} = 0$$
$$\Sigma F_y = -H_y + E_y = 0$$

Solving these equations, we get

$$F_{BD} = 1.2W \qquad H_x = -1.2W \qquad H_y = 1.5W \qquad \textbf{\textit{Answer}}$$

The net forces acting upon member AC are summarized in Figure 3.21(d).

The equilibrium equations for member AC can be used as a check. This step will be left as an exercise. It will also be left as an exercise to show that the same results would be obtained if the forces C_x and C_y were considered to act on member EC. The values of H_x and H_y will be different, but the net forces on each of the bars at C will be the same.

PROBLEMS

3.56 Two identical smooth cylinders are supported by smooth surfaces as shown. Determine all the forces acting upon the bottom cylinder.

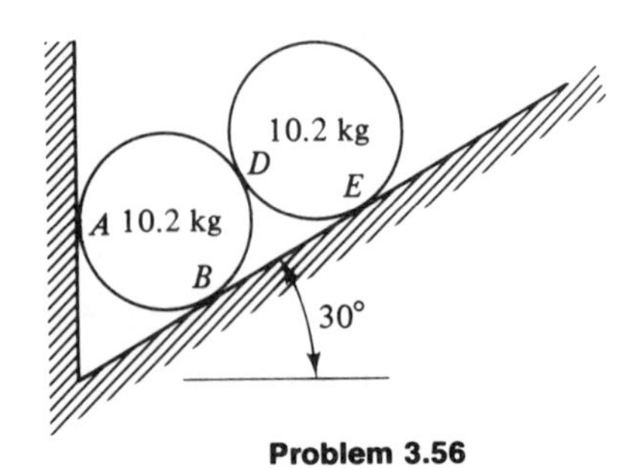

Problem 3.56

3.57 For the pulley arrangement shown, determine the pull P necessary for equilibrium. Neglect friction. (See Figure on p. 99.)

3.58 The top unit of the differential chain hoist shown consists of two grooved pulleys with diameters d_1 and d_2 keyed together to

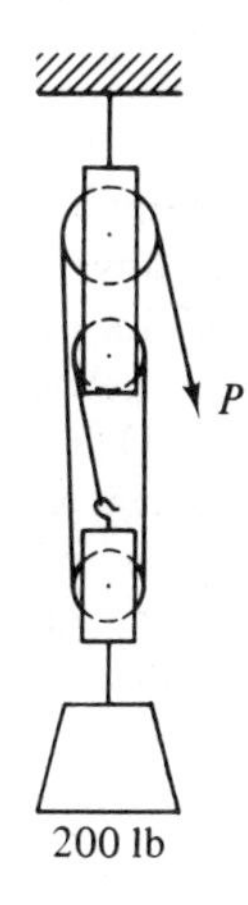

Problem 3.57

rotate as a unit. The diameter of the lower pulley is $\frac{1}{2}(d_1 + d_2)$. A continuous chain passes over the pulleys as shown. Assume that there is no tension on the slack side of the chain. Show that the force F necessary to lift an object with weight W is $F = W(d_2 - d_1)/(2d_2)$.

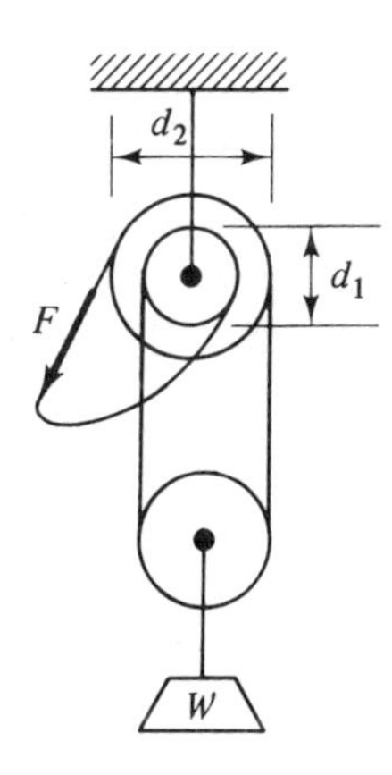

Problem 3.58

3.59 Determine the magnitude F of the forces that must be applied perpendicular to the handles of the pliers shown to generate a gripping force of 50 N on the bolt B. The gripping force acts perpendicular to the jaws of the pliers.

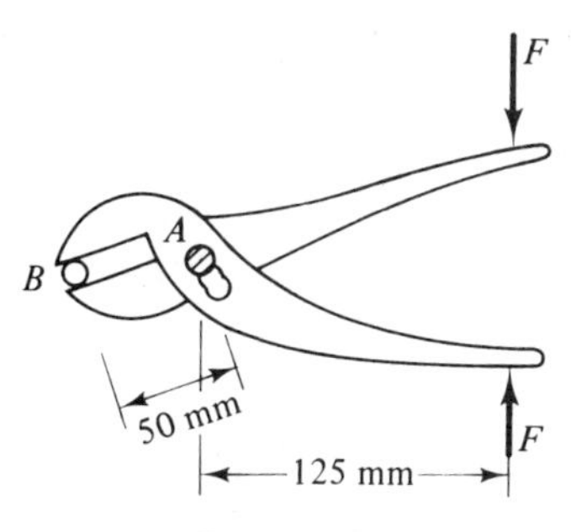

Problem 3.59

3.60 The length of the hacksaw shown can be adjusted by engaging pin A attached to member ABC in different circular notches in member DE. For the position shown, determine the forces acting upon member DE if the tension on the blade has a magnitude of 50 lb. Assume that the two members make contact only at pin A and at point D.

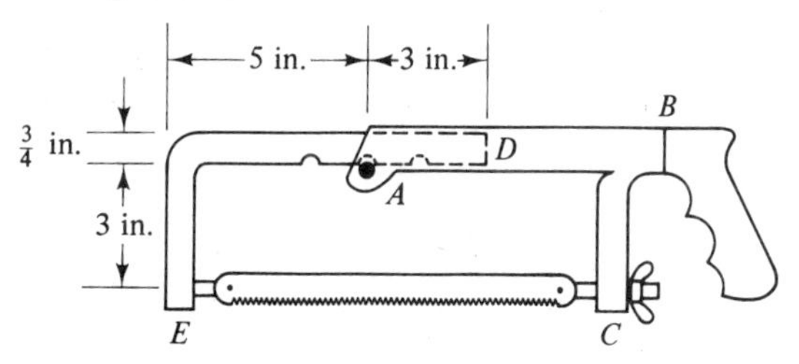

Problem 3.60

3.61 The piston in an engine is subjected to a 200-N force as shown. Find (a) the couple C necessary for equilibrium and (b) the reactions at B. Neglect friction and the weights of the members.

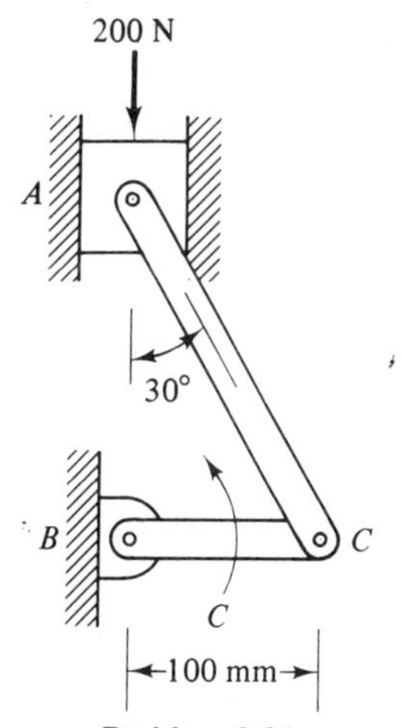

Problem 3.61

3.62 Determine the horizontal and vertical components of the reactions on the frame shown at B and C. Neglect the weights of the members.

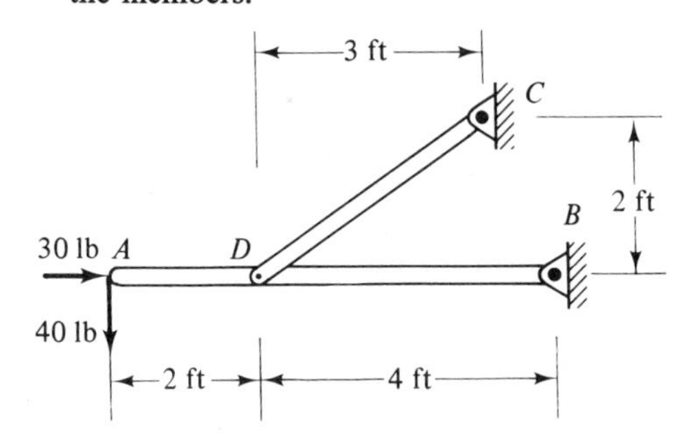

Problem 3.62

3.63 Same as Problem 3.62 except that member AB weighs 60 lb.

3.64 Determine the reactions at A and C for the structure shown. All surfaces are smooth.

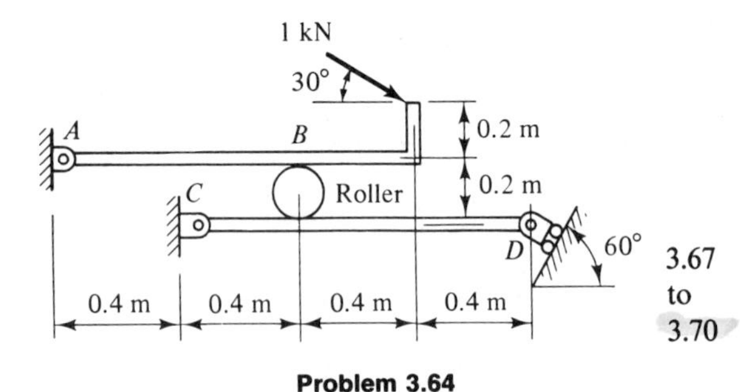

Problem 3.64

3.65 The folding camp chair shown consists of two U-shaped tubular legs hinged together at their midpoints and connected at the top with a canvas seat. When supporting a load of 100 lb, the left and right edges of the seat are inclined at an angle θ from the horizontal. Determine the forces acting upon each leg of the chair. Assume that the floor is smooth.

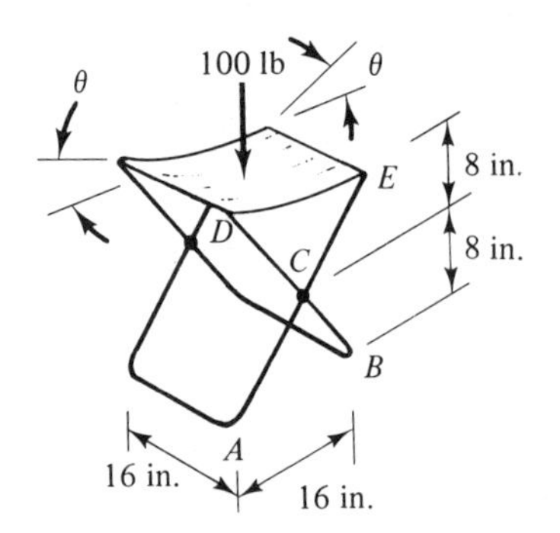

Problem 3.65

3.66 Determine the reactions at A and B and the forces acting upon member AED of the frame shown.

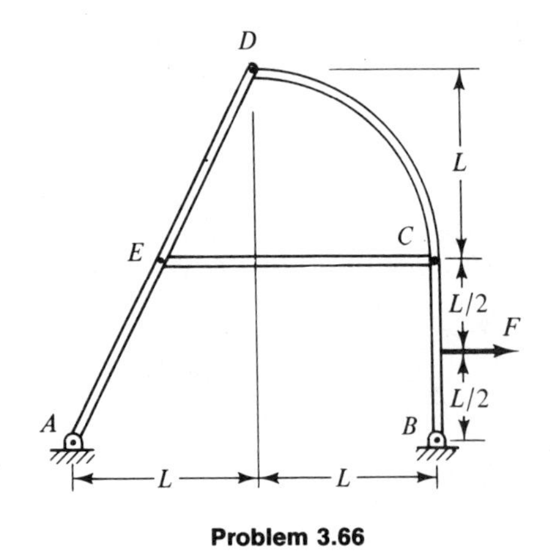

Problem 3.66

3.67 to 3.70 For the pin-connected frames shown, determine the horizontal and vertical components of the forces acting upon each member.

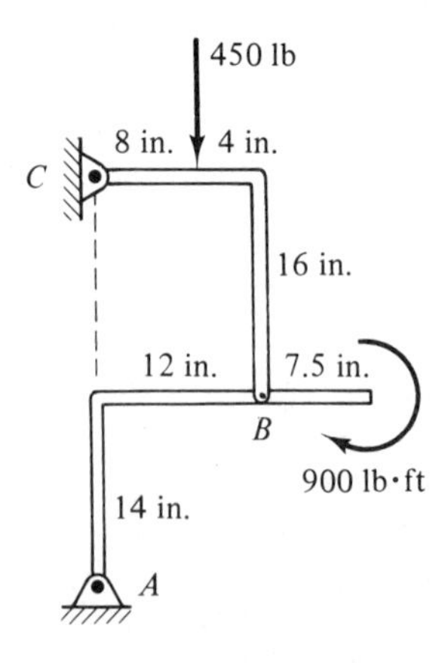

Problem 3.67

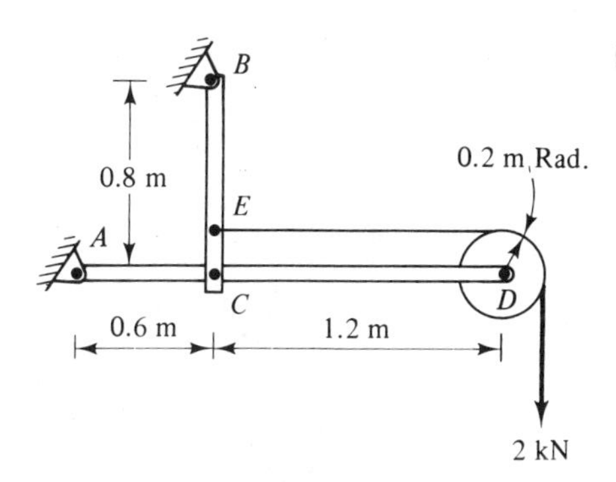

Problem 3.68

3.71 A trailer with a mass of 0.5 tonne is attached to an automobile with a mass of 1.5 tonne by a ball and socket trailer hitch. The locations of the centers of gravity are as shown. Determine the reactions at each axle when the system is at rest.

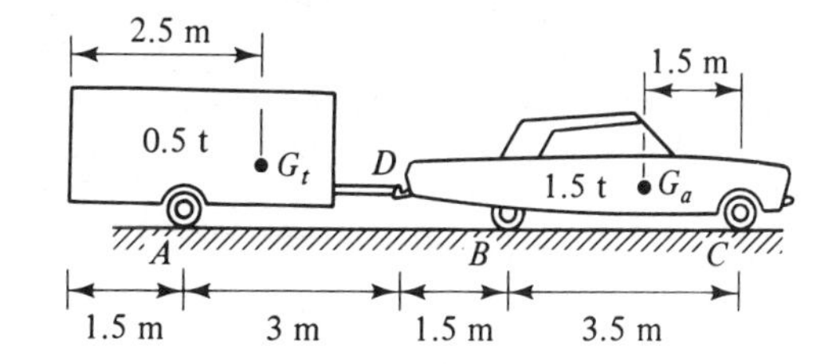

Problem 3.71

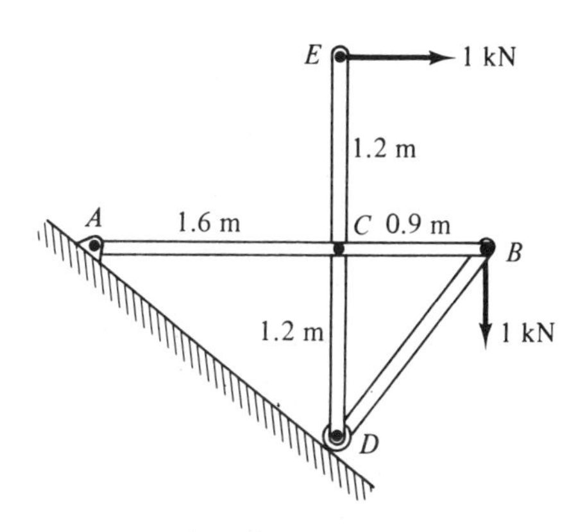

Problem 3.69

3.72 The figure shows a loader for handling pipe. The arm BCG is connected to linkages at B and C and is raised and lowered by the hydraulic cylinder DF, which is pinned at each end. The loader is symmetrical about a central vertical plane in the fore-and-aft direction, with an arm on each side. For the particular position shown, portion CG of the arm is horizontal. Determine the force that the hydraulic cylinder must exert on the arm to hold it in this position and find the force on link CE. The pipe section weighs 5 kips, and the weights of the various members are negligible compared to the forces they support.

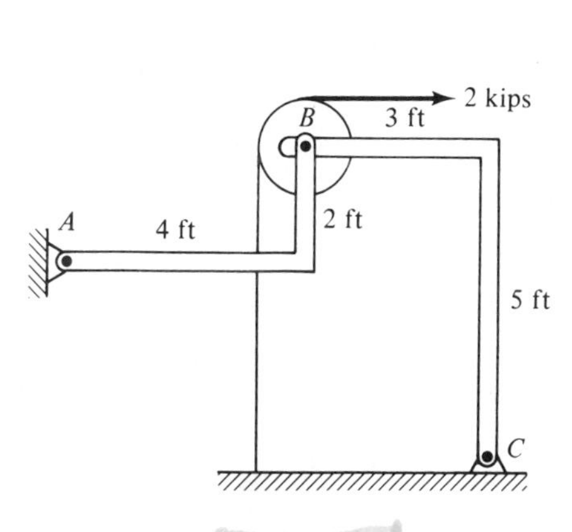

Problem 3.70

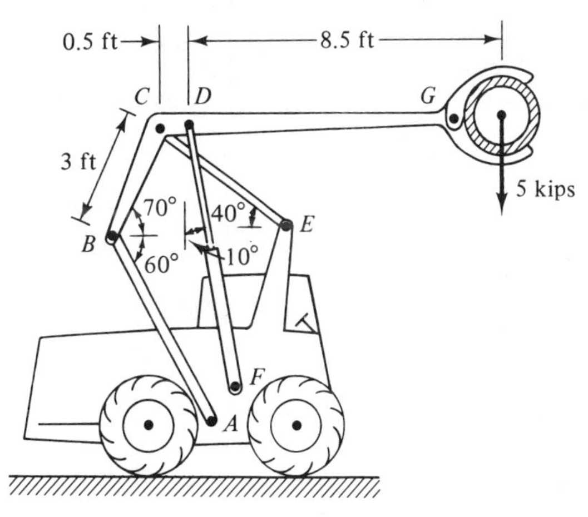

Problem 3.72

3.73 A folding table sits on a smooth floor and is loaded as shown. The legs are hinged at the top and are held in place by braces that lock into position. The two halves of the table top are connected by hinges at the bottom surface on each side at A and B and their top surfaces bear against each other, as shown in the enlargement. Determine the forces acting upon the hinge at A and the force acting upon brace CD.

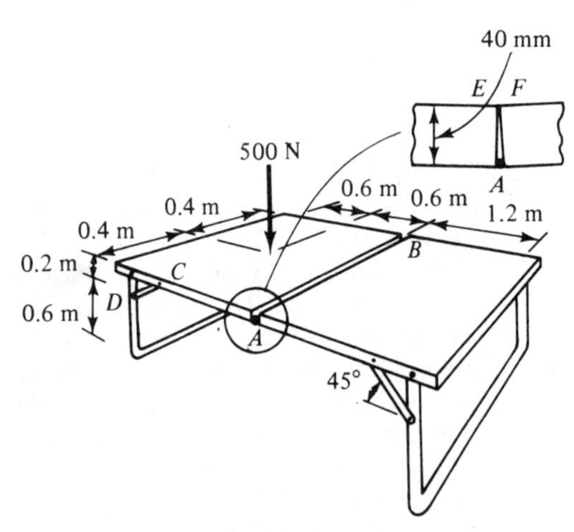

Problem 3.73

3.74 In the assembly shown, the smaller gear has a pitch diameter of 4 in. and the larger gear a pitch diameter of 8 in. If a 400-lb·in. counterclockwise couple is applied to the smaller gear, determine the couple acting upon the larger gear necessary for equilibrium.

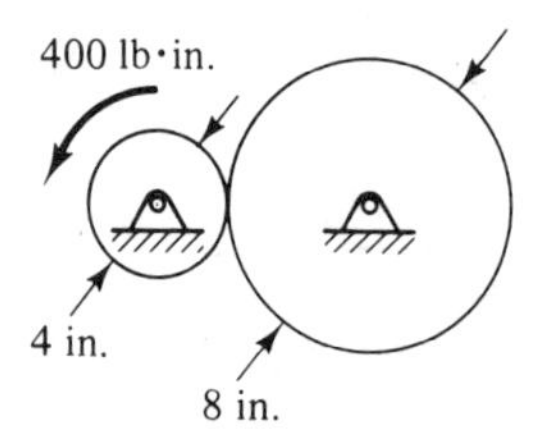

Problem 3.74

3.75 Bevel gears A and B shown have pitch diameters of 3 in. and 8 in., respectively. Determine the couple C necessary for equilibrium.

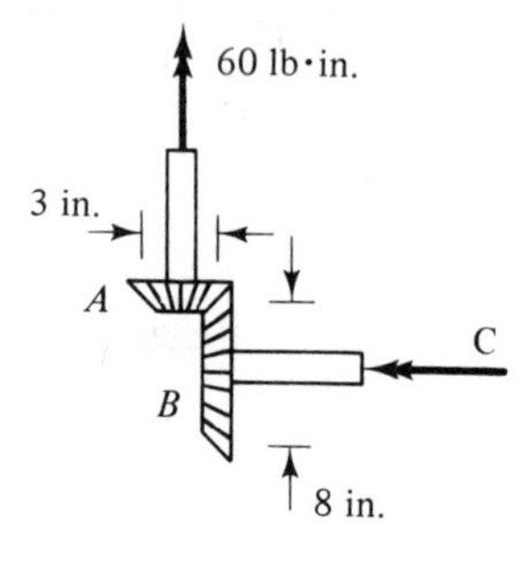

Problem 3.75

3.76 For the gear train shown, determine the couple acting upon the shaft at D necessary for equilibrium. The diameters given are the pitch diameters.

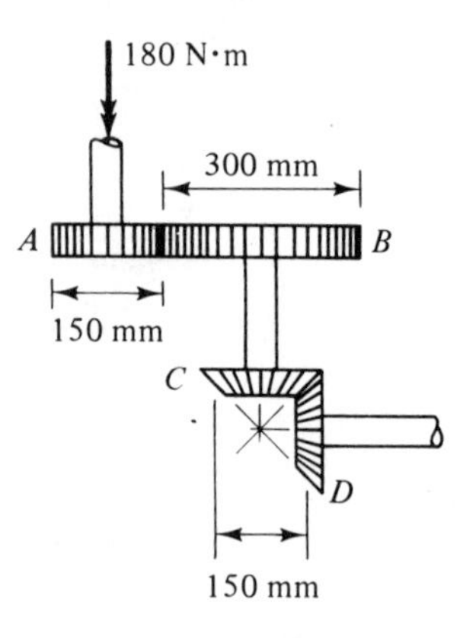

Problem 3.76

3.8 FORCE ANALYSIS OF TRUSSES

A *truss* is a structure consisting of a number of straight, slender members connected at their ends to form a rigid unit. By rigid, we mean that the structure will not collapse under a small applied load. For example, the structure shown in Figure 3.22(a) is nonrigid, and it will collapse if given a slight push. However, it can be made rigid by adding a brace, as in Figure 3.22(b). Note that the resulting structure forms two connected triangles. The triangle is the simplest rigid unit, and most trusses consist of a number of connected triangular elements.

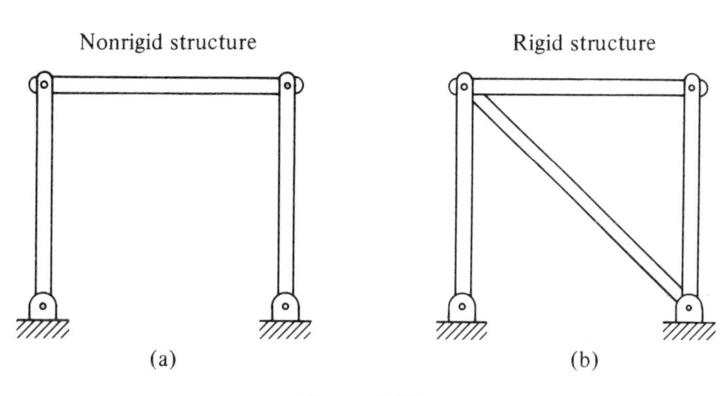

Figure 3.22

Trusses are efficient structures because they are lightweight and can support relatively large loads. They are commonly used in bridges, in the roofs of buildings, and in numerous other applications. Schematics of two common types of trusses are shown in Figure 3.23. These trusses are examples of so-called *plane trusses*, in which the individual members all lie in one plane. Trusses whose members do not all lie in one plane are called *space trusses.*

The force analysis of trusses is based upon the assumption that all members can be treated as two-force members. This feature distinguishes trusses from the frames and other complex systems considered in Section 3.7. The assumption of two-force members is reasonable if:

1. All connections are the equivalent of smooth pins (or ball and socket joints in the case of space trusses).

2. The weights of the members are negligible compared to the applied loads.

3. All loads and reactions consist only of forces (no couples) and act only at the joints.

Most trusses are designed so that the preceding conditions are met. For example, bridge decks are attached to the supportive side trusses only at the joints; consequently, the applied loads are transmitted to the trusses only at these locations. The connections in a truss are usually arranged so that the centerlines of the members joined together are concurrent, or nearly so. In this case, the joints

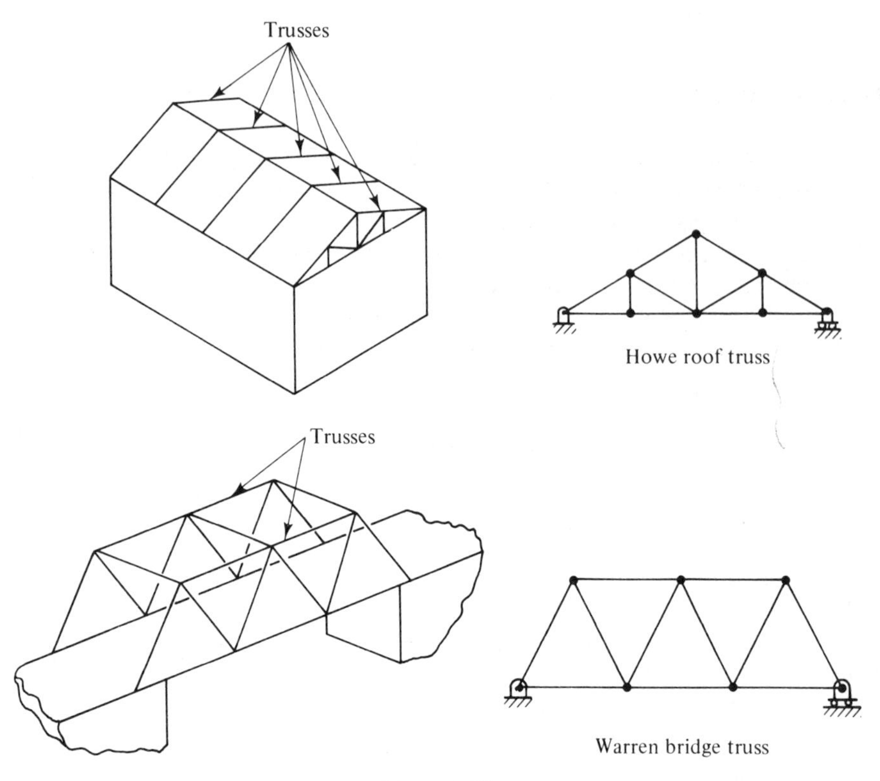

Figure 3.23

behave like pinned connections to a good degree of approximation, even though the members may actually be joined by bolts, rivets, welds, nails, adhesives, or other types of connectors. If the centerlines of the members are not concurrent at the connections, couples of appreciable magnitude can be exerted on their ends and they can no longer be treated as two-force members.

If the weight of a truss member is significant, it can be accounted for approximately by replacing the weight W with loads equal to $W/2$ acting at each end of the member. This procedure makes it possible to continue to treat the members as two-force members, and it gives the correct average tension or compression acting upon them. However, any effects due to bending of the members are not accounted for.

The object of a force analysis of a truss is to determine the tensile or compressive forces acting upon the various members. The procedure for doing this is basically the same as that used for frames and other complex systems in Section 3.7. We first determine the reactions by considering the equilibrium of the entire structure (not always necessary), and then we dismember the structure and determine the forces on the individual members. However, since trusses consist only of two-force members, the procedures for carrying out the second of these two steps are more specific than in the case of frames. Two common methods for determining the forces on the members of a truss are discussed in the following.

Method of Joints. In the *method of joints*, the forces acting upon the members of the truss are determined by considering the equilibrium of the connecting pins at each of the joints. To illustrate the procedure, let us consider the simple truss shown in Figure 3.24(a). We shall assume that the reactions at A and D have already been determined.

The free-body diagrams of the members and connecting pins are shown in Figure 3.24(b). Note that all reactions and applied forces are shown as acting directly upon the pins. Also, all members have been assumed to be in tension, in which case they tend to "pull" on the pins. This assumption simplifies the interpretation of the final results. If the value of a force turns out to be positive, we know immediately that the member upon which it acts is in tension; if the force is negative, the member is in compression.

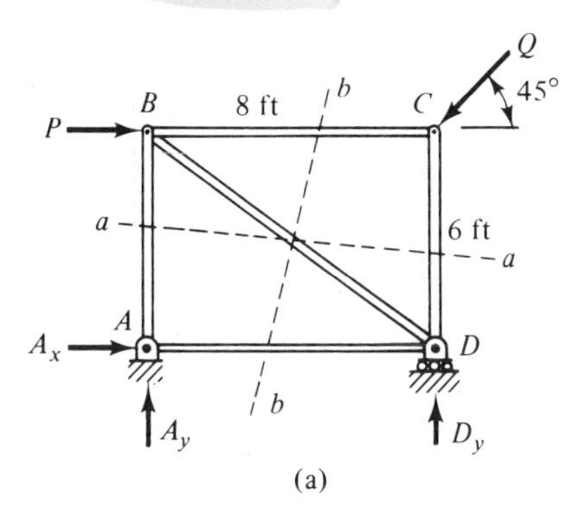

(a)

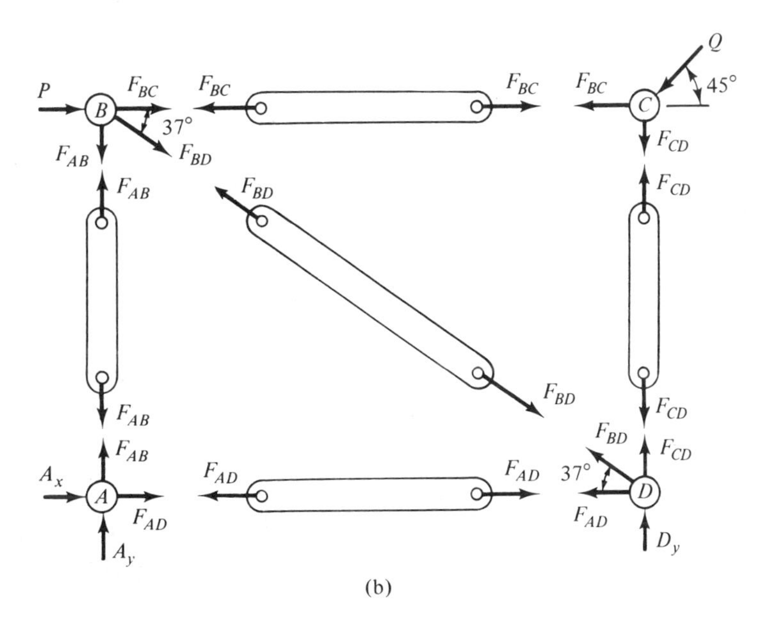

(b)

Figure 3.24

From the free-body diagrams of Figure 3.24(b), it is clear that the forces acting upon the members can be determined from the equations of equilibrium ($\Sigma F_x = 0$ and $\Sigma F_y = 0$) for the pins. (For space trusses we also have $\Sigma F_z = 0$.) We start at a pin at which there are no more than two unknowns, such as pins A or C in Figure 3.24(b). Once we have determined the unknowns at the first pin, we proceed to another. The process is repeated until all the unknowns have been determined. The order in which the pins are considered is immaterial; the only restriction is that the pin under consideration not involve more unknowns than equations of equilibrium.

For example, suppose we start at pin A and determine F_{AB} and F_{AD}. Once F_{AB} and F_{AD} are known, we can go to pin B or pin D, or we can go to pin C. Suppose we go to pin B and solve for F_{BC} and F_{BD} and then go to pin C and solve for the remaining unknown F_{CD}. The equilibrium equations for pin D are not needed; they can be used as a check.

It is not necessary to draw free-body diagrams of the truss members when using the method of joints. These diagrams were included here only to illustrate more clearly the force interactions between the members and the connecting pins. Only the free-body diagrams of the pins are needed.

Method of sections. In the *method of sections*, the forces on the members of the truss are determined by imagining that the truss is cut into sections and then considering the equilibrium of the resulting pieces. To illustrate the procedure, let us again consider the truss of Figure 3.24(a).

If the truss is sectioned along line a-a, the free-body diagrams of the resulting pieces are as shown in Figure 3.25. Again, it is convenient to assume that all members are in tension. Since the forces acting upon each section of the truss are coplanar, there are three equilibrium equations for each section (there are six equations for space trusses). Thus, the forces F_{AB}, F_{BD}, and F_{CD} can be determined by considering the equilibrium of either the top or bottom piece of the truss.

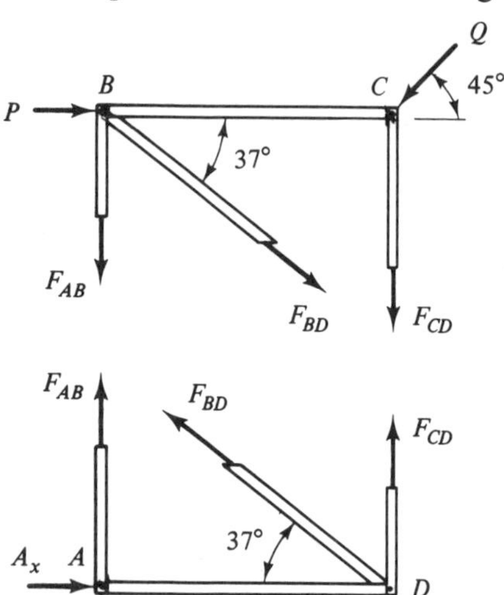

Figure 3.25

In order to determine the forces on the rest of the members, it is necessary to section the truss in a different way. For example, the remaining unknowns F_{BC} and F_{AD} can be determined by sectioning the truss along line b-b [Figure 3.24(a)] and considering the equilibrium of either the left or right portions.

There is no definite procedure for determining where or how to section a truss, but the preceding example does reveal two general guidelines: (1) The section must pass through the member on which the force is to be determined and (2) the resulting free-body diagram should not involve more unknowns than equations of equilibrium. The first condition guarantees that the force of interest will appear on the free-body diagram and enter into the equilibrium equations, and the second condition guarantees that the force can be determined from these equations.

As a rule, the method of sections involves less work than the method of joints, but it requires more ingenuity to apply. The method of sections is particularly convenient when the forces on only a few members in the interior of a truss are to be determined. In many problems it is advantageous to use a combination of the method of sections and method of joints.

Zero-force members. Under any one given loading, the forces on some members of a truss may be zero. Member BD of the truss shown in Figure 3.26 is an example of such a *zero-force member*, as can be seen from the free-body diagram of pin D. Of course, a zero-force member under one loading may not be a zero-force member under a different loading. For instance, member BD would not be a zero-force member if the load were applied at joint D instead of joint B.

Even though they support no load, zero-force members have an important function. They serve to stabilize a truss and to hold in position the members that do carry a load. For example, if the zero-force member BD were removed from the truss shown in Figure 3.26, joint D would be free to move up or down, and any small lateral load or disturbance applied to member AD or DC would cause the truss to fold.

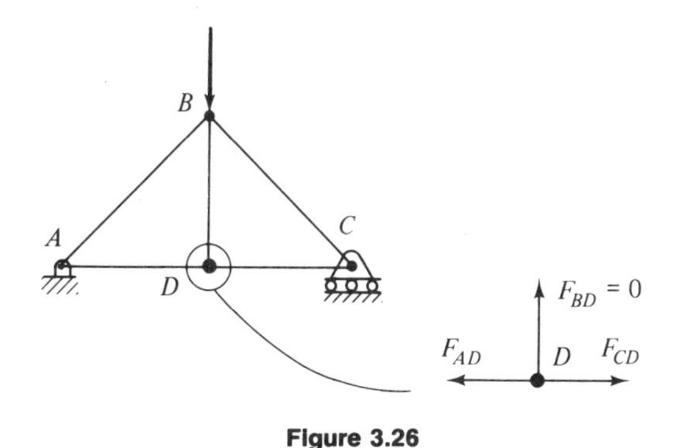

Figure 3.26

Example 3.12. Force Analysis of a Truss. Determine the forces on members BC and DE of the truss shown in Figure 3.27(a) by using (a) the method of joints and (b) the method of sections.

Solution. (a) The forces on members BC and CE can be determined by considering joint C; the force on member DE can then be obtained by considering joint E. If it is assumed that all members are in tension, the FBDs of pins C and D

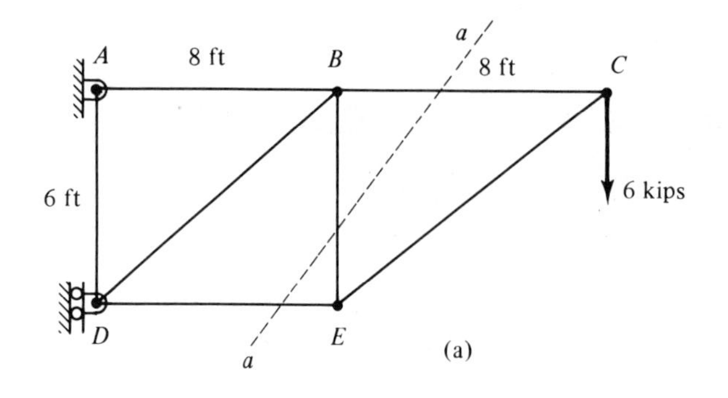

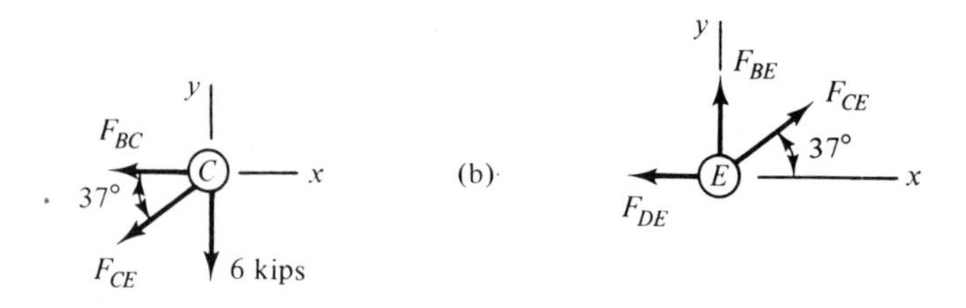

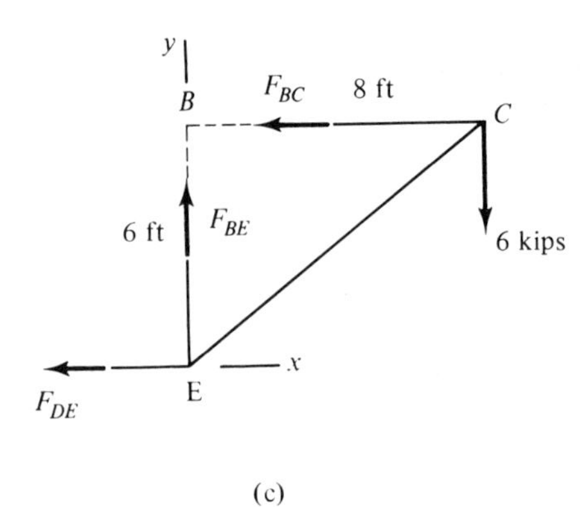

Figure 3.27

are as shown in Figure 3.27(b). For equilibrium, we have:

PIN C:
$$\Sigma F_x = -F_{BC} - F_{CE}\cos 37^\circ = 0$$
$$\Sigma F_y = -6 \text{ kips} - F_{CE}\sin 37^\circ = 0$$
$$F_{CE} = -10 \text{ kips or } 10 \text{ kips (compression)}$$ ***Answer***
$$F_{BC} = +8 \text{ kips or } 8 \text{ kips (tension)}$$

PIN E:
$$\Sigma F_x = F_{CE}\cos 37^\circ - F_{DE} = 0$$
$$F_{DE} = -8 \text{ kips or } 8 \text{ kips (compression)}$$ ***Answer***

(b) Sectioning the truss along line *a-a* and considering the portion to the right, we obtain the FBD shown in Figure 3.27(c). Again, all members are assumed to be in tension. For equilibrium

$$\overset{+}{\curvearrowleft}\Sigma M_E = F_{BC}\,(6\text{ ft}) - 6\text{ kips}(8\text{ ft}) = 0$$

$$\Sigma F_x = -F_{DE} - F_{BC} = 0$$

from which we obtain the same results as before:

$$F_{BC} = 8\text{ kips (tension)}$$

$$F_{DE} = 8\text{ kips (compression)}$$

Note that the reactions did not enter into the computations in this problem; thus, there was no need to determine them. This is not usually the case, however.

PROBLEMS

3.77 Determine the forces acting on each member of the truss shown. Use the method of joints.

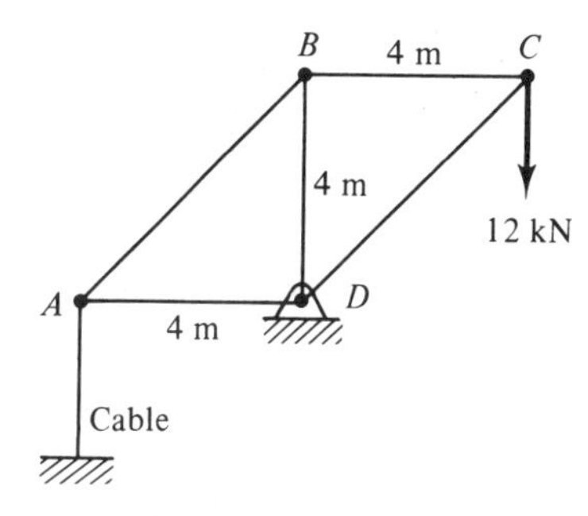

Problem 3.77

3.78 Each member of the truss shown is a uniform 6-ft bar weighing 40 lb/ft. Determine the average tension or compression on each member due to the weights of the members. Use the method of joints.

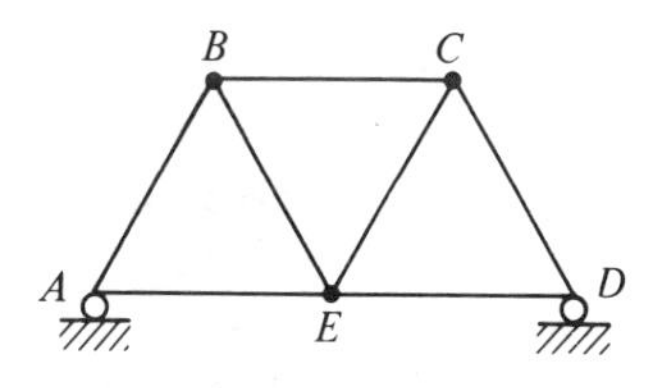

Problem 3.78

3.79 A 900-lb reel of electrical cable is supported as shown. If the tension on the cable has a magnitude of 260 lb, determine the force on each member of the supporting truss ABC. The members weigh 40 lb each and are of equal length. If the weights of the truss members are neglected, what is the percent error in the resulting values of the forces?

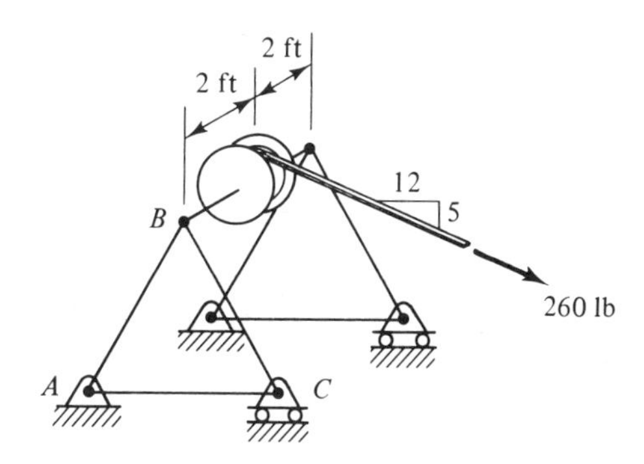

Problem 3.79

3.80 to 3.82 For the trusses shown, determine the forces acting upon the members indicated by using the method of sections.

3.80 Members BC and BD.

3.81 Members DC and DE.

3.82 Members DE and EI.

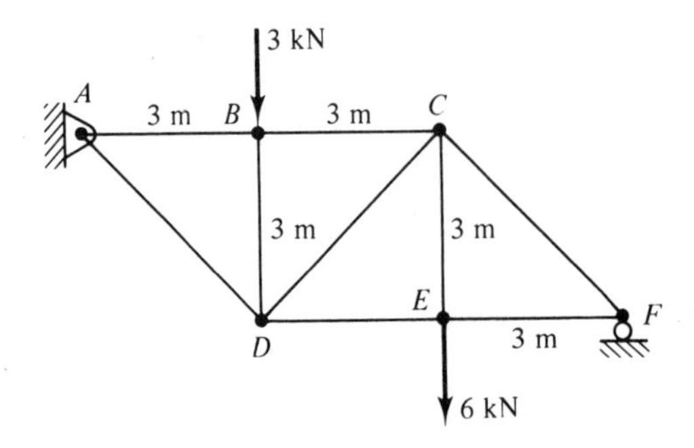

Problem 3.80

3.83 Members AB, BC, and CE.
3.84 Members CD, DF, FG, and CG.
3.85 All members.

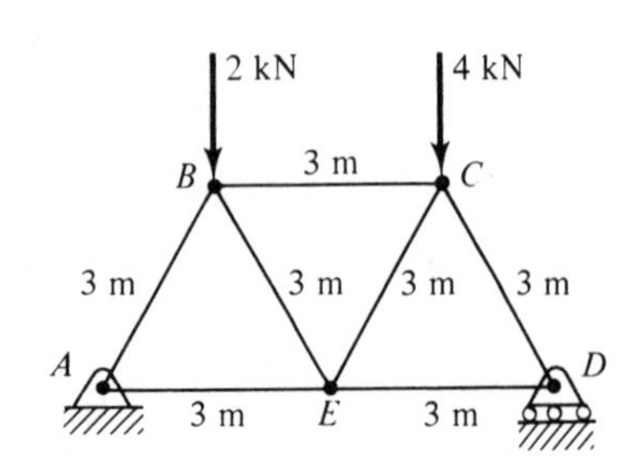

Problem 3.83

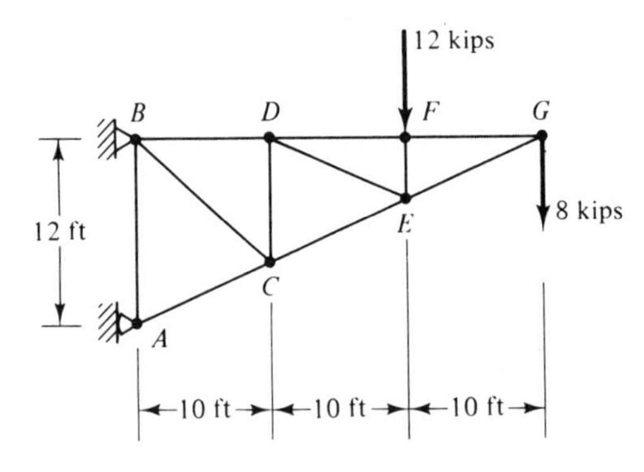

Problem 3.81

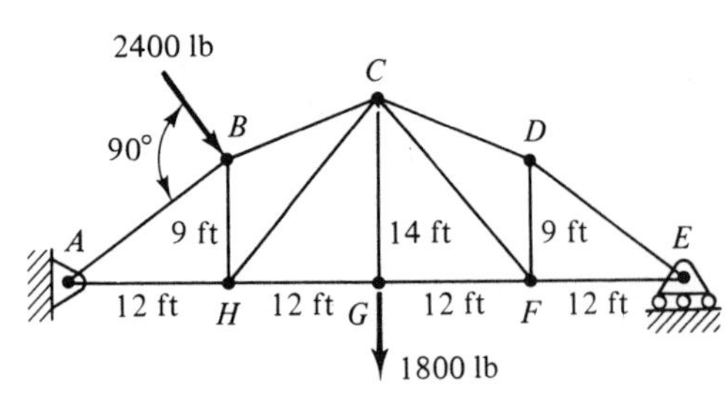

Problem 3.84

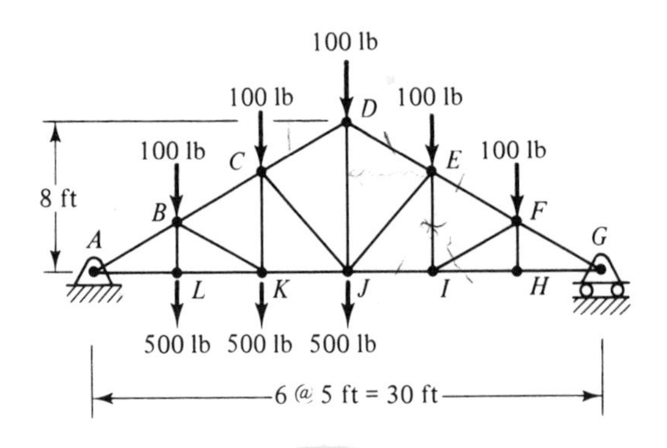

Problem 3.82

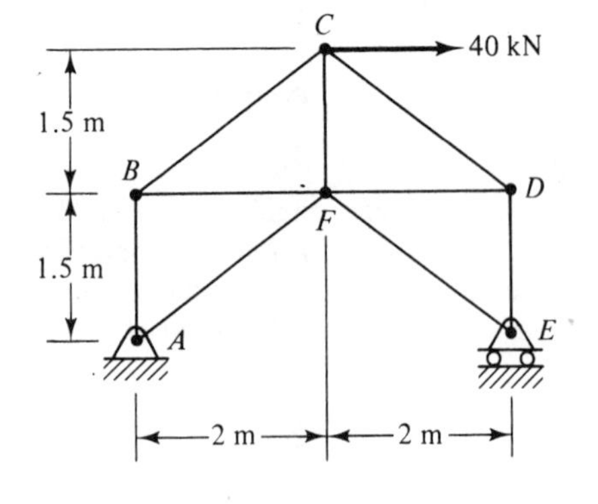

Problem 3.85

3.83 to 3.85 For the trusses shown, determine the forces acting upon the members indicated.

3.86 and 3.87 For the trusses shown, identify the zero-force members, if any, and determine the forces acting upon the members indicated.

3.86 Cable AB and members DF, DG, FG, IK, and JK.

3.87 Members FG, JK, and the four members that intersect at joint E.

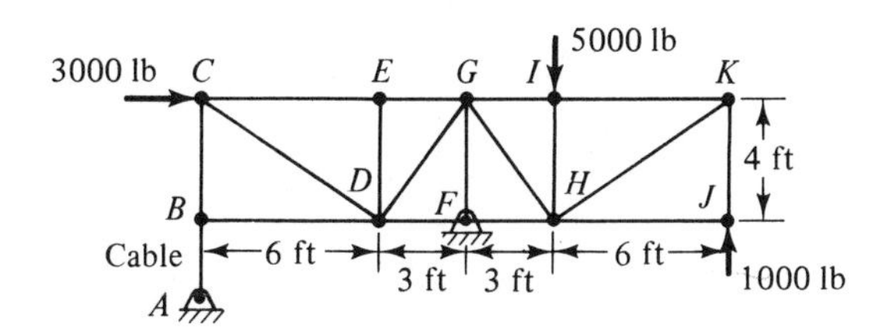

Problem 3.86

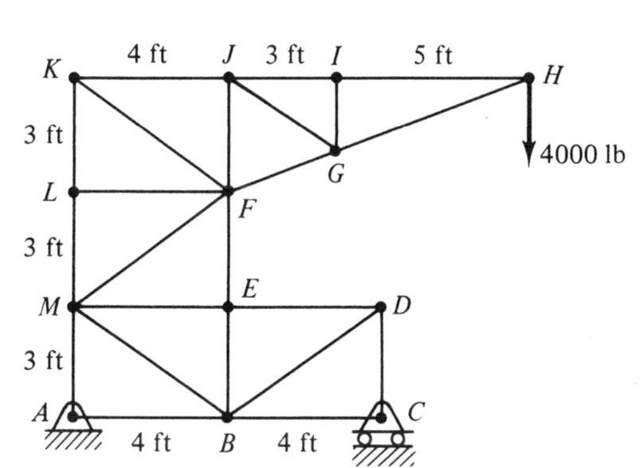

Problem 3.87

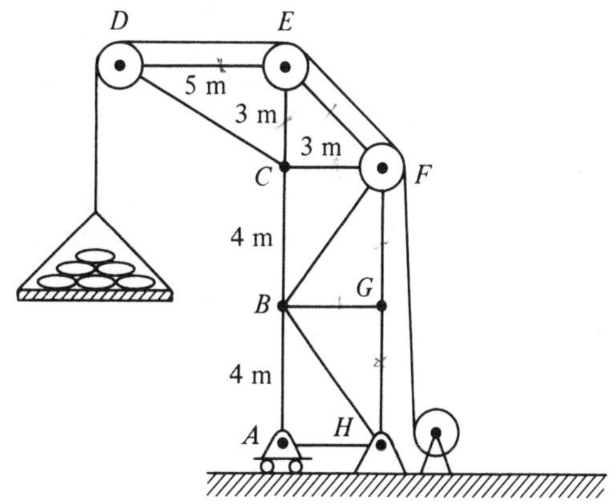

Problem 3.88

3.88 Building materials with a total mass of 3.26 tonne are being hoisted to the roof of a building with the arrangement shown. Determine the forces on the truss members that intersect at joints E, F, and G.

3.89 The mouth of a purse seine is held open by separating the tow cables with a "spreader" BCD as shown. The spreader consists of three bars, each 2 m long, pinned together at the ends. If it takes a force of 20 kN to tow the seine through the water, what are the forces acting upon members BC, BD, and CD of the spreader?

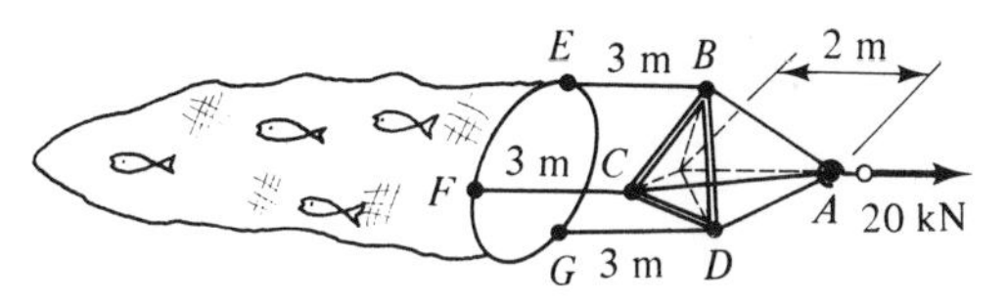

Problem 3.89

3.90 During erection, the support structure for a playground swing is subjected to a 450-N force acting vertically downward at D as shown. Determine the forces on members AB, AC, and AD. All connections may be assumed to be equivalent to ball and socket joints; the weights of all members are negligible.

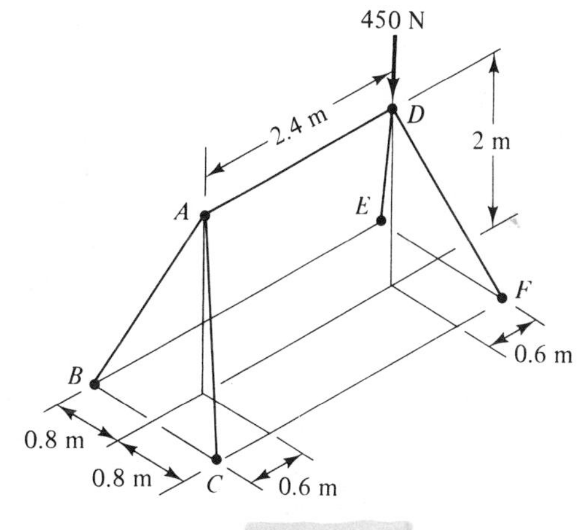

Problem 3.90

3.91 The figure shows an observation tower for field sports. The weight of the tower and occupants is equivalent to a 6000-lb force acting downward at M. Determine the forces on members HL, HI, EH, EF, FI, and FJ. Neglect the weights of the individual members involved, and assume all connections are equivalent to ball and socket joints.

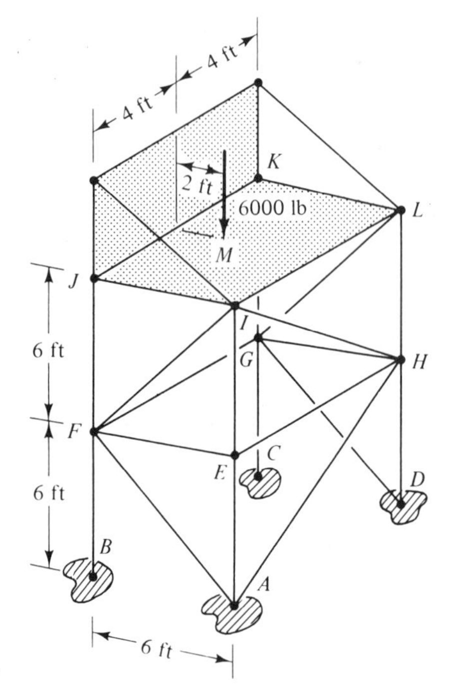

Problem 3.91

3.9 FORCE ANALYSES INVOLVING FRICTION

As we have already mentioned in Section 3.4, there is a friction force developed between rough surfaces in contact when one surface tends to slide relative to the other. This force is tangent to the plane of contact, and it always tends to oppose the motion of the body upon which it acts. In this section we shall consider the friction force in more detail.

Friction is somewhat of a paradox because it is both a tremendous asset and a liability. For instance, friction contributes to the wear in the engine of our automobile, but the vehicle could not move were it not for the friction between the tires and the road. There are several types of friction, but we shall consider only the so-called *dry*, or *Coulomb, friction* that occurs between nonlubricated surfaces.

The unique thing about the friction force is that there is a definite limit to its magnitude. Consider, for example, a block sitting on a rough surface and subjected to a horizontal force F [(Figure 3.28(a)]. A free-body diagram of the block is shown in Figure 3.28(b). Experiments show that (a) the block will remain in equilibrium if F is sufficiently small ($F = f$ in this case), (b) the block will start to move if F is increased to a certain value, and (c) the force required to keep the block in motion is less than that required to start it moving. The existence of these various types of responses is easily demonstrated. For example, try pushing a book or other fairly

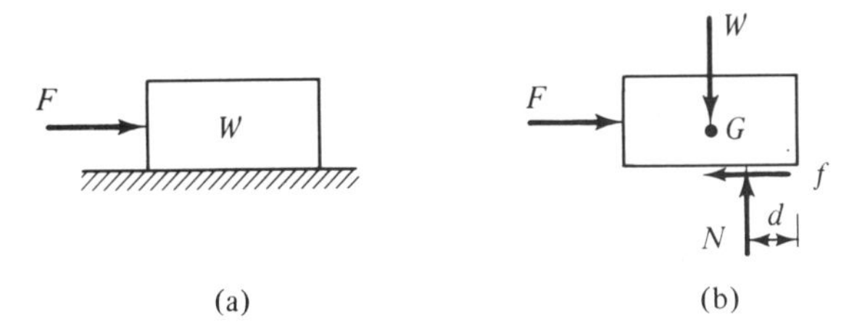

Figure 3.28

heavy object over a desk or table top. You will be able to sense with your hand the relative changes in F as you push harder upon the object and it starts to move.

The behavior of the block shown in Figure 3.28(a) is conveniently displayed by plotting the friction force f versus the applied force F, as in Figure 3.29. As F increases, f also increases until it reaches a maximum value f_{max}. At this instant the block is in a state of *impending motion*; it is still in equilibrium, but it is on the verge of moving. If F is at all increased beyond this point, the friction force will suddenly decrease. As a result, F will be greater than f and the block will move.

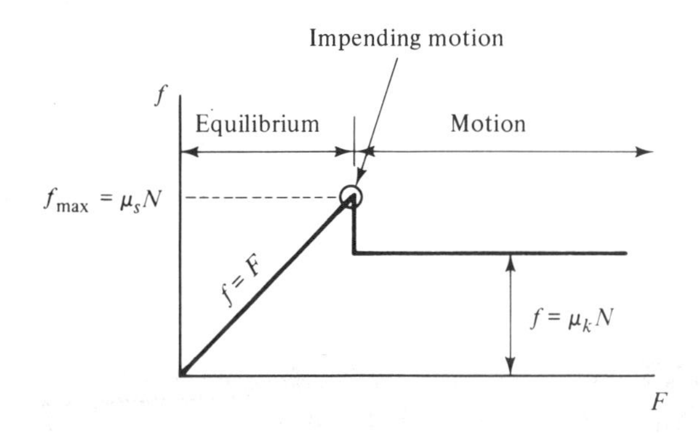

Figure 3.29

The maximum friction force that can be developed between any two given surfaces depends upon a number of factors, many of which are not entirely understood. However, experiments show that f_{max} is proportional to the normal force N between the contacting surfaces. Thus, we can write

$$f_{max} = \mu_s N \tag{3.7}$$

The constant of proportionality μ_s is called the *coefficient of static friction*. The friction force is also proportional to the normal force when there is relative motion between the contacting surfaces. In this case,

$$f = \mu_k N \tag{3.8}$$

where μ_k is the *coefficient of kinetic friction*. The laws of friction expressed in Eqs. (3.7) and (3.8) are based upon the experiments of the French physicist C. A. Coulomb, the results of which were published in 1781.

The coefficients of static and kinetic friction vary greatly for different materials and for different conditions of their surfaces. They also depend upon environmental factors such as temperature and humidity. Representative values of the coefficients of friction for various materials are given in Table 3.1. Note that $\mu_k < \mu_s$, which is in keeping with the fact that the friction force decreases when motion occurs.

Table 3.1 Coefficients of Static and Kinetic Friction

Materials	μ_s	μ_k
Metal on metal	0.7	0.5
Metal on wood	0.6	0.4
Metal on stone	0.7	0.4
Metal on ice	0.03	0.02
Wood on wood	0.6	0.5
Rubber tires on dry pavement	0.9	0.8

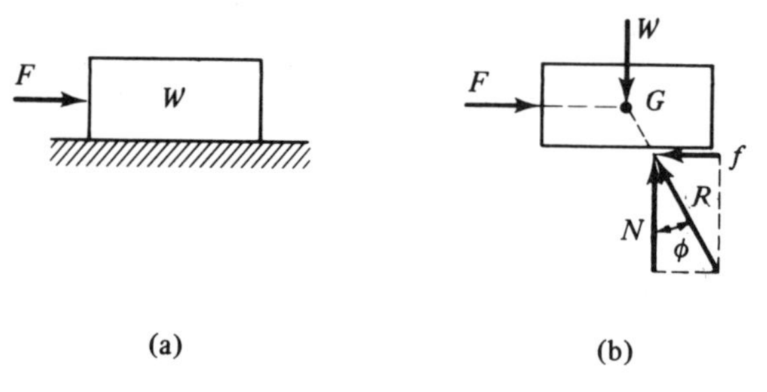

Figure 3.30

The frictional characteristics of two contacting surfaces can also be expressed in terms of the angle of inclination ϕ of the resultant contact force R (Figure 3.30). From this figure, we have

$$\tan \phi = \frac{f}{N}$$

The angle ϕ is zero for frictionless surfaces, and it attains its maximum value ϕ_s when motion is impending.

For impending motion, $f = \mu_s N$; so

$$\tan \phi_s = \mu_s \tag{3.9}$$

When there is relative motion between the two surfaces, Eq. (3.8) applies, and we have

$$\tan \phi_k = \mu_k \tag{3.10}$$

The angles ϕ_s and ϕ_k are called the *angles of static* and *kinetic friction*, respectively. In some fields of engineering it is common practice to describe the friction forces in terms of these angles instead of the coefficients of friction.

The procedure for handling problems involving friction depends upon what is known about the problem. If it's known that motion is impending, then $f = f_{max} = \mu_s N$. In this case, the friction force is a known quantity (once N is determined) and, therefore, it must be shown on the free-body diagram with the proper sense. The friction force on a body always points in the direction opposite to that in which the body tends to move. The preceding comments also apply when the body is in motion, except that in this case $f = \mu_k N$.

If the system is in equilibrium and motion is not impending, the friction force is an unknown that must be determined from the equations of equilibrium. In this case, the friction force is just like any other unknown force, and its sense can be chosen arbitrarily on the free-body diagram.

In some problems it is not known beforehand whether the system will move or remain in equilibrium under the action of the applied forces. The following procedure is convenient for problems of this type:

1. Assume the system is in equilibrium and solve for f; f will be negative if its sense is chosen incorrectly on the FBD.

2. Compare $|f|$, which is the magnitude of the friction force required to maintain equilibrium, with $f_{max} = \mu_s N$, which is the maximum friction force available.

3. If $|f| \leqslant f_{max}$, the system is in equilibrium as assumed, and the friction force is as determined in step 1. In other words, the surfaces can generate a friction force of sufficient magnitude to maintain equilibrium.

4. If $|f| > f_{max}$, the surfaces cannot generate a friction force large enough to maintain equilibrium. The system will move, and $|f| = \mu_k N$. The sense of f is as determined in step 1, and the direction of motion is opposite to the sense of f.

Example 3.13. Force Analysis of a Block. The metal block shown in Figure 3.31(a) has a weight of 70 lb, and the incline is made of wood. Determine the force F required to (a) prevent the block from sliding down the incline and (b) start it moving up the incline.

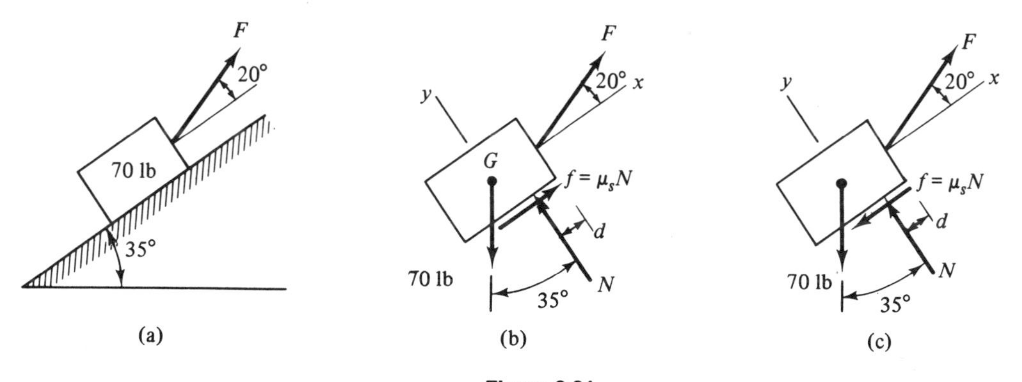

Figure 3.31

Solution. (a) When the block is on the verge of sliding down the incline, the FBD of the block is as shown in Figure 3.31(b). The friction force has magnitude $\mu_s N$ and in directed up the plane, opposite the direction in which the block tends to move. For equilibrium,

$$\Sigma F_x = F \cos 20^\circ + \mu_s N - (70 \text{ lb})\sin 35^\circ = 0$$

$$\Sigma F_y = F \sin 20^\circ + N - (70 \text{ lb})\cos 35^\circ = 0$$

Solving these equations by using the value $\mu_s = 0.6$ obtained from Table 3.1 for metal on wood, we get

$$N = 54.7 \text{ lb} \qquad F = 7.8 \text{ lb} \qquad \textit{Answer}$$

(b) When motion is impending up the incline, the FBD of the block is as shown in Figure 3.31(c). The only difference from the previous case is that the sense of the friction force is reversed. For equilibrium,

$$\Sigma F_x = F \cos 20^\circ - \mu_s N - (70 \text{ lb})\sin 35^\circ = 0$$

$$\Sigma F_y = F \sin 20^\circ + N - (70 \text{ lb})\cos 35^\circ = 0$$

from which we obtain

$$N = 35.1 \text{ lb} \qquad F = 65.1 \text{ lb} \qquad \textit{Answer}$$

To summarize, the block will slide down the incline if $F < 7.8$ lb, it will slide up the incline if $F > 65.1$ lb, and for $7.8 \text{ lb} \leqslant F \leqslant 65.1$ lb it will remain at rest.

Example 3.14. Determination of Friction Force. Determine the braking (friction) force acting upon the wheel shown in Figure 3.32(a) if the force F developed in the hydraulic cylinder that actuates the brake is (a) 2 kN and (b)1 kN. The coefficients of friction are $\mu_s = 0.4$ and $\mu_k = 0.3$.

Solution. Since we do not know beforehand whether the wheel will rotate or remain at rest under the action of the applied couple, we assume it is in equilibrium. In this case, the friction force f is an unknown. Referring to the FBDs of Figure 3.32(b), we have for the wheel

$$\overset{+}{\curvearrowleft}\Sigma M_D = f(0.25 \text{ m}) - 500 \text{ N·m} = 0$$

For the brake arm,

$$\overset{+}{\curvearrowleft}\Sigma M_A = N(0.35 \text{ m}) - F(0.65 \text{ m}) - f(0.25 \text{ m}) = 0$$

(a) Solving the equilibrium equations with $F = 2$ kN, we obtain

$$f = 2.00 \text{ kN} \qquad N = 5.14 \text{ kN} \qquad f_{max} = \mu_s N = 2.06 \text{ kN}$$

Since the friction force needed to maintain equilibrium is smaller than the maximum available friction force ($|f| < f_{max}$), the wheel is in equilibrium as assumed.

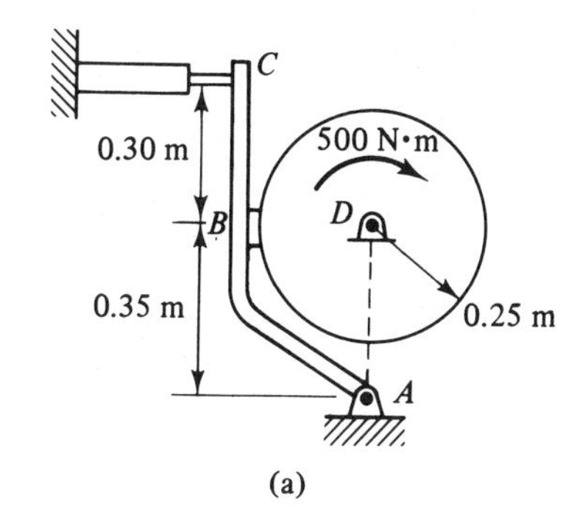

(a)

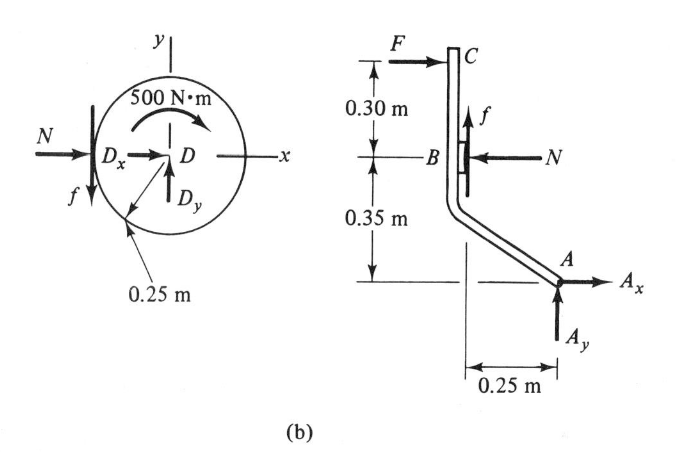

(b)

Figure 3.32

Thus, the friction force is

$$f = 2.00 \text{ kN} \qquad \textbf{\textit{Answer}}$$

The positive value of f indicates that its sense is as shown on the FBDs.

(b) For $F = 1$ kN, we find

$$f = 2.00 \text{ kN} \qquad N = 3.29 \text{ kN} \qquad f_{\max} = \mu_s N = 1.32 \text{ kN}$$

Since $|f| > f_{\max}$, the surfaces cannot generate a friction force large enough to maintain equilibrium. Thus, the wheel will rotate, and

$$f = \mu_k N = 0.3(3.29 \text{ kN}) = 0.99 \text{ kN} \qquad \textbf{\textit{Answer}}$$

Example 3.15. Force Analysis of Blocks. In Figure 3.33(a) the uniform slab A weighs 5000 lb and block B weighs 1000 lb. What minimum force F is required to disturb equilibrium of the system? The coefficients of friction are as indicated in the figure.

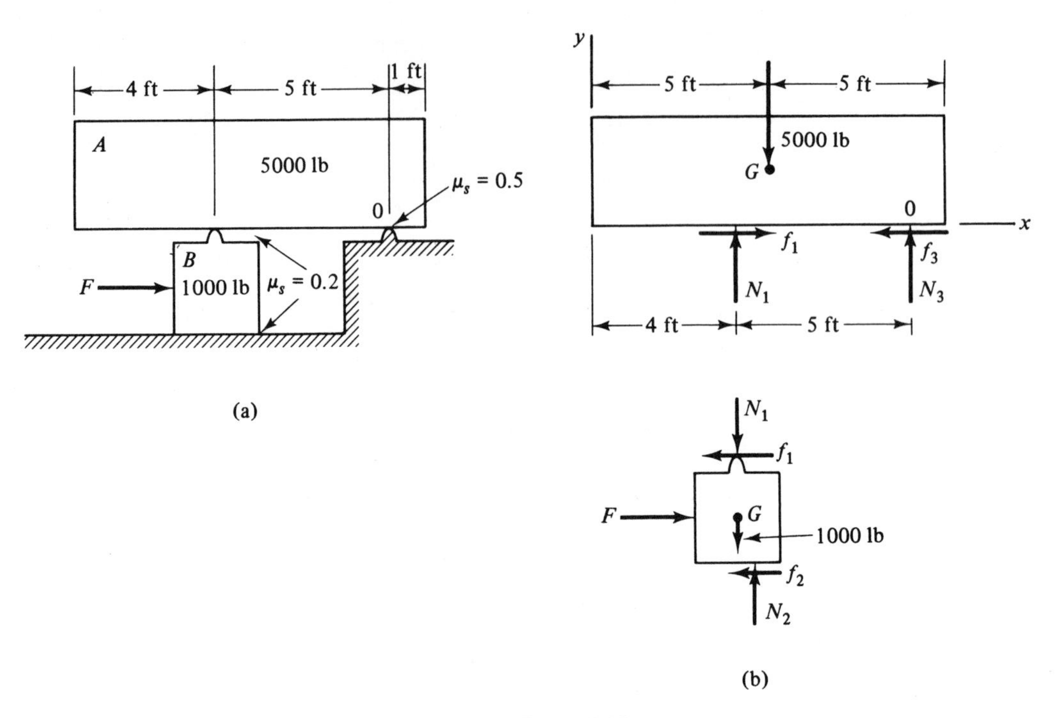

Figure 3.33

Solution. This is an impending motion problem, but there is more than one way the system can start to move: (a) B can start to slide while A remains at rest or (b) B and A can both start to slide as a unit. The most straightforward way to proceed is to treat each of these cases as a separate problem and solve for F. The case that gives the smallest value of F is the one that actually occurs.

Case a. The FBDs are shown in Figure 3.33(b). For this case, $f_1 = 0.2N_1$ and $f_2 = 0.2N_2$, while the friction force f_3 is unknown. For block B, we have

$$\Sigma F_x = F - 0.2N_1 - 0.2N_2 = 0$$
$$\Sigma F_y = N_2 - N_1 - 1000 \text{ lb} = 0$$

For slab A,

$$\overset{+}{\curvearrowleft} \Sigma M_0 = 5000 \text{ lb}(4 \text{ ft}) - N_1(5 \text{ ft}) = 0$$

Solving these equations, we get

$$F = 1800 \text{ lb}$$

Case b. The only difference from the previous case is that now $f_3 = 0.5N_3$, while f_1 is unknown.

For block B, we have

$$\Sigma F_x = F - f_1 - 0.2N_2 = 0$$
$$\Sigma F_y = N_2 - N_1 - 1000 \text{ lb} = 0$$

For slab A,

$$\overset{+}{\curvearrowleft}\Sigma M_0 = 5000 \text{ lb}(4 \text{ ft}) - N_1(5 \text{ ft}) = 0$$
$$\Sigma F_y = N_1 + N_3 - 5000 \text{ lb} = 0$$
$$\Sigma F_x = f_1 - 0.5N_3 = 0$$

Solving these equations, we obtain

$$F = 1500 \text{ lb}$$

Case (b) gives the smallest value of F. Therefore, both blocks start to slide as a unit, and

$$F_{min} = 1500 \text{ lb} \qquad \textbf{\textit{Answer}}$$

PROBLEMS

3.92 Three boards are held together with a C-clamp as shown. If the clamping force has a magnitude of 100 lb, determine the force F required to separate the boards; $\mu_s = 0.6$.

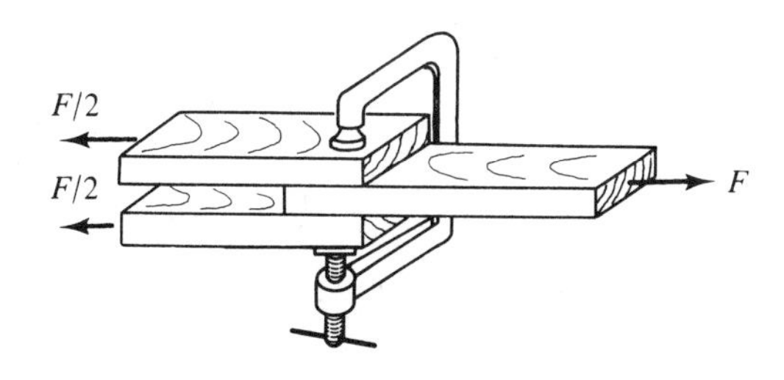

Problem 3.92

3.93 A boy is pulling a girl on a sled as shown. If the sled weighs 10 lb and the girl weighs 110 lb, how hard must the boy pull to (a) prevent the sled from sliding down the slope, (b) start the sled moving up the slope, and (c) keep the sled moving up the slope at constant speed? The coefficients of static and kinetic friction between the sled runners and snow are 0.1 and 0.05, respectively.

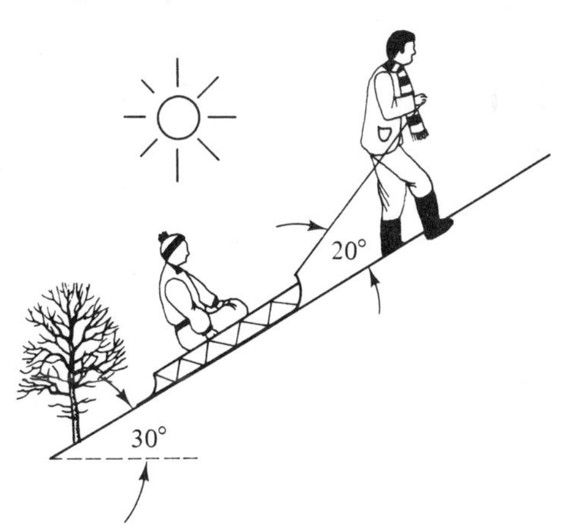

Problem 3.93

3.94 If $\mu_s = 0.5$, for what range of values of the force F will the block shown remain in equilibrium on the incline?

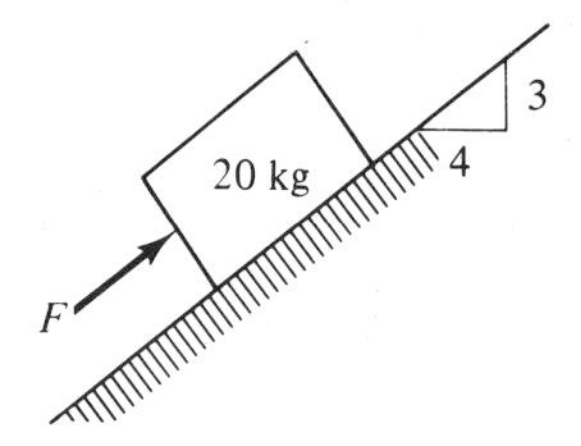

Problem 3.94

3.95 A uniform ladder weighing 40 lb rests against a wall as shown. The coefficient of static friction at both A and B is 0.3. (a) Determine the height h that a 178-lb man can climb before the ladder starts to slip. (b) If the ladder has negligible weight, is the value of h increased or decreased? Verify your answer.

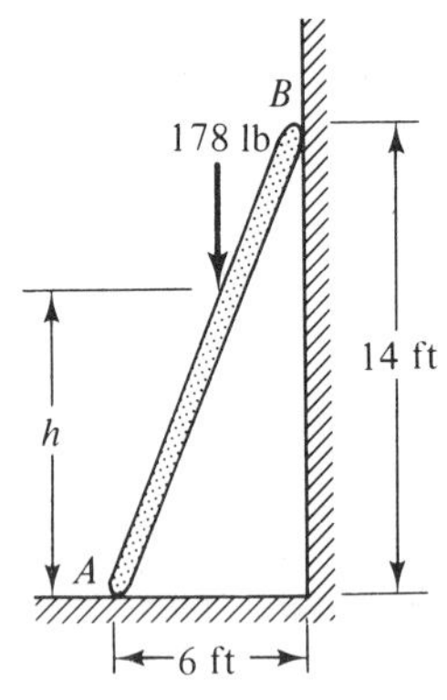

Problem 3.95

3.96 A uniform cylinder is held in equilibrium on a rough incline by the arrangement shown. Determine (a) the angle θ and (b) the minimum required value of the coefficient of static friction between the cylinder and incline. The mass of the cylinder is twice that of the block.

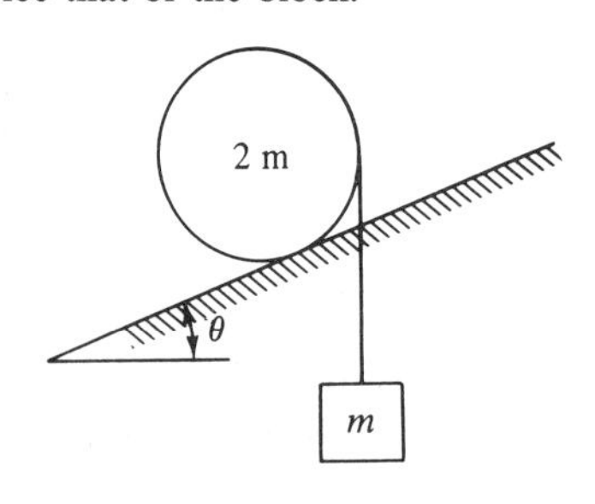

Problem 3.96

3.97 A boom with negligible mass supports a 2-tonne load as shown. The bottom of the boom rests on a rough surface and the top is held in place by cable AC. (a) Determine the reactions at B and the tension on the cable. (b) What minimum coefficient of static friction at B is required to prevent the boom from slipping?

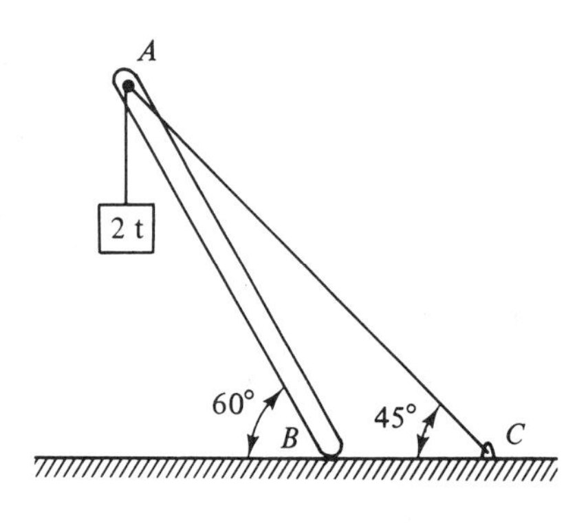

Problem 3.97

3.98 One end of a heavy machine with its center of gravity at G is to be raised with a wedge as shown. The machine weighs 4050 lb and is made of metal. The wedge and floor are made of wood. Estimate the force F required to raise the machine (see Table 3.1 for values of μ_s). Surface AB remains very nearly horizontal, and a board nailed to the floor prevents slippage at A.

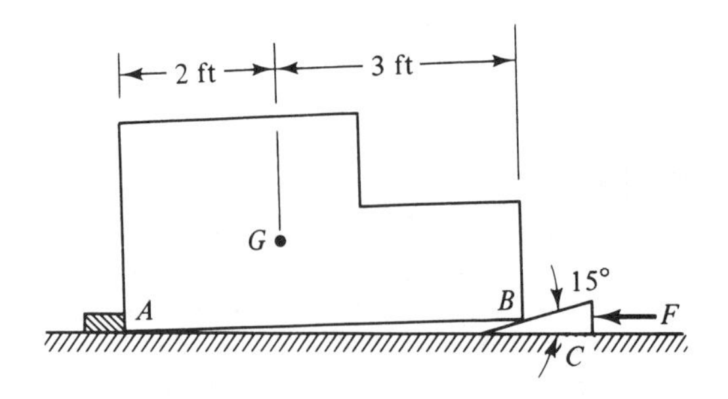

Problem 3.98

3.99 A door is held in position during installation by a wedge placed beneath it as shown. What should be the value of the angle θ so that the wedge will not slip out from under the door? Neglect the weight of the wedge, and assume equal coefficients of friction μ_s at all surfaces.

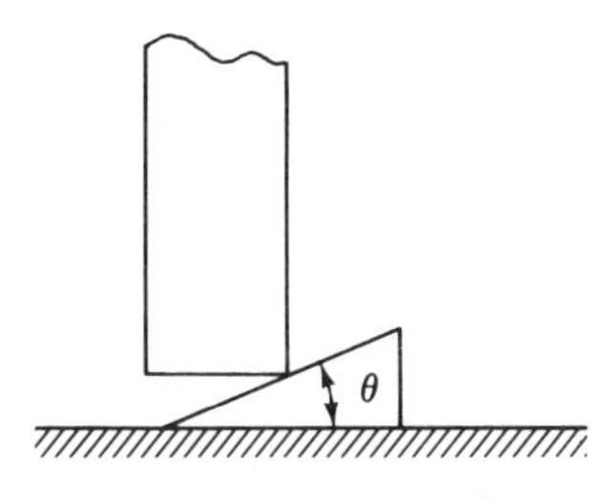

Problem 3.99

3.100 What is the angle of inclination θ of the steepest incline the vehicle shown can start up without slipping the driving wheels if it has (a) rear-wheel drive, (b) front-wheel drive, and (c) four-wheel drive? Assume rubber tires and dry pavement (see Table 3.1).

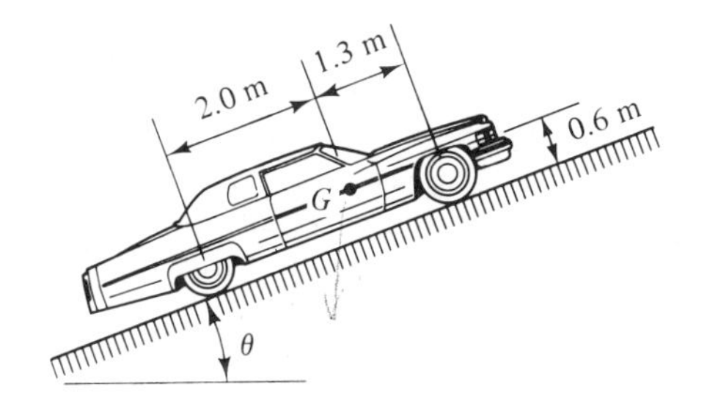

Problem 3.100

3.101 and 3.102 Determine whether or not the systems shown are in equilibrium, and find the friction forces at all contact surfaces. The coefficients of friction are as indicated in the figures.

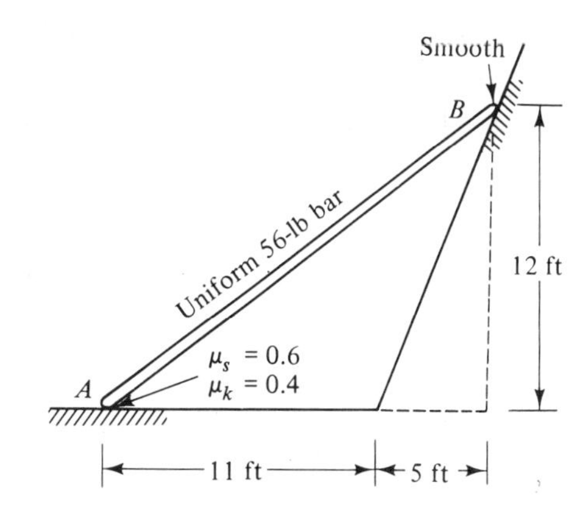

Problem 3.101

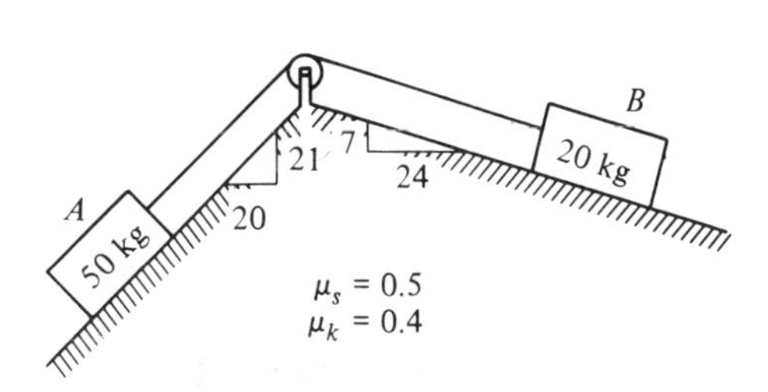

Problem 3.102

3.103 What is the smallest force F that will disturb equilibrium of the uniform crate shown if (a) $\mu_s = 1/4$ and (b) $\mu_s = 2/3$? Does the crate slide or tip?

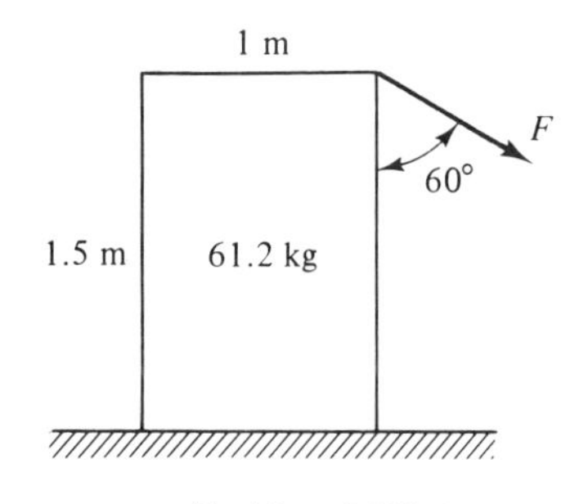

Problem 3.103

3.104 and 3.105 Determine the smallest force F and couple C, respectively, that will disturb equilibrium of the systems shown and describe clearly the manner in which they start to move. The coefficients of friction are as indicated in the figures.

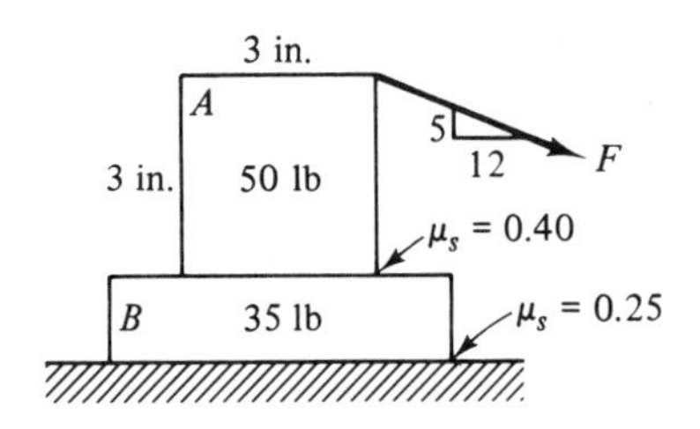

Problem 3.104

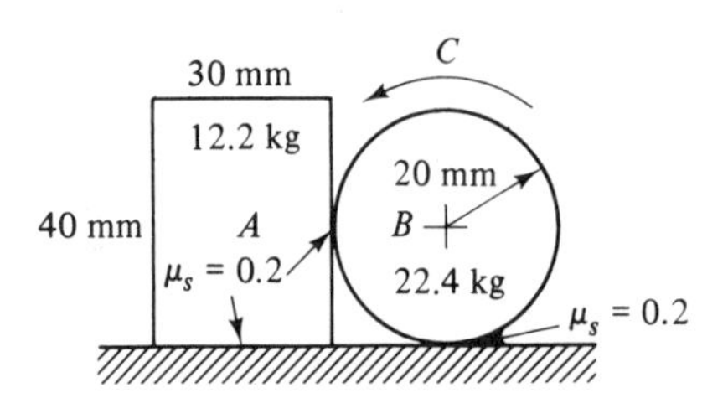

Problem 3.105

3.106 A sheet of tracing paper is held against a smooth glass surface by spring clips A and B as shown. Each spring clip exerts a normal force of 10 N on the paper. What is the smallest force F that will cause the sheet to move? Assume $\mu_s = 0.5$ between the clips and paper.

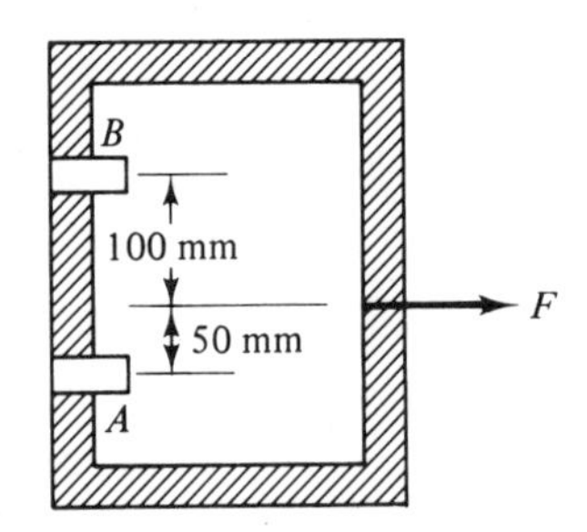

Problem 3.106

3.107 A can of roofing cement weighing 10 lb is sitting on a roof as shown. What is the smallest side force F that will cause the can to start to slide? Use $\mu_s = 0.6$.

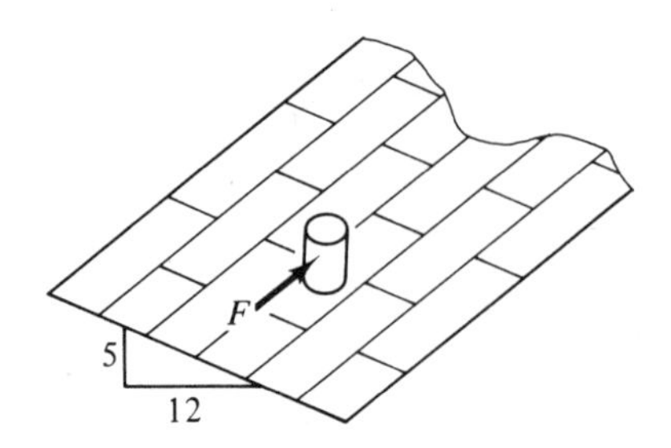

Problem 3.107

3.108 Collar A shown weighs 6 lb and is constrained to move along a horizontal rod. If an 8-lb horizontal side force acts upon the collar, find the force F required to start it moving along the rod. Use $\mu_s = 0.7$.

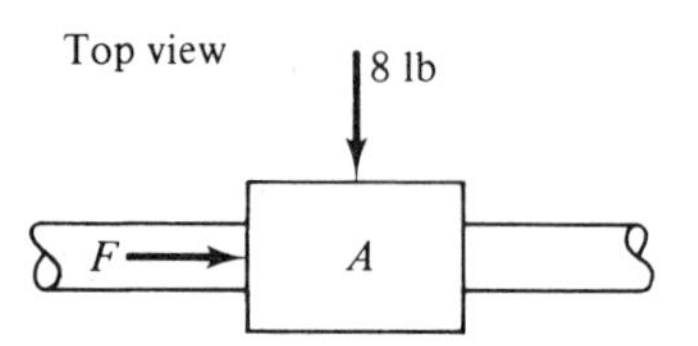

Problem 3.108

3.109 The drive wheel A for a phonograph turntable B is attached to an arm pivoted at C and held in position by a spring attached to the chassis at D as shown. Wheel A is turned by a small electric motor (not shown). (a) If the frictional effects at the bearing O are equivalent to a couple with magnitude 0.54 N·m, what is the required spring tension if the drive wheel is to rotate the turntable at constant speed without slipping? Use $\mu_s = 0.6$. (b) Would a stronger or weaker spring be required if the direction of rotation of the drive wheel were reversed? Verify your answer.

3.110 Same as Problem 3.109, except that the frictional couple at the bearing O has magnitude 0.72 N·m.

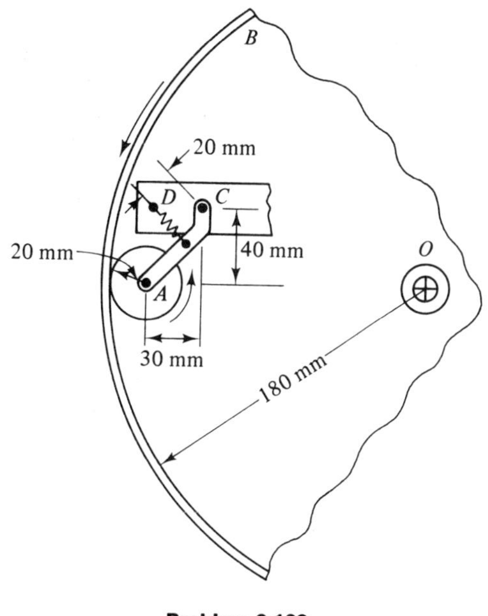

Problem 3.109

3.10 THE RIGID-BODY APPROXIMATION

Now that we have used the rigid-body assumption in the solution of a number of problems, let us consider the nature of this approximation in a bit more detail. In particular, we may ask: What do we gain by neglecting the deformations of a body when performing a force analysis, and how large are the errors introduced by this approximation? To answer these questions, let us consider the following problem.

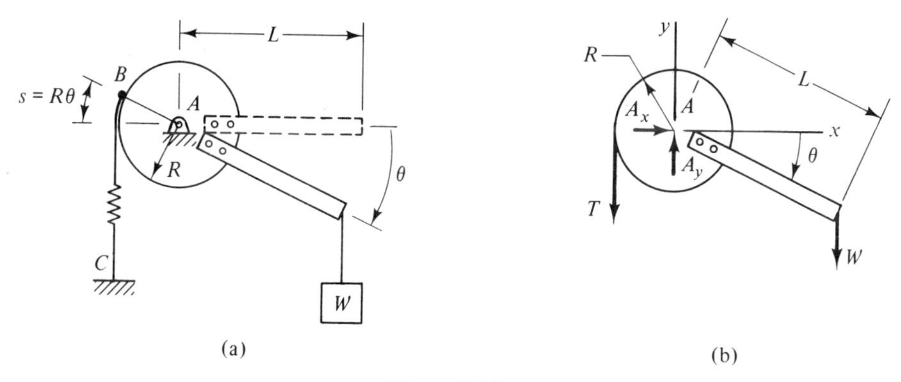

Figure 3.34

Figure 3.34(a) shows a block with weight W suspended from the end of a rigid arm attached to a disc. The disc and arm have negligible weight, and the assembly is supported by a smooth bearing at A and a spring BC. Before the block is attached, the arm is horizontal and the spring is unstretched. After the block is attached and the system has come to rest, the spring is stretched and the arm is inclined at an angle θ from the horizontal, as shown. We wish to determine the tension T on the spring. The dimensions L and R and the weight W of the block are assumed known.

From the free-body diagram of Figure 3.34(b), we have

$$\overset{+}{\curvearrowleft}\ \Sigma M_A = TR - WL\cos\theta = 0$$

or

$$T = \frac{WL}{R}\cos\theta \qquad (3.11)$$

If the spring stretches only a small amount, θ will be approximately equal to zero (the rigid-body approximation), and we get

$$T_{\text{rigid}} = \frac{WL}{R} \qquad (3.12)$$

However, if the spring stretches an appreciable amount, the assumption $\theta \approx 0$ is unreasonable. In this case, both θ and T are unknowns, and the problem cannot be solved from the equilibrium equations alone; it is statically indeterminate.

The additional information needed to solve the indeterminate problem is the resistive behavior of the spring. Suppose that the spring is tested by subjecting it to different tensions and measuring how much it stretches. The results of such an experiment performed on a typical coil spring are shown in Figure 3.35. Note that the amount of stretch s is proportional to the applied tension T. Thus, the data can be represented by the equation

$$T = ks \qquad (3.13)$$

where k is the so-called *spring constant*. A spring that behaves according to Eq. (3.13) is said to be *linear*. For the spring of Figure 3.35, $k = 50$ lb/in.

From the geometry of Figure 3.34(a), we see that the spring will stretch an amount

$$s = R\theta \qquad (3.14)$$

when the arm rotates through an angle θ. Combining this result with Eqs. (3.11) and (3.13), we obtain

$$kR^2\theta = WL\cos\theta \qquad (3.15)$$

This equation can be solved for θ, provided the values of the other quantities are known. The solution can be obtained either by trial and error or by plotting both sides of the equation versus θ and determining the points of intersection of the two

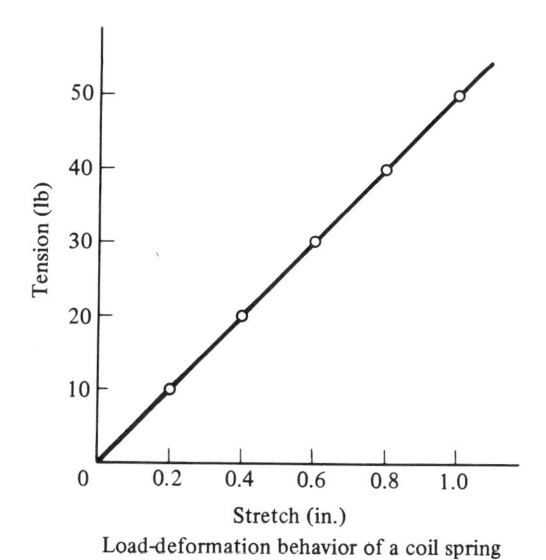

Load-deformation behavior of a coil spring

Figure 3.35

resulting curves. Once θ has been determined, the tension T can be found from Eq. (3.11).

The advantages of the rigid-body approximation should now be obvious. The exact solution for the tension not only requires a knowledge of the spring characteristics, but it also involves considerably more work than does the solution for T_{rigid}.

The error introduced by the rigid-body approximation can best be demonstrated by considering a specific case. A trial-and-error solution of Eq. (3.15) with $W = 50$ lb, $k = 50$ lb/in., $L = 3$ ft, and $R = 1$ ft gives $\theta = 13.904°$. From Eq. (3.11), we then find $T = 145.61$ lb. The value of T_{rigid} obtained from Eq. (3.12) is 150.00 lb, which exceeds the true value of T by 3%. An error of this magnitude would be acceptable for most engineering purposes. In fact, values of the parameters W, L, R, and k may not be known to this degree of accuracy.

An increase in the stiffness of the system decreases the angle of rotation of the arm and improves the accuracy of the rigid-body approximation. For example, if the spring is replaced by a $\frac{1}{16}$ in. diameter steel wire 3 feet long ($k = 2560$ lb/in.), the error in the tension is only 0.001%! Clearly, the rigid-body approximation gives accurate results when the deformations are small, as they are in most typical engineering structures.

In later chapters we shall be concerned with the determination of the deformations of structures; therefore, before leaving this section let us illustrate how the rigid-body approximation can be used to advantage in problems of this type.

Suppose that we wish to determine the stretch s of the spring, which can be computed from Eq. (3.11) once the tension is known. As we have seen, there is considerable work involved in obtaining the exact value of T. However, an approximate value of s can be easily determined by using the value of T_{rigid}: $s = T/k$ and $s_{\text{approx.}} = T_{\text{rigid}}/k$. For the particular system considered here, the error in the value of s is the same as the error in the value of the tension (3% for $k = 50$ lb/in.).

The preceding result suggests that we can assume a body is rigid while performing the force analysis and then use the values obtained for the forces and couples in finding the deformations of the body. This procedure greatly reduces the amount of work involved in determining the deformations, and we shall make considerable use of it in later chapters.

To summarize, the rigid-body approximation greatly reduces the amount of work involved in the force analysis, and it introduces little error if the deformations are small. It can be used for all the problems we shall consider, with the exception of the buckling problems discussed in Chapter 10.

3.11 CLOSURE

Our work in this chapter has been concerned with equilibrium—the single most important topic we shall consider, and one of the most important in all science and engineering.

The general equilibrium equations, $\Sigma \mathbf{F} = \mathbf{0}$ and $\Sigma \mathbf{M}_p = \mathbf{0}$, are a mathematical statement of the physical requirement that the total force and moment acting upon a system must be zero if the system is to remain at rest. These equations are used in force analyses to determine the unknown forces and couples that may be acting upon a body. This step is a prerequisite for determining the response of the body to applied loads or for properly designing it to withstand a given loading.

As we have seen, the key to a successful force analysis is the construction of proper free-body diagrams. For most problems, the force analysis can be based upon the dimensions of the system before the loads are applied (the rigid-body approximation). This enables us to separate the problem of determining the forces and couples acting upon the body from the problem of determining the deformations they produce, which greatly simplifies the force analysis.

chapter four

DISTRIBUTED FORCES AND RELATED CONCEPTS

4.1 INTRODUCTION

In the preceding chapters we dealt only with so-called *concentrated forces* (also called *discrete* or *point forces*) acting at a specific point. However, the concept of a concentrated force is an idealization. In reality, a force never acts at a point; it always exists as a *distributed force* acting over some area or volume. For example, if two spheres are pressed together, they will flatten slightly, and the contact force between them will be distributed over a small area. Similarly, the tension on a string is distributed over its cross section, and the gravitational attraction of the earth is distributed throughout the volume of the body upon which it acts. This being the case, what is our justification for working with concentrated forces?

It is reasonable to treat a force as being concentrated if the area or volume over which it is distributed is small enough compared to the other dimensions of the body to be considered a point. Typical examples are the force exerted on a telephone pole by a support cable and the reactions at the ends of a ladder due to the support surfaces. However, this argument does not justify the use of concentrated forces in every instance. It does not explain, for example, why the weight of a body can be treated as a concentrated force acting at the center of gravity or why the force distributed over the bottom surface of a block sitting on a smooth surface can be treated as a concentrated normal force.

The justification for working with concentrated forces in the aforementioned cases lies in the fact that distributed forces can be replaced by "equivalent" concentrated forces for the purpose of performing a force analysis. Our goal in this chapter is to determine the nature of this equivalence and to learn how to use it to advantage in the solution of problems. We shall begin by considering systems of discrete forces, with which we are already familiar. The results obtained will then be applied to distributed force systems. Certain geometric quantities associated with distributed force problems will also be considered.

4.2 STATICALLY EQUIVALENT FORCE SYSTEMS

Consider a body subjected to two different systems of forces, I and II (Figure 4.1). If the systems have the same total force and the same total moment about a common point, they will have the same effect upon the equilibrium of the body, and one system can be replaced by the other when performing a force analysis. Such force systems are said to be *statically equivalent*, or *equipollent*, a condition which we shall denote by the symbol $\sim$.

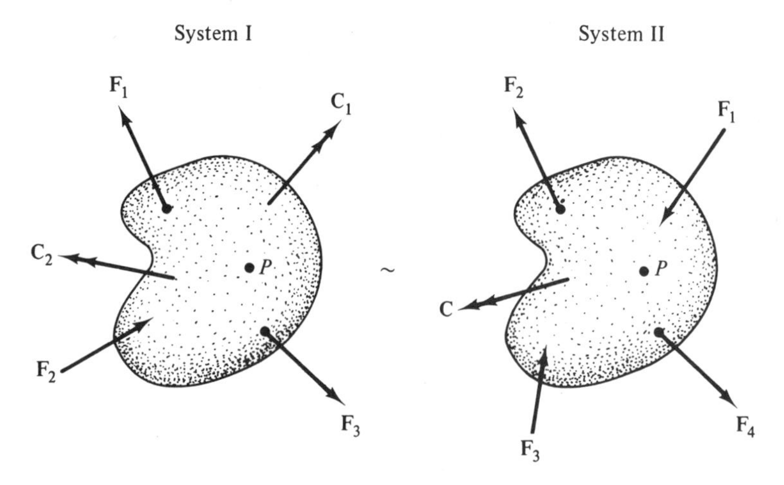

Figure 4.1

In the case of discrete forces, the conditions for static equivalence can be written as

$$(\Sigma\,\mathbf{F})_{\mathrm{I}} = (\Sigma\,\mathbf{F})_{\mathrm{II}}$$
$$(\Sigma\,\mathbf{M}_P)_{\mathrm{I}} = (\Sigma\,\mathbf{M}_P)_{\mathrm{II}} \tag{4.1}$$

where the point P about which moments are computed is arbitrary. These conditions also hold if there are distributed forces. The only differences are the procedures for computing the total force and moment, which will be discussed in later sections.

When expressed in component form, Eqs. (4.1) yield six scalar equations:

$$\begin{array}{lll} (\Sigma F_x)_{\text{I}} = (\Sigma F_x)_{\text{II}} & (\Sigma F_y)_{\text{I}} = (\Sigma F_y)_{\text{II}} & (\Sigma F_z)_{\text{I}} = (\Sigma F_z)_{\text{II}} \\ (\Sigma M_{Px})_{\text{I}} = (\Sigma M_{Px})_{\text{II}} & (\Sigma M_{Py})_{\text{I}} = (\Sigma M_{Py})_{\text{II}} & (\Sigma M_{Pz})_{\text{I}} = (\Sigma M_{Pz})_{\text{II}} \end{array} \tag{4.2}$$

All six of these equations must be satisfied in order to have static equivalence. However, for two-dimensional and other special force systems, certain of these conditions will be satisfied automatically.

Several general results of practical interest can be deduced from Eqs. (4.1). First, we see that two couples are statically equivalent if they produce equal moments. Since the moment of a couple is the same about every point, it makes no difference where the couple is applied to a body insofar as static equivalence is concerned.

Second, two forces are statically equivalent if they are equal and have the same line of action. They need not have the same point of application. Two equal forces with different lines of action are not statically equivalent. In order to satisfy the moment condition in Eqs. (4.1), a couple with the appropriate moment must be added to one force system or the other, as in Figures 4.2(a) and 4.2(b).

It is essential to recognize that equipollent force systems are equivalent only in the sense that they have the same effect upon the equilibrium of a body. (In dynamics you will learn that equipollent force systems are also dynamically equivalent in the sense that they produce the same motion when applied to a given rigid body; equilibrium is just a special case of motion.) Equipollent force systems

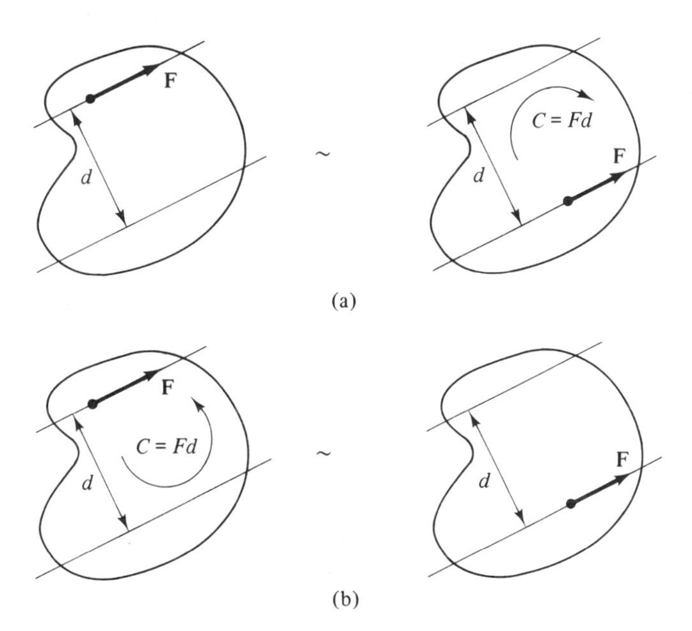

Figure 4.2

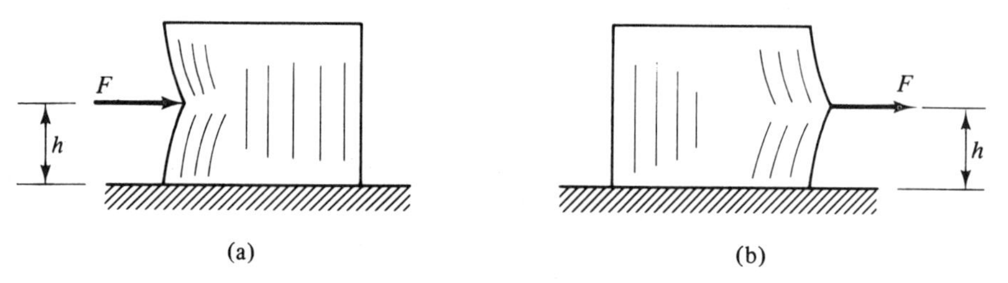

Figure 4.3

are not equivalent in every sense. For example, Figure 4.3 shows a rubber block subjected to two different forces. The two forces are statically equivalent, but they produce completely different deformations of the body.

As a second example, let us consider the truss shown in Figure 4.4. The equilibrium of the truss is not affected if the load Q is applied at joint F instead of at joint B, since the two loadings are equipollent. However, the forces acting upon the individual members of the truss will be different. The force on member BF, for example, is zero when the load is applied at B, and it is a tension with magnitude Q when the load is applied at F.

A still different type of situation is illustrated in Figure 4.5. Insofar as equilibrium is concerned, it makes no difference whether the stick is suspended from a cord or balanced on the end of our finger, since the forces exerted on the stick in each case are statically equivalent ($\mathbf{T} \sim \mathbf{N}$). However, the nature of the equilibrium is completely different. When the stick is suspended from a cord, the equilibrium position is stable. If the stick is disturbed slightly, it will return to its original position after the resulting oscillations have died out. The stick is in an unstable equilibrium position when it is balanced on our finger. Any slight bump will cause it to fall off, and it will not return to its original equilibrium position. (Stability will be discussed in Chapter 10.)

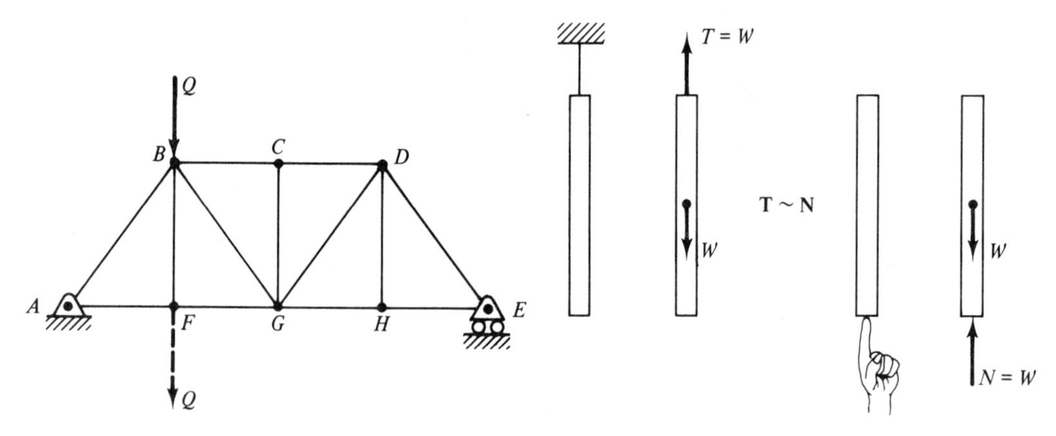

Figure 4.4

Figure 4.5

To summarize, replacement of a system of forces by an equipollent system does not affect the equilibrium of a body, but the deformations and forces developed within it may be altered. Furthermore, the stability of the equilibrium can be affected. We shall need to keep these characteristics of equipollent force systems in mind when determining the response of bodies to applied loadings in later chapters.

Example 4.1. Statically Equivalent Couples. Replace the forces acting upon the block shown in Figure 4.6(a) by another statically equivalent force system.

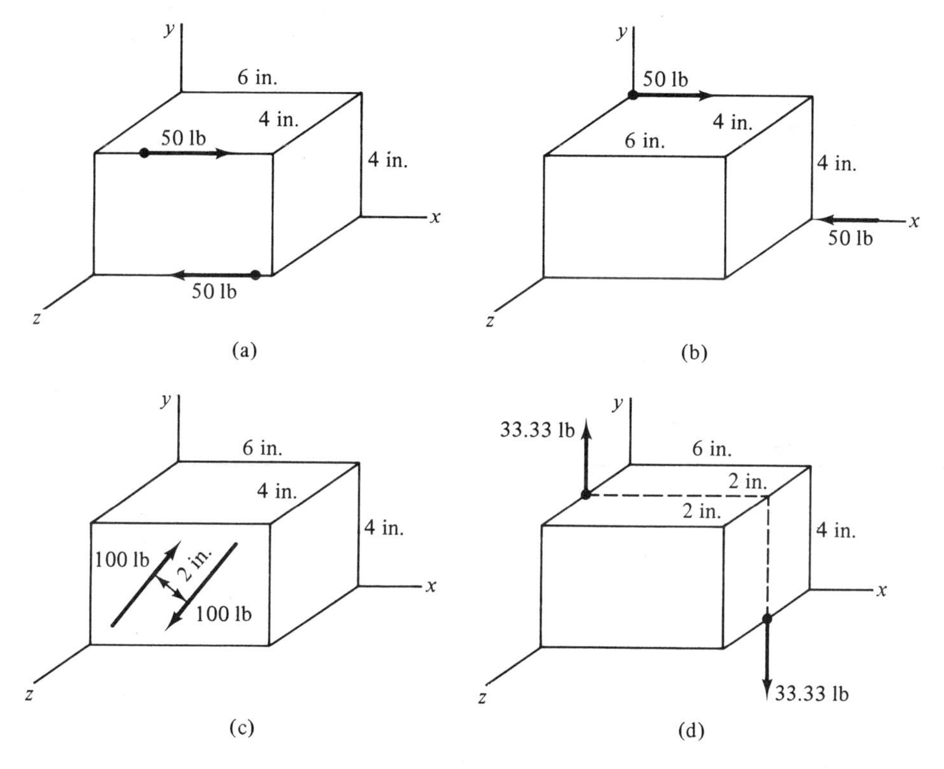

Figure 4.6

Solution. The two forces form a couple

$$\mathbf{C} = -200\mathbf{k} \text{ lb·in.}$$

The conditions for static equivalence, Eqs. (4.1), will be satisfied if this couple is replaced by any other couple with the same moment. There are an infinite number of ways to form such a couple. Several possibilities are shown in Figures 4.6(b)–4.6(d). What other possibilities can you think of?

Example 4.2. Statically Equivalent Force Systems. Replace the system of forces and couples acting upon the plate shown in Figure 4.7(a) by a statically equivalent system consisting of a vertical force acting along side AC and a horizontal force.

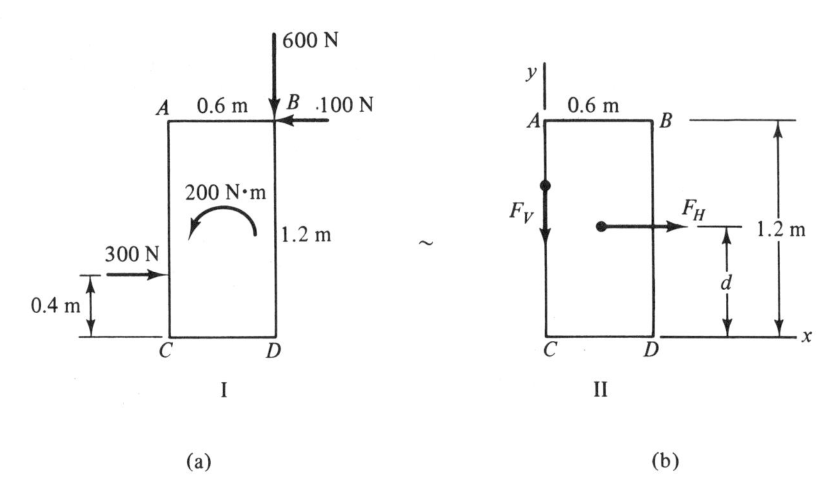

Figure 4.7

Solution. Let I denote the given force system and II the desired equipollent system. We now draw a sketch of system II showing all the known information about the forces and all the information that must be determined [Figure 4.7(b)]. The location of the line of action of the horizontal force is not known; therefore, it is shown an unknown distance d from the bottom of the plate. The sense of the unknown forces is chosen arbitrarily.

Applying the conditions for static equivalence, Eqs. (4.2), we obtain

$$(\Sigma F_x)_{\text{I}} = (\Sigma F_x)_{\text{II}}: \quad 300 \text{ N} - 100 \text{ N} = F_H$$

$$(\Sigma F_y)_{\text{I}} = (\Sigma F_y)_{\text{II}}: \quad -600 \text{ N} = -F_V$$

$$(\overset{+}{\curvearrowleft}\Sigma M_C)_{\text{I}} = (\overset{+}{\curvearrowleft}\Sigma M_C)_{\text{II}}: \quad -300 \text{ N } (0.4 \text{ m}) - 600 \text{ N } (0.6 \text{ m}) + 100 \text{ N } (1.2 \text{ m}) + 200 \text{ N·m} = -F_H d$$

Solving these equations, we get

$$F_H = 200 \text{ N} \qquad F_V = 600 \text{ N} \qquad d = 0.8 \text{ m} \qquad \textit{Answer}$$

Since F_H and F_V are positive, their sense is as assumed in Figure 4.7(b). Note that these forces do not have a definite point of application; they can be acting anywhere along their line of action. The fact that d is positive indicates that the line of action of F_H lies above side CD of the plate, as assumed.

Example 4.3. Statically Equivalent Force Systems. Determine the horizontal force acting at D that must be applied to the plate shown in Figure 4.8(a) in order for the total force system to be statically equivalent to a 40-lb upward force at B and a couple.

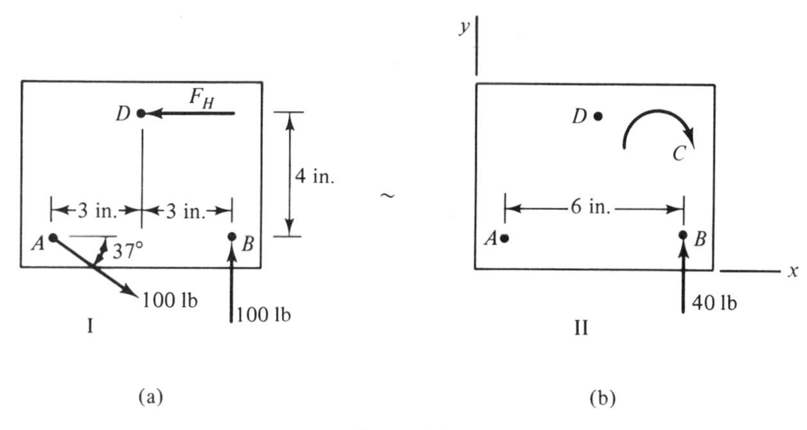

Figure 4.8

Solution. Neither force system is known completely. The unknown horizontal force at D is shown in Figure 4.8(a), and force system II is shown in Figure 4.8(b). The sense of the unknown force and couple is chosen arbitrarily.

For static equivalence, we have

$$(\Sigma F_x)_{\text{I}} = (\Sigma F_x)_{\text{II}}: \qquad (100 \text{ lb}) \cos 37^\circ - F_H = 0$$

$$(\Sigma F_y)_{\text{I}} = (\Sigma F_y)_{\text{II}}: \qquad -(100 \text{ lb}) \sin 37^\circ + 100 \text{ lb} = 40 \text{ lb}$$

$$\overset{+}{\curvearrowleft}(\Sigma M_A)_{\text{I}} = \overset{+}{\curvearrowleft}(\Sigma M_A)_{\text{II}}: \qquad F_H(4 \text{ in.}) + 100 \text{ lb } (6 \text{ in.}) = -C + 40 \text{ lb } (6 \text{ in.})$$

Since the second equation is satisfied identically, it yields no information; it serves only as a check. If this equation were not satisfied, it would be impossible to make the two force systems statically equivalent in the form stated. Solving for F_H and C from the first and third equations, we get

$$F_H = 80 \text{ lb} \qquad C = -680 \text{ lb·in.} \qquad \textit{Answer}$$

The negative value for C indicates that the sense of the couple is opposite to that assumed in Figure 4.8(b). Note that the couple can be shown as acting anywhere upon the plate because its moment is the same about any point. Also, the couple need not consist of any particular set of forces. Only the moment of the couple matters here.

PROBLEMS

4.1 to 4.3 Determine whether or not the force systems I and II shown are statically equivalent. If not, can they be made equivalent by adding to system I (a) a force, (b) a couple, or (c) a force and a couple?

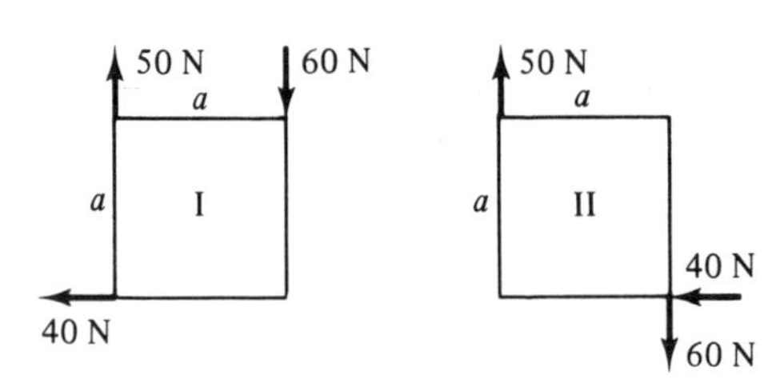

Problem 4.1

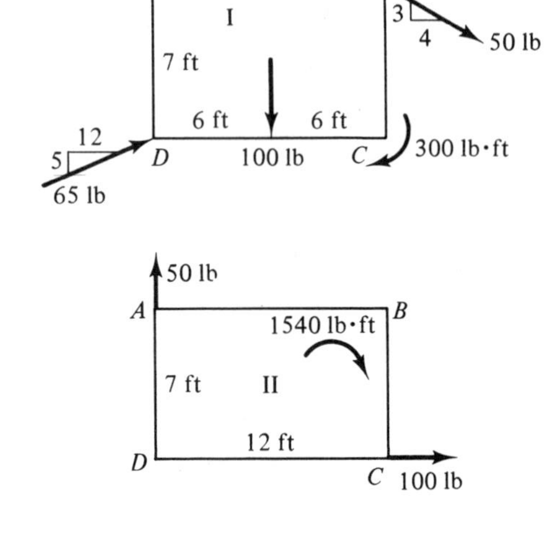

Problem 4.3

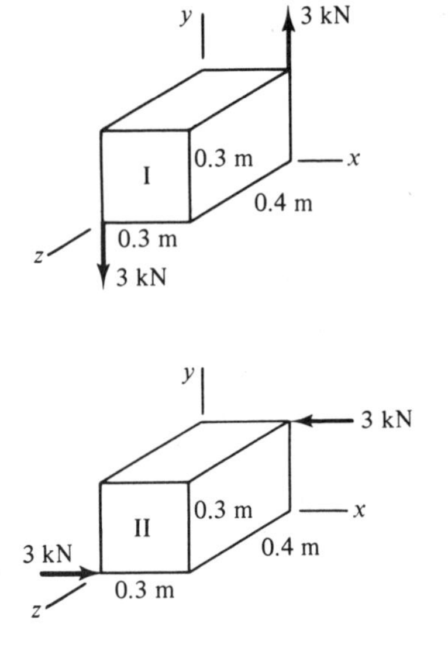

Problem 4.2

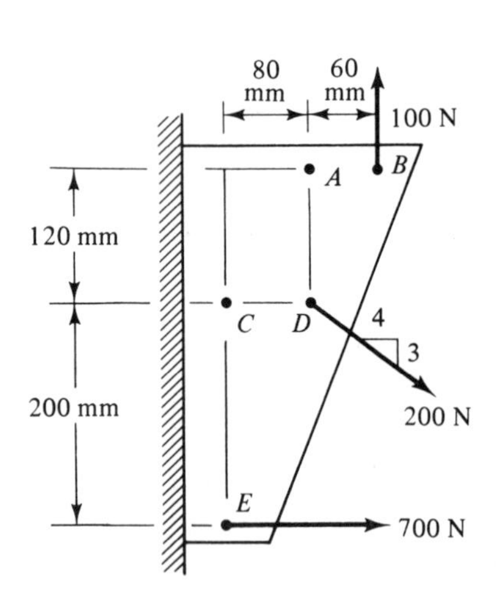

Problem 4.4

4.4 Transform the given force system into an equivalent system consisting of a single force at A and a couple.

4.5 Same as Problem 4.4, except that the single force is to act at point C.

4.6 Replace the 200-N force acting at the corner of the bracket shown by an equivalent system consisting of (a) a force at A and a couple and (b) a force at A and a 450-N force at B.

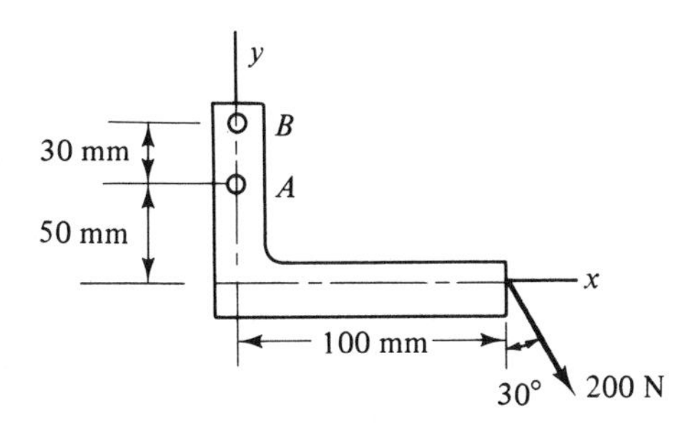

Problem 4.6

4.7 Replace the force system shown by an equivalent system consisting of a horizontal and vertical force acting at a point along side *BD* of the rectangle.

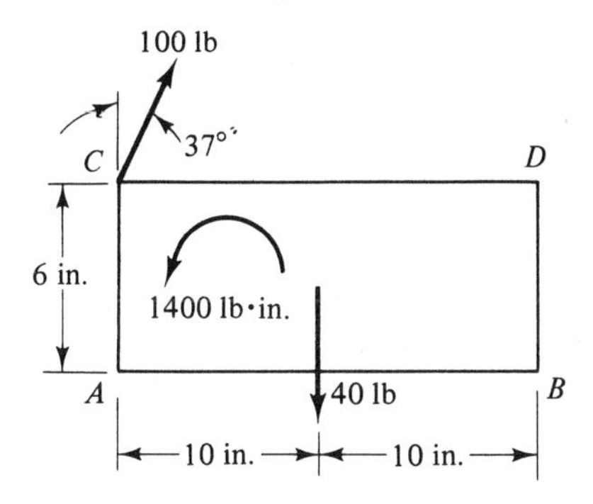

Problems 4.7 and 4.8

4.8 Determine completely the vertical force **F** that must be applied to the plate shown if the total force system is to be equivalent to a single 60-lb force acting to the right along side *AB*.

4.9 A force is applied at the end of a rigid arm attached to a circular shaft as shown. Replace the force by an equivalent system consisting of a force at *A* and a couple. What is the couple tending to twist the shaft about its longitudinal axis?

4.10 Replace the force acting upon the column shown by an equivalent system consisting of a force at the center *O* of the cross section and a couple. What is the couple tending to bend the column about (a) the *x* axis and (b) the *y* axis? Is there any tendency for the column to twist about the *z* axis?

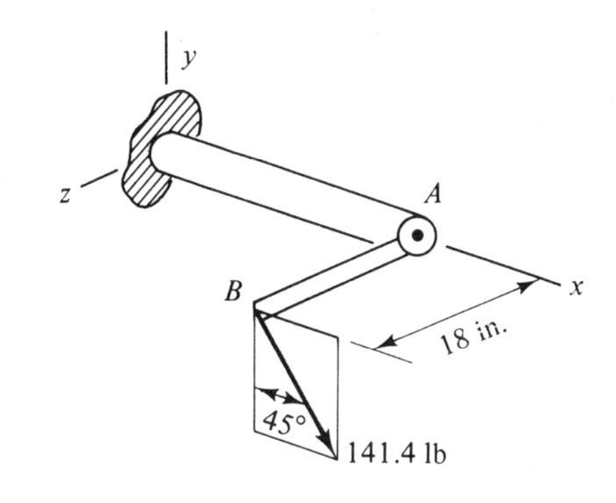

Problem 4.9

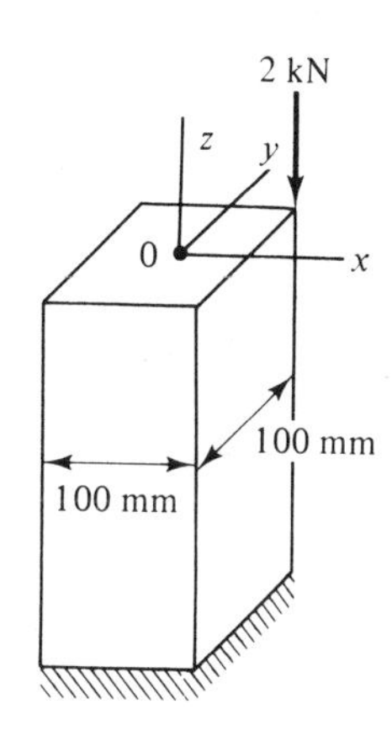

Problem 4.10

4.11 The outboard motor shown develops a horizontal thrust of 800 N. Replace this force by an equivalent system consisting of a force acting at the center of gravity *G* of the boat and a couple. The propeller is 0.3 m below *G*; the positive *y* axis is upward. What is the couple tending to turn the boat about the *y* axis? Does the thrust tend to raise or lower the bow of the boat?

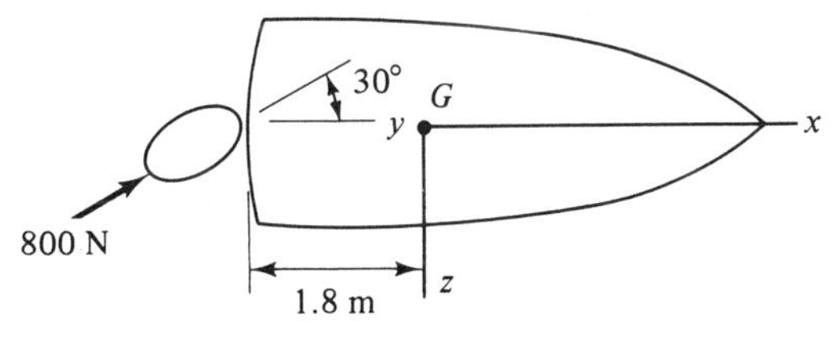

Problem 4.11

4.3 RESULTANTS OF FORCE SYSTEMS

When replacing one force system by another, it is usually desired to make the new system as simple as possible. The simplest force system equipollent to a given system is called the *resultant* of that system, and it can be determined from the conditions for static equivalence if we have some idea of its form. As we shall show in the following, the resultant is generally a force, a couple, or a force and a couple. We shall also develop guidelines to help determine a priori just which of these forms the resultant will take.

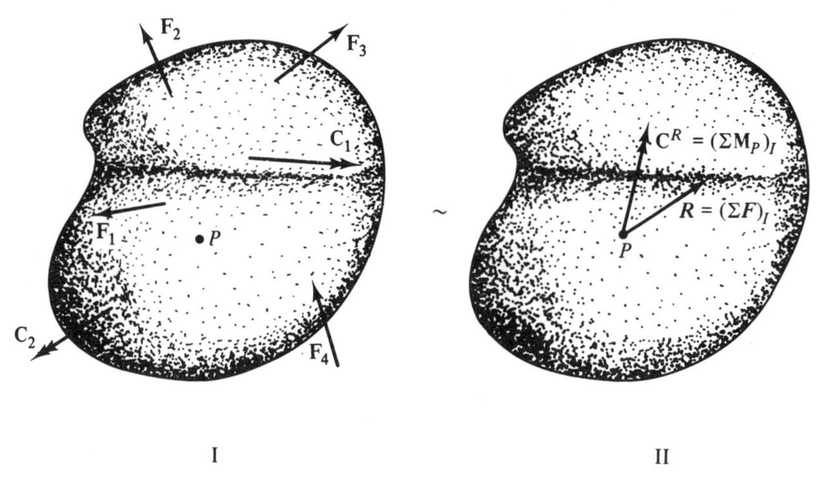

Figure 4.9

We begin by noting that any system of forces can be reduced to an equipollent system consisting of a single force **R** acting at an arbitrary point P and a single couple $\mathbf{C}^R$ (Figure 4.9). We shall call **R** the *resultant force* and $\mathbf{C}^R$ the *resultant couple*. Applying the conditions for static equivalence to the force systems of Figure 4.9, we have

$$\begin{aligned} (\Sigma\,\mathbf{F})_{\mathrm{I}} = (\Sigma\,\mathbf{F})_{\mathrm{II}}: &\qquad (\Sigma\,\mathbf{F})_{\mathrm{I}} = \mathbf{R} \\ (\Sigma\,\mathbf{M}_P)_{\mathrm{I}} = (\Sigma\,\mathbf{M}_P)_{\mathrm{II}}: &\qquad (\Sigma\,\mathbf{M}_P)_{\mathrm{I}} = \mathbf{C}^R \end{aligned} \tag{4.3}$$

Thus, the resultant force **R** is simply the total force of the original system, and the resultant couple $\mathbf{C}^R$ is equal to the total moment of the system about point P. The couple can be shown as acting at any point, but it is common practice to place it at the point of application of **R**, as in Figure 4.9.

For any one particular force system, **R**, $\mathbf{C}^R$, or both may be zero. If **R** is zero, the resultant is a couple; if $\mathbf{C}^R$ is zero, it is a single force. If both **R** and $\mathbf{C}^R$ are zero, the force system has a *null resultant*. When all the forces and couples acting upon a body have a null resultant, the body is in equilibrium.

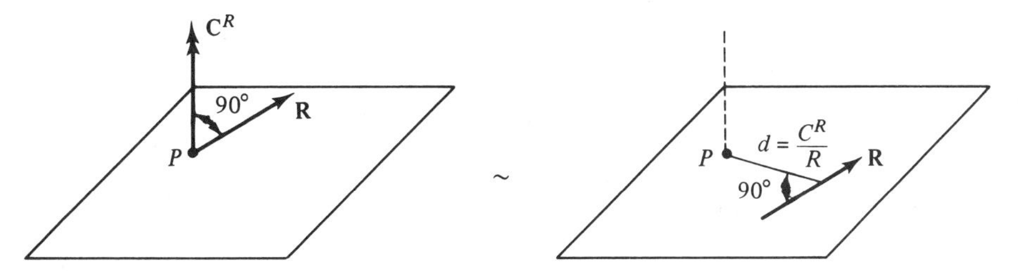

Figure 4.10

If the resultant force and couple are perpendicular, and $\mathbf{R} \neq \mathbf{0}$, the force system can be further reduced to a single force. As illustrated in Figure 4.10, this reduction requires that **R** be located such that it provides the same moment about point P as does the couple $\mathbf{C}^R$, thus assuring that static equivalence is maintained.

With the preceding general results in mind, let us now consider some specific force systems.

Concurrent force systems. If the forces are concurrent, their lines of action will intersect at some point P (Figure 4.11). Taking moments about P, we see that the resultant will be a single force **R** whose line of action passes through the point of concurrency. If $\mathbf{R} = \mathbf{0}$, the system has a null resultant.

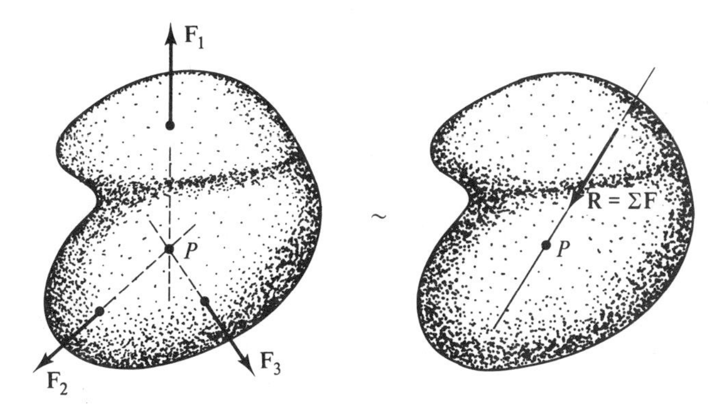

Figure 4.11

Coplanar and parallel force systems. Consider a system of coplanar forces (Figure 4.12). Using Eqs. (4.3) and taking moments about a point P in the plane of the forces, we see that $\mathbf{C}^R$ will be perpendicular to **R**. Thus, the system can be further reduced to a single force, provided $\mathbf{R} \neq \mathbf{0}$. This single force is the resultant of the system, and its line of action can be located by using the moment condition for static equivalence. If $\mathbf{R} = \mathbf{0}$, either the resultant will be a couple or the system will have a null resultant.

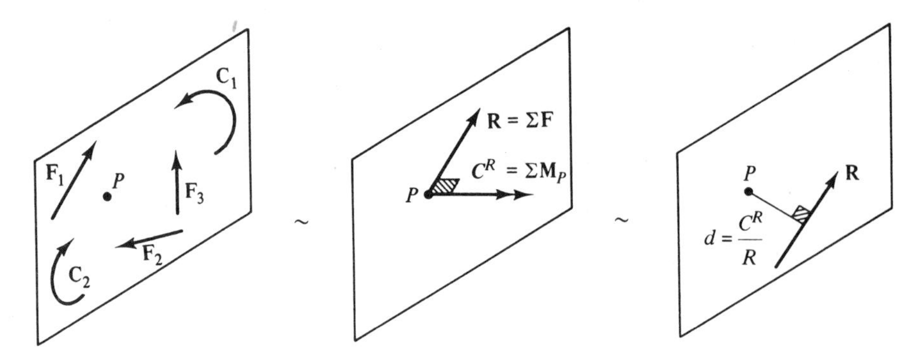

Figure 4.12

The preceding results also hold for systems of parallel forces (Figure 4.13). Since the moment of each force about a point P is perpendicular to the force, $\mathbf{R}$ and $\mathbf{C}^R$ will be perpendicular. Thus, the resultant will have the same form as in the case of coplanar forces. It is a single force $\mathbf{R}$ parallel to the original forces, provided $\mathbf{R} \neq \mathbf{0}$. For $\mathbf{R} = \mathbf{0}$, either the resultant is a couple or the system has a null resultant.

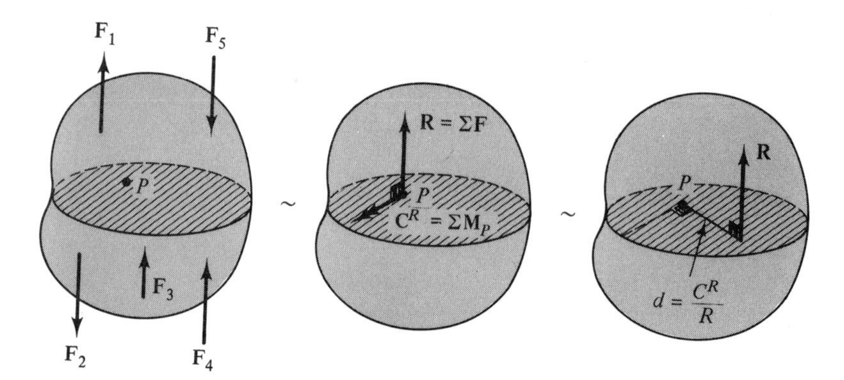

Figure 4.13

When determining the resultant of systems of parallel or coplanar forces, it is not necessary to first reduce the system to a force at a point and a couple. This was done here only to increase clarity. We can proceed directly from the given force system to the resultant, as illustrated in Examples 4.5 and 4.6 at the end of this section.

General force systems. For a general three-dimensional force system, it is not possible to determine a priori just what form the resultant will take. It will usually be a force and a couple. The component of the couple perpendicular to the force can be eliminated by moving the force to an appropriate point, as explained earlier. The resulting force system, consisting of a single force **R** and a couple parallel to it, is called a *wrench*. Reduction of a force system to a wrench has little practical application and will not be discussed further. It is usually most convenient to reduce the system to a single force and a couple and stop at that point.

Application to reactions. In the force analyses of Chapter 3 the reactions due to various supports and connections were treated as concentrated forces and couples, even though the forces acting were often distributed. This procedure can now be justified by using the properties of the resultants of force systems.

Consider the block shown in Figure 4.14. The normal force on the block is distributed over its bottom surface. Since the forces acting upon each element of area are parallel, their resultant will be a single vertical force. Thus, the distributed normal force can be replaced by a statically equivalent concentrated force, as indicated.

As a second example, consider a beam supported at one end by a pinned connection (Figure 4.15). Assuming that the contact surface between the pin and

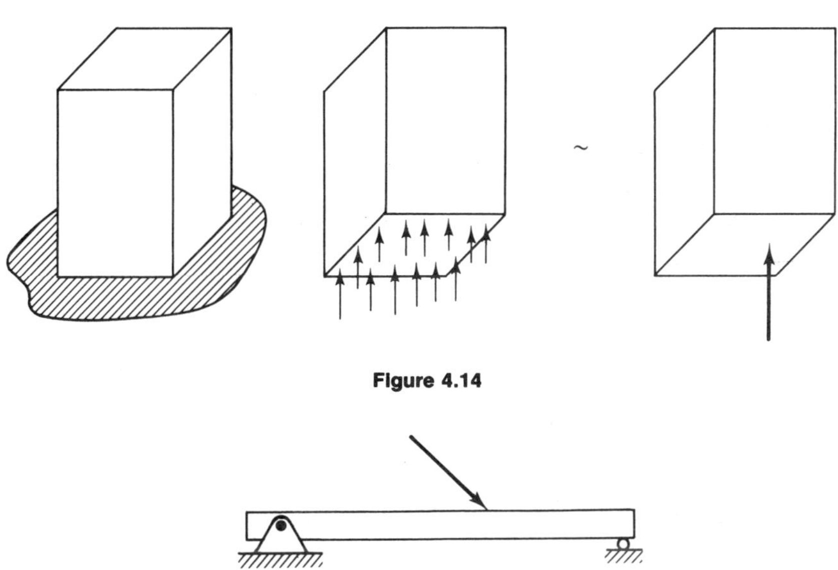

Figure 4.14

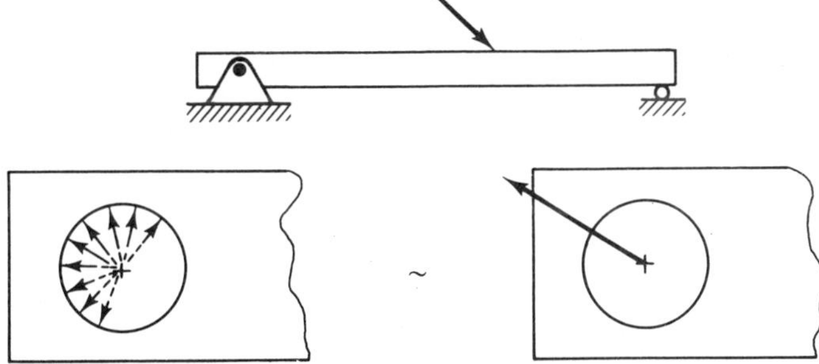

Figure 4.15

the beam is smooth and circular, it follows that the forces exerted on each element of the surface by the pin must lie along radii of the hole in the beam. Since these forces are concurrent, they can be replaced by a single equivalent force acting at the center of the hole.

Our final example involves a beam with a fixed support (Figure 4.16). The wall exerts some distribution of force on the end of the beam, the details of which are generally unknown. However, any system of forces is equipollent to a force and a couple. Thus, the reactions due to the wall can be treated as a concentrated force and a couple, as shown.

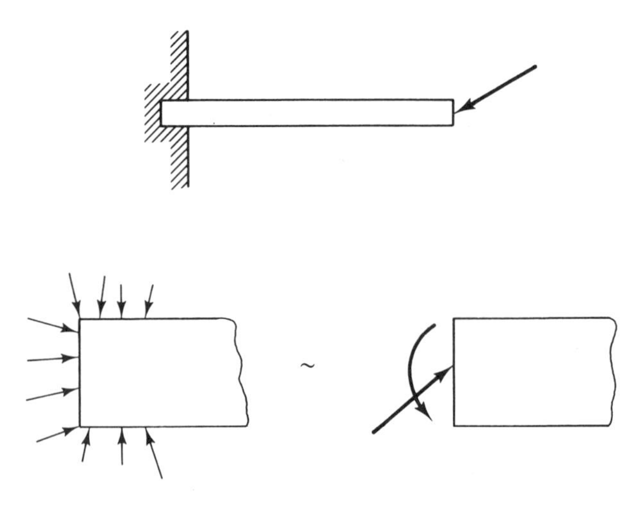

Figure 4.16

It is clear from the preceding examples that the quantities that we called the reactions in Chapter 3 are actually the resultant force and resultant couple of whatever system of forces is acting at the support or connection. It is these resultants that are determined in a force analysis; the actual force distribution cannot be determined from statics alone.

Example 4.4. Reduction of a Force System to a Force and a Couple. Reduce the force system shown in Figure 4.17(a) to a force **R** at the origin and a couple $\mathbf{C}^R$. Can the system be further reduced to a single force?

Solution. We first sketch the desired equipollent system [Figure 4.17(b)]. Applying the conditions for static equivalence, Eqs. (4.1), and determining the moments of the forces by inspection, we obtain

$$(\Sigma\, \mathbf{F})_{\mathrm{I}} = (\Sigma\, \mathbf{F}_{\mathrm{II}}): \quad -50\mathbf{i} + 40\mathbf{j} + 60\mathbf{j}\ N = \mathbf{R}$$

$$\mathbf{R} = -50\mathbf{i} + 100\mathbf{j}\ \mathrm{N} \qquad \textit{Answer}$$

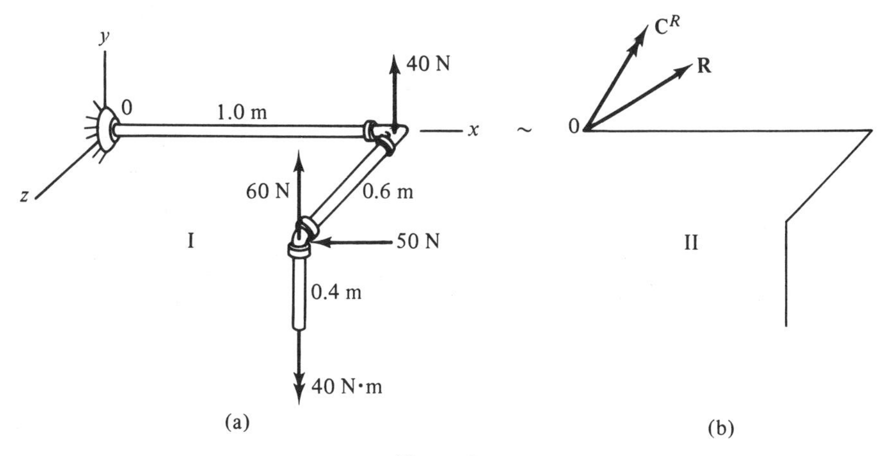

Figure 4.17

and

$$(\Sigma\, \mathbf{M}_0)_{\mathrm{I}} = (\Sigma\, \mathbf{M}_0)_{\mathrm{II}}: \quad (40\text{ N})(1.0\text{ m})\mathbf{k} - 60\text{ N }(0.6\text{ m})\mathbf{i} + 60\text{ N }(1.0\text{ m})\mathbf{k}$$

$$-50\text{ N }(0.6\text{ m})\mathbf{j} - 40\mathbf{j}\text{ N·m} = \mathbf{C}^R$$

$$\mathbf{C}^R = -36\mathbf{i} - 60\mathbf{j} + 60\mathbf{k}\text{ N·m} \qquad \textit{Answer}$$

The force system can be further reduced to a single force only if $\mathbf{C}^R$ is perpendicular to **R**. Taking the dot product of the two vectors, we have

$$\mathbf{C}^R \cdot \mathbf{R} = (-36)(-50) + (-60)(100) = -4200$$

Hence, the two vectors are not perpendicular, and the force system cannot be reduced to a single force.

Example 4.5. Resultant of a Coplanar Force System. Determine the resultant of the forces acting upon the bracket shown in Figure 4.18(a) if (a) $F = 100$ lb and (b) $F = 200$ lb.

Solution. Since the forces are coplanar, we try a force **R** as the resultant [Figure 4.18(b)]. The magnitude and direction of **R** and the location of its line of action are unknown and are chosen arbitrarily in the figure. The resultant is placed at point E, although it can be acting at any other point along its line of action. The location of the line of action is defined by the distance d, but it can be defined in other ways. For example, it could be defined by the perpendicular distance between it and some known point, such as a corner of the bracket. For convenience, we work with **R** in component form. Applying the conditions for static equivalence, Eqs. (4.2), we have

$$(\Sigma\, F_x)_{\mathrm{I}} = (\Sigma\, F_x)_{\mathrm{II}}: \quad -120\text{ lb} + F\cos 53° = R_x$$

$$(\Sigma\, F_y)_{\mathrm{I}} = (\Sigma\, F_y)_{\mathrm{II}}: \quad F\sin 53° - 40\text{ lb} - 120\text{ lb} = R_y$$

$$\overset{+}{(\circlearrowleft \Sigma M_B)_{\mathrm{I}}} = \overset{+}{(\circlearrowleft \Sigma M_B)_{\mathrm{II}}}: \quad 40\text{ lb}(10\text{ in.}) = -R_y d$$

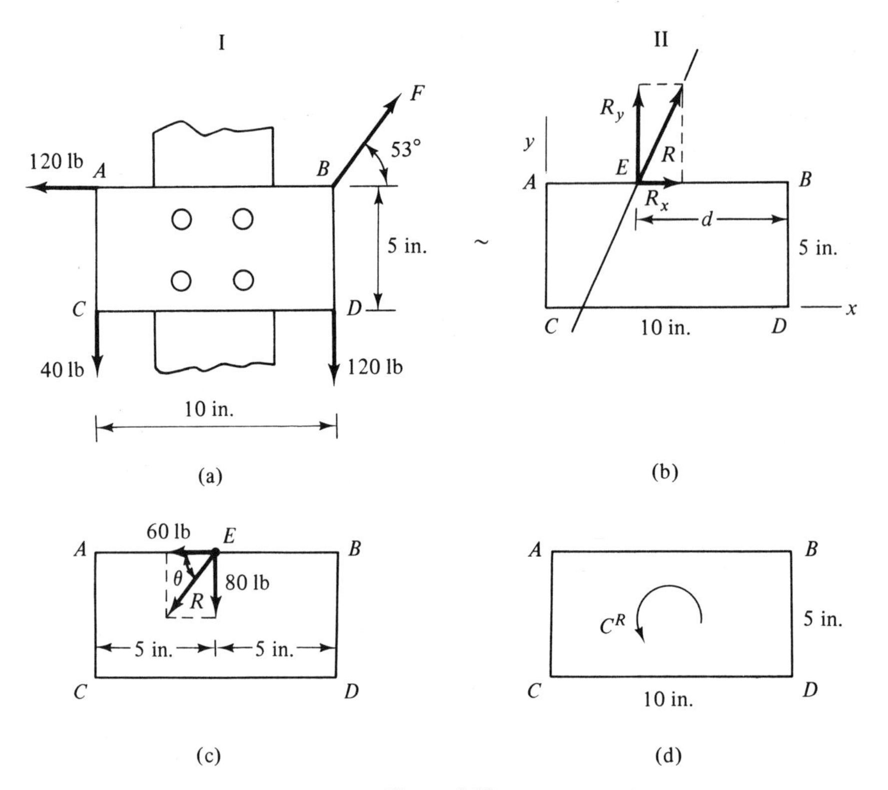

Figure 4.18

(a) Solving the preceding equations for $F = 100$ lb, we obtain

$$R_x = -60 \text{ lb} \qquad R_y = -80 \text{ lb} \qquad d = 5.0 \text{ in.} \qquad \textbf{Answer}$$

Since R_x and R_y are negative, the resultant is as shown in Figure 4.18(c). It is usually sufficient to leave the resultant in component form, but its magnitude and orientation can be easily determined. From Figure 4.18(c), we have

$$R = \sqrt{(60 \text{ lb})^2 + (80 \text{ lb})^2} = 100 \text{ lb}$$

$$\tan\theta = \frac{80}{60} = 1.33$$

$$\theta = 53°$$

(b) Solving the force equations for static equivalence with $F = 200$ lb, we get

$$R_x = 0 \qquad R_y = 0$$

Thus, the resultant will not be a force; therefore, we try a couple [Figure 4.18(d)]. From the moment condition for static equivalence, we have

$$(\overset{+}{\curvearrowleft} M_B)_{\text{I}} = (\overset{+}{\curvearrowleft} M_B)_{\text{II}}: \quad 40 \text{ lb } (10 \text{ in.}) = C^R$$

$$C^R = 400 \text{ lb·in.} \qquad \textbf{Answer}$$

Example 4.6. Resultant of a Parallel Force System. The forces exerted on a raft by its four occupants are as shown in Figure 4.19(a). Determine the resultant of these forces.

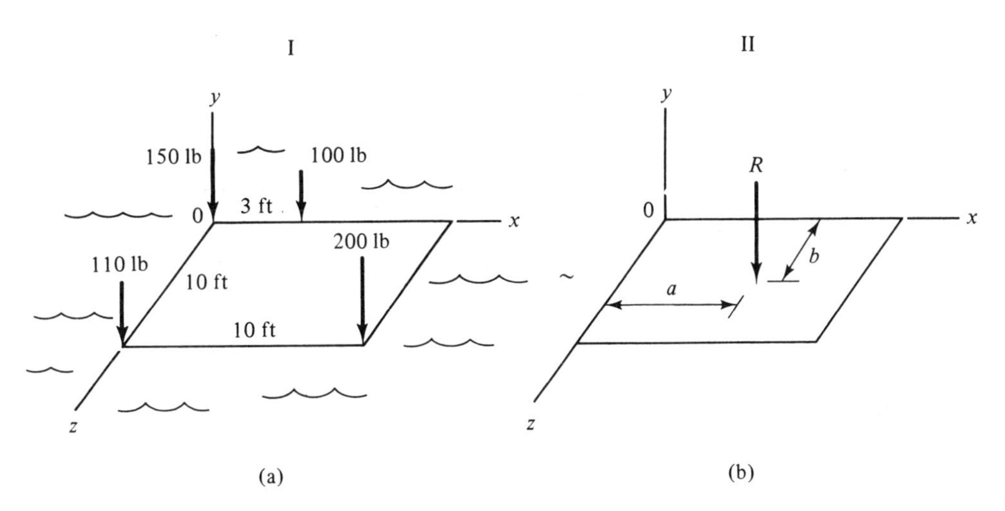

Figure 4.19

Solution. Since the forces are parallel and vertical, we try a vertical force **R** as the resultant [Figure 4.19(b)]. The location of the line of action of **R** is defined by the unknown distances a and b.

From the conditions for static equivalence, we have

$$(\Sigma F_y)_{\mathrm{I}} = (\Sigma F_y)_{\mathrm{II}}: \quad -150 \text{ lb} - 100 \text{ lb} - 110 \text{ lb} - 200 \text{ lb} = -R$$

$$R = 560 \text{ lb} \qquad \textbf{\textit{Answer}}$$

and

$$(\Sigma \mathbf{M}_0)_{\mathrm{I}} = (\Sigma \mathbf{M}_0)_{\mathrm{II}}: \quad -100 \text{ lb}(3 \text{ ft})\mathbf{k} + 200 \text{ lb}(10 \text{ ft})\mathbf{i} - 200 \text{ lb}(10 \text{ ft})\mathbf{k}$$
$$+ 110 \text{ lb}(10 \text{ ft})\mathbf{i} = Rb\mathbf{i} - Ra\mathbf{k}$$

or

$$3100\mathbf{i} - 2300\mathbf{k} \text{ lb·ft} = Rb\mathbf{i} - Ra\mathbf{k}$$

where the moments have been determined by inspection. Solving for a and b from the moment equation, we obtain

$$a = \frac{2300 \text{ lb·ft}}{R} = \frac{2300 \text{ lb·ft}}{560 \text{ lb}} = 4.11 \text{ ft}$$

$$b = \frac{3100 \text{ lb·ft}}{R} = \frac{3100 \text{ lb·ft}}{560 \text{ lb}} = 5.54 \text{ ft} \qquad \textbf{\textit{Answer}}$$

PROBLEMS

4.12 and **4.13** Reduce the force systems shown to a single force **R** acting at the origin and a couple $\mathbf{C}^R$. Can the systems be further reduced to a single force?

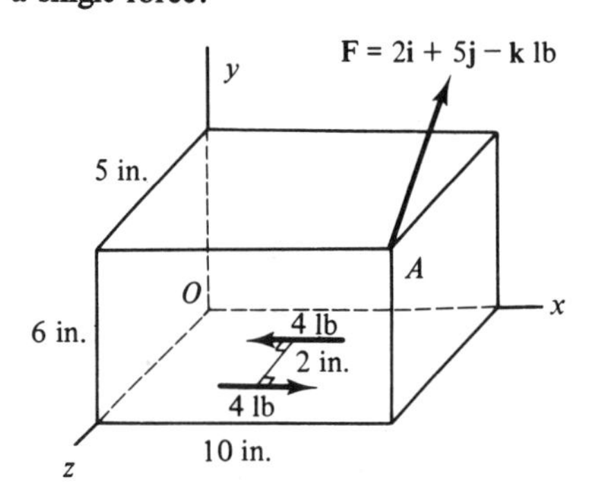

Problem 4.12

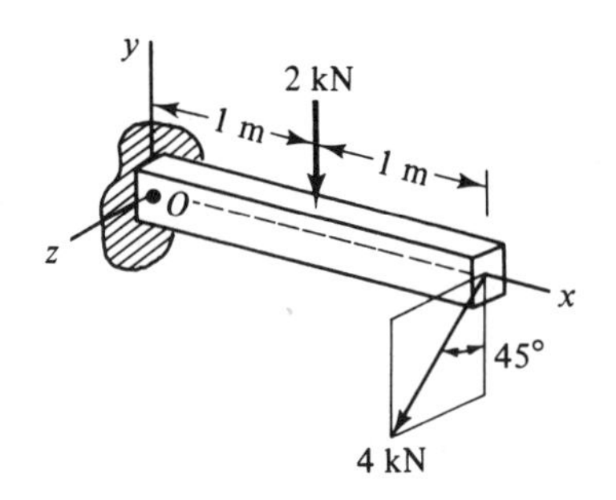

Problem 4.13

4.14 to **4.20** Determine completely and show on a sketch the resultant of the force systems shown.

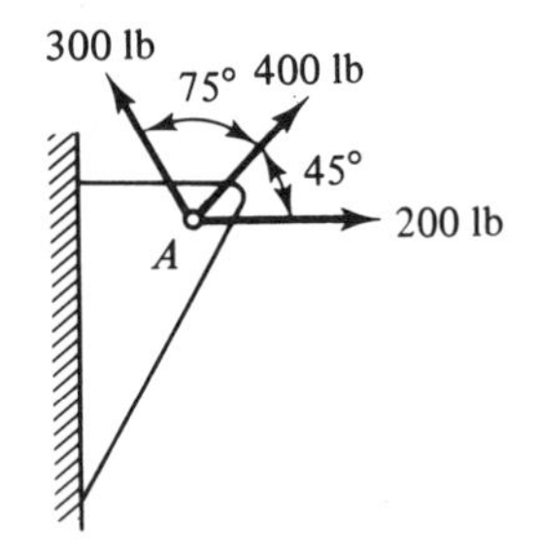

Problem 4.14

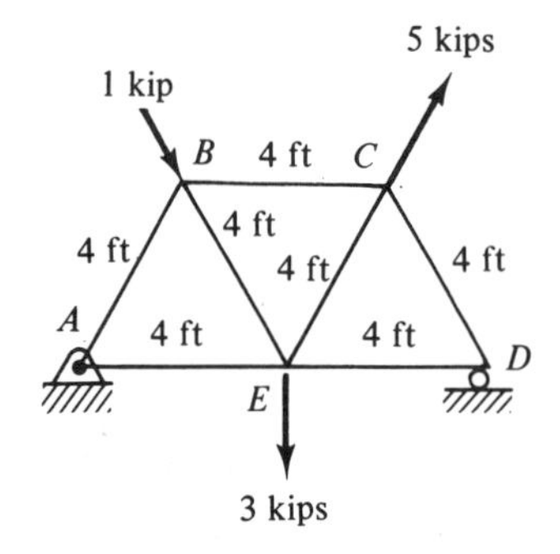

Problem 4.15

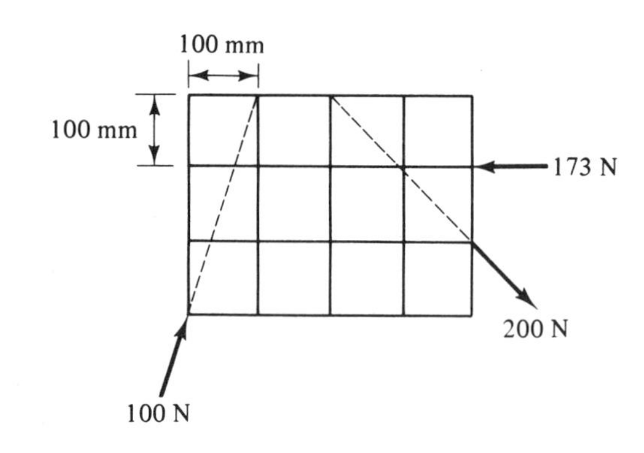

Problem 4.16

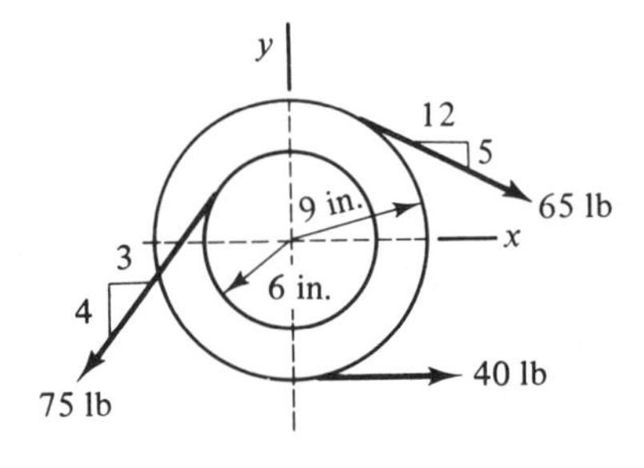

Problem 4.17

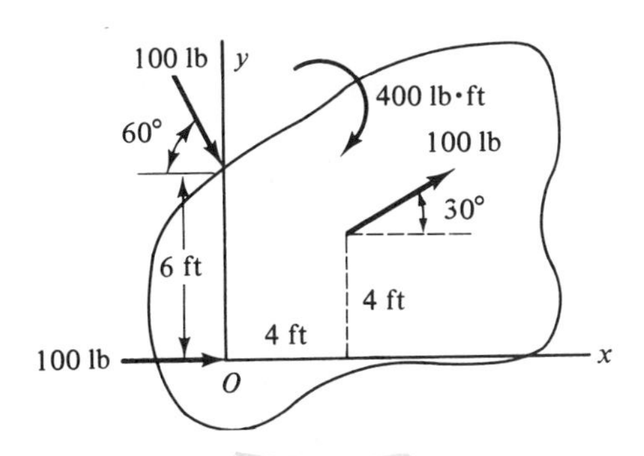

Problem 4.18

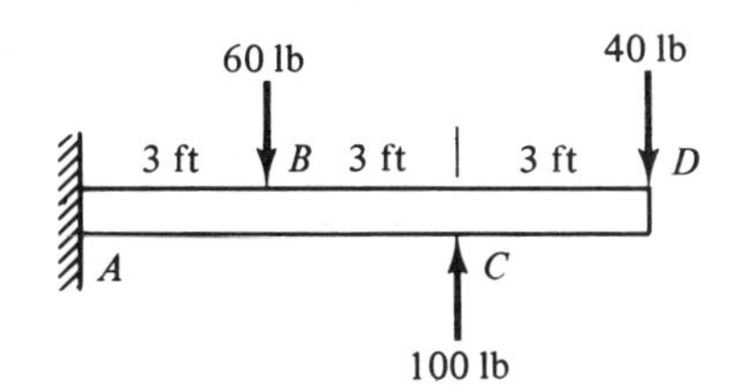

Problem 4.19

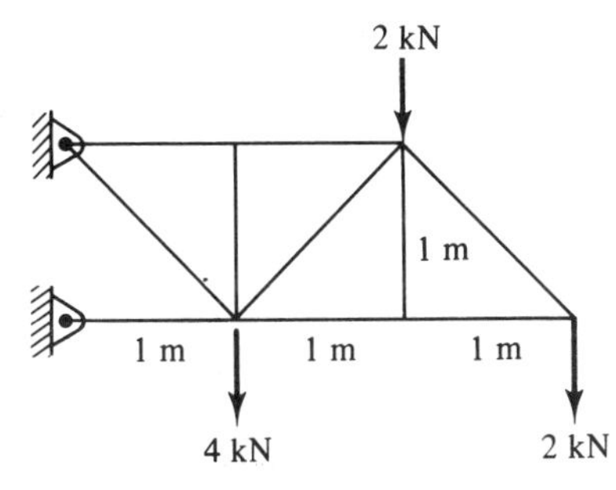

Problem 4.20

4.21 The forces exerted on a bridge deck by a tractor-trailer truck are as shown. Determine completely the resultant load on the bridge.

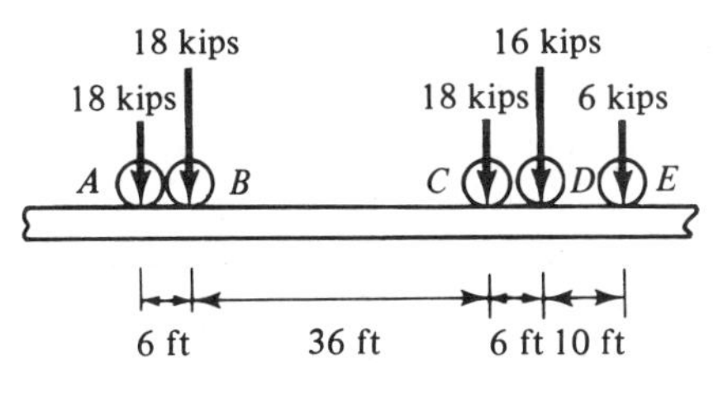

Problem 4.21

4.22 The resultant of the five forces shown is a couple with a magnitude of 12 kN·m. Determine the magnitudes of the three unknown forces and the sense of the resultant couple.

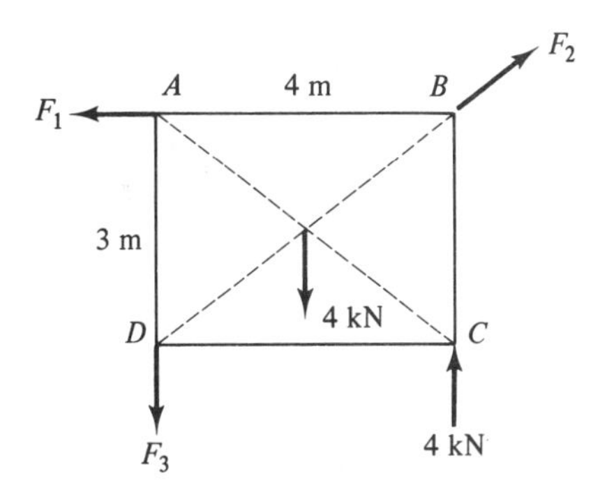

Problem 4.22

4.23 Determine the vertical force that must be applied to the foot pedal at point *D* in order that the resultant of the total force system pass through point *C*. What is the resultant force?

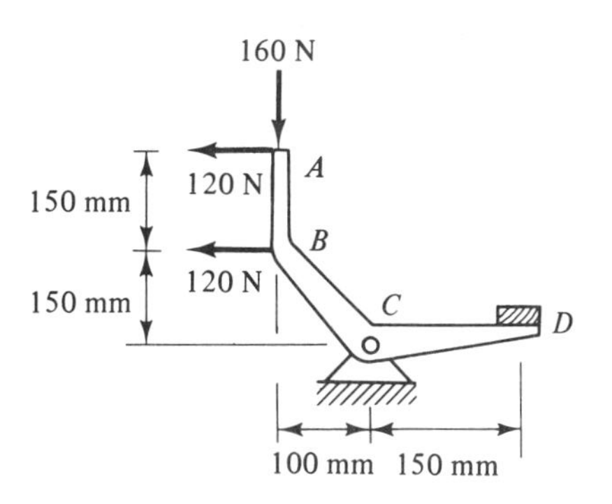

Problem 4.23

4.24 The support columns in a building exert the forces shown on the floor slab. Determine completely the resultant load on the slab.

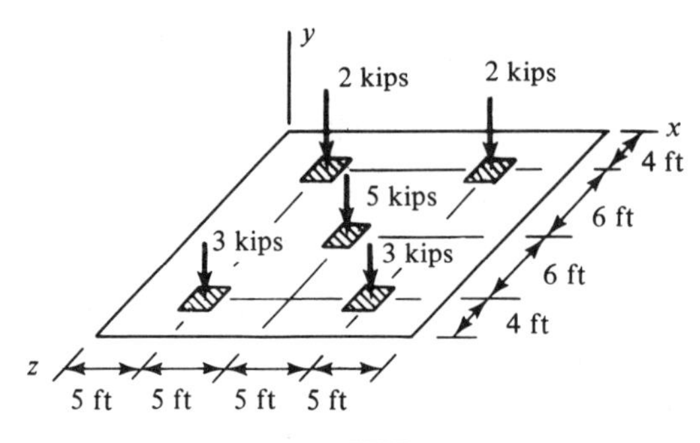

Problem 4.24

4.25 A crane supported by a floating platform as shown weighs 4 tons and is lifting an object that weighs 1 ton. Where should the 5-ton counterweight A be placed if the resultant of the three weights is to be a force acting through the geometric center of the platform? What is the resultant force?

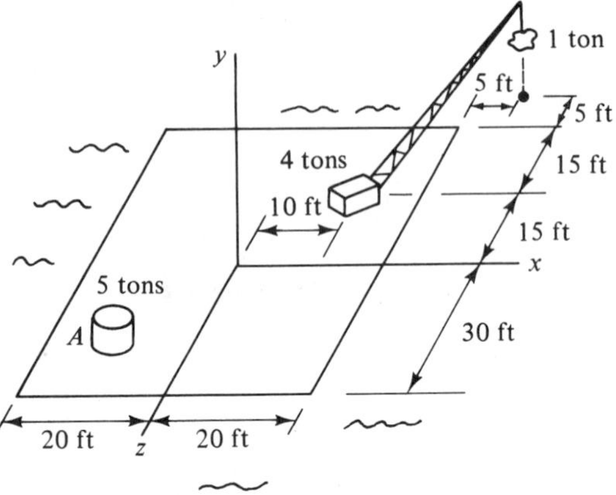

Problem 4.25

4.4 WEIGHT, CENTER OF GRAVITY, AND CENTER OF MASS

In the force analyses of Chapter 3 the gravitational attraction of the earth was treated as a single force acting at the center of gravity. We are now in a position to justify this step and to develop procedures for locating the center of gravity of bodies of arbitrary shape.

Consider two bodies with mass m and M separated a distance r. According to *Newton's law of gravitation*, the bodies attract each other with equal and opposite forces that lie along a line connecting their centers and that have magnitude

$$F = \frac{GMm}{r^2} \tag{4.4}$$

where G is the so-called *universal constant of gravitation*; $G = 3.442 \times 10^{-8}$ $\text{ft}^4/(\text{lb·s}^4)$ or $G = 6.658 \times 10^{-11}$ $\text{m}^3/(\text{kg·s}^2)$. This law, which is analogous to Coulomb's law of attraction between electric charges, was formulated by Sir Isaac Newton during his studies of the motions of the planets. Strictly speaking, the law of gravitation applies only for particles. However, it has also been shown to hold for spherical bodies of finite size whose density (mass per unit volume) is constant or varies only in the radial direction. The earth, being very nearly a stratified sphere, satisfies these conditions to a reasonable degree of approximation.

The attractive forces of the earth on a finite body of arbitrary shape can be determined by considering the body to consist of an infinite number of infinitesimal elements, or particles, with volume dV and mass $dm = \rho dV$, where ρ is the density. According to the law of gravitation, each of these elements is subjected to a force directed toward the earth's center and with magnitude defined by Eq. (4.4). Denoting the magnitude of the force acting upon a typical element by dW, we have

$$dW = \left(\frac{GM_{\text{earth}}}{r^2} \right) \rho dV$$

This expression can be rewritten as

$$dW = g\rho dV = \gamma dV$$

where

$$g = \frac{GM_{\text{earth}}}{r^2} \tag{4.5}$$

is the so-called *acceleration of gravity* and $\gamma = \rho g$ is the weight per unit volume, or *specific weight*, of the body. (The acceleration of gravity due to the attraction of the moon or other celestial bodies can also be obtained from Eq. (4.5) by replacing M_{earth} with the mass of the particular body of interest.)

Unless a body is exceptionally large, the forces acting upon its various elements can be assumed to be parallel. Thus, the earth's gravitational attraction is statically equivalent to a single force. The magnitude of the resultant attractive force is known as the *weight* of the body and its point of application is called the *center of gravity*, or c.g. for short. Both the weight and location of the center of gravity can be determined from the conditions for static equivalence, as we shall now show.

Let $dW = \gamma dV$ denote the weight of an infinitesimal element of a body and $(\bar{x}_{\text{el}}, \bar{y}_{\text{el}}, \bar{z}_{\text{el}})$ denote the coordinates of its center of gravity G_{el} [Figure 4.20(a)]. Similarly, let $(\bar{x}, \bar{y}, \bar{z})$ denote the coordinates of the center of gravity G of the entire

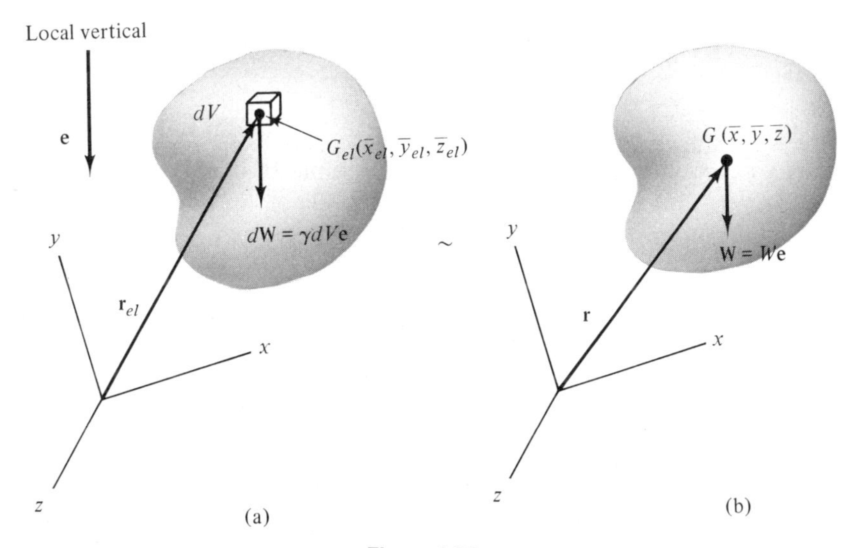

Figure 4.20

body and W denote its total weight [Figure 4.20(b)]. The locations of the centers of gravity G_{el} and G can also be defined by their respective position vectors $\mathbf{r}_{el}$ and $\mathbf{r}$.

If $\mathbf{e}$ is a unit vector in the direction of the local vertical, the force condition for static equivalence can be expressed as

$$(\Sigma\, \mathbf{F})_{\mathrm{I}} = (\Sigma\, \mathbf{F})_{\mathrm{II}}: \quad \int \gamma dV \mathbf{e} = W\mathbf{e}$$

or

$$W = \int \gamma dV \tag{4.6}$$

Note that for system I, the summation becomes an integration since the forces are infinitesimal and infinite in number. For constant γ, Eq. (4.6) reduces to

$$W = \gamma V = \rho g V = mg \tag{4.7}$$

where V is the total volume of the body and m is its total mass. This is the same result presented in Eq. (1.1).

The moment condition for static equivalence is

$$(\Sigma\, \mathbf{M}_0)_{\mathrm{I}} = (\Sigma\, \mathbf{M}_0)_{\mathrm{II}}: \quad \int \left[\mathbf{r}_{el} \times (\gamma dV \mathbf{e})\right] = \mathbf{r} \times (W\mathbf{e})$$

Using the distributive property of the cross product, we can rewrite this expression as

$$\left(\int \mathbf{r}_{el} \gamma dV\right) \times \mathbf{e} = W\mathbf{r} \times \mathbf{e}$$

Equating the terms on the left side of the cross products and introducing Eq. (4.6), we obtain for the position vector of the center of gravity

$$\mathbf{r} = \frac{\int \mathbf{r}_{el} \gamma dV}{\int \gamma dV} \tag{4.8}$$

Expressions for the coordinates of the center of gravity can be obtained from Eq. (4.8) by expressing the various position vectors in component form and equating the coefficients of the base vectors:

$$\bar{x} = \frac{\int \bar{x}_{el} \gamma dV}{\int \gamma dV} \qquad \bar{y} = \frac{\int \bar{y}_{el} \gamma dV}{\int \gamma dV} \qquad \bar{z} = \frac{\int \bar{z}_{el} \gamma dV}{\int \gamma dV} \tag{4.9}$$

In these equations, and in Eqs. (4.6) and (4.8), the integration is to be carried out over the entire volume of the body.

Equation (4.8) can also be expressed in terms of the density by introducing the relation $\gamma = \rho g$. For a uniform gravitational field, g will be a constant in both the

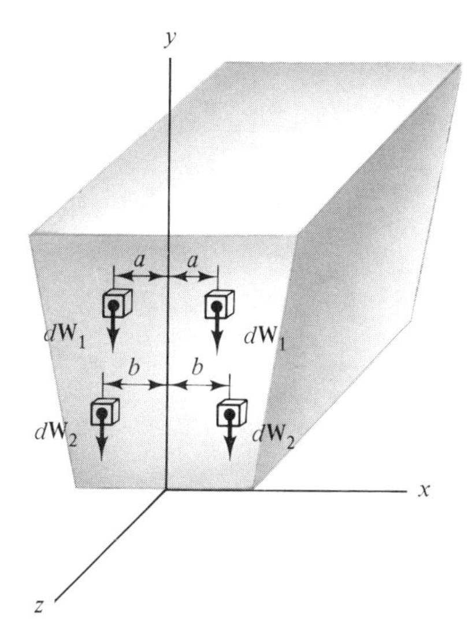

Figure 4.21

numerator and denominator of the equation and it will cancel. The resulting expression

$$\mathbf{r} = \frac{\int \mathbf{r}_{el} \rho dV}{\int \rho dV} \tag{4.10}$$

defines the position of the *center of mass*.

Since the center of gravity and the center of mass coincide for constant g, the two terms are often used interchangeably. However, the term center of gravity is properly associated with the effect of gravitational forces on a body, and center of mass with the manner in which the mass of a body is distributed. The mass center is an important concept in dynamics.

The evaluation of the integrals in the preceding equations will be considered in detail in Section 4.5. However, if the weight distribution of the body is symmetrical, one or more coordinates of the center of gravity can be determined by inspection.

The weight distribution of a body is said to have a *plane of symmetry* if for every element located a certain perpendicular distance to one side of the plane there is an element with equal weight located the same perpendicular distance to the other side. This condition is illustrated in Figure 4.21, where the y-z plane is the plane of symmetry. Expressing the first of Eqs. (4.9) in terms of weights, we have for the x coordinate of the center of gravity of the body

$$\bar{x} = \frac{\int \bar{x}_{el} dW}{W} = \frac{dW_1(a) + dW_1(-a) + dW_2(b) + dW_2(-b) + \cdots}{W} = 0$$

Since $\bar{x} = 0$, it follows that the center of gravity lies in the plane of symmetry. This is true in general, not just for this specific example.

If a body has two planes of symmetry, its center of gravity must lie in both planes. Consequently, it must lie along their line of intersection. For example, the center of gravity of a homogeneous right circular cone lies along its axis. If there are three planes of symmetry that intersect at a point, the center of gravity coincides with their point of intersection. Thus, the center of gravity of many uniform bodies with simple shapes, such as spheres, cylinders, ellipsoids, and rectangular parallelepipeds, can be located completely by inspection.

The weight distribution of a body is said to have an *axis of symmetry* if for every element located a certain perpendicular distance to one side of the axis there is an element with equal weight located the same perpendicular distance to the other side. For example, the z axis is an axis of symmetry for the Z-shaped wire

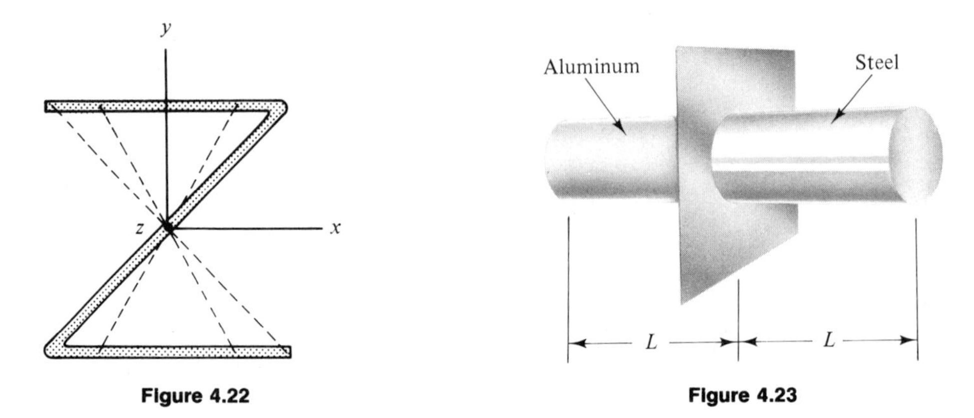

Figure 4.22

Figure 4.23

shown in Figure 4.22. If a body has an axis of symmetry, the center of gravity must lie along that axis. This can be readily shown by using the same procedure as in the case of a plane of symmetry.

When using symmetry conditions to locate the center of gravity, it is essential to remember that it is symmetry of the weight distribution that is required. This is not to be confused with geometric symmetry. Geometric symmetry implies symmetry of the weight distribution only if the body is homogeneous and is located in a uniform gravitational field. For example, the composite shaft shown in Figure 4.23 has geometric symmetry with respect to the plane through its center, but it does not have weight symmetry.

In concluding this section, we note that the location of the center of gravity can also be determined experimentally. Usually, this requires that the body be supported in such a way that the reactions can be measured. The weight and location of the center of gravity are then determined from the equilibrium equations, as illustrated in Example 4.8.

Example 4.7. Location of the Center of Gravity. Locate the center of gravity of the homogeneous body shown in Figure 4.24.

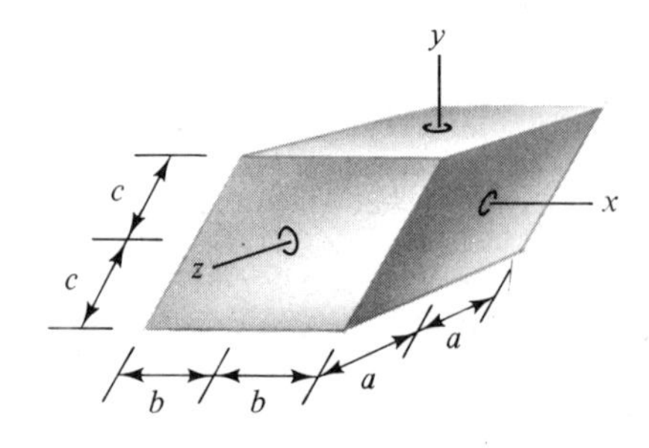

Figure 4.24

Solution. The x-y plane is a plane of symmetry and the z axis is an axis of symmetry. Therefore, the c.g. is at the origin of the coordinate system shown; $\bar{x} = \bar{y} = \bar{z} = 0$. Note that only geometric symmetry is required because the body is homogeneous.

Example 4.8. Experimental Determination of Center of Gravity. A connecting rod from an engine is suspended from wires attached to a spring scale at each end, as shown in Figure 4.25(a). When the rod is horizontal, scale A reads 12 N and scale B reads 6 N. Determine the weight and location of the center of gravity of the connecting rod (assume that it is homogeneous).

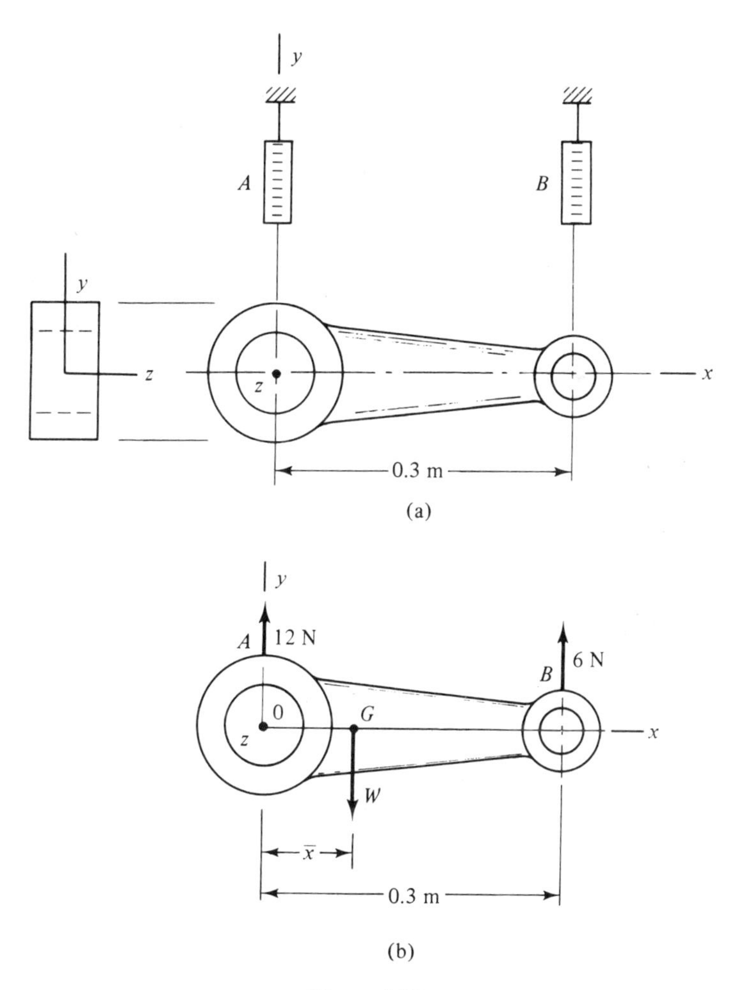

Figure 4.25

Solution. By symmetry, $\bar{y} = \bar{z} = 0$. From the FBD of the rod [Figure 4.25(b)], we have

$$\Sigma F_y = 12N + 6N - W = 0$$

$$\overset{+}{\curvearrowleft}\Sigma M_0 = 6N\,(0.3 \text{ m}) - W\bar{x} = 0$$

from which we get

$$W = 18N \qquad \bar{x} = 0.1 \text{ m} \qquad \textbf{\textit{Answer}}$$

PROBLEMS

4.26 Determine the specific weight γ of a uniform solid body weighing 23 kN and occupying a volume of 0.3 m^3. What is the density ρ of the body?

4.27 What is the force of mutual attraction between two 1-ft diameter solid steel spheres placed such that they are touching? How does this force compare in magnitude with the gravitational attraction of the earth on each sphere? The specific weight of steel is 490 lb/ft^3.

4.28 If the acceleration of gravity is 32.2 ft/s^2 at the earth's surface, what is its value at the surface of the moon? The radius and mass of the moon are 0.273 times and 0.0123 times that of the earth, respectively.

4.29 If the mass of the moon is 0.0123 times that of the earth, at what fraction of the total distance d between them will a spacecraft experience an equal attractive force from each planet?

4.30 to 4.33 Locate the center of gravity of the uniform bodies shown.

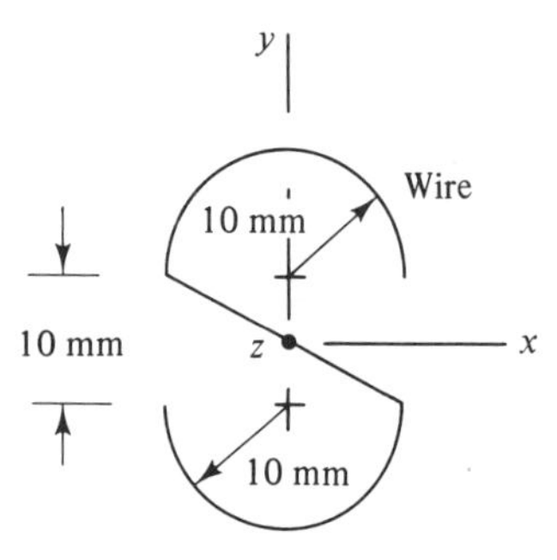

Problem 4.30

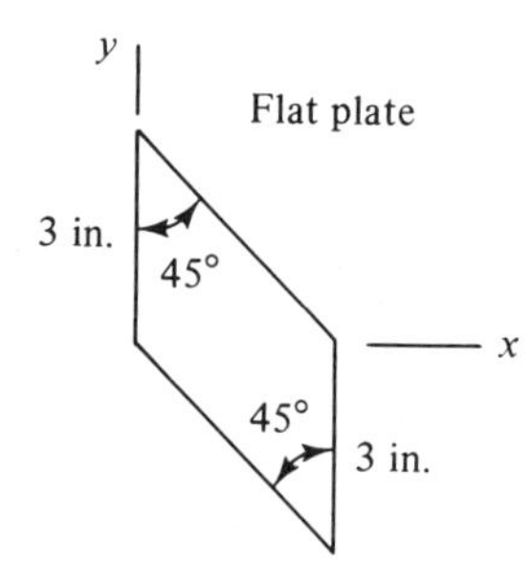

Problem 4.31

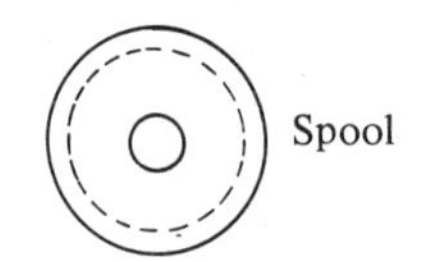

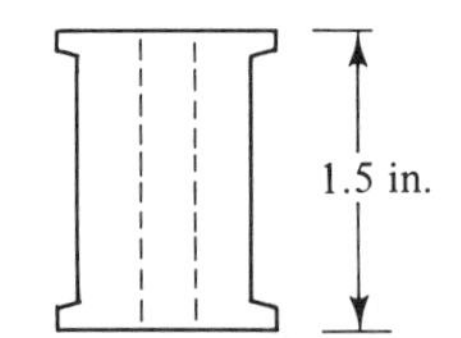

Problem 4.32

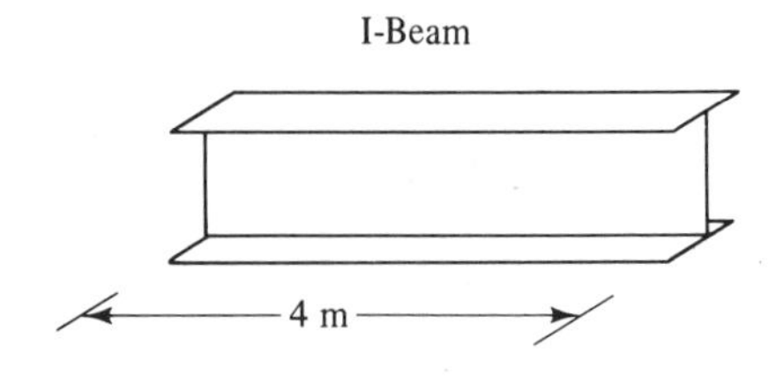

Problem 4.33

4.34 Locate the mass center of a thin homogeneous plate in the shape of an equilateral triangle.

4.35 Three circular holes of the same size are punched in a uniform circular plate as shown. Determine the angle θ for which the mass center will be at the origin.

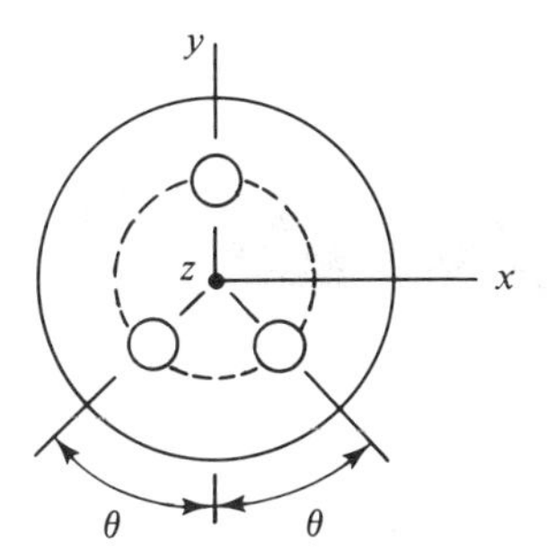

Problem 4.35

4.36 Show that if a body is suspended from a cord or frictionless pivot, its center of gravity will lie directly below the point of suspension. Thus, show that the center of gravity of an irregular-shaped plate can be found by suspending it from two or more different points and locating the point of intersection of vertical lines drawn through the points of suspension. Cut an irregular shape from cardboard and use this procedure to locate its center of gravity.

4.37 When the front wheels of a truck are placed on a scale, it registers 6000 lb. When the rear wheels are placed on the scale, it registers 10 000 lb. How much does the truck weigh and how far forward from the rear axle is its center of gravity located? The wheelbase is 10 ft.

4.38 It takes a force F of 400 lb to raise end A of the heavy machine shown and a force of 300 lb to raise end D by using the same lever arrangement. Determine the weight of the machine and the x coordinate of its center of gravity. Can $\bar{y}$ be determined from the information given? Assume negligible friction between the machine and lever.

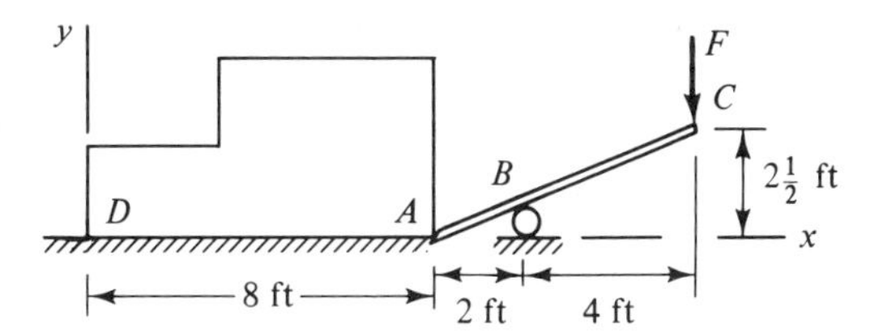

Problem 4.38

4.39 An irregular-shaped plate is suspended from three wires as shown. When the plate is horizontal, the tensions on the wires are as indicated. Determine the coordinates $\bar{x}$, $\bar{y}$, and $\bar{z}$ of the mass center.

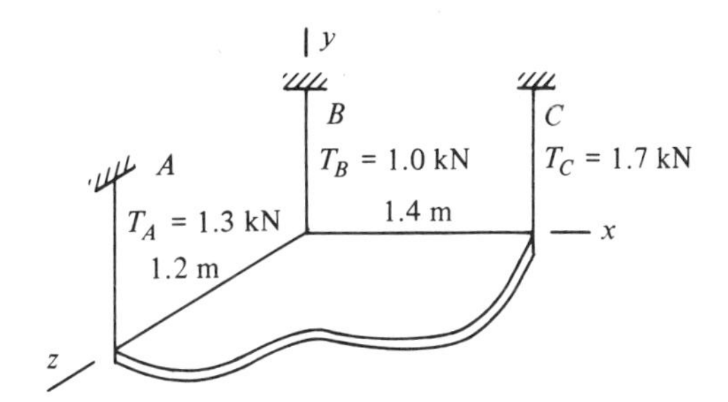

Problem 4.39

4.40 The center of gravity of a 12-ft long rowboat is to be determined by suspending it from two ropes and measuring the tensions on the ropes as shown. Scales A and B read 90 lb and 60 lb, respectively, when the boat is horizontal. When the bow is lowered 1 ft, scale B reads 59 lb. Determine the coordinates $\bar{x}$, $\bar{y}$, $\bar{z}$ of the center of gravity.

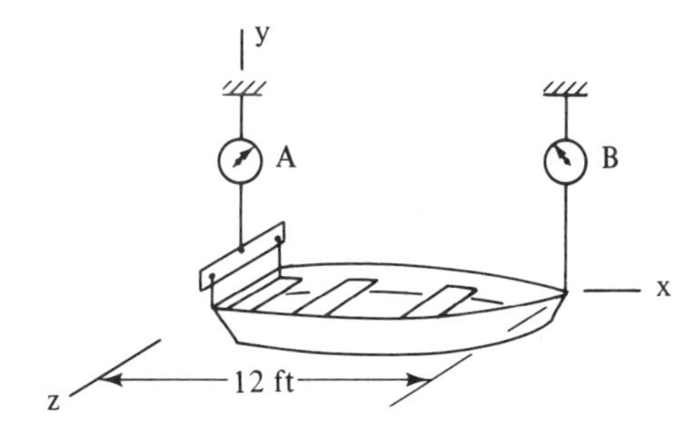

Problem 4.40

4.5. CENTERS OF GRAVITY BY INTEGRATION—CENTROIDS

In this section we shall consider the procedure for locating the center of gravity of a body by formal evaluation of the integrals in Eqs. (4.9). From a computational standpoint, this approach is practical only for bodies with simple shapes whose boundaries can be defined in equation form. More complicated shapes can be handled by methods of approximation or by evaluating the integrals numerically.

Many bodies are uniform throughout, at least to a reasonable degree of approximation. For such homogeneous bodies, γ is a constant, and Eqs. (4.9) reduce to

$$\bar{x} = \frac{\int \bar{x}_{el} dV}{V} \qquad \bar{y} = \frac{\int \bar{y}_{el} dV}{V} \qquad \bar{z} = \frac{\int \bar{z}_{el} dV}{V} \tag{4.11}$$

where

$$V = \int dV$$

is the total volume of the body. These equations define the *centroid* C *of a volume.*

Although the centroid and center of gravity coincide for homogeneous bodies, they involve completely different physical concepts. The center of gravity is associated with the distribution of gravitational forces on a body. The centroid is associated with geometric shapes. We shall use $(\bar{x}, \bar{y}, \bar{z})$ to denote the coordinates of both centers of gravity and centroids. This should present no problem if we keep in mind the physical concepts associated with each. The symmetry conditions of Section 4.4 also apply to the location of centroids. However, only geometric symmetry is required in this case.

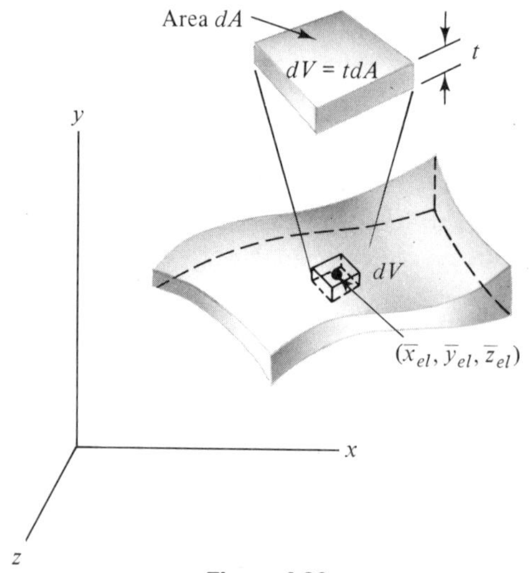

Figure 4.26

For thin, shell-like shapes that approximate a surface area (Figure 4.26), the volume of an element can be expressed as $dV = tdA$, where dA is the area of the element and t is its thickness. If t is a constant, Eqs. (4.11) become

$$\bar{x} = \frac{\int \bar{x}_{el} dA}{A} \qquad \bar{y} = \frac{\int \bar{y}_{el} dA}{A} \qquad \bar{z} = \frac{\int \bar{z}_{el} dA}{A} \tag{4.12}$$

where

$$A = \int dA$$

is the total surface area and ($\bar{x}_{el}$, $\bar{y}_{el}$, $\bar{z}_{el}$) are the coordinates of the centroid of the element of area dA. These equations define the *centroid of an area*.

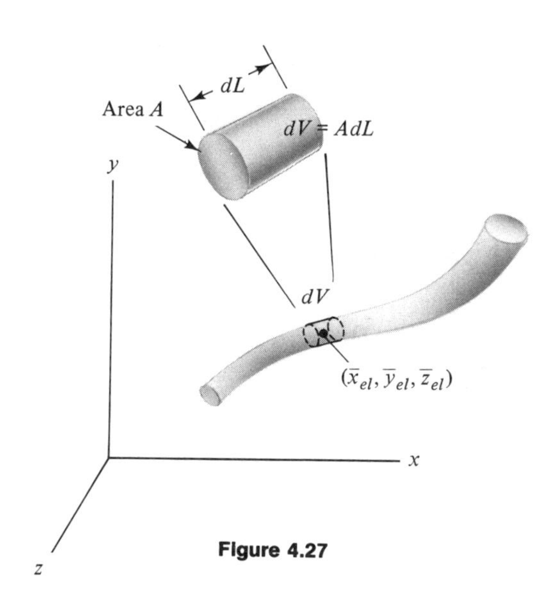

Figure 4.27

The volume of an element of a thin, wire-like shape that approximates a line (Figure 4.27) can be expressed as $dV = AdL$, where A is the cross-sectional area of the element and dL is its length. If A is a constant, Eqs. (4.11) can be written as

$$\bar{x} = \frac{\int \bar{x}_{el} dL}{L} \qquad \bar{y} = \frac{\int \bar{y}_{el} dL}{L} \qquad \bar{z} = \frac{\int \bar{z}_{el} dL}{L} \tag{4.13}$$

where

$$L = \int dL$$

is the total length of the shape and ($\bar{x}_{el}$, $\bar{y}_{el}$, $\bar{z}_{el}$) are the coordinates of the centroid of the element of length dL. These equations define the *centroid of a line*.

At this point, it is important to note that the preceding equations define the location of the centroid of a shape in terms of the locations of the centroids of its constituent elements. In other words, $(\bar{x}_{el}, \bar{y}_{el}, \bar{z}_{el})$ must be known before we can determine $(\bar{x}, \bar{y}, \bar{z})$. This presents no problem. The elements can always be chosen with simple shapes such that the locations of their centroids are obvious or can be determined from symmetry conditions.

Once an infinitesimal element has been selected, expressions for its area, length, or volume and location of its centroid can be readily determined from simple geometry, and the necessary integrals can be evaluated to locate the centroid of the shape. For centers of gravity, the variation of the specific weight γ throughout the body must also be known. Most problems can be handled by using first-order elements (infinitesimal in one direction only) and single integrals. However, double and triple integrals can be used in many instances, if desired. An appropriate choice of coordinate systems can also simplify the computations. For example, polar coordinates are usually most convenient for problems involving circular shapes.

The locations of the centroids of many shapes are tabulated in engineering and mathematical handbooks. For reference purposes, results for several common shapes are listed in Table A.1 of the Appendix.

Example 4.9. Centroid of an Area. Locate the centroid of the area shown in Figure 4.28(a).

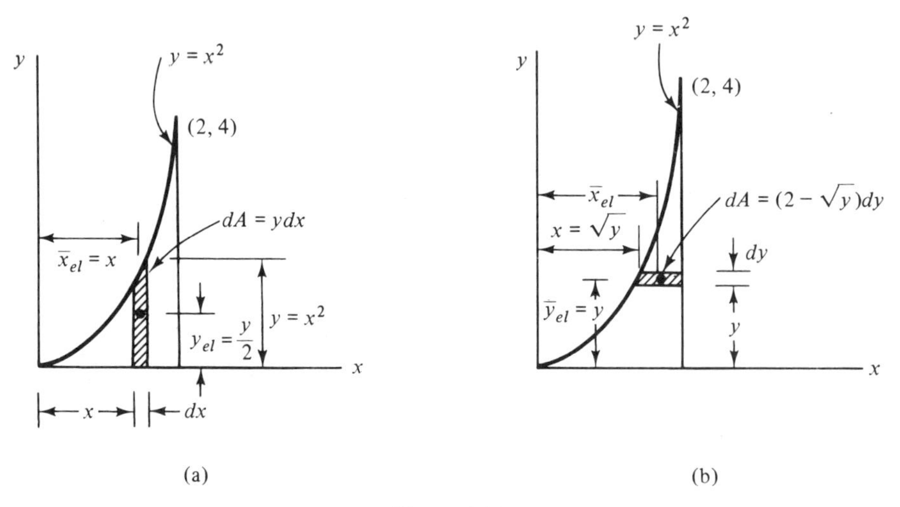

Figure 4.28

Solution. We take a vertical strip of width dx located an arbitrary distance x from the origin as the element of area. The height y of the element is given by the equation of the boundary, $y = x^2$. By symmetry, the centroid of the element is at its geometric center. Thus,

$$\bar{x}_{el} = x \qquad \bar{y}_{el} = \frac{y}{2} = \frac{x^2}{2} \qquad dA = ydx = x^2dx$$

To cover the entire area, all elements that lie between $x = 0$ and $x = 2$ must be considered. This establishes the limits of integration.

Substituting the expressions for $\bar{x}_{el}$, $\bar{y}_{el}$, and dA into Eqs. (4.12) and integrating, we obtain

$$A = \int dA = \int_0^2 x^2dx = \frac{x^3}{3}\Bigg|_0^2 = \frac{8}{3}$$

$$A\bar{x} = \int \bar{x}_{el}dA = \int_0^2 (x)x^2dx = \frac{x^4}{4}\Bigg|_0^2 = 4$$

$$A\bar{y} = \int \bar{y}_{el}dA = \int_0^2 \left(\frac{x^2}{2}\right)x^2dx = \frac{x^5}{10}\Bigg|_0^2 = \frac{16}{5}$$

Solving for $\bar{x}$ and $\bar{y}$, we have

$$\bar{x} = \frac{4}{(8/3)} = 1.5 \qquad \bar{y} = \frac{(16/5)}{(8/3)} = 1.2 \qquad \textit{Answer}$$

Alternate Solution. A horizontal strip can also be used as the element of area [Figure 4.28(b)]. The element has width dy and is located an arbitrary distance y from the origin. Since the centroid of the element lies at its midpoint, $\bar{x}_{el}$ is equal to the average of the horizontal coordinates of the two ends of the strip. The horizontal coordinate x of the left end is obtained from the equation of the boundary ($x = \sqrt{y}$), and the coordinate of the right end is given as $x = 2$. Thus,

$$\bar{x}_{el} = \frac{2 + \sqrt{y}}{2} \qquad \bar{y}_{el} = y \qquad dA = (2 - \sqrt{y})dy$$

Substituting these expressions into Eqs. (4.12) and integrating from $y = 0$ to $y = 4$, we get the same results as before.

The computations with the horizontal element are a bit lengthier than those with the vertical element. (The reverse may be true for other problems.) This is because the horizontal element runs from one curve ($x = \sqrt{y}$) to another ($x = 2$), while the vertical element runs from a coordinate axis ($y = 0$) to a curve ($y = x^2$). It is usually convenient to choose the element with one end on a coordinate axis, if possible.

Example 4.10. Centroid of a Volume. Locate the centroid of the pyramidal volume shown in Figure 4.29. The pyramid had height h and a square base with side b.

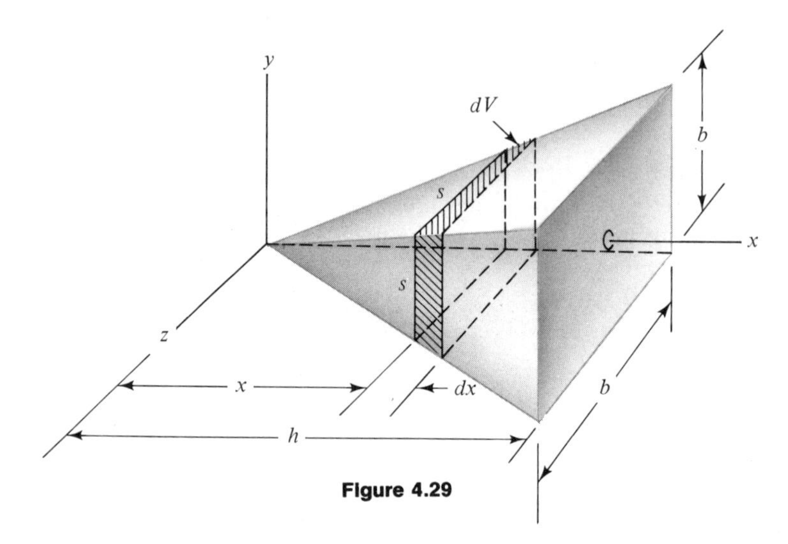

Figure 4.29

Solution. By symmetry, the centroid must lie along the x axis. Thus, $\bar{y} = \bar{z} = 0$, and only $\bar{x}$ remains to be determined. The element of volume is taken to be a square plate of thickness dx located an arbitrary distance x from the origin. By similar triangles, the length s of a side of the element is $s = (b/h)x$. Thus,

$$\bar{x}_{el} = x \qquad dV = s^2 dx = (b/h)^2 x^2 dx$$

The integration is with respect to x and ranges from $x = 0$ to $x = h$. From Eqs. (4.11), we obtain

$$V = \int dV = \int_0^h \left(\frac{b}{h}\right)^2 x^2 dx = \left(\frac{b}{h}\right)^2 \frac{x^3}{3}\bigg|_0^h = \frac{b^2 h}{3}$$

$$V\bar{x} = \int \bar{x}_{el} dV = \int_0^h x \left(\frac{b}{h}\right)^2 x^2 dx = \left(\frac{b}{h}\right)^2 \frac{x^4}{4}\bigg|_0^h = \frac{b^2 h^2}{4}$$

$$\bar{x} = \frac{3}{4}h \qquad \bar{y} = 0 \qquad \bar{z} = 0 \qquad \textbf{\textit{Answer}}$$

Example 4.11. Center of Gravity of a Wire. A thin, homogeneous wire is bent into the shape of a circular arc (Figure 4.30). Locate its center of gravity.

Solution. Since the wire is homogeneous, the center of gravity coincides with the centroid. By symmetry, $\bar{y} = 0$. Using polar coordinates and choosing an element of length dL, we have from the geometry of the figure

$$\bar{x}_{el} = R\cos\theta \qquad dL = R\,d\theta$$

The integration is with respect to θ and ranges from $\theta = -\alpha$ to $\theta = +\alpha$. Thus, we

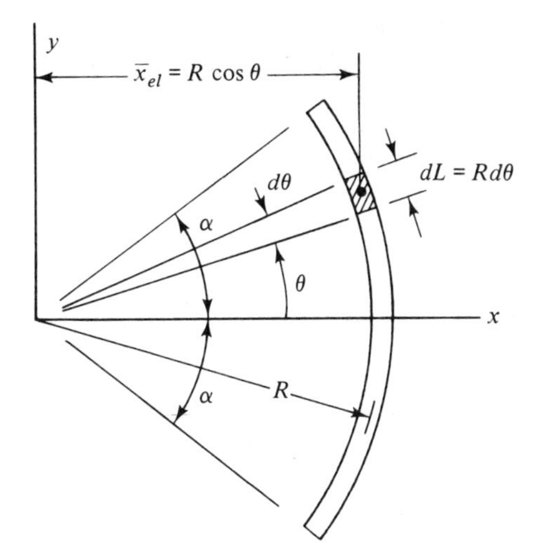

Figure 4.30

have from Eqs. (4.13)

$$L = \int dL = \int_{-\alpha}^{\alpha} R d\theta = R\theta \Big|_{-\alpha}^{\alpha} = 2R\alpha$$

$$L\bar{x} = \int \bar{x}_{\text{el}} dL = \int_{-\alpha}^{\alpha} (R\cos\theta) R d\theta = R^2 \sin\theta \Big|_{-\alpha}^{\alpha} = 2R^2 \sin\alpha$$

$$\bar{x} = \frac{R\sin\alpha}{\alpha} \qquad \bar{y} = 0 \qquad \textit{Answer}$$

Example 4.12. Centroid of an Area. Locate the centroid of the circular segment shown in Figure 4.31(a).

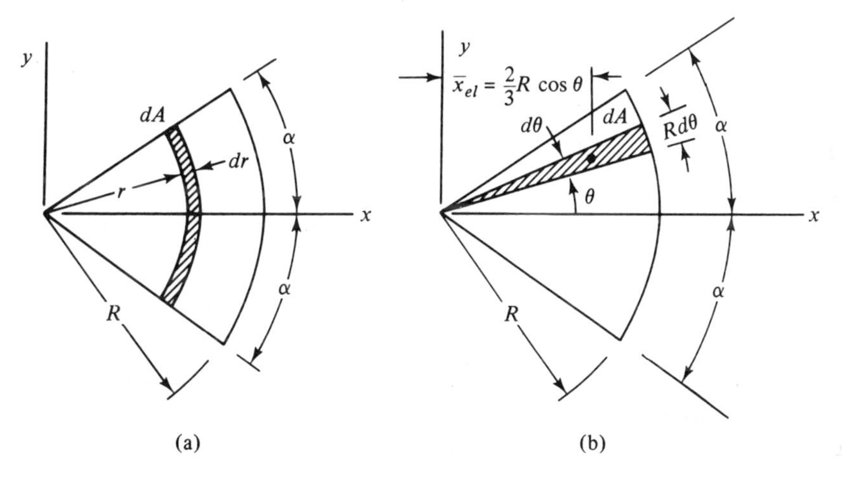

Figure 4.31

Solution. By symmetry, $\bar{y} = 0$. The element of area can be taken as a circular strip of width dr [Figure 4.31(a)] or as a triangle with included angle $d\theta$ [Figure 4.31(b)]. The centroid of these elements cannot be located by symmetry. However, the location of the centroid of the circular strip can be determined from the results of Example 4.11, and Table A.1 of the Appendix gives the necessary information for the triangular element.

Using the triangular element, we have

$$\bar{x}_{\text{el}} = \frac{2}{3} R \cos \theta \qquad dA = \frac{R^2}{2}\, d\theta$$

Thus,

$$A = \int dA = \int_{-\alpha}^{\alpha} \frac{R^2}{2}\, d\theta = \frac{R^2}{2}\, \theta \Big|_{-\alpha}^{\alpha} = R^2 \alpha$$

$$A\bar{x} = \int \bar{x}_{\text{el}} dA = \int_{-\alpha}^{\alpha} \left(\frac{2}{3} R \cos \theta \right) \frac{R^2}{2}\, d\theta = \frac{R^3}{3} \sin \theta \Big|_{-\alpha}^{\alpha} = \frac{2R^3 \sin \alpha}{3}$$

$$\bar{x} = \frac{2R \sin \alpha}{3\alpha} \qquad \bar{y} = 0 \qquad \textit{Answer}$$

It is left as an exercise to confirm that use of the circular element leads to the same results. Note that the location of the centroid of various other circular segments can be obtained by letting α take on the appropriate values. For example, $\alpha = \pi/2$ corresponds to a semicircle.

PROBLEMS

4.41 to 4.46 Determine by integration the coordinates $\bar{x}$ and $\bar{y}$ of the centroid of the areas shown.

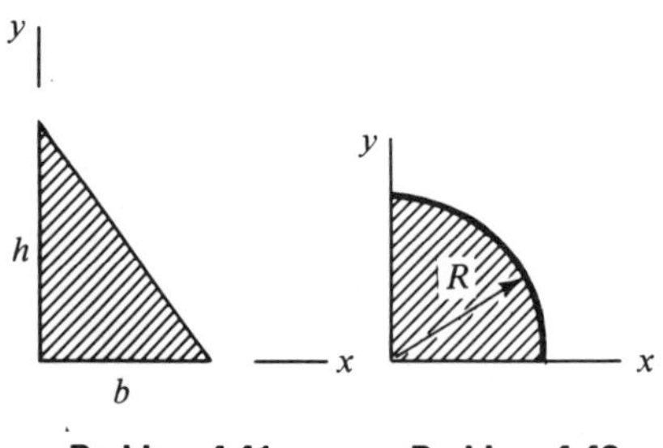

Problem 4.41 Problem 4.42

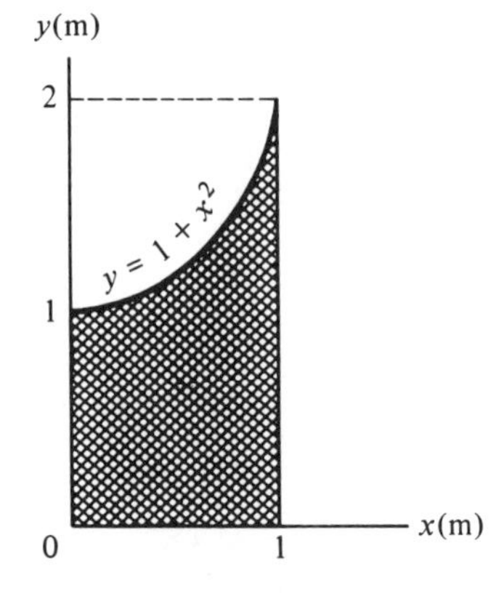

Problem 4.43

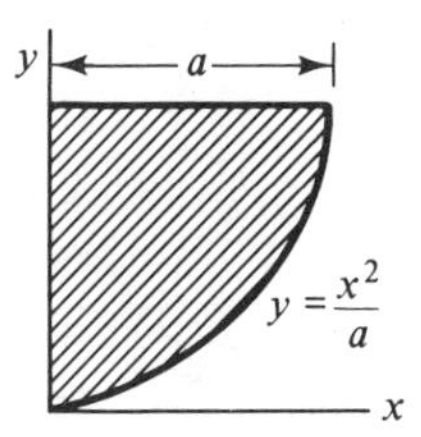

Problem 4.44

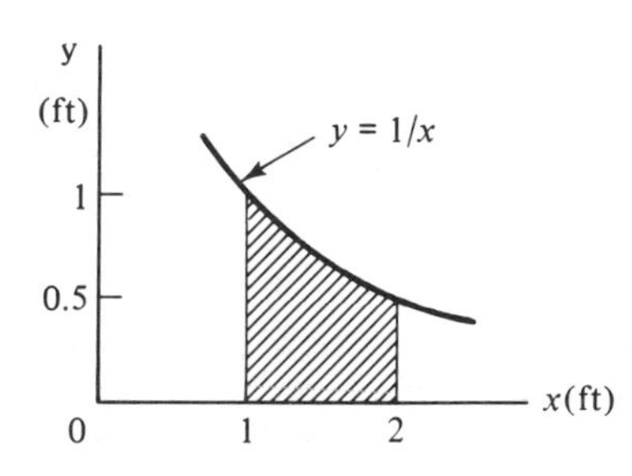

Problem 4.45

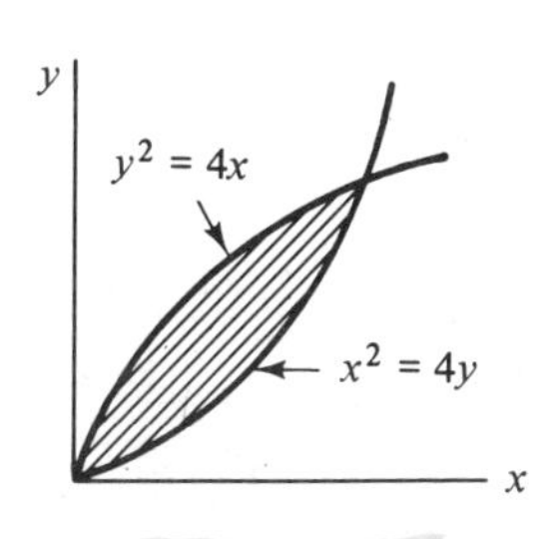

Problem 4.46

4.47 and 4.48 For the lines shown, determine the coordinates $\bar{x}$ and $\bar{y}$ of the centroid. Note that from the Pythagorean theorem, $dL = [(dy/dx)^2 + 1]^{1/2}dx = [(dx/dy)^2 + 1]^{1/2}dy$.

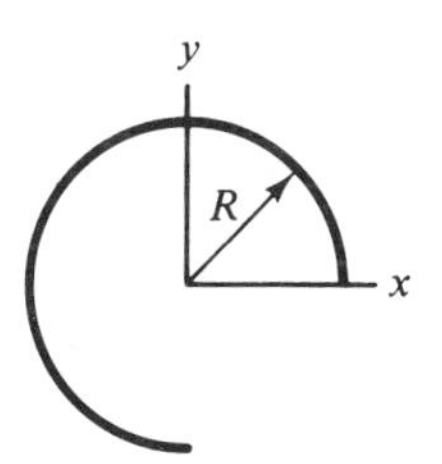

Problem 4.47

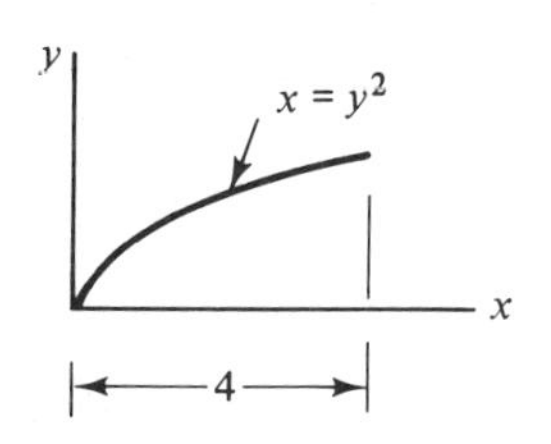

Problem 4.48

4.49 Locate the center of gravity of a thin conical shell with height h and a circular base with radius R.

4.50 Locate the center of gravity of a thin-walled hemispherical dome.

4.51 A homogeneous sheet-metal part has the form of a pyramid, open at the bottom, with height h and a square base with side $2b$. Locate the mass center.

4.52 Locate the center of gravity of a solid homogeneous cone with height h and a circular base with radius R.

4.53 Locate the centroid of the volume formed by rotating the area in Problem 4.45 about the x axis.

4.54 Same as Problem 4.53, except that the area is rotated about the y axis.

4.55 and 4.56 Locate the mass center of the solid uniform bodies shown.

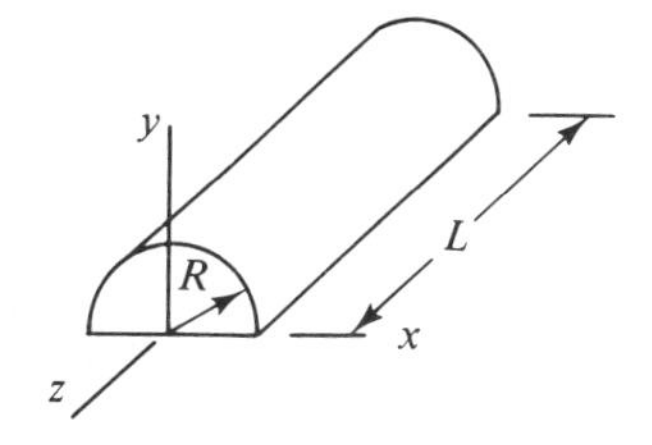

Problem 4.55

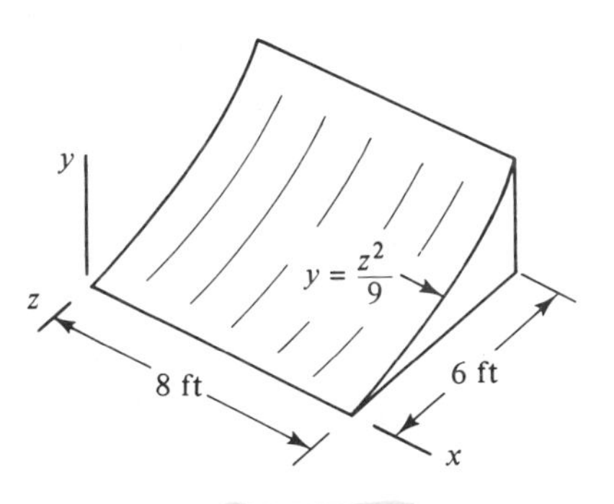

Problem 4.56

4.57 A surface of revolution may be generated by rotating a plane curve about a fixed axis, and a body of revolution may be generated by rotating a plane area about a fixed axis. Show that (a) the area of a surface of revolution is equal to the length of the generating curve times the distance its centroid travels while the surface is being generated and (b) the volume of a body of revolution is equal to the generating area times the distance its centroid travels while the body is being generated. These results are known as the Pappus—Guldinus theorems.

4.58 Use the results of Problem 4.57 to determine the surface area and volume of a sphere.

4.59 Determine the volume of the ring segment shown. Use the results of Problem 4.57.

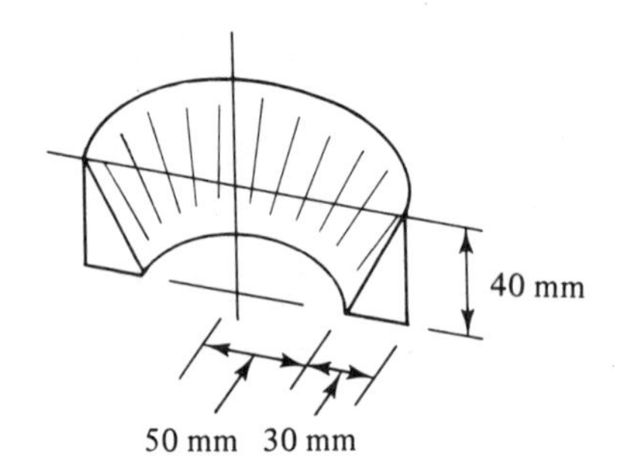

Problem 4.59

4.60 Determine the surface area and volume of the body of revolution generated by rotating the area in Problem 4.45 about the y axis. Use the results of Problem 4.57

4.6. CENTERS OF GRAVITY AND CENTROIDS OF COMPOSITE BODIES AND SHAPES

Many of the bodies encountered in engineering practice can be divided into finite-sized pieces of simple shape whose centers of gravity can be easily located either from symmetry, integration, or from tables. For such composite bodies, the center of gravity can be located with a minimum of effort.

Consider a composite body consisting of pieces with volumes $V_1, V_2, \ldots, V_n$. Integrals over the total volume can be expressed as the sum of the integrals over each of these subvolumes. In particular, Eq. (4.8) for the position vector of the center of gravity can be written as

$$\mathbf{r} = \frac{\int_V \mathbf{r}_{el}\gamma dV}{\int_V \gamma dV} = \frac{\int_{V_1} \mathbf{r}_{el}\gamma dV + \int_{V_2} \mathbf{r}_{el}\gamma dV + \cdots + \int_{V_n} \mathbf{r}_{el}\gamma dV}{\int_{V_1} \gamma dV + \int_{V_2} \gamma dV + \cdots + \int_{V_n} \gamma dV}$$

From Eqs. (4.6) and (4.8), we recognize that

$$\int_{V_i} \gamma dV = W_i \qquad \int_{V_i} \mathbf{r}_{el}\gamma dV = W_i\mathbf{r}_i$$

where W_i is the weight of the ith piece of the body and $\mathbf{r}_i = \bar{x}_i\mathbf{i} + \bar{y}_i\mathbf{j} + \bar{z}_i\mathbf{k}$ is the position vector of its center of gravity. Thus, we have

$$\mathbf{r} = \frac{W_1\mathbf{r}_1 + W_2\mathbf{r}_2 + \cdots + W_n\mathbf{r}_n}{W_1 + W_2 + \cdots + W_n} = \frac{\Sigma W_i\mathbf{r}_i}{\Sigma W_i} \tag{4.14}$$

or in component form,

$$\bar{x} = \frac{\Sigma W_i \bar{x}_i}{\Sigma W_i} \qquad \bar{y} = \frac{\Sigma W_i \bar{y}_i}{\Sigma W_i} \qquad \bar{z} = \frac{\Sigma W_i \bar{z}_i}{\Sigma W_i} \tag{4.15}$$

This same procedure yields for the coordinates of the centroid of a composite volume

$$\bar{x} = \frac{\Sigma V_i \bar{x}_i}{\Sigma V_i} \qquad \bar{y} = \frac{\Sigma V_i \bar{y}_i}{\Sigma V_i} \qquad \bar{z} = \frac{\Sigma V_i \bar{z}_i}{\Sigma V_i} \tag{4.16}$$

where V_i is the volume of the ith piece and $(\bar{x}_i, \bar{y}_i, \bar{z}_i)$ are the coordinates of its centroid. Similar relations hold for composite areas and lines and can be obtained from Eqs. (4.16) by replacing the Vs by As and Ls, respectively. For example, for a composite area,

$$\bar{x} = \frac{\Sigma A_i \bar{x}_i}{\Sigma A_i} \qquad \bar{y} = \frac{\Sigma A_i \bar{y}_i}{\Sigma A_i} \qquad \bar{z} = \frac{\Sigma A_i \bar{z}_i}{\Sigma A_i} \tag{4.17}$$

where A_i is the area of the ith piece and $(\bar{x}_i, \bar{y}_i, \bar{z}_i)$ are the coordinates of its centroid.

Bodies with holes or cavities can be thought of as solid bodies minus one or more pieces. In this case, the weights of the pieces removed are considered to be negative quantities in Eqs. (4.14) and (4.15). The same procedure applies for centroids. The volume, area, or length of a piece removed from a shape is simply considered to be negative.

Example 4.13. Center of Gravity of a Shaft. Locate the center of gravity of the composite shaft shown in Figure 4.32. The specific weight of steel is 77 kN/m^3 and the specific weight of aluminum is 27 kN/m^3. Each portion of the shaft is homogeneous.

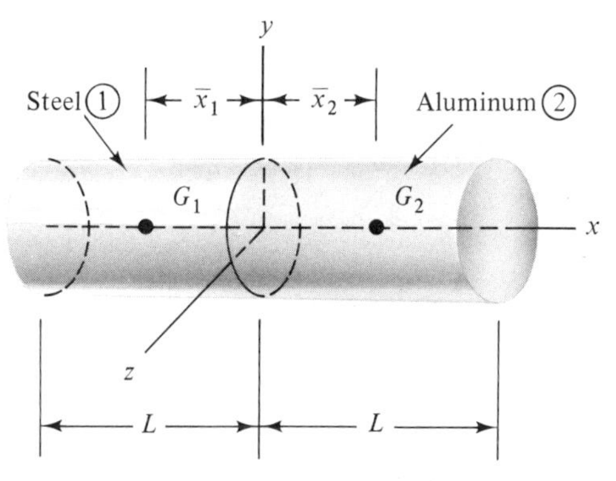

Figure 4.32

Solution. We choose coordinates and number the pieces as shown. By symmetry, $\bar{y} = \bar{z} = 0$, and $\bar{x}_1 = -L/2$ and $\bar{x}_2 = L/2$. The weights of the pieces are

$W_1 = \gamma_1 V_1$ and $W_2 = \gamma_2 V_2$. Since the pieces have the same volume,

$$W_1 = \left(\frac{\gamma_1}{\gamma_2}\right) W_2 = \frac{77 \text{ kN/m}^3}{27 \text{ kN/m}^3} W_2 = 2.85 W_2$$

Applying the first of Eqs. (4.15), we have

$$\bar{x} = \frac{W_1 \bar{x}_1 + W_2 \bar{x}_2}{W_1 + W_2} = \frac{2.85 W_2(-L/2) + W_2(L/2)}{2.85 W_2 + W_2} = -0.24L \qquad \textit{Answer}$$

Thus, the center of gravity lies along the centerline a distance $0.24L$ to the left of the center of the shaft. Note that is was not necessary to calculate the actual weights of the pieces in this example.

Example 4.14. ***Center of Gravity of a Sheet-Metal Part.*** The sheet-metal part shown in Figure 4.33 is homogeneous and has constant thickness. Locate its center of gravity.

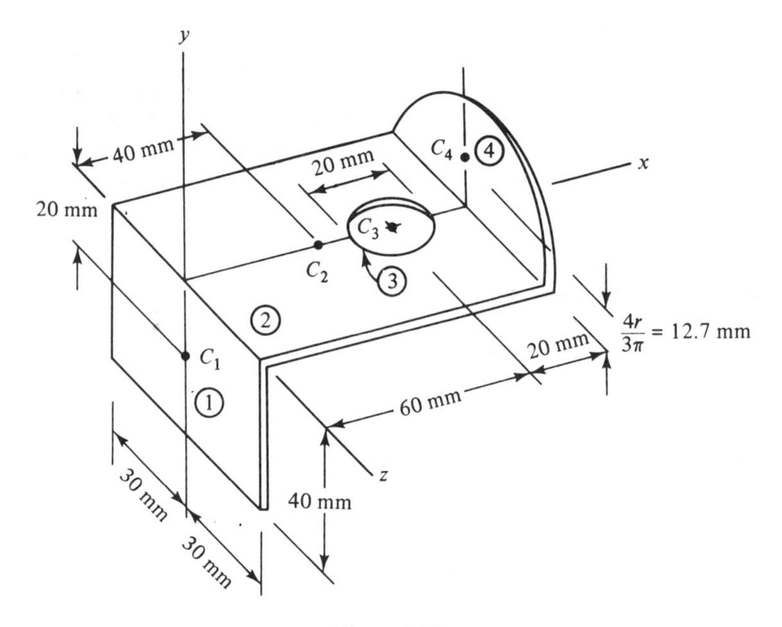

Figure 4.33

Solution. Since the body is homogeneous, its center of gravity coincides with the centroid of the area. We choose coordinates and divide the shape into pieces as indicated. Piece 2 is a solid rectangle and piece 3 is the circular area removed from it to form the hole. By symmetry, $\bar{z} = 0$. The location of the centroids of the various pieces can be determined from symmetry and from Table A.1 of the

Appendix. Since there are a number of pieces, it is convenient to tabulate the data necessary for making the computations indicated in Eqs. (4.17):

No.	A_i(mm²)	$\bar{x}_i$(mm)	$\bar{y}_i$(mm)	$A_i\bar{x}_i$(mm³)	$A_i\bar{y}_i$(mm³)
1	2.40×10^3	0.0	−20.0	0.0	-48.0×10^3
2	4.80×10^3	40.0	0.0	192.0×10^3	0.0
3	-0.31×10^3	60.0	0.0	-18.6×10^3	0.0
4	1.41×10^3	80.0	12.7	112.8×10^3	18.0×10^3
Σ	8.30×10^3			286.2×10^3	-30.0×10^3

Note that A_3 is considered negative because it corresponds to an area removed from the shape. Using Eqs. (4.17) and the sums found in the table, we have

$$\bar{x} = \frac{286.2 \times 10^3 \text{ mm}^3}{8.3 \times 10^3 \text{ mm}^2} = 34.5 \text{ mm} \qquad \bar{y} = \frac{-30.0 \times 10^3 \text{ mm}^3}{8.3 \times 10^3 \text{ mm}^2} = -3.6 \text{ mm}$$

Answer

$$\bar{z} = 0$$

Example 4.15. Centroid of an Area. Locate the centroid of the cross-sectional area of the structural member shown in Figure 4.34(a).

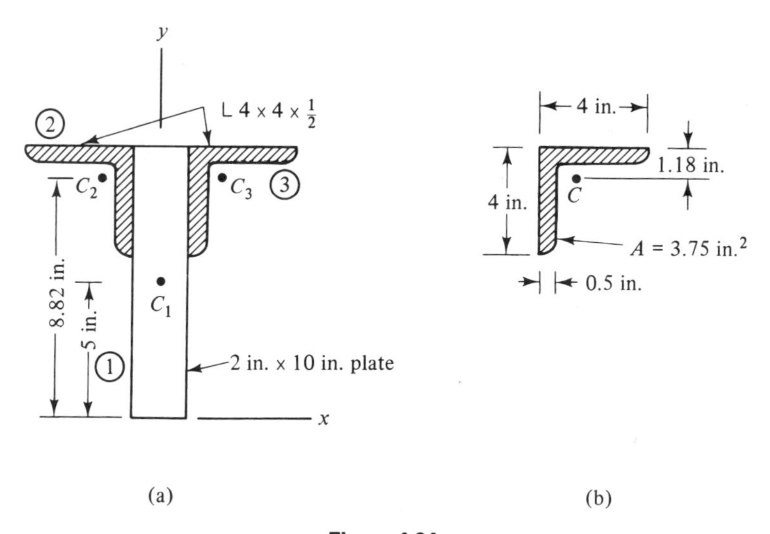

Figure 4.34

Solution. We choose coordinates and divide the area into elements as shown. By symmetry, $\bar{x} = 0$. The geometric properties of the angle sections are obtained from Table A.5 of the Appendix. These data are listed in Figure 4.34(b). Using a tabular format and Eqs. (4.17), we have

No.	A_i(in.^{2})	$\bar{y}_i$(in.)	$A_i\bar{y}_i$(in.^{3})
1	20.00	5.00	100.00
2	3.75	8.82	33.08
3	3.75	8.82	33.08
Σ	27.50		166.16

$$\bar{y} = \frac{166.16 \text{ in.}^3}{27.50 \text{ in.}^2} = 6.04 \text{ in.} \qquad \bar{x} = 0 \qquad \textbf{\textit{Answer}}$$

PROBLEMS

4.61 to 4.69 Locate the centroid of the shapes shown.

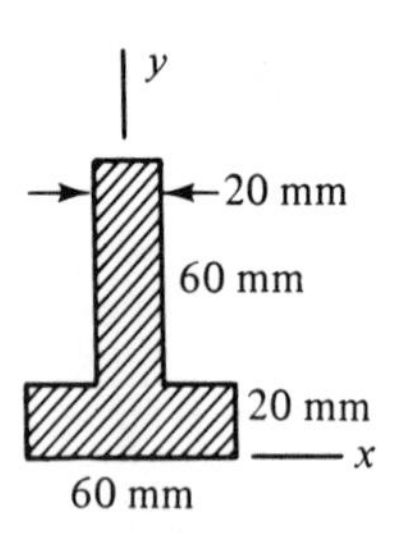

Problem 4.61

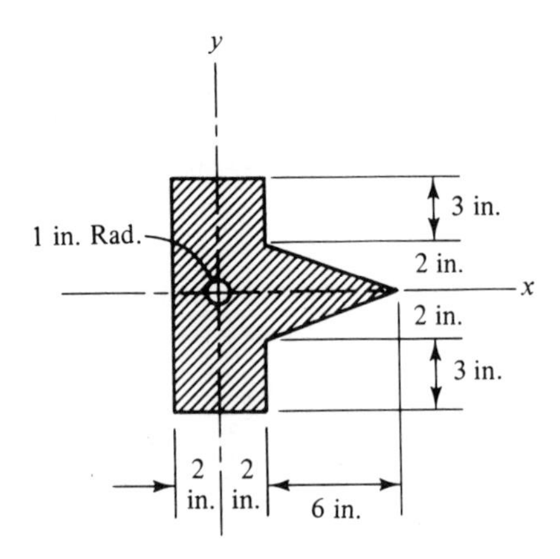

Problem 4.63

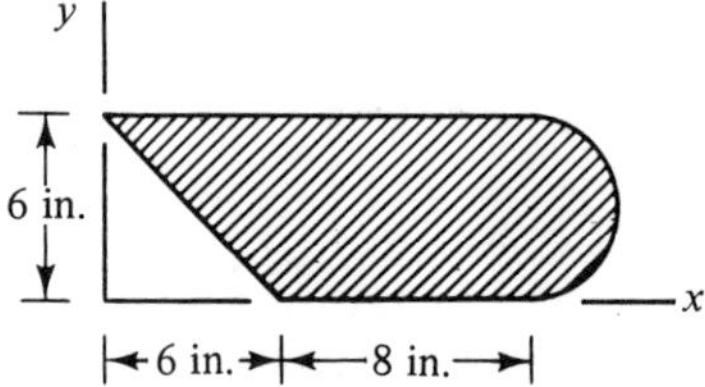

Problem 4.62

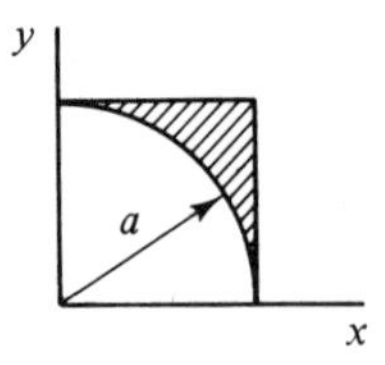

Problem 4.64

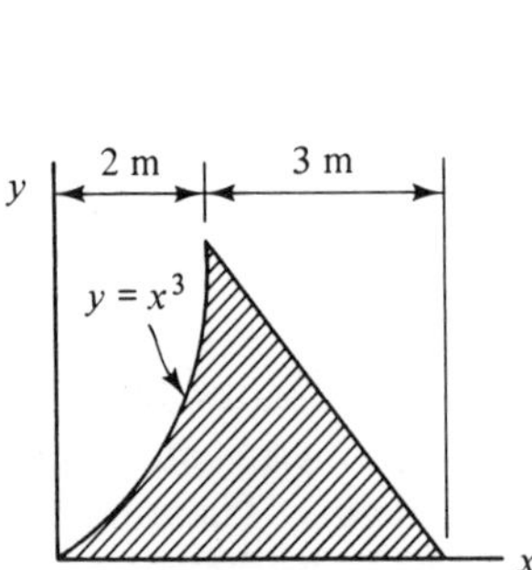

Problem 4.65

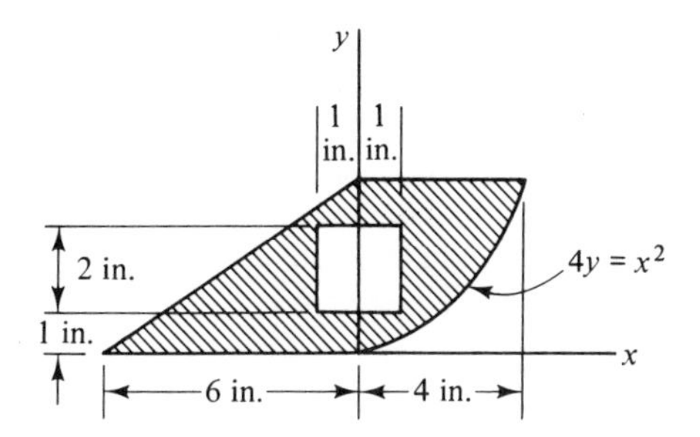

Problem 4.66

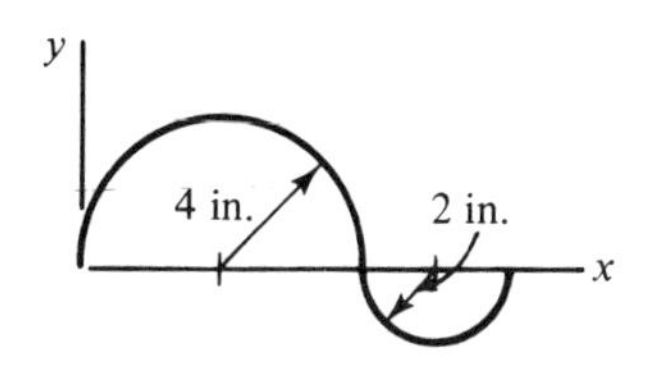

Problem 4.67

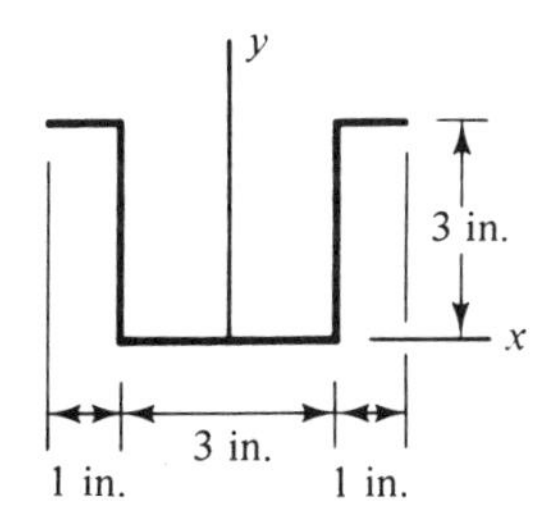

Problem 4.68

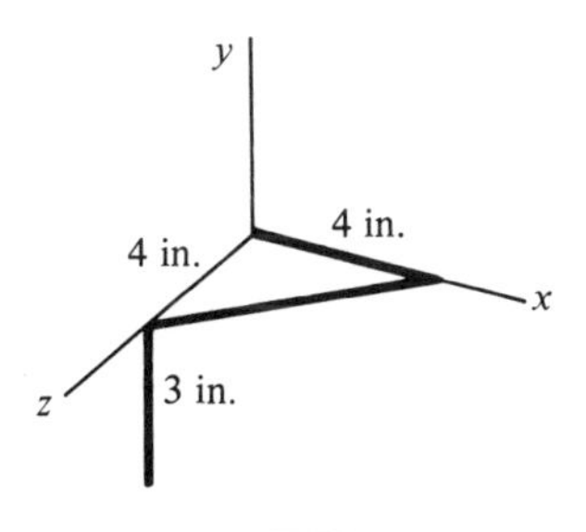

Problem 4.69

4.70 The location of the center of gravity of a triangular plate is to be adjusted by drilling a hole at the location shown. What is the required hole diameter so that $\bar{x} = 6.0$ in.? What is the corresponding value of $\bar{y}$?

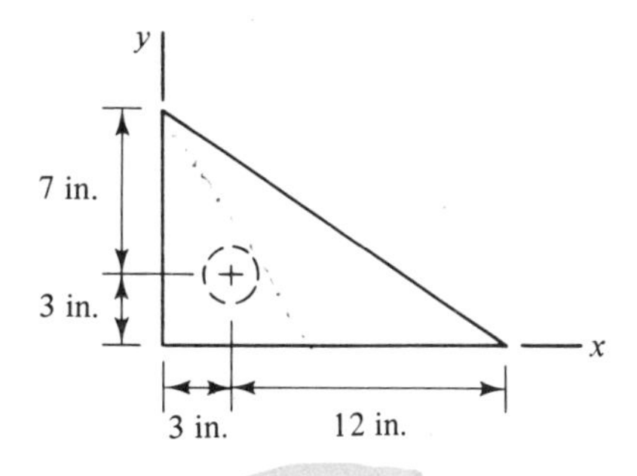

Problem 4.70

4.71 Three timbers are nailed together to form a symmetrical section with the nominal dimensions shown. Locate the centroid of the section by using the geometric properties of timbers given in Table A.12 of the Appendix.

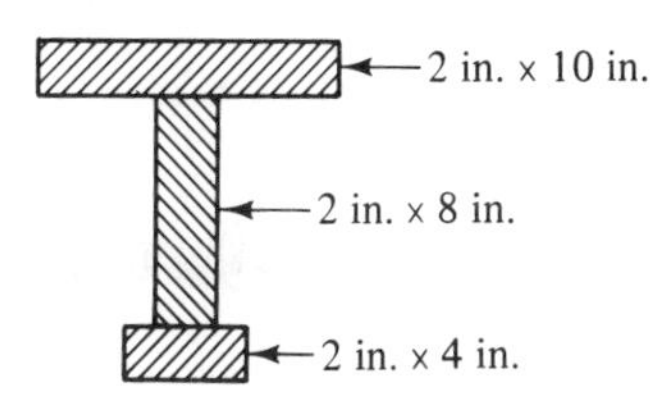

Problem 4.71

4.72 A channel section is connected to a W-shape beam to form a structural member with the symmetrical cross section shown. Locate the centroid of the cross section by using the geometric properties for the members given in Tables A.3 and A.4 of the Appendix.

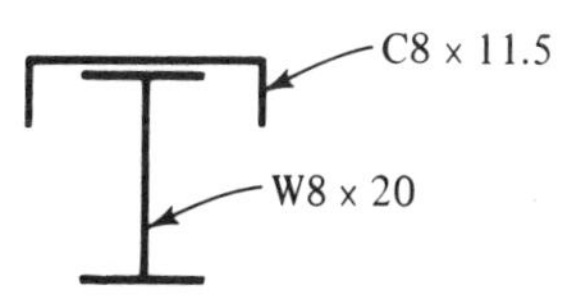

Problem 4.72

4.73 Locate the mass center of the sheet-metal part shown.

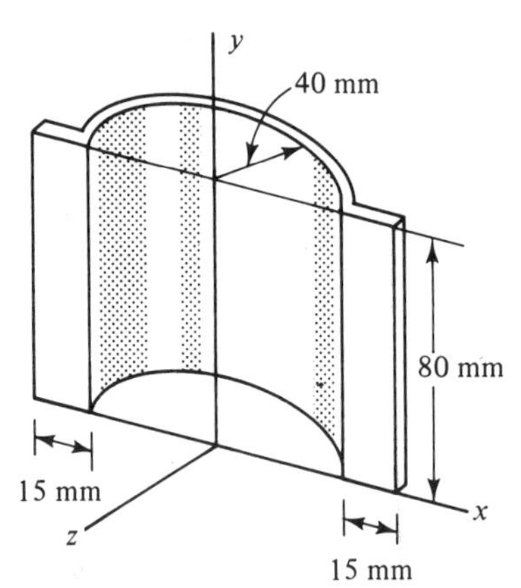

Problem 4.73

4.74 A sheet-metal part has the shape shown. Locate its mass center.

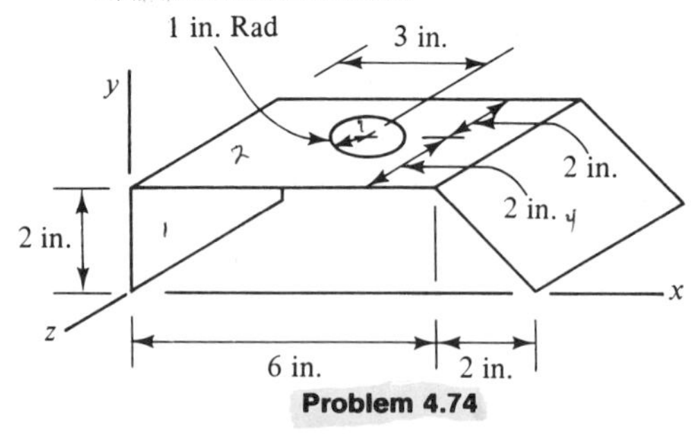

Problem 4.74

4.75 A solid circular cone is joined to a solid cylinder as shown. Locate the centroid of the combined volume.

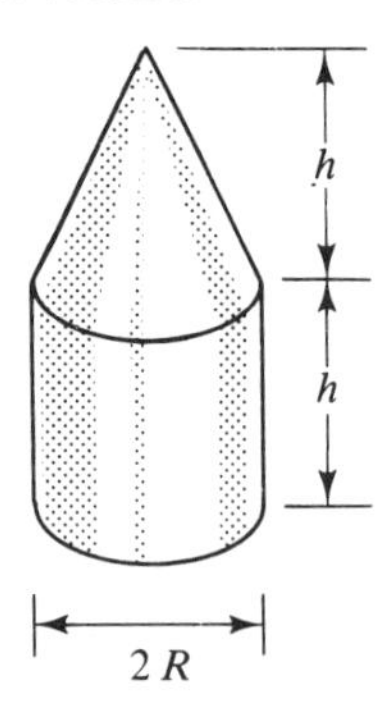

Problem 4.75

4.76 A 4-in. diameter hub is fitted onto a 1.5-in. diameter steel shaft as shown. Locate the center of gravity of the assembly if (a) the hub is made of steel and (b) the hub is made of aluminum. The specific weights of steel and aluminum are 490 lb/ft^3 and 170 lb/ft^3, respectively.

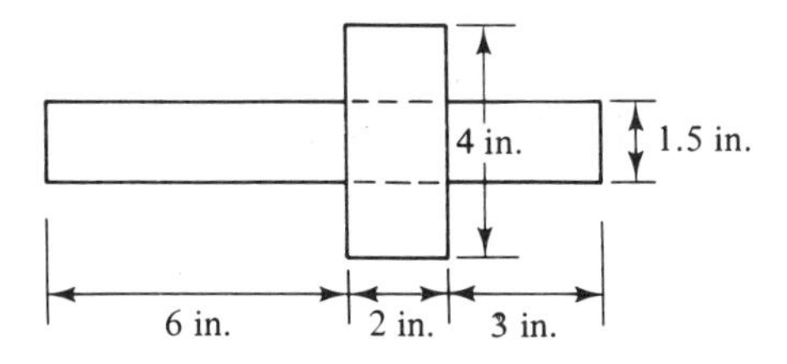

Problem 4.76

4.77 A 10-in. nominal diameter steel pipe (see Table A.7 of Appendix) of length L is stood on end and partially filled with concrete. What is the required depth d of concrete if the center of gravity of the assembly is to be located a distance 0.4 L above the base? Concrete weighs 150 lb/ft^3.

4.78 A cast iron pulley has an outer diameter of 250 mm and the rim has the cross section shown. Use the results of Problem 4.57 (Section 4.5) to determine the mass of the pulley. The density of cast iron is 7.20 Mg/m^3.

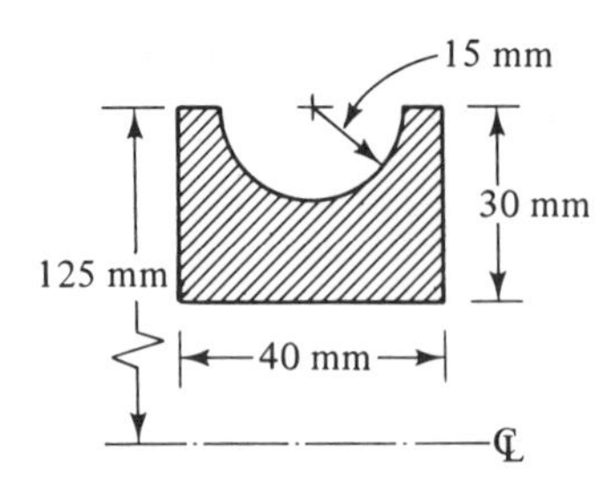

Problem 4.78

4.7 SURFACE FORCES

Forces that are distributed over a surface area are called *surface forces*. Common examples are the forces between solid bodies in contact and the pressure loadings exerted on a solid body by a fluid. Some common types of surface forces will be considered in this section, along with the procedures for determining their resultants.

Surface and line forces. Surface forces are defined in terms of the force acting per unit of area. If the force acting upon an element of area ΔA is $\Delta \mathbf{F}$ (Figure 4.35), the average force per unit area is $\Delta \mathbf{F}/\Delta A$. The *surface force at a point*, which we shall denote by $\mathbf{s}$, is defined as the limiting value of this ratio as the area surrounding the point approaches zero. Stated mathematically,

$$\mathbf{s} = \lim_{A \to 0} \frac{\Delta \mathbf{F}}{\Delta A} = \frac{d\mathbf{F}}{dA} \tag{4.18}$$

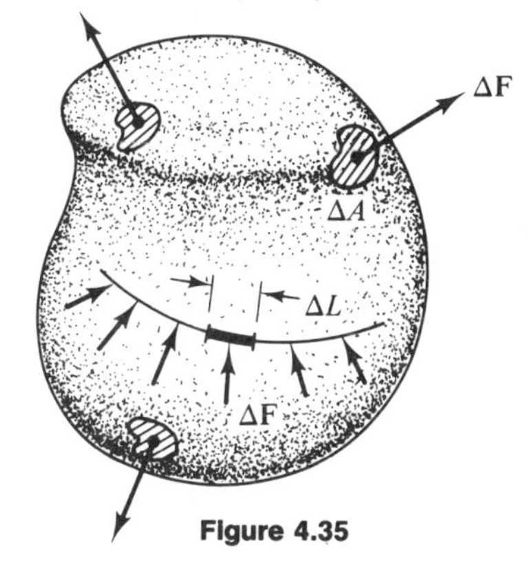

Figure 4.35

Both the magnitude and direction of $\mathbf{s}$ can vary from point to point over the surface. Common units for the magnitude of a surface force are lb/ft^2 and lb/in.^2(psi). In SI, the units are N/m^2, or pascals.

If the surface area is long and narrow, the distributed force can usually be assumed to vary only along its length. Such forces are called *line forces*. An example would be loading that material stacked on a floor would impart to the top surface of a floor beam. Line forces are expressed in terms of the force acting per unit of length of the line. Thus, the *line force at a point*, $\mathbf{q}$, is defined as

$$\mathbf{q} = \lim_{\Delta L \to 0} \frac{\Delta \mathbf{F}}{\Delta L} = \frac{d\mathbf{F}}{dL} \tag{4.19}$$

where $\Delta \mathbf{F}$ is the force acting upon a line element of length ΔL (Figure 4.35). Common units for the magnitude of a line force are lb/ft, lb/in., and N/m.

Surface forces can often be represented graphically by means of a *loading diagram*, which is a plot of the variation of $\mathbf{s}$ over the surface. For example, if identical cartons measuring 1 ft on a side and weighing 50 lb are stacked four deep on a warehouse floor (Figure 4.36), each unit of area of the floor is subjected to the same surface force of $200\ \text{lb/ft}^2$. The corresponding loading diagram would be as shown. The height of the diagram represents the magnitude of the surface force, and the arrows indicate its direction. If the cartons are stacked to a variable depth, the loading diagram would have a variable height.

Loading diagrams can also be used to represent line forces. For example, suppose that the floor shown in Figure 4.36 is supported by beams spaced 2 ft apart. Each beam supports the load on a 2-ft wide strip of the floor, or 400 lb/ft of length. The corresponding loading diagram is as shown in Figure 4.37. The height

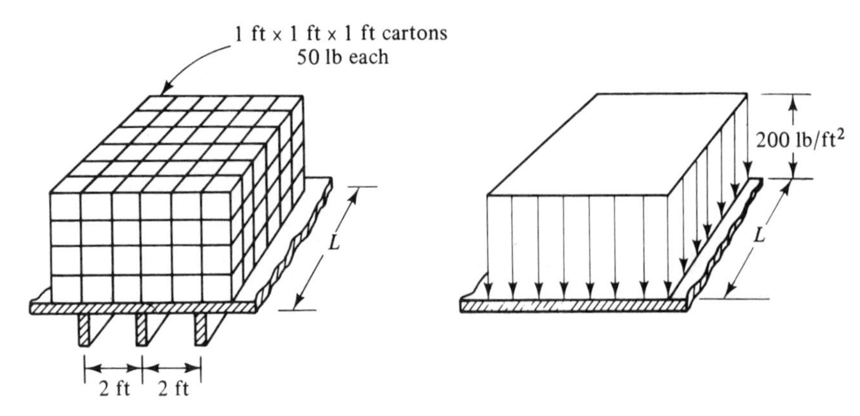

Figure 4.36

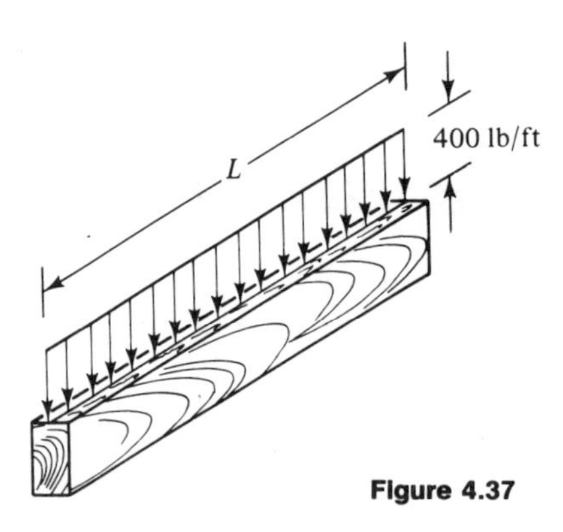

Figure 4.37

of the diagram represents the magnitude of the line force. Note that the loading diagram forms an area for line forces and a volume for surface forces.

Pressure loadings. The magnitude of the surface force exerted by a fluid is called *pressure*. Pressure loadings are characterized by the fact that they are always directed toward the surface upon which they act and are perpendicular to it. Also, the pressure at any point in a fluid is the same in all directions.

The pressure in a liquid at rest, the so-called *hydrostatic pressure*, is due to the weight of the liquid lying above the particular point of interest. Consider a point situated at a depth h (Figure 4.38). The weight of a column of liquid with cross-sectional area dA above this point is $dW = \gamma h dA$, where γ is the specific weight of the liquid. The pressure at the point, which is the force per unit area, is $p = dF/dA = dW/dA$, or

$$p = \gamma h \tag{4.20}$$

Thus, the hydrostatic pressure increases linearly with depth. This relationship can also be used to estimate the pressure loadings due to loose, granular materials, such as dry sand and grain.

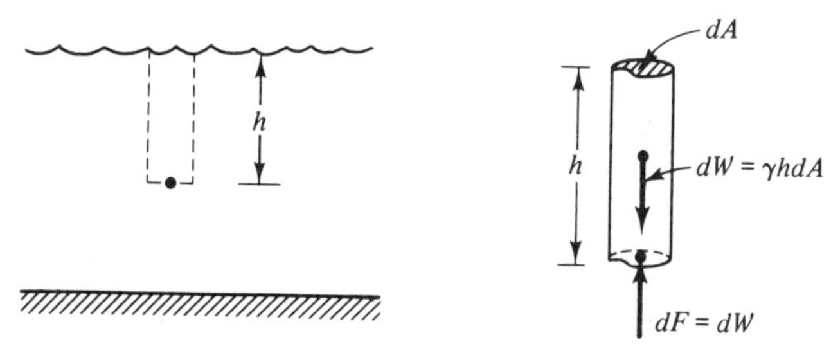

Figure 4.38

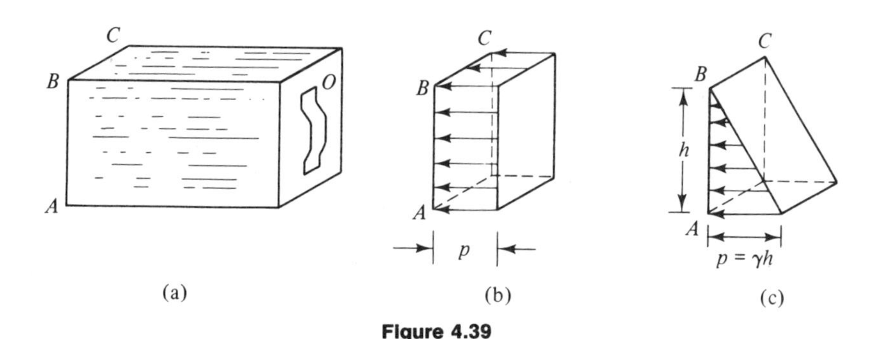

Figure 4.39

Gasses usually have negligible weight. Thus, they exert a uniform pressure over the surface upon which they act. For example, if the can shown in Figure 4.39(a) is filled with air at pressure p, the surface force on the end of the can is as shown in Figure 4.39(b). Figure 4.39(c) shows the loading if the can is filled with a liquid with specific weight γ.

All bodies situated at the earth's surface are subjected to an atmospheric pressure of approximately 14.7 lb/in.^2 ($10.1 \times 10^4\ \text{N/m}^2$) at sea level due to the weight of the overlying atmosphere. If all surfaces of the body are exposed to the atmosphere, this loading has a null resultant and need not be considered in a force analysis. Consequently, it is usually the *gage pressure*, or difference between the total pressure and atmospheric pressure, which is of interest. Equation 4.20 defines the gage pressure in a liquid.

Resultants of parallel surface and line forces. Figure 4.40(a) shows a loading diagram for a general system of parallel surface forces. According to Eq. (4.18), each element of area dA is subjected to a force $d\mathbf{F} = sdA\mathbf{e}$, where $\mathbf{e}$ is a unit vector in the direction of the surface force. Since these forces are all parallel, their resultant will be a single force $\mathbf{R}$ [Figure 4.40(b)]. From the conditions for static equivalence, we have

$$(\Sigma\,\mathbf{F})_{\text{I}} = (\Sigma\,\mathbf{F})_{\text{II}}:\quad \int dF\mathbf{e} = R\mathbf{e}$$

$$(\Sigma\,\mathbf{M}_P)_{\text{I}} = (\Sigma\,\mathbf{M}_P)_{\text{II}}:\quad \int (\mathbf{r}_{\text{el}} \times dF\mathbf{e}) = \mathbf{r} \times R\mathbf{e}$$

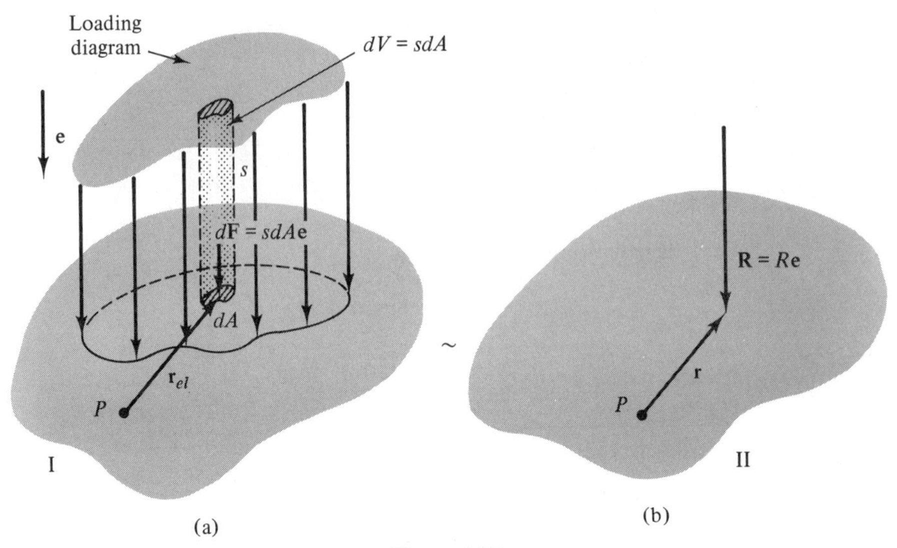

Figure 4.40

where $\mathbf{r}$ and $\mathbf{r}_{el}$ are the respective position vectors of $\mathbf{R}$ and $d\mathbf{F}$ relative to point P. Rewriting the moment condition as

$$\left(\int \mathbf{r}_{el} dF\right) \times \mathbf{e} = (\mathbf{r}R) \times \mathbf{e}$$

and solving for R and $\mathbf{r}$, we obtain

$$R = \int dF = \int s dA \tag{4.21}$$

$$\mathbf{r} = \frac{\int \mathbf{r}_{el} dF}{\int dF} = \frac{\int \mathbf{r}_{el} s dA}{\int s dA} \tag{4.22}$$

If the variation of s over the surface is known, these integrals can be evaluated to determine the resultant force and its location. However, the integrals have a geometric interpretation that makes their formal evaluation unnecessary in many cases.

Referring to Figure 4.40(a), we see that sdA is equal to the volume dV of a column under the loading diagram with cross-sectional area dA. If the quantity sdA is replaced by dV in Eqs. (4.21) and (4.22), they correspond to the equations for the total volume of a shape and the position vector of its centroid (see Section 4.5). Interpreted physically, this means that the resultant of a parallel surface force has magnitude equal to the total volume under the loading diagram and its line of action passes through the centroid of that volume. This result makes it possible to determine the resultant of many simple loadings without actually performing an integration. For example, we can immediately say that the resultant of the uniform pressure loading shown in Figure 4.41 has magnitude $R = pA$ and acts at the center of the plate.

Equations (4.21) and (4.22) can also be used to determine the resultants of parallel line forces. For a line force, $dF = qdL$ [Eq. (4.19)]. Now qdL is the area dA of an element with width dL under the loading diagram [Figure 4.42(a)]; therefore, Eqs. (4.21) and (4.22) correspond to the equations for the total area of a shape and position vector of its centroid. Thus, the resultant of a parallel line force has magnitude equal to the area under the loading diagram and its line of action passes through the centroid of that area [Figure 4.42(b)].

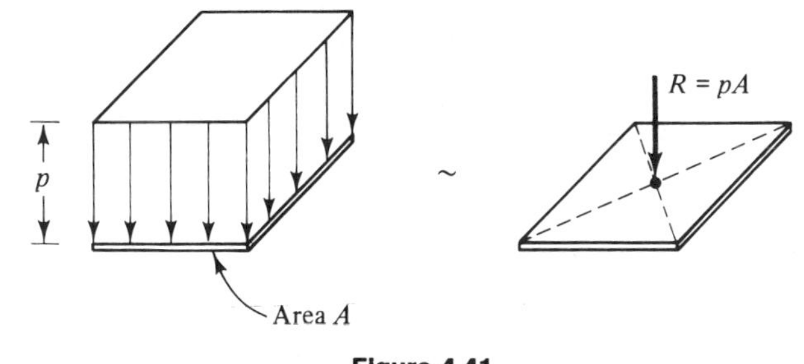

Figure 4.41

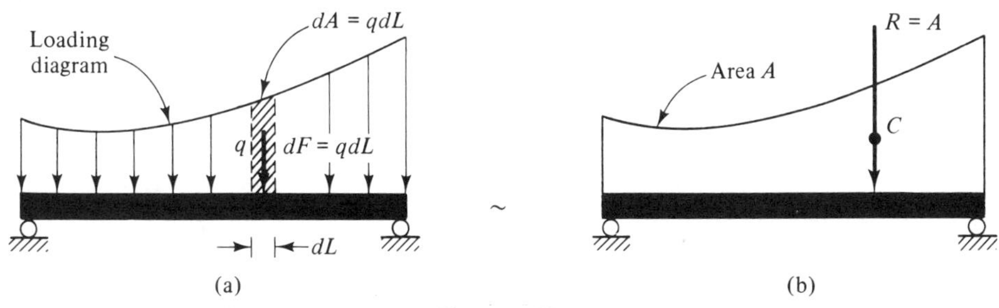

Figure 4.42

Pressure loadings on curved surfaces. Figure 4.43(a) shows an end view of the pressure distribution on the curved upstream face of a dam. If we attempt to determine the resultant force on the dam directly from the surface force distribution, we find that the integrals involved are very

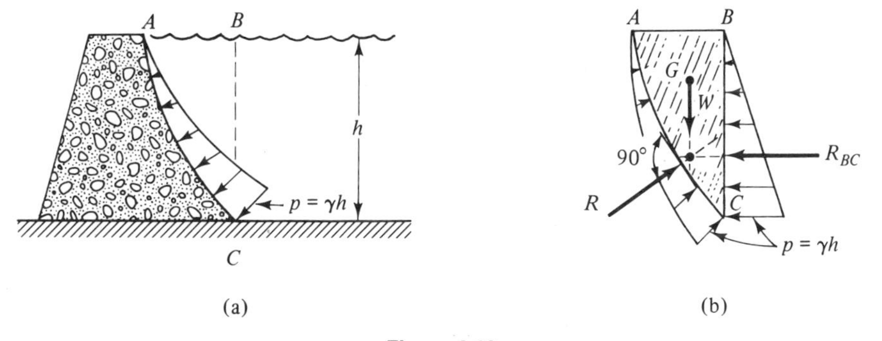

Figure 4.43

complicated because both the magnitude and the orientation of the loading vary over the surface. This complication can be avoided by using an alternate procedure wherein the resultant is determined indirectly from a force analysis of a portion of the fluid.

For example, suppose we wish to determine the resultant force on the face of the dam. To do this, we consider the equilibrium of the volume of fluid bounded by the dam and the vertical plane *BC*. The free-body diagram is shown in Figure 4.43(b). Since the pressure loading on plane *BC* involves parallel surface forces, its resultant $\mathbf{R}_{BC}$ can be easily determined. Once the weight of the fluid has been computed and its center of gravity located, the only remaining unknown is the resultant **R** of the pressure acting upon the curved surface *AC* of the fluid volume. This force can be determined from the equations of equilibrium, with the resultant force on the dam obtained from Newton's third law. Uniform pressure loadings due to gasses can be handled in the same way.

Example 4.16. Snow Load on a Roof. Snow accumulates to a variable depth on a flat roof, as shown in Figure 4.44(a). Determine the pressure distribution and total load on the roof if the snow weighs 15 lb/ft³.

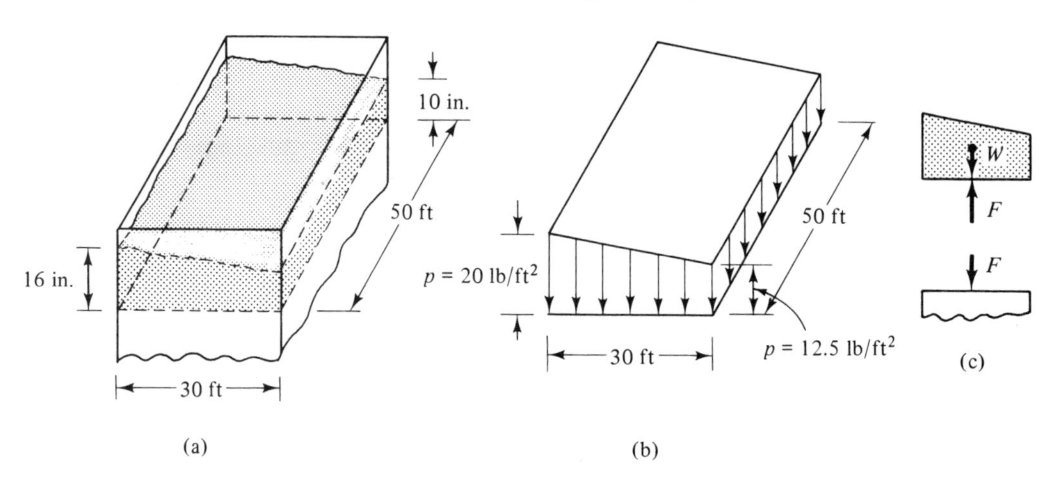

Figure 4.44

Solution. Assuming that the snow can be treated as a liquid, we have for the pressure on the roof

$$p = \gamma h$$

where h is the depth of the snow and γ is its specific weight. The loading diagram is as shown in Figure 4.44(b). At the left edge of the roof

$$p = (15 \text{ lb/ft}^3)\frac{(16 \text{ in.})}{(12 \text{ in./ft})} = 20 \text{ lb/ft}^2$$

and at the right edge

$$p = (15\ \text{lb/ft}^3)\,\frac{(10\ \text{in.})}{(12\ \text{in./ft})} = 12.5\ \text{lb/ft}^2$$

A simple force analysis [Figure 4.44(c)] shows that the total load F on the roof is equal to the total weight of the snow (it is also equal to the volume under the loading diagram):

$$F = W = \gamma V = (15\ \text{lb/ft}^3)\,\frac{(16\ \text{in.} + 10\ \text{in.})}{2(12\ \text{in./ft})}(30\ \text{ft})(50\ \text{ft}) = 24\ 375\ \text{lb} \qquad \textit{Answer}$$

Example 4.17 Line Loads on a Beam. A beam supports the distributed loads shown in Figure 4.45(a). Determine the reactions at the supports.

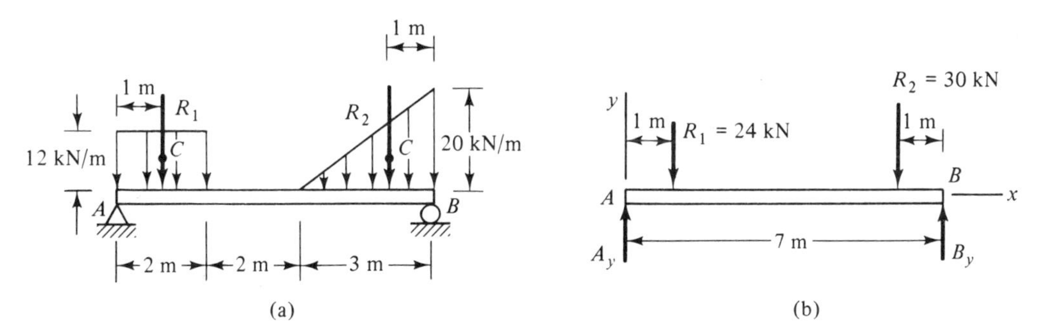

Figure 4.45

Solution. We first replace the distributed loads by their resultants $\mathbf{R}_1$ and $\mathbf{R}_2$. Each resultant has a magnitude equal to the area under the corresponding loading diagram and acts through the centroid of that area. Thus,

$$R_1 = (12\ \text{kN/m})(2\ \text{m}) = 24\ \text{kN}$$

$$R_2 = \frac{1}{2}(20\ \text{kN/m})(3\ \text{m}) = 30\ \text{kN}$$

From the FBD of Figure 4.45(b), we have

$$\overset{+}{\curvearrowleft}\ \Sigma M_A = B_y(7\ \text{m}) - 24\ \text{kN}(1\ \text{m}) - 30\ \text{kN}(6\ \text{m}) = 0$$

$$\Sigma F_y = A_y + B_y - 24\ \text{kN} - 30\ \text{kN} = 0$$

or

$$B_y = 29.1\ \text{kN} \qquad A_y = 24.9\ \text{kN} \qquad \textit{Answer}$$

Example 4.18. Resultant Force on a Wall. Determine the resultant force **R** acting upon the curved portion AC of the pool wall shown in Figure 4.46(a). The wall is 30 ft long, and the specific weight of water is 62.4 lb/ft³.

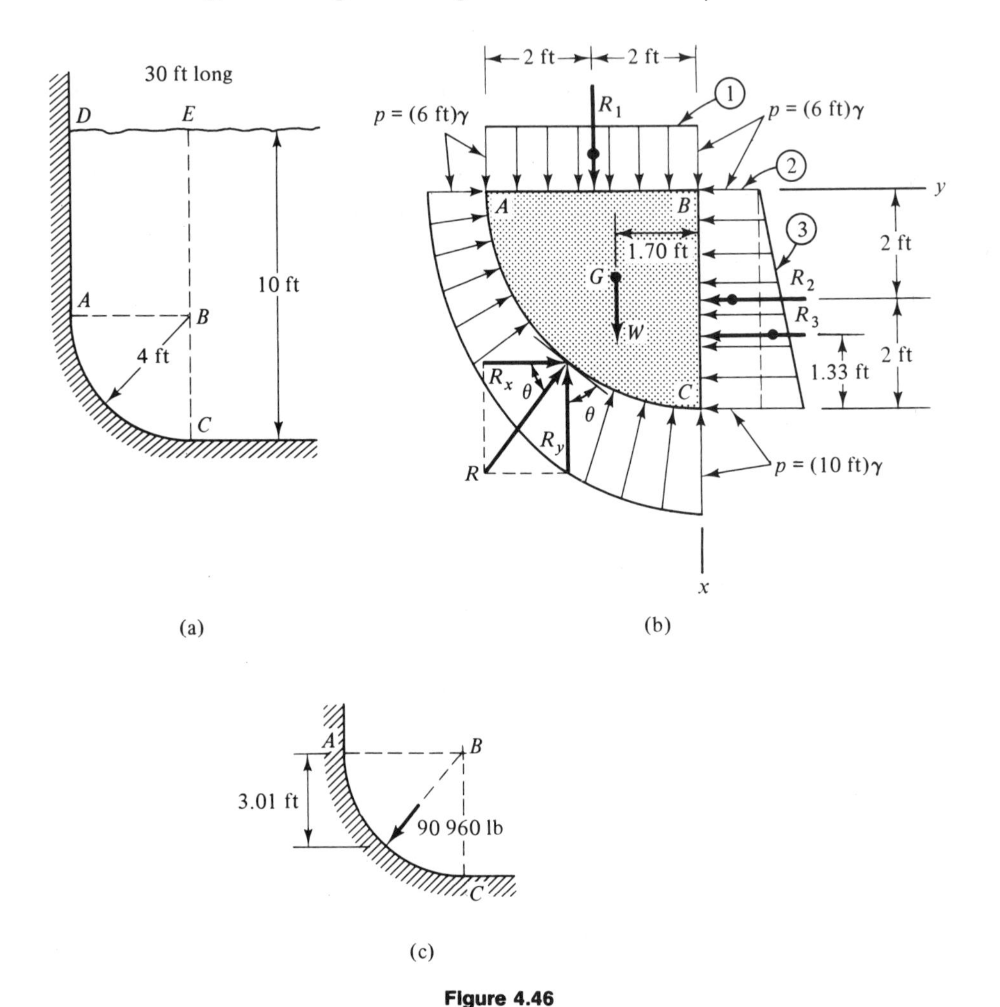

Figure 4.46

Solution. To determine **R**, we consider the equilibrium of the volume of water bounded by the curved surface and planes AB and BC. (Volume $ADEBC$ could also be used.) Since the pressure varies linearly with depth, the FBD is as shown in Figure 4.46(b). We now replace all distributed forces by their resultants. In doing this, it is convenient to break the loading on plane BC into two parts, as indicated. The resultant of each of the parallel distributed forces is equal to the volume under the corresponding loading diagram and acts through the centroid of that volume. Since none of the loadings varies along the length of the wall, all the resultants lie in a plane located at its midpoint. Carrying out the indicated computations, we obtain

$$W = \gamma V = \left(62.4 \frac{\text{lb}}{\text{ft}^3}\right) \frac{\pi}{4} (4 \text{ ft})^2 (30 \text{ ft}) = 23\ 520 \text{ lb}$$

$$R_1 = 6 \text{ ft}\left(62.4 \frac{\text{lb}}{\text{ft}^3}\right)(4 \text{ ft})(30 \text{ ft}) = 44\ 930 \text{ lb}$$

$$R_2 = 6 \text{ ft}\left(62.4 \frac{\text{lb}}{\text{ft}^3}\right)(4 \text{ ft})(30 \text{ ft}) = 44\ 930 \text{ lb}$$

$$R_3 = \frac{1}{2} (4 \text{ ft})\left(62.4 \frac{\text{lb}}{\text{ft}^3}\right)(4 \text{ ft})(30 \text{ ft}) = 14\ 980 \text{ lb}$$

For equilibrium,

$$\Sigma F_x = R_x - R_2 - R_3 = 0$$
$$\Sigma F_y = R_y - R_1 - W = 0$$

from which we obtain

$$R_x = 59\ 910 \text{ lb} \qquad R_y = 68\ 450 \text{ lb}$$

$$R = \sqrt{R_x^2 + R_y^2} = 90\ 960 \text{ lb} \qquad \textbf{\textit{Answer}}$$

To locate the resultant force, we note that it must be perpendicular to surface AC, which is defined by the equation $x^2 + y^2 = (4 \text{ ft})^2$. Using the coordinates shown in Figure 4.46(b), we have

$$\tan \theta = \frac{dy}{dx} = \frac{R_y}{R_x}$$

or

$$-\frac{x}{y} = \frac{68\ 450 \text{ lb}}{59\ 910 \text{ lb}} = 1.14$$

Substituting this result back into the equation of the surface AC, we obtain

$$x = 3.01 \text{ ft}$$

Thus, the location of the resultant force is as shown in Figure 4.46(c). The resultant can also be located by combining the moment equation of equilibrium for the fluid volume with the equation of the curved surface.

Example 4.19. Resultant Force on a Gate. Determine the resultant water force on the triangular gate in the irrigation dam shown in Figure 4.47(a).

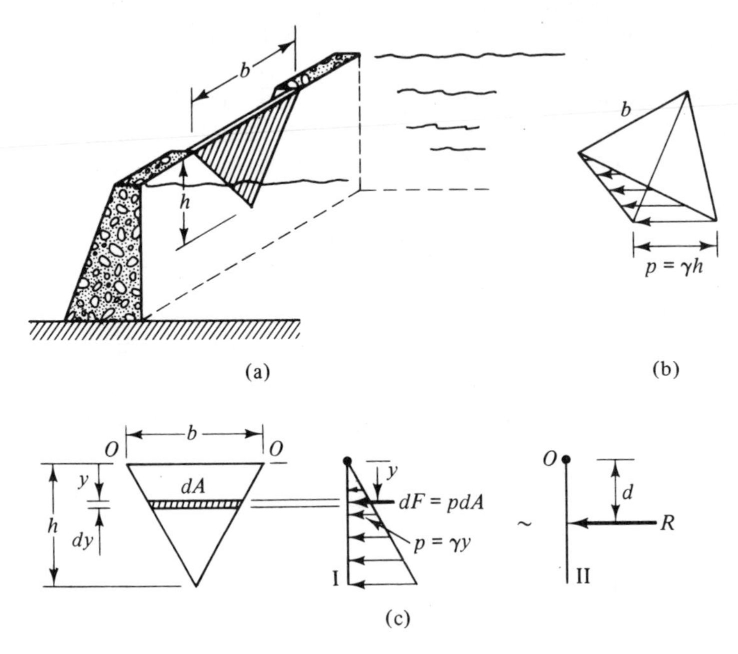

Figure 4.47

Solution. The resultant force on the gate is equal to the volume under the loading diagram [Figure 4.47(b)] and acts through its centroid. However, the geometry of the diagram is sufficiently complicated in this case that it is simpler to start from first principles. Since we do know from the symmetry of the diagram that the resultant acts through a point on the centerline of the gate, only its vertical location need be determined.

We select a horizontal strip with width dy located a distance y down from the top of the gate as the element of the area [Figure 4.47(c)]. The force acting upon this element is

$$dF = pdA = \gamma y dA$$

Denoting the resultant of the distributed forces by **R** and applying the conditions for static equivalence, we have

$$(\Sigma \mathbf{F})_{\mathrm{I}} = (\Sigma \mathbf{F})_{\mathrm{II}}: \quad \int \gamma y dA = R$$

$$(\overset{+}{\curvearrowleft}\Sigma M_0)_{\mathrm{I}} = (\overset{+}{\curvearrowleft}\Sigma M_0)_{\mathrm{II}}: \quad -\int y(\gamma y dA) = -Rd$$

From the definition of the centroid of an area, Eqs. (4.12), we see that

$$\int y dA = A\bar{y}$$

where A is the area of the gate and $\bar{y}$ is the distance to its centroid. Thus,

$$R = \cdot \quad \bar{y} = \gamma\left(\frac{1}{2}bh\right)\left(\frac{h}{3}\right) = \frac{1}{6}\gamma bh^2 \qquad \textbf{\textit{Answer}}$$

Solving for the distance d, we get

$$d = \frac{\gamma \int y^2 dA}{R}$$

By similar triangles, the width of the element of area dA is $b(1 - y/h)$; so

$$dA = b\left(\frac{1 - y}{h}\right)dy$$

$$\int y^2 dA = \int_0^h y^2 b\left(\frac{1 - y}{h}\right)dy = \frac{1}{12}bh^3$$

$$d = \frac{\gamma\left(\frac{bh^3}{12}\right)}{\gamma \frac{bh^2}{6}} = \frac{h}{2} \qquad \textbf{\textit{Answer}}$$

The integral $\int y^2 dA$ in the numerator of the expression for d is called the moment of inertia of the gate area. Moments of inertia will be discussed in Section 4.8.

PROBLEMS

4.79 The vertical walls of a concrete form are held in place by equally spaced posts imbedded into the ground as shown. Determine the pressure distribution on the walls of the form and the distribution of loading on each of the posts. Fresh concrete weighs approximately 24 kN/m^3.

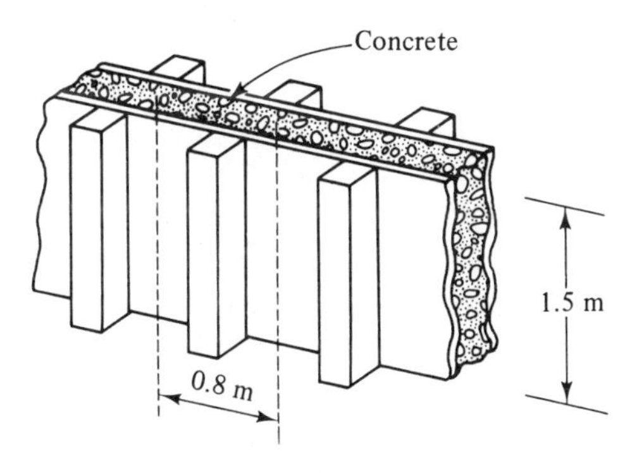

Problem 4.79

4.80 The two halves of a flat roof have a slope of 1/100 toward the center as shown in order to facilitate drainage. If the drain becomes clogged and water collects until it is even with the sides of the roof, what is the pressure distribution and total load on the roof? The density of water is 10^3kg/m^3.

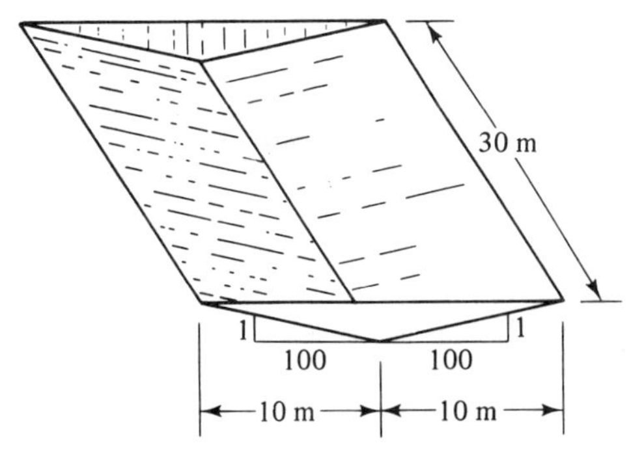

Problem 4.80

4.81 Three identical cylindrical tanks with radius R contain a gas at gage pressure p. One tank has flat ends, another has hemispherical ends, and the third has ends in the shape of elliptical domes. Compare the net longitudinal forces acting upon the ends of each of the tanks. Explain your results.

4.82 to 4.84 Determine the reactions at the supports for the beams shown.

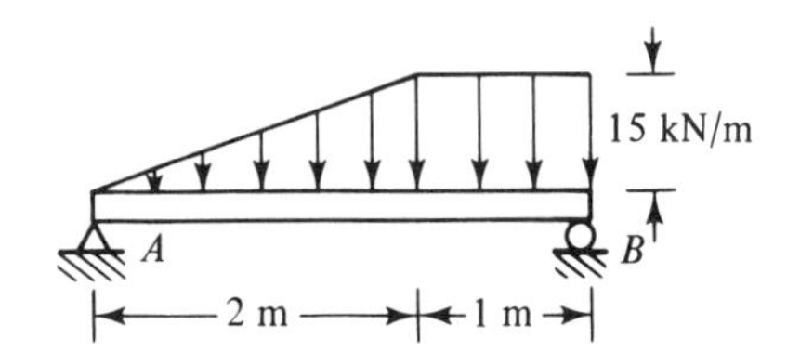

Problem 4.82

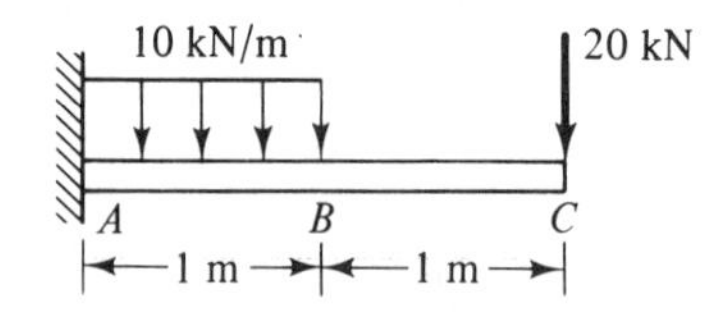

Problem 4.83

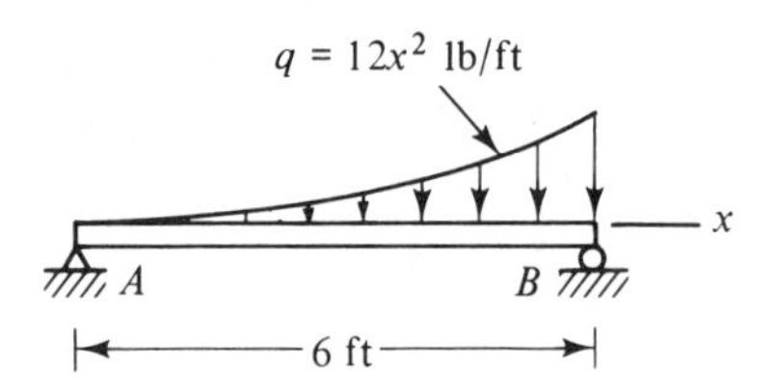

Problem 4.84

4.85 Determine the forces acting upon the horizontal member of the frame shown at points A and B.

4.86 The aquarium shown is filled with sea water, for which $\gamma = 64\ \text{lb/ft}^3$. Determine the magnitude and location of the net force acting upon the window.

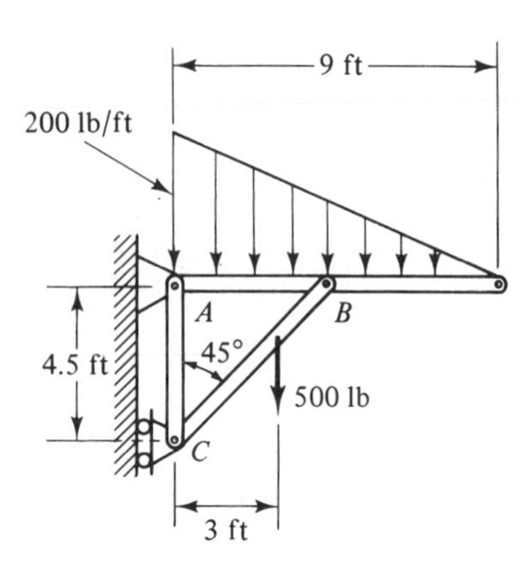

Problem 4.85

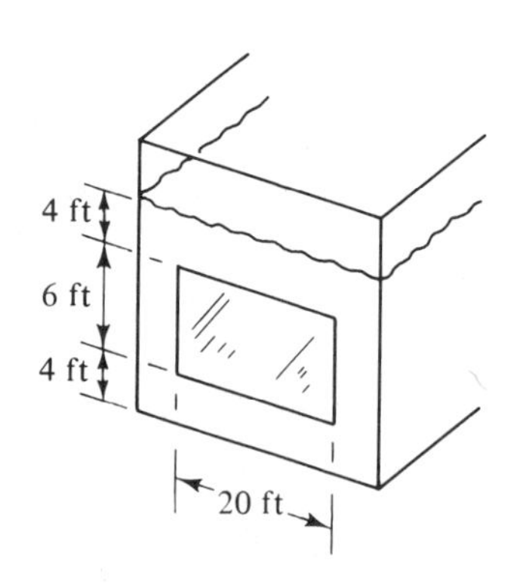

Problem 4.86

4.87 Due to wave action, the pressure on the 20-ft long portion of concrete sea wall shown is twice that due to a liquid at rest. What is the required width w of the wall if it is not to overturn? The specific weight of concrete is $150\ \text{lb/ft}^3$; sea water weighs $64\ \text{lb/ft}^3$.

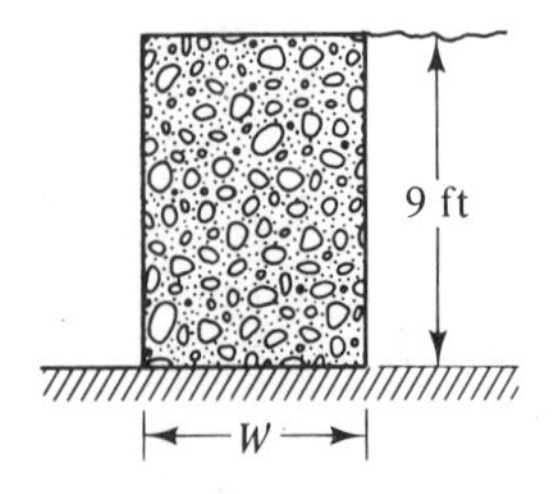

Problem 4.87

4.88 The reservoir behind a dam is filled with silt to a depth of 1 m as shown. The water depth is 2 m. Determine the magnitude and location of the resultant force on the face of the dam per metre of length. The specific weight of water is 9.8 kN/m^3 and the silt weighs 17.3 kN/m^3.

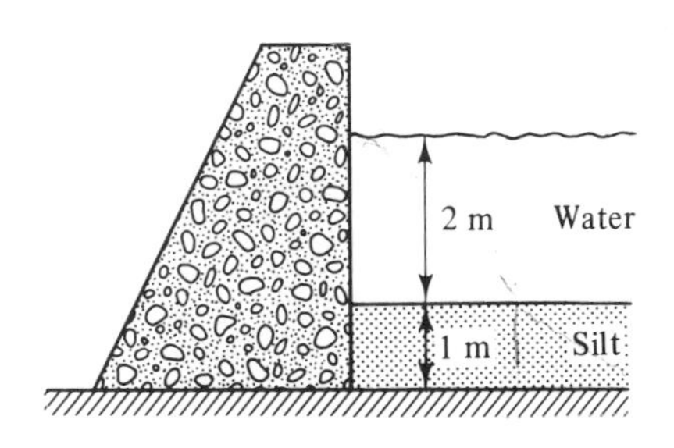

Problem 4.88

4.89 The dam shown holds back water in a 140 ft-wide river that has a depth of 24 ft. Find the resultant force of the water against the upstream face of the dam. To prevent seepage under the dam, the resultant force exerted by the foundation must lie within the central one-third of the base. Is this condition satisfied?

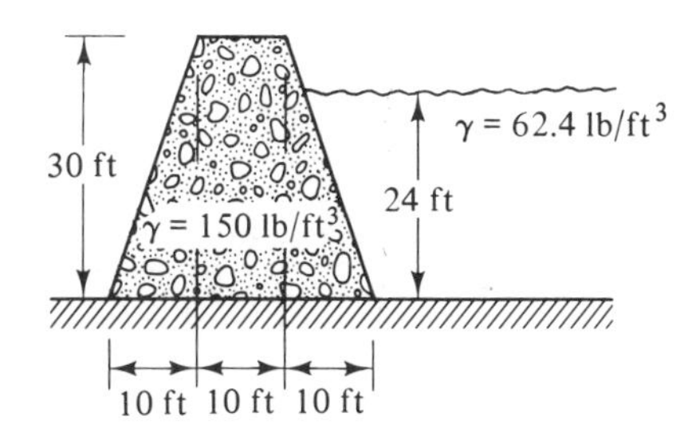

Problem 4.89

4.90 The gate in an irrigation ditch is 2 m long and is held shut by a counterweight as shown. What is the required mass of the counterweight if the gate is to open when the water reaches a depth of 1.2 m?

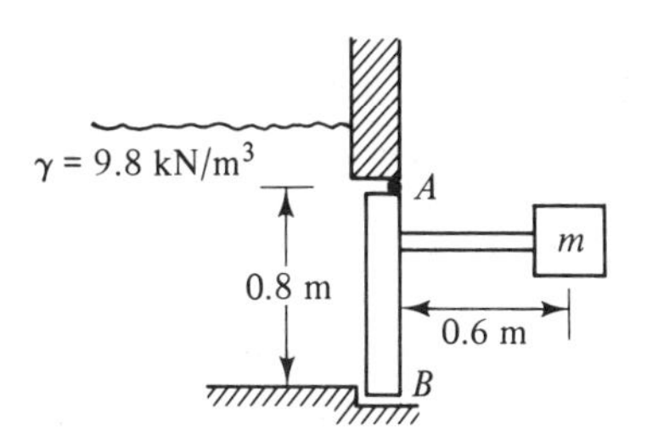

Problem 4.90

4.91 A grain bin 4 m long has the cross section shown. If the density of the grain is 770 kg/m^3, what is the magnitude of the resultant force on each of the wall segments AB and BC and on the bottom CD of the bin? What is the magnitude of the resultant force on the vertical ends of the bin?

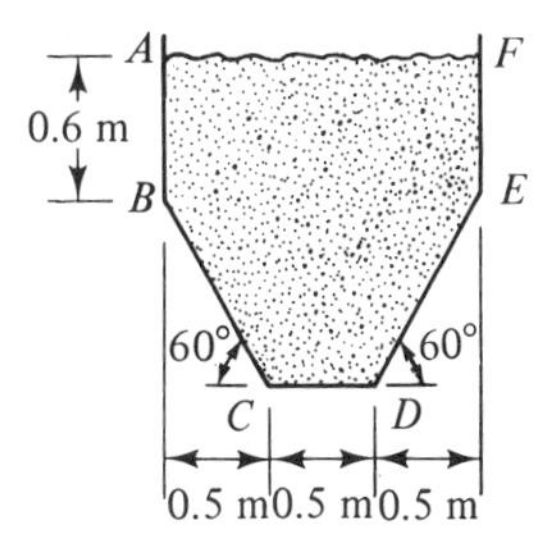

Problem 4.91

4.92 Determine the magnitude and location of the resultant force per foot of length acting upon the upstream face of the dam shown.

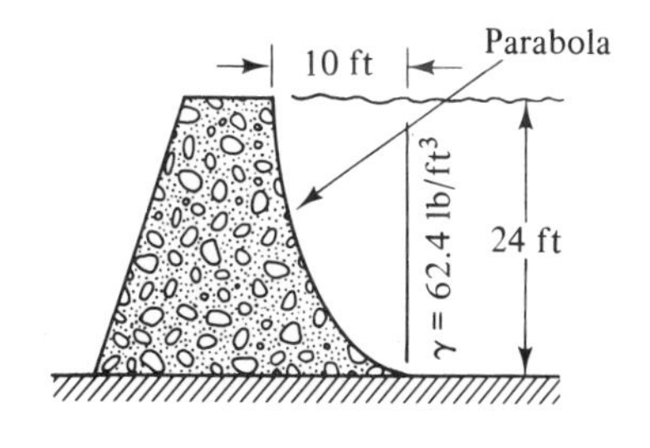

Problem 4.92

4.93 Because of a nearby explosion, one side of a cylinder of length L and radius a experiences the pressure distribution shown. What is the net force exerted on the cylinder?

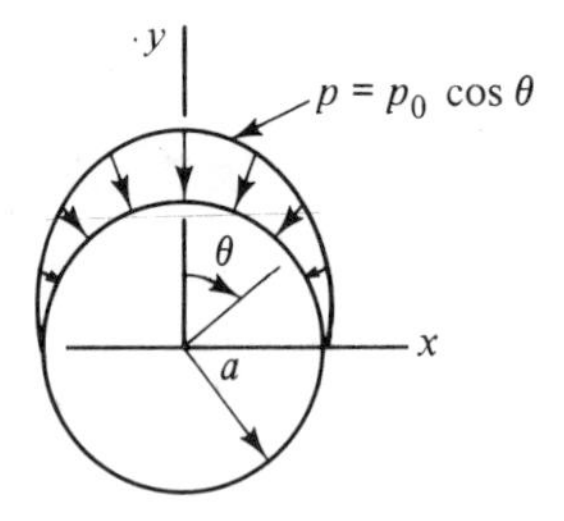

Problem 4.93

4.94 An access hole in the inclined side of a tank is covered with a circular plate as shown. Obtain an expression for the magnitude R of the resultant force acting upon the plate and for the distance d that defines its location.

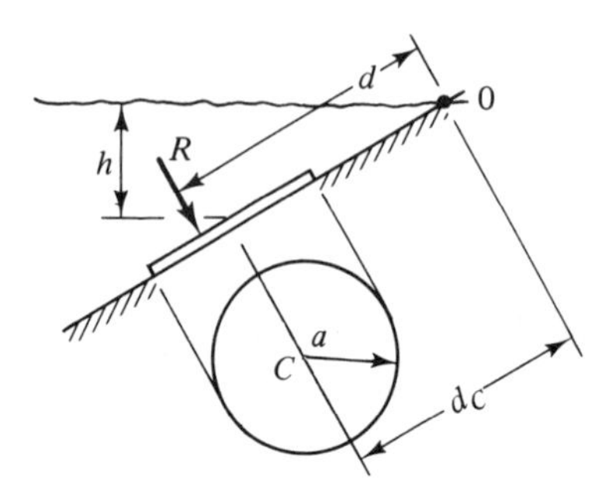

Problem 4.94

4.95 Show that the buoyant force on a submerged body has magnitude equal to the weight of the volume of the fluid displaced and that its line of action passes through the centroid of that volume. *Hint*: The net force on the top portion of the body can be obtained from a force analysis of the fluid volume $ACDEB$ shown, and the force on the bottom surface can be obtained from a force analysis of the volume of fluid $ACFEB$ formed if the body is removed. The difference between these forces is the buoyant force.

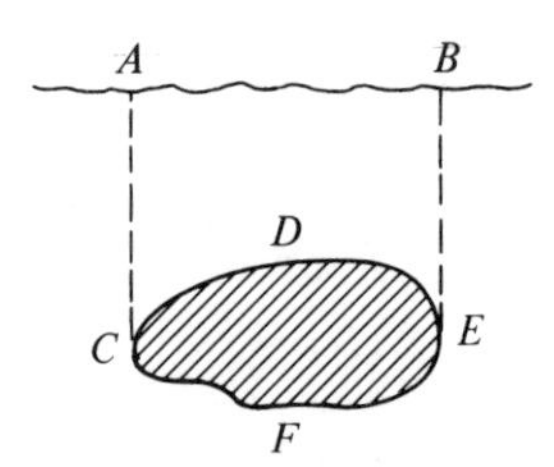

Problem 4.95

4.96 A spherical diving bell with a radius of 1 m and a mass of 8 tonne is lowered into the ocean. If the density of sea water is 1.03×10^3 kg/m^3, what is the tension on the supporting cable? See Problem 4.95.

4.97 If wood weighs 30 lb/ft^3 and water weighs 62.4 lb/ft^3, at what angle θ will the 4 in.$\times$4 in. square timber shown float when in equilibrium? See Problem 4.95.

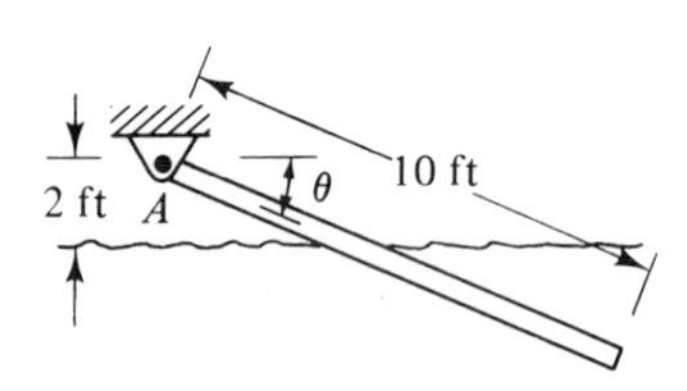

Problem 4.97

4.98 If the density of ice is 90% of that of sea water, what percentage of the volume of an iceberg remains below the surface? See Problem 4.95.

4.8 AREA MOMENTS OF INERTIA

When locating the resultant of the pressure loading in Example 4.19, we encountered an integral quantity called the *area moment of inertia*. Integrals of this type arise whenever the magnitude of the surface force varies linearly with distance, as in the case of a pressure loading. Since these surface forces will be encountered on numerous occasions in the following chapters, we shall pause here to consider some useful properties of area moments of inertia and the procedures for computing them.

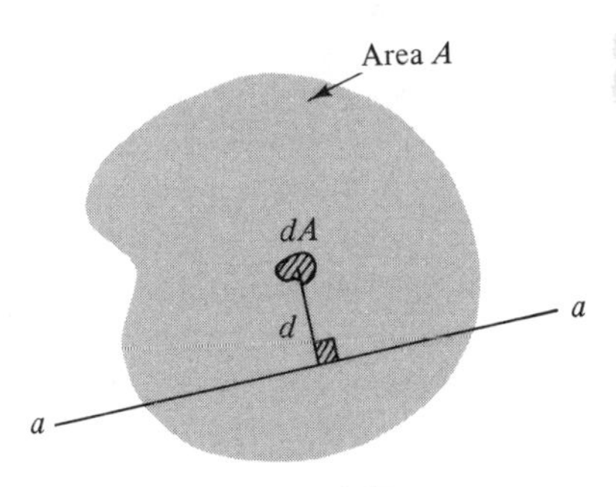

Figure 4.48

By definition, the moment of inertia dI_a of an element of area dA about a line or axis aa within the plane of the area [Figure 4.48(a)] is

$$dI_a = d^2(dA) \qquad (4.23)$$

where d is the perpendicular distance from the axis to the element. The moment of inertia for the total area is the sum of the moments of inertia for each of its constituent parts:

$$I_a = \int dI_a = \int d^2(dA) \qquad (4.24)$$

The moment of inertia is a relative measure of the manner in which an area is distributed with respect to the axis of interest. The further the area is away from the axis, the larger its moment of inertia. From the definitions in Eqs. (4.23) and (4.24), it is clear that area moments of inertia are always positive and have the dimensions of length to the fourth power.

In many problems it is the values of I about the coordinate axes that are of interest. Referring to Figure 4.49 and applying the definition in Eq. (4.24), we have

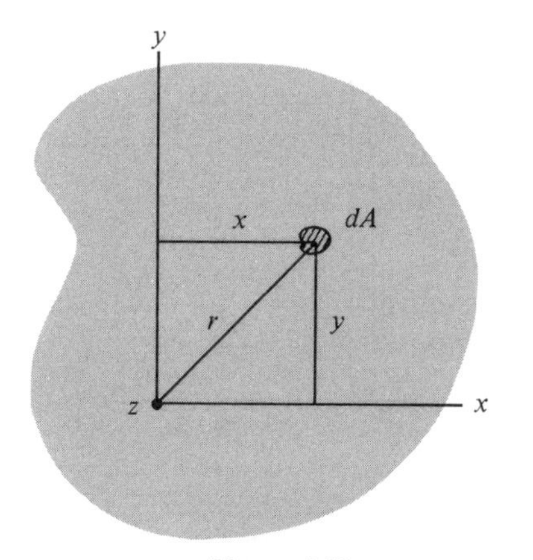

Figure 4.49

$$I_x = \int y^2 dA$$

$$I_y = \int x^2 dA$$

The moment of inertia about an axis perpendicular to the area is called the *polar moment of inertia* and is commonly denoted by J. Thus, the polar moment of inertia about the z axis for the area shown in Figure 4.49 is

$$J_z = \int r^2 dA = \int (x^2 + y^2) dA$$

or

$$J_z = I_x + I_y \qquad (4.25)$$

This equation states that the sum of the moments of inertia about two perpendicular axes within the plane of the area is equal to the moment of inertia about a third concurrent axis perpendicular to the plane. If the moments of inertia about any two of the coordinate axes are known, the value about the third axis can be determined directly from Eq. (4.25).

The distribution of an area with respect to a given axis can also be described in terms of the so-called *radius of gyration*, r_a. By definition,

$$I_a \,(\text{or } J_a) = r_a^2 A \tag{4.26}$$

where A is the total area of the shape. The radius of gyration has no particular physical significance. It is probably best thought of as a fictitious dimension, which, when squared and multiplied by the area, yields the moment of inertia. Equations are sometimes expressed in terms of the radius of gyration instead of the moment of inertia as a matter of convenience. We shall do this in Chapter 10, for example.

As an illustration of the use of the preceding definitions, let us determine the moments of inertia of a rectangle with height h and base b about axes parrallel to its sides and passing through its centroid [Figure 4.50(a)]. Here, and in the sequel, values of I and J about centroidal axes will be denoted by $\bar{I}$ and $\bar{J}$.

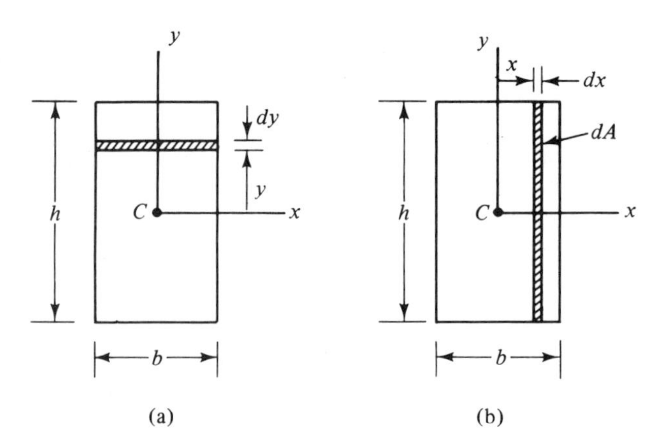

Figure 4.50

To compute $\bar{I}_x$, we select a horizontal strip as the element of area. Referring to Figure 4.50(a), we have

$$\bar{I}_x = \int y^2 dA = \int_{-h/2}^{h/2} y^2 (b dy) = \frac{bh^3}{12}$$

To compute $\bar{I}_y$, we select a vertical element of area [Figure 4.50(b)]. The horizontal element cannot be used directly because each portion of it lies at a different distance from the y axis. We have

$$\bar{I}_y = \int x^2 dA = \int_{-b/2}^{b/2} x^2 (h dx) = \frac{hb^3}{12}$$

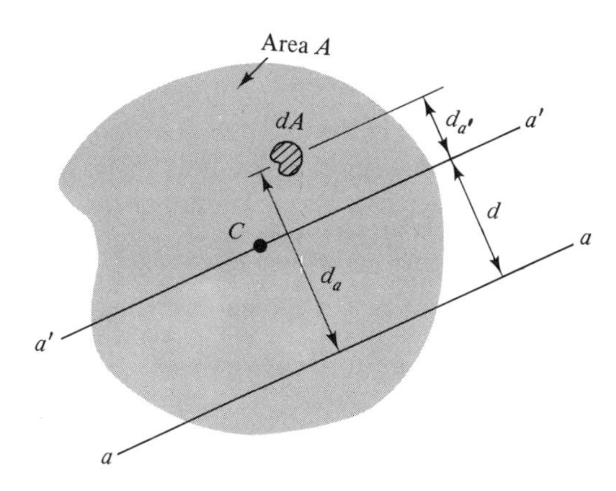

Figure 4.51

This result can also be obtained from the expression for $\bar{I}_x$ simply by interchanging the x and y dimensions.

Values of the moment of inertia and radius of gyration of many common shapes have been computed and tabulated in engineering and mathematical handbooks. Results for some common geometrical and structural shapes are given in Tables A.1 and A.3 through A.12 of the Appendix. Usually, only the values of I and J about centroidal axes are given. However, if these values are known, the values about any other parallel axis can be readily determined, as we shall now show.

Consider the area shown in Figure 4.51. Let $a'a'$ be an axis through the centroid and aa be any other axis parallel to it. The distance between the axes is d. From the definition of moment of inertia, Eq. (4.24), we have

$$I_a = \int d_a^2 (dA) = \int (d_{a'} + d)^2 dA$$

$$= \int d_{a'}^2 (dA) + 2d \int d_{a'} (dA) + d^2 \int dA$$

where d_a is the distance from axis aa to the element of area and $d_{a'}$ is the distance to the element from the centroidal axis $a'a'$.

The first term in the preceding expression is recognized to be the moment of inertia, $\bar{I}_{a'}$, about the centroidal axis. Using the definition of the centroid of an area, Eqs. (4.12), the integral in the second term can be expressed as

$$\int d_a' (dA) = A\bar{d}$$

where $\bar{d}$ is the distance from the centroidal axis to the centroid. Since this distance is zero, the integral expression must also be zero. Thus, we have

$$I_a = \bar{I}_{a'} + Ad^2 \tag{4.27}$$

This result, which is called the *parallel axis theorem*, relates the moment of inertia about a centroidal axis to that about any other parallel axis. The theorem also applies for polar moments of inertia.

As in the case of centroids, the greatest difficulty in determining moments of inertia by integration lies in the formulation of the integrals. Again, the first, and most important, step is the selection of the element of area. The element should be selected parallel to the axis of interest, if possible. Otherwise, each portion of the element is a different distance from the axis, and the definitions in Eqs. (4.23) and (4.24) cannot be used directly. In this case, the parallel axis theorem must be used to determine dI for the element about the desired axis, and the resulting expression integrated to determine I for the entire area (see Example 4.20). The other alternative is to use double integration.

An approach similar to that used for centroids can be used to compute the moment of inertia of composite areas. The values of the moments of inertia of simple shapes about centroidal axes can be determined from tables, and the parallel axis theorem can be used to transfer them to the desired axis. The moments of inertia of each element are then summed to determine I for the whole area. Cutouts and holes are handled by considering their moments of inertia and areas to be negative.

Example 4.20. Moment of Inertia of an Area. Determine the moments of inertia I_x, I_y, and J_z for the area shown in Figure 4.52.

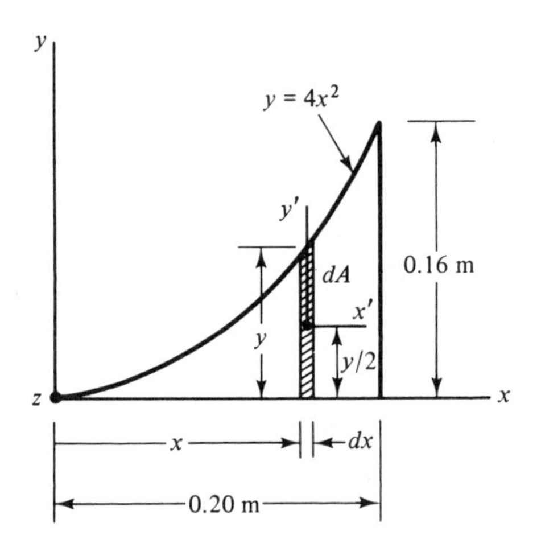

Figure 4.52

Solution. Taking a vertical strip as the element of area, we have

$$dA = ydx = 4x^2dx$$

Since the entire element is the same distance from the y axis, I_y can be computed directly from the definition of I [Eq. (4.24)]:

$$I_y = \int x^2dA = \int_0^{0.2} x^2(4x^2)dx = \frac{4x^5}{5}\Bigg|_0^{0.2} = 25.6 \times 10^{-5}\ \text{m}^4 \qquad \textbf{\textit{Answer}}$$

To compute I_x, we must either select another element parallel to the x axis or use the parallel axis theorem. Using the latter approach, we have for the rectangular element

$$d\bar{I}_{x'} = \frac{bh^3}{12} = \frac{1}{12}y^3dx$$

$$dI_x = d\bar{I}_{x'} + Ad^2 = \frac{1}{12}y^3dx + ydx\left(\frac{y}{2}\right)^2 = \frac{1}{3}y^3dx = \frac{64}{3}x^6dx$$

(This result could also have been obtained by using the equation $bh^3/3$ from Table

A.1 for the moment of inertia of a rectangle about its base.) We now integrate dI_x over the entire area to determine I_x:

$$I_x = \int_0^{0.2} \frac{64}{3} x^6 dx = \frac{64}{21} x^7 \Big|_0^{0.2} = 3.9 \times 10^{-5}\ \text{m}^4 \qquad \textit{Answer}$$

From Eq. (4.25), the polar moment of inertia is

$$J_z = I_x + I_y = 29.5 \times 10^{-5}\ \text{m}^4 \qquad \textit{Answer}$$

Example 4.21. Moment of Inertia of a Composite Area. A triangular bracket with a circular hole is welded to an angle section to form the shape shown in Figure 4.53(a). Determine the moment of inertia and radius of gyration about axis aa.

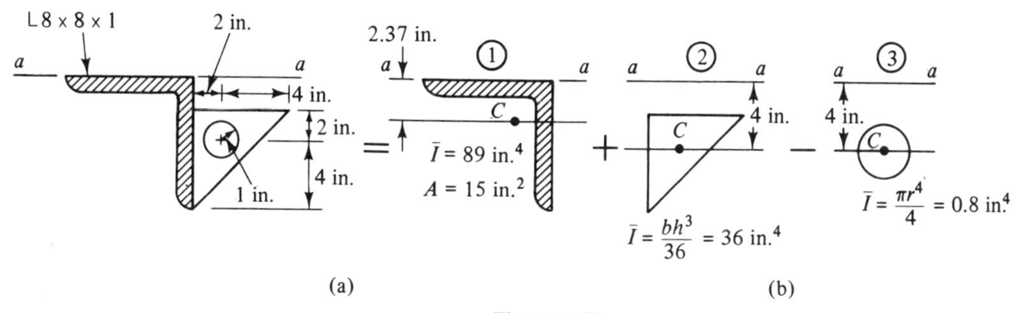

Figure 4.53

Solution. The area is considered to consist of three elements [Figure 4.53(b)]. Their centroidal moments of inertia, obtained from Tables A.1 and A.5 of the Appendix, are as indicated. The parallel axis theorem is then used to determine their moments of inertia about axis aa. Carrying out the indicated computations in tabular form, we have

No.	$\bar{I}_i$ (in.^{4})	A_i (in.^{2})	d_i (in.)	$A_i d_i^2$ (in.^{4})
1	89.0	15.00	2.37	84.3
2	36.0	18.00	4.00	288.0
3	−0.8	−3.14	4.00	−50.3
Σ	124.2	29.86		322.0

Note that $\bar{I}$ and A for the circle are considered negative because this piece is removed to form the hole. Using the sums found in the table, we obtain

$$I_a = \Sigma \bar{I}_i + \Sigma A_i d_i^2 = 124.2 + 322.0 = 446.2\ \text{in.}^4$$

Answer

$$r_a = \sqrt{\frac{I_a}{A}} = \sqrt{\frac{446.2\ \text{in.}^4}{29.9\ \text{in.}^2}} = 3.9\ \text{in.}$$

PROBLEMS

4.99 to 4.102 Determine by integration the moments of inertia I_x and I_y for the areas shown. Also determine J_z.

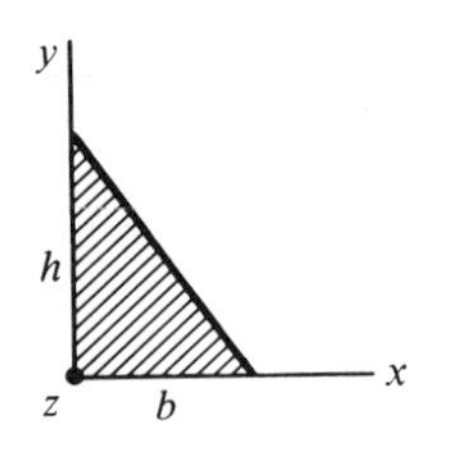

Problem 4.99

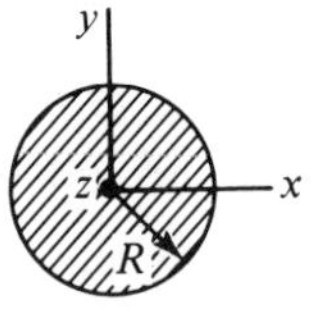

Problem 4.100

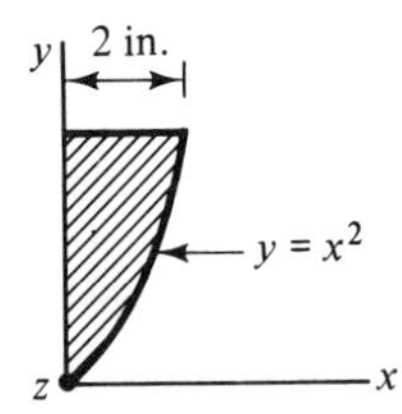

Problem 4.101

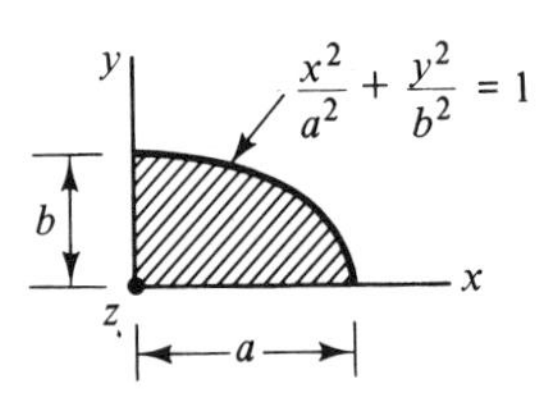

Problem 4.102

4.103 For the area shown, determine the moments of inertia and the radii of gyration about the x and y axes.

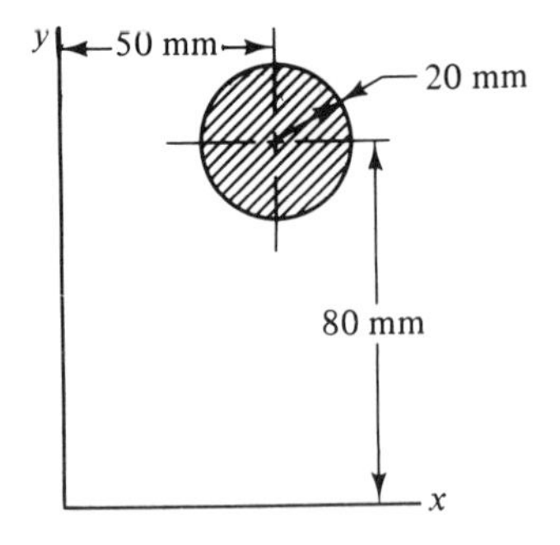

Problem 4.103

4.104 Given that $I_a = 108 \text{ in.}^4$ for the area shown, find I_b.

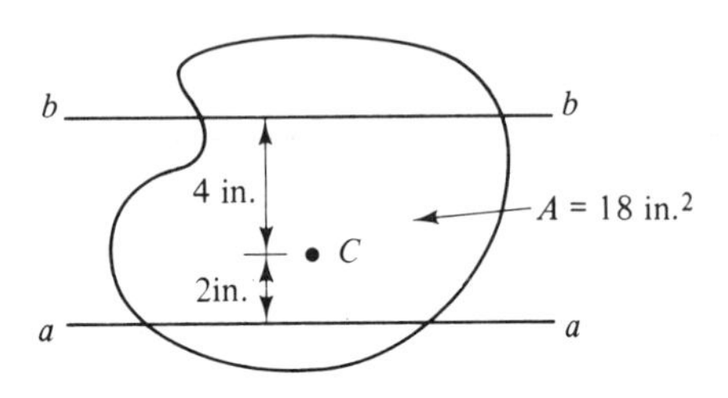

Problem 4.104

4.105 and 4.106 Determine the moments of inertia of the areas shown about the x and y axes.

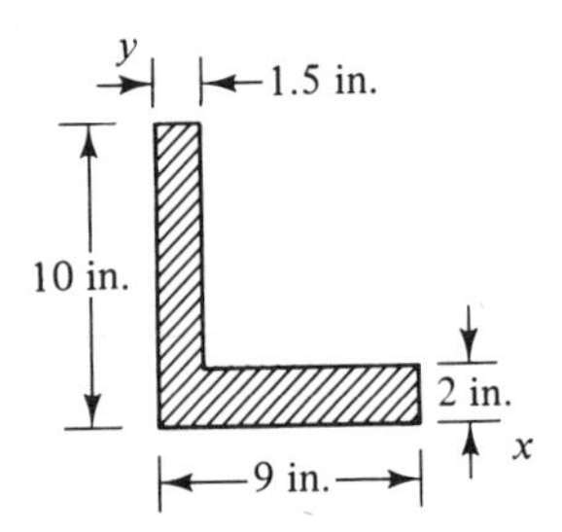

Problem 4.105

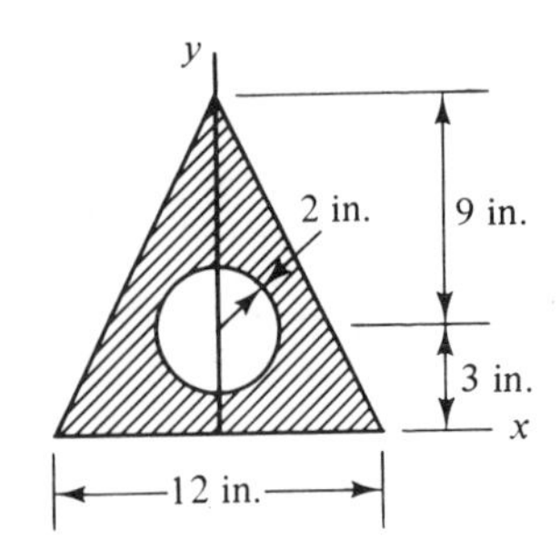

Problem 4.106

4.107 Determine the dimension b such that $I_x = 2000\ \text{in.}^4$ for the area shown. What is the corresponding value of I_y?

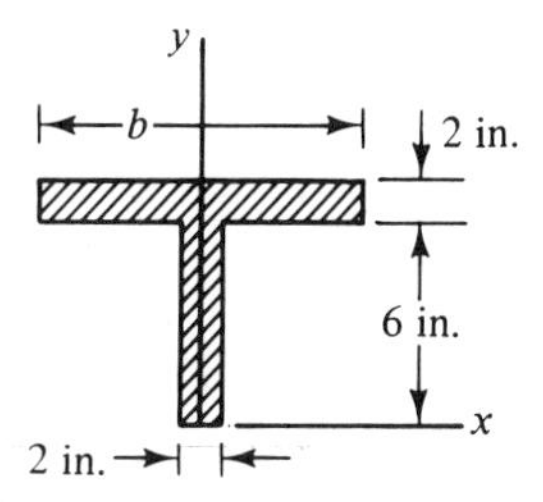

Problem 4.107

4.108 A precast concrete roof slab has the cross section shown. Determine the moment of inertia about a horizontal axis through the centroid.

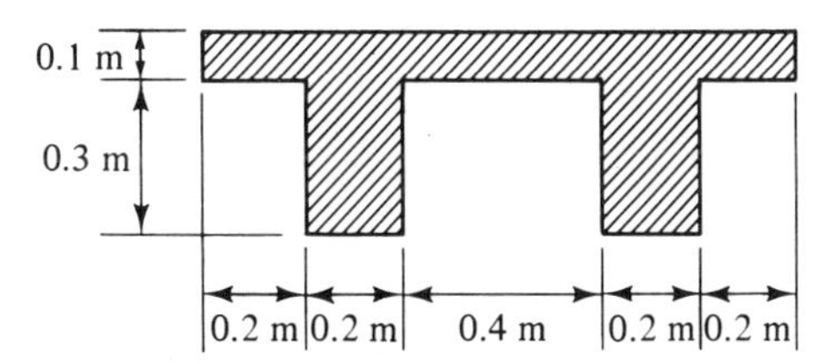

Problem 4.108

4.109 A flat steel plate is attached to the top surface of a 254 × 102 mm Universal Beam with a mass of 28 kg/m (see Table A.8) to form a structural member with the cross section shown. Determine the moment of inertia about a horizontal axis through the centroid.

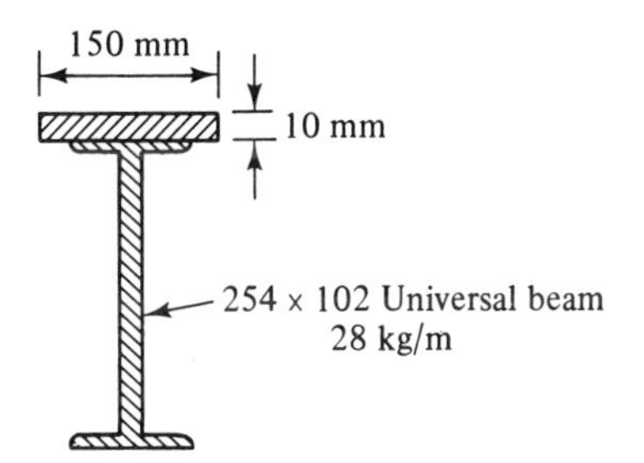

Problem 4.109

4.110 A structural member is formed from two channel sections (see Table A.4) connected by two $\frac{1}{2}$-in. thick steel plates as shown. Determine the spacing d between the channels for which $I_x = I_y$.

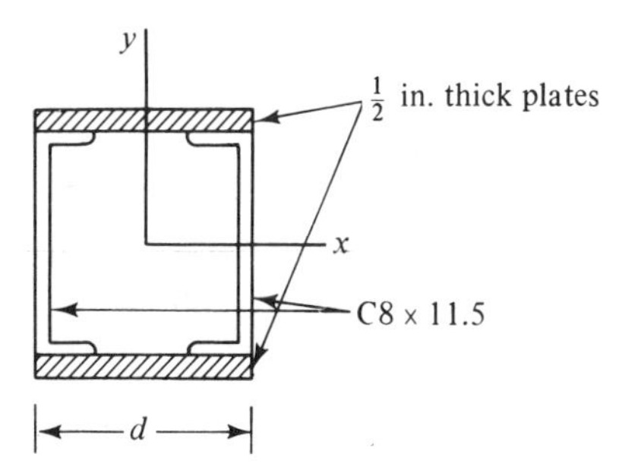

Problem 4.110

4.111 Four equal angle sections (see Table A.10) are riveted together to form a member with the cross section shown. Determine the radius of gyration about the centroidal x axis.

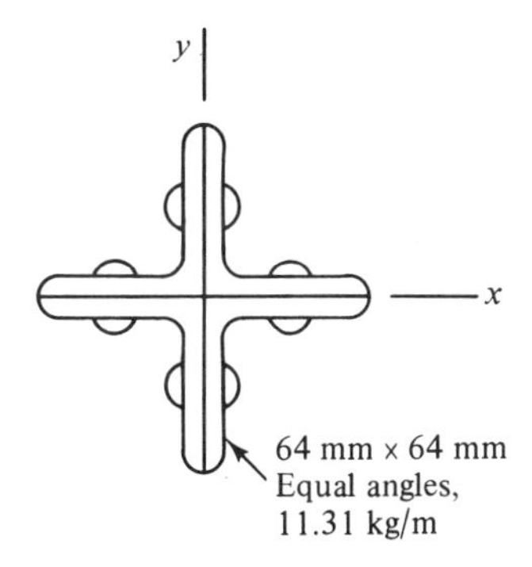

Problem 4.111

4.112 Determine the polar moment of inertia about the longitudinal axis of a hollow tube with inside radius R_i and outside radius R_o. What is the radius of gyration about this axis? Show that if the walls are thin, $J \approx 2\pi R_{avg}^3 t$, where R_{avg} is the average radius and t is the wall thickness.

4.9. CLOSURE

In this chapter we have considered the conditions for the static equivalence of two force systems. These conditions apply to force systems in general, but they are particularly useful when dealing with distributed forces. They were used in Section 4.3 to show that the reactions due to various supports and connections can be treated as concentrated forces and couples in a force analysis. They were also used to locate the center of gravity of a body and to develop procedures for handling distributed applied loads. We shall use these conditions again in later chapters when determining the stresses developed within load-carrying members.

It is essential that we not forget the limitations on static equivalence. Static equivalencc of two force systems implies only that they have the same effect upon the equilibrium of a body. If one of the force systems is replaced by the other, the deformations and forces developed within the body may be altered and the stability of the equilibrium may be affected.

chapter five

MECHANICS OF DEFORMABLE BODIES

5.1 INTRODUCTION

Our discussions in the previous chapters have been concerned primarily with the problem of determining the forces and couples acting upon various bodies loaded and supported in different ways. This was done without regard to the response of the bodies to the applied loadings, other than assuming that the resulting deformations are small (the rigid-body approximation).

Clearly, the ability of a load-carrying member to fulfill its intended purpose requires that the member not break under the applied loadings. The "strength" of a member depends in large measure upon the stresses, or intensity of the forces developed within it. Deformations of a member can also dictate the limits of its usefulness. A floor that sags an excessive amount is a "failure," even though it may be in no danger of breaking.

The effects of forces and couples upon load-carrying members will be investigated in this chapter and in those to follow. Our goal is to develop relationships between the applied loadings and the stresses and deformations they produce. These relationships are needed for analyzing members to see if they can adequately perform their job and for designing (sizing) members to carry a given loading.

Once the force analysis of a body has been completed, the rigid-body approximation must be abandoned if its response is to be determined. Obviously, we cannot determine the deformations of a body if we insist on assuming that the body is rigid. This is one of the major points of departure from our work in earlier chapters. Thus, we shall now be considering the *mechanics of deformable bodies*.

The mechanics of deformable bodies is more involved than the mechanics of rigid bodies, but not unduly so. Actually, we have already performed an analysis of a deformable system; namely, the example of the rigid arm supported by a spring considered in Section 3.10. A review of this example shows that, in order to obtain a solution, we had to consider equilibrium, the geometry of the deformations, and the resistive properties of the material (spring). These three items are the basic ingredients in the mechanics of deformable bodies.

We have already considered equilibrium. The geometry of deformations and resistive properties of materials will be discussed in this chapter, along with the application of all three concepts to some simple, but important, problems. Structural elements such as rods, shafts, and beams under simple loadings will be considered in Chapters 6 through 8. These results will be combined in Chapter 9 in a study of more complicated members and loadings. The buckling of columns under compressive loads will be considered in Chapter 10.

It was mentioned in Section 3.4 that most engineering structures are designed such that the deformations are small. Accordingly, *small deformations will be assumed throughout* all our work. As a result, the rigid-body approximation can still be used in force analyses, even though the values of the forces and couples so obtained are then used to determine the deformations. The justification for this procedure, which greatly simplifies the analysis, was given in Section 3.10. This approach can be used for all but buckling problems, for which the equilibrium equations must be based upon the deformed geometry of the body.

In the way of historical perspective, we note that the systematic study of the response of load-carrying members dates back at least to Leonardo da Vinci (1452–1519), who performed experiments on the strength of wires and beams. The advent of railroads in the late 1800's provided the impetus for much of the basic work in this area, and problems associated with the design of aircraft, space vehicles, and nuclear reactors have led to extensive studies of the more advanced aspects of the subject.

5.2 STRESSES AND STRESS RESULTANTS

Loadings applied to a body are transmitted through the body to the supports, which provide the reactions necessary to hold the entire body in equilibrium. In the process, forces are developed within the body that resist the applied loads and maintain equilibrium of each of its parts. These resistive forces are distributed over surfaces within the body, and their magnitude per unit area is called *stress*. The resultants of these forces are called *stress resultants* (also called internal reactions or internal forces).

As an illustration, consider a bar made of several blocks glued together end to end [Figure 5.1(a)]. If we pull on the end of the bar with a small force **F**, it is clear

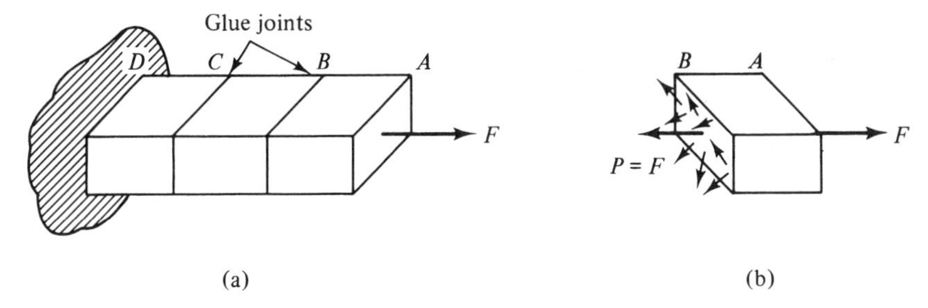

Figure 5.1

that the glue joints will resist the tendency for the blocks to be pulled apart. The joints provide resistive forces that hold the various pieces of the bar together in equilibrium. They also serve to transmit the applied load from block to block to the wall, which supports the entire member.

The resultant force transmitted by a joint can be determined by passing an imaginary cutting plane through it and then considering the equilibrium of one of the resulting pieces of the bar. (Note that this is the method of sections used in Chapter 3 to determine the forces in the members of a truss.) For example, sectioning the bar along joint B and considering the equilibrium of the piece to the right [Figure 5.1(b)], we see that the stress resultant, denoted by **P**, must be equal and opposite to the applied force **F**. If we push on the bar instead of pull on it, the sense of the stress resultant is reversed. In both cases, the resultant is a force perpendicular to the cut surface. This type of stress resultant is called a *normal force*, and is associated with the tendency of the member to elongate or shorten.

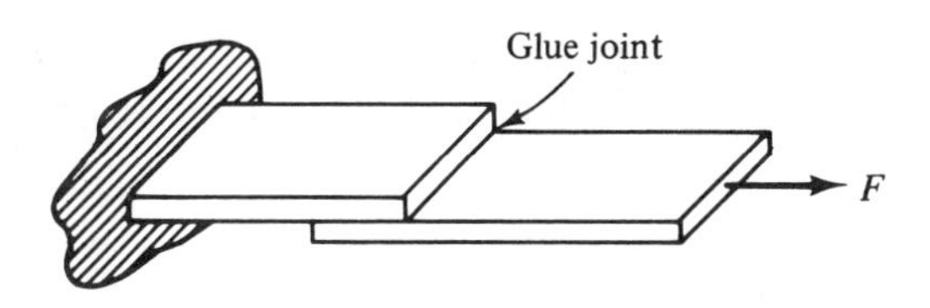

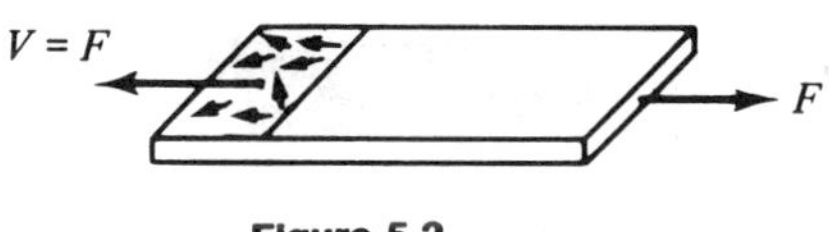

Figure 5.2

A different situation is illustrated in Figure 5.2. In this case, the stress resultant is a force parallel to the joint and is associated with the tendency of one part of the member to slide with respect to the other. This tendency is called *shear*, and the stress resultant is called a *shear force*. Shear forces will be denoted by **V**.

The stress resultant can also be a couple. If the couple is perpendicular to the cutting plane [Figure 5.3(a)], the stress resultant, denoted by **T**, is called a *torque*. Torques are associated with the tendency of the member to twist. Couples that lie in the cutting plane [Figure 5.3(b)] are associated with the tendency of the member to bend and are called *bending moments*. Bending moments will be denoted by **M**. Under more complex loading conditions, the various types of stress resultants can occur simultaneously.

The preceding ideas apply directly to bodies of one piece. The only difference is that they are held together by cohesive forces instead of glue joints. Any body

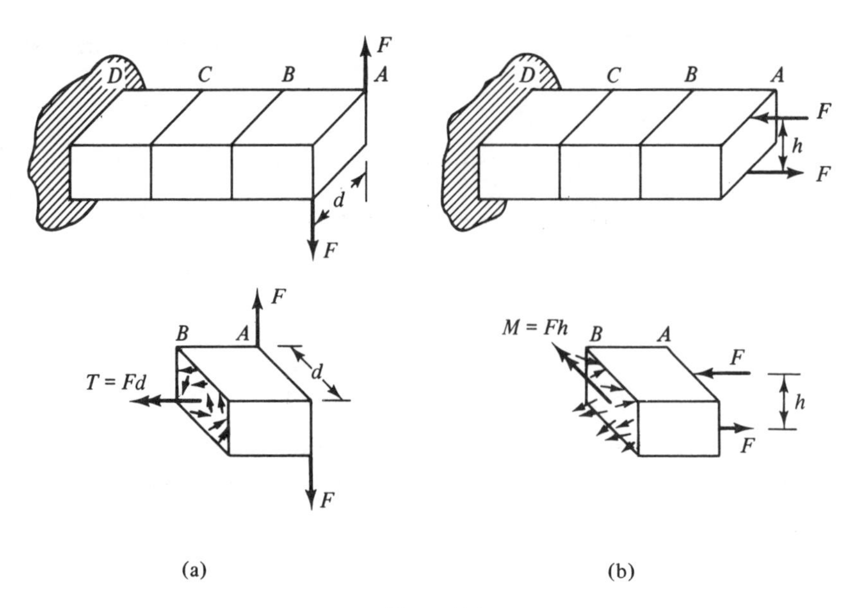

Figure 5.3

can be thought of as consisting of two pieces, one on either side of an imaginary plane [Figure 5.4(a)]. Each piece exerts forces on the other piece. These forces are transmitted across the plane, just as the adjoining blocks in our previous examples exert forces on each other. In this sense, the cutting plane is analogous to a glue joint.

Figure 5.4(b) shows the resistive forces acting upon a plane through a body. The resultant force $\Delta\mathbf{R}$ acting upon an element of area ΔA within the plane can be resolved into a normal force $\Delta\mathbf{P}$ and a shear force $\Delta\mathbf{V}$. The magnitudes of the normal force and shear force per unit area are called the *normal stress* and *shear stress*, respectively, and will be denoted by the Greek letters σ and τ. The *average stresses* are

$$\sigma_{\text{avg}} = \frac{\Delta P}{\Delta A} \qquad \tau_{\text{avg}} = \frac{\Delta V}{\Delta A} \tag{5.1}$$

and, from the definition of a surface force [Eq. (4.18)], the *normal* and *shear stresses at a point* in the plane are

$$\sigma = \lim_{\Delta A \to 0} \frac{\Delta P}{\Delta A} = \frac{dP}{dA} \qquad \tau = \lim_{\Delta A \to 0} \frac{\Delta V}{\Delta A} = \frac{dV}{dA} \tag{5.2}$$

If the stresses are uniformly distributed over the plane, the average stresses and the stresses at a point are the same. Common units for stress are lb/in^2 (psi) or N/m^2 (pascals). Another common unit is ksi (1 ksi = 1000 psi).

Normal stresses (and forces) directed away from the cut section are called *tensile* and will be considered positive. When directed toward the section, they are called *compressive* and will be considered negative. There is no universal sign convention for shear stress. Usually, only its magnitude is of importance.

Under general loading conditions, the stress resultant acting upon any given plane within a body will be a force and a couple [Figure 5.4(c)]. For convenience, the resultant force will always be placed at the centroid of the cut section. (Recall from Section 4.3 that the resultant couple depends upon the location of the resultant force.) The resultant force can be resolved into a normal force **P** perpendicular to the plane and a shear force **V** parallel to the plane. Similarly, the resultant couple can be resolved into a torque **T** normal to the plane and a bending moment **M** parallel to the plane. If desired, the shear force and bending moment can be further resolved into components within the cut section.

Both the stresses and stress resultants generally vary throughout the body. They also depend upon the orientation of the plane on which they act. Thus, the orientation of the cutting plane must always be specified in order for stated values of the stresses and stress resultants to have meaning.

The stresses are related to the stress resultants via the conditions for static equivalence. As shown earlier, the stress resultants can be determined by using the method of sections. Unfortunately, since there are any number of possible stress distributions that have the same resultant, the stresses cannot be determined

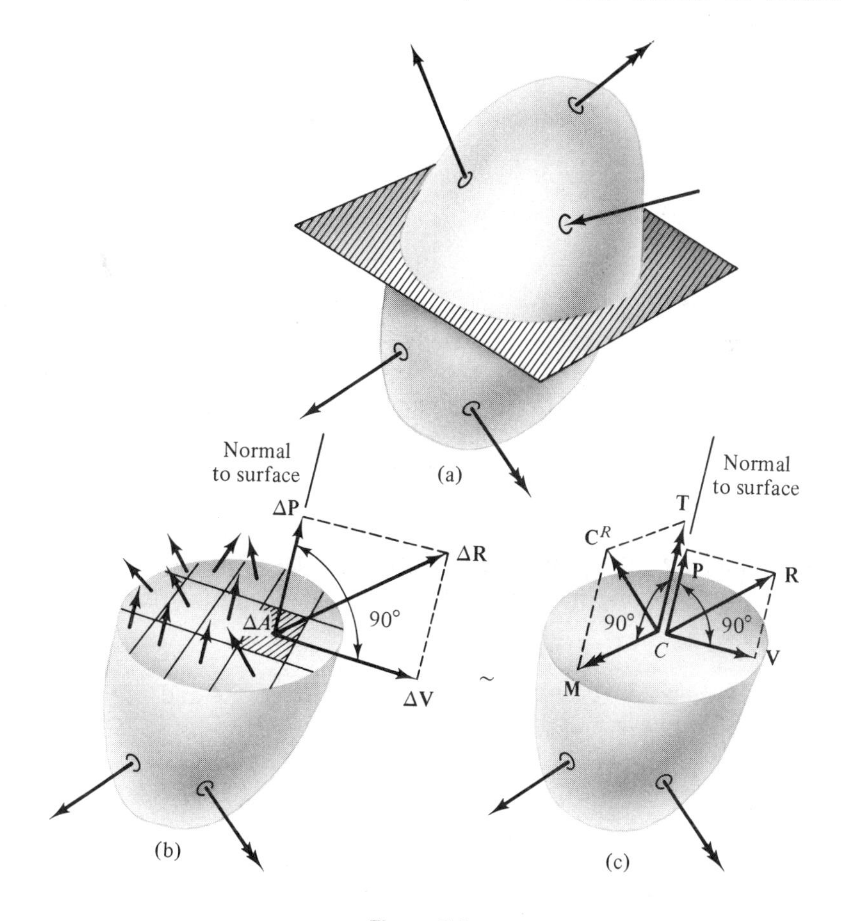

Figure 5.4

without having some additional information. But is it necessary to know the stresses and not just the stress resultants? Yes, and for two very important reasons.

First, the strength of a material is best expressed in terms of stresses instead of forces. For example, the force that a glue joint in the bar shown in Figure 5.1(a) can support is not a good measure of the strength of the glue. Obviously, joints with a larger area can support a larger force, even though the properties of the glue remain the same. A better measure of the material capabilities is the force per unit area, or stress, that it can withstand.

Second, the deformations produced by the applied loads are intimately related to the stress distribution. This is illustrated in Figure 5.5, which shows two different distributions of normal stress acting upon identical rubber blocks. Each stress distribution has the same resultant, but it produces a different pattern of deformation. Clearly, it is not enough to know only the stress resultants if the deformations are to be determined.

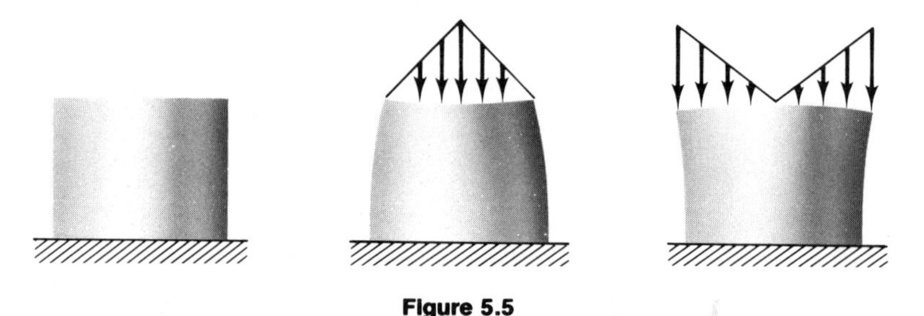

Figure 5.5

At this point, we are in no position to outline a general procedure for determining the stresses in a body. This must await our study of the geometry of the deformations and the resistive properties of materials. However, there are problems for which average stresses have significance and can be readily computed. Some of these problems will be considered in Sections 5.3 and 5.4. The following examples further illustrate the procedures for determining the stress resultants.

Example 5.1. Stress Resultants in a Beam. Determine the normal and shear forces and bending moment at sections *a-a* and *b-b* in the beam shown in Figure 5.6(a).

Solution. We first determine the reactions at A and B. From the FBD [Figure 5.6(b)], we have

$$\Sigma F_x = A_x + B_x = 0$$

$$\Sigma F_y = B_y - 6000 \text{ lb} = 0$$

$$\overset{+}{\curvearrowleft}\Sigma M_B = 6000 \text{ lb } (5 \text{ ft}) - A_x(4 \text{ ft}) = 0$$

$$A_x = 7500 \text{ lb} \qquad B_x = -7500 \text{ lb} \qquad B_y = 6000 \text{ lb}$$

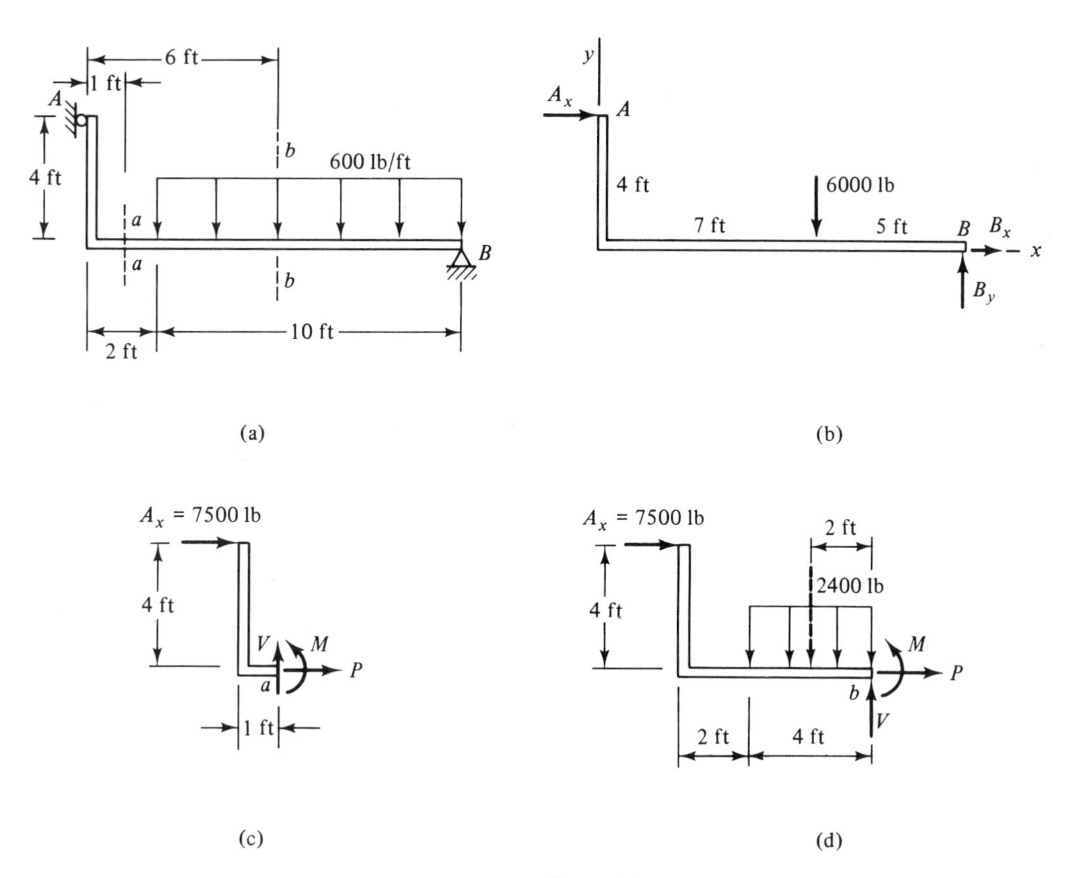

Figure 5.6

Section a-a. Sectioning the beam at *a-a* and considering the piece to the left, we obtain the FBD shown in Figure 5.6(c). For equilibrium,

$$\Sigma F_x = P + 7500 \text{ lb} = 0$$

$$\Sigma F_y = V = 0$$

$$\overset{+}{\curvearrowleft} \Sigma M_a = M - 7500 \text{ lb } (4 \text{ ft}) = 0$$

so

$$P = -7500 \text{ lb or } 7500 \text{ lb (compression)} \qquad V = 0$$

$$M = 30\ 000 \text{ lb·ft}$$

Answer

The portion of the beam to the right of *a-a* could also have been used. The stress resultants on this segment are equal and opposite to those on the left-hand segment, by Newton's third law.

Section b-b. Sectioning through the beam and the distributed load at *b-b* and proceeding as before, we have from Figure 5.6(d)

$$\Sigma F_x = P + 7500 \text{ lb} = 0$$

$$\Sigma F_y = V - 2400 \text{ lb} = 0$$

$$\overset{+}{\curvearrowleft} \Sigma M_b = M + 2400 \text{ lb } (2 \text{ ft}) - 7500 \text{ lb } (4 \text{ ft}) = 0$$

or

$$P = -7500 \text{ lb or } 7500 \text{ lb (compression)} \qquad V = 2400 \text{ lb}$$

$$M = 25\ 200 \text{ lb·ft}$$

Answer

Note that the distributed load was replaced by its resultant only after the member was sectioned. If this were done before sectioning the member, the forces within the body could be altered (see Section 4.2), leading to incorrect values for the stress resultants.

Example 5.2. Stress Resultants In a Rod. Find the stress resultants acting upon section *a-a* of the rod shown in Figure 5.7(a). Identify the shear and normal forces, bending moments, and torque. Describe the responses associated with each.

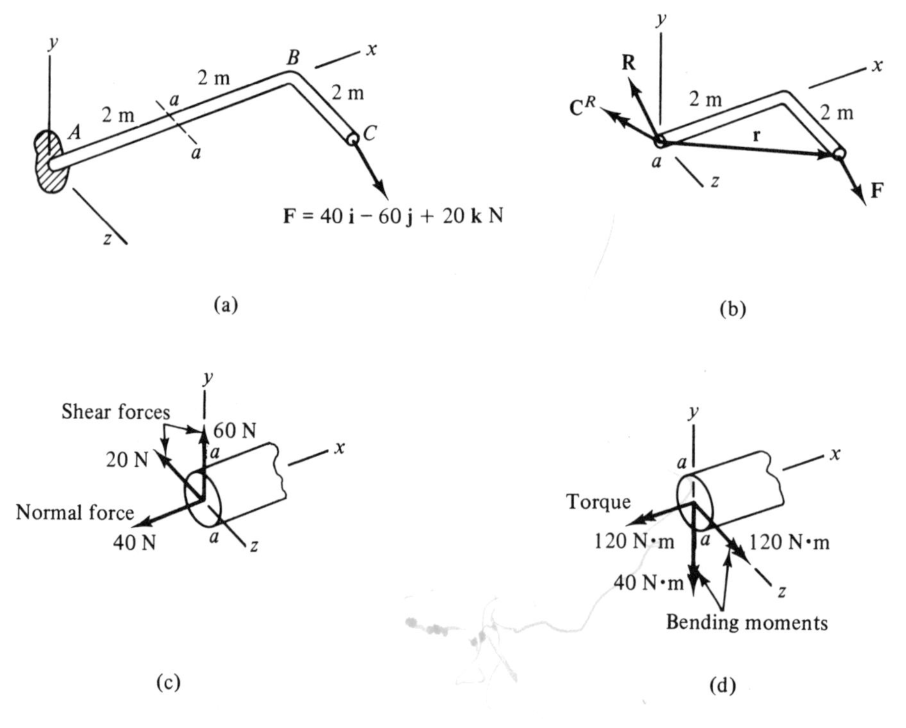

Figure 5.7

Solution. We section the bar at *a-a* and consider the portion to the right [Figure 5.7(b)]. Note that this choice eliminates the need to determine the reactions at the wall. The stress resultant is shown as a single force **R** acting at the centroid of the cross section and a couple $\mathbf{C}^R$. For equilibrium:

$$\Sigma \mathbf{F} = \mathbf{F} + \mathbf{R} = \mathbf{0}$$

$$\mathbf{R} = -\mathbf{F} = -40\mathbf{i} + 60\mathbf{j} - 20\mathbf{k}\ \text{N} \qquad \textit{Answer}$$

$$\Sigma \mathbf{M}_a = \mathbf{C}^R + \mathbf{r} \times \mathbf{F} = \mathbf{0}$$

$$\mathbf{C}^R + (2\mathbf{i} + 2\mathbf{k}) \times (40\mathbf{i} - 60\mathbf{j} + 20\mathbf{k})\ \text{N·m} = \mathbf{0}$$

$$\mathbf{C}^R = -120\mathbf{i} - 40\mathbf{j} + 120\mathbf{k}\ \text{N·m} \qquad \textit{Answer}$$

The components of **R** are shown in Figure 5.7(c). The normal force (perpendicular to the cut section) is 40 N (tension). It is associated with a stretching of portion *AB* of the member (the stress resultants counteract the effects of the applied loads). The 60-N and 20-N forces are the components of the shear force within the plane. They are associated with the tendency of part *aBC* of the bar to move downward and in the positive *z* direction relative to the piece attached to the wall. The components of $\mathbf{C}^R$ are shown in Figure 5.7(d). The torque (normal to the cut section) is associated with a tendency of portion *AB* of the member to twist about the *x* axis. The other two components are bending moments. The 40-N·m moment is associated with a bending of the bar in the *x-z* plane and the 120-N·m moment with bending in the *x-y* plane.

PROBLEMS

5.1 to 5.11 For the members or structures shown, determine the stress resultants on the sections indicated. Identify the normal and shear forces, bending moments, and torques where applicable.

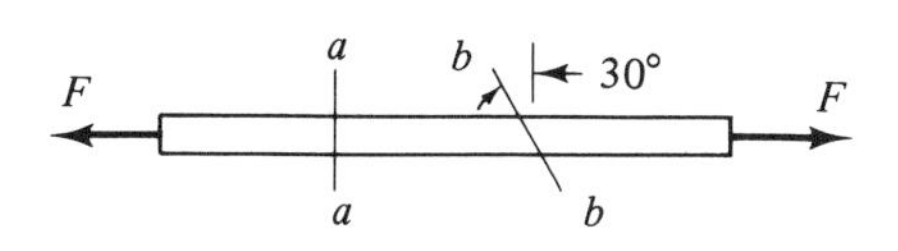

Problem 5.1

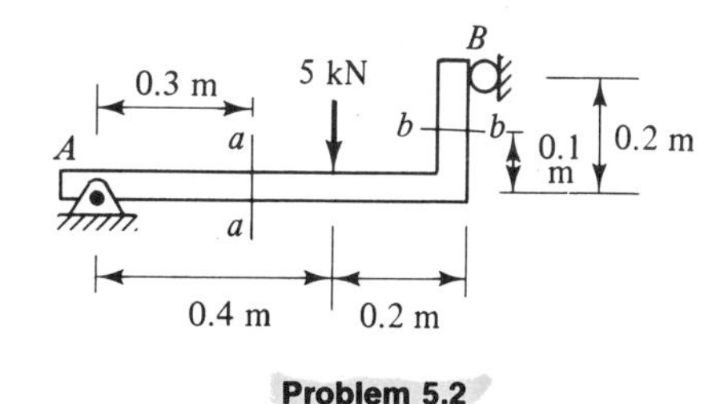

Problem 5.2

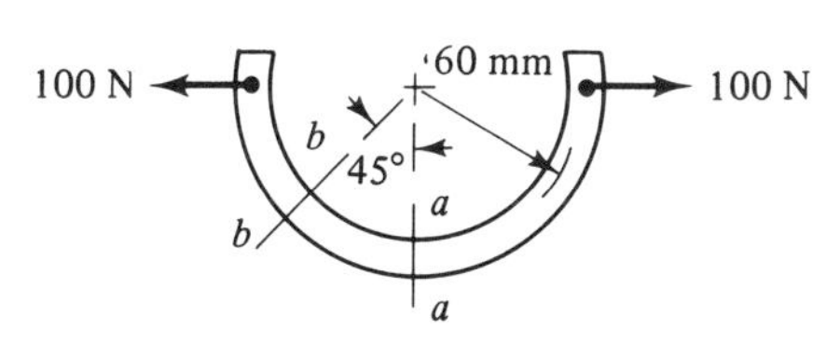

Problem 5.3

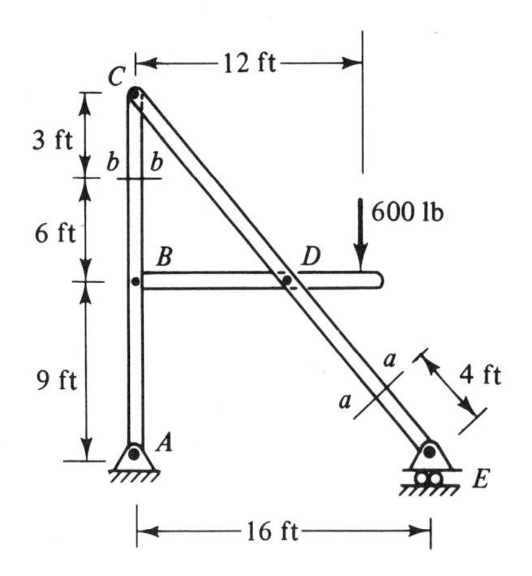

Problem 5.4

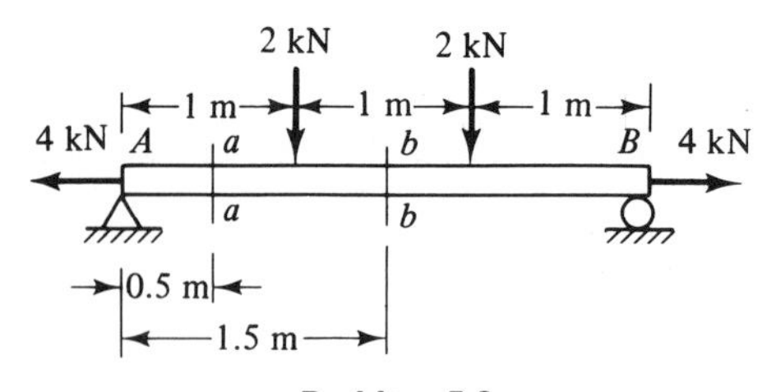

Problem 5.8

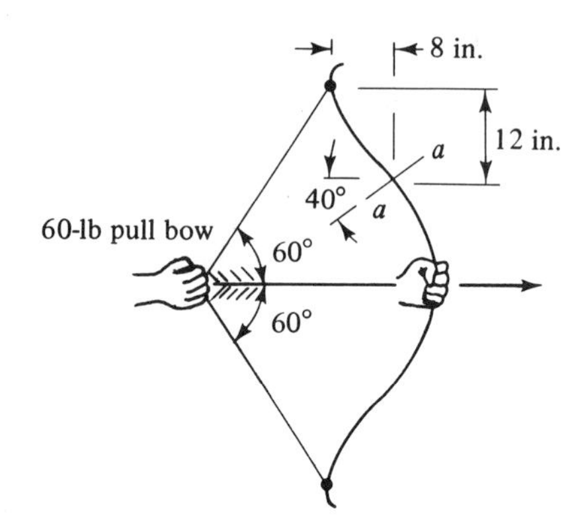

Problem 5.9

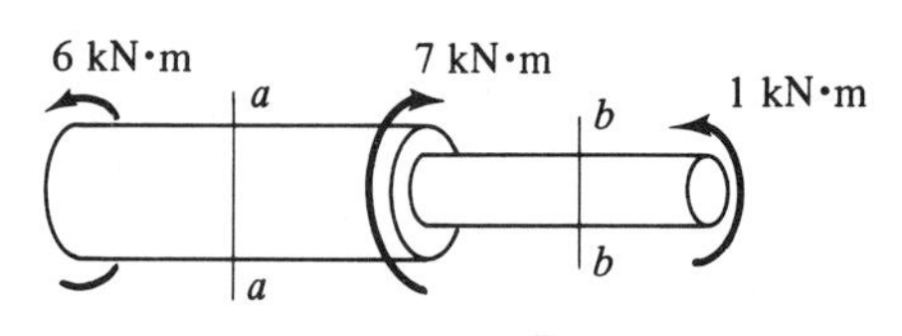

Problem 5.5

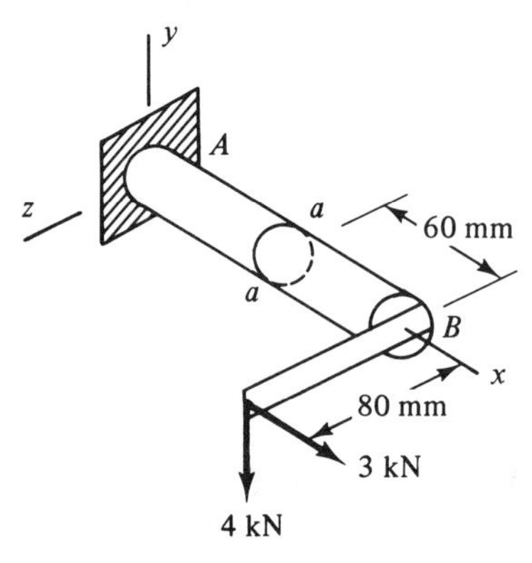

Problem 5.10

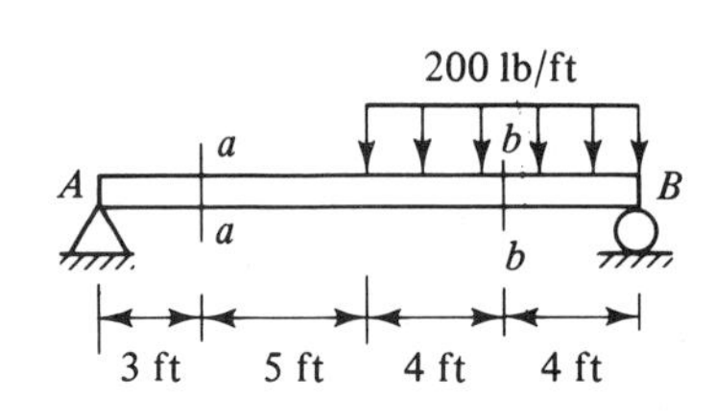

Problem 5.6

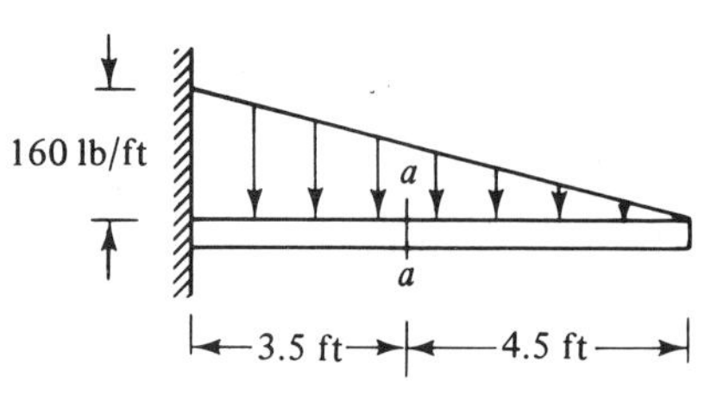

Problem 5.7

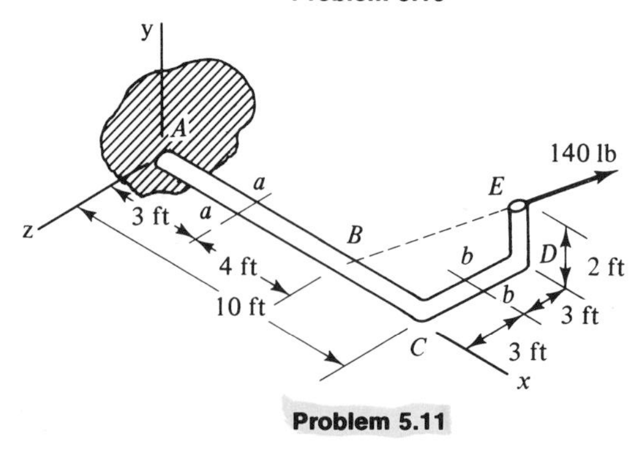

Problem 5.11

5.3. STRESSES DUE TO AXIAL LOADING

Straight, slender members subjected to forces that lie along the longitudinal axis passing through the centroid of the cross section are said to be *axially loaded*. Such members are one of the more common types of structural elements. Typical examples are support wires and cables and the members of a truss. The stress resultant in an axially loaded member consists only of a force; therefore, the average stresses can be computed directly from their definitions. As we shall show, the actual stresses in axially loaded members are often uniformly distributed, in which case they are equal to the average stresses and can also be determined.

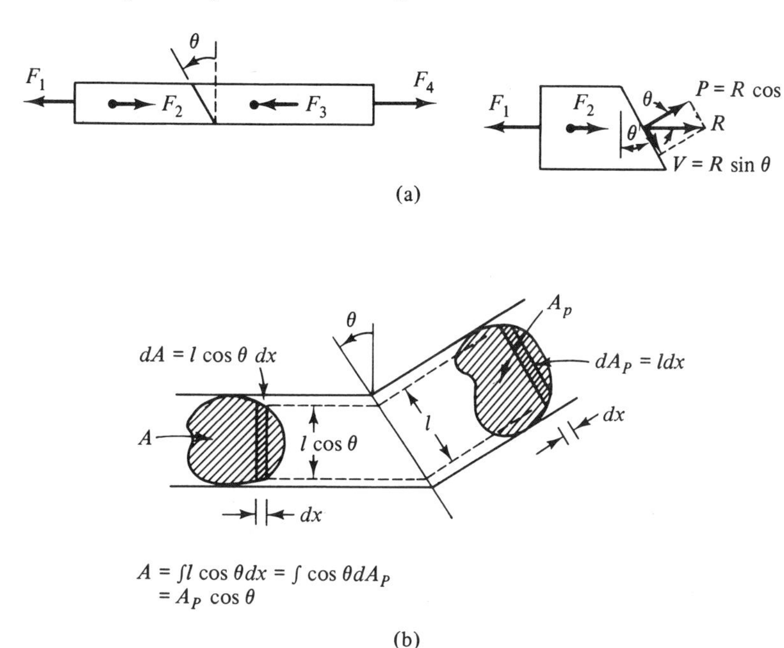

Figure 5.8

Consider an axially loaded member with arbitrary cross section in equilibrium under the action of two or more forces [Figure 5.8(a)]. The normal and shear forces on a plane oriented at an arbitrary angle θ, measured as shown, are

$$P = R \cos \theta \qquad V = R \sin \theta$$

where R is the magnitude of the stress resultant. The area A_P of the plane over which these forces act is

$$A_P = \frac{A}{\cos \theta}$$

where A is the cross-sectional area of the member. This result is easily derived, as shown in Figure 5.8(b).

We now make the assumption that the actual stresses are uniformly distributed, in which case they are equal to the average stresses. From the definitions of the average stresses, Eqs. (5.1), we have for the normal stress σ_θ on the plane

$$\sigma_\theta = \sigma_{avg} = \frac{P}{A_P} = \frac{R}{A}\cos^2\theta \tag{5.3}$$

and for the shear stress τ_θ

$$\tau_\theta = \tau_{avg} = \frac{V}{A_P} = \frac{R}{A}\sin\theta\cos\theta = \frac{R}{2A}\sin 2\theta \tag{5.4}$$

These equations relate the stresses to the stress resultant, which, in turn, is related to the applied loads.

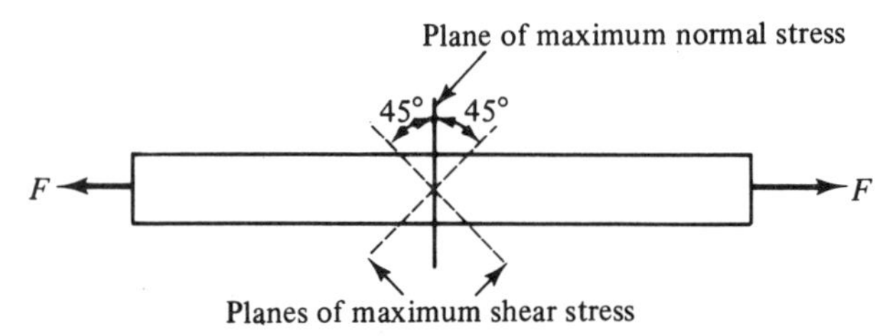

Figure 5.9

From Eq. (5.3), we see that the normal stress is a maximum on planes perpendicular to the axis of the member, for which $\theta = 0$ and $\cos\theta = 1$ (Figure 5.9). Since $P = R$ on these planes, the maximum value of normal stress is simply the normal force divided by the cross-sectional area of the member:

$$\sigma_{max} = \frac{R}{A} = \frac{P}{A} \tag{5.5}$$

The maximum shear stress occurs on planes oriented at $\pm 45°$ with respect to the longitudinal axis (Figure 5.9), as can be seen from Eq. (5.4). Thus,

$$|\tau_{max}| = \frac{1}{2}\frac{|R|}{A} = \frac{1}{2}|\sigma_{max}| \tag{5.6}$$

The fact that the magnitude of the maximum shear stress is only one-half that of the maximum normal stress does not mean that shear stresses are unimportant. Some materials are much weaker in shear than in tension or compression.

The assumption of uniformly distributed stresses is correct, and Eqs. (5.3) through (5.6) valid, for axially loaded members made of a homogeneous material, except near points of load application and sudden changes in the cross section, such as notches or holes. The member may be tapered slightly, and the applied loads may vary along its length. For compressive loads, the member must not be so slender that it buckles. As a rule of thumb, a member will not buckle if its length is no more than ten times its least cross-sectional dimension.

The conditions for the validity of the load–stress relations given in Eqs. (5.3) through (5.6) will be examined in more detail in Chapter 6. However, the need for

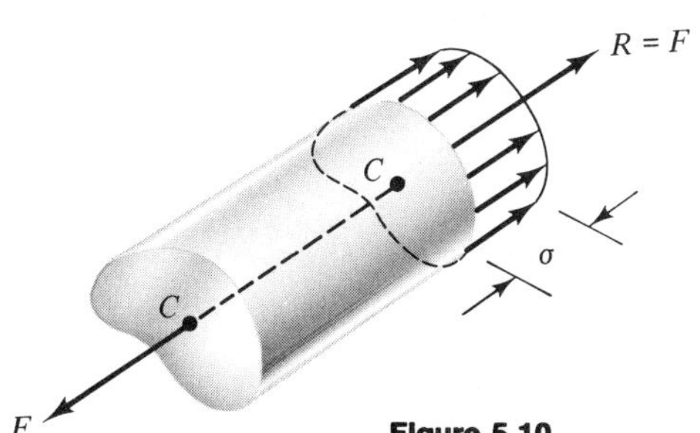

Figure 5.10

axial loading can be easily demonstrated here by considering the stresses on a plane perpendicular to the axis of the member. If the stress distribution is uniform, as has been assumed, the stress resultant will lie along the longitudinal axis passing through the centroid of the cross section (Figure 5.10). (Recall from Section 4.7 that the resultant of a surface force passes through the centroid of the volume under the loading diagram.) Consequently, the applied loads must also lie along the centroidal axis in order to maintain equilibrium. If they don't, the stresses cannot be uniformly distributed, and Eqs. (5.3) through (5.6) do not apply.

There is no need to memorize any of the equations presented in this section. The important thing to remember is that the stresses on any given plane are simply the normal and shear forces acting upon that plane divided by the area over which they act.

Example 5.3. Stresses in a Plate. A broken plate is welded back together, as shown in Figure 5.11(a). (a) What is the maximum load F that can be applied if the tensile stress in the weld cannot exceed 5000 psi? (b) What are the maximum normal and shear stresses in the plate for this value of load and on what planes do they occur? Assume axial loading.

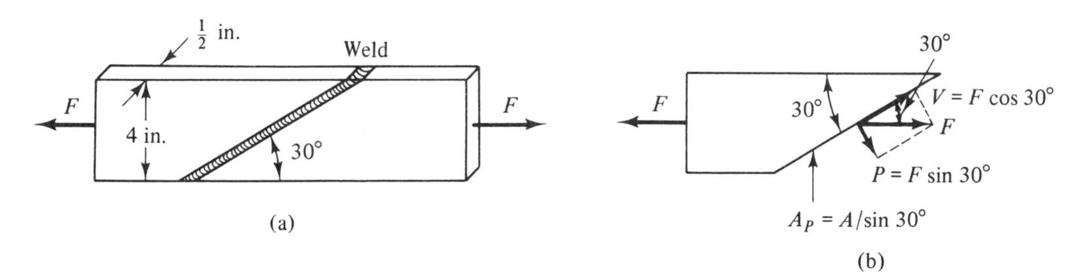

Figure 5.11

Solution. (a) Sectioning the plate along the weld [Figure 5.11(b)], we find that the tensile force is

$$P = F \sin 30° = \frac{F}{2}$$

The area over which this force acts is

$$A_P = \frac{A}{\sin 30°} = 2\left(\frac{1}{2} \text{ in.}\right)(4 \text{ in.}) = 4 \text{ in.}^2$$

Thus, the condition on the tensile stress is

$$\sigma = \frac{P}{A_P} = \frac{(F/2)}{4\ \text{in.}^2} \leqslant 5000\ \frac{\text{lb}}{\text{in.}^2}$$

from which we obtain for the allowable load

$$F \leqslant 40\ 000\ \text{lb} \qquad \textit{Answer}$$

(b) The maximum normal stress occurs on planes perpendicular to the axis. On these planes, $P = F$; so

$$\sigma_{\text{max}} = \frac{40\ 000\ \text{lb}}{\left(\frac{1}{2}\ \text{in.}\right)(4\ \text{in.})} = 20\ 000\ \text{psi(tension)} \qquad \textit{Answer}$$

The maximum shear stress is one-half the maximum normal stress [Eq. (5.6)]:

$$|\tau_{\text{max}}| = \tfrac{1}{2}|\sigma_{\text{max}}| = 10\ 000\ \text{psi} \qquad \textit{Answer}$$

It occurs on planes oriented $\pm 45°$ from the axis of the member.

Example 5.4. Design of a Column. A column in a building supports the loads shown in Figure 5.12(a). What is the required cross-sectional area of the member if the maximum allowable compressive stress is 50 MN/m^2? Assume the member does not buckle.

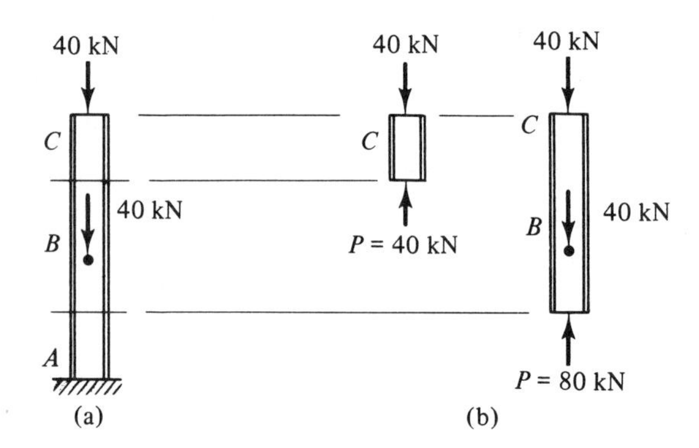

Figure 5.12

Solution. From the FBDs of Figure 5.12(b), we see that the largest compressive force occurs in portion AB of the member and is equal to 80 kN. Thus, the condition on the compressive stress is

$$\sigma_{\text{max}} = \frac{P}{A} = \frac{80 \times 10^3\ \text{N}}{A} \leqslant 50 \times 10^6\ \text{N/m}^2$$

from which we obtain

$$A \geqslant 1.6 \times 10^{-3}\ \text{m}^2 \qquad \textit{Answer}$$

PROBLEMS

5.12 What diameter rope is required to support an 800-kg object if the allowable tensile stress in the rope is 12 MN/m^2?

5.13 The uniform beam AB shown has a mass of 81.6 kg. What is the maximum tensile stress in the support cable BC if its diameter is 10 mm?

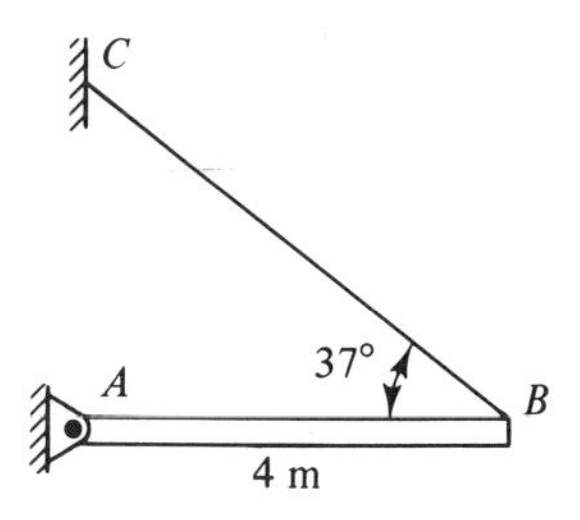

Problem 5.13

5.14 The bar shown is 2 in. wide and $\frac{1}{2}$ in. thick. Determine the normal and shear stresses on (a) section a-a and (b) section b-b.

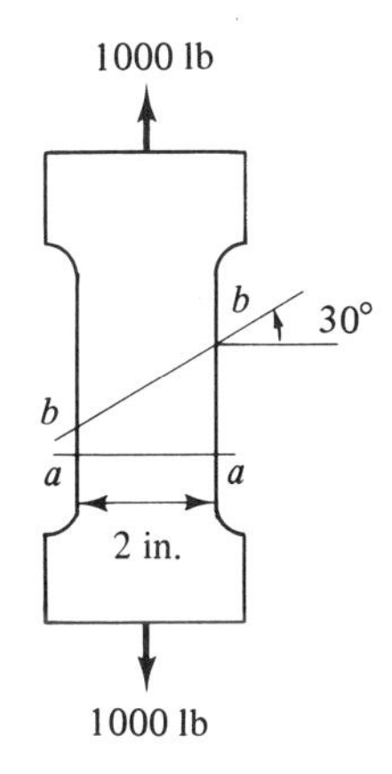

Problem 5.14

5.15 Two 2 in. × 2 in. square bars are glued together to form the member shown. The allowable tensile and shear stresses in the glue are 1000 psi and 500 psi, respectively. For what joint angle θ and applied load F will the glue be simultaneously stressed to its limit in tension and shear?

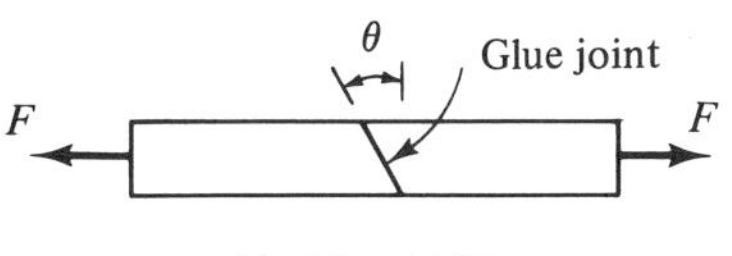

Problem 5.15

5.16 For the wood block shown, the allowable shear stress parallel to the grain is 150 psi and the maximum allowable compressive stress in any direction is 600 psi. What is the maximum compressive load F the block can support?

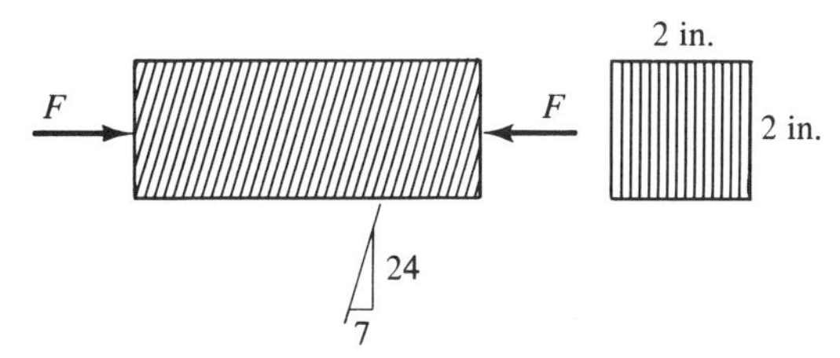

Problem 5.16

5.17 Determine the required cross-sectional areas of members DE and EG of the truss shown if the allowable stresses are 140 MN/m^2 in tension and 70 MN/m^2 in compression. Assume that compression members do not buckle.

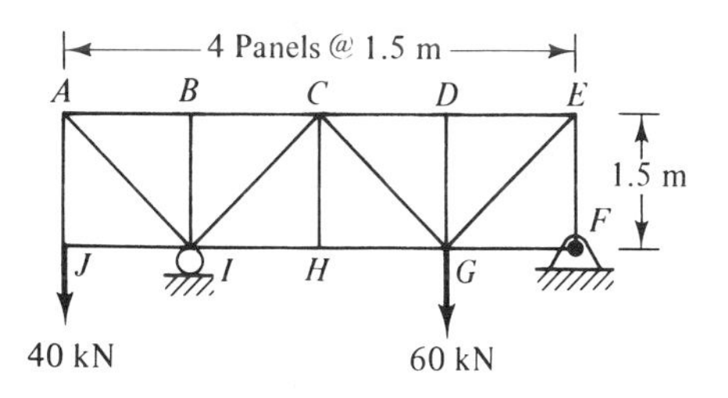

Problem 5.17

5.18 Same as Problem 5.17, except that the members of interest are *BC* and *CI*.

5.19 For what ratio of the cross-sectional areas of members *AC* and *BC* of the structure shown will both members be stressed to their limit if the allowable compressive stress is two-thirds of the allowable tensile stress? Assume member *BC* does not buckle.

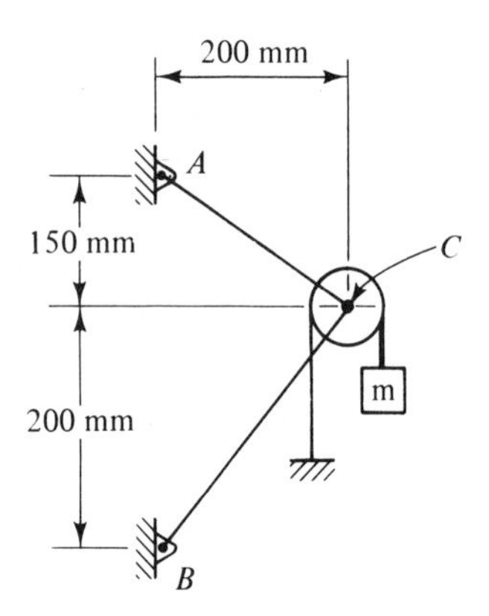

Problem 5.19

5.20 The heavy door shown is hinged along side *OA* and is held in a horizontal position by the steel rod *BD*. If the mass of the door is 500 kg, what is the required diameter of the rod? The allowable tensile stress is 150 MN/m^2.

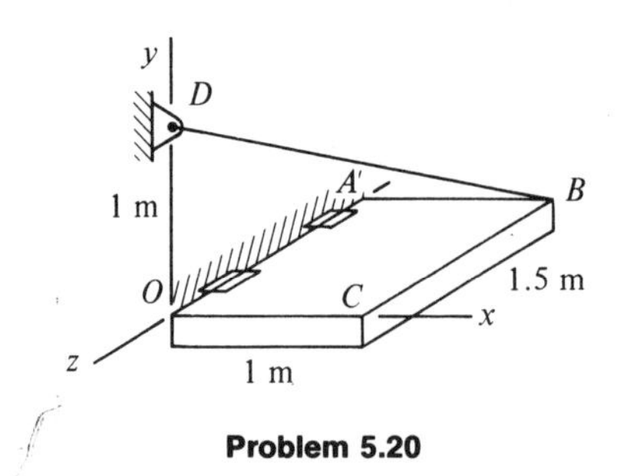

Problem 5.20

5.21 What is the maximum load *F* that can be applied to the structure shown if the allowable stress in guy wires *AD* and *BC* is 500 MN/m^2 and each has a cross-sectional area of 60 mm^2? Boom *CDE* is supported by a ball and socket joint at *E*, and the applied load *F* is parallel to the *z* axis.

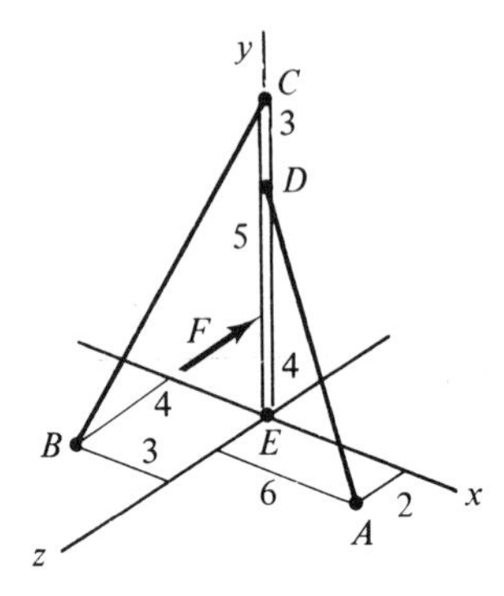

Problem 5.21

5.22 A short pedestal is made of a length of 3-in. nominal diameter steel pipe (see Table A.7) placed atop a length of 5-in. nominal diameter pipe as shown. If the allowable compressive stress is 15 ksi, what is the maximum allowable value of the load *F*?

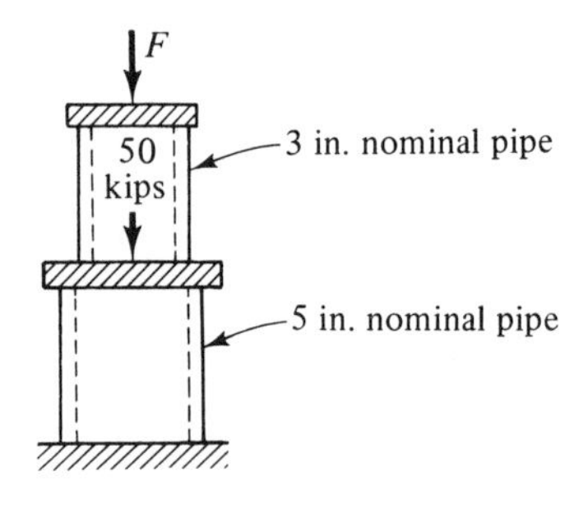

Problem 5.22

5.23 What is the lightest weight W-shape (see Table A.3) that will support the loading shown if the allowable stress is 20 ksi and (a) the weight of the member is neglected and (b) the weight of the member is included?

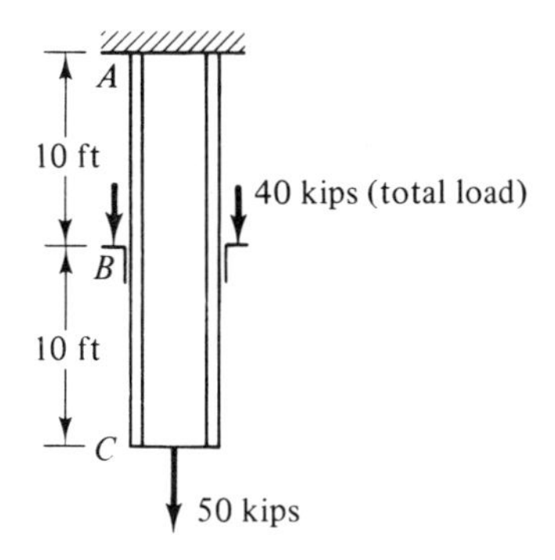

Problem 5.23

5.24 Determine the maximum normal and shear stresses in the member shown. The cross-sectional area is 0.5 in.^{2}.

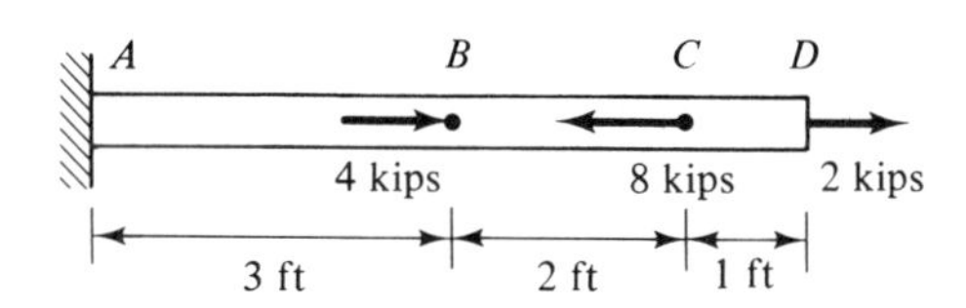

Problem 5.24

5.25 The tapered column shown is 1-ft. thick and is made of concrete with a specific weight of 150 lb/ft^3. Determine the compressive stress over the cross section as a function of the distance y from the top surface. Include the effect of the weight of the member.

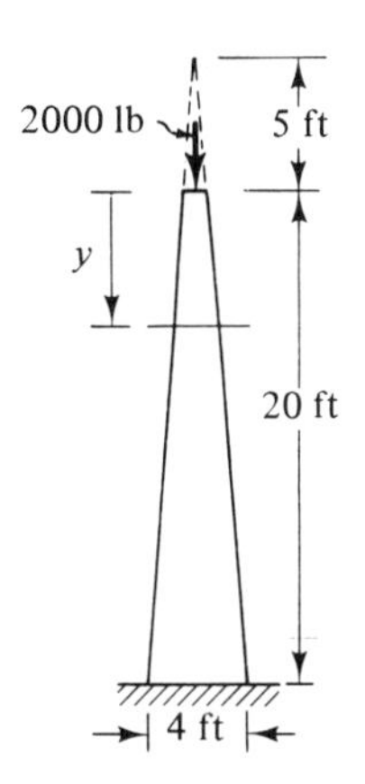

Problem 5.25

5.26 A body of revolution supports a compressive load F as shown. If the radius of the top surface is r_0 and the specific weight of the material is γ, how should the radius r vary with the distance y in order that the compressive stress at all cross sections be the same? Include the effect of the weight of the member.

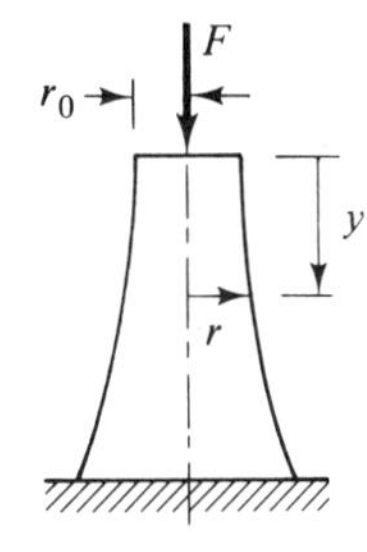

Problem 5.26

5.4 AVERAGE NORMAL, SHEAR, AND BEARING STRESSES

In many problems the actual stresses are difficult, if not impossible, to determine; therefore, it is necessary to work with average stresses. This situation is commonly encountered in structural joints where loads are transferred from one member to another. Average stresses do not necessarily provide a close estimate of the true stresses in such problems, but this can be accounted for by reducing the maximum allowable loads in order to compensate for inaccuracies in the analysis. This point will be considered in more detail in Section 5.10. The procedure for determining the average stresses is the same as that used for axially loaded members, as we shall show in the following.

For example, suppose that we wish to determine the significant stresses in the various members of the assembly shown in Figure 5.13(a). The assembly consists of

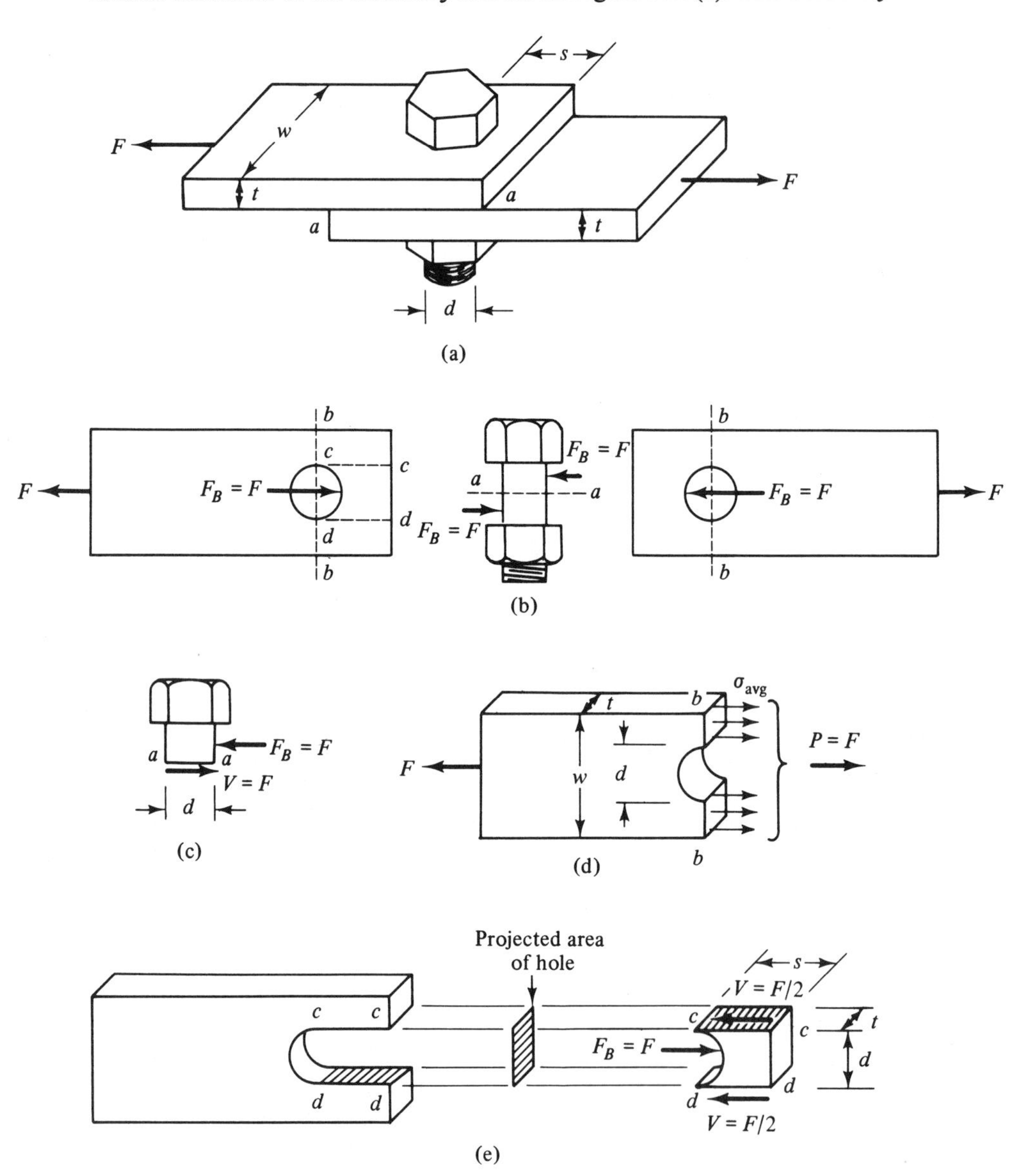

Figure 5.13

two plates connected by a bolt and loaded in tension. The most important step in the analysis of problems of this type is to draw free-body diagrams of each of the members so that the forces acting upon them can be clearly identified. Figure 5.13(b) shows the major forces acting upon the plates and upon the bolt. Other forces that may be acting, such as those due to friction or tightening of the bolt, are usually of secondary importance and can be neglected. This is fortunate, for they are very difficult to determine accurately.

Once the forces acting upon the various members have been determined, the next step is to decide just which stresses are of interest and on what planes. Obviously, the stresses that are likely to cause a failure of the joint are of primary concern. For example, since there is a possibility of shearing the bolt off along the interface between the two plates (plane *a-a*), the shear stress in the bolt on this plane is of interest.

Sectioning the bolt along plane *a-a* and considering the equilibrium of the top piece [Figure 5.13(c)], we obtain the shear force $V = F$. Thus, the average shear stress in the bolt on this plane is

$$\tau_{avg} = \frac{V}{A_{aa}} = \frac{F}{A_B}$$

where A_B is the cross-sectional area of the bolt.

In addition to shearing off the bolt, there are several other ways the connection can fail. For example, the plates may break because of the tensile stresses developed. These stresses are largest on plane *b-b* because the area available to carry the load is smallest there. Sectioning the plate along this plane [Figure 5.13(d)], we find that the average tensile stress is

$$\sigma_{avg} = \frac{P}{A_{bb}} = \frac{F}{t(w - d)}$$

where the dimensions t, w, and d are as defined in the figure. Since both plates in this example have the same dimensions, the stress will be the same in each.

It is also possible for a piece of a plate to shear out along planes *c-c* and *d-d* [Figure 5.13(e)], which would allow the plates to separate. Considering the piece removed, we see that the average shear stress on these planes is

$$\tau_{avg} = \frac{V}{A_{cc}} = \frac{V}{A_{dd}} = \frac{(F/2)}{ts}$$

where s is the distance from the center of the hole to the end of the plate.

Another, and less obvious, possible mode of failure is excessive deformation of the plates due to the contact pressure between them and the bolt. This causes the holes to elongate, and it can result in a loose joint. In some applications this would not matter so long as the joint remained intact, but in others it would. For example, if this were a rivet joint in a pressure vessel, elongation of the holes could cause the vessel to leak.

The contact pressure generated when one solid body bears against another is called the *bearing stress*, which we shall denote by σ_B. The average bearing stress is the bearing force divided by the area over which it acts:

$$\sigma_{B_{\text{avg}}} = \frac{F_{\text{Bearing}}}{A_{\text{Bearing}}} \tag{5.7}$$

The actual bearing stresses in a connection depend upon many factors, including how snugly the bolt or rivet fits the hole. Thus, it is common practice to work with an average value of bearing stress obtained by dividing the bearing force by the projected area of the hole. Referring to Figure 5.13(e), we have

$$\sigma_{B_{\text{avg}}} = \frac{F_B}{A_{\text{proj}}} = \frac{F}{td}$$

where d is the diameter of the hole and t is the thickness of the plate.

This completes our analysis of the connection. The equations obtained relate the various stresses to the applied load and dimensions of the members. Note carefully the procedure used, for it applies to all average stress problems. If the stresses and planes of interest are not specified, they can usually be identified by examining the likely modes of failure. The procedure is further illustrated in the following examples.

Example 5.5. Stresses in a Glue Joint. Three 20-mm by 100-mm boards are glued together to form the member shown in Figure 5.14(a). For an applied load of 15 kN, determine (a) the average shear stress in the glue and (b) the average bearing stress between the member and the floor.

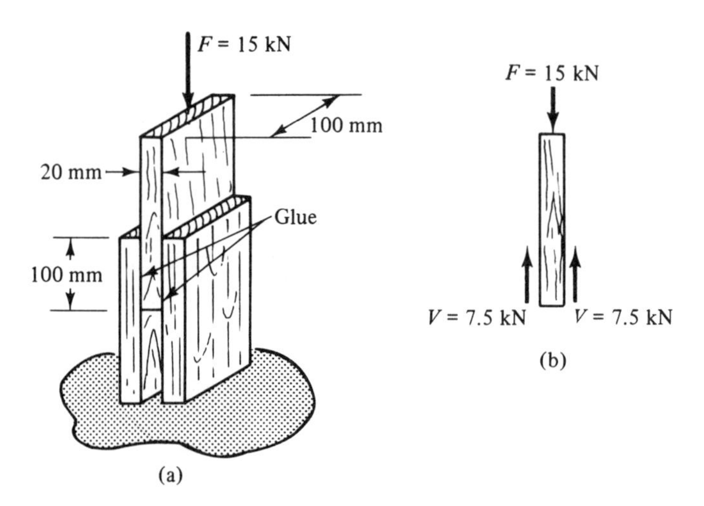

Figure 5.14

Solution. (a) From the FBD of the top board [Figure 5.14(b)], we see that the shear force in each joint is $V = 7.5$ kN, acting over a square area 100 mm on a side. Thus, the average shear stress in the glue is

$$\tau_{avg} = \frac{V}{A} = \frac{7.5\ \text{kN}}{(0.1\ \text{m})^2} = 750\ \text{kN/m}^2 \qquad \textbf{\textit{Answer}}$$

The shear stress can also be computed by dividing the total shear force acting upon both joints, $2V$, by the total joint area, $2A$.

(b) A force analysis of the entire member shows that the total contact force on the floor is 15 kN acting over an area equal to the cross-sectional area of both bottom boards. Thus,

$$\sigma_{B_{avg}} = \frac{15\ \text{kN}}{2(0.02\ \text{m})(0.10\ \text{m})} = 3.75 \times 10^3\ \text{kN/m}^2 \text{ or } 3.75\ \text{MN/m}^2 \qquad \textbf{\textit{Answer}}$$

Example 5.6. Stresses in a Connection. The steel truss member shown in Figure 5.15(a) supports a tension of 20 kips and is connected to a gusset plate by two rivets. (a) What diameter rivets should be used if the allowable average shear and bearing stresses are 15 ksi and 27 ksi, respectively? (b) What is the maximum average tensile stress in the member for this size rivet?

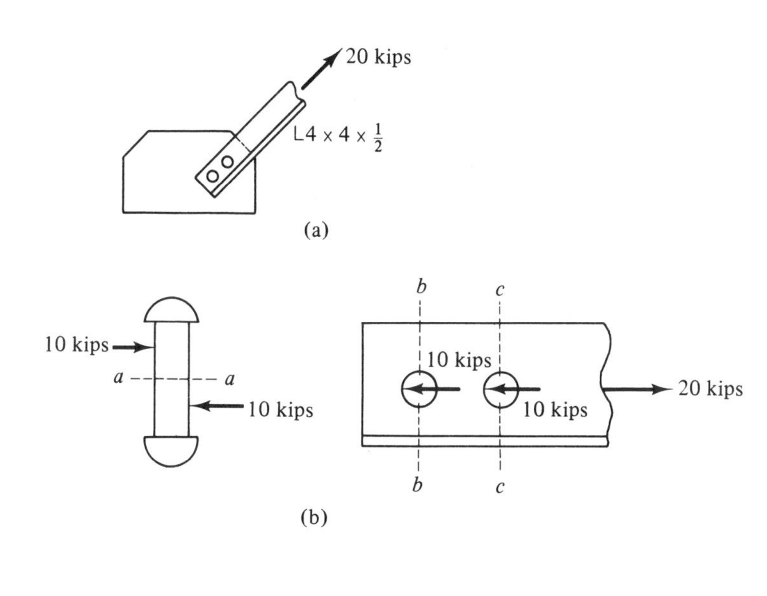

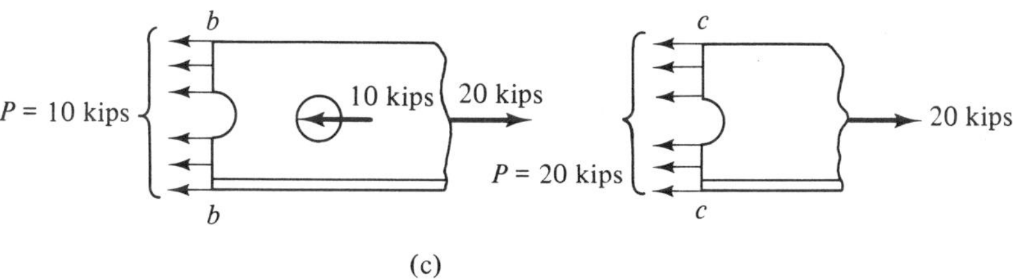

Figure 5.15

Solution. (a) When the rivets or other connecting elements are arranged in a row parallel to the direction of loading, as in this example, it is customary to assume that each supports an equal share of the load. Thus, the forces acting upon the angle and the rivets are as shown in Figure 5.15(b).

The rivets tend to shear off at the interface between the two members (plane a-a). The average shear stress in the rivets on this plane is

$$\tau_{avg} = \frac{V}{A_{rivet}} = \frac{10\ 000 \text{ lb}}{(\pi d^2/4) \text{ in.}^2} \leqslant 15\ 000 \text{ lb/in.}^2$$

from which we obtain for the diameter of the rivets

$$d \geqslant 0.92 \text{ in.}$$

The average bearing stress is the bearing force divided by the projected area of the rivet hole, which is the hole diameter times the thickness of the angle:

$$\sigma_{B_{avg}} = \frac{10\ 000 \text{ lb}}{(0.5 \text{ in.})d} \leqslant 27\ 000 \text{ lb/in.}^2$$

$$d \geqslant 0.74 \text{ in.}$$

Since both stress conditions must be satisfied, the required diameter is the larger of these two values (0.92 in.). Since rivets do not come in odd sizes, a 1-in. diameter would be used.

(b) The maximum tensile stress will occur at one or the other of the rivet holes, since the cross-section area is smallest there. The normal force is largest at the first hole [Figure 5.15(c)]; therefore, that is the location of the largest stress. From Table A.5, the cross-sectional area of the solid angle is 3.75 in.^{2} and its thickness is 0.5 in. Thus, the net area is

$$A_{net} \doteq 3.75 \text{ in.}^2 - (1 \text{ in.})(0.5 \text{ in.}) = 3.25 \text{ in.}^2$$

and

$$\sigma_{avg} = \frac{P}{A_{net}} = \frac{20\ 000 \text{ lb}}{3.25 \text{ in.}^2} = 6150 \text{ psi (tension)} \qquad \textit{Answer}$$

PROBLEMS

5.27 Two boards 100 mm wide are joined as shown with an adhesive with a shear strength of 500 kN/m^2. What is the required overlap L if the member is to support a tensile load of 5 kN?

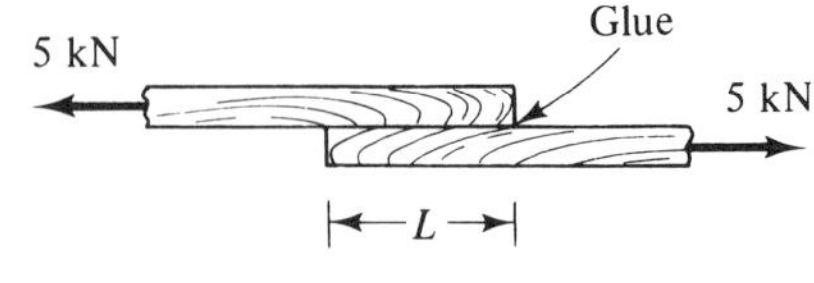

Problem 5.27

5.28 A column is fabricated by slipping a solid circular bar part way into a hollow tube and joining them by welding around the circumference as shown. What is the maximum load F that can be supported if the allowable shear stress in the weld is 60 MN/m^2 and the allowable bearing stress between the column and the concrete floor is 20 MN/m^2?

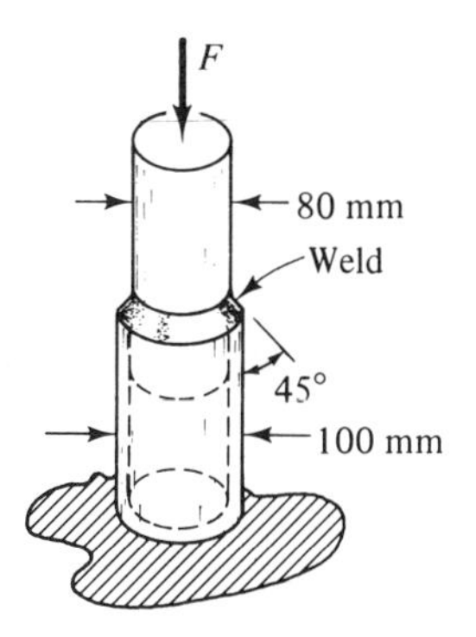

Problem 5.28

5.29 For the assembly shown, determine the lengths L_1 and L_2 of fillet weld needed so that the connection is as strong as the angle section, for which the allowable tensile stress is 20 ksi. Each weld can support a shear force of 3600 lb/in.

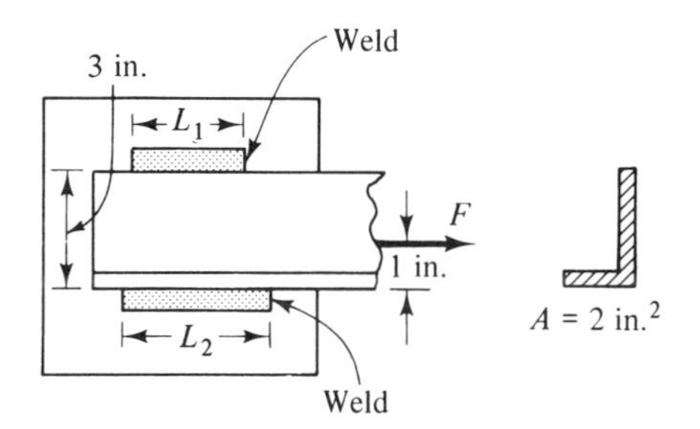

Problem 5.29

5.30 A car for elevating people up and down a steep incline is raised and lowered by a cable attached to the car with a clevis arrangement and connected to an electric winch as shown. If the mass of the car and occupants does not exceed 2 tonne and if the allowable tensile stress in the cable and shear stress in the clevis pin are 50 MN/m^2 and 35 MN/m^2, respectively, what are the required cable and pin diameters? To account for dynamic effects, double all loads.

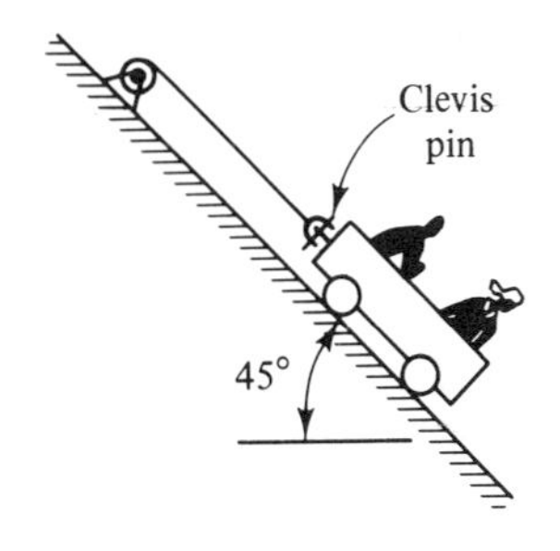

Problem 5.30

5.31 A rigid bar is supported by pins at A and B as shown. Compute the maximum shear stress in the pins and the maximum bearing stress between the pins and bar, and indicate at which pin they occur.

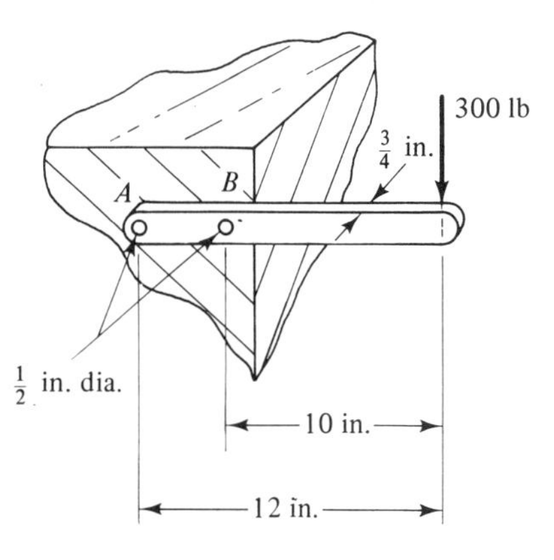

Problem 5.31

5.32 The truss shown is made of 4 in. × 4 in. nominal timbers (see Table A.12) connected at A and C by $\frac{1}{2}$ in. diameter bolts.

Determine the bearing stress between the members at these locations and the shear stress in the bolts.

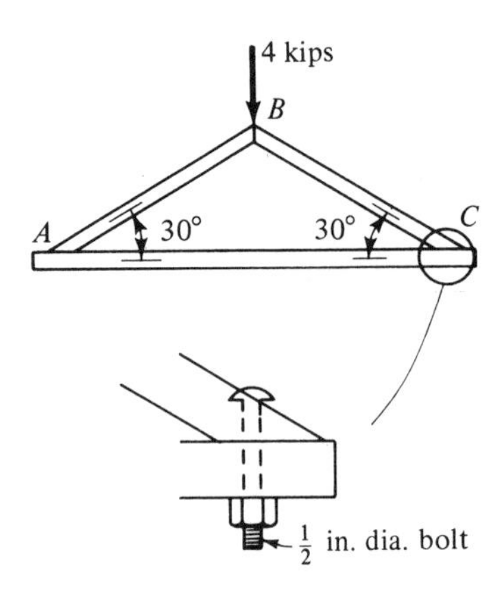

Problem 5.32

5.33 A 1-in. thick wooden bar is tested by applying loads at the ends as shown. If the ultimate tensile strength of the wood is 5000 psi and the shear strength parallel to the grain is 300 psi, what is the required length L of the end tabs if they are not to shear off before the bar breaks?

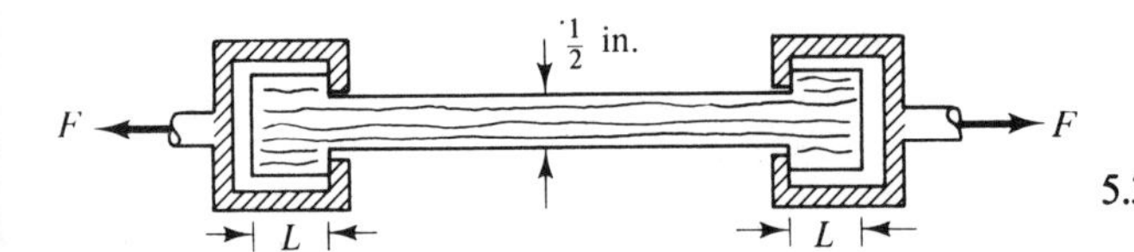

Problem 5.33

5.34 A flat bar is attached to a rigid support as shown by two $\frac{1}{2}$-in. diameter bolts that fit snugly into the holes. If the bar supports a load of 12 000 lb, determine (a) the shear stress in the bolts, (b) the maximum average tensile stress in the bar, and (c) the bearing stress between the bolts and bar.

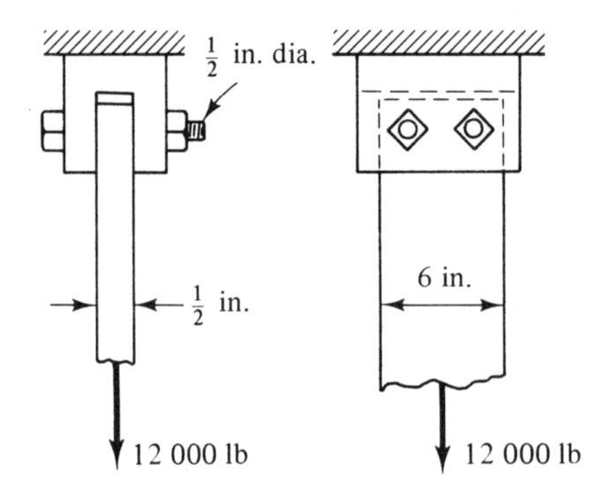

Problem 5.34

5.35 Determine the maximum load F that can be supported by the member shown. The rivets are 6 mm in diameter, and the allowable stresses are 120 MN/m^2 in tension and 60 MN/m^2 in bearing and shear.

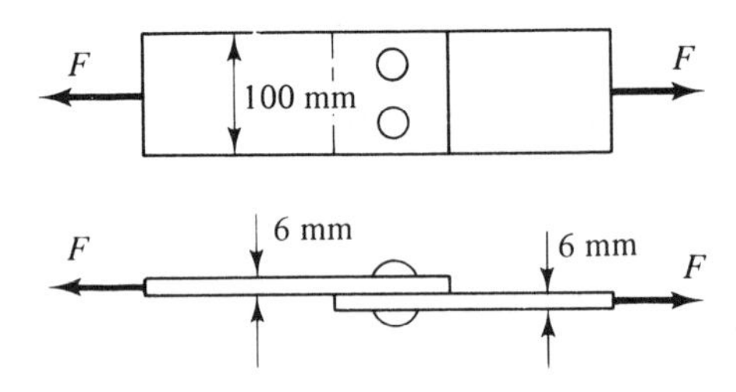

Problem 5.35

5.36 Same as Problem 5.35, except that the rivets are arranged in a longitudinal row.

5.37 A W 10 × 19 beam (see Table A.3) is attached to a wall as shown by two brackets, one on each side, made of 4 × 4 × $\frac{1}{2}$ angle sections. If the connection must support a total vertical load of 9000 lb, how many $\frac{1}{2}$-in. diameter rivets through the beam are required? The allowable shear stress is 10 000 psi. For this number of rivets, what is the bearing stress between (a) the rivets and beam and (b) between the rivets and bracket?

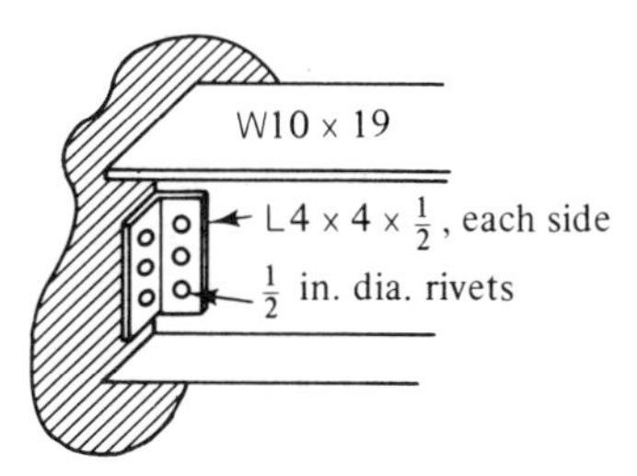

Problem 5.37

5.38 For satisfactory performance, the bar shown must not break or shear out at the pin or fail in bearing. If the allowable shear and bearing stresses are one-half the allowable tensile stress, determine the required ratios of dimensions h/d and a/d.

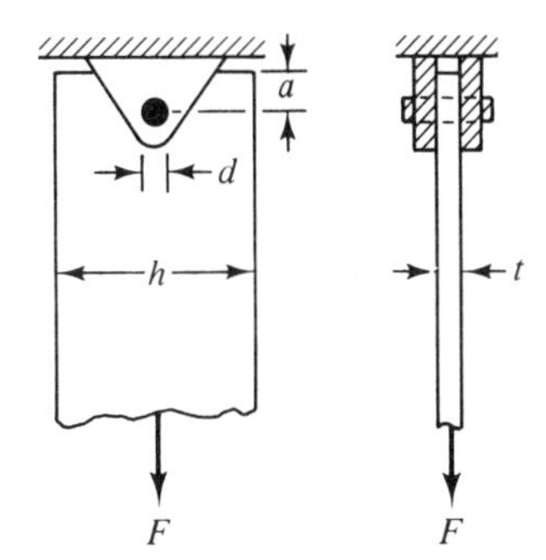

Problem 5.38

5.39 The assembly shown is advertised as a 2-ton capacity jack stand. It consists of an inner tube that telescopes into a larger tube and is held in various positions by a pin passing through both members. The inner tube, which has a wall thickness of $\frac{1}{4}$-in. and a cross-sectional area of 1.59 in.^{2}, and the pin, which is $\frac{1}{8}$-in. in diameter, are made of structural steel for which the failure stresses in compression and shear are 36 ksi and 21 ksi, respectively. Are the advertising claims justified? What simple changes could you recommend to improve the safety of the stand?

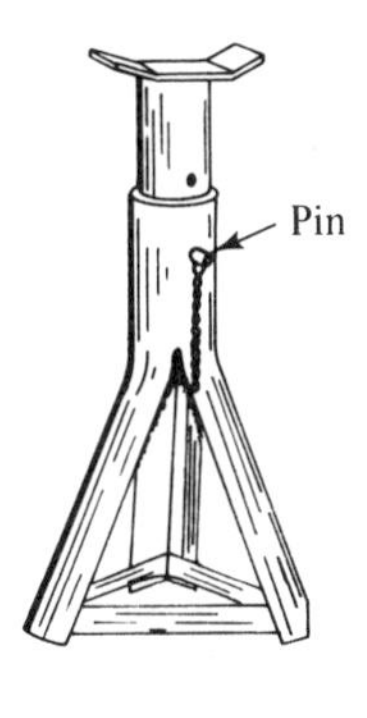

Problem 5.39

5.5 DEFORMATION AND STRAIN

Now that stress has been defined and we have developed some feel for it by considering some simple problems, we turn our attention to the *deformations* produced by the applied loads. By deformation, we mean a change in the geometry of a body. This can be a change in size, a change in shape, or both.

All bodies deform when loaded: A rubber band elongates when pulled, a diving board deflects downward when we stand at its end, a rubber ball flattens when squeezed, and a floor sags when we walk across it. The degree of deformation depends upon the resistive properties of the material and the magnitude and distribution of the applied loads. The deformations may be relatively large and visible to the naked eye, as in the case of a rubber band, or they may be very small, as in the case of a floor. For members made of very stiff materials like steel or concrete, the deformations may be so small that they can be detected only with sensitive instruments. As stated in the introduction to this chapter, we shall consider only those problems for which the deformations are small.

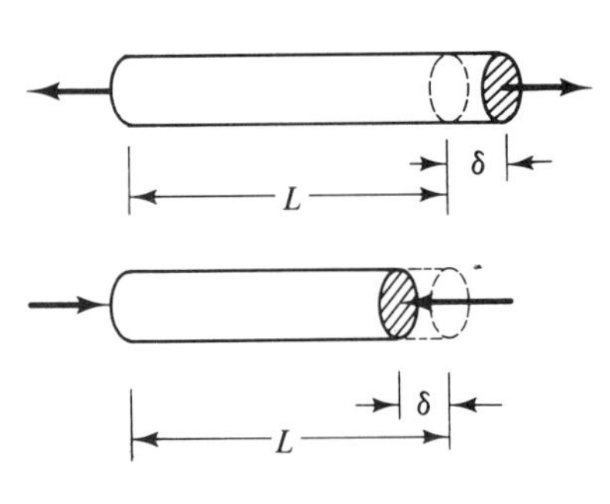

Figure 5.16

We begin our discussion by considering a rod subjected to an axial load. Under a tensile load, the member will elongate, and under a compressive load, it will shorten (Figure 5.16). The amount of elongation or shortening is not really the best measure of the tendency of the applied load to produce deformation, however, because it depends upon the original length of the member. A long rod will elongate more than a short rod of the same size and material under the same loading, as can be easily demonstrated by using rubber bands of different lengths. A better measure of the tendency of the force to produce deformation is the change in length it causes per unit of length of the rod. This ratio, which is called the *average normal strain*, is independent of the length of the member.

Denoting the elongation and normal strain by the Greek letters δ and ε, respectively, we have

$$\varepsilon_{\text{avg}} = \frac{\delta}{L} = \frac{L_{\text{final}} - L_{\text{initial}}}{L_{\text{initial}}} \tag{5.8}$$

where L is the length of the member. If the member increases in length, ε is positive and is called a *tensile strain*; if the member shortens, ε is negative and is called a *compressive strain*. This sign convention is consistent with that stated earlier for normal stresses, in that a tensile stress is associated with a tensile strain and a compressive stress is associated with a compressive strain. Physically, normal strains provide a measure of the change in size of a body when it is loaded.

Now let us consider the deformation due to a shear force. The assembly shown in Figure 5.17(a) consists of a metal plate bonded between two rubber pads attached to fixed surfaces. This arrangement is frequently used in machinery and

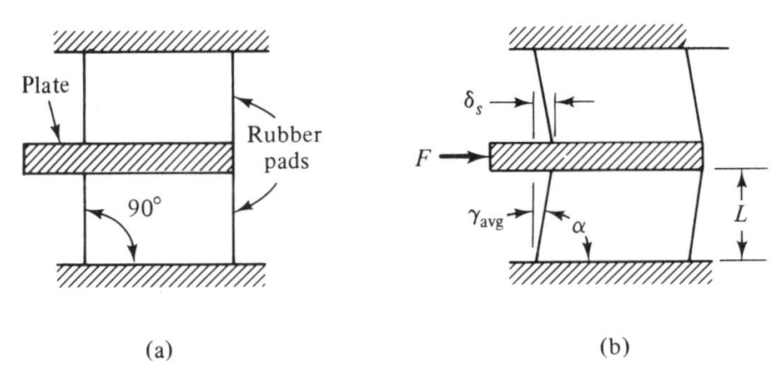

Figure 5.17

engine mounts. When a horizontal force is applied to the plate, a shear force is exerted on the pads and the pads deform approximately as indicated.

The distance δ_s that one surface of a pad moves relative to the other [Figure 5.17(b)] is called the *shear deformation*. As in the case of the axially loaded rod, the amount of deformation depends upon the dimensions of the body. In this particular example, δ_s depends upon the dimension L. However, the angle of inclination of the side of the pad is independent of L and, therefore, provides a better measure of the tendency of the shear force to produce shear deformation.

Accordingly, we define the *average shear strain* as the change in angle between the two originally perpendicular sides of the pad, measured in radians. Denoting the shear strain by the Greek letter γ, we have

$$\gamma_{avg} = \frac{\pi}{2} - \alpha \tag{5.9}$$

where α is the angle between the two sides after loading [Figure 5.17(b)]. The average shear strain can also be defined in terms of the shear deformation. Referring to Figure 5.17(b) and recognizing that the change in angle will be small for small deformations, we obtain

$$\gamma_{avg} \approx \tan(\gamma_{avg}) = \frac{\delta_s}{L} \tag{5.10}$$

Shear strains are associated with shear stresses, and they are a measure of the change in shape, or distortion, of a body due to the applied loads. They are considered positive if the angle between the two perpendicular sides decreases and negative if it increases.

The concepts of normal and shear strain developed from the preceding examples apply in general. To show this, consider two perpendicular directions n and t at a point P within a body before it is loaded [Figure 5.18(a)]. Let PQ and PR be short line segments along these directions with lengths ΔL_n and ΔL_t, respec-

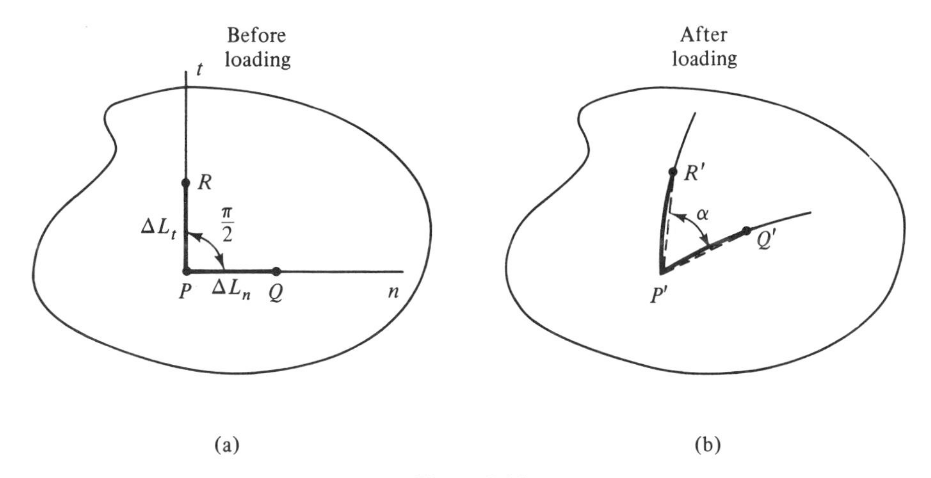

Figure 5.18

tively. The individual segments are analogous to the rod in our previous example (Figure 5.16), and the pair of perpendicular segments is analogous to the sides of the shear pad shown in Figure 5.17(a).

Upon loading, the body will deform and points P, Q, and R will move to new positions P', Q', and R' [Figure 5.18(b)]. In general, the length of the line segments will increase or decrease and the angle between them will change. The deformations that may occur can be easily demonstrated by drawing lines on a balloon and then inflating it. Notice that the original straight lines may become curved, but segments $P'Q'$ and $P'R'$ can still be considered straight because they are short. The deformations have been shown greatly exaggerated in the figure.

If we let $\Delta\delta_n$ be the change in length of line segment PQ, the average normal strain at the point P associated with the direction n is

$$\varepsilon_n(P)_{\text{avg}} = \frac{P'Q' - PQ}{PQ} = \frac{\Delta\delta_n}{\Delta L_n} \tag{5.11}$$

The *normal strain at a point* in the original direction n is defined as the limit of this ratio as ΔL_n tends to zero:

$$\varepsilon_n(P) = \lim_{\Delta L_n \to 0} \frac{\Delta\delta_n}{\Delta L_n} = \frac{d\delta_n}{dL_n} \tag{5.12}$$

The average normal strain is the same as the strain at a point, if the latter does not vary over the length of the line segment. Equations (5.11) and (5.12) hold for any direction because the choice of n is arbitrary.

The average shear strain at P associated with the perpendicular directions n and t is

$$\gamma_{nt}(P) = \frac{\pi}{2} - \alpha \tag{5.13}$$

where α is the angle between the chords $P'Q'$ and $P'R'$ [Figure 5.18(b)]. The *shear strain at a point* is then defined as

$$\gamma_{nt}(P) = \frac{\pi}{2} - \lim_{\substack{\Delta L_n \to 0 \\ \Delta L_t \to 0}} \alpha \tag{5.14}$$

Average shear strains are the same as shear strains at a point if straight lines before loading remain straight after loading. In this case, the angle α in Eqs. (5.13) and (5.14) is independent of the length of the line segments considered.

It is important to note that normal strains are always measured in a particular direction and shear strains are always measured between two perpendicular directions. Accordingly, the values of the strains at a point depend upon the directions considered, just as the values of the stresses at a point depend upon the orientation of the plane considered. Of course, the strains may also vary from point to point within the body. The directions associated with the strains are indicated by n and t, or other appropriate subscripts. For example, ε_{AB} signifies the strain of a particular line element AB.

Strains are dimensionless quantities and have the same value no matter what units of length are used. However, it is customary to express them in terms of the units in./in. or m/m. Since strains are usually very small quantities, the units μ in./in. or μ m/m are also widely used (1 μ in./in.= 10^{-6} in./in.). Strains can also be expressed as a percentage.

Example 5.7. Normal Strain of a Wire. The frame shown in Figure 5.19(a) consists of three rigid bars hinged together at the ends and braced with a wire AC. Upon loading, member BC moves 50 mm to the right. Determine the average normal strain of the wire.

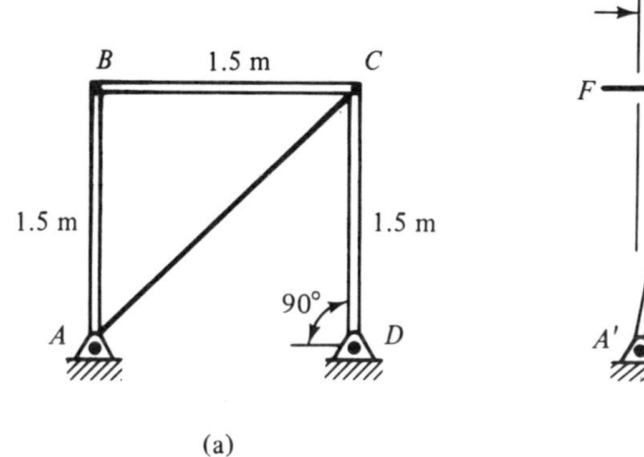

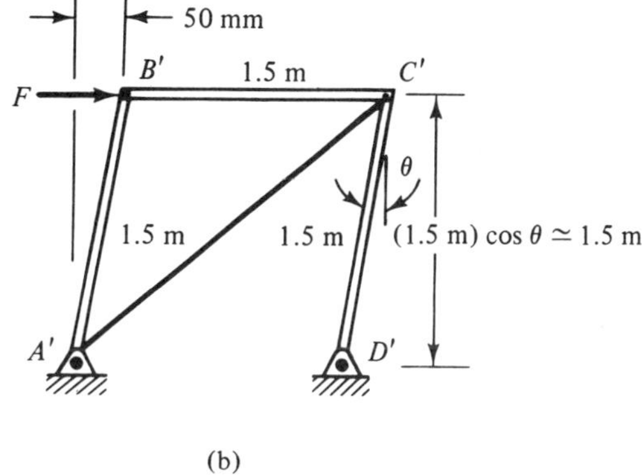

Figure 5.19

Solution. Referring to Figure 5.19(b), we have for the final length of the wire

$$A'C' = \sqrt{(1550 \text{ mm})^2 + (1500 \text{ mm})^2} = 2157 \text{ mm}$$

In this calculation the vertical coordinate of point C' is taken equal to the length of bar DC. This is a good approximation, since the angle of inclination θ of the bar is very small. Of course, the actual coordinate of C' can be computed; it is 1499.2 mm. The initial length of the wire is

$$AC = 1500\sqrt{2} \text{ mm} = 2121 \text{ mm}$$

so the average normal strain is

$$\varepsilon_{AC} = \frac{A'C' - AC}{AC} = \frac{2157 \text{ mm} - 2121 \text{ mm}}{2121 \text{ mm}}$$

$$= 0.017 \text{ or } 0.017 \text{ m/m (tension)} \qquad \textit{Answer}$$

Example 5.8. Shear Strain in a Plate. A rectangular plate [Figure 5.20(a)] distorts into a parallelogram [Figure 5.20(b)] when loaded. Determine the shear strain

γ_{xy} at points A and C. Are the values computed average shear strains or the actual shear strains at these points?

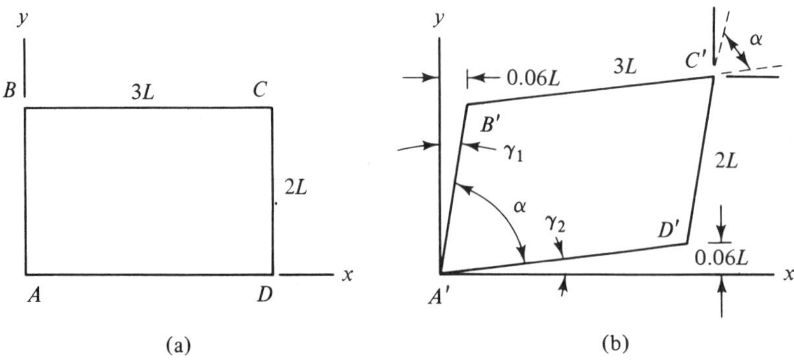

Figure 5.20

Solution. $\gamma_{xy}(A)$ is the change in angle between the positive x and y directions at A, or

$$\gamma_{xy}(A) = \gamma_1 + \gamma_2$$

Since the deformations are small,

$$\gamma_1 \approx \sin \gamma_1 = \frac{0.06L}{2L} = 0.03$$

$$\gamma_2 \approx \sin \gamma_2 = \frac{0.06L}{3L} = 0.02$$

and

$$\gamma_{xy}(A) = 0.05 \text{ radians or } 0.05 \text{ in./in.} \qquad \textbf{\textit{Answer}}$$

The shear strain at C is the change in angle between the positive x and y directions there. From Figure 5.20(b), it can be seen that

$$\gamma_{xy}(C) = \gamma_{xy}(A) \qquad \textbf{\textit{Answer}}$$

In fact, γ_{xy} has the same value at all points.

Since the sides of the plates remain straight, the angles γ_1 and γ_2 do not depend upon the lengths of the line segments used to compute them. Thus, the values given are the actual shear strains.

PROBLEMS

5.40 What is the average normal strain of a wire 4 m long if it is stretched 1 mm?

5.41 What is the change in length of a bar originally 2 m long if the average longitudinal strain is (a) 0.002 and (b) −0.003?

5.42 A spherical balloon with a radius of 4 in. is inflated until the radius is 6 in. What is the average normal strain in the circumferential direction?

5.43 In the dark of night, a burglar runs into a

clothesline and stretches it 1 ft laterally at its center as shown. Determine the average normal strain of the line.

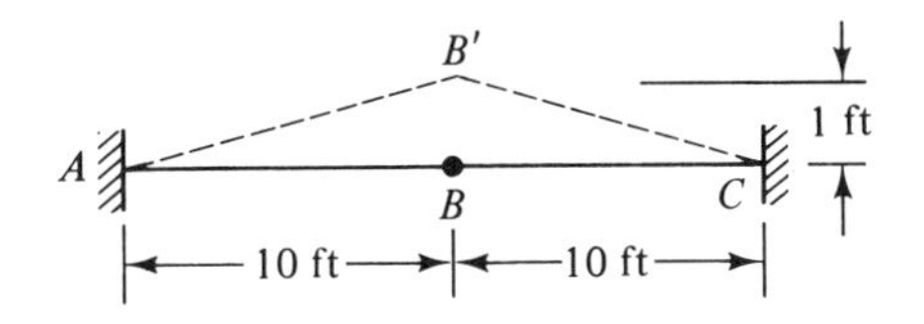

Problem 5.43

5.44 Determine the change in volume of a cube of steel 10 in. on a side which is heated until the normal strain along each side is 0.0004 in./in.

5.45 A vertical wire BD is attached to the midpoint of a stretched horizontal cable AC as shown. When a block is suspended from end D of the wire, B moves to B' and D moves to D'. What is the displacement DD' of the block if the strain of the horizontal cable is 0.005 m/m and the strain of the vertical wire is 0.010 m/m?

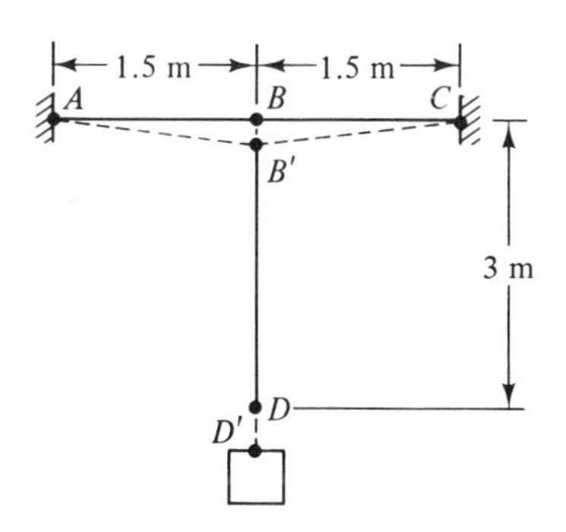

Problem 5.45

5.46 The weight of the rigid slab shown is such that the average normal strains of the outer posts are 500×10^{-6} m/m. What is the strain of the middle post if it was originally 0.2 mm shorter than the other two, which were 1 m long?

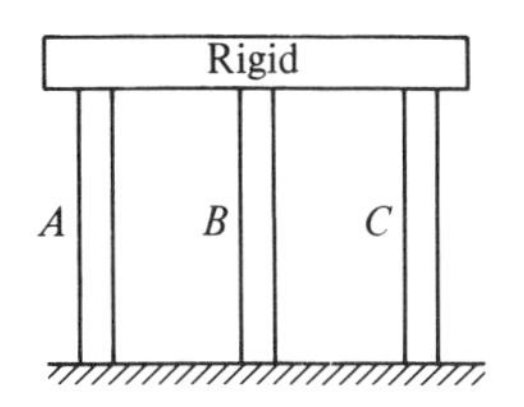

Problem 5.46

5.47 If the rigid bar shown rotates downward through a small angle θ when loaded, determine the ratio of the average normal strains of the support wires BD and CE.

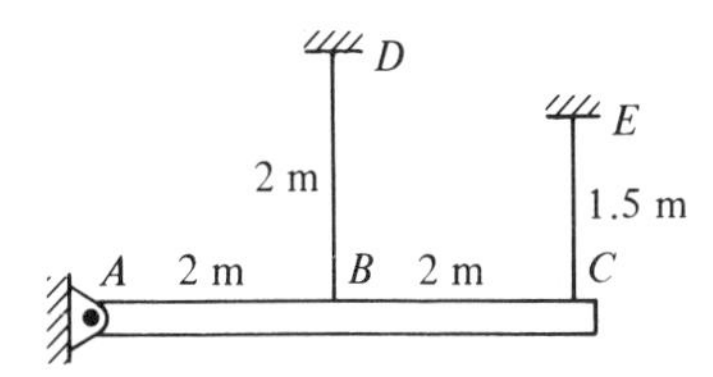

Problem 5.47

5.48 A motor mount made of a rectangular pad of rubber deforms as shown when loaded. What is the shear strain γ_{xy}?

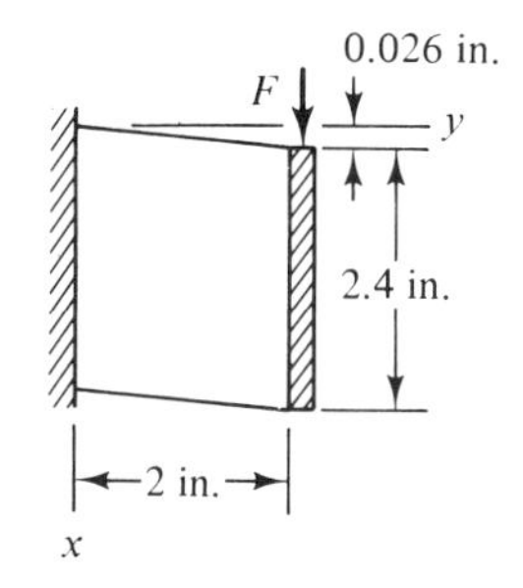

Problem 5.48

5.49 A thin triangular plate deforms as shown when loaded. Determine the shear strain γ_{nt} and the normal strain ε_n.

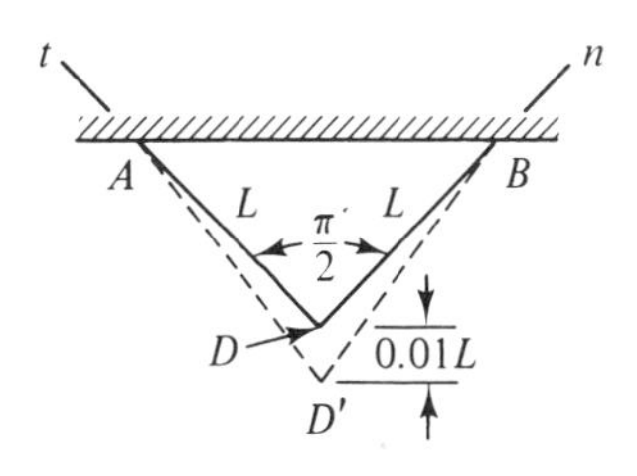

Problem 5.49

5.50 When a rectangular block is subjected to a shear force, the originally straight sides become curves described by the equation $x = ay^2/h^2$, where a and h are as defined in the figure. Determine the shear strain γ_{xy} at (a) any point x, y in the block, (b) the origin $x = y = 0$, and (c) the center of the block $x = b/2$, $y = h/2$.

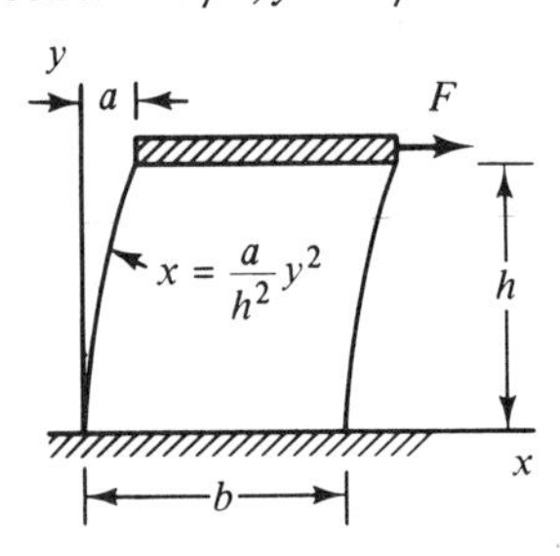

Problem 5.50

5.51 A 0.50-in. diameter circle drawn on a beam before it is loaded distorts into an ellipse after loading, as shown. If the major and minor axes of the ellipse are 0.52 in.

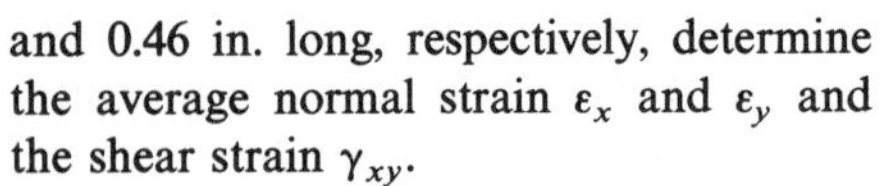
and 0.46 in. long, respectively, determine the average normal strain ε_x and ε_y and the shear strain γ_{xy}.

5.52 When loaded, the square plate shown deforms in such a way that diagonals AC and BD remain perpendicular and side AB remains horizontal. It is observed that diagonal AC contracts 0.04 in. and that diagonal BD elongates 0.02 in. Determine the average values of the strains ε_x, ε_y, and γ_{xy}.

5.53 A rectangle 15.000 mm long × 5.000 mm wide was scribed on the surface of an unloaded member. When the member was loaded, the long sides of the rectangle rotated counterclockwise through an angle of 0.067° and lengthened 0.009 mm, while the short sides rotated clockwise through an angle of 0.033° and shortened 0.002 mm. If the x axis is along the original long side of the rectangle and the y axis is along the short side, what are the average values of the strains ε_x, ε_y, and γ_{xy}?

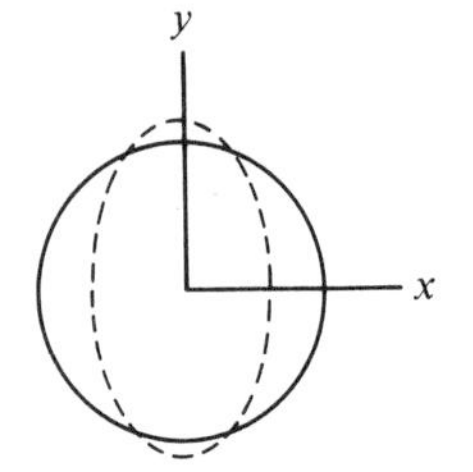

Problem 5.51

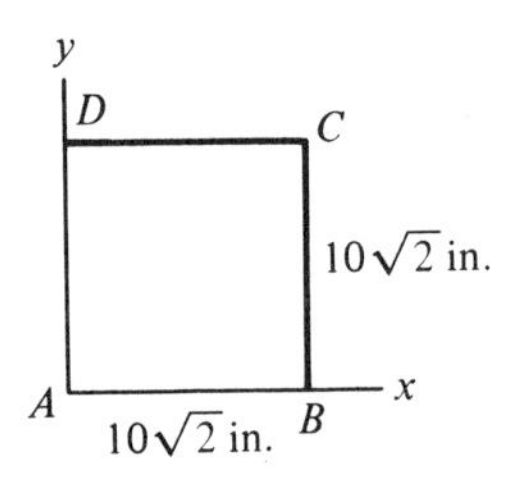

Problem 5.52

5.6 MATERIAL TESTS AND DATA REPRESENTATION

The remaining basic ingredient in our study of the mechanics of deformable bodies is the resistive properties of materials. These properties relate the stresses to the strains and can only be determined by experiment. In this and the following two sections, we shall consider the behavior of some common engineering materials under tensile, compressive, and shear loadings and the tests for determining this behavior. Our primary concern will be the description of observed material responses. The metallurgical and other physical and chemical factors associated with these responses will not be considered in any detail. They are discussed in texts on materials science.

Tensile and compressive tests. One of the simplest tests for determining mechanical properties of a material is the so-called *tensile test*. In this test, a load is applied along the longitudinal axis of a rod or bar, usually with rectangular or circular cross section. The applied load and the resulting elongation of the member are measured. The load is then increased and the process repeated until the desired load levels are reached or the member breaks. The procedure used in *compressive tests* is basically the same, except that the member is loaded in compression.

There are many different testing machines available for performing tensile and compressive tests. All of them involve a fixed and a movable surface between which the specimen can be stretched or compressed. The testing machine shown in Figure 5.21 is typical. The top and bottom crossheads are driven upward as a unit by a hydraulic cylinder; the middle crosshead remains stationary. Thus, a specimen

Figure 5.21. Tinius Olsen 60,000 lb Standard Super "L" Universal Testing Machine. (*Courtesy* Tinius Olsen Testing Machine Co., Inc.)

gripped between the top and middle heads will be loaded in tension; one placed between the central and bottom heads will be compressed. The load applied to the specimen is determined by components internal to the machine and is displayed on a load dial. Many of the testing machines in use today are electronically controlled and can subject a specimen to arbitrary loading or deformation histories. Automatic data recording systems are also widely used.

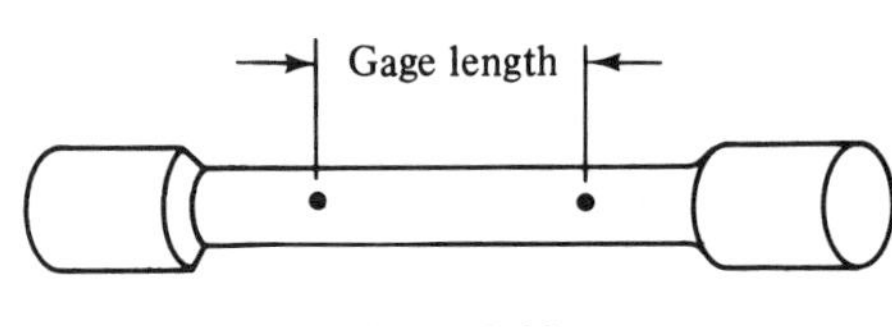

Figure 5.22

Standards governing specimen dimensions and test procedures have been developed by the American Society for Testing Materials (ASTM) and other agencies to assure some degree of uniformity between data obtained in different laboratories. A typical tensile specimen is shown in Figure 5.22. It has an enlarged portion on each end for gripping in the testing machine and a uniform central test section of smaller size. The test section is designed as an axially loaded member; therefore, the stresses and strains within it can be easily computed. Before the specimen is loaded, a known distance called the *gage length* is marked off in the center of the test section. It is the elongation or shortening of this gage length which is measured. One-inch and 2-inch gage lengths are common for metal specimens, but other lengths can be used.

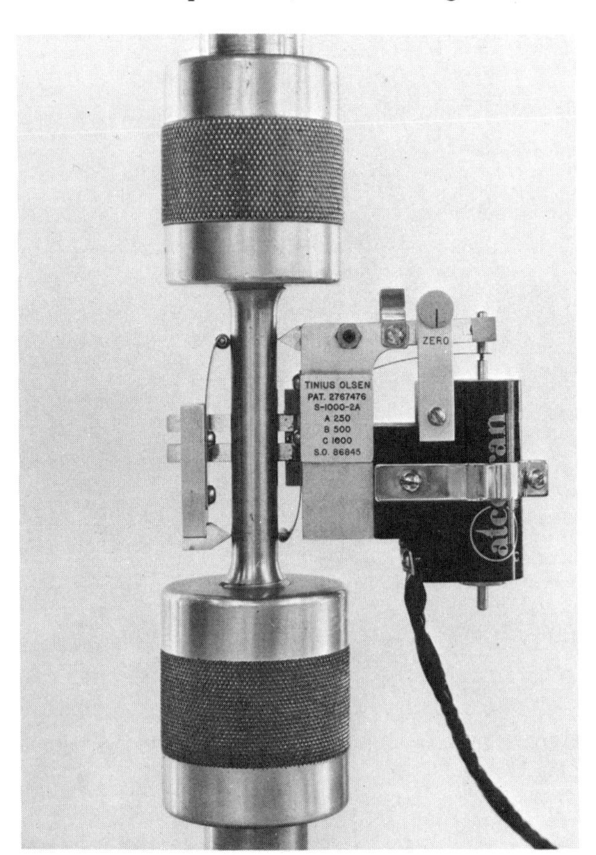

Figure 5.23. Tinius Olsen Model S-1000 Averaging Type Extensometer shown in place on a standard specimen. (*Courtesy* Tinius Olsen Testing Machine Co., Inc.)

The deformations are usually very small, particularly at low load levels, and must be magnified in order to be measured accurately. There are numerous instruments available for making these measurements. The instrument shown in Figure 5.23 is held in place by leaf springs, and it has two knife edges 2 in. apart that establish the gage length. One knife edge remains fixed and the other moves along with the specimen as it deforms. The deformations are magnified electronically. The movement of the knife edge moves the core of the differential transformer, which creates an electrical signal proportional to the deformation. This signal can be used for automatic recording of the load-deformation data. Other devices and techniques for measuring deformations will be discussed in Chapter 11.

Tension and compression members also undergo changes in cross-sectional dimensions when loaded. Tension members contract; compression members expand. This phenomenon is called the *Poisson effect* in honor of S. D. Poisson (1781–1840), who studied it analytically. This effect can be easily demonstrated by stretching a large rubber band or squeezing a rubber ball. For constructional materials, the changes in the transverse dimensions are relatively small. They can be measured with a micrometer or devices similar to those used to determine the longitudinal deformations.

Stress-strain diagrams. Load-deformation data obtained from tensile or compressive tests do not give a direct indication of the material behavior, because they also depend upon the specimen geometry. The load required to produce a certain amount of deformation obviously will depend upon the size of the member. Similarly, the change in the gage length at a given load will depend upon the gage length used; the longer the gage length, the greater the change. This dependence upon specimen geometry is evident in Figure 5.24(a), which shows a portion of the load-deformation curves for aluminum tensile specimens with different cross-sectional areas and gage lengths. All specimens were taken from the same bar of material.

Figure 5.24(b) shows the same data with the loads and deformations converted to stresses and strains via the relationships

$$\sigma = \frac{F}{A} \qquad \varepsilon = \frac{\delta}{L}$$

Here, σ is the normal stress on a plane perpendicular to the longitudinal axis of the specimen, ε is the normal strain in the longitudinal direction, F is the applied load, A is the original cross-sectional area, δ is the change in the gage length, and L is the original gage length.

The fact that the data now fall along a single curve indicates that this conversion eliminates the effects of specimen size and gage length. Consequently, the resulting *stress-strain curve*, or *diagram*, gives a direct indication of the material properties. This result is of no little consequence. It confirms our earlier statements that force per unit area (stress) and deformation per unit length (strain) are the appropriate measures for expressing the effects of applied loadings upon material bodies.

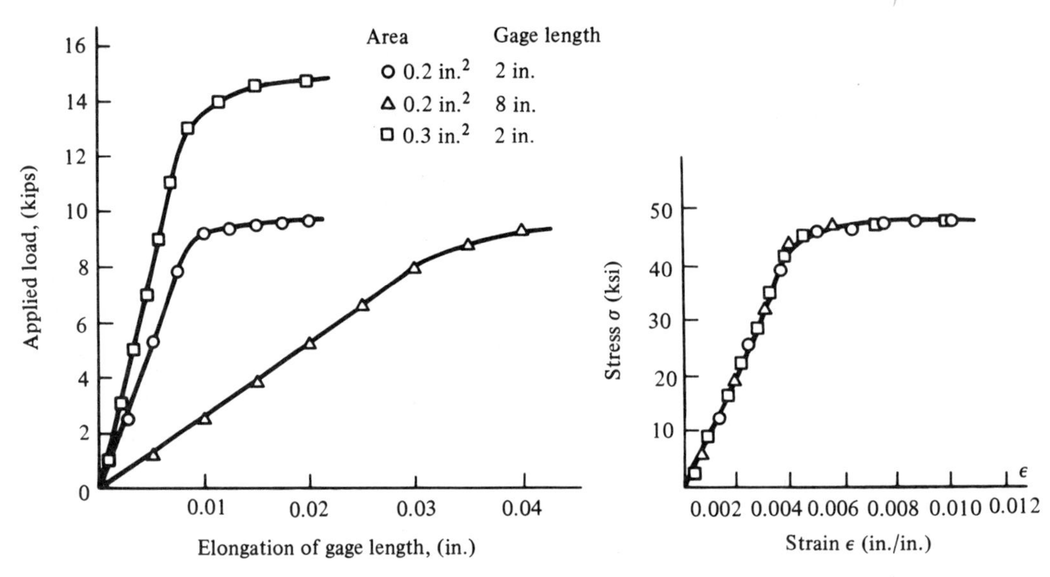

Figure 5.24. Load–Deformation Relations. and (a) Stress–Strain Relation (b) For 2017-T451 Aluminum Alloy

Note that the stress-strain diagram is based upon the original cross-sectional area and gage length, even though these quantities change continuously as the test progresses. These changes have a negligible effect, except possibly during the final stages of the test before the specimen breaks. Moreover, since the job of design engineers is to determine the necessary original dimensions of members, material data based upon the so-called *engineering stress* and *engineering strain* computed by using the original specimen dimensions are of the most use to them. Material scientists concerned with the detailed responses of materials need to know the precise values of stress and strain to which the material is subjected at each instant. Accordingly, they usually work with the so-called *true stress* and *true strain* based upon the instantaneous values of area and gage length. All of the problems considered in this text involve only engineering stress and strain.

Shear tests. Material behavior in shear can be determined by twisting a specimen in the form of a thin-walled tube. By measuring the applied couple and the resulting angle of twist, the shear stresses and strains can be determined and a shear stress-shear strain diagram can be constructed. A more detailed discussion of this test must await our study of torsion in Chapter 7.

5.7. RESISTIVE BEHAVIOR OF MATERIALS

Different materials exhibit different responses, and these responses are reflected in the shapes of their stress-strain diagrams. The responses typical of common structural materials subjected to slowly applied tensile, compressive, and shear loadings are discussed in this section. Emphasis will be placed upon tensile behavior, for the behavior in compression and shear is similar in many respects.

Tensile behavior. Figure 5.25 shows complete tensile stress-strain curves for structural steel, Plexiglas, and cast iron. These curves, and those for most other structural materials loaded in tension, fall into one of the three general categories shown in Figure 5.26.

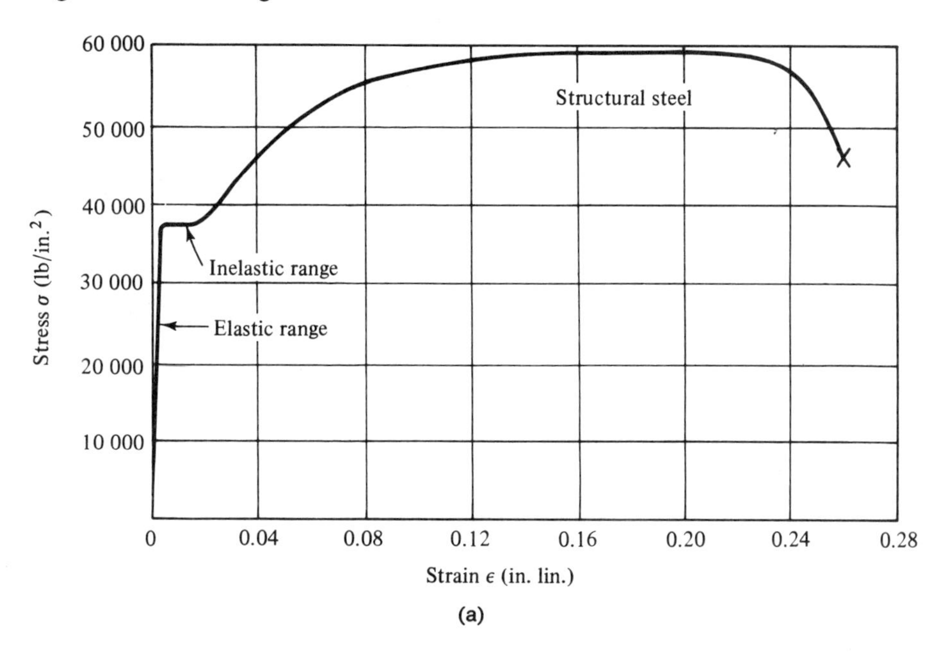

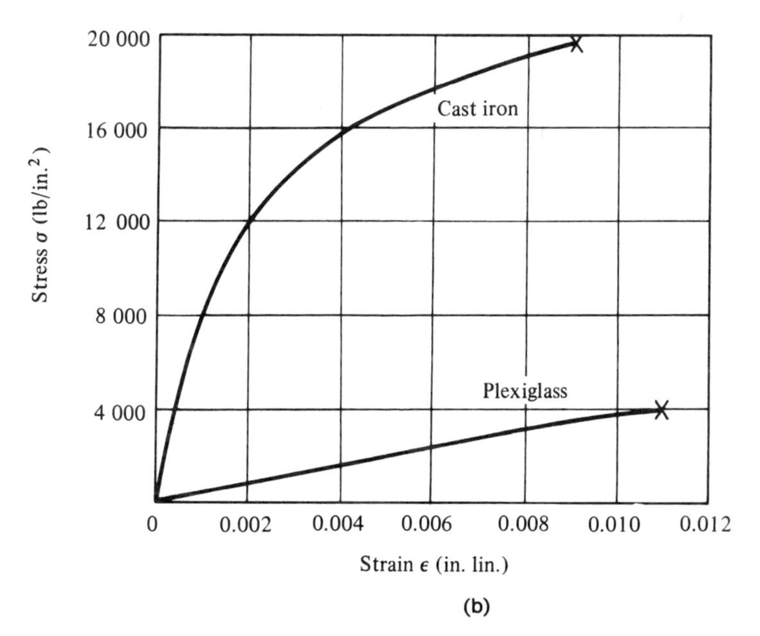

Figure 5.25. Behavior of Structural Steel (a) Plexiglass and Cast Iron (b) In Tension

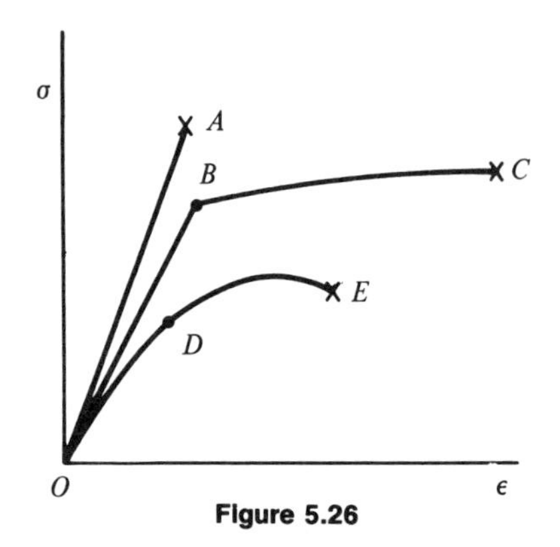

Figure 5.26

Diagram OA in Figure 5.26 is very nearly linear up to the stress at which the specimen breaks (denoted by ×). Glass and certain plastics have stress-strain curves like this one. Diagram OBC is typical of metals such as steel, aluminum, and brass. The stress is proportional to strain up to point B, at which the material is said to *yield*. Beyond this point, the strain increases much more rapidly than the stress. Materials such as cast iron and concrete (in compression) have curves with the shape ODE. The stress-strain diagrams for these materials are characterized by the fact that they have no well-defined linear region.

There are two distinct ranges of material response: the *elastic range* and the *inelastic*, or *plastic*, *range*. In the elastic range (OA, OB, or OD in Figure 5.26), the transmission of force from point to point within the material is reversible. Ideally speaking, if a material is loaded within the elastic range and then unloaded, each atom will regain its original equilibrium position and the entire member will return to its original dimensions. As can be seen from Figure 5.25(a), the elastic range corresponds to relatively small strains. Nevertheless, most members are designed so that this range is not exceeded under normal circumstances.

The inelastic range (BC or DE in Figure 5.26) is caused by a disruption of the bonds between atoms in some localized regions within the material. When this happens, shear deformations and slip between atoms occur, and some new bonds are established. This phenomenon is not completely reversible, and the member will not return to its original dimensions when unloaded. This behavior is clearly illustrated in Figure 5.27, which shows the loading and unloading behavior of an aluminum alloy in tension. The strain that remains after the load is removed is called the *permanent strain*. Experiments show that for most metals the stress-strain

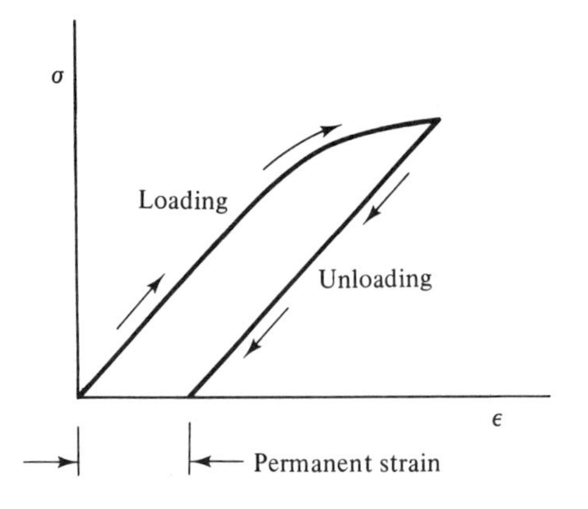

Figure 5.27. Loading-Unloading Behavior of an Aluminum Alloy in Tension.

curve for unloading is approximately linear and parallel to the initial straight line portion of the curve for loading, although not always to the degree indicated in Figure 5.27.

If the load on the specimen is continously increased, the phenomenon of disruption and reestablishment of atomic bonds is eventually propagated throughout the member. A limit of inelastic deformation is reached when no new bonds are established, with the result that the specimen fractures or separates. These limits are shown as points A, C, or E in Figure 5.26.

Material behavior can also be categorized according to the amount of inelastic deformation that occurs prior to fracture. Materials such as glass, plaster, and cast iron which exhibit very little inelastic deformation are said to be *brittle*. Steel, aluminum, and other materials which undergo comparatively large inelastic deformations are called *ductile*.

Brittle materials tend to fail suddenly with little warning when they are overloaded. In contrast, ductile materials yield and continue to support the load while undergoing inelastic deformation. This ability to deform inelastically helps prevent catastrophic accidents, and it is one of the reasons metals such as steel and aluminum are so outstanding as construction materials. Ductility is also an important factor in metalforming operations.

The fracture of brittle materials is associated with tensile stresses, but ductile materials tend to fail as a result of the shear stresses that develop. This is illustrated in Figure 5.28, which shows the fracture patterns for a brittle and a ductile material. The brittle material [Figure 5.28(a)] breaks normal to the axis of the specimen because this is the plane of maximum tensile stress. Ductile materials

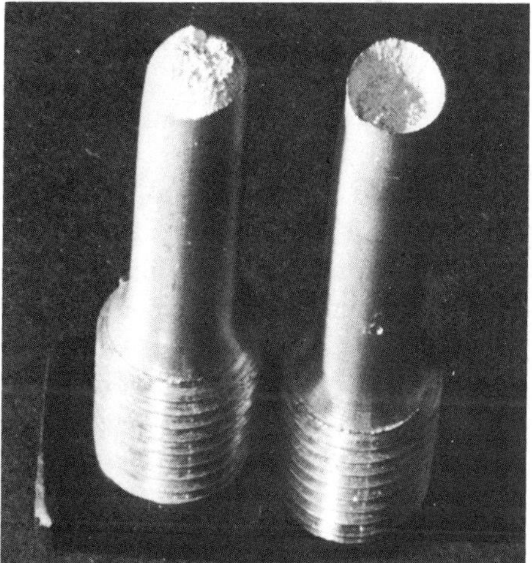

Figure 5.28. Typical Fracture Patterns of (a) Brittle and (b) Ductile Materials in Tension.

usually exhibit a cup-and-cone type fracture [Figure 5.28(b)]. The sides of the cup-and-cone portions are inclined at approximately 45° to the specimen axis, which corresponds to the planes of maximum shear stress.

The reduced cross section of the specimen shown in Figure 5.28(b) in the vicinity of the fracture is due to a phenomenon called *necking*. This localized reduction in area is responsible for the decrease in stress beyond the maximum point on the stress-strain diagram [see Figure 5.25(a), for example]. The load required to continue the deformation decreases as the specimen necks down because of the smaller cross-sectional area. Since the stress-strain curve is based upon the original area of the specimen, it also reflects this decrease.

Compressive behavior. Materials that behave in a ductile fashion in tension usually exhibit approximately the same response in compression, well into the inelastic range, but the later stages of the tensile and compressive curves differ significantly. Since compression specimens expand instead of necking down, the compressive stress-strain curve continues to rise instead of reaching a maximum and then dropping off. Materials that are brittle in tension usually exhibit ductile behavior in compression. Most structural materials, brittle as well as ductile, behave approximately the same in tension and in compression over the elastic range.

Behavior in shear. The shear stress-shear strain curve for most materials closely resembles the tensile curve. Of course, the magnitudes of the stresses and strains involved are different. The magnitude of the shear stresses is generally approximately one-half that of the corresponding tensile stresses. For example, a material that yields at a certain stress in tension will yield at approximately one-half that stress in shear.

Environmental and other factors. In concluding this section, it is important that our comments on material behavior be put in proper perspective. Material behavior is not absolute; it depends upon many factors. Temperature, rate of loading, number of repetitions of the load, and the state of stress all influence material behavior. Other possible factors are irradiation or exposure to corrosive environments.

Materials may also exhibit certain directional effects. For example, the behavior of wood is different when loaded parallel to the grain than when loaded normal to the grain. Materials whose properties depend upon direction are said to be *anisotropic*; materials whose properties are the same in all directions are called *isotropic*.

Most materials exhibit directional effects if the specimens tested are sufficiently small. For example, the individual crystals of a metal are anisotropic. However, these effects tend to cancel in bodies of larger size because of the large number of randomly oriented crystals involved. As a result, the material may behave anisotropically on a microscopic scale and isotropically on a macroscopic scale. Anisotropy can also be produced by fabrication processes such as forging and rolling, which tend to give a preferred alignment to the grains of metals. Anisotropy is often introduced intentionally to enhance the material performance, as in fiber reinforced composite materials and steel reinforced concrete.

Nonhomogeneous materials have properties that vary from point to point. Most structural materials are nonhomogeneous on a microscopic scale, but they are reasonably homogeneous on a macroscopic scale.

Although it is important to be aware of the existence of effects such as those mentioned in the preceding, their incorporation into the analysis of the response of load-carrying members is beyond the scope of this book. Unless stated otherwise, we shall assume that (1) the materials considered are homogeneous and isotropic, at least to a reasonable degree of approximation; (2) all loads are slowly applied and nonrepetitive; and (3) only moderate temperatures are involved. The results presented will apply in a large percentage of cases.

5.8 MATERIAL PROPERTIES AND STRESS-STRAIN RELATIONS

Material responses can be described quantitatively in terms of certain mechanical properties and by equations that represent the stress-strain curves. This quantitative description, which is necessary for computational purposes, is discussed in this section.

Material properties. Most of the mechanical properties of interest can be determined directly from the stress-strain diagram. In discussing these properties we shall refer to Figure 5.29, which shows stress-strain curves typical of those for mild steel and aluminum in tension. Except where noted, the definitions given also apply for compression and shear. The symbols used to denote the various quantities, if any, are given in parentheses after the name of the property.

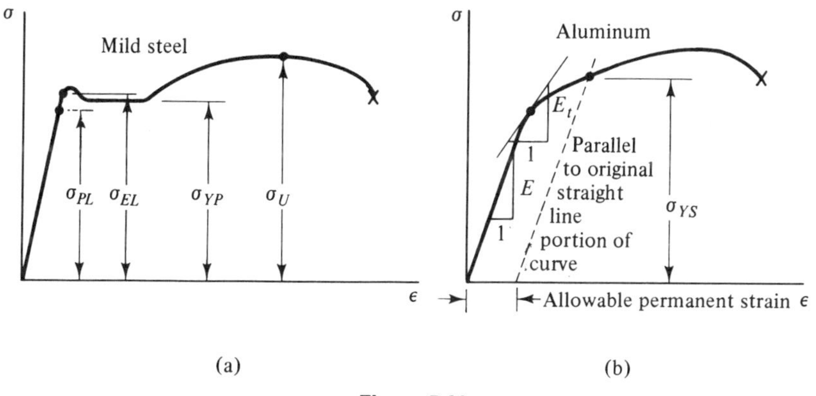

Figure 5.29

Among the properties of interest are the following:

1. *Proportional Limit* (σ_{PL} or τ_{PL}). The stress above which stress is no longer proportional to strain.

2. *Modulus of Elasticity* (E). Slope of the initial linear portion of the stress-strain diagram. This quantity has the same units as stress and is a measure of the material stiffness; the larger the value of E, the larger the stress required to produce a given strain. For the common structural metals, the value of E ranges from approximately 10×10^6 psi (69×10^9 N/m^2) for aluminum to 30×10^6 psi (207×10^9 N/m^2) for steel.

3. *Tangent Modulus* (E_t). Slope of the stress-strain curve beyond the proportional limit. There is no one single value of E_t; it varies with the amount of strain. The value of E_t at the origin is used as the modulus of elasticity for materials such as cast iron whose stress-strain curves have no well-defined linear range.

4. *Shear Modulus* (G). Slope of the initial linear portion of the shear stress-shear strain curve. This quantity is a measure of the material stiffness in shear and has the same units as stress. Typical values of G are 4×10^6 psi (28×10^9 N/m^2) for aluminum and 12×10^6 psi (83×10^9 N/m^2) for steel.

5. *Elastic Limit* (σ_{EL} or τ_{EL}). The maximum stress that can be applied without resulting in permanent deformation upon unloading. The elastic limit is difficult to determine experimentally; the proportional limit is usually used in place of the elastic limit because they have very nearly the same values for most materials.

6. *Yield Point* (σ_{YP} or τ_{YP}). The stress at which there are large increases in strain for little or no increase in stress. Among the common structural materials, only steel exhibits this type of response.

7. *Yield Strength* (σ_{YS} or τ_{YS}). The maximum stress that can be applied without exceeding a specified value of permanent strain (usually 0.2%) upon unloading. The procedure for determining the yield strength is illustrated in Figure 5.29(b). The yield strength is not really a material property because it depends upon the amount of permanent strain allowed. It is used for materials that do not have a yield point to define a stress above which excessive inelastic deformations will occur.

8. *Ultimate Strength* (σ_U or τ_U). The maximum stress the material will withstand.

9. *Percent Elongation*. The strain at fracture in tension, expressed as a percentage. This is a measure of ductility. The larger the percent elongation, the more ductile the material. If the complete stress-strain curve is not available, the percent elongation can be computed from the definition of average strain:

$$\text{Percent elongation} = \frac{\text{final gage length} - \text{initial gage length}}{\text{initial gage length}} \times 100 \tag{5.15}$$

The broken specimen must be pieced back together in order to measure the final gage length.

10. *Percent Reduction of Area*. The reduction in cross-sectional area of a tensile specimen at fracture, expressed as a percentage. This quantity is another measure of ductility. It is defined by the relationship

$$\text{Percent reduction area} = \frac{A_{\text{initial}} - A_{\text{final}}}{A_{\text{initial}}} \times 100 \tag{5.16}$$

The final area is measured at the fracture site where the cross section is smallest. The percent reduction of area cannot be determined from the stress-strain diagram.

11. *Poisson's Ratio* (ν). The negative of the ratio of the strain in the transverse direction (ε_t) to that in the longitudinal direction (ε_L) for a tension or compression specimen:

$$\nu = -\frac{\varepsilon_t}{\varepsilon_L} \tag{5.17}$$

This ratio which is a measure of the strain due to the Poisson effect is a positive constant for stresses below the elastic limit. The negative sign is required for consistency with the experimental observation that when a member is stretched ($\varepsilon_L > 0$) it contracts ($\varepsilon_t < 0$) and that when it is compressed ($\varepsilon_L < 0$) it expands ($\varepsilon_t > 0$). The value of ν cannot be determined from the stress-strain diagram. A separate measurement of the transverse strain is required. The value of ν for the common structural metals lies in the range from 0.27 to 0.35 for stresses below the elastic limit.

Relationship between E, G, and ν. For isotropic materials, the modulus of elasticity, shear modulus, and Poisson's ratio are related via the expression

$$G = \frac{E}{2(1 + \nu)} \tag{5.18}$$

Proof of this relationship will be given in Chapter 9 (Section 9.7).

Statistical nature of material properties. Supposedly identical tests do not necessarily yield the same values for the material properties. This may be because there are natural variations in the material from specimen to specimen, differences in the loading or environmental conditions from test to test, inaccuracies introduced in reading values off the stress-strain curves, and a myriad other reasons. If there is only a small variation in the values obtained for a particular quantity, it is sufficient to use an average value for it based upon the results of several tests. However, if the values vary widely, a larger number of tests must be conducted and the data analyzed by using statistical methods.

Representative values for several important properties are listed for a number of materials in Table A.2 of the Appendix. More detailed information and information on other materials can be found in engineering handbooks, manufacturers' literature, and other references.

Stress-strain relations. Below the proportional limit the stress-strain curve is a straight line completely defined by its slope. Thus, we have the stress-strain relations

$$\sigma = E\varepsilon \qquad (\sigma \leqslant \sigma_{PL}) \tag{5.19}$$

for normal stresses and normal strains and

$$\tau = G\gamma \qquad (\tau \leqslant \tau_{PL}) \tag{5.20}$$

for shear stresses and shear strains, where E and G are the modulus of elasticity and the shear modulus, respectively. Materials that obey linear relations of this type are said to be *linearly elastic*. Equation (5.19) is commonly called *Hooke's law* in honor of Robert Hooke (1635–1703), who carried out experiments on wires and observed a linear relationship between the applied load and resulting elongation. Equation (5.20) is the counterpart of Hooke's law for shear.

Problems involving brittle materials are often such that an equation describing the complete stress-strain curve is required. Highly brittle materials, such as glass and Plexiglas in tension, obey Hooke's law to a good degree of approximation all the way to fracture [see Figure 5.25(b), for example]. Nonlinear stress-strain curves, such as that for cast iron in tension [Figure 5.25(b)], can usually be represented by a power law relationship of the form

$$\sigma = K\varepsilon^n$$

where K and n are constants obtained by fitting the equation to the data.

In most problems involving ductile materials, the strains are sufficiently small that only the elastic range and first part of the inelastic range need be described. For materials with a yield point, the resulting stress-strain relations are simple. Hooke's law holds up to the proportional limit, beyond which the stress has a constant value equal to the yield point of the material (Figure 5.30):

$$\begin{aligned} \sigma &= E\varepsilon \qquad (\sigma \leqslant \sigma_{YP}) \\ \sigma &= \sigma_{YP} \qquad (\sigma > \sigma_{YP}) \end{aligned} \tag{5.21}$$

Materials that exhibit responses described by Eqs. (5.21) are called *elastic-perfectly plastic*.

The behavior of ductile materials without a yield point is more difficult to describe. However, it is usually sufficiently accurate to treat the material as being elastic-perfectly plastic, with the yield point equal to the actual yield strength. The nature of this approximation is illustrated in Figure 5.31. It is widely used in structural engineering and machine design, and it will be used almost exclusively in this book. Note that, for this representation, the yield point, yield strength, and proportional limit all have the same values.

Obviously, more accurate representations of the stress-strain curves are possible, but only at the expense of increased mathematical complexity. After an

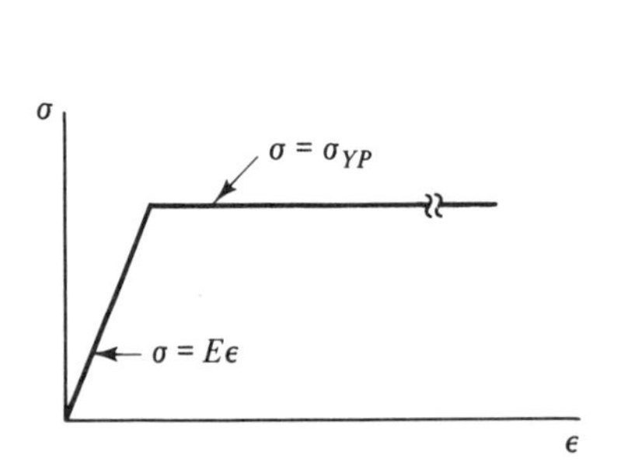

Figure 5.30

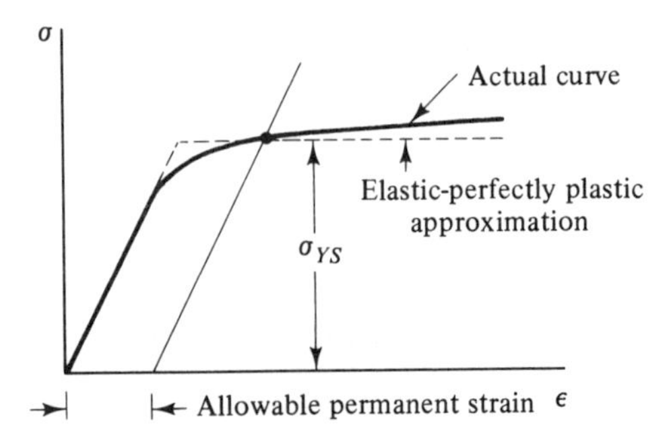

Figure 5.31

appropriate change of symbols, all of the stress-strain relations discussed here also apply for shear.

Example 5.9. Tensile Test. During a tensile test the original 50.000-mm gage length is observed to have increased to 50.030 mm and the original diameter of 12.000 mm to have decreased to 11.998 mm. The applied load is 4.68 kN. Determine the modulus of elasticity and Poisson's ratio. Assume that the stress is below the proportional limit.

Solution. The stress is

$$\sigma = \frac{F}{A} = \frac{4.68 \times 10^3 \text{ N}}{(\pi/4)(12)^2(10^{-6}) \text{ m}^2} = 41.38 \times 10^6 \text{ N/m}^2$$

and the strains in the longitudinal and transverse directions are

$$\varepsilon_L = \frac{50.030 \text{ mm} - 50.000 \text{ mm}}{50.000 \text{ mm}} = 0.0006$$

$$\varepsilon_t = \frac{11.998 \text{ mm} - 12.000 \text{ mm}}{12.000 \text{ mm}} = -0.0002$$

Hooke's law applies because the stresses are below the proportional limit; therefore,

$$E = \frac{\sigma}{\varepsilon} = \frac{41.38 \times 10^6 \text{ N/m}^2}{6 \times 10^{-4}} = 69.0 \times 10^9 \text{ N/m}^2 \text{ or } 69.0 \text{ GN/m}^2 \qquad \textit{Answer}$$

From the definition of Poisson's ratio, Eq. (5.17), we obtain

$$\nu = -\frac{\varepsilon_t}{\varepsilon_L} = \frac{-(-0.0002)}{0.0006} = 0.33 \qquad \textit{Answer}$$

PROBLEMS

5.54 Diagonals AC and BD of the rhombus $ABCD$ shown were originally 2.0 mm long. Determine the modulus of elasticity E and Poisson's ratio ν for this material. Assume linearly elastic material behavior.

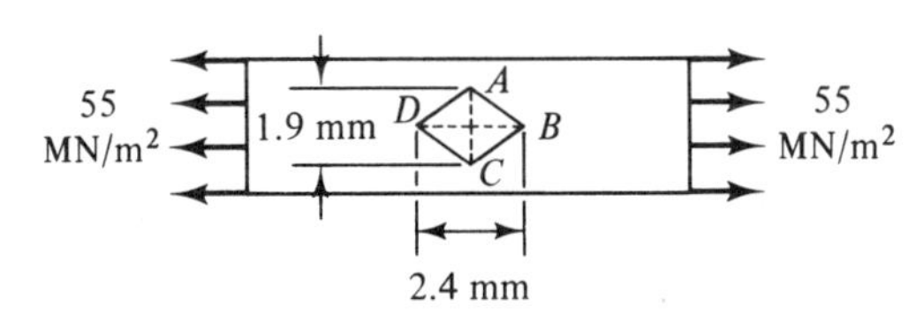

Problem 5.54

5.55 A tensile specimen is loaded until the transverse strain is -0.00045. If the material is elastic with $E = 55$ MN/m^2 and $\nu = 0.3$, what is the maximum tensile stress in the member?

5.56 When a 2-in. × 2-in. square aluminum tensile specimen is subjected to a load of 32 kips, the original 12-in. gage length is observed to have increased by 0.00938 in. and the original 2-in. lateral dimension to have decreased by 0.00052 in. Determine (a) the modulus of elasticity E, (b) Poisson's ratio ν, and (c) the shear modulus G. Assume that the stresses are below the proportional limit.

5.57 What is the shear modulus G for an isotropic elastic material with $E = 12 \times 10^6$ psi and $\nu = 0.35$?

5.58 A portion of the tensile stress-strain diagram for 2017-T451 Aluminum is shown in Figure 5.24(b). For this material, determine (a) the modulus of elasticity, (b) the proportional limit, and (c) the yield strength for 0.2% allowable permanent strain.

5.59 Referring to Figure 5.25(b), compare the tensile modulus of elasticity of cast iron with that of Plexiglas.

5.60 Determine the yield point and ultimate tensile strength for a structural steel specimen with the stress-strain curve shown in Figure 5.25(a). What is the percent elongation of the specimen?

5.61 For a cast-iron tensile specimen with the stress-strain curve shown in Figure 5.25(b), determine the tangent modulus at (a) $\sigma = 0$ and (b) $\sigma = 12$ ksi. Also determine the yield strength for 0.2% allowable permanent strain and the ultimate tensile strength.

5.62 The following data are for a steel tensile specimen: initial diameter = 0.606 in., final diameter = 0.432 in., initial gage length = 8.0 in., final gage length = 10.2 in. Determine the percent reduction of area and percent elongation.

5.63 The stress-strain curve for a material, indicated by the dashed curve shown, is approximated by two straight lines. Using these straight-line approximations, determine (a) the modulus of elasticity, (b) the tangent modulus, (c) the yield strength for 0.2% allowable permanent strain, and (d) the permanent strain if the member is loaded to 30 ksi and then unloaded.

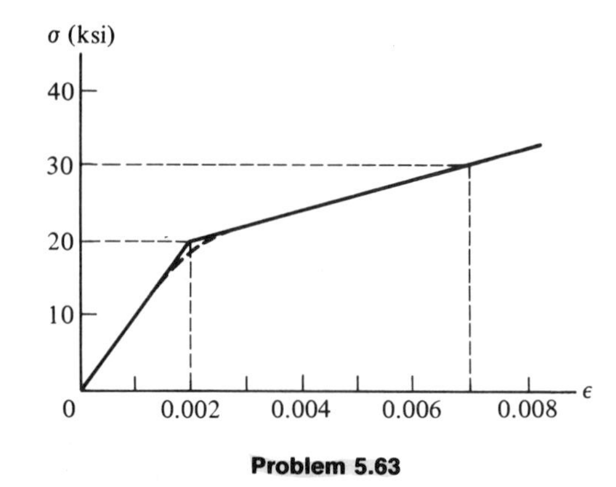

Problem 5.63

5.64 The following data are from a compression test on a square white-pine specimen 1.8 in. on a side and 12 in. long. The gage length is 8 in., and the extensometer used has a magnification factor of 5. Plot the stress-strain diagram and determine the modulus of elasticity and ultimate compressive stress.

Load (kips)	*Extensometer Reading (in.)*
0	0.000
2	0.006
4	0.017
6	0.029
8	0.041
10	0.054
12	0.066
14	0.079
16	0.092
18	0.106
20	0.119
22	0.132
24	0.148 (fracture)

5.9 LOAD-STRESS AND LOAD-DEFORMATION RELATIONS

We are now in a position to outline a general procedure for obtaining equations that relate the applied loadings to the stresses and deformations they produce. Recall from Section 5.2 that the actual stresses cannot be determined from the equations of statics alone. Additional information about the manner in which the stresses are distributed is required.

One possible approach is to make assumptions about the stress distribution and then check to see if the results obtained are reasonable. A more systematic approach is to relate the stress distribution to some quantity that can be readily observed and/or measured. But what quantity would this be? For obvious reasons, quantities occurring within the body cannot be easily observed and measured, if at all. The one quantity that meets our requirements is the deformations occurring on the surface. Observation of the pattern of these deformations is the key to the determination of the stresses and strains throughout the body. The procedure is as follows:

1. Observe the geometry of the deformation pattern produced by the loading.
2. Obtain the distribution of strain from the equations relating strains to deformations (Sections 5.5). At this point, only the manner in which the strains are distributed can be determined; their magnitudes remain unknown.
3. Obtain the stress distribution from the strain distribution by using the stress-strain relations (Section 5.8).
4. Relate the stresses to the stress resultants by using the conditions for static equivalence (Section 4.7).
5. Relate the stress resultants to the applied loadings by using the method of sections (Section 5.2).

These steps, which need not be carried out in the exact order listed, yield the relation between the stresses and the applied loadings. To relate the deformations to the loads, we go back to steps 2 and 3. Step 3 provides the actual strains, since the stresses are now known. Actual deformations can then be obtained from step 2 by solving the strain-deformation relations. This procedure will be used extensively in the following chapters to determine the load-stress and load-deformation relations for common structural members.

5.10 DESIGN CRITERIA AND SAFETY FACTORS

Load-stress and load-deformation relations can be used to check to see if a given member will perform adequately under a given loading (analysis) or to determine the material and size of member necessary to support a particular load (design). The design criteria by which the performance of a member or adequacy of its design are judged depend, in large measure, upon the intended purpose of the member. As we have mentioned several times before, a member need not break in order to fail.

There are four general types of structural failure: fracture, general yielding, excessive deformations, and buckling. The first three will be considered in this section, along with the significant material properties and design criterion associated with each. Buckling will be discussed in Chapter 10.

Fracture. When a member breaks, its usefulness as a load-carrying member obviously ceases. Most members made of brittle material fail in this manner. Unfortunately, since the deformations are usually very small, there is little warning of the impending failure. Fracture of materials is associated with the ultimate stress; therefore, the appropriate design criterion for this mode of failure is that the stresses be kept below the ultimate strength of material. Members made of ductile materials also break, but they usually fail first because of general yielding or excessive deformations.

General yielding. Members made of ductile materials can undergo very large inelastic deformations before breaking. As a result of these large deformations, the geometry of the structure can change so drastically that it literally collapses. The design criterion for this type of failure is obvious—the applied load must be kept below the load at which collapse occurs. The collapse load of simple members and structures will be considered in the following chapters. Suffice it to say at this point that the collapse load is intimately related to the yield strength of the material, for this is the stress at which the inelastic deformations start to become significant.

Excessive deformations. This type of failure occurs when the deformations, even though they may be small, exceed some allowable value. For example, deformations in machine tools must be kept small so that small tolerances on the dimensions of the workpiece can be maintained. The amount of deformation that can be tolerated in any given problem is often a matter of engineering judgment. Excessive deformation is a result of insufficient "stiffness" of the member or structure. For many problems in which deformations are the limiting factor, the stresses are below the proportional limit. In these cases, the significant material parameters are the elastic modulus, E, or shear modulus, G.

Safety factors. It should be apparent by now that there is a degree of inexactness involved in determining the response of load-carrying members and that it would be rather presumptuous to think that one could design right to the limit of the capability of the material without encountering difficulties. For example, various approximations are involved in the analysis, and the loads, dimensions, and material properties may not be known precisely. In addition, the structure may not be built according to specifications, or it may later be used in a manner not anticipated by the designer.

The necessary margin for error is usually provided by introducing a *safety factor* into the design. The safety factor is defined as the ratio of the load to produce failure (the design load) to the anticipated load to which the member will be subjected (the working load):

$$\text{Safety factor } (SF) = \frac{\text{failure load}}{\text{working load}} \tag{5.22}$$

For example, a safety factor of 3 means that the member is designed to withstand three times the expected applied load. A safety factor can also be introduced by reducing the allowable stresses or deformations. However, we shall use the definition given here.

The value of the safety factor is often set by building codes or other specifications and is based upon the consequences of a failure, economics, the degree of confidence in the accuracy of the analysis, and other factors. For example, since failure of a large building would involve potential loss of life, a substantial safety factor would be used in its design. Aircraft also require a high degree of safety, but an airplane could not get off the ground if it were designed with a large safety factor because of the extra material and weight involved. A small safety factor must be used, but this is compensated for by making the analysis highly accurate.

5.11 CLOSURE

The function of a load-carrying member is to support or transmit applied loads, and its success or failure is governed by the magnitudes of the resulting stresses and deformations. In this chapter we outlined the basic procedures for obtaining the load-stress and load-deformation relations and for using them in design and analysis. Some simple problems were considered, but most of the applications will come in the following chapters.

A review of this chapter will show that only a few new concepts have been introduced, a fact easily lost in the deluge of new definitions and terminology. Many of the points discussed undoubtedly remain a bit vague, but they should fall into focus when they are used over and over again in the applications to follow. There are several key points that apply to all our work and they are of sufficient importance to merit repeating here:

1. The three basic ingredients in the mechanics of deformable bodies are equilibrium, the geometry of the deformations, and the resistive properties of materials.

2. We consider only small deformation problems. A large majority of the problems encountered in practice fall into this category.

3. Since the deformations are small, the geometry of the body before loading can be used in all force analyses, except for buckling problems. This is an approximation without which few problems of practical interest could be solved. The accuracy of this approximation was demonstrated in Section 3.10.

chapter six

AXIAL LOADING

6.1 INTRODUCTION

The stresses in members subjected to simple axial loadings were discussed in Section 5.3. In this chapter we shall consider the stresses that result from more complex axial loadings, as well as the resulting deformations. Some structural and design problems involving axially loaded members will also be considered. Our first step will be to examine more carefully the conditions under which the load-stress relations developed in Section 5.3 are valid.

6.2 VALIDITY OF LOAD-STRESS RELATIONS

Equations (5.3) through (5.6) derived in Section 5.3 define the actual stresses in an axially loaded member provided they are uniformly distributed over the section. Thus, it is only necessary to determine the conditions for a uniform stress distribution in order to establish the range of validity of these equations. This can be done by using the procedure outlined in Section 5.9.

Uniform axially loaded members deform in such a way that planes perpendicular to the axis before loading remain plane and perpendicular to the axis after loading. This is illustrated in Figure 6.1, which shows the deformation pattern in an axially loaded rubber sheet. Note that the deformed grids are distorted in the

(a)Before Loading

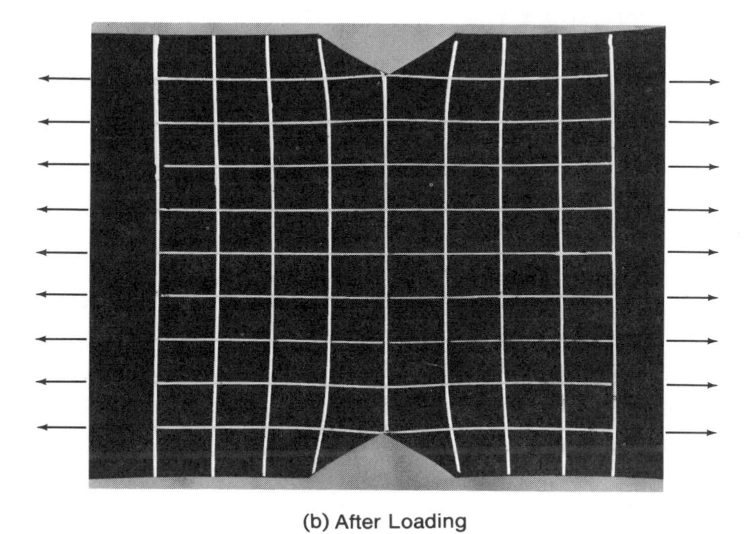

(b) After Loading

Figure 6.1. Deformation of an Axially Loaded Rubber Sheet as Indicated by Grid Patterns

vicinity of the notches, but are uniform some distance away. Similar distortions in the deformation pattern may also occur near supports, connections, and points of load application.

Now that we have observed the deformations in the sheet (step 1 in Section 5.9), we consider the strains associated with them. In doing so, it will be helpful to think of the member as consisting of a number of parallel fibers, or line elements, laid side by side.

Concentrating for the moment on the regions away from the notches, we have the situation illustrated in Figure 6.2(a). This figure shows the deformation of a typical row of squares across the sheet. Since each line element has the same original length, L, and elongates the same amount, δ, the strain is the same for each. Thus, the normal strain in the longitudinal direction is uniformly distributed across the sheet, as shown in Figure 6.2(b). This determines the strain distribution (step 2 in Section 5.9).

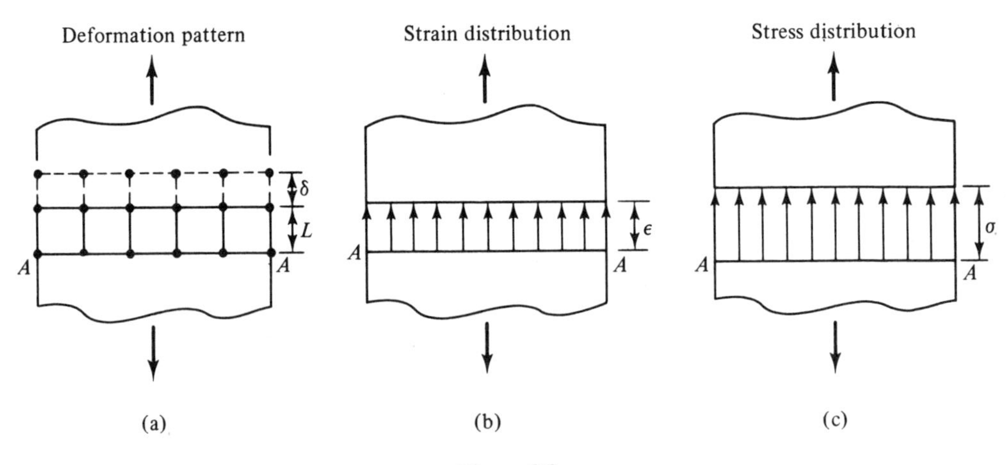

Figure 6.2

If the material is uniform across the sheet, each fiber will have the same properties (stress-strain diagram). Accordingly, each will experience the same stress because the strain is the same for each. Thus, the stress is also uniformly distributed across the sheet [Figure 6.2(c)]. This will not be the case near the notches, or at any other locations where the deformations are nonuniform. The stresses and deformations near the points of load application depend upon how the load is applied. For example, it is not too difficult to visualize that a load uniformly distributed over the end of the sheet would produce very different deformations locally than would a statically equivalent load applied as a concentrated force at its center. Members that have significant taper or are curved also exhibit nonuniform deformations.

The conditions for a uniform stress distribution and the validity of the load-stress relations given in Eqs. (5.3) through (5.6) can be summarized as follows:

1. The member must be axially loaded (this was shown in Section 5.3). Another way of saying this is that the stress resultant must consist only of a force acting at the centroid of the cut section. If there is any torque or bending moment, the stresses cannot be uniformly distributed.

2. For compressive loads, the member must not be so slender that it buckles. Once the member bends, it is no longer axially loaded.

3. The member must be straight, or nearly so, for some distance on either side of the plane of interest.

4. The member must have constant cross-sectional area, or nearly so, for some distance on either side of the plane of interest.

5. The plane of interest must be some distance away from connections, supports, and points of load application.

6. The material must be uniform across the width of the member. Its properties can vary along the length, however, as in the case of two bars of different materials joined end to end to form a single composite member.

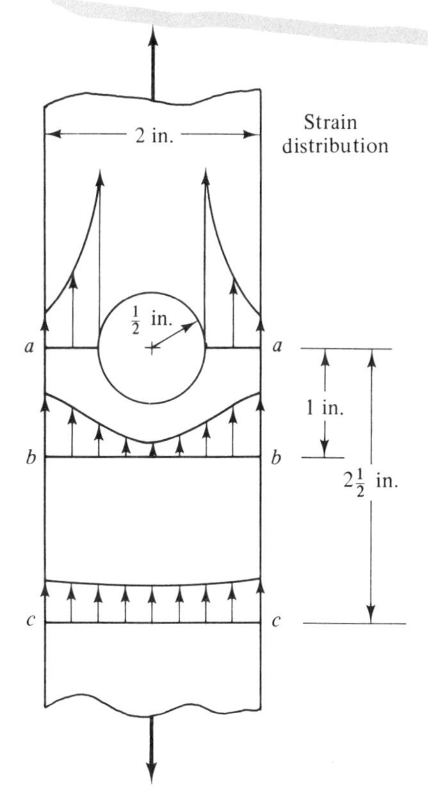

Figure 6.3

The regions of a member over which the load-stress relations are not strictly valid tend to be highly localized. As a rule of thumb, they extend only over a distance approximately equal to the width of the member. This is illustrated in Figure 6.3, which shows the experimentally determined strain distribution in an axially loaded bar containing a hole. The strains are significantly affected near the hole, but $1\frac{1}{4}$ bar-widths away at section *c-c* the strains are practically the same as if the hole weren't there.

This localizing effect, which is also evident near the notches in the sheet shown in Figure 6.1(b), occurs in all types of members under all types of loadings and is known as *St. Venant's principle*. This principle is very important, for without it there would be little hope of determining the stresses and deformations analytically. It makes it possible to develop simple load-stress and load-deformation relations that apply throughout most of the body. Special techniques can then be developed for handling the remaining regions involving the localized effects.

The maximum normal stress at holes, notches, or other geometrical discontinuities can be obtained by applying a correction factor to the values computed from the relationship $\sigma = P/A$. These correction factors, which are usually determined experimentally, will be discussed in Section 6.6. The situation near supports, connections, and points of load application is not so easily handled. There, the usual procedure is to work with the average stresses and to account for any errors by using a safety factor. In some cases, empirical methods of analysis have been devised.

PROBLEM

6.1 For each of the members shown, indicate whether or not the equation $\sigma = P/A$ is valid at the section indicated. If it is not valid, which of the necessary conditions are violated?

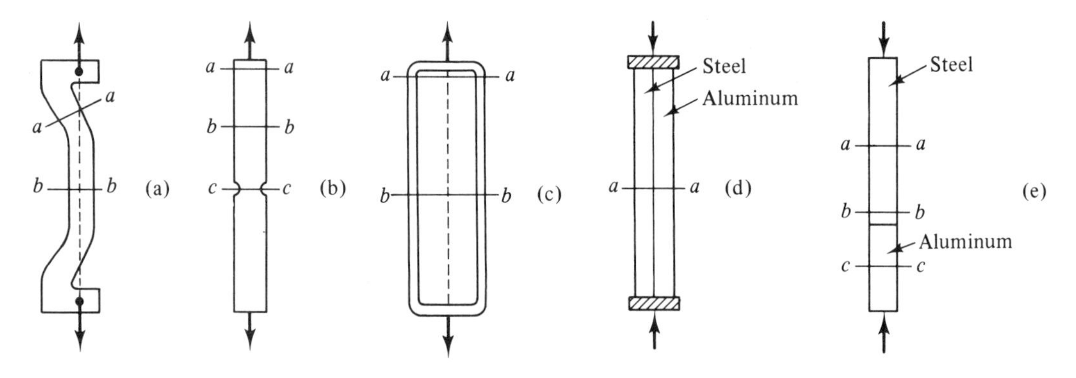

Problem 6.1

6.3 DEFORMATIONS

The elongation or shortening of an axially loaded member can be determined by combining the expression for the normal stress with the definition of normal strain and the stress-strain relation for the material. For the present, we shall assume that the material is linearly elastic.

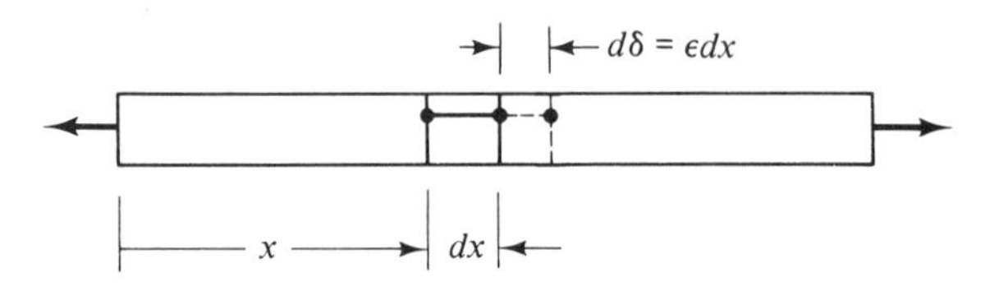

Figure 6.4

Consider a line element of length dx located an arbitrary distance x from the end of the member (Figure 6.4). From the definition of normal strain, Eq. (5.12), the elongation of the element is

$$d\delta = \varepsilon dx$$

From Hooke's law and the load-stress relation, we obtain for the strain

$$\varepsilon = \frac{\sigma}{E} = \frac{P}{AE}$$

where P is the normal force on planes perpendicular to the axis of the member, A is the cross-sectional area, and E is the modulus of elasticity. Combining these expressions and integrating over the length of the member, we obtain

$$\delta = \int_0^L \frac{P}{AE}\, dx \qquad (\sigma \leqslant \sigma_{PL}) \tag{6.1}$$

The quantities P, A, and E must be expressed as functions of x before the integral can be evaluated. If they are constant over the length of the member, Eq. (6.1) can be written as

$$\delta = \frac{PL}{AE} \qquad (\sigma \leqslant \sigma_{PL}) \tag{6.2}$$

The sign of δ is the same as that of P; it is positive if P is tensile and negative if it is compressive.

The preceding equations define the total deformation of the entire member. However, the deformation of any particular segment of it can be determined by carrying out the integration in Eq. (6.1) over the length of the segment or by using the segment length instead of the total length in Eq. (6.2).

Equations (6.1) and (6.2) apply only if the stresses are below the proportional limit, since Hooke's law was used in their derivation. For other types of material behavior, it is only necessary to introduce the appropriate stress-strain relation and proceed as before. Since the load-deformation equations are also based upon the relation $\sigma = P/A$ there are regions of the member where they are not strictly valid. However, this has a negligible influence upon the total deformation of the member. These regions are relatively short (St. Venant's principle) and, therefore, do not contribute appreciably to the value of the integral in Eq. (6.1).

If the member is loaded at several points along its length, the normal force P will vary from section to section. In this case, it is helpful to plot the variation of P along the length of the member so that its value at any location can be readily determined (Figure 6.5). This plot, which is called the *normal force diagram*, is useful in determining both the stresses and the deformations. In accordance with our sign convention, tensile forces are plotted as positive and compressive forces as negative.

Once the value of P in each portion of the member has been determined, the deformation of each segment can be computed and the results summed to determine the total deformation. In other words, the member is treated as a series of

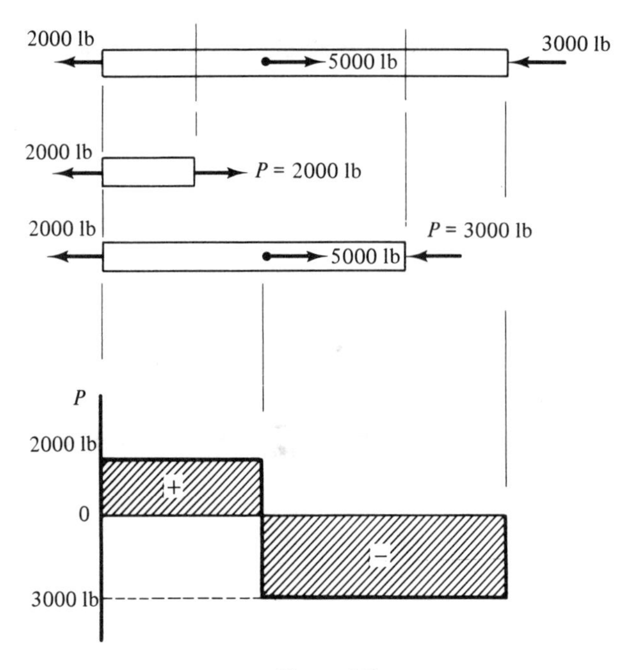

Figure 6.5

shorter members placed end to end, each with constant normal force. The deformation of the entire member is then equal to the sum of the deformations of its parts. This approach can also be used when different segments of the member have different cross-sectional areas or modulus of elasticity.

Example 6.1. Deformation of Truss Members. Members AB and AE of the truss shown in Figure 6.6(a) are made of structural steel and have cross-sectional areas of $2.4 \times 10^{-3}\text{m}^2$ and $1.2 \times 10^{-3}\text{m}^2$, respectively. How much will these members elongate or shorten as a result of the 400-kN applied load? Assume that compression members do not buckle.

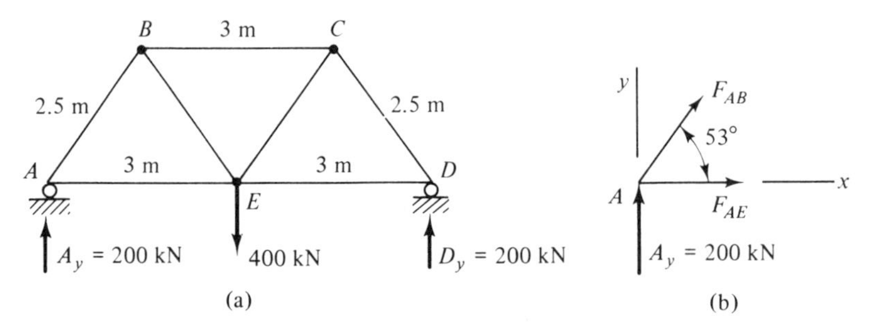

Figure 6.6

Solution. The first step is to determine the forces on the members of interest. A force analysis of the entire truss (details not given here) shows that the reactions at A and D are 200-kN upward forces. From the FBD of joint A [Figure 6.6(b)], we then have

$$\Sigma Fy = F_{AB} \sin 53° + 200 \text{ kN} = 0$$

$$\Sigma F_x = F_{AE} + F_{AB} \cos 53° = 0$$

$$F_{AB} = -250 \text{ kN or } 250 \text{ kN (compression)}$$

$$F_{AE} = 150 \text{ kN (tension)}$$

We next check the stresses to see if they are below the proportional limit:

$$\sigma_{AB} = \left(\frac{P}{A}\right)_{AB} = \frac{250 \times 10^3 \text{ N}}{2.4 \times 10^{-3} \text{ m}^2} = 104 \times 10^6 \text{ N/m}^2 \text{ (compression)}$$

$$\sigma_{AE} = \left(\frac{P}{A}\right)_{AE} = \frac{150 \times 10^3 \text{ N}}{1.2 \times 10^{-3} \text{ m}^2} = 125 \times 10^6 \text{ N/m}^2 \text{ (tension)}$$

Both of these values are well below the proportional limit of structural steel (see Table A.2); therefore, Eq. (6.2) applies:

$$\delta_{AB} = \left(\frac{PL}{AE}\right)_{AB} = \frac{(-250 \times 10^3 \text{ N})(2.5 \text{ m})}{(2.4 \times 10^{-3} \text{ m}^2)(207 \times 10^9 \text{ N/m}^2)}$$

$$= -1.26 \times 10^{-3} \text{ m or } 1.26 \text{ mm (shortening)} \qquad \textit{Answer}$$

$$\delta_{AE} = \left(\frac{PL}{AE}\right)_{AE} = \frac{(150 \times 10^3 \text{ N})(3 \text{ m})}{(1.2 \times 10^{-3} \text{ m}^2)(207 \times 10^9 \text{ N/m}^2)}$$

$$= 1.81 \times 10^{-3} \text{ m or } 1.81 \text{ mm (elongation)} \qquad \textit{Answer}$$

Example 6.2. Deformation of a Stepped Bar. (a) What load F can be applied to the aluminum bar shown in Figure 6.7(a) if the maximum normal stress cannot exceed 24 ksi? Use a safety factor of 3. (b) How much will the bar deform under this loading? Assume linearly elastic behavior.

Solution. (a) We first determine the reaction at A and the normal force diagram [Figure 6.7(b)]. It is clear from this diagram that the largest normal stress occurs in BC. (The normal force in BC is only one-half that in AB, but the area is six times smaller.) Thus, we have

$$\sigma_{BC} = \left(\frac{P}{A}\right)_{BC} = \frac{F}{0.5 \text{ in.}^2} \leqslant 24\ 000 \text{ lb/in.}^2$$

$$F \leqslant 12\ 000 \text{ lb}$$

This defines the failure load. The actual load that can be applied to the bar (the working load) is this value divided by the safety factor [Eq. (5.22)]:

$$F_{\text{working}} = \frac{F_{\text{failure}}}{SF} = \frac{12\ 000 \text{ lb}}{3} = 4000 \text{ lb} \qquad \textit{Answer}$$

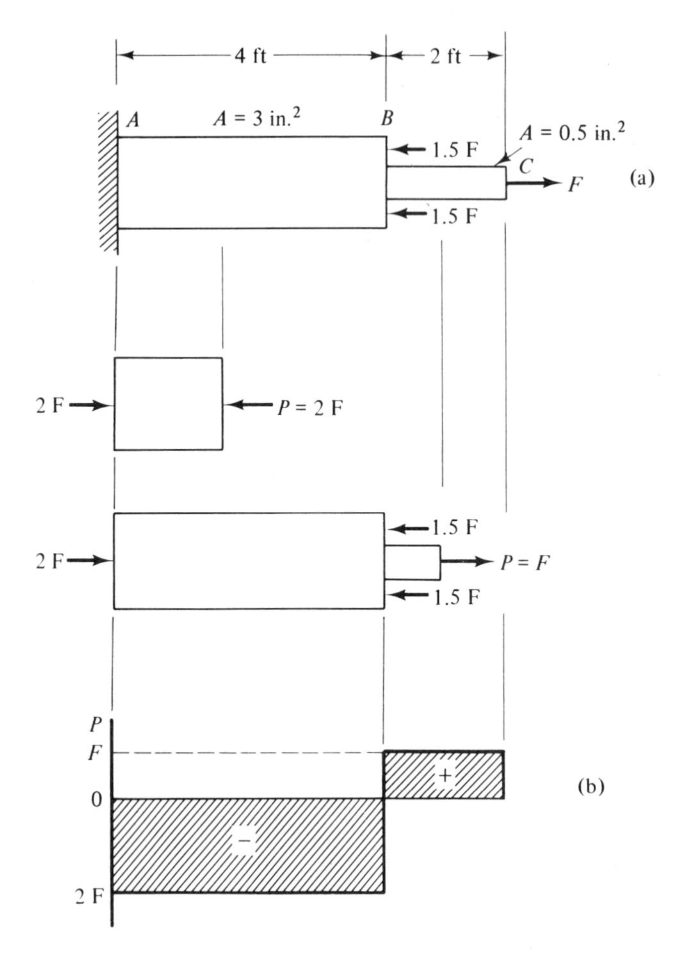

Figure 6.7

(b) The total deformation of the member is the sum of the deformations of each part. The working load is used in the computations because it is the load actually applied to the bar. We have

$$\delta = \delta_{AB} + \delta_{BC} = \left(\frac{PL}{AE}\right)_{AB} + \left(\frac{PL}{AE}\right)_{BC}$$

or

$$\delta = \frac{(-8000\text{ lb})(48\text{ in.})}{(3\text{ in.}^2)(10 \times 10^6\text{ lb/in.}^2)} + \frac{(4000\text{ lb})(24\text{ in.})}{(0.5\text{ in.}^2)(10 \times 10^6\text{ lb/in.}^2)}$$

$$= -0.0128\text{ in.} + 0.0192\text{ in.} = 0.0064\text{ in. (elongation)} \qquad \textbf{\textit{Answer}}$$

Units must be watched carefully when computing δ because A and E are usually given in terms of square inches and pounds per square inch, respectively, while L is often given in feet.

Example 6.3. Elongation of a Bar Due to Its Own Weight. A bar with cross-sectional area A, specific weight γ, and length L is suspended vertically from one end [Figure 6.8(a)]. How much will the bar elongate under the action of its own weight? Assume that Hooke's law applies.

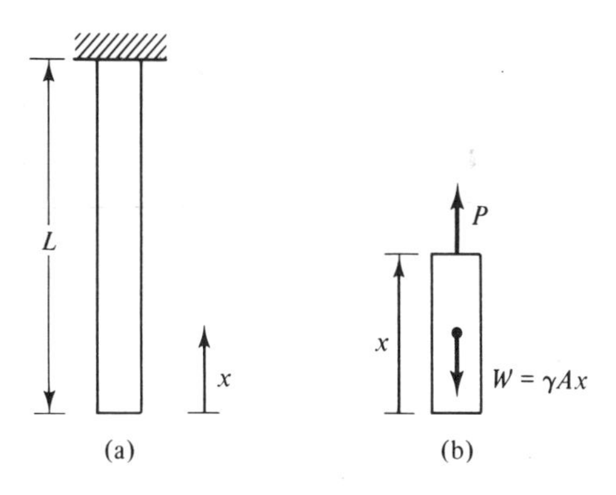

Figure 6.8

Solution. For convenience, we take the origin of coordinates at the free end of the bar. Sectioning the bar at an arbitrary location and considering the equilibrium of the resulting piece [Figure 6.8(b)], we obtain for the normal force

$$P = \gamma A x$$

Since P varies continuously along the length of the member, it is necessary to integrate in order to determine δ. From Eq. (6.1), we have

$$\delta = \int_0^L \frac{P}{AE}\,dx = \int_0^L \frac{\gamma x}{E}\,dx = \frac{\gamma L^2}{2E} \qquad \textit{Answer}$$

The elongation can also be expressed in terms of the total weight of the member, which is $W = \gamma AL$:

$$\delta = \frac{WL}{2AE}$$

PROBLEMS

6.2 What is the maximum allowable length of a 20-mm × 20-mm square bar of 2024-T3 aluminum if the elongation is not to exceed 1 mm for an axial tensile load of 100 kN?

6.3 If the nut at the end of a steel spoke in a bicycle wheel is advanced one-quarter turn beyond that required to take up any slack, what is the tension induced in the spoke? The spoke is 0.3 m long and the nut has 1 thread/mm. Use $E = 207$ GN/m^2 and assume that the rim of the wheel does not deform.

6.4 The members of the truss shown are made of 64-mm × 64-mm equal angles with a mass of 5.96 kg/m (see Table A.10). Determine the elongation or shortening of members AB and BD. Use $E = 207$ GN/m^2 and neglect the weights of the

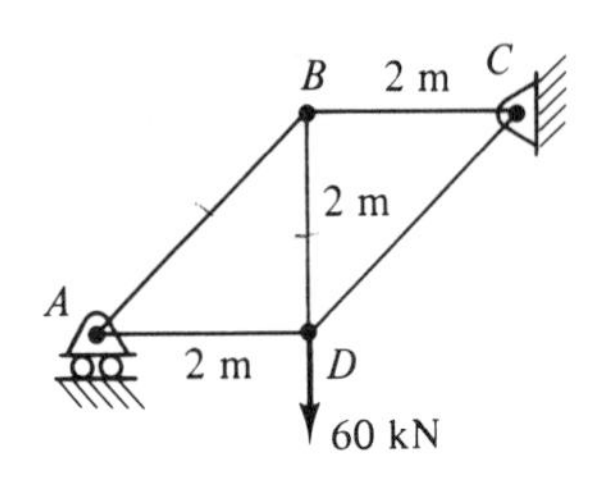

Problem 6.4

members. Assume that compression members do not buckle.

6.5 A steel ($E = 30 \times 10^6$ psi) bar with a cross-sectional area of 0.5 in.^{2} is loaded as shown. Determine the maximum normal stress in the bar and the total change in its length.

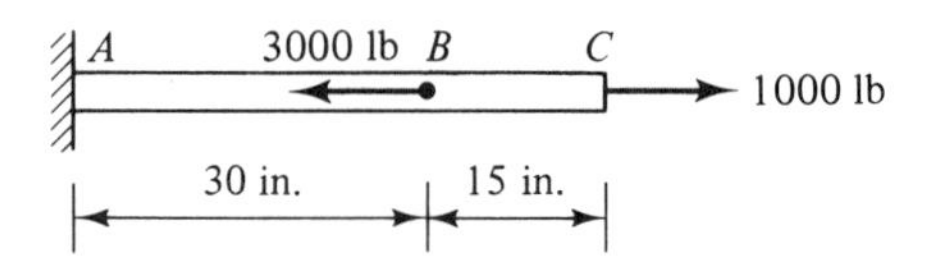

Problem 6.5

6.6 If the tensile stress in portion *AB* of the rod shown is not to exceed 20 ksi, what is the maximum allowable value of the load *F*? For this value of *F*, determine the change in length of portion *BC* of the rod and the total change in its length. $A = 0.5$ in.^{2} and $E = 30 \times 10^6$ psi.

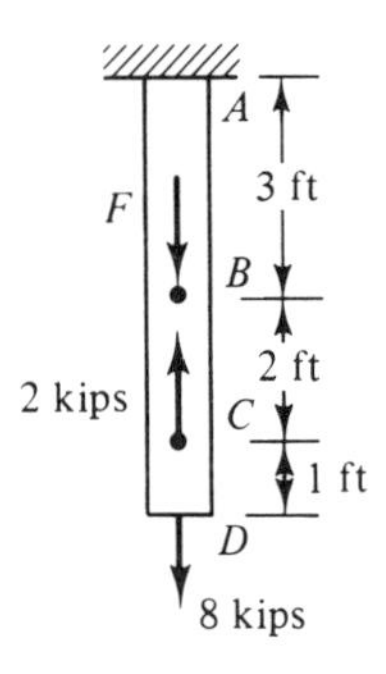

Problem 6.6

6.7 What is the total elongation of the composite bar shown when subjected to an axial tensile force of 50 kN?

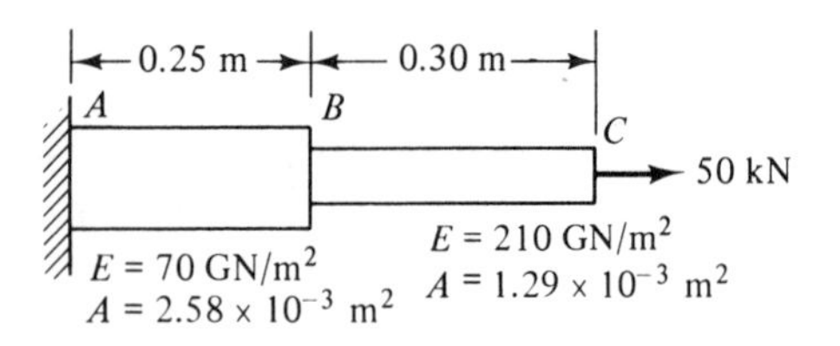

Problem 6.7

6.8 A wood post made of a 4-in.×4-in. timber rests on a concrete pier as shown and is used to support a roof. The top of the post is observed to have settled $\frac{1}{4}$ in. after the roof is constructed. Determine the load *F* on the post and the bearing stress between the post and concrete. $E_W = 1.8 \times 10^6$ psi and $E_C = 3 \times 10^6$ psi.

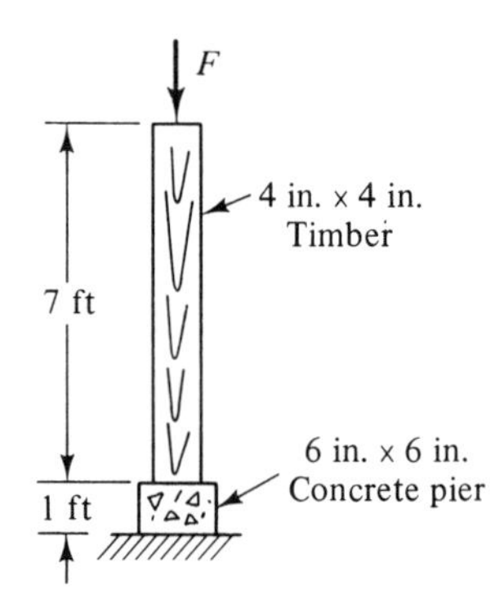

Problem 6.8

6.9 If the overall change in length of the composite rod shown must not exceed 0.001 in., what is the maximum allowable compressive load *F*?

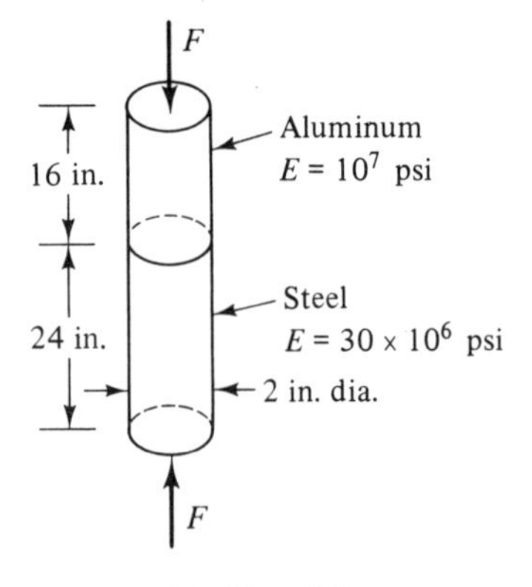

Problem 6.9

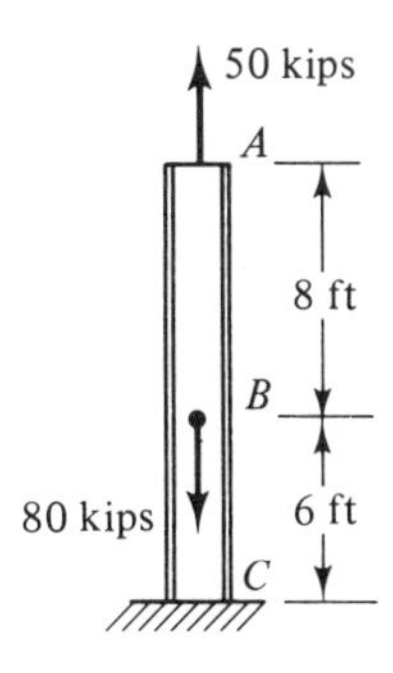

Problem 6.10

6.10 Select the lightest weight W-shape (see Table A.3) that will support the loads shown if the overall change in length is not to exceed 0.02 in. $E = 30 \times 10^6$ psi. Assume that the member does not buckle.

6.11 Determine the minimum required cross-sectional area of the steel ($E = 207\ \text{GN/m}^2$) column shown if the maximum compressive stress cannot exceed $200\ \text{MN/m}^2$ and the free end A cannot settle more than 2.5 mm. How far does the point of load application B settle? Assume that the member does not buckle.

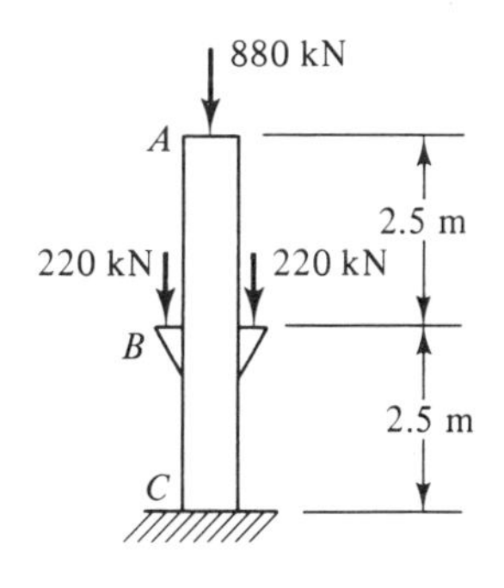

Problem 6.11

6.12 A 10-in. × 10-in. square concrete pillar with a specific weight of $150\ \text{lb/ft}^3$ is loaded as shown. What is the overall shortening of the member if (a) its weight is neglected and (b) its weight is included? $E = 3 \times 10^6$ psi.

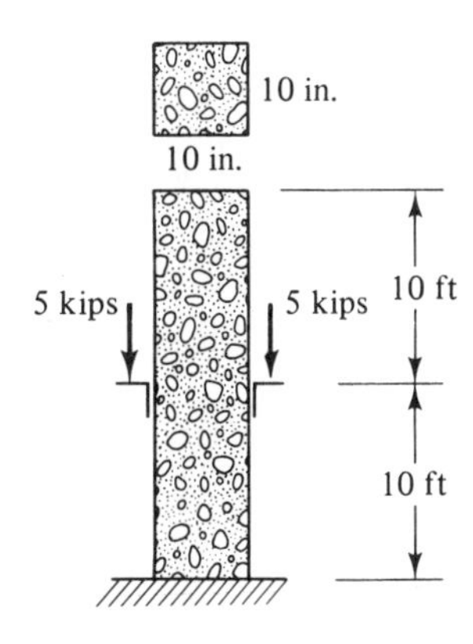

Problem 6.12

6.13 Obtain an expression for the change in length of the slightly tapered bar shown due to its own weight. Express the results in terms of the modulus of elasticity E, the specific weight γ, and the given dimensions.

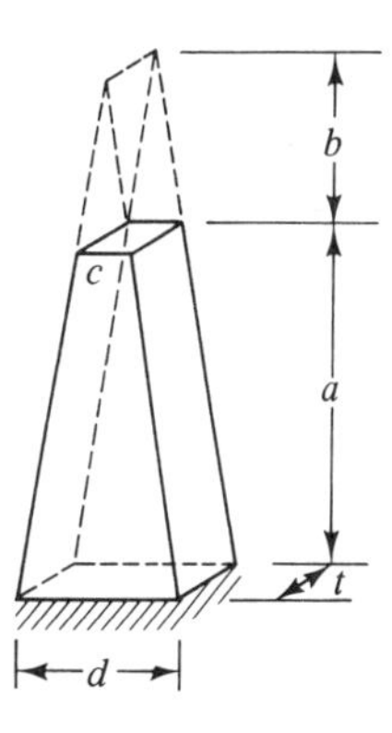

Problem 6.13

6.14 A 20-in. long steel rod with a cross-sectional area of $0.4\ \text{in.}^2$ is subjected to an axial tensile load F. If the resulting longitudinal strain is 0.003 and the material yield point is $\sigma_{YP} = 30$ ksi, determine (a) the applied load, (b) the total change in length of the rod, and (c) the final length of the rod if it is unloaded.

6.15 A round bar with a cross-sectional area of 0.5 in.^{2} is loaded as shown. The strain gage on the bar indicates an axial strain of 1400 μ in./in. ($\mu = 10^{-6}$) and the stress-strain curve for the material is as indicated. Determine F and the total elongation of the member.

6.16 A uniform bar made of a material with specific weight γ and whose stress-strain curve is described by the equation $\sigma = k\varepsilon^n$, where k and n are constants, is suspended from one end. Determine the total elongation of the member due to its own weight.

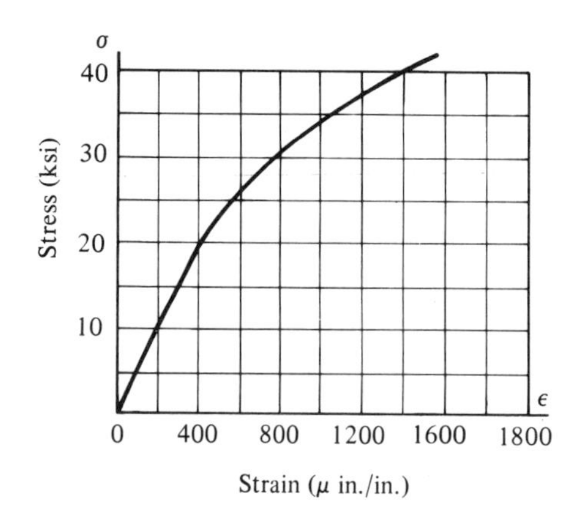

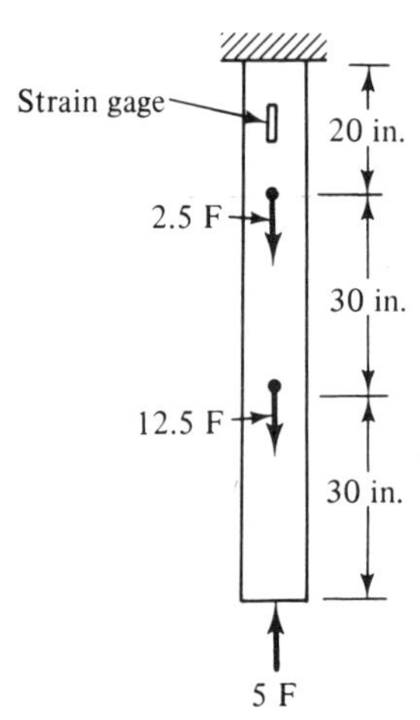

Problem 6.15

6.4 STATICALLY INDETERMINATE PROBLEMS

Consider the problem of determining the forces acting upon the rigid bar shown in Figure 6.9(a). The bar is hinged at the left end and supported by cables at B and C. From the free-body diagram [Figure 6.9(b)], it is evident that there are four unknowns and only three equations from which to determine them. Thus, the forces acting cannot be determined solely from the equations of equilibrium. This is an example of what was referred to in Section 3.3 as a *statically indeterminate* problem.

Indeterminate situations arise because there are more constraints than necessary to maintain equilibrium. In this problem, for instance, the bar could be held in place by a single cable. Fortunately, the additional constraints also place restrictions upon the deformations, which, when expressed analytically, provide the additional equations necessary to obtain a solution.

Referring to Figure 6.9(c), we see that the cables cannot elongate independently of one another when the bar is loaded. Instead, their elongations must

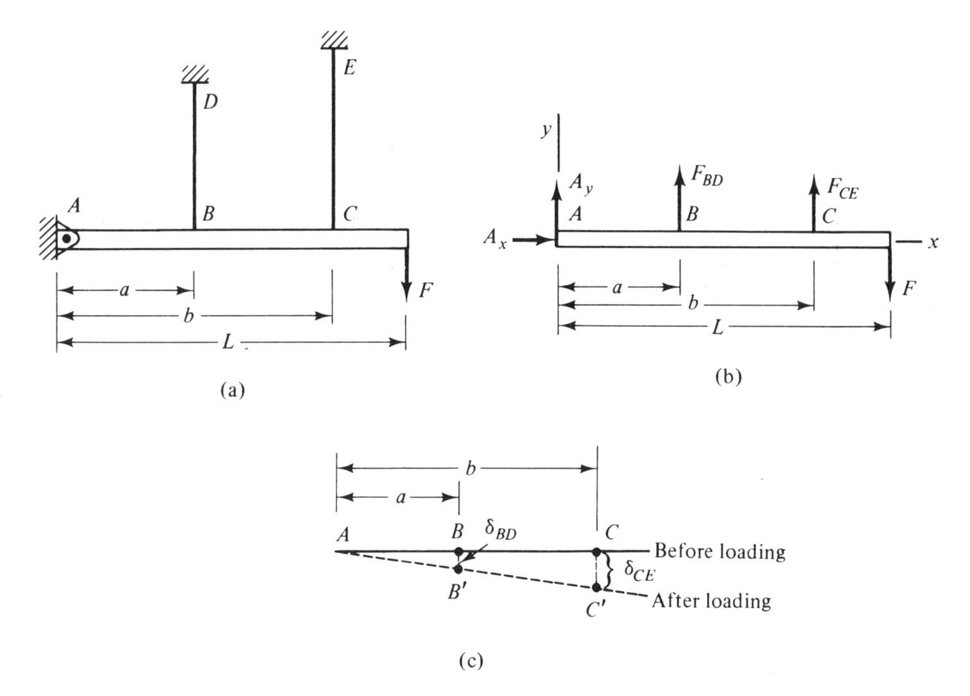

Figure 6.9

satisfy a definite relationship. From the geometry of the figure, we have

$$\frac{\delta_{BD}}{a} = \frac{\delta_{CE}}{b}$$

or

$$\delta_{BD} = \left(\frac{a}{b}\right)\delta_{CE}$$

This relationship is called the *compatibility condition*, for it states the condition that the deformations must satisfy in order to be compatible with the physical situation.

The fourth equation needed for the force analysis of the bar is obtained by substituting the appropriate load-deformation relations into the compatibility condition. Assuming that the stresses in the cables are below the proportional limit, we obtain

$$\left(\frac{FL}{AE}\right)_{BD} = \left(\frac{a}{b}\right)\left(\frac{FL}{AE}\right)_{CE}$$

This equation and the three equations of equilibrium are sufficient for determining all the forces acting upon the bar and upon the cables. Once the forces on the cables have been determined, the assumption concerning the stresses within them can be checked.

The most important thing to notice about this example is that the deformations had to be considered in order to perform the force analysis, even though they were of no interest otherwise. This is characteristic of statically indeterminate problems.

The compatibility conditions are generally different for each different problem, but a clue to their form can usually be obtained from the problem statement. Sketches of the expected deformations, as in Figure 6.9(c), are an invaluable aid in determining the necessary relationships between them. Further illustrations are given in the following examples. The basic procedures outlined here will also be used in later chapters for statically indeterminate torsion and bending problems.

Example 6.4 Forces in a Statically Indeterminate Bar. A bar with cross-sectional area A and length L is attached to immovable walls at each end and loaded as shown in Figure 6.10(a). (a) What are the reactions at each end of the bar and (b) how far does the point of load application move? Assume that the stresses are below the proportional limit.

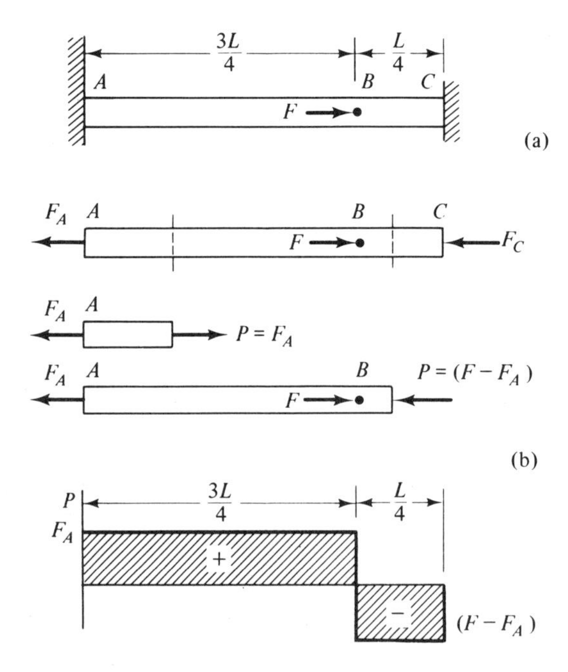

Figure 6.10

Solution. (a) From the FBD of the bar [Figure 6.10(b)], we obtain the equilibrium condition

$$F_A + F_C = F$$

There are two unknown reactions and only one equilibrium equation; therefore, the problem is statically indeterminate. Consequently, we must also consider the deformations.

Since the walls are immovable, the total elongation of the bar must be zero. Thus, the compatibility condition is

$$\delta = \delta_{AB} + \delta_{BC} = 0$$

or

$$\left(\frac{PL}{AE}\right)_{AB} + \left(\frac{PL}{AE}\right)_{BC} = 0$$

Using the values of P obtained from the normal force diagram [Figure 6.10b] and recognizing that the quantity AE cancels out, we have

$$F_A \frac{3L}{4} + \left[-(F - F_A)\frac{L}{4}\right] = 0$$

$$F_A = \frac{F}{4}$$

Thus, the reactions are

$$F_A = \frac{F}{4} \qquad \textbf{\textit{Answer}}$$

$$F_C = \frac{3F}{4}$$

This problem can also be approached from the viewpoint that the amount of elongation of portion AB of the bar must equal the amount of shortening of portion BC. In this case, the compatibility condition is written as

$$|\delta_{AB}| = |\delta_{BC}|$$

The results obtained are the same, but there is more chance of making sign errors by using this approach.

(b) The point of load application moves to the right a distance equal to the elongation of portion AB of the bar, which is

$$\delta_{AB} = \left(\frac{PL}{AE}\right)_{AB} = \frac{(F/4)(3L/4)}{AE} = \frac{3}{16}\frac{FL}{AE} \qquad \textbf{\textit{Answer}}$$

This distance is also equal to the amount of shortening of portion BC.

Example 6.5. Forces in a Composite Column. A column consisting of a 10-in. standard steel pipe filled with high-strength concrete supports a total load of 100 kips applied through a rigid loading plate [Figure 6.11(a)]. What portion of the load is carried by each material if (a) the concrete is initially level with the top of the pipe and (b) the level of the concrete is initially 0.01 in. below the top of the pipe? $E_S = 30 \times 10^6$ psi and $E_C = 3 \times 10^6$ psi.

Solution. (a) Both parts of the member are obviously in compression and will shorten. However, it is easier to avoid sign errors if we assume that they are in

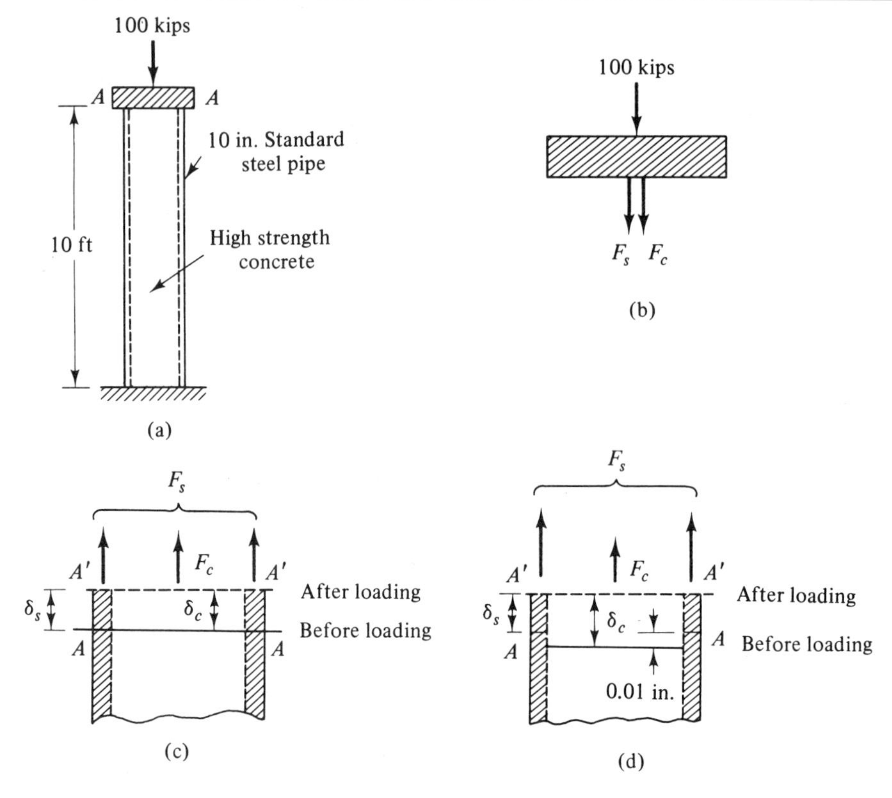

Figure 6.11

tension and elongate. (An alternate approach is to introduce a new sign convention wherein compressive normal forces and shortenings are considered positive.) We shall also assume linearly elastic material behavior.

Let F_S and F_C denote the forces acting upon the steel and concrete, respectively. From the FBD of the loading plate [Figure 6.11(b)], we have

$$F_S + F_C = -100 \text{ kips}$$

Since the problem is obviously statically indeterminate, the deformations must also be considered.

Since the concrete is level with the top of the pipe, both materials must deform the same amount [Figure 6.11(c)]. Thus, the compatibility condition is

$$\delta_S = \delta_C$$

or

$$\left(\frac{FL}{AE}\right)_S = \left(\frac{FL}{AE}\right)_C$$

From Table A.7, the cross-sectional area of the pipe is $A_S = 11.9 \text{ in.}^2$ and its inside diameter is 10.020 in. Thus,

$$A_C = \frac{\pi d^2}{4} = \frac{\pi(10.02 \text{ in.})^2}{4} = 78.9 \text{ in.}^2$$

Substituting the values of A and E into the compatability condition and recognizing that the lengths cancel, we have

$$F_S = \frac{(EA)_S}{(EA)_C} F_C = \frac{(30 \times 10^6 \text{ lb/in.}^2)(11.9 \text{ in.}^2)}{(3 \times 10^6 \text{ lb/in.}^2)(78.9 \text{ in.}^2)} F_C = 1.5 F_C$$

Combining this expression with the equilibrium equation, we obtain

$$F_S = -60 \text{ kips or } 60 \text{ kips (compression)}$$
$$F_C = -40 \text{ kips or } 40 \text{ kips (compression)}$$

Answer

The forces on both materials are compressive, which is physically correct.

We now check the assumption of linearly elastic material behavior. The maximum compressive stresses in the steel and in the concrete are

$$\sigma_S = \left(\frac{P}{A}\right)_S = \frac{60 \text{ kips}}{11.9 \text{ in.}^2} = 5.0 \text{ ksi}$$

$$\sigma_C = \left(\frac{P}{A}\right)_C = \frac{40 \text{ kips}}{78.9 \text{ in.}^2} = 0.5 \text{ ksi}$$

The stress-strain diagram for concrete does not have a definite linear range. However, the stress σ_C is well below the ultimate strength of 5 ksi (see Table A.2); therefore, it is reasonable to approximate the stress-strain curve by a straight line over this range. Since the stress in the steel is also below the proportional limit, the solution is valid.

(b) The question here is, does the steel deform enough so that the loading plate comes into contact with the concrete? We assume that it does. If the load on the concrete turns out to be compressive, the assumption is confirmed; if it turns out to be tensile, which is physically impossible, the assumption is incorrect and all the load is carried by the pipe.

Referring to Figure 6.11(d), we see that the compatibility condition is

$$\delta_S = \delta_C - 0.01 \text{ in.}$$

Otherwise, the solution is the same as in part (a). We have

$$\left(\frac{FL}{AE}\right)_S = \left(\frac{FL}{AE}\right)_C - 0.01 \text{ in.}$$

$$F_S = \frac{(EA)_S}{(EA)_C} F_C - \frac{(AE)_S}{L_S} (0.01 \text{ in.})$$

The coefficient of F_C was computed in part (a). Thus,

$$F_S = 1.5 F_C - \frac{(11.9 \text{ in.}^2)(30 \times 10^6 \text{ lb/in.}^2)(0.01 \text{ in.})}{120 \text{ in.}}$$
$$= 1.5 F_C - 29\ 800 \text{ lb}$$

Combining this equation with the equilibrium equation, we obtain

$$F_C = -28 \text{ kips or } 28 \text{ kips (conpression)}$$
$$F_S = -72 \text{ kips or } 72 \text{ kips (compression)}$$

Answer

Both forces are compressive; therefore, our assumption that the loading plate makes contact with the concrete is correct.

The stress in the steel is now

$$\sigma_S = \left(\frac{P}{A}\right)_S = \frac{72 \text{ kips}}{11.9 \text{ in.}^2} = 6.1 \text{ ksi}$$

which is still well below the proportional limit. Since the stress in the concrete is less than in part (a), it is also below the proportional limit. Thus, our solution is valid.

PROBLEMS

6.17 Determine the stresses in each of the wires supporting the rigid bar shown if $F = 5$ kips.

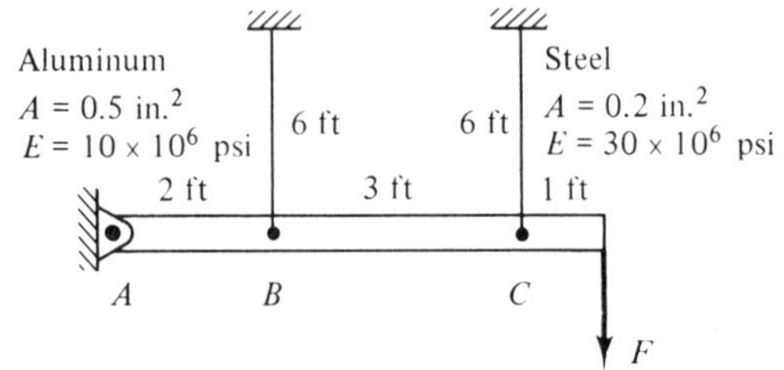

Problem 6.17

6.18 What is the maximum load F that can be applied to the rigid horizontal bar shown in Problem 6.17 if the allowable stress in the steel wire is 30 ksi?

6.19 In the assembly shown, bar C is rigid and the nuts on the rods A and B are tightened until they are just snug. Determine the stress induced in rod B by advancing the nut at its end by one-half turn. There are 10 threads/in.

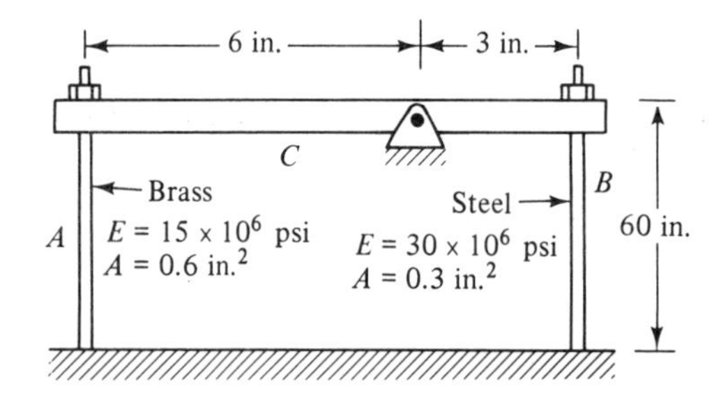

Problem 6.19

6.20 A rigid floor slab with a mass of 3200 kg is supported by three columns as shown. What is the compressive force on each of the members?

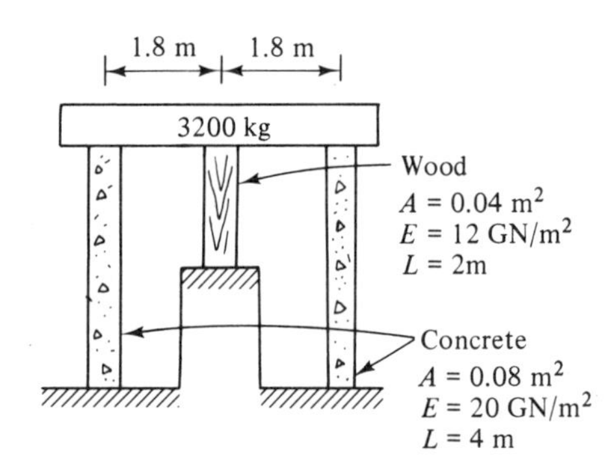

Problem 6.20

6.21 Same as Problem 6.20, except that there is an initial gap of 0.02 mm between the slab and center column.

6.22 A concrete pedestal 250 mm high has a 100-mm × 100-mm square cross section and is reinforced by five steel rods with a diameter of 14 mm and a yield strength of 690 MN/m^2. The concrete has an ultimate compressive strength of 35 MN/m^2. What is the largest load F that can be supported without yielding the steel or crushing the concrete? Use a safety factor of 2. How does this compare with the load that could be supported by the pedestal if it had no reinforcing? $E_S = 207\ \text{GN/m}^2$ and $E_C = 21\ \text{GN/m}^2$.

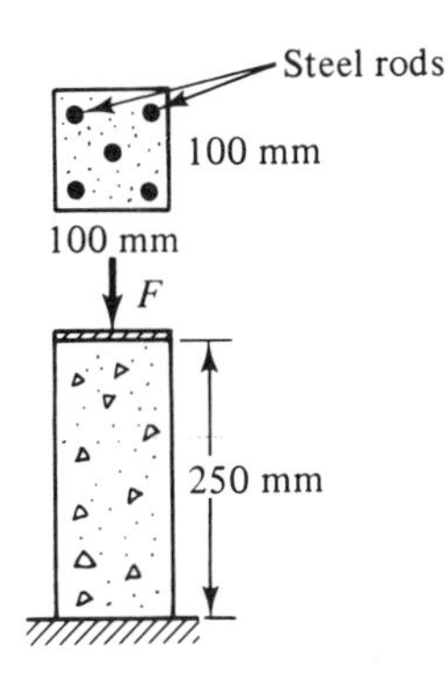

Problem 6.22

6.23 In an attempt to increase the load-carrying capacity of a 150-mm × 150-mm square oak column, 6 mm thick × 150 mm wide steel plates are attached to two sides as shown. By what percentage does the presence of the plates change the load-carrying capacity of the column? The allowable stresses are 40 MN/m^2 for the oak and 150 MN/m^2 for the steel. $E_S = 207\ \text{GN/m}^2$ and $E_W = 12.4\ \text{GN/m}^2$.

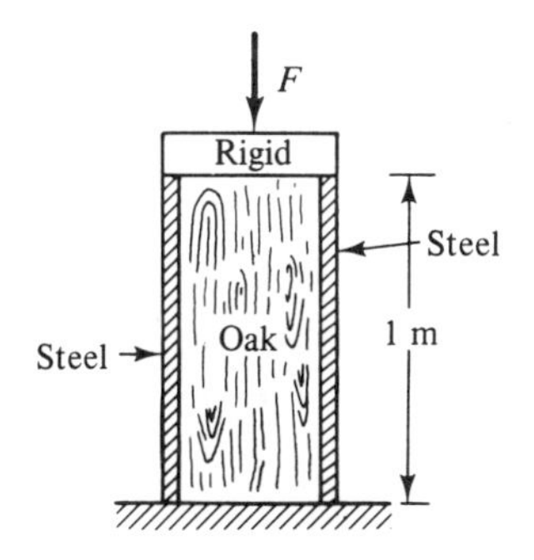

Problem 6.23

6.24 A 4000-lb object is supported by two rods as shown. The aluminum rod is cut 0.04 in. too long; thus, it is initially slack. Determine the tensile stress in each rod after the load is applied. What is the elongation of the steel rod?

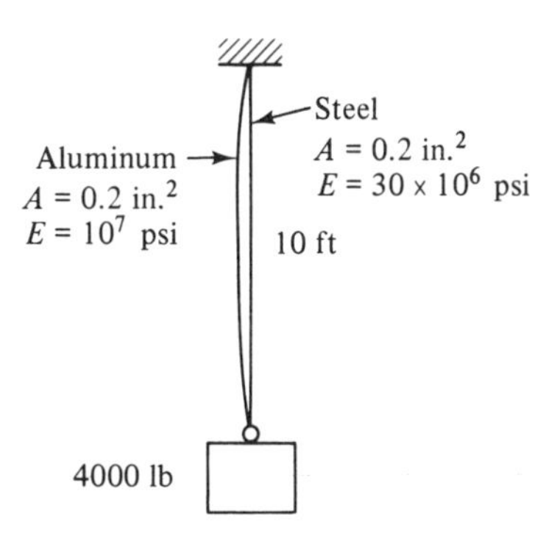

Problem 6.24

6.25 Three steel cables support a load F as shown. The cross-sectional area of the middle cable is three-fourths that of the outer cables. Determine the angle θ such that all cables carry the same load.

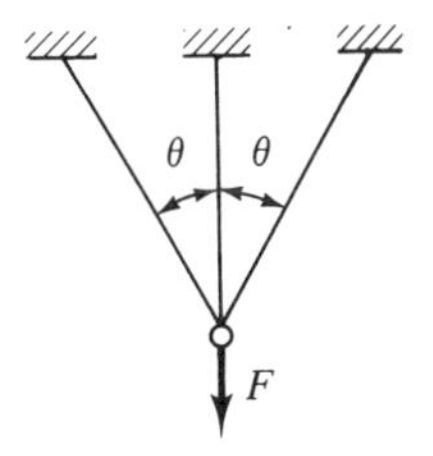

Problem 6.25

6.26 A 254-mm × 102-mm Universal Beam with a mass of 22 kg/m (see Table A.8) is used as a column in a building. The member is attached to rigid supports at the top and bottom and is loaded as shown. What are the reactions at the supports if the weight of the member is neglected?

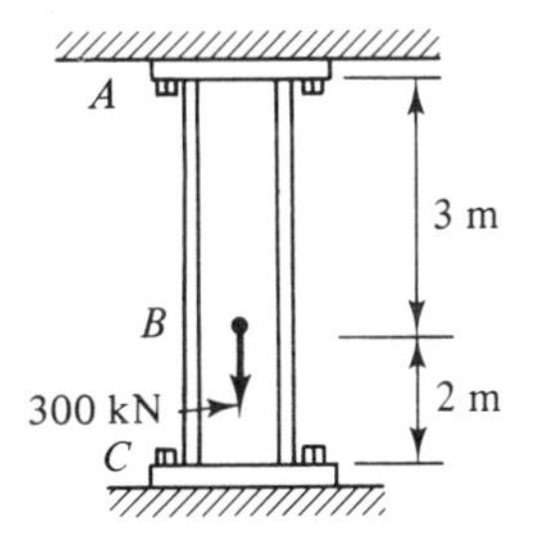

Problem 6.26

6.27 Determine the stress in each portion of the composite rod shown. The member is initially stress free.

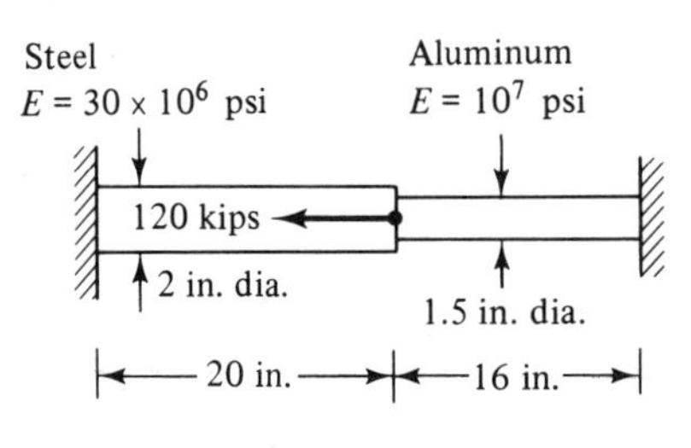

Problem 6.27

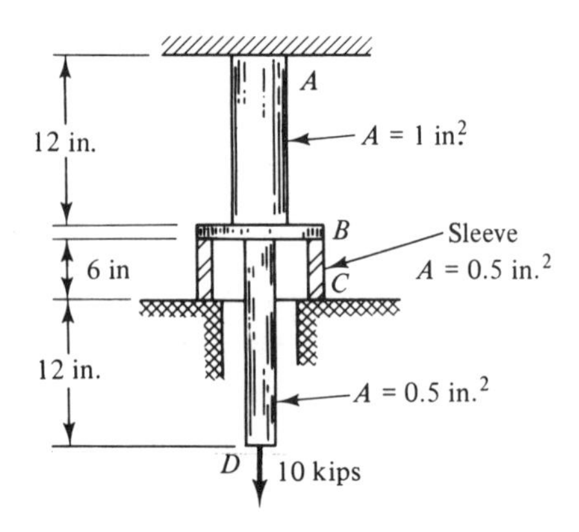

Problem 6.29

6.28 The 1-in. × 1-in. square steel bar shown is initially 0.001 in. shorter than the distance between the rigid walls. Determine the stress in each portion of the member due to the applied load.

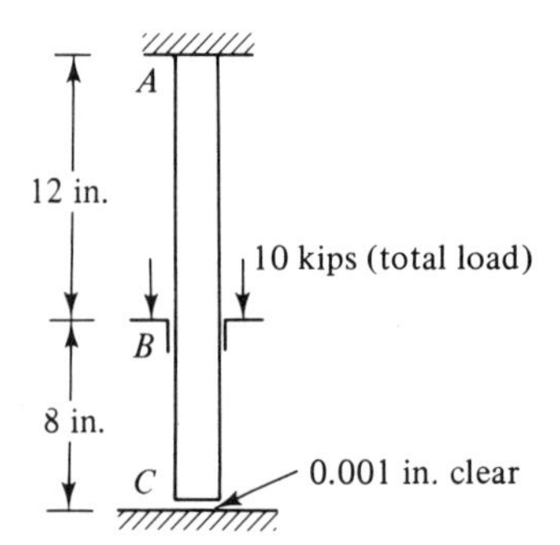

Problem 6.28

6.29 The vertical steel rod ABD shown is fastened at its upper end A and is supported at B by a loose-fitting steel sleeve. Determine the displacement of end D of the rod and the bearing stress between the sleeve and the support. Assume the sleeve is just snug before the load is applied. $E = 30 \times 10^6$ psi.

6.30 A uniform bar with a cross-sectional area of 2 in.^{2} is fixed at each end and loaded as shown. The stress-strain curve for the material is as indicated and is the same in tension and compression. If the compressive strain in portion BC of the bar is 0.0016 in./in., determine the stress in each section of the bar and the applied load F.

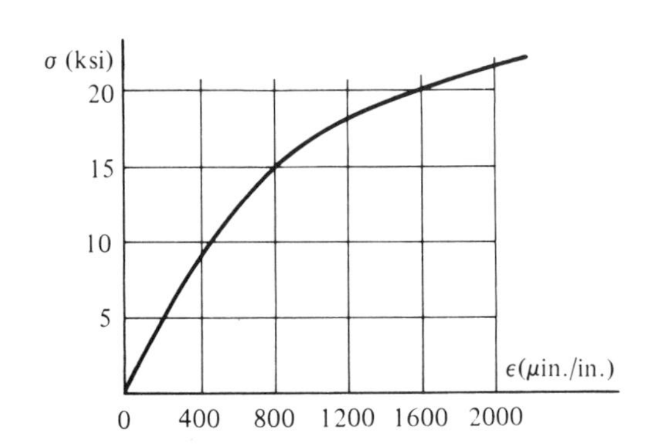

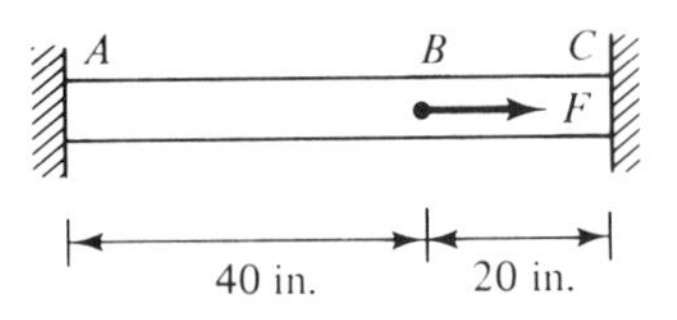

Problem 6.30

6.5 THERMAL STRAINS AND STRESSES

When a material is heated or cooled, it expands or contracts. For example, consider a dimension L_0 of a body at some reference temperature T_0. Under a uniform temperature change $\Delta T = T - T_0$, this dimension will change by an amount

$$\delta = L - L_0 = \alpha L_0 \Delta T \tag{6.3}$$

This relation is based upon experimental observations and is valid for moderate changes in temperature. The constant of proportionality, α, is called the *coefficient of thermal expansion.* It has the units of $(°\text{F})^{-1}$ or $(°\text{C})^{-1}$. Typical values of α for some common materials are given in Table A.2 of the Appendix.

Equation (6.3) applies to all linear dimensions of a body subjected to a uniform temperature change. For example, it applies to the change in diameter of a rod as well as to the change in length. If ΔT varies from point to point within the body, the situation is not so simple. In this case, Eq. (6.3) must be applied to infinitesimal line elements and the results integrated to determine the total dimensional change. Problems involving nonuniform temperature changes will not be considered in this text.

The thermally induced change in dimension gives rise to a normal strain

$$\varepsilon_T = \frac{\delta}{L_0} = \alpha \Delta T \tag{6.4}$$

This strain, denoted by ε_T, is called the *thermal strain* to distinguish it from the *mechanical strain* due to the applied loads. The total strain and deformation is the sum of those due to the applied loads and the temperature change. Note that the thermal strain and deformation are positive if the temperature increases and negative if it decreases.

Thermal strains due to a uniform temperature change differ from mechanical strains in that there is no stress associated with them unless the expansion or contraction of the body is restricted. Then, very large stresses can be generated. These so-called *thermal stresses* are responsible for the buckling of pavements on hot summer days. They are also a problem in large-scale structures, such as bridges, piping systems, and railroad tracks, where expansion joints or some other means of accommodating the thermal deformations must be provided. By their very nature, thermal stress problems are statically indeterminate.

Example 6.6. Stresses and Deformations in a Cooled Rod. A brass rod [$\alpha = 19 \times 10^{-6}\,(°\text{C})^{-1}$] has a length of 3 m and a cross-sectional area of $1.2 \times 10^{-3}\text{m}^2$ at 20°C. The rod is then subjected to an axial tensile loading of 24 kN and the temperature is decreased to 0°C. Determine (a) the maximum normal stress within the member and (b) the change in its length; $E = 83\ \text{GN/m}^2$.

Solution. (a) Since the thermal contraction is not restricted in any way, the only stress is that due to the applied load. The change in cross-sectional area due to the temperature change is proportional to the square of α and is considered negligible. Thus,

$$\sigma = \frac{P}{A} = \frac{24 \times 10^3 \text{ N}}{1.2 \times 10^{-3} \text{ m}^2} = 20 \times 10^6 \text{ N/m}^2 \text{ or } 20 \text{ MN/m}^2 \text{ (tension)} \qquad \textbf{Answer}$$

(b) The total change in length is the sum of the changes due to the temperature decrease and the applied load. Since the stress in the rod is below the proportional limit (see Table A.2), we have

$$\begin{aligned}\delta_{\text{total}} &= \frac{PL}{AE} + \alpha L \Delta T \\ &= \frac{(24 \times 10^3 \text{ N})(3 \text{ m})}{(1.2 \times 10^{-3} \text{ m}^2)(83 \times 10^9 \text{ N/m}^2)} \\ &\quad + \left[19 \times 10^{-6} (^\circ\text{C})^{-1}\right](3 \text{ m})(0^\circ\text{C} - 20^\circ\text{C}) \\ &= -0.42 \times 10^{-3} \text{ m or } 0.42 \text{ mm (shortening)} \qquad \textbf{Answer}\end{aligned}$$

Here, the contraction due to the decrease in temperature exceeds the elongation due to the applied load.

Example 6.7. *Thermal Stresses in a Bar.* A 6-ft long steel bar with a cross-sectional area of 3 in.^{2} is placed between two walls and the temperature is increased by 70°F [Figure 6.12(a)]. If we know that the bar does not buckle, what is the maximum normal stress within it if (a) the walls are immovable and (b) the walls move apart 0.01 in.?

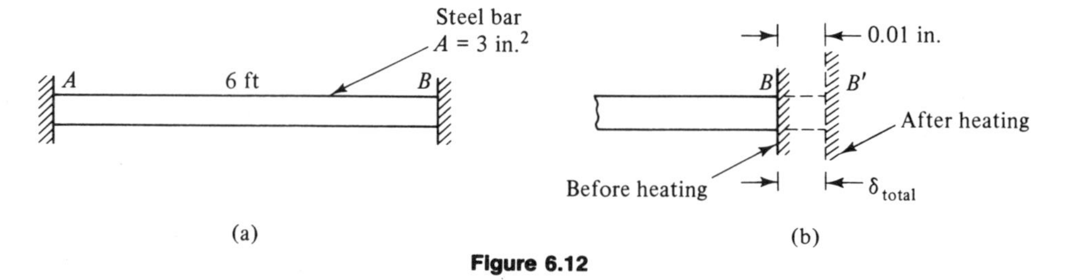

Figure 6.12

Solution. (a) As the bar tends to expand, the walls exert compressive forces of unknown magnitude F on the bar and prevent it from changing length. The compatibility condition is

$$\delta_{\text{total}} = \frac{-FL}{AE} + \alpha L \Delta T = 0$$

from which we obtain

$$\begin{aligned}F = \alpha E \Delta T &= \left[6.5 \times 10^{-6} (^\circ\text{F})^{-1}\right](3 \text{ in.}^2)(30 \times 10^6 \text{ lb/in.}^2)(70^\circ\text{F}) \\ &= 40\ 950 \text{ lb}\end{aligned}$$

The maximum normal stress is

$$\sigma = \frac{P}{A} = \frac{40\ 950 \text{ lb}}{3 \text{ in.}^2} = 13\ 650 \text{ psi (compression)} \qquad \textbf{\textit{Answer}}$$

Since this stress is well below the proportional limit of steel, the solution is valid.

(b) In this case, the bar can expand an amount equal to the distance the walls move apart [Figure 6.12(b)]. Thus, the compatibility condition is

$$\delta_{\text{total}} = \frac{-FL}{AE} + \alpha L \Delta T = 0.01 \text{ in.}$$

from which we obtain

$$F = \alpha A \Delta E T - \frac{AE}{L}(0.01 \text{ in.})$$

The first term on the right-hand side of the equation is simply the value of F computed in part (a). Thus,

$$F = 40\ 950 \text{ lb} - \frac{(3 \text{ in.}^2)(30 \times 10^6 \text{ lb/in.}^2)(0.01 \text{ in.})}{72 \text{ in.}} = 28\ 450 \text{ lb}$$

$$\sigma = \frac{P}{A} = \frac{28\ 450 \text{ lb}}{3 \text{ in.}^2} = 9480 \text{ psi (compression)} \qquad \textbf{\textit{Answer}}$$

A comparison of this result with that obtained in part (a) shows that the thermal stress can be reduced considerably by providing for even a relatively small amount of expansion or contraction.

This problem can also be approached in a slightly different way by imagining that one wall is removed so that the bar can expand freely and then computing the force F required to compress it back to its required length. It is left to the reader to show that this approach leads to the same results presented here.

PROBLEMS

6.31 A section of welded-steel railroad rail is 1 mi long. What would be the variation in its length from the cold of winter ($-20°$F) to the heat of summer ($+140°$F) if it were not attached to the ties and were free to expand and contract? $\alpha = 6.5 \times 10^{-6}$ $(°\text{F})^{-1}$.

6.32 To prevent overstressing of the piping in an oil refinery, expansion joints are placed at regular intervals. What is the required spacing of the joints if each can accommodate an expansion of 1 in.? The pipe is made of steel [$\alpha = 6.5 \times 10^{-6}$ $(°\text{F})^{-1}$] and operates at a temperature of 1400°F above ambient.

6.33 An aluminum [$\alpha = 23 \times 10^{-6}$ $(°\text{C})^{-1}$] plate 1.00 m long, 0.50 m wide, and 0.02 m thick contains a 0.2-m diameter hole at the center. What are the dimensions of the plate and diameter of the hole if the temperature is raised 100°C? What is the change in area of the hole?

6.34 A 2-mm diameter copper wire 20 m long is subjected to a tensile loading of 400 N while undergoing a temperature decrease of 40°C. Determine the changes in length and diameter of the wire. $E = 120$ GN/m^2, $\nu = 1/3$, and $\alpha = 17 \times 10^{-6}$ (°C)$^{-1}$.

6.35 A 100-mm wide × 20-mm thick 2024-T3 aluminum bar 2 m long is attached to rigid walls at each end by smooth pins as shown. Determine the required diameter d of the pins if the assembly is to withstand a temperature decrease of 30°C. The allowable average normal stress in the bar is 150 MN/m^2 and the allowable average shear stress in the pins is 75 MN/m^2.

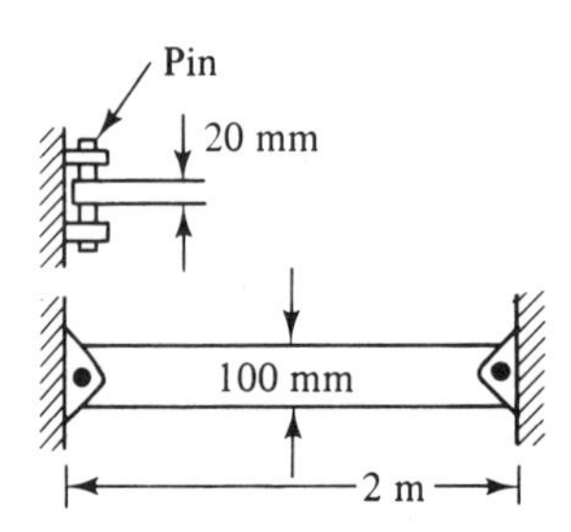

Problem 6.35

6.36 A steel rod attached to rigid walls at each end experiences a tensile force of 1200 lb at 70°F. If the stress is not to exceed 18 ksi at 0°F, what is the required cross-sectional area of the rod? At what temperature will the rod be stress-free? $E = 30 \times 10^6$ psi and $\alpha = 6.5 \times 10^{-6}$ (°F)$^{-1}$.

6.37 A concrete cylinder 0.5 m long is placed between rigid surfaces with an initial gap between them of 0.5 mm. What temperature increase can be tolerated without crushing the concrete if its ultimate compressive strength is 30 MN/m^2? Use a safety factor of 3. $\alpha = 10.8 \times 10^{-6}$ (°C)$^{-1}$ and $E = 21$ GN/m^2.

6.38 A composite bar is attached to walls at each end and loaded as shown. When the temperature is decreased 100°F, the walls are observed to move 0.015 in. closer together. What is the normal stress in each segment of the bar?

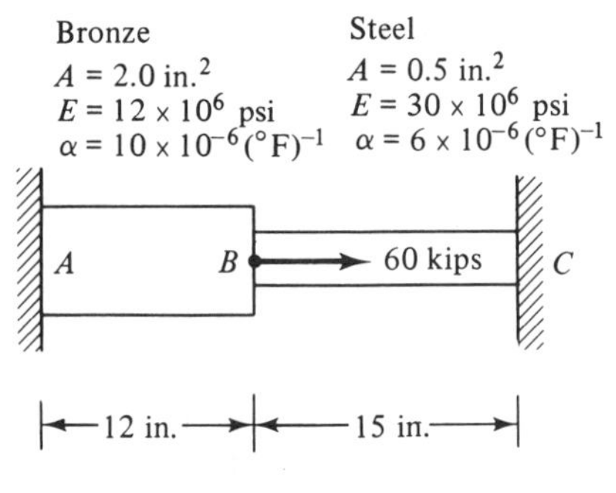

Problem 6.38

6.39 A rigid block supported by a 20-ft long steel wire that passes through a hole in a rigid floor has the position shown when the temperature is 90°F. If the tensile stress in the wire is 18 ksi when the temperature is 10°F, what is the weight W of the block? $E = 30 \times 10^6$ psi and $\alpha = 6.5 \times 10^{-6}$ (°F)$^{-1}$.

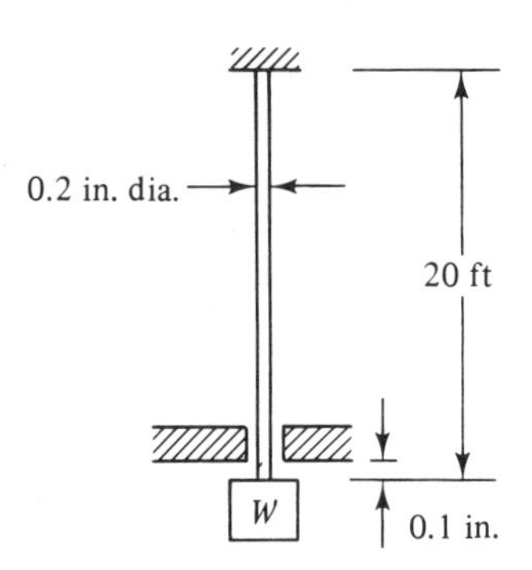

Problem 6.39

6.40 A rigid bar is hinged at the center and supported by a wire at each end as shown. If the temperature is decreased 40°C, what are the resulting tensile stresses in the wires? Through what angle θ will the bar rotate? (See Figure on p. 265).

6.41 Determine the vertical force F that must be applied at end A to make the initially horizontal rigid bar ABC in Problem

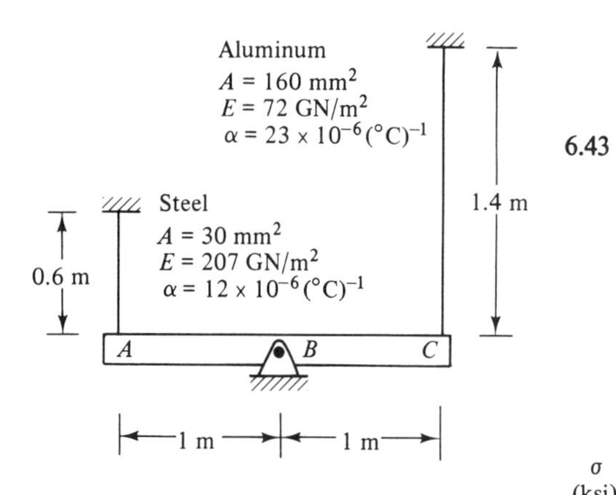

Problem 6.40

6.40 return to horizontal after a temperature drop of 20°C. What are the resulting stresses in the support wires?

6.42 A brass sleeve 16 in. long with $\frac{3}{4}$ in. ID and 1 in. OD is slipped over a $\frac{1}{2}$-in. diameter high-strength steel bolt and the nut is advanced one-eighth turn beyond the point where the ends of the members first come into contact. If the bolt has 16 threads/in., what is the stress induced in each of the members? If the part is assembled at 70°F, at what temperature will the sleeve become loose on the bolt?

6.43 A square bar with a cross-sectional area of 2 in.^{2} is attached to rigid walls at each end. At 70°F, the tensile force in the bar is 20 kips. If the stress-strain diagram for the material is as shown, what is the minimum temperature to which the bar can be subjected if the maximum stress is not to exceed 20 ksi? $\alpha = 12 \times 10^{-6}$ (°F)$^{-1}$.

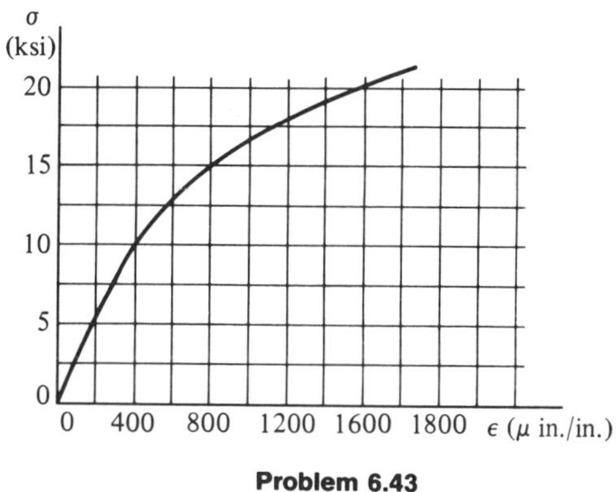

Problem 6.43

6.6. STRAIN AND STRESS CONCENTRATIONS

The strains in a member depend upon its geometry and the applied loads. Discontinuities in the geometry, such as holes and notches, give rise to localized areas of high strain. These effects, which are evident in Figures 6.1 and 6.3, are called *strain concentrations*. Strain concentrations may also occur at other discontinuities, such as voids and rigid inclusions in the material, and at points of load application.

The effect of geometrical discontinuities on the strains can be described in terms of a *strain concentration factor*, K_ε, defined as the ratio of the maximum strain with the discontinuity to the value the strain would have if the discontinuity were not present. For an axially loaded member, this definition can be expressed as

$$K_\varepsilon = \frac{\varepsilon_{max}}{\varepsilon_{avg}} \tag{6.5}$$

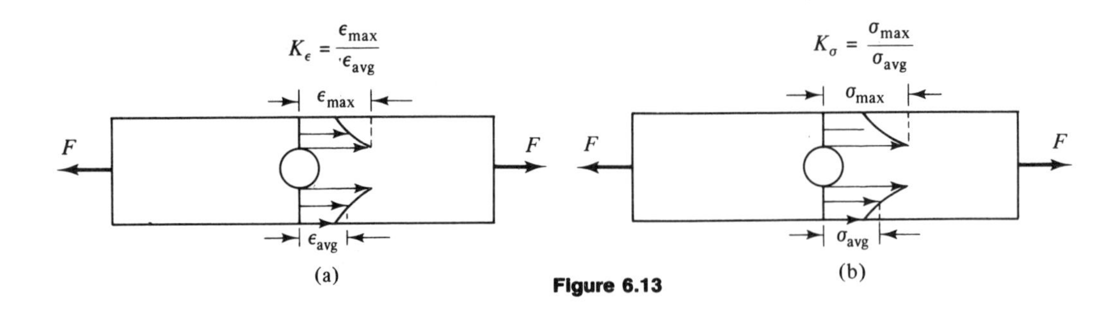

Figure 6.13

where ε_{avg} is the average strain over the cross section and ε_{max} is the maximum strain [Figure 6.13(a)]. The value of K_ε is usually determined experimentally.

Geometrical discontinuities can also result in localized areas of high stress, or *stress concentrations*. The increase in the maximum value of stress due to a discontinuity can be accounted for by a *stress concentration factor*, K_σ. For an axially loaded member,

$$K_\sigma = \frac{\sigma_{max}}{\sigma_{avg}} \tag{6.6}$$

where σ_{max} is the maximum stress and σ_{avg} is the average stress over the cross section [Figure 6.13(b)]. Thus, the maximum stress can be determined by multiplying the value obtained from the relation $\sigma = P/A$ by K_σ.

The value of K_σ depends upon the strain distribution (i.e., upon K_ε) and the stress-strain relation for the material. For linearly elastic material behavior, the stress is directly proportional to strain, and $K_\sigma = K_\varepsilon$. The corresponding values of K_σ for axially loaded flat bars containing circular holes, semicircular grooves, and fillets are given in Figure 6.14. These results are based upon average stresses computed by using the net cross-sectional area of the member. For members with fillets, the area of the smaller portion is used. Note that the more drastic the change in cross section, the larger the value of K_σ. Stress concentration factors for other discontinuities, members, and loadings are available in engineering handbooks and other references.

Materials that exhibit brittle behavior usually have stress-strain diagrams that are very nearly linear all the way to fracture. Consequently, the values of K_σ given in Figure 6.14 can be used to determine the failure load for members made of such materials. The load to produce fracture is determined by setting the maximum stress at the discontinuity equal to the ultimate strength of the material.

If the material can undergo inelastic deformations, the stress distribution tends to become uniform once the yield strength is exceeded. As a result, the stress concentration factor is lower than the values given in Figure 6.14 ($K_\sigma < K_\varepsilon$). This is illustrated in Figure 6.15, which shows the stress distribution at a hole in a bar made of an elastic-perfectly plastic material for three different load levels. The conditions at locations *a*, *b*, and *c* on the cross section are denoted by the corresponding points on the stress-strain diagram.

As the load is increased from zero, the maximum strain increases and so does the maximum stress, until it is just equal to the yield strength of the material. This

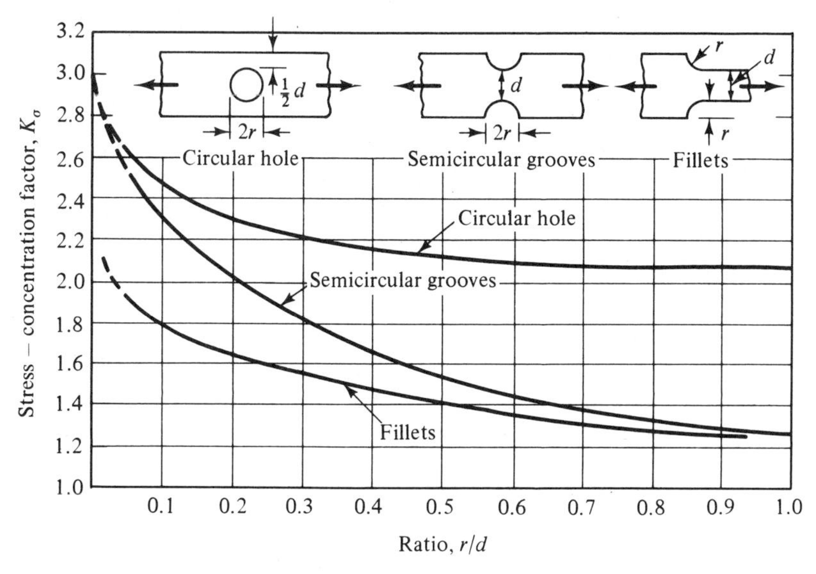

Figure 6.14. Stress Concentration Factors for Axially Loaded Flat Bars.

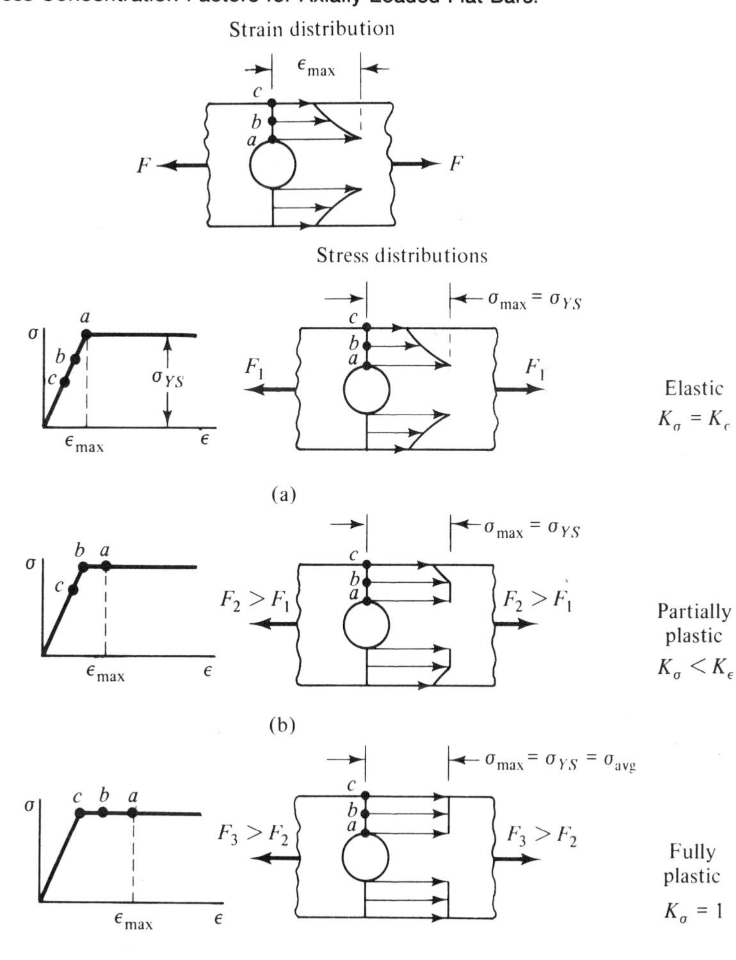

Figure 6.15

is the situation shown in Figure 6.15(a). Up to this load, $K_\sigma = K_\varepsilon$, and the values given in Figure 6.14 apply. If the load is increased further, the maximum strain increases, but the maximum stress remains equal to the yield strength. Strains at some distance away from the hole are now at, or above, the proportional limit strain; therefore, the stress distribution becomes more nearly uniform [Figure 6.15(b)]. This causes K_σ to decrease, since the average stress increases while the maximum stress remains the same. Thus, $K_\sigma < K_\varepsilon$.

If the load continues to increase, a situation is eventually reached where the strains are at, or above, the proportional limit strain everywhere over the cross section. The stresses are then uniformly distributed, as shown in Figure 6.15(c). This is referred to as the *fully plastic condition*, and the load at which it occurs is called the *fully plastic load*, F_{fp}. The maximum stress is equal to the average stress in this case; therefore, $K_\sigma = 1$. In other words, the discontinuity has no effect other than to decrease the cross-sectional area available to carry the load. All other effects are wiped out by the ability of the material to undergo inelastic deformations. This will not be the case for all types of loadings, however. Materials that are ductile under static loadings tend to behave in a brittle fashion when subjected to cyclic loadings, such as occur in moving machine parts or vibrating structures. For such loadings, the stress concentration remains.

A ductile member does not necessarily fail when the fully plastic condition is reached. Experiments show that the region of large strains is still very much localized near the discontinuity; therefore, the overall deformations of the member will be relatively small. Gross yielding of the member and the accompanying large deformations do not result until the stresses at sections away from the discontinuity reach the yield strength of the material. However, the member may fracture at the reduced section before this happens. It is usually sufficiently accurate to take $K_\sigma = 1$ when computing the fracture load for ductile materials.

It should be clear from the preceding discussion that strain concentrations are most serious in members made of materials that behave in a brittle manner. The discontinuities result in greatly increased stresses, which can easily lead to fracture. For statically loaded members that behave in a ductile manner, the material can yield in the localized regions of high stress and continue to carry the load.

Example 6.8. Design of a Member with a Hole. What thickness t is required if the axially loaded member shown in Figure 6.16 is to support a load of 10 kips with a safety factor of 2 and (a) it is made of a brittle material with an ultimate strength of 40 ksi and (b) it is made of a ductile material with a yield strength of 36 ksi and an ultimate strength of 65 ksi? For satisfactory performance, the member must not break or undergo gross yielding.

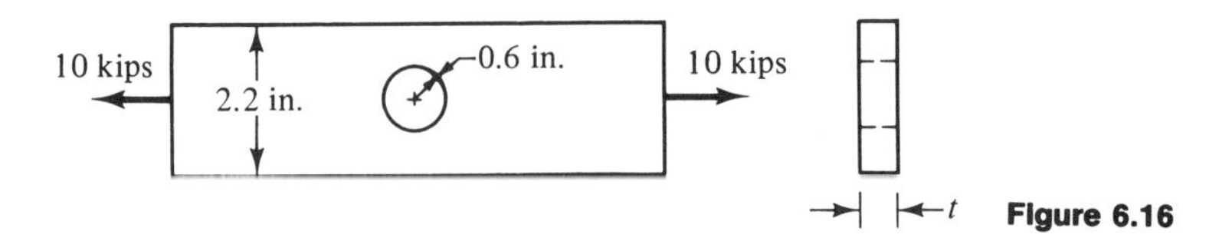

Figure 6.16

Solution. (a) The brittle material can only fail by fracture. The design (failure) load is equal to the safety factor times the working load, or 20 kips. Referring to Figure 6.14, we have $r/d = 0.6$ and $K_\sigma = 2.1$. Now

$$\sigma_{\max} = K_\sigma \sigma_{\text{avg}} = K_\sigma \frac{P}{A}$$

where A is the net cross-sectional area at the hole. Fracture occurs when the maximum stress is equal to the ultimate strength of the material. Thus,

$$\sigma_{\max} = \frac{2.1(20\ 000\ \text{lb})}{(1\ \text{in.})t} \leqslant 40\ 000\ \text{lb/in.}^2$$

from which we obtain

$$t \geqslant 1.05\ \text{in.} \qquad \textbf{\textit{Answer}}$$

(b) Fracture at the hole occurs when the maximum stress there is equal to the ultimate strength of the material. We take $K_\sigma = 1$. Thus,

$$\sigma = \frac{P}{A} = \frac{20\ 000\ \text{lb}}{(1\ \text{in.})t} \leqslant 65\ 000\ \text{lb/in.}^2$$

or

$$t \geqslant 0.31\ \text{in.}$$

Gross yielding occurs when the stress away from the hole reaches the material yield strength:

$$\sigma = \frac{P}{A} = \frac{20\ 000\ \text{lb}}{(2.2\ \text{in.})t} \leqslant 36\ 000\ \text{lb/in.}^2$$

or

$$t \geqslant 0.25\ \text{in.}$$

Failure by fracture governs, and the minimum required thickness is 0.31 in.

PROBLEMS

6.44 The strain distribution in an axially loaded notched bar of constant thickness is as shown. Determine (a) the strain concentration factor K_ϵ, (b) the stress concentration factor K_σ if the material is linearly elastic, and (c) the value of K_σ if the material has the stress-strain diagram shown.

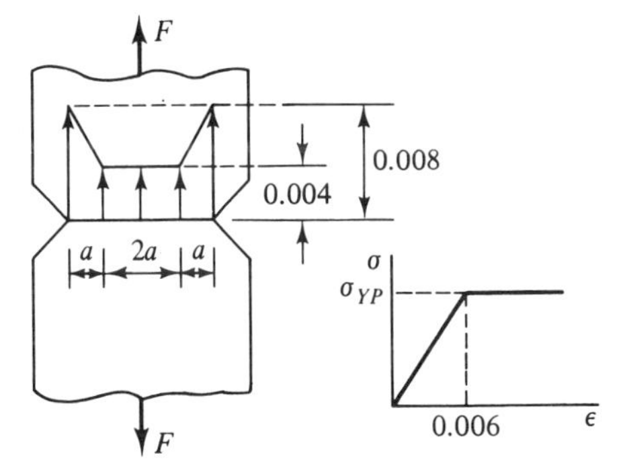

Problem 6.44

6.45 What is the maximum stress in a bar 50 mm wide × 10 mm thick if it has a 10-mm radius semicircular groove in each side and is subjected to a tensile load of 15 kN? Assume the material is linearly elastic.

6.46 A 2-in. wide × $\frac{1}{2}$-in. thick steel bar containing a 1-in. diameter hole at its center is subjected to a tensile load F. The yield point of the material is 35 ksi and the ultimate tensile strength is 60 ksi. Determine (a) the load F at which yielding first occurs, (b) the fully plastic load F_{fp}, and (c) the maximum load that can be supported with a safety factor of 3 without fracture or gross yielding.

6.47 If the bar shown is $\frac{1}{4}$ in. thick and is made of a ductile material with a yield strength of 40 ksi and an ultimate strength of 68 ksi, determine (a) the load F at which yielding first occurs, (b) the fully plastic load F_{fp}, and (c) the maximum load F that can be supported without fracture or gross yielding.

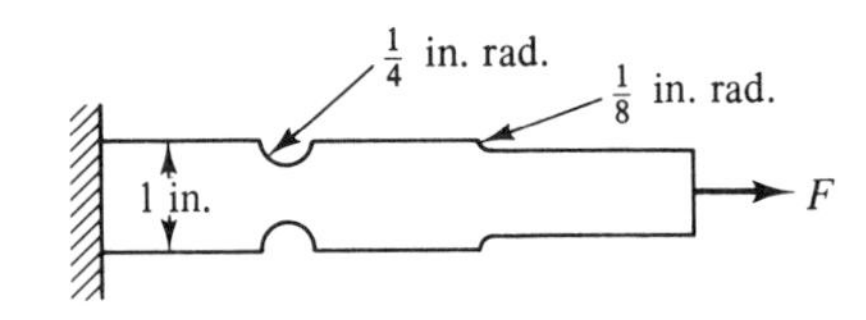

Problem 6.47

6.48 If the bar in Problem 6.47 is made of a brittle material with an ultimate tensile strength of 10 ksi, what is its required thickness t if it is to support a tensile load of 500 lb with a safety factor of 2?

6.49 The bar shown is 20 mm thick and is made of a brittle material with an ultimate tensile strength of 140 MN/m^2. If the bar is to support a tensile load of 20 kN with a safety factor of 3, what is the minimum allowable width d?

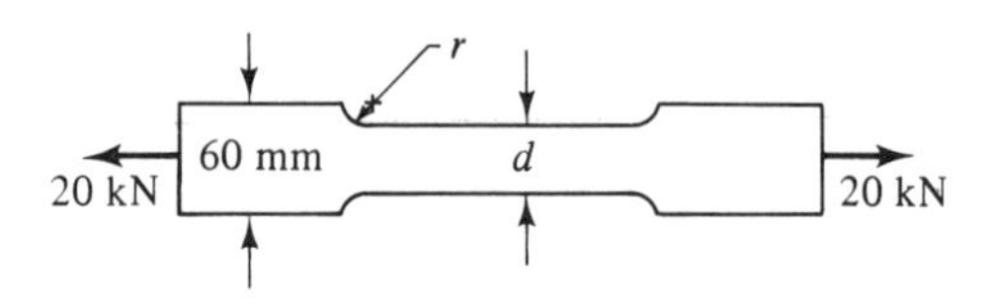

Problem 6.49

6.50 A 30-mm diameter circular hole is to be drilled in a stepped bar as shown to provide a passageway for electrical cables. By what percentage does the presence of the hole reduce the load-carrying capacity of the bar if it is (a) made of a brittle material and (b) made of a ductile material with yield point equal to two-thirds of the ultimate strength?

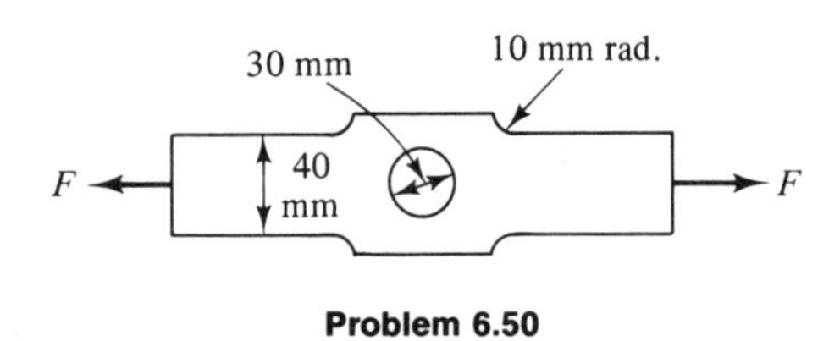

Problem 6.50

6.7 INELASTIC BEHAVIOR OF STATICALLY INDETERMINATE STRUCTURES

When the stresses in an axially loaded member made of a ductile material reach the yield strength, failure by general yielding usually occurs for little or no increase in load. However, if the member is part of a statically indeterminate structure, its deformations are restricted by the other members and additional load can be supported until they, too, yield.

To illustrate this, let us consider the example of a rigid bar hinged at one end and supported by two cables [Figure 6.17(a)]. This is the same example considered in Section 6.4. For simplicity, we assume that both cables have the same cross-sec-

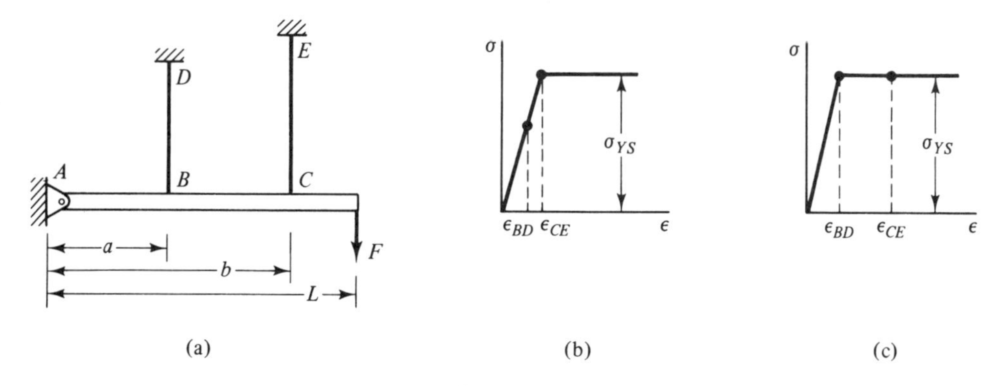

Figure 6.17

tional area and length are made of the same material. We further assume that the material behavior is essentially elastic-perfectly plastic.

An elastic analysis shows that the loads on the cables are

$$F_{BD} = \frac{FLa}{(a^2 + b^2)} \qquad F_{CE} = \frac{FLb}{(a^2 + b^2)}$$

where F is the applied load and the dimensions a, b, and L are as defined in Figure 6.17(a). Cable CE is the most highly stressed because it has the same area as cable BD and supports a larger load.

As the applied load is increased from zero, cable CE will yield first. The stress in cable BD will still be within the elastic range when this happens [Figure 6.17(b)]. Consequently, its elongation will be small and the rotation of the bar will be limited. Any further increases in load must be supported entirely by cable BD because the stress in cable CE will remain equal to the yield strength. Eventually, the second cable will also yield [Figure 6.17(c)]. Both cables will be free to undergo large deformations when this happens, allowing the structure to literally collapse. The load at which this occurs is called the *limit load* for the structure.

Since the stresses in the cables are known to be equal to the yield strength at the limit load, the forces acting upon them can be determined by multiplying the stresses by the cross-sectional areas. Once these forces are known, the problem becomes statically determinate, and the limit load can be obtained directly from the equilibrium equations. Thus, a fully plastic analysis of an indeterminate structure is much simpler than an elastic analysis.

Example 6.9. Limit Load for a Simple Structure. A rigid bar is supported by three wires and is loaded as shown in Figure 6.18(a). The outer wires have a cross-sectional area of $1 \times 10^{-3} m^2$ and are made of a steel alloy with a yield strength of 690 MN/m^2. The center wire has a cross-sectional area of $2 \times 10^{-3} m^2$ and is made of aluminum (σ_{YS} = 480 MN/m^2). What is the limit load for this structure?

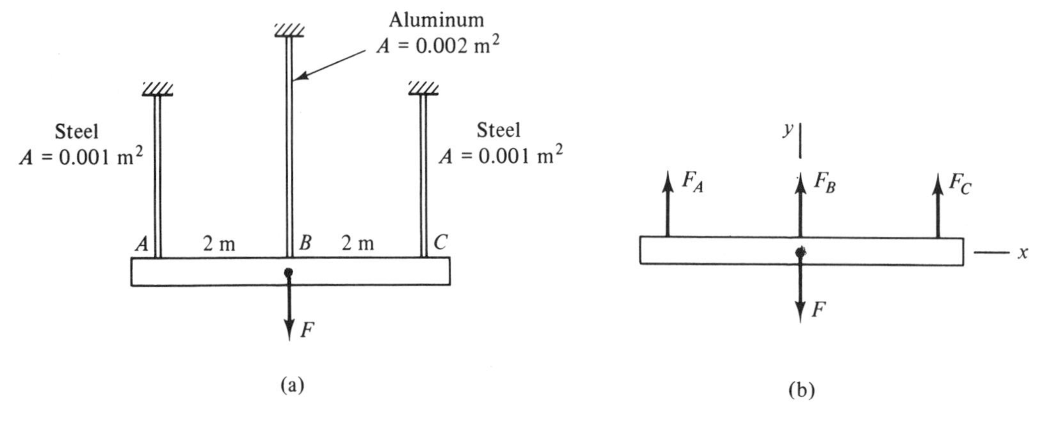

Figure 6.18

Solution. The limit load is reached when all three wires have yielded. The force on each wire at this load is equal to the material yield strength times the cross-sectional area:

$$F_A = F_C = F_{\text{steel}} = (690 \times 10^6 \text{ N/m}^2)(10^{-3} \text{ m}^2) = 690 \times 10^3 \text{ N}$$
$$F_B = F_{\text{alum}} = (480 \times 10^6 \text{ N/m}^2)(2 \times 10^{-3} \text{ m}^2) = 960 \times 10^3 \text{ N}$$

From the FBD of the bar [Figure 6.18(b)], we have

$$\Sigma F_y = F_A + F_B + F_C - F = 0$$

Thus, the fully plastic load is

$$F_{fp} = 2F_A + F_B = 2(690 \times 10^3 \text{ N}) + 960 \times 10^3 \text{ N}$$
$$= 2340 \times 10^3 \text{N or } 2.34 \text{ MN} \qquad \textit{Answer}$$

Notice that the lengths of the wires have no bearing upon the limit load.

PROBLEMS

6.51 to 6.60 Determine the limit load for the structures shown. The cross-sectional dimensions and material properties are as indicated in the figures; otherwise, express the results in terms of the cross-sectional area A and yield strength σ_{YS}.

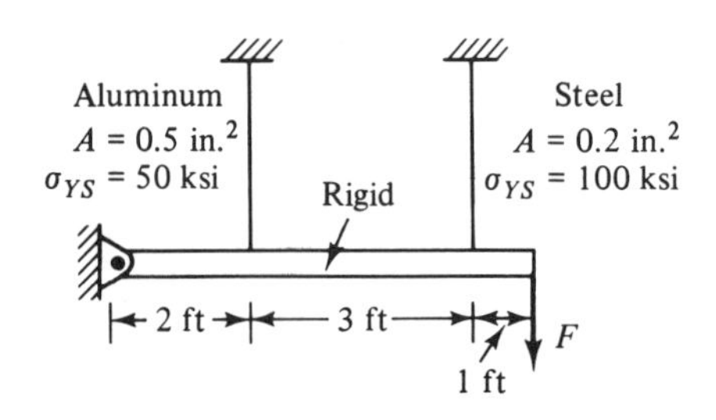

Problem 6.51

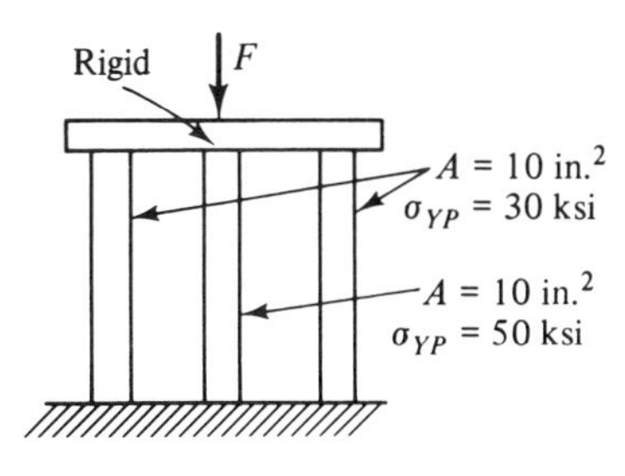

Problem 6.52

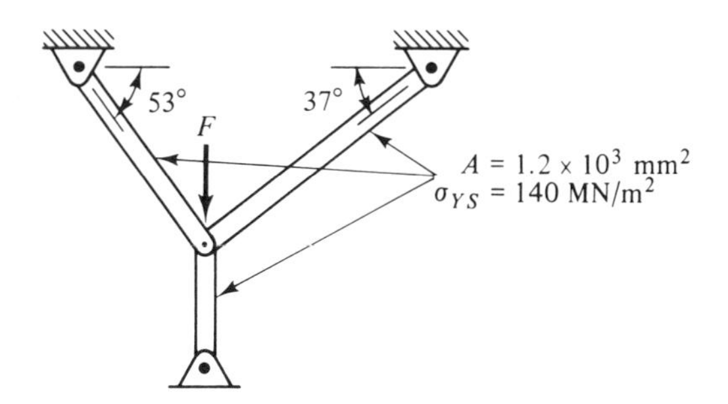

Problem 6.53

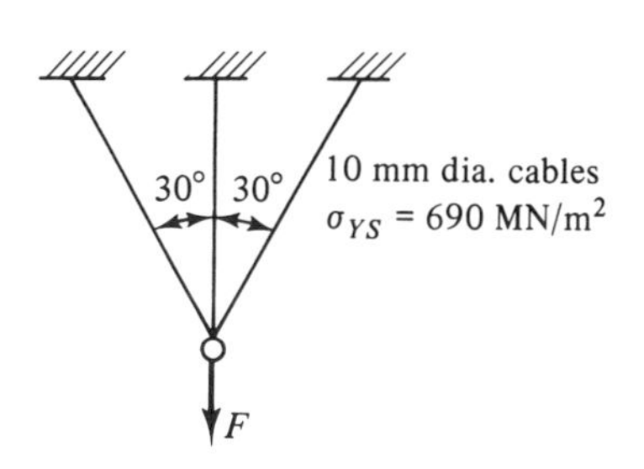

Problem 6.54

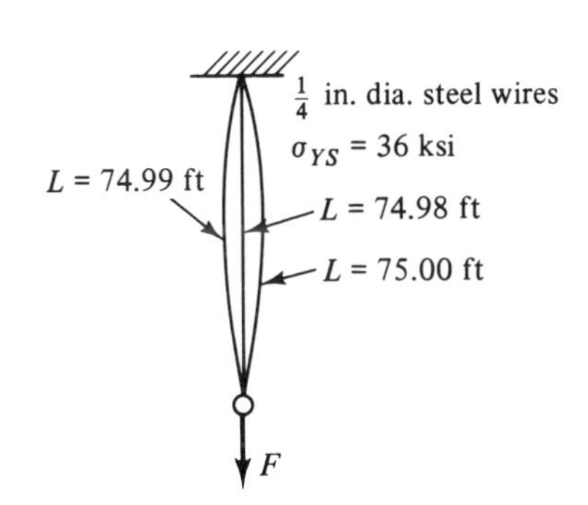

Problem 6.55

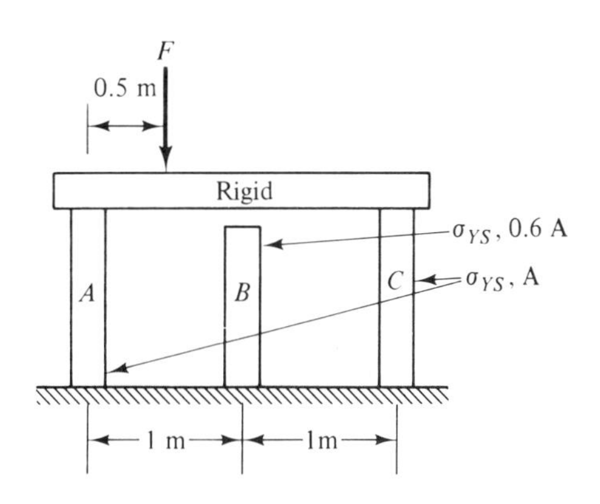

Problem 6.56

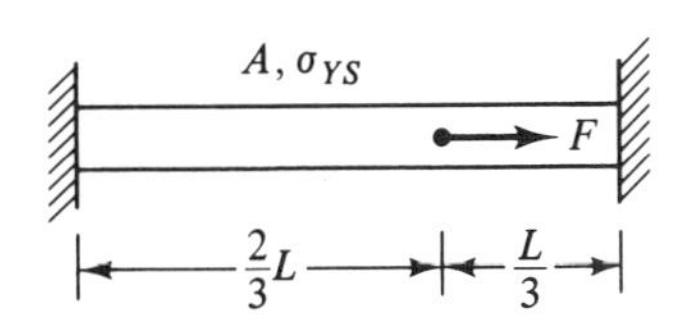

Problem 6.57

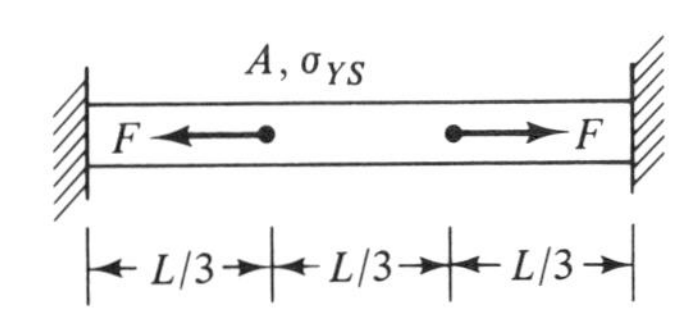

Problem 6.58

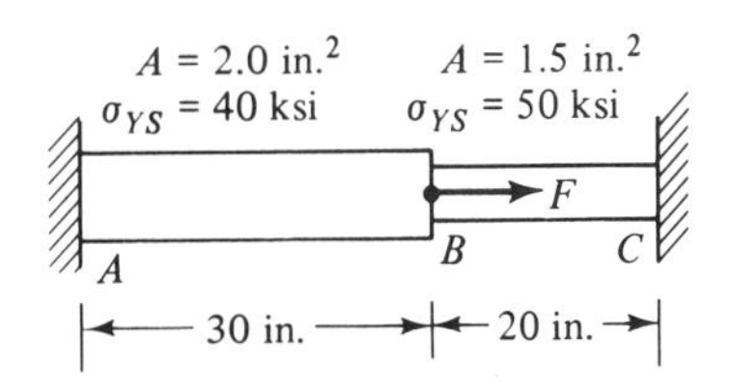

Problem 6.59

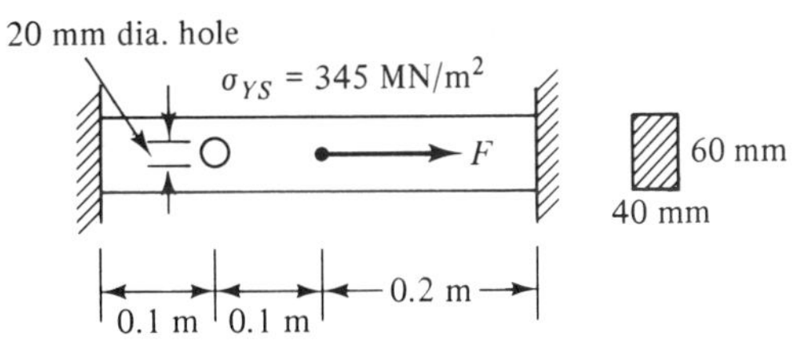

Problem 6.60

6.8 STRESSES IN THIN-WALLED CYLINDERS AND SPHERES

Pressurized cylinders and spheres are commonly encountered in practice, often in the form of piping systems and storage vessels. If the walls of the member are thin (as a rule of thumb, one-tenth of the radius or less), the stresses within it can be readily determined.

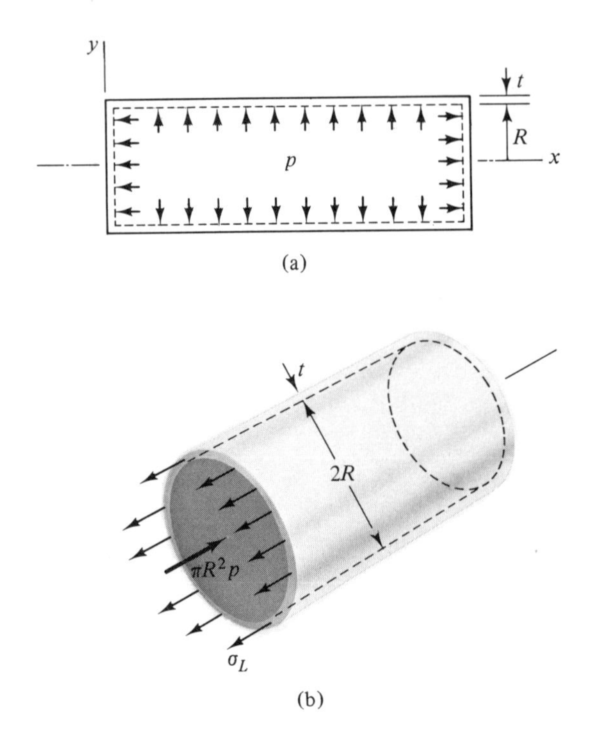

Figure 6.19

Consider a thin-walled circular cylinder with closed ends subjected to a uniform internal gage pressure p [Figure 6.19(a)]. The cylinder tends to elongate when pressurized, giving rise to normal stresses acting upon planes perpendicular to the axis. Sectioning the cylinder and the gas or fluid within it along such a plane, we obtain the free-body diagram shown in Figure 6.19(b). The resultant force due to the internal pressure lies along the centerline of the cylinder; therefore, the conditions for axial loading are met. Consequently, the normal stress σ_L acting in the longitudinal direction will be uniformly distributed over the cross section if the material is homogeneous.

Referring to Figure 6.19(b), we have for equilibrium of the segment of the cylinder

$$\Sigma F_x = \pi R^2 p - 2\pi R t \sigma_L = 0$$

Thus, the longitudinal stress is

$$\sigma_L = \frac{pR}{2t} \tag{6.7}$$

where p is the gage pressure and R and t are the radius and wall thickness of the cylinder, respectively. No distinction is made between the inside and outside radius because they are very nearly equal for thin walls.

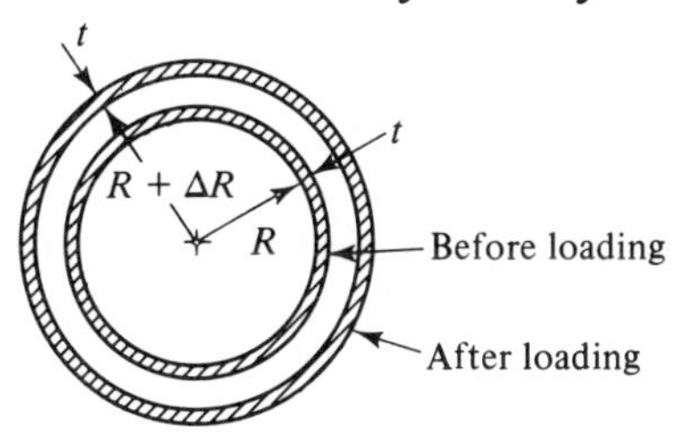

Figure 6.20

Cylinders also tend to expand in the radial direction when pressurized, causing the walls to stretch circumferentially. This is illustrated in Figure 6.20, where the change in radius, ΔR, has been shown greatly exaggerated for clarity. As a result of the radial expansion, there are normal stresses σ_C acting in the circumferential direction on planes parallel to the cylinder axis. These stresses can be determined by using the procedure outlined in Section 5.10.

The strain in the circumferential direction is the change in circumference divided by the original circumference. Referring to Figure 6.20, we have at the inside of the cylinder

$$\varepsilon_{\text{inside}} = \frac{2\pi(R + \Delta R) - 2\pi R}{2\pi R} = \frac{\Delta R}{R}$$

while at the outside

$$\varepsilon_{\text{outside}} = \frac{2\pi(R + \Delta R + t) - 2\pi(R + t)}{2\pi(R + t)} = \frac{\Delta R}{R + t}$$

If the wall thickness is small compared to the radius, the latter expression reduces to

$$\varepsilon_{\text{outside}} \approx \frac{\Delta R}{R} = \varepsilon_{\text{inside}}$$

Thus, the circumferential strains are approximately the same at the inside and outside of the cylinder and, consequently, at all points in between. As a result, the circumferential stresses will be very nearly uniform through the wall thickness if the material is homogeneous.

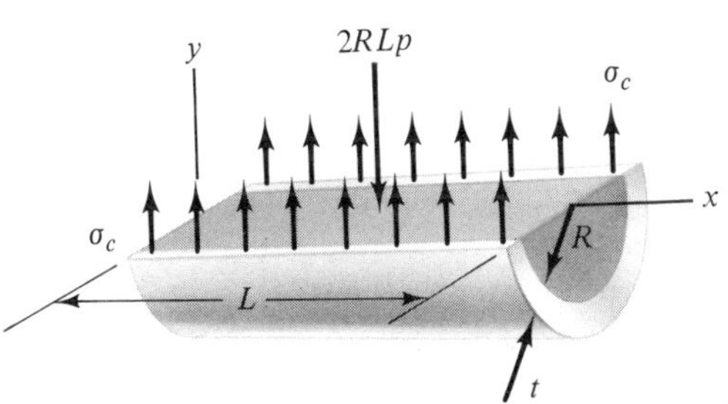

Figure 6.21

The circumferential stresses can be related to the internal pressure by considering the equilibrium of a segment of the cylinder. Figure 6.21 shows the free-body diagram of a segment obtained by removing the ends and then sectioning the cylinder and the fluid within it lengthwise. For equilibrium,

$$\Sigma F_y = 2Lt\sigma_C - 2RLp = 0$$

from which we obtain for the circumferential stress

$$\sigma_C = \frac{pR}{t} \tag{6.8}$$

A comparison of Eqs. (6.7) and (6.8) shows that the circumferential stresses are twice as large as the longitudinal stresses. Equation (6.8) also holds for thin-walled hoops, bands, and rings subjected to radial pressure.

Spheres. A sphere subjected to a uniform internal pressure expands into a larger sphere. From this observation it can be shown that the stresses are uniformly distributed through the wall thickness if the walls are thin and the material is homogeneous. Since the procedure is the same as that used for the circumferential stresses in cylinders, it will not be repeated here.

Sectioning through the sphere and the fluid or gas within it (Figure 6.22), we obtain the equilibrium condition

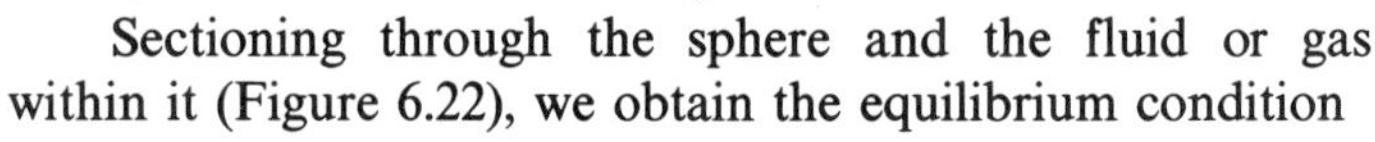

$$\Sigma F_x = \pi R^2 p - 2\pi R t \sigma = 0$$

or

$$\sigma = \frac{pR}{2t} \tag{6.9}$$

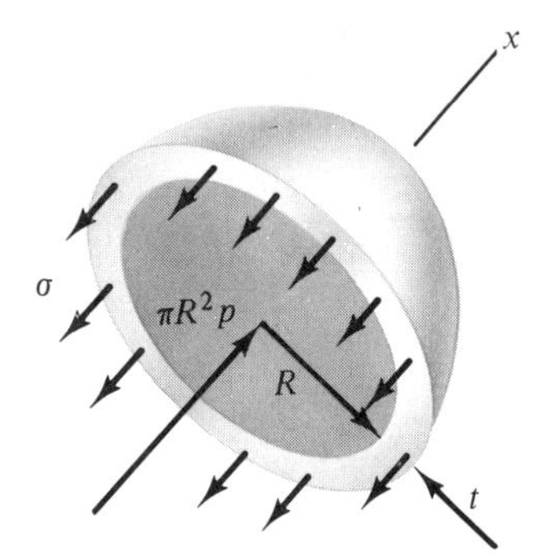

Figure 6.22

This is the same expression obtained for the longitudinal stress in a cylinder. Since cutting planes with different orientation yield free-body diagrams identical to that shown in Figure 6.22, the stresses in a sphere are the same in all directions.

Validity of pressure-stress relations. Although an internal pressure was assumed in the preceding derivation's, Eqs. (6.7) through (6.9) also apply for external pressures if the member does not buckle and lose its cylindrical or spherical shape. The pressure to produce buckling is usually relatively small compared to the internal pressure that can be withstood. For external pressures, the stresses in the member are compressive.

Equations (6.7) and (6.8) do not ordinarily apply near the ends of a cylinder where the radial expansion may be restricted by the end elements. This restriction causes localized shearing and bending effects that are not accounted for in these equations. Similar effects arise in cylinders and spheres at points of attachment to other members and at the junction between wall segments of different thickness. Also, the pressure-stress relations do not apply near holes or other sources of stress concentration. The determination of the stresses in locations such as these is complicated and is beyond the scope of an introductory text such as this.

The assumption of a uniform pressure loading implies that the pressure is imparted by a gas of negligible weight. If imparted by a fluid, the pressure varies linearly with depth. In this case, the pressure-stress relations in Eqs. (6.7) through (6.9) are not strictly valid, but they can be used to obtain an estimate of the stresses. The accuracy of this estimate depends upon the nature of the particular loading of interest.

Representation of stresses. The various stresses acting at a point in a cylinder can be conveniently displayed on a small element of material containing the point and with sides coinciding with the planes upon which the stresses act. To display the stresses acting upon these planes, we need only show them as acting upon the corresponding sides of the element. This representation is illustrated in Figure 6.23(a) in, where the element is shown greatly enlarged for clarity. Note that the normal stresses are shown as being equal on opposite sides of the element. This is a result of the equilibrium requirement that the sum of the forces acting upon the element be zero.

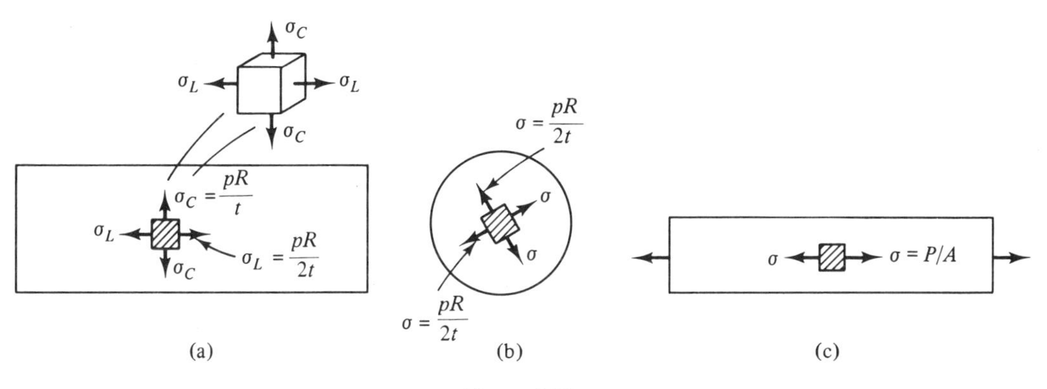

Figure 6.23

This same representation can be used for stresses in a sphere [Figure 6.23(b)]. It can also be used for axially loaded members [Figure 6.23(c)], although it is not so advantageous in this case. We shall have occasion to use this representation many times in later chapters.

Deformations. Deformations of cylinders and spheres are not as easily determined as deformations for axially loaded members because the material is simultaneously stretched or compressed in two directions. However, for rings and similar members in which only the circumferential stress is acting, deformations and strains can be determined by using the methods developed in Section 6.3 (see Example 6.11). Further discussion of the deformations of cylinders and spheres will be deferred until Chapter 9.

Example 6.10. Stresses in an Air Tank. A cylindrical tank with hemispherical ends contains air at a gage pressure of 120 psi [Figure 6.24(a)]. The radius of the tank is 8 in. and the wall thickness is $\frac{1}{4}$ in. Determine the stresses in each portion of the tank.

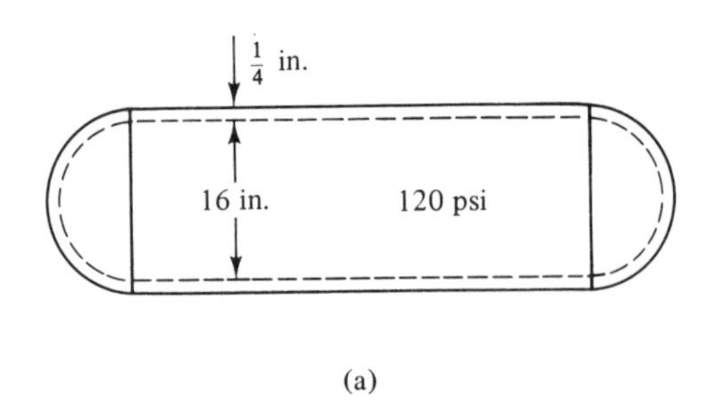

(a)

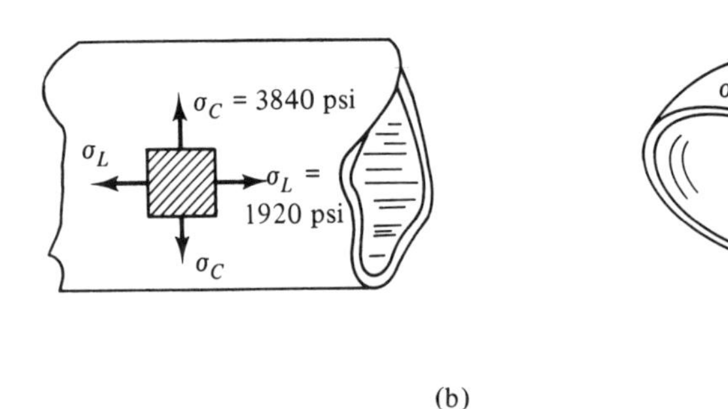

(b)

Figure 6.24

Solution. The longitudinal stresses in the cylindrical portion of the tank and the stresses in the hemispherical ends have the same values. From Eq. (6.7) or Eq. (6.9), we have

$$\sigma_L = \frac{pR}{2t} = \frac{(120\ \text{lb/in.}^2)(8\ \text{in.})}{2(\frac{1}{4}\ \text{in.})} = 1920\ \text{psi} \qquad \textbf{\textit{Answer}}$$

The circumferential stress in the cylindrical portion of the tank is [Eq. (6.8)]

$$\sigma_C = \frac{pR}{t} = 2\sigma_L = 3840\ \text{psi} \qquad \textbf{\textit{Answer}}$$

These stresses are shown on the elements in Figure 6.24(b). The orientation of the element in the hemispherical ends is arbitrary, since the stresses there are the same in all directions.

Example 6.11. Stresses in a Ring. A thin ring with an inside diameter of 500 mm and a wall thickness of 12 mm is fitted onto a shaft with a diameter of 503 mm by heating the ring until it slips over the shaft and then allowing it to cool to its original temperature. What is the resulting circumferential stress in the ring and what is the contact pressure between the ring and the shaft? Assume that the shaft deforms a negligible amount. The ring is made of cast iron with the stress-strain curve shown in Figure 5.25(b).

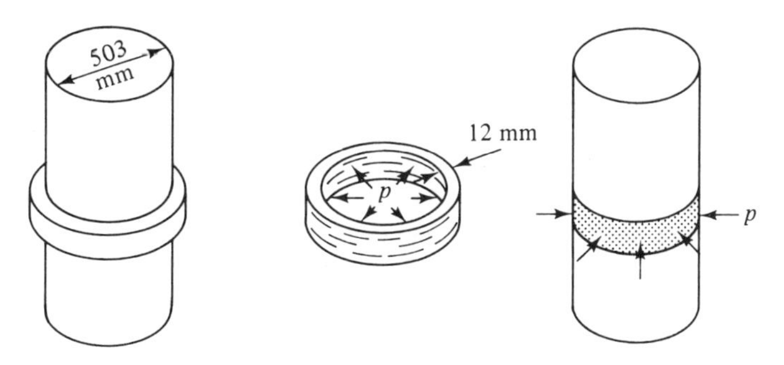

Figure 6.25

Solution. The shrinking of the ring onto the shaft has the same effect as subjecting it to a radial pressure sufficient to expand it from its original diameter of 500 mm to its final diameter of 503 mm (Figure 6.25). This pressure is exerted by the shaft and is assumed to be uniformly distributed over the contact surface.

The circumferential strain is equal to the change in circumference of the ring divided by the original circumference:

$$\varepsilon_C = \frac{\pi(503 \text{ mm}) - \pi(500 \text{ mm})}{\pi(500 \text{ mm})} = 0.006$$

From the stress-strain curve in Figure 5.25, we find that the stress at this value of strain is approximately

$$\sigma_C = 18\ 000 \text{ psi} \quad \text{or} \quad 124 \text{ MN/m}^2 \qquad \textit{Answer}$$

The contact pressure is now determined from the equation for the circumferential stress, Eq. (6.8):

$$\sigma_C = \frac{pR}{t}$$

so

$$p = \frac{\sigma_C t}{R} = \frac{(124 \times 10^6 \text{ N/m}^2)(12 \times 10^{-3} \text{ m})}{(0.5 \text{ m})}$$

$$= 2980 \times 10^3 \text{ N/m}^2 \text{ or } 2.98 \text{ MN/m}^2 \qquad \textit{Answer}$$

PROBLEMS

6.61 A cylindrical pressure vessel 1 m in diameter with a wall thickness of 6 mm is subjected to an internal gage pressure of 600 kN/m^2. Determine the longitudinal and circumferential stresses and show them acting upon a small material element.

6.62 A cylindrical tank made of 6-mm thick steel plate is to be subjected to an internal gage pressure of 1.2 MN/m^2. If the circumferential stress is not to exceed 120 MN/m^2, what is the maximum allowable tank diameter?

6.63 A 20-in. diameter pipe with $\frac{1}{2}$-in. wall thickness carries steam at 500 psi and 1200°F. If the material yield strength at this temperature is 20 ksi, what is the safety factor against failure by the onset of yielding? Treat the pipe as a cylinder with open ends, i.e. $\sigma_L = 0$.

6.64 What is the required wall thickness t of a 20-ft diameter spherical tank that is to be subjected to an internal pressure of 200 psi if the maximum allowable tensile stress is 18 ksi?

6.65 A 20-in. diameter cylindrical tank subjected to an internal pressure of 500 psi has $\frac{1}{8}$-in. thick hemispherical ends. What is the required thickness t of the cylindrical portion of the tank if it is to be stressed to the same level as the ends? What is the maximum stress in the tank? Neglect any stress concentration at the junctures between the tank segments.

6.66 Estimate the maximum tensile stress in a vertical cylindrical tank 6 m in diameter filled with water to a depth of 4 m. The tank has a wall thickness of 30 mm and is open at the top. The specific weight of water is 9.8 kN/m^3.

6.67 A form for circular concrete columns consists of two half-cylinders hinged together along one side and latched together along the other with $\frac{1}{2}$-in. diameter bolts spaced 1 ft apart. The form is 18 in. in diameter and stands 12 ft high. If fresh concrete weighs 150 lb/ft^3, what is the tensile stress in the bolts when the form is filled to the top? Base your answer on the maximum pressure existing within the form.

6.68 The tank shown is fabricated from 4 mm thick steel plate. Determine the maximum longitudinal and circumferential stresses caused by an internal pressure of 860 kN/m^2. Neglect localized end effects.

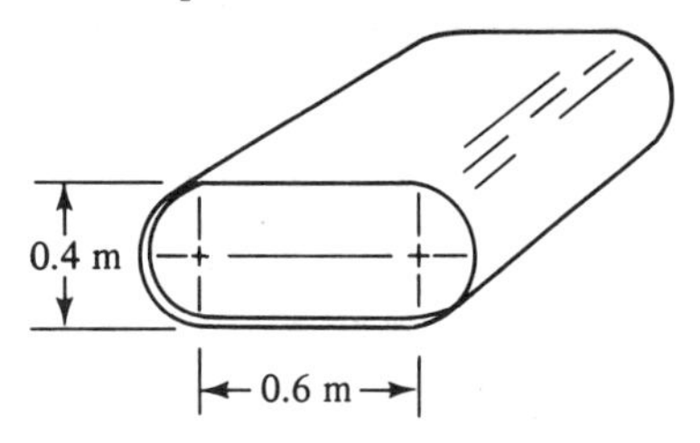

Problem 6.68

6.69 A spherical pressure vessel is fabricated by bolting two hemispheres together along rigid diametral flanges with 50 equally spaced bolts as shown. The sphere has an internal diameter of 6 ft and the flanges are 3 in. wide. The tank is to be subjected to an internal gage pressure of 100 psi, and, to prevent leakage, the bolts are initially tightened so that the final contact pressure between the flanges is also 100 psi. If the allowable stresses are 20 ksi in the bolts and 6 ksi in the hemispheres, what is the required bolt diameter d and vessel wall thickness t? Neglect localized effects near the flanges.

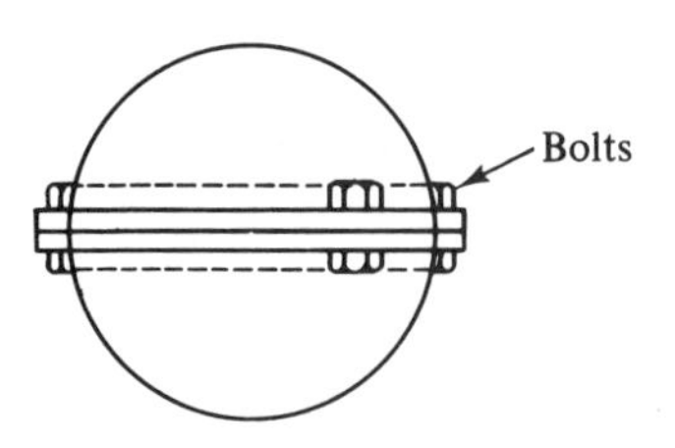

Problem 6.69

6.70 A 2-mm thick band with $E = 2\ GN/m^2$ is placed snugly, but without stress, around the trunk of a tree 300 mm in diameter. What is the stress induced in the band if the tree grows to a diameter of 310 mm? What is the resulting contact pressure between the band and the tree? Assume the tree does not deform.

6.71 A 40-mm wide clamp consisting of two bands made of 1.5-mm thick sheet steel is placed around a 100-mm diameter pipe as shown and the screws at each side are tightened until they are subjected to a tensile force of 500 N. Determine the contact pressure between the clamp and pipe and the circumferential stress in the bands. Assume the pipe doesn't deform.

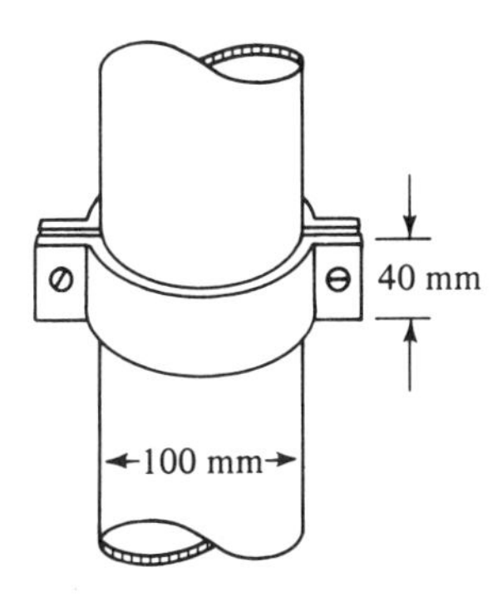

Problem 6.71

6.72 The two rings shown are 0.2 in. thick and just fit together snugly at 70°F. If the temperature is increased to 150°F, what is the circumferential stress in each ring and the contact pressure between them?

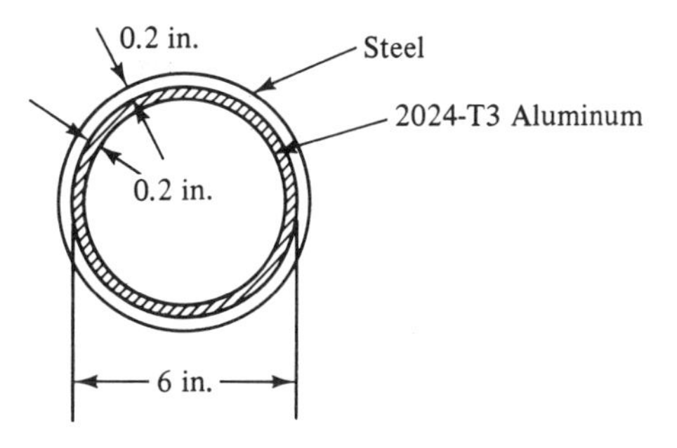

Problem 6.72

6.9 CLOSURE

In this chapter we have shown that the actual stresses in axially loaded members and in pressurized cylinders and spheres are generally uniformly distributed. Consequently, they have the same values as the average stresses and can be computed by dividing the normal and shear forces by the areas over which they act. The load-stress relations so obtained stem directly from the equations of equilibrium and, therefore, apply regardless of the material behavior. This property is unique to these types of members and loadings. In contrast, the load-deformation relations depend upon the material stiffness and have different forms for different ranges of stress.

We have also introduced the concept of using a small material element to display the stresses acting upon the various planes at a point within a body. This representation will be of great help in later chapters in visualizing and analyzing stresses due to complex loadings.

chapter seven TORSION

7.1 INTRODUCTION

In this chapter we shall consider the *torsion* of circular shafts. More precisely, we shall consider circular members loaded in such a way that the stress resultant is a couple that lies along the longitudinal axis and whose response consists of a twisting about that axis. Such members are commonly used as drive shafts in power transmission systems and in other mechanical and structural applications. Our goal is to determine the load-stress and load-deformation relations, both for the linearly elastic and inelastic cases, and to learn how to apply them.

As we shall see, there is a direct analogy between all aspects of the torsion problem and the analysis of axially loaded members. The only significant difference is that shear stresses and rotational effects are involved instead of normal stresses and elongations or contractions. Accordingly, the basic procedures used in Chapters 5 and 6 for axially loaded members still apply. This is true for both statically determinate and statically indeterminate problems.

Only members in equilibrium will be considered. Thus, the results obtained will apply to shafts at rest and to shafts rotating with constant angular velocity.

7.2 DEFORMATION PATTERN AND STRAINS

Following the procedure outlined in Section 5.10, we first consider the nature of the deformations in a twisted shaft. Experiments show that a circular shaft made of a homogeneous and isotropic material deforms in such a way that planes perpendicular to the axis before loading remain plane and perpendicular to the axis after loading, and radial lines in the cross section remain radial. Furthermore, the length does not change appreciably. In other words, the shaft behaves as a series of thin discs that rotate slightly with respect to one another when the shaft is twisted. As a result, lines originally parallel to the axis of the shaft distort into helixes. This pattern of deformation, which also follows from symmetry, is illustrated in Figure 7.1 and is confirmed by Figure 7.2.

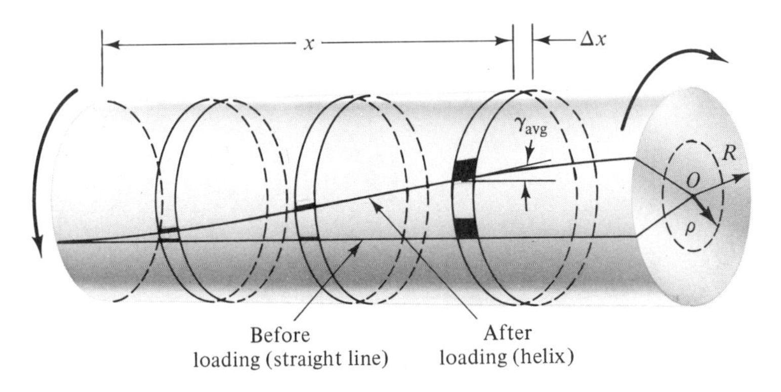

Figure 7.1

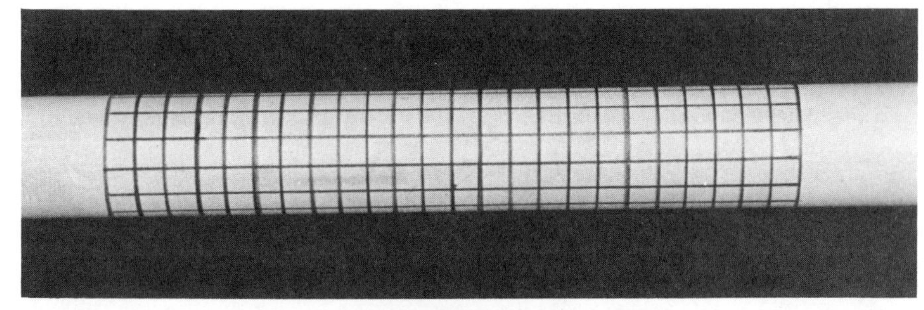

(a) Before Loading

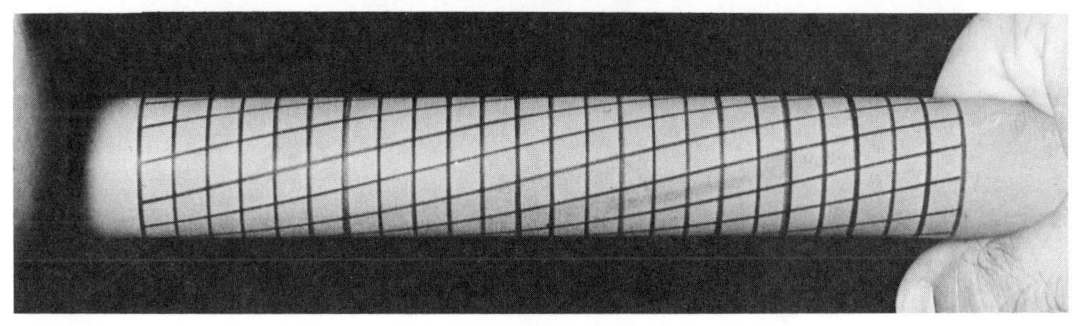

(b) After Loading.

Figure 7.2. Deformation of a Rubber Shaft as Indicated by Grid Patterns

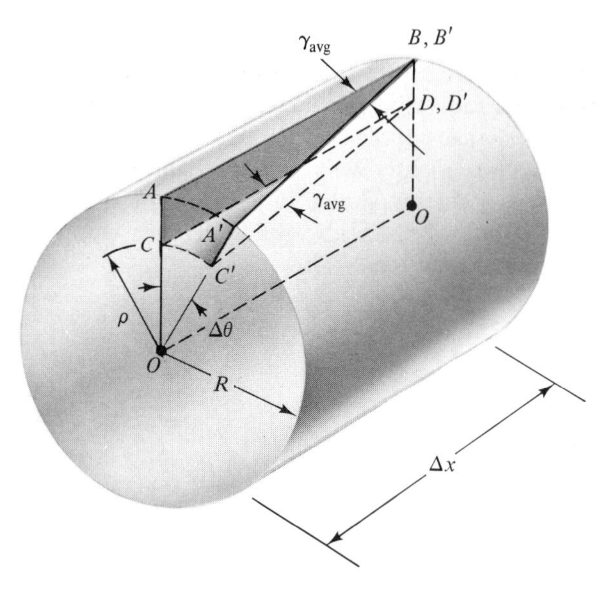

Figure 7.3

The deformations can be related to the strains by considering a short segment of the shaft with length Δx (Figure 7.3). Lines AB and CD before loading become lines $A'B'$ and $C'D'$ after loading. They remain very nearly straight because they are short. The angle between the original and final line elements is the average shear strain, γ_{avg}. This is most easily seen from Figure 7.1. It is clear from Figure 7.3 that this angle varies with the radius ρ.

Denoting the amount of rotation of one end of the segment with respect to the other by $\Delta\theta$ and assuming that the deformations are small, we have from the geometry of Figure 7.3

$$\gamma_{avg} \approx \tan \gamma_{avg} = \frac{CC'}{\Delta x} = \frac{\rho\Delta\theta}{\Delta x}$$

The shear strain at a point is obtained by taking the limit as $\Delta x \to 0$:

$$\gamma = \lim_{\Delta x \to 0} \frac{\rho\Delta\theta}{\Delta x} = \frac{\rho d\theta}{dx} \tag{7.1}$$

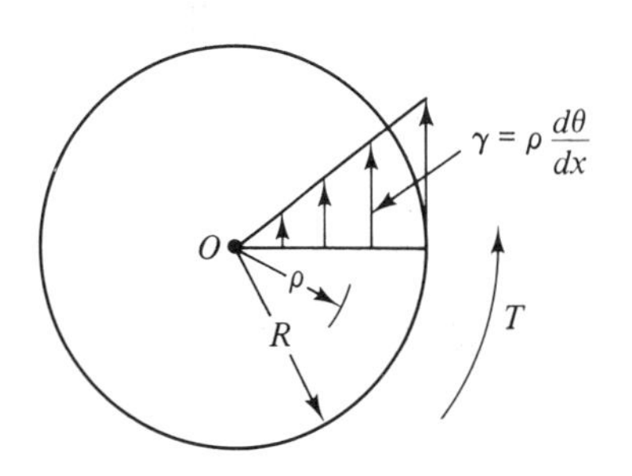

Figure 7.4

Equation (7.1) shows that the shear strain varies linearly with the radius and is a maximum at the outside of the shaft where $\rho = R$. This is illustrated in Figure 7.4, where the height of the diagram denotes the relative magnitude of γ.

The strain distribution in a homogeneous and isotropic shaft depends only upon the geometry of the member and the manner in which it is loaded. It is independent of the resistive properties of the material. However, there are several conditions, in addition to those already mentioned, that must be satisfied before

Eq. (7.1) is valid. For convenience, all of these requirements are summarized below:

1. The member must be straight and circular, or nearly so, for some distance on either side of the cross section of interest.

2. The loading must be such that the stress resultant is a couple that lies along the longitudinal axis.

3. The cross-sectional area must be constant, or nearly so, for some distance on either side of the section of interest.

4. The section of interest must be some distance away from connections, supports, and points of load application.

5. The material must be homogeneous and isotropic.

The fifth condition can be relaxed somewhat, but it is sufficiently general for our purposes. If the conditions on the geometry of the shaft are not met, there will be distortions in the deformation pattern and Eq. (7.1) does not apply. However, these effects tend to be localized (St. Venant's principle), and they generally die out over a distance approximately equal to the shaft diameter.

If the shaft is uniform and the torque doesn't vary along its length, then neither will the shear strain. In this case, Eq. (7.1) can be integrated directly to obtain

$$\theta = \int_0^L \frac{\gamma}{\rho}\,dx = \frac{\gamma L}{\rho} \tag{7.2}$$

where L is the length of the member and θ is the *angle of twist* (measured in radians) through which one end rotates relative to the other.

7.3 SHEAR STRESSES—LINEARLY ELASTIC CASE

Continuing with the procedure outlined in Section 5.10, we now determine the stress distribution and torque-stress relation for a shaft in torsion by combining the strain distribution with the stress-strain relation and the conditions for static equivalence. Linearly elastic material behavior will be assumed for the present. The inelastic behavior of shafts will be considered in Section 7.6.

Torque-stress relation. If the material is homogeneous and isotropic and the stresses are below the proportional limit, the stress-strain relation $\tau = G\gamma$ applies throughout the shaft. Combining this expression with the strain-deformation relation of Eq. (7.1), we obtain for the stress distribution

$$\tau = G\gamma = G\rho\,\frac{d\theta}{dx} \tag{7.3}$$

Thus, the shear stresses also vary linearly with the radius, ρ, as shown in Figure 7.5. They are oriented in the tangential direction and act over the entire cross section.

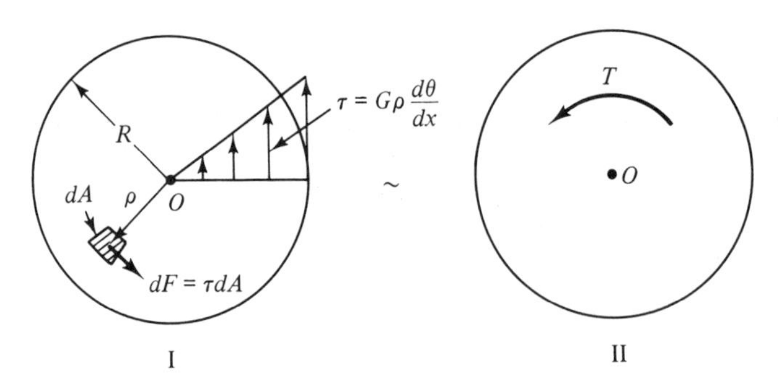

Figure 7.5

The shear stresses can be related to the stress resultant (torque), T, by using the conditions for static equivalence. Referring to Figure 7.5, we see that the force dF acting upon an element of area dA of the cross section has a moment about the center of the shaft of magnitude $dM = \rho dF = \rho\tau dA$. Thus, we have

$$(\overset{+}{\curvearrowleft}\Sigma M_0)_I = (\overset{+}{\curvearrowleft}\Sigma M_0)_{II}: \quad \int \rho\tau dA = T \tag{7.4}$$

Substituting Eq. (7.3) into this relation and recognizing that G and $d\theta/dx$ do not vary over the cross section, we obtain

$$T = \int \rho\left(G\rho \frac{d\theta}{dx}\right)dA = G\frac{d\theta}{dx}\int \rho^2 dA$$

The latter integral is the polar moment of inertia, J, of the cross-sectional area; therefore, this expression can be rewritten as

$$T = GJ\frac{d\theta}{dx} \tag{7.5}$$

Combining Eqs. (7.3) and (7.5), we obtain the torque-stress relation

$$\tau = \frac{T\rho}{J} \qquad (\tau \leqslant \tau_{PL}) \tag{7.6}$$

Here, T is the torque acting upon the particular cross section of interest, and J is the polar moment of inertia of that cross section. For a shaft with radius R or diameter D,

$$J = \frac{\pi R^4}{2} = \frac{\pi D^4}{32} \tag{7.7}$$

Equation (7.6) defines the shear stresses acting upon planes perpendicular to the axis, which are also the maximum shear stresses in a shaft. Proof of this will be given in the following section. The sense of the shear stress is the same as that of the torque, as indicated in Figure 7.5.

The connection between the stresses and the applied loads is completed by relating the torque to the loads by using the method of sections, as discussed in Section 5.2. If the torque varies along the length of the shaft, as when couples are applied at intermediate points as well as at the ends, it is convenient to construct a *torque diagram* showing the value of the torque at every location along the member.

Validity of torque-stress relation. The torque-stress relation given in Eq. (7.6) applies only if the stresses are below the proportional limit, since the relation $\tau = G\gamma$ was used in its derivation. For other types of material behavior, it is only necessary to use the appropriate stress-strain relation in Eq. (7.3) and then proceed as before. Equation (7.6) is also based upon the strain-deformation relation of Eq. (7.1); therefore, the five conditions stated in Section 7.2 concerning the geometry and loading of the shaft must also be met.

Many shafts contain oil holes, keyways, fillets, and other geometrical discontinuities, the effects of which can often be accounted for by multiplying the maximum value of shear stress obtained from Eq. (7.6) by a stress concentration factor, K_τ. Thus,

$$\tau_{\max} = K_\tau \frac{TR}{J} \tag{7.8}$$

Values of K_τ for shafts with fillets are given in Figure 7.6 for the linearly elastic case. These values apply to the stresses in the smaller portion of the member.

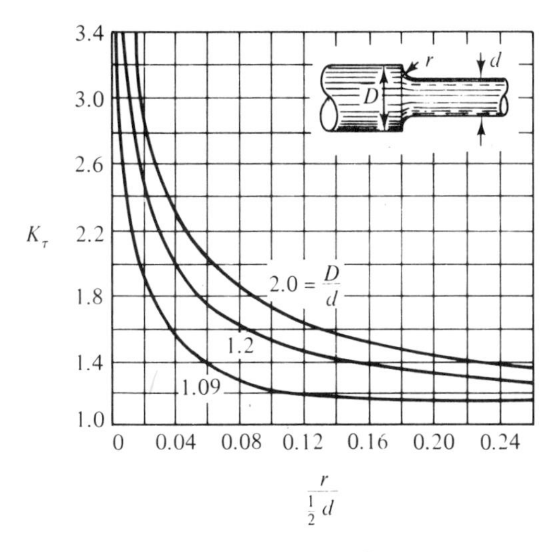

Figure 7.6. Torsional Stress-Concentration Factors for Stepped Circular Shafts

As in the case of axially loaded members, geometrical discontinuities are most serious if the material behaves in a brittle manner. Shafts made of ductile materials can yield and continue to transmit the torque.

Hollow shafts. From Figure 7.5, it can be seen that the shear stresses are smallest near the center of a shaft and that the moment arms of the forces due to these stresses are also small. Consequently, the material near the center contributes little to the torque-carrying capacity of a member, and considerable economy can be achieved by eliminating it. This is why many shafts are hollow. For example, a hollow shaft with an inside radius equal to one-half the outside radius can transmit 94% of the torque carried by a solid shaft with the same outside radius, but it contains only 75% as much material.

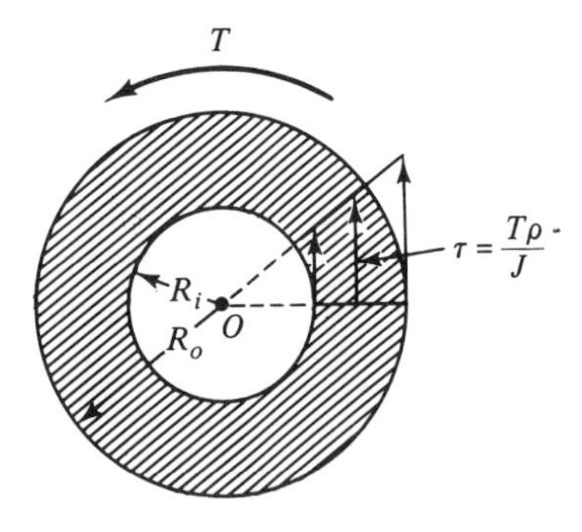

Figure 7.7

The torque-stress relation of Eq. (7.6) also applies to hollow shafts, in which case J is the polar moment of inertia of the net cross-sectional area:

$$J = \frac{\pi}{2}(R_o^4 - R_i^4) = \frac{\pi}{32}(D_o^4 - D_i^4) \tag{7.9}$$

The subscripts i and o denote inside and outside, respectively. The stress distribution in a hollow shaft loaded within the elastic range is shown in Figure 7.7.

Thin-walled circular shafts. For a hollow shaft with thin walls, the shear strain is very nearly uniform through the wall thickness [Figure 7.8(a)]. Accordingly, the stress distribution [Figure 7.8(b)], will also be nearly uniform. This is true regardless of the material properties.

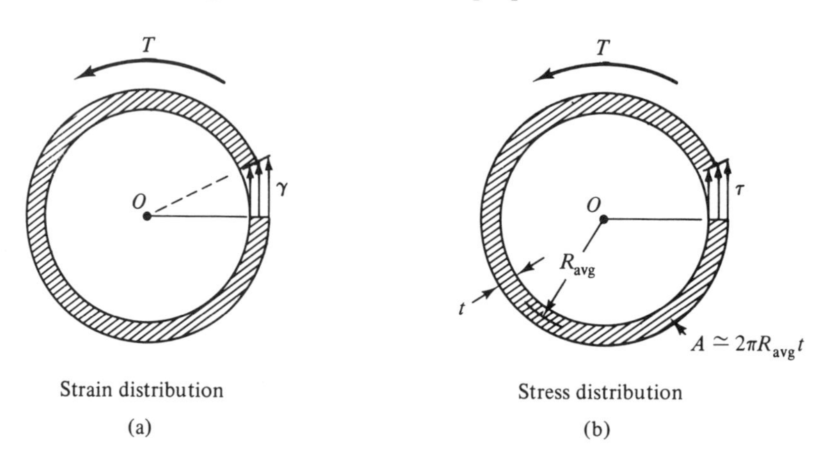

Figure 7.8

If we assume that τ does not vary over the cross section and that all of the material is located at approximately the same radius $\rho = R_{avg}$, the torque-stress relation of Eq. (7.4) becomes

$$T = \int \rho\tau dA \approx \tau R_{avg} \int dA = \tau R_{avg} A$$

where A is the cross-sectional area of the shaft and R_{avg} is the radius to the centerline of this area [Figure 7.8(b)]. Now $A \approx 2\pi R_{avg}t$, where t is the wall thickness; therefore, this expression can be rewritten as

$$\tau \approx \frac{T}{2\pi t R_{avg}^2} \tag{7.10}$$

The accuracy of this relation increases with decreasing wall thickness. However, shafts with very thin walls tend to buckle, or wrinkle, in which case Eq. (7.10) is no longer valid.

Equations (7.10) and (7.2) provide the basis for the experimental determination of the properties of a material in shear. If couples of known magnitude are applied to a thin-walled specimen with given dimensions and if the resulting angles of twist are measured, or vice versa, the shear stress and shear strain can be determined from these equations and the stress-strain curve can be constructed. This experiment is not easy to perform, however, because of the tendency of the member to buckle.

Example 7.1. Shear Stress in a Stepped Shaft. A stepped shaft made of 2024-T3 aluminum alloy is loaded as shown in Figure 7.9(a). Determine the maximum shear stress in each section of the shaft and at the fillet.

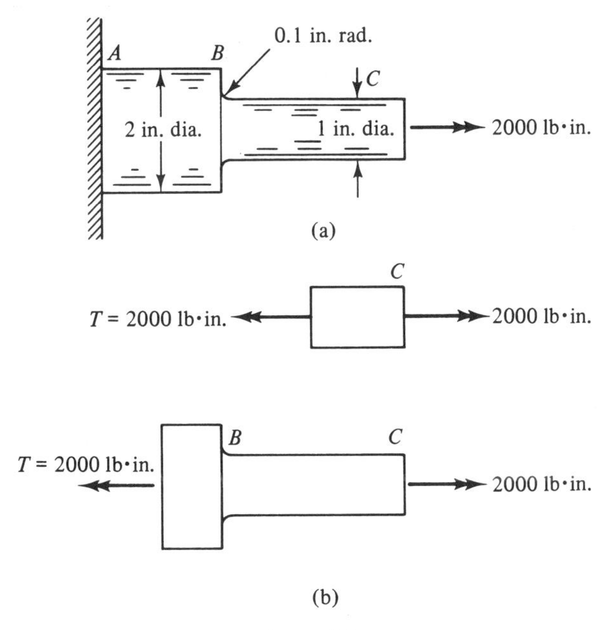

Figure 7.9

Solution. The torque is the same in both portions of the shaft and is equal in magnitude to the applied couple [Figure 7.9(b)]. From the torque-stress relation of

Eq. (7.6), we have

$$\tau_{max} = \frac{TR}{J} = \frac{TR}{(\pi/2)R^4} = \frac{2T}{\pi R^3}$$

For section AB of the shaft,

$$\tau_{max} = \frac{2(2000 \text{ lb·in.})}{\pi(1 \text{ in.})^3} = 1270 \text{ psi} \qquad \textit{Answer}$$

while for section BC,

$$\tau_{max} = \frac{2(2000 \text{ lb·in.})}{\pi(0.5 \text{ in.})^3} = 10\ 190 \text{ psi} \qquad \textit{Answer}$$

Referring to Figure 7.6, we have $r/(d/2) = 0.2$, $D/d = 2$, and $K_\tau = 1.45$. Applying this factor to the stress in the small portion of the shaft, we obtain for the stress at the fillet

$$\tau_{max} = 1.45(10\ 190 \text{ psi}) = 14\ 780 \text{ psi} \qquad \textit{Answer}$$

We now check to see if the stresses are below the proportional limit. If they aren't, Eq. (7.6) used in the computations isn't valid and the value $K_\tau = 1.45$ doesn't apply. From Table A.2, the yield strength of the material in shear is found to be 30 ksi. Since all stresses are well below this value, the results are valid.

Example 7.2. Allowable Couple on a Drive Shaft. A motor rotating at constant speed exerts a couple, C, on a drive shaft that distributes two-thirds of it to a machine at the left end and one-third of it to another machine at the right end [Figure 7.10(a)]. How large a couple can be applied without exceeding the shear yield strength of 210 MN/m^2? The shaft is hollow with an outside diameter of 50 mm and an inside diameter of 40 mm. Use a safety factor of 1.5.

Solution. We first obtain the torque diagram, as shown in Figure 7.10(b). Here, we have arbitrarily chosen torque vectors directed to the right to be positive. Actually, the sign of the torque has little significance insofar as the shear stresses are concerned. Since the shaft is uniform, the stress will be largest where the torque is largest. The torque diagram indicates that this is in portion AB of the shaft, where $T = 2/3C$.

The polar moment of inertia of the shaft is

$$J = \frac{\pi}{32}\left(D_0^4 - D_i^4\right) = \frac{\pi}{32}\left[(50 \text{ mm})^4 - (40 \text{ mm})^4\right] = 36.23 \times 10^4 \text{ mm}^4$$

From Eq. (7.6), we have

$$\tau_{max} = \frac{TR}{J} = \frac{(2/3C)(25 \times 10^{-3} \text{ m})}{36.23 \times 10^{-8} \text{ m}^4} \leqslant 210 \times 10^6 \text{ N/m}^2$$

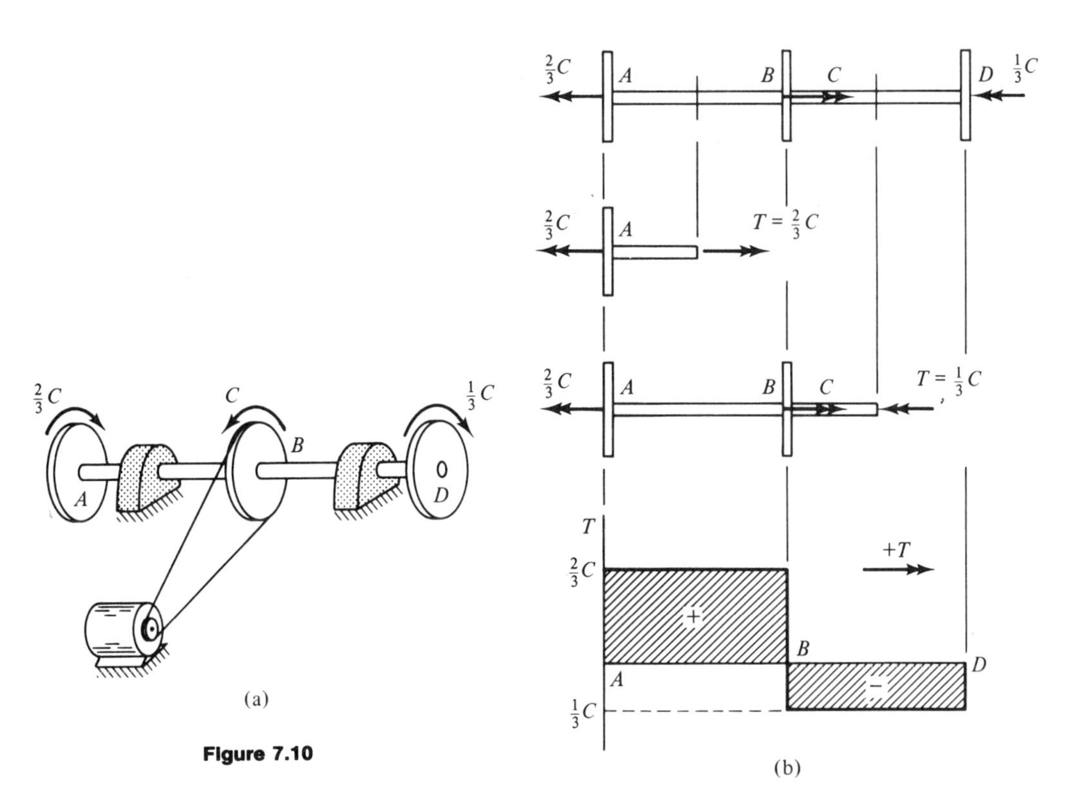

Figure 7.10

or

$$C \leqslant 4570 \text{ N}\cdot\text{m}$$

The working couple is obtained by dividing this value by the safety factor:

$$C_{\text{working}} = \frac{4570 \text{ N}\cdot\text{m}}{1.5} = 3050 \text{ N}\cdot\text{m} \qquad \textbf{\textit{Answer}}$$

PROBLEMS

7.1 Determine the maximum shear stress in the shaft shown if $C = 400$ lb·ft and indicate where this stress occurs. The fillet radius is $\frac{1}{16}$ in.

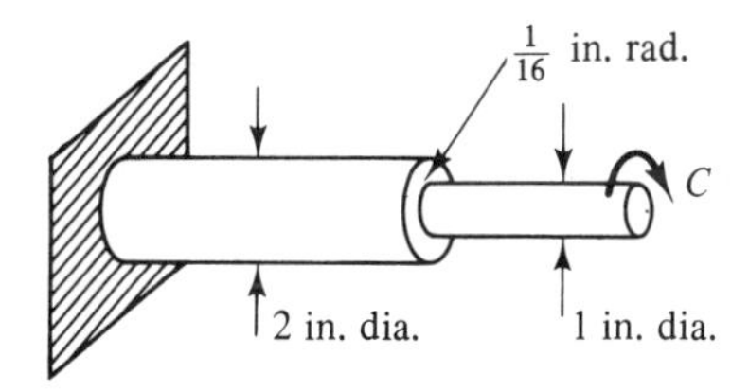

Problem 7.1

7.2 What is the largest couple C that can be applied to the shaft in Problem 7.1 if the maximum allowable shear stress is 10 ksi? Use a safety factor of 1.5.

7.3 Verify the statement in the text that a hollow shaft with inside radius equal to one-half the outside radius can transmit 94% of the torque carried by a solid shaft with the same outside radius but contains only 75% as much material. What are the corresponding figures for a hollow shaft with inside radius equal to three-fourths the outside radius?

7.4 A thin-walled tube with an inside diameter of 100 mm and a wall thickness of 5 mm transmits a torque of 8 kN·m. What diameter solid shaft would transmit the same torque at the same stress? What is the ratio of the masses of the two shafts if they are made of the same material and have the same length?

7.5 The horsepower transmitted by a rotating shaft is given by the relation hp = $Tn/63\ 000$ where T is the torque in lb·in. and n is the rotational speed in revolutions per minute. What horsepower is transmitted by a hollow shaft with ID = 2 in. and OD = 4 in. rotating at 275 rpm if the maximum shear stress is 8000 psi?

7.6 A solid shaft is to transmit 200 hp at 1800 rpm. What is the required shaft diameter if the maximum allowable shear stress is 5000 psi? See Problem 7.5.

7.7 A hollow shaft made of structural steel is connected to a solid shaft made of 2024-T3 aluminum by a bolt as shown. What is the largest couple C that can be applied to the assembly without causing either shaft to yield? What is the required diameter d of the structural steel bolt if it is not to shear off?

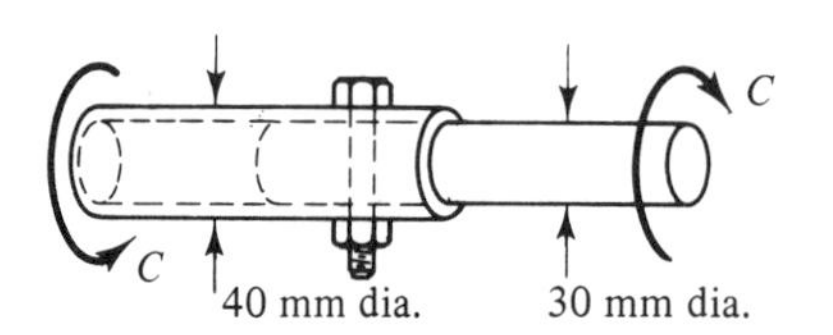

Problem 7.7

7.8 Determine the maximum shear stress in the stepped shaft shown if $C_1 = 700$ N·m and $C_2 = 100$ N·m. Ignore stress concentrations.

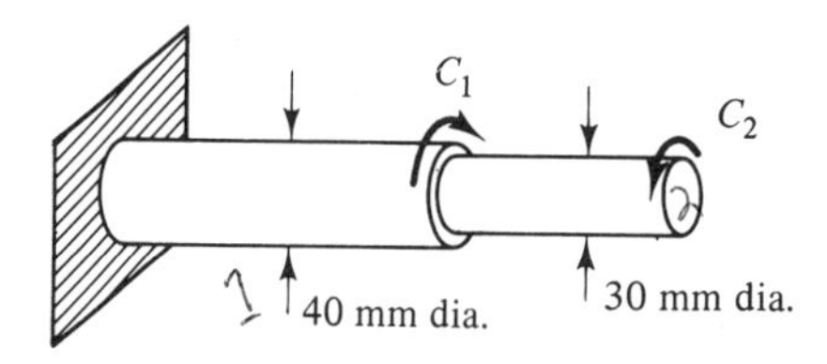

Problem 7.8

7.9 For what values of the applied couples C_1 and C_2 will each portion of the aluminum shaft shown in Problem 7.8 be stressed to the shear yield strength $\tau_{YS} = 200\ \text{MN/m}^2$? Ignore stress concentrations.

7.10 A hollow drive shaft with 4 in. OD and 2 in. ID is in equilibrium under the action of the couples shown. Determine the maximum shear stress in the shaft and indicate where it occurs.

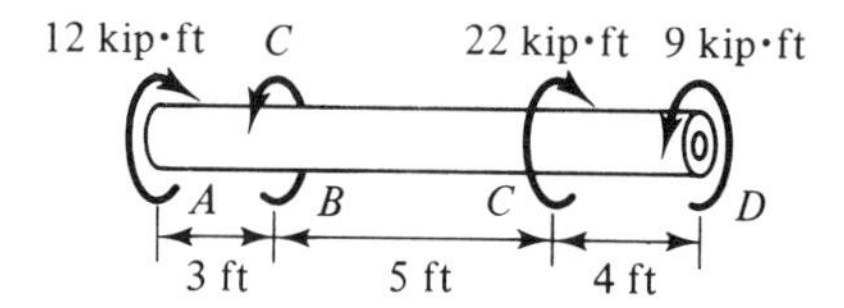

Problem 7.10

7.11 Two steel shafts are connected by gears as shown. What is the largest couple C that can be applied to shaft DE if the maximum allowable torsional shear stress in each member is 145 MN/m²? Use a safety factor of 2.

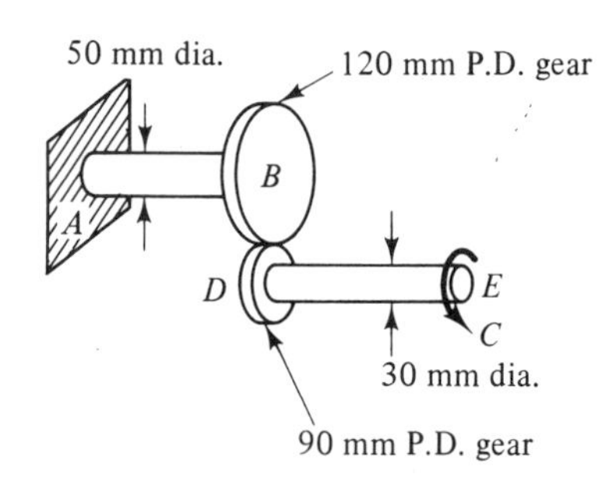

Problem 7.11

7.12 A small pipe is slipped inside a larger pipe and the two pipes are joined by welding around the circumference as shown. If the pipes can withstand a shear stress of 80 MN/m² and the weld can withstand a shear stress of 60 MN/m² on the interface with the smaller tube, what is the maximum applied couple C the assembly can support?

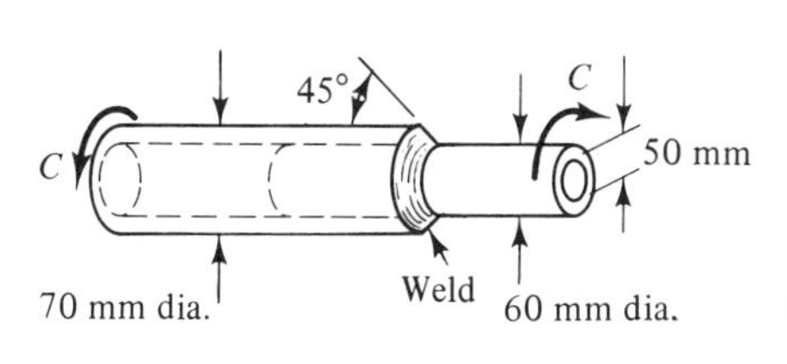

Problem 7.12

7.13 For the drive assembly shown, what is the required diameter of the bolts if the coupling is to be as strong as the main shaft? Assume the maximum allowable shear stress for the bolts is the same as for the shaft.

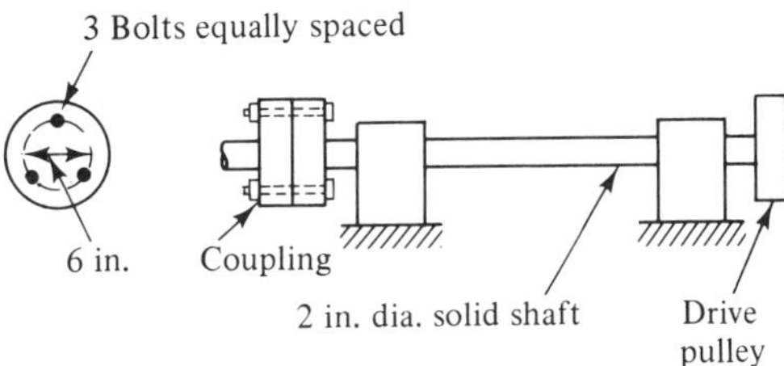

Problem 7.13

7.4 LONGITUDINAL SHEAR STRESSES AND NORMAL STRESSES

Equation (7.6), or its equivalent for materials with a stress-strain relation other than $\tau = G\gamma$, defines the shear stresses acting upon planes perpendicular to the axis of a shaft. However, there are also other stresses acting that are of significance. In particular, there are longitudinal shear stresses acting upon planes parallel to the shaft axis and normal stresses acting upon planes inclined to the axis. These stresses will be considered in this section.

Stresses on a material element. We begin by considering the stresses acting upon an infinitesimal element of material with dimensions dx, dy, and dz oriented with sides parallel and perpendicular to the shaft axis [Figure 7.11(a)]. Since one side of the element lies in the shaft cross section, the

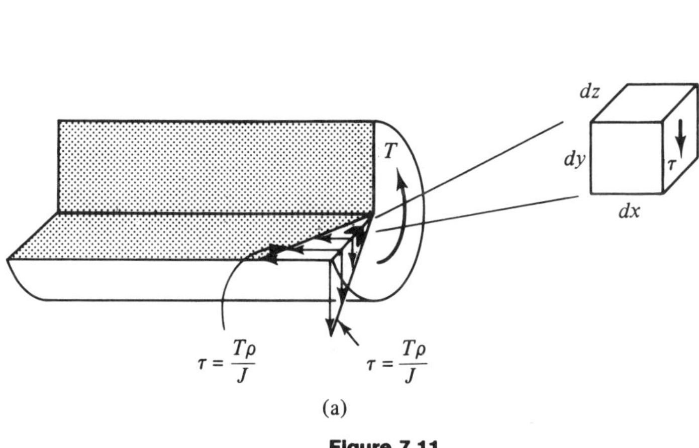

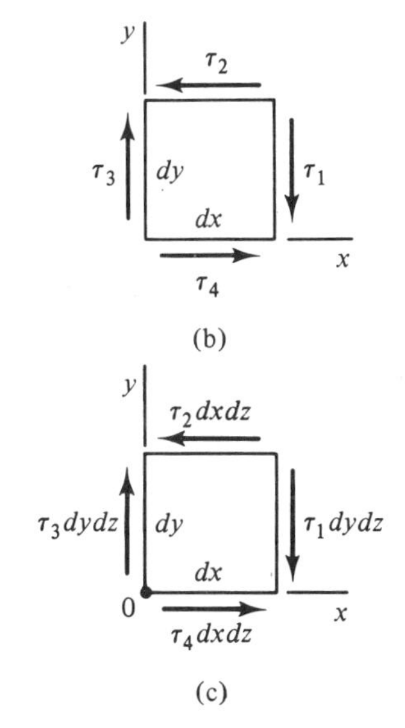

Figure 7.11

shear stress acting upon it is given by Eq. (7.6), or its equivalent. This stress has the same sense as the torque, as indicated in the figure.

The shear stress acting upon the right-hand face of the element gives rise to a shear force with magnitude equal to the stress times the area over which it acts. It is obvious that the element cannot be in equilibrium under the action of this force alone. To maintain equilibrium, there must also be shear stresses of equal magnitude acting upon the top, bottom, and left-hand faces of the element with the sense shown in Figure 7.11(b). This is easily proved, as follows.

Let us denote the shear stresses acting upon the various faces of the element by τ_1, τ_2, τ_3, and τ_4. Converting these stresses to forces by multiplying by the respective areas, we obtain the free-body diagram shown in Figure 7.11(c). The equations of equilibrium are

$$\Sigma F_x = \tau_4(dxdz) - \tau_2(dxdz) = 0$$

$$\Sigma F_y = -\tau_1(dydz) + \tau_3(dydz) = 0$$

$$\overset{+}{\curvearrowleft}\Sigma M_0 = (\tau_2 dxdz)dy - (\tau_1 dydz)dx = 0$$

from which we obtain

$$\tau_1 = \tau_2 = \tau_3 = \tau_4$$

Thus, the shear stresses must all have the same magnitude and their sense must be such that they meet tip to tip and tail to tail, as shown in Figure 7.11(b). If the sense of the torque is reversed, the sense of all four of the shear stresses is also reversed. This leads us to the following important result, which holds for all bodies and loadings:

> *If there is a shear stress acting upon a plane at a point in a body, there must also be a shear stress of equal magnitude acting at the point on a plane perpendicular to the first one. The sense of these stresses is such that they meet tip to tip or tail to tail.*

Longitudinal shear stresses. According to the preceding result, there must be shear stresses acting upon planes parallel to the shaft axis which have the same magnitude as those acting over the cross section and which vary with the radius in the same manner. This is illustrated in Figure 7.11(a) for the linearly elastic case. The longitudinal shear stresses are particularly important in wooden shafts. Since wood is relatively weak in shear parallel to the grain, these stresses tend to cause the shaft to split along its length when twisted.

Stresses on other planes. Stresses on planes other than those parallel and perpendicular to the axis of the shaft can be determined by sectioning the material element along the desired plane and considering the equilibrium of one of the resulting pieces.

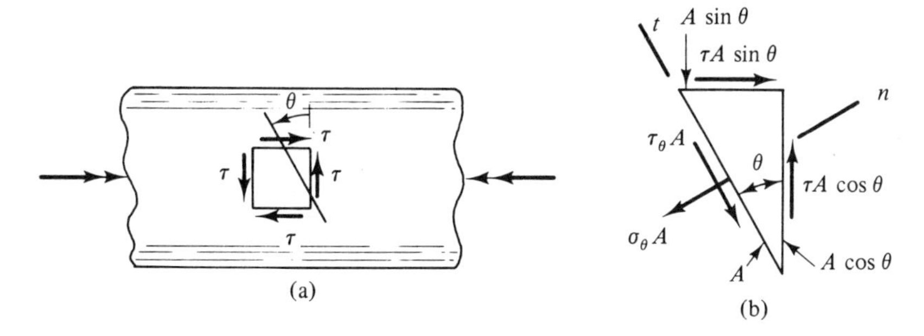

Figure 7.12

Applying this procedure to the element shown in Figure 7.12(a) and then converting the stresses to forces, we obtain the free-body diagram shown in Figure 7.12(b). In this figure, σ_θ and τ_θ denote the stresses on a plane oriented at an angle θ with respect to the shaft cross section, and A is the area over which these stresses act. As before, τ is the shear stress on planes parallel and perpendicular to the longitudinal axis.

The equations of equilibrium are

$$\Sigma F_n = -\sigma_\theta A + (\tau A \cos\theta)\sin\theta + (\tau A \sin\theta)\cos\theta = 0$$
$$\Sigma F_t = -\tau_\theta A + (\tau A \cos\theta)\cos\theta - (\tau A \sin\theta)\sin\theta = 0$$

from which we obtain

$$\sigma_\theta = 2\tau \sin\theta \cos\theta = \tau \sin 2\theta \tag{7.11}$$

and

$$\tau_\theta = \tau(\cos^2\theta - \sin^2\theta) = \tau \cos 2\theta \tag{7.12}$$

These equations indicate that both normal and shear stresses exist in a twisted shaft.

From Eq. (7.12), it can be seen that τ_θ is largest when $\cos 2\theta = 1$, which corresponds to $\theta = 0°$ or $90°$. Thus, the maximum shear stress occurs on planes parallel and perpendicular to the axis of the shaft and is equal in magnitude to τ:

$$|\tau_\theta|_{max} = |\tau| \tag{7.13}$$

This confirms our earlier statement to this effect in Section 7.3.

The normal stress is a maximum when $\sin 2\theta = 1$, or $\theta = \pm 45°$. For $\theta = +45°$, σ_θ is positive (tension), and for $\theta = -45°$, it is negative (compression). In both cases, it is equal in magnitude to τ:

$$|\sigma_\theta|_{max} = |\tau| \tag{7.14}$$

These maximum shear and normal stresses and the planes upon which they act are shown in Figure 7.13. If the sense of the shear stress is reversed, the planes of maximum tension and compression also reverse. Notice that the shear stresses appear to pull the element in the direction of the maximum tension. This observa-

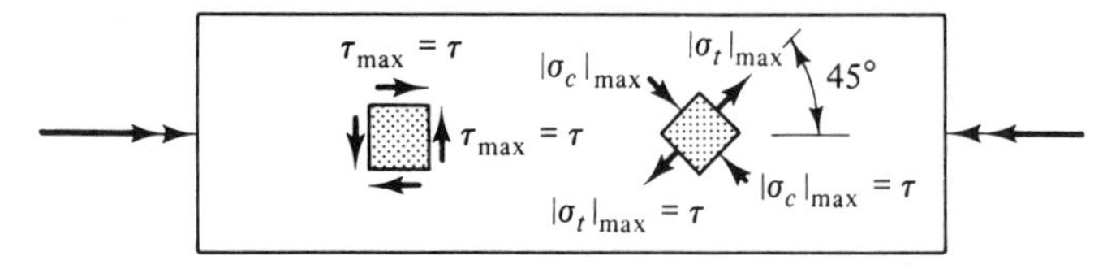

Figure 7.13

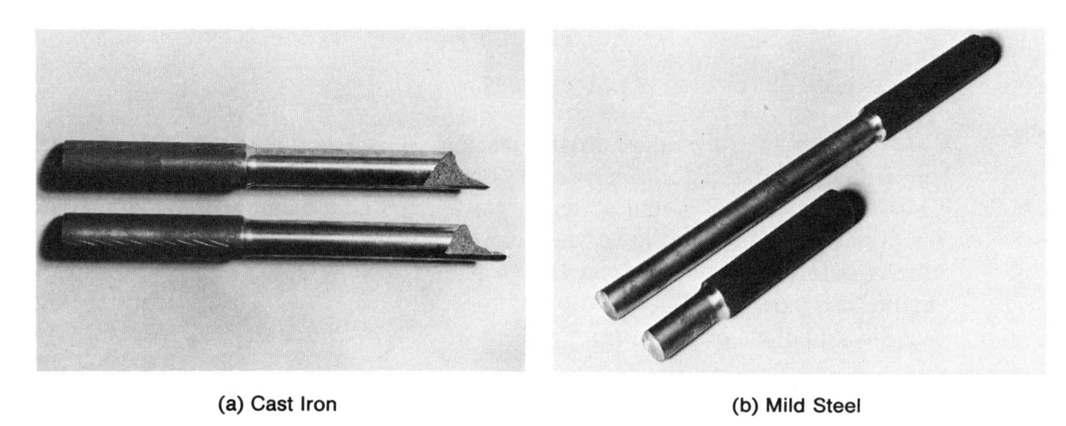

(a) Cast Iron

(b) Mild Steel

Figure 7.14. Fracture Patterns for Brittle and Ductile Shafts

tion can be used to readily determine which of the maximum normal stresses is tension and which is compression.

The maximum tensile stress is usually the most significant stress in shafts made of brittle material because it is the one that tends to cause fracture. This is illustrated in Figure 7.14(a), which shows the fracture surface of a cast-iron shaft. The surface is oriented at 45° with respect to the axis, which corresponds to the plane of maximum tensile stress. This type of failure can be readily demonstrated by twisting a piece of chalk.

Figure 7.14(b) shows the fracture surface for a steel shaft. As mentioned previously, ductile materials generally fail due to the shear stresses developed, which, in this case, are largest on planes perpendicular to the axis. Accordingly, the shaft breaks along one of these planes.

The compressive stresses developed can cause buckling, or wrinkling, of thin-walled hollow shafts. Experiments show that the wrinkle is oriented at 45° with respect to the axis, along the plane of maximum compressive stress.

Example 7.3. Design of a Cast-Iron Shaft. What diameter shaft is required to transmit a couple of 4000 lb·in. with a safety factor of 2 if the material is cast iron with an ultimate tensile strength of 20 ksi?

Solution. The design couple is the safety factor times the applied couple, or 8000 lb·in. The maximum tensile stress is equal in magnitude to the maximum torsional shear stress [Eq. (7.14)]:

$$\sigma_{max} = \tau_{max} = \frac{TR}{J} = \frac{16T}{\pi D^3}$$

Thus, we have

$$\sigma_{max} = \frac{16(8000 \text{ lb·in})}{\pi D^3} \leqslant 20\ 000 \text{ lb/in.}^2$$

or

$$D \geqslant 1.26 \text{ in.} \qquad \textbf{\textit{Answer}}$$

PROBLEMS

7.14 What diameter wooden shaft is required to transmit a torque of 100 lb·in. with a safety factor of 2 if the allowable longitudinal shear stress is 100 psi?

7.15 A 1-in. diameter shaft is made of an acrylic for which the allowable stresses are 4000 psi in shear and 5000 psi in tension. Determine the maximum torque T that can be transmitted by the shaft.

7.16 The stepped shaft shown is made of cast iron with an ultimate tensile strength of 140 MN/m^2. What is the maximum allowable value of the couple C? Use a safety factor of 3 and ignore stress concentrations.

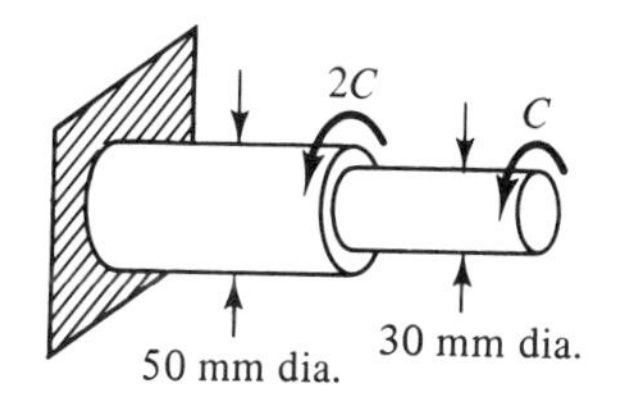

Problem 7.16

7.17 Experiments show that a thin-walled hollow shaft made of a certain material will buckle if the maximum compressive stress exceeds 35 MN/m^2. What is the required wall thickness t of a shaft made of this material if it has an average radius of 50 mm and is to transmit a torque of 3 kN·m without buckling?

7.18 If the maximum shear stress in a shaft is 60 MN/m^2, what are the magnitudes of the normal and shear stresses acting upon a plane oriented 30° counterclockwise from the longitudinal axis?

7.19 A hollow shaft is constructed by rolling a thin sheet into a cylindrical shape and cementing the edges together along the helical seams as shown. If the allowable shear stress in the cement is 20% of the allowable compressive stress in the sheet, what should be the angle of inclination θ of the seams so that both the cement and sheet are stressed to their respective limits?

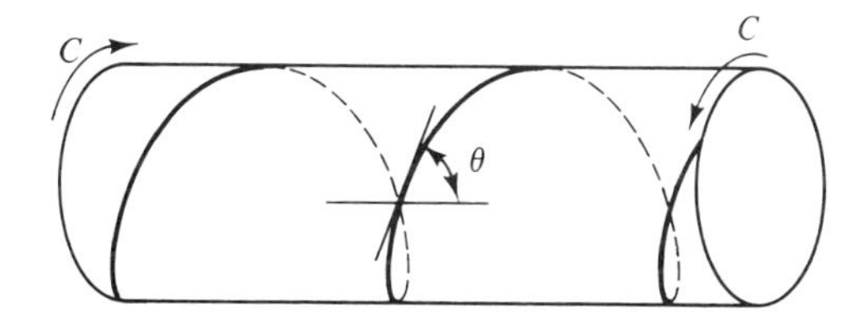

Problem 7.19

7.5 ANGLE OF TWIST

As indicated in Section 7.2, the angle θ through which one cross section of a shaft rotates with respect to another is called the *angle of twist*. If the stresses are below the proportional limit, this angle can be determined by integrating Eq. (7.5). We have

$$\theta = \int_0^L \frac{T}{JG}\,dx \qquad (\tau \leqslant \tau_{PL}) \tag{7.15}$$

where L is the length of the shaft and θ is measured in radians. The sense of θ is the same as that of the torque T. Accordingly, the shaft twists in the direction indicated by the fingers on the right hand when the thumb is placed along the torque vector.

The quantities T, J, and G must be expressed as functions of the distance x along the shaft before the preceding integral can be evaluated. If they are constant over the length of the member, Eq. (7.15) reduces to

$$\theta = \frac{TL}{JG} \qquad (\tau \leqslant \tau_{PL}) \tag{7.16}$$

Note the close similarity between this equation and the expression $\delta = PL/AE$ for the elongation of an axially loaded member.

Equations (7.15) and (7.16) define the total angle of twist between one end of the shaft and the other. To determine the twist of any particular portion of the shaft, it is only necessary to carry out the integration in Eq. (7.15) over the length of this portion or to use this length in place of the total length in Eq. (7.16).

Note that Eqs. (7.15) and (7.16) apply only if the shear stress is below the proportional limit because the stress-strain relation $\tau = G\gamma$ was used in their derivation. For other types of material behavior, it is only necessary to combine the appropriate stress-strain relation with the strain-deformation relation of Eq. (7.1) and the torque-stress relation of Eq. (7.4) and proceed as before.

There are regions of the shaft near the supports, points of load application, and geometrical discontinuities where Eq. (7.5) is not strictly valid. However, this has a negligible effect upon the total twist of the member. These regions are relatively short (St. Venant's principle) and, therefore, contribute little to the value of the integral in Eq. (7.15).

If the torque, polar moment of inertia, of shear modulus vary in steps along the length, the shaft can be treated as a series of shorter members placed end to end within which T, J, and G are constant. The twist of each portion can then be computed from Eq. (7.16) and the results summed to determine the twist of the entire member. In doing this, the sign convention for θ is the same as that for T, which can be chosen arbitrarily. The torque diagram is an invaluable aid in keeping track of the signs of T and θ in the various portions of the shaft.

Statically indeterminate torsion problems are handled in the same way as problems involving statically indeterminate axially loaded members. The only difference is that the compatibility condition involves angles of twist instead of elongations.

Our discussion of the deformations of shafts has purposely been brief because the procedures for determining them are basically the same as for axially loaded members. This is further illustrated in the following examples.

Example 7.4 Angle of Twist of a Shaft. Through what angle can a solid shaft with a diameter of 30 mm and a length of 2 m be twisted without exceeding the shear yield strength of 145 MN/m^2 if $G = 83\ \text{GN/m}^2$?

Solution. We first determine the torque from the torque-stress relation, Eq. (7.6):

$$T = \tau_{\max} \frac{J}{R}$$

Substituting this expression into the torque-twist relation of Eq. (7.16), we have

$$\theta = \frac{TL}{JG} = \tau_{\max} \frac{J}{R}\left(\frac{L}{JG}\right) = \frac{\tau_{\max} L}{RG}$$

or

$$\theta = \frac{(145 \times 10^6\ \text{N/m}^2)(2\ \text{m})}{(30 \times 10^{-3}\ \text{m})(83 \times 10^9\ \text{N/m})} = 0.12 \text{ radians or } 6.9^\circ \qquad \textit{Answer}$$

Example 7.5. Twist of a Composite Shaft. A steel pipe is joined to a solid aluminum rod to form the composite shaft shown in Figure 7.15(a). Through what angle will the free end of the shaft rotate under the loading shown?

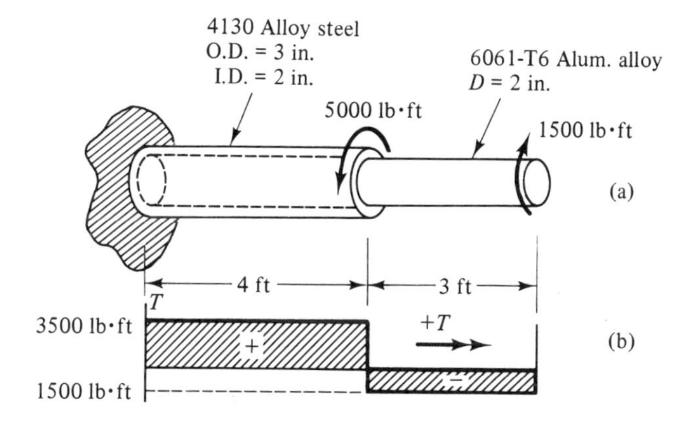

Figure 7.15

Solution. We first determine the reaction at the wall and the torque diagram [Figure 7.15(b)]. Here, we have used the sign convention that torque vectors directed to the right are positive. According to this convention, θ will be positive if the shaft twists in the counterclockwise direction when viewed from the right end.

The material properties are obtained from Table A.2:

$$\text{Steel:} \quad G = 12 \times 10^6 \text{ psi} \qquad \text{Alum:} \quad G = 3.8 \times 10^6 \text{ psi}$$
$$\tau_{YS} = 60 \text{ ksi} \qquad \qquad \tau_{YS} = 21 \text{ ksi}$$

The polar moments of inertia are

$$J_S = \frac{\pi}{32}\left[(3 \text{ in.})^4 - (2 \text{ in.})^4\right] = 6.38 \text{ in.}^4$$

$$J_A = \frac{\pi}{32}(2 \text{ in.})^4 = 1.57 \text{ in.}^4$$

We next check the stresses to see if they are below the proportional limit:

$$(\tau_{\max})_S = \left(\frac{TR}{J}\right)_S = \frac{(3500 \text{ lb·ft})(12 \text{ in./ft})(1.5 \text{ in.})}{6.38 \text{ in.}^4} = 9870 \text{ psi}$$

$$(\tau_{\max})_A = \left(\frac{TR}{J}\right)_A = \frac{(1500 \text{ lb·ft})(12 \text{ in./ft})(1 \text{ in.})}{1.57 \text{ in.}^4} = 11\ 460 \text{ psi}$$

Since both of these values are well below the respective yields strengths, Eq. (7.14) applies.

The total angle of twist of the shaft is the sum of the angles of twist of each part:

$$\theta = \theta_S + \theta_A = \left(\frac{TL}{JG}\right)_S + \left(\frac{TL}{JG}\right)_A$$

or

$$\theta = \frac{(3500 \text{ lb·ft})(12 \text{ in./ft})(48 \text{ in.})}{(6.38 \text{ in.}^4)(12 \times 10^6 \text{ lb/in.}^2)} + \frac{(-1500 \text{ lb·ft})(12 \text{ in./ft})(36 \text{ in.})}{(1.57 \text{ in.}^4)(3.8 \times 10^6 \text{ lb/in.}^2)}$$

$$= 0.026 - 0.109 = -0.083 \text{ radians} \qquad \textit{Answer}$$

Since θ_S is positive and θ_A is negative, the steel pipe twists in the counterclockwise direction (when viewed from the right end of the shaft) while the aluminum rod twists in the clockwise direction. Since the aluminum twists more than the steel, the net twist of the free end is clockwise. Note that units must be watched carefully when computing θ.

Example 7.6. Torques in an Indeterminate Shaft. An undersized shaft is "strengthened" by slipping a hollow tube over it and connecting the two shafts at the ends so that they twist as a single unit [Figure 7.16(a)]. By how much will this procedure reduce the shear stresses in the solid shaft and its angle of twist? Both shafts have the same length, and the quantity JG for the outer shaft is twice that of the inner shaft. Assume linearly elastic behavior.

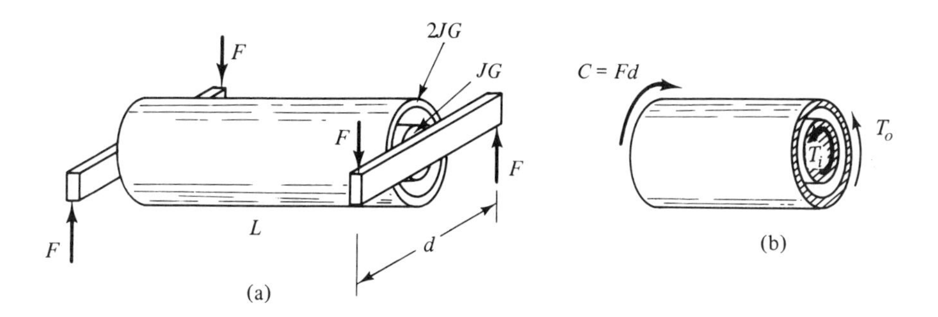

Figure 7.16

Solution. Let T_o and T_i denote the torques in the outer and inner shafts, respectively. Sectioning the composite shaft perpendicular to its axis and considering the equilibrium of the resulting piece [Figure 7.16(b)], we have

$$T_o + T_i = C$$

where $C = Fd$ is the magnitude of the applied couple. The problem is statically indeterminate because there are two unknowns and only one equilibrium equation from which to determine them.

The angle of twist of each shaft must be the same because they are connected at the ends. Thus, the compatibility condition is

$$\theta_o = \theta_i$$

Substituting the torque-twist relation of Eq. (7.16) into this expression, we obtain

$$\left(\frac{TL}{JG}\right)_o = \left(\frac{TL}{JG}\right)_i$$

or

$$T_o = \frac{2(JG)_i}{(JG)_i} T_i = 2T_i$$

Combining this result with the equilibrium equation, we get

$$T_i = \frac{C}{3}$$

Thus, the presence of the outer shaft reduces the torque in the inner shaft by two-thirds (the applied couple is assumed to be the same in each case). Since the shear stress and angle of twist are directly proportional to the torque, they are reduced by the same amount.

PROBLEMS

7.20 What must be the length of a steel rod 5 mm in diameter so that it can be twisted through one complete revolution without exceeding the shear yield strength of 400 MN/m^2? $G = 83$ GN/m^2.

7.21 A solid circular shaft 3 m long with a diameter of 150 mm is loaded in pure torsion. What is the angle of twist θ of the shaft when the maximum shear stress is 125 MN/m^2? $G = 83$ GN/m^2.

7.22 What is the maximum torque T that can be transmitted by a cast-iron shaft 80 mm in diameter if the ultimate tensile strength is 140 MN/m^2 and the allowable angle of twist per metre of length is 0.5°? Use $G =$ 43 GN/m^2.

7.23 What is the angle of twist of the free end of the composite shaft shown? Through what angle does the steel portion twist?

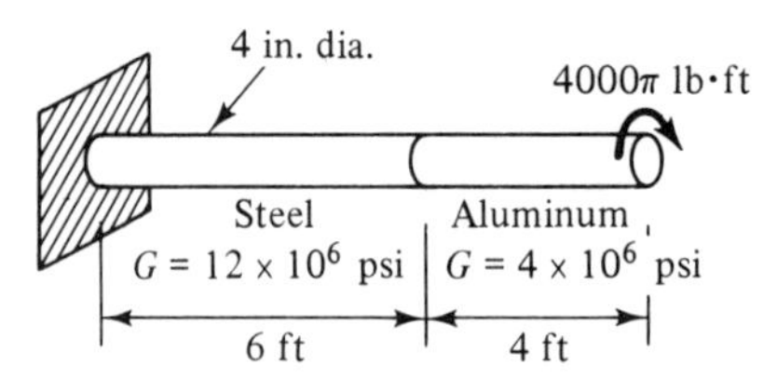

Problem 7.23

7.24 Determine the inner and outer diameters of a hollow shaft with $G = 4 \times 10^6$ psi to transmit a torque of 3000π lb·ft with a maximum shear stress of 9600 psi and an angle of twist of 5.5 degrees in an 80 in. length.

7.25 A sign post is made by welding a length of 3-in. nominal diameter standard steel pipe to a length of 6-in. nominal diameter pipe as shown. During a high wind the top of the post is observed to rotate 5°. What is the maximum torsional shear stress in the post and where does it occur?

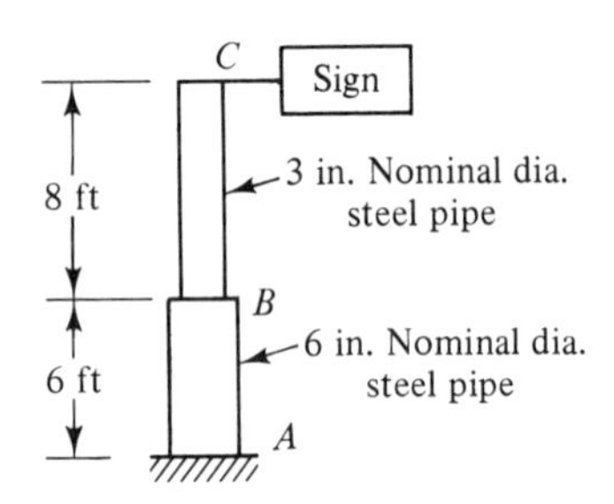

Problem 7.25

7.26 An aluminum shaft with $G = 4 \times 10^6$ psi is loaded as shown. What is the maximum angle of twist in the shaft and where does it occur?

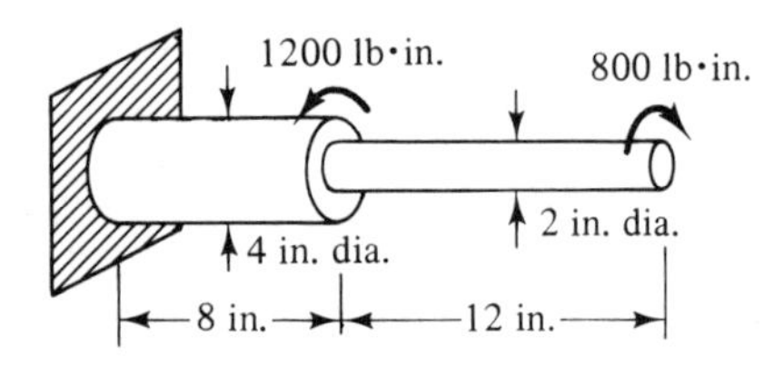

Problem 7.26

7.27 A 100-mm diameter steel drive shaft with $G = 83$ GN/m^2 is in equilibrium under the action of the couples shown. Determine the relative angle of twist between (a) gears A and D and (b) gears B and C.

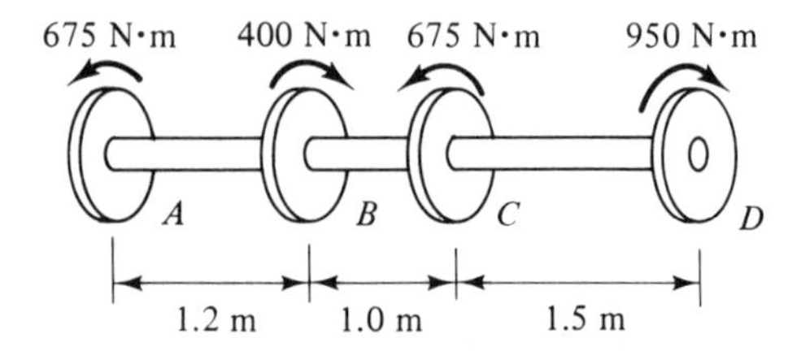

Problem 7.27

7.28 A 50-mm diameter aluminum shaft with $G = 28\ \text{GN/m}^2$ is to be lightened by boring a 25-mm diameter hole along the centerline as shown. To what depth d can the hole be bored if the total angle of twist is not to exceed 0.074 radian?

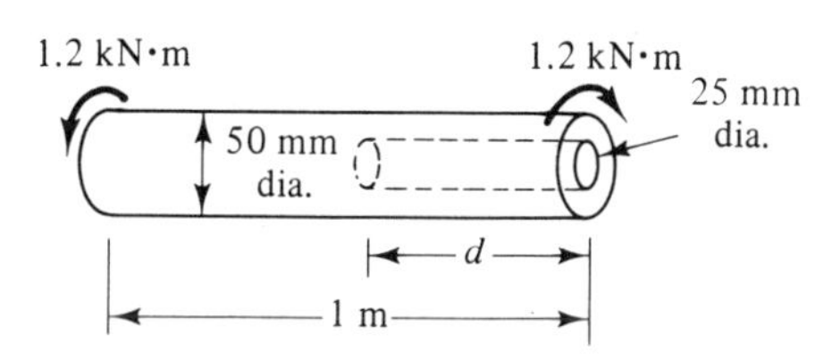

Problem 7.28

7.29 A composite shaft is constructed by slipping a hollow steel shaft 10 ft long with 3 in. OD and 2 in. ID over a solid steel shaft 8 ft long with a diameter of 2 in. The shafts overlap for 3 ft. Determine the total angle of twist of the assembly when it transmits a torque of 3000 lb·ft if the shafts are (a) bonded together over the entire 3 ft of overlap and (b) welded together only at the end of the hollow shaft. Use $G = 12 \times 10^6$ psi.

7.30 Two shafts are connected by gears as shown. If $C = 15\ 000$ lb·in., through what angle will end A rotate and what is the maximum torsional shear stress developed in each of the shafts? Use $G = 12 \times 10^6$ psi and assume there are sufficient bearings to prevent bending.

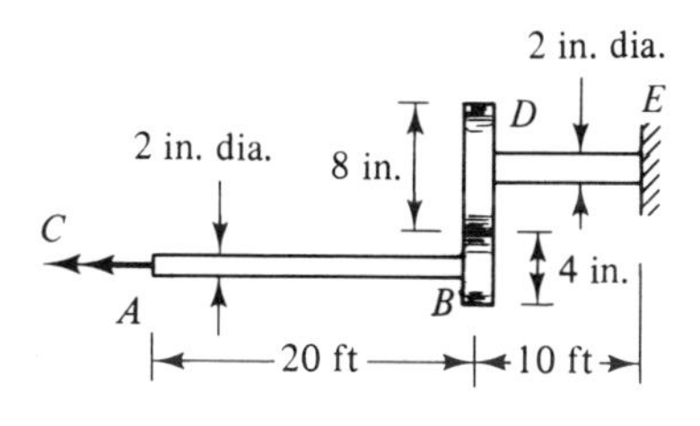

Problem 7.30

7.31 If end A of the shaft in Problem 7.30 rotates through an angle of 0.1 radian, determine the magnitude C of the applied couple and the maximum shear stress in each of the shafts.

7.32 The ends of the uniform shaft shown are fixed so that they cannot rotate. Determine the reactions at the walls. Assume elastic behavior.

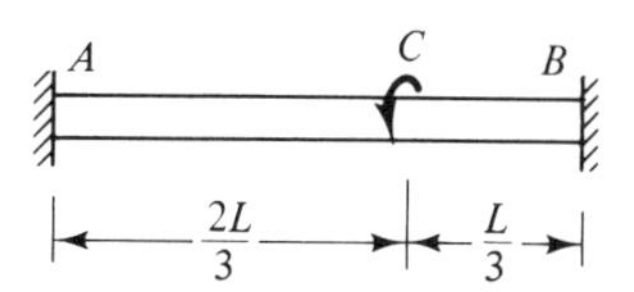

Problem 7.32

7.33 The solid composite shaft shown is attached to rigid supports at the ends. The allowable shear stress for bronze is 8 ksi and for steel it is 12 ksi. Determine the ratio of lengths b/a so that each material will be stressed to its permissible limit. What couple C is required to produce these stresses?

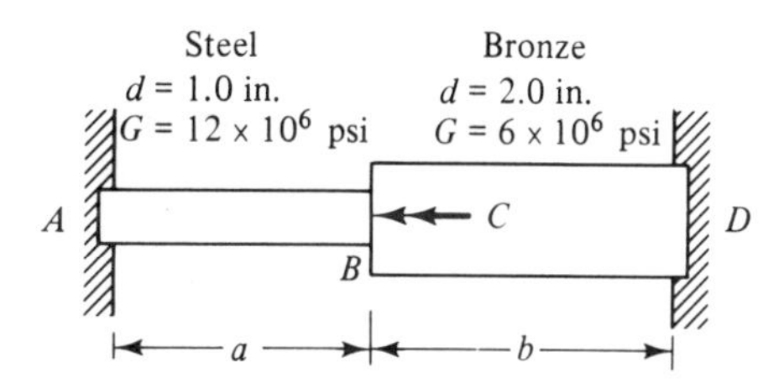

Problem 7.33

7.34 Upon application of the couple to the shaft shown in Problem 7.33, end D was observed to turn 0.0406 radian in its socket. Determine the maximum shear stress in each segment of the member if $C = 1000$ lb·ft, $a = 168$ in., and $b = 192$ in.

7.35 When a steel shaft 50 mm in diameter and 2 m long transmits a torque of 1 kN·m, the twist is found to exceed the allowable value of 0.025 radian. The situation is to be corrected by slipping a hollow steel sleeve with 70 mm OD and 50 mm ID over the shaft and welding it in place at each end. What is the required length L of the sleeve? Does it make any difference where the sleeve is located along the shaft?

7.36 A rigid bar AB is attached to the end of a 20-mm diameter aluminum shaft as shown. If the initial gap between the bar and the rigid stop D is 10 mm, what is the maximum shear stress in the shaft? $G = 28$ GN/m^2.

7.37 Derive the torque-stress and torque-twist relations for a circular shaft of radius R and length L made of a material whose shear stress-shear strain curve is described by the equation $\tau = k\gamma^{1/3}$, where k is a material constant.

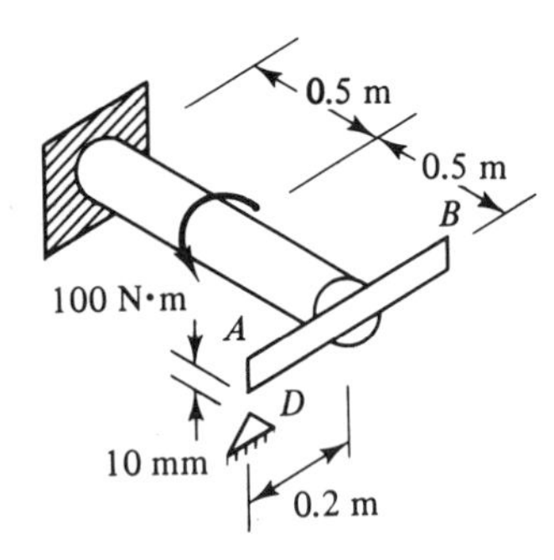

Problem 7.36

7.38 Same as Problem 7.37, except that the stress-strain relation is $\tau = k\gamma^{1/2}$.

7.6 INELASTIC BEHAVIOR

In this section we shall consider the response of torsional members strained into the inelastic range. All of our preceding assumptions concerning the geometry and loading of the shaft still apply, plus the additional assumption that the material behavior is essentially elastic-perfectly plastic.

Stress distribution. As the shaft undergoes inelastic deformations, the stress distribution tends to become uniform over the cross section, and the torque-stress and torque-twist relations derived in the preceding sections no longer apply. The situation is illustrated in Figure 7.17, which shows the shear stress distribution in a shaft made of an elastic-perfectly plastic material for three different angles of twist. The conditions at locations a, b, and c on the cross section are denoted by the corresponding points on the stress-strain diagram.

The shear strain in a shaft is defined by Eq. (7.1) and is proportional to the angle of twist regardless of the stress level. As the twist increases from zero, the maximum shear strain increases, and so does the maximum shear stress, until it is just equal to the yield strength of the material. This is the situation shown in Figure 7.17(a). Up to this point, the torque-stress and torque-twist relations given in the preceding sections are valid.

If the angle of twist is further increased, the maximum shear strain increases, but the maximum shear stress remains equal to the yield strength. Since strains at points in the interior of the shaft are now at or above the proportional limit strain, the stress distribution becomes more nearly uniform [Figure 7.17(b)]. This is referred to as the *partially plastic condition* because the strains are within the inelastic, or plastic, range only over a portion of the cross section.

Strain distribution

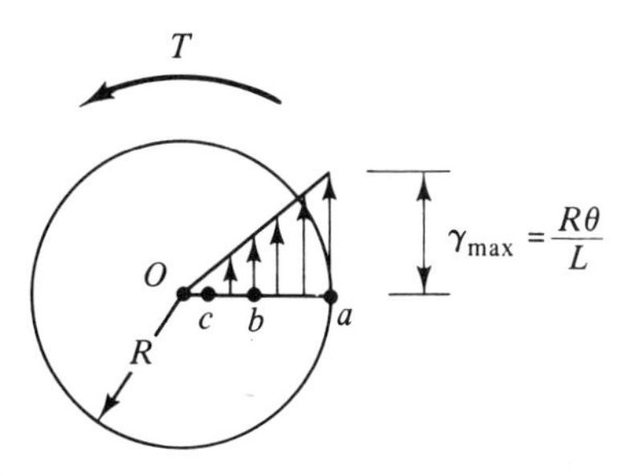

Stress distributions

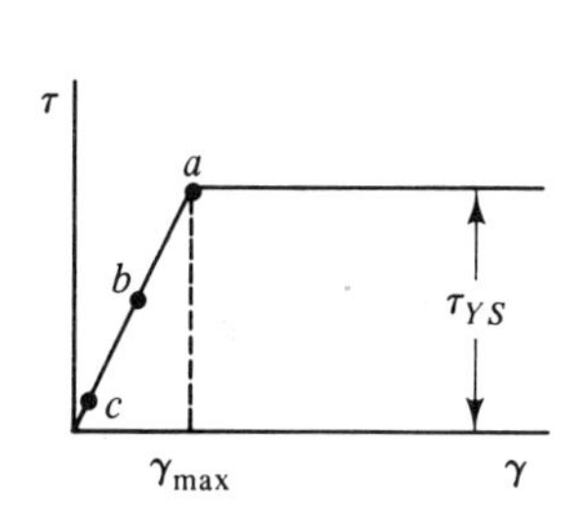

(a)

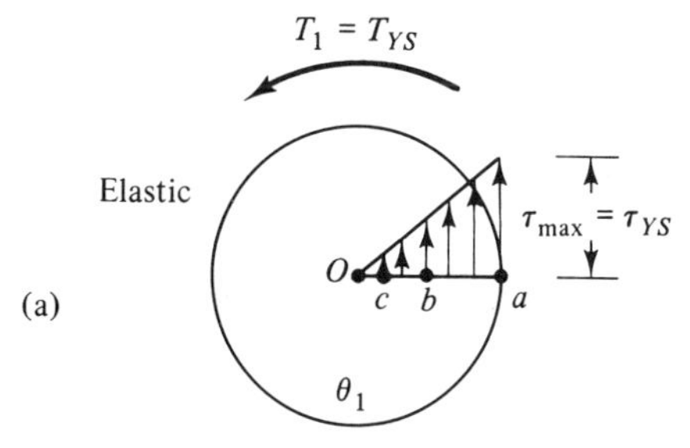

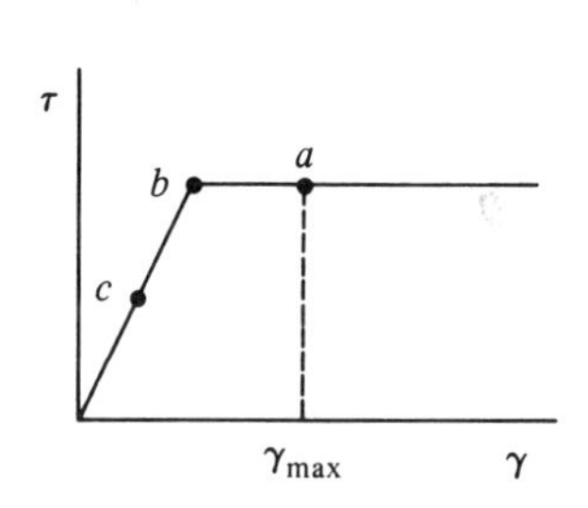

(b)

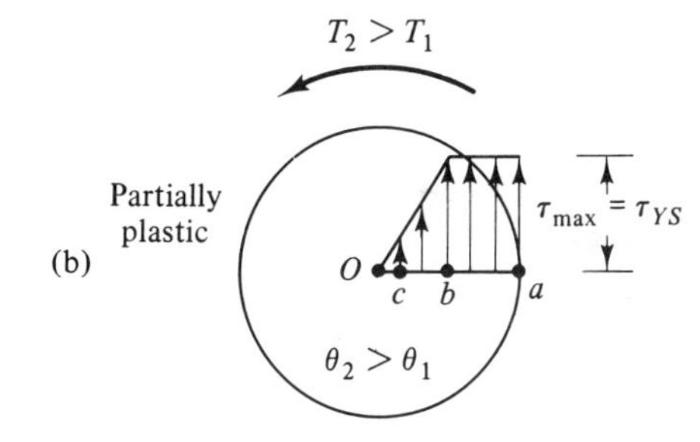

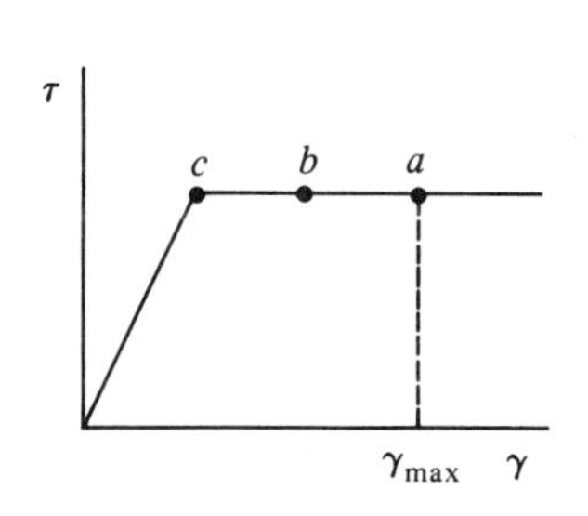

(c)

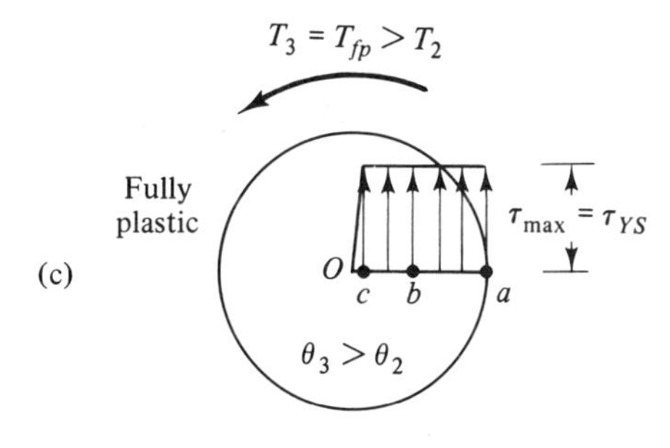

Figure 7.17

As the angle of twist continues to increase, a situation is eventually reached in which the strains are at or above the proportional limit strain everywhere over the cross section, except near the center where the strain is always zero. The stress distribution is then very nearly uniform, as shown in Figure 7.17(c). This is referred to as the *fully plastic condition*, and the torque at which it occurs is called the *fully plastic torque*, T_{fp}.

Fully plastic torque. The torque-stress relation for the fully plastic case is obtained from Eq. (7.4):

$$T = \int \rho \tau dA$$

Referring to Figure 7.18, we take a ring with radius ρ and thickness $d\rho$ as the element of area dA. The shear stress is equal to the yield strength, τ_{YS}, and is constant over the cross section. Thus, we have

$$T_{fp} = \int_0^R \rho \tau_{YS} (2\pi \rho d\rho) = \frac{2}{3} \pi \tau_{YS} R^3 \tag{7.17}$$

or, expressed in terms of the polar moment of inertia,

$$T_{fp} = \frac{4}{3} \frac{\tau_{YS} J}{R} \tag{7.18}$$

It is also informative to express the fully plastic torque in terms of the *yield torque*, T_{YS}, at which the maximum shear stress first reaches the yield strength. From Eq. (7.6), we have

$$T_{YS} = \frac{\tau_{YS} J}{R} \tag{7.19}$$

therefore, Eq. (7.18) can be written as

$$T_{fp} = \frac{4}{3} T_{YS} \tag{7.20}$$

Thus, the fully plastic torque is four-thirds of the torque at which the shaft starts to yield.

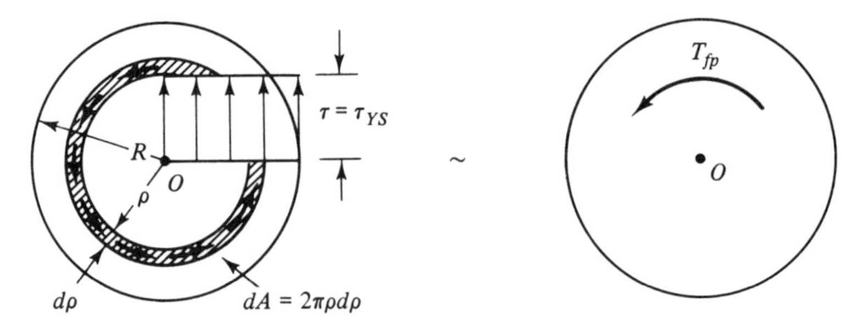

Figure 7.18

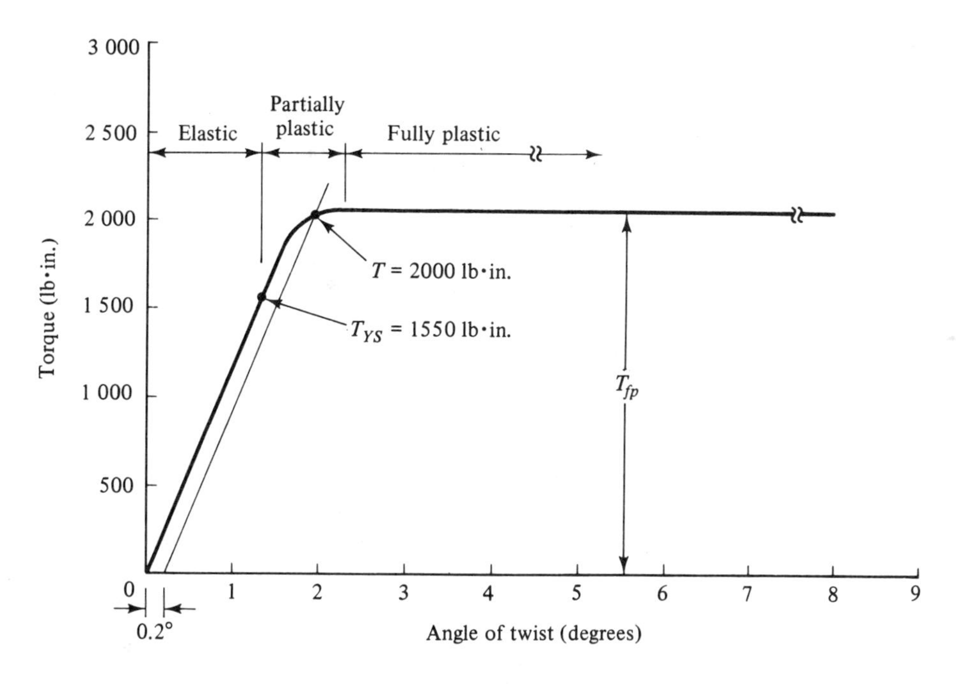

Figure 7.19. Torsional Response of a Mild Steel Shaft (Initial Portion Only)

These various torques and the different ranges of behavior associated with them are illustrated in Figure 7.19, which shows a portion of the experimentally determined torque-twist curve for a mild steel shaft. The point at which the curve first starts to deviate from a straight line corresponds to the onset of yielding. This point is difficult to locate accurately because the deviation from linearity is initially very slight. The horizontal portion of the curve corresponds to the fully plastic condition.

Notice that in the fully plastic range the angle of twist can increase to very large values with little or no increase in torque. Thus, the shaft usually fails by general yielding once the fully plastic condition is reached, unless the twisting is somehow restricted.

Figure 7.19 also shows the advantage of allowing some permanent deformations whenever possible. If no permanent deformations are allowed, the maximum torque that can be supported is T_{YS}, which for this particular shaft is 1550 lb·in. However, if a permanent twist of 0.2° is permitted, a torque of 2000 lb·in. can be supported, which is an increase of 29%. As illustrated in the figure, this torque is determined in the same manner as the yield strength of a material.

Ultimate torque. If the stress-strain diagram is reasonably flat in the vicinity of the ultimate strength, the stress distribution will be very nearly uniform, as in the fully plastic case. Thus, the torque-stress relation is the same as that given in Eq. (7.17), except that the yield strength is replaced by the ultimate strength, τ_U, and the fully plastic torque is replaced by the *ultimate torque*, T_U:

$$T_U = \frac{2}{3}\pi\tau_U R^3 \tag{7.21}$$

or

$$T_U = \frac{4}{3}\frac{\tau_U J}{R} \tag{7.22}$$

These equations define the maximum torque that can be applied to a shaft made of a ductile material without it breaking.

Hollow shafts. The inelastic behavior of hollow shafts is the same as for solid shafts. However, the torque-stress relations cannot be obtained by simply replacing J in the various equations by that for a hollow shaft, as in the linearly elastic case. These relations must be derived independently. This can be done by carrying out the integration in Eq. (7.17) from the inside radius R_i of the shaft to the outside radius R_o. For the fully plastic torque, we obtain

$$T_{fp} = \frac{2}{3}\pi\tau_{YS}\left(R_o^3 - R_i^3\right) = \frac{4}{3}\tau_{YS}\left(\frac{J_o}{R_o} - \frac{J_i}{R_i}\right) \tag{7.23}$$

where J_o and J_i are the polar moments of inertia of areas with radius R_o and R_i, respectively. Similarly, the ultimate torque is

$$T_U = \frac{2}{3}\pi\tau_U\left(R_o^3 - R_i^3\right) = \frac{4}{3}\tau_U\left(\frac{J_o}{R_o} - \frac{J_i}{R_i}\right) \tag{7.24}$$

Example 7.7 Fully Plastic and Ultimate Torques for a Hollow Shaft. Determine the fully plastic and ultimate torques for a hollow shaft with an outside diameter of 80 mm and an inside diameter of 50 mm made of a ductile material for which $\tau_{YS} = 400\ \text{MN/m}^2$ and $\tau_U = 550\ \text{MN/m}^2$.

Solution. The fully plastic torque for a hollow shaft is defined by Eq. (7.23):

$$T_{fp} = \frac{2}{3}\pi\tau_{YS}\left(R_o^3 - R_i^3\right)$$

or

$$T_{fp} = \frac{2}{3}\pi(400 \times 10^6\ \text{N/m}^2)\left[(0.040\ \text{m})^3 - (0.025\ \text{m})^3\right]$$

$$= 40.5 \times 10^3\ \text{N·m} \quad \text{or} \quad 40.5\ \text{kN·m} \qquad \textit{Answer}$$

The ultimate torque is given by the same expression, but with the yield strength replaced by the ultimate strength [Eq. (7.24)]:

$$T_U = \frac{2}{3}\pi\tau_U\left(R_o^3 - R_i^3\right) = \frac{\tau_U}{\tau_{YS}}T_{fp}$$

Thus,

$$T_U = \frac{(550 \text{ MN/m}^2)(40.5 \text{ kN·m})}{(400 \text{ MN/m}^2)} = 55.7 \text{ kN·m} \qquad \textit{Answer}$$

Example 7.8. Inelastic Response of a Shaft. (a) What applied couple C is required to twist the shaft shown in Figure 7.20(a) through an angle of 11°? The stress-strain diagram for the material is given in Figure 7.20(b). (b) What is the maximum couple that can be applied without the shaft failing due to excessive inelastic deformations (general yielding)?

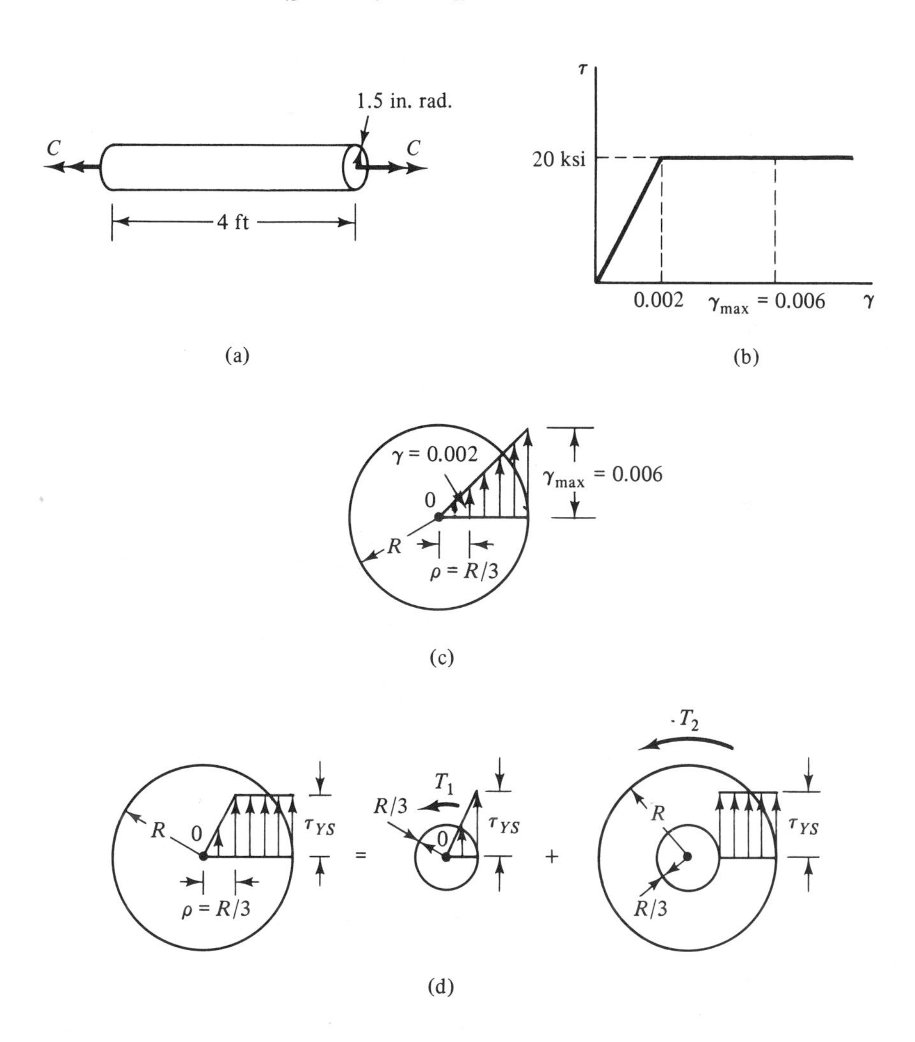

Figure 7.20

Solution. (a) We first check to see whether the strains are in the linearly elastic or inelastic range. From Eq. (7.2), the maximum strain is

$$\gamma_{max} = \frac{\theta R}{L} = \frac{(11°)(\pi \text{rad}/180°)(1.5 \text{ in.})}{48 \text{ in.}} = 0.006$$

Since this value exceeds the strain at the proportional limit [Figure 7.20(b)], the shaft is strained into the inelastic range. The strain distribution is as shown in Figure 7.20(c). By similar triangles, the strain is found to be equal to the proportional limit strain at a radius of $\rho = R/3$.

The corresponding stress distribution is shown in Figure 7.20(d). As indicated in the figure, the shaft can be thought of as consisting of two parts: an inner elastic core with radius $R/3$ stressed to the yield strength and an outer fully plastic part with inner radius $R/3$ and outer radius R. The torques T_1 and T_2 carried by each of these parts can be obtained from Eqs. (7.19) and (7.23), respectively:

$$T_1 = \frac{\tau_{YS}(\pi/2)(R/3)^4}{(R/3)} = \frac{\pi}{54}\tau_{YS}R^3$$

$$T_2 = \frac{2}{3}\pi\tau_{YS}\left[R^3 - (R/3)^3\right] = \frac{52}{81}\pi\tau_{YS}R^3$$

The total torque, which is equal in magnitude to the applied couple, is

$$T = T_1 + T_2 = \frac{53.5}{81}\pi\tau_{YS}R^3 \qquad \textit{Answer}$$

or

$$T = \frac{53.5\pi}{81}(20\ 000 \text{ lb/in.}^2)(1.5 \text{ in.})^3 = 14.01 \times 10^4 \text{ lb·in.} \qquad \textit{Answer}$$

Note that only a small fraction of the torque is carried by the elastic core.

(b) General yielding occurs when the fully plastic condition is reached. Thus, the maximum couple that can be transmitted is equal in magnitude to the fully plastic torque. From Eq. (7.15), we have

$$T_{fp} = \frac{2}{3}\pi\tau_{YS}R^3 = \frac{2\pi}{3}(20\ 000 \text{ lb/in.}^2)(1.5 \text{ in.})^3 = 14.14 \times 10^4 \text{ lb·in.} \qquad \textit{Answer}$$

which is only slightly larger than the value obtained in part (a).

PROBLEMS

7.39 Compare the fully plastic and ultimate torques for a hollow shaft with an outside diameter of 4 in. and an inside diameter of 2 in. with those for a solid shaft with the same outside diameter. The shafts are made of steel with a yield point in shear of 21 ksi and an ultimate shear strength of 45 ksi.

7.40 The nut on a $\frac{1}{2}$-in. diameter steel bolt is "frozen" by rust and will not turn. If the stress concentration factor due to the threads is $K_\tau = 3$, what is the maximum couple that can be exerted on the nut without causing the bolt to (a) yield and (b) twist off? $\tau_{YS} = 60$ ksi and $\tau_U = 80$ ksi.

7.41 Determine the ratio of the fully plastic torque to the yield torque for a hollow circular shaft whose outer radius is twice the inner radius.

7.42 A 50-mm diameter shaft 1 m long made of a material with the stress-strain curve shown is twisted through an angle of 0.12 radian. Sketch the distribution of shear stress and shear strain over the cross section and compute the torque T.

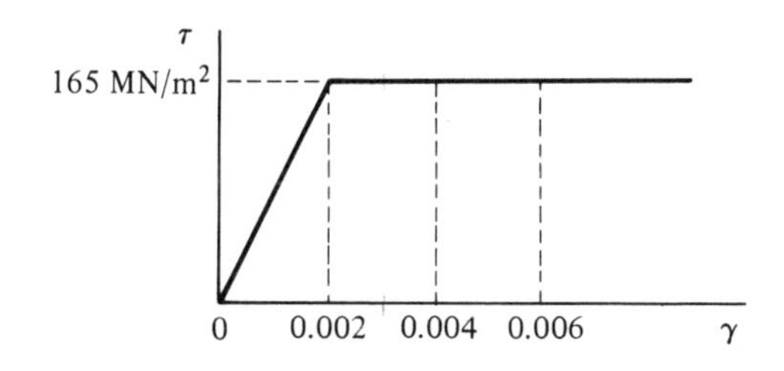

Problem 7.42

7.43 A 24-in. long solid circular shaft with a radius of 1.5 in. is twisted until the maximum shear strain is 0.01. The shear modulus G of the material is 10^7 psi and the yield point in shear is 25 ksi. Determine the angle of twist θ and the torque T.

7.44 A composite shaft is constructed by inserting a 2-in. diameter solid shaft into a hollow shaft with 2 in. ID and 4 in. OD and bonding them together over their entire length. What is the maximum torque T that the composite shaft can transmit without causing yielding in the inner member? The stress-strain curves for the two shafts are as shown. Sketch the distribution of shear stress over the cross section for this value of torque.

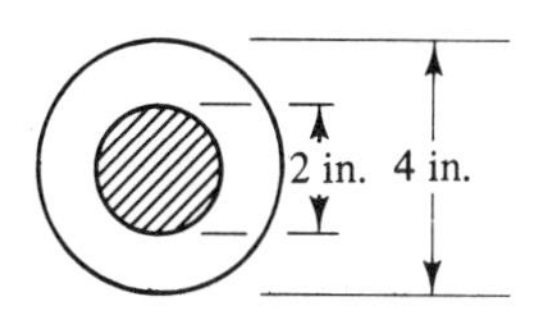

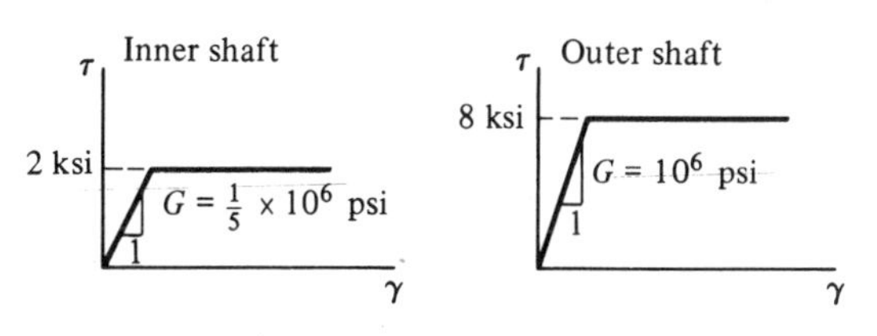

Problem 7.44

7.45 What is the fully plastic torque for the composite shaft in Problem 7.44?

7.46 The stepped shaft shown is fixed at each end and is made of steel with shear modulus $G = 83$ GN/m^2 and shear yield strength $\tau_{YS} = 414$ MN/m^2. If the angle of twist at the intersection of the two segments is 0.12 radian, what is the applied couple C?

7.47 A solid circular shaft 3 in. in diameter and 4 ft long carries a torque of 12 kip·ft. If the material is 6061-T6 aluminum, what is the radius ρ_E of the elastic core and what is the angle of twist θ?

7.48 Same as Problem 7.47, except that the material is brass.

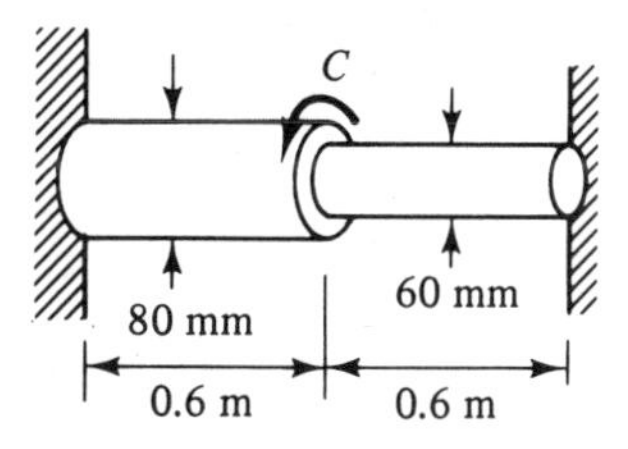

Problem 7.46

7.7 LIMIT LOADS

When the fully plastic condition is reached in a shaft made of a ductile material, failure by general yielding and the associated large angles of twist usually occurs for little or no increase in load. If, however, the member is statically indeterminate, the angle of twist is restricted by the additional support conditions and further loading is possible before general yielding occurs.

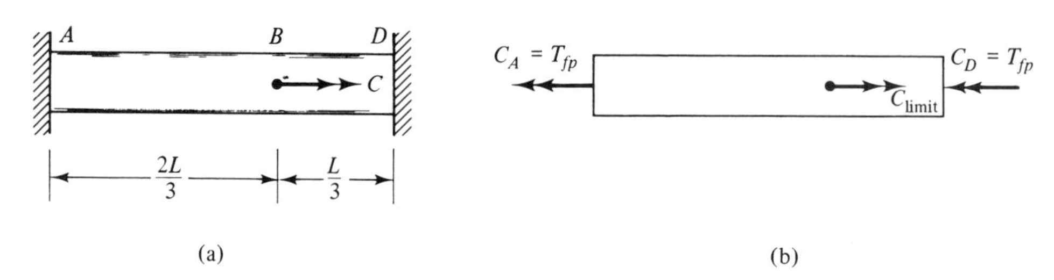

Figure 7.21

To illustrate this, let us consider the example of a shaft attached to rigid walls at each end and subjected to a couple C applied at an intermediate point along its length [Figure 7.21(a)]. The material behavior is assumed to be elastic-perfectly plastic. An elastic analysis shows that the torques in the two portions of the shaft are

$$T_{AB} = \frac{C}{3} \qquad T_{BD} = \frac{2C}{3}$$

Thus, portion BD is the most highly stressed.

As the applied couple is increased from zero, the fully plastic condition is first reached in portion BD of the shaft. However, its angle of twist will be restricted by the adjoining portion, AB, which remains elastic, or possibly partially plastic. Any further increases in load must be supported entirely by portion AB because the torque in portion BD will remain equal to the fully plastic torque. Eventually, portion AB will also become fully plastic. When this happens, both parts of the shaft will be free to undergo large angles of twist, and its usefulness as a load-carrying member will generally be ended. As in the axial loading case, the load at which this occurs is called the *limit load.*

At the limit load, the torques are equal to the fully plastic torques, and the problem becomes statically determinate. Thus, the limit load can be determined solely from the equations of equilibrium, as is evident from the free-body diagram of the member [Figure 7.21(b)]. For this particular example, the limit load (couple) is

$$C_{\text{limit}} = 2T_{fp}$$

where the fully plastic torque is defined by Eq. (7.17).

Example 7.9. Limit Load of a Stepped Shaft. Derive an expression for the limit load, C_{limit}, for the stepped shaft shown in Figure 7.22(a). Express the result in terms of the material yield strength, τ_{YS}, and the diameter, D, of the smaller portion of the shaft.

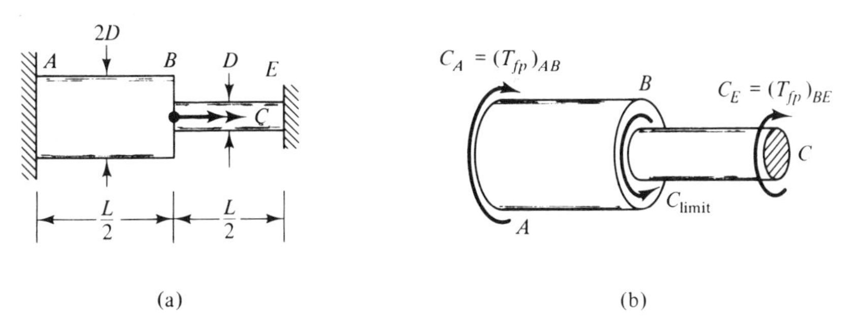

Figure 7.22

Solution. General yielding cannot occur until both portions of the shaft become fully plastic. The free-body diagram for this case is shown in Figure 7.22(b). For equilibrium, the sum of the moments about the longitudinal axis must be zero, or

$$C_{\text{limit}} = (T_{fp})_{AB} + (T_{fp})_{BE}$$

From Eq. (7.17), we have

$$(T_{fp})_{BE} = \frac{2\pi}{3}\tau_{YS}\left(\frac{D}{2}\right)^3 = \frac{\pi}{12}\tau_{YS}D^3$$

$$(T_{fp})_{AB} = \frac{2}{3}\pi\tau_{YS}D^3$$

so

$$C_{\text{limit}} = \frac{3}{4}\pi\tau_{YS}D^3 \qquad \textit{Answer}$$

Note that the lengths of the segments don't enter into the computations.

PROBLEMS

7.49 to 7.53 Determine the limit load (couple) for the members shown. The shaft diameters and material shear yield strengths are as indicated; otherwise, express the results in terms of the fully plastic torque T_{fp} for the members.

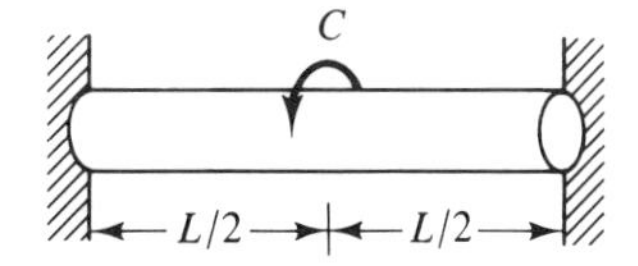

Problem 7.49

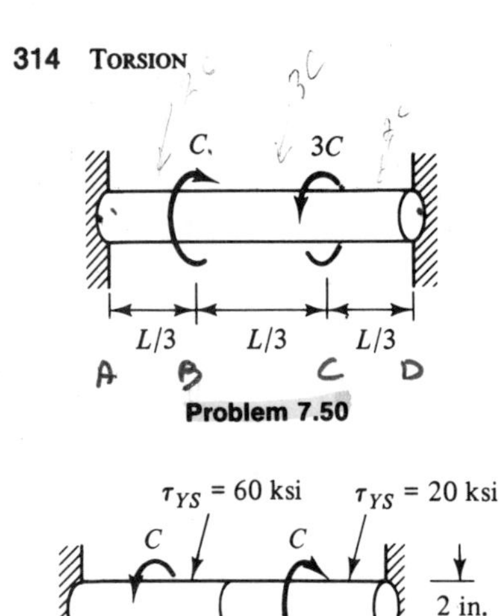

Problem 7.50

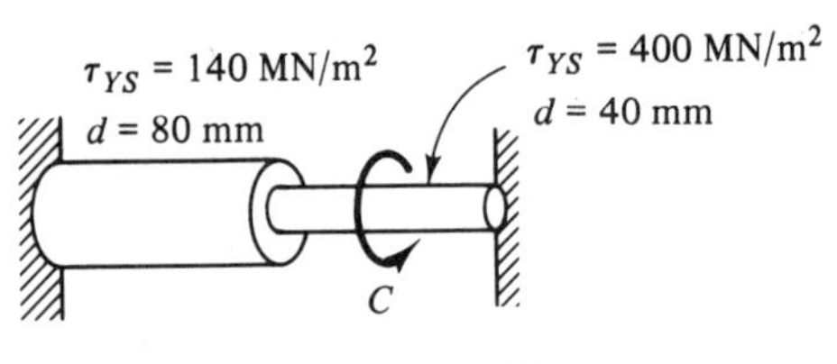

Problem 7.52

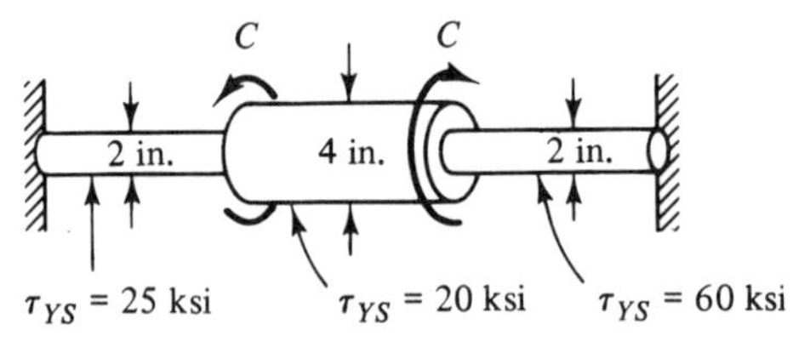

Problem 7.53

Problem 7.51

7.54 A composite shaft is made by slipping a 3-in. nominal diameter steel pipe inside a 4-in. nominal diameter steel pipe and connecting them together at the ends. What is the maximum torque that can be transmitted without failure by general yielding if τ_{YS} = 20 ksi?

7.8 CLOSURE

This chapter was devoted to a study of the torsional response of solid and hollow circular shafts. The stress distribution in a shaft depends upon the stress-strain relation for the material and, consequently, so do both the torque-stress and torque-twist relations. Most torsional members are designed to operate within the linearly elastic range, but the inelastic response is also of importance. Accordingly, it was considered in some detail for the case of an elastic-perfectly plastic material.

The couples applied to a shaft are often referred to as torques, but here we used this term only to denote the axial component of the couple in the stress resultant. This was done in an attempt to avoid any confusion about which of these quantities is involved in the torque-stress and torque-twist relations. In many cases, but not in all, the torque is equal in magnitude to the applied couples. The distinction between the two should always be kept in mind, for many errors can be avoided by doing so.

As mentioned in Section 7.1, all the results presented in this chapter apply for circular shafts at rest and for circular shafts rotating with constant angular velocity. The torsion of shafts with noncircular cross sections is considerably more complicated than that of circular shafts and is treated in more advanced texts.

chapter eight BENDING

8.1 INTRODUCTION

To this point, we have considered the response of slender members to loadings acting along the longitudinal axis. These loadings were forces in the case of axially loaded members and couples in the case of torsional members. In this chapter we shall consider the response of straight, slender members to forces and couples acting perpendicular to the longitudinal axis. Such members are called *beams* and are probably the most common type of structural element.

The loadings applied to a beam cause its originally straight axis to deform into a curve (Figure 8.1). This type of deformation is referred to as *bending*, or *flexure*. Our goal in this chapter is to develop the load-stress and load-deformation relations for the bending response of beams and to learn how to apply them. Again, the general procedure outlined in Section 5.10 will be used, and the deformations will be assumed to be small.

Beams are usually classified according to their support conditions. For example, a beam supported by rollers, or the equivalent, at each end is called a *simply supported beam* or a *simple beam* [Figure 8.1(a)]. A *cantilever beam* is one that is fixed at one end and free at the other [Figure 8.1(b)]. If the free end is placed on a roller, as in Figure 8.1(c), it becomes a *propped cantilever*. Beams clamped on both ends are usually called *fixed-fixed beams* [Figure 8.1(d)], and those that extend

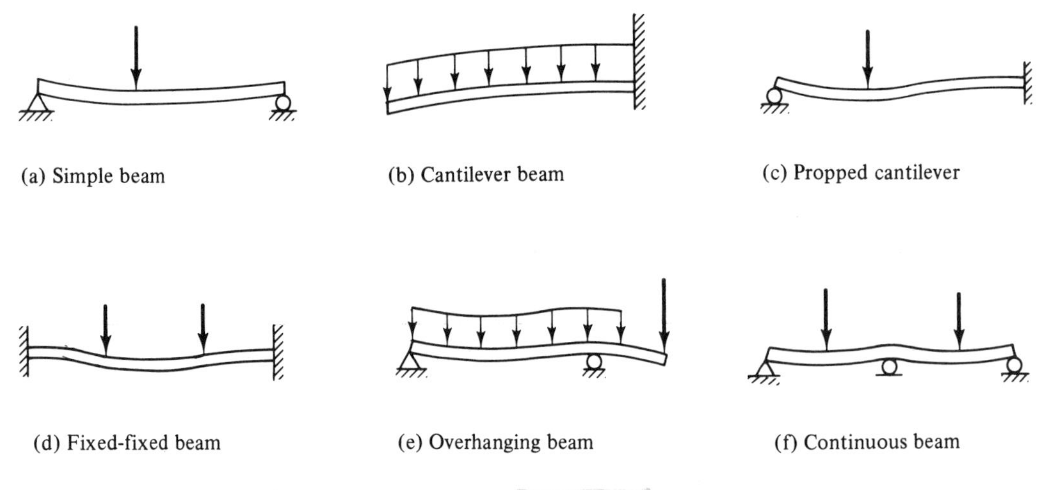

(a) Simple beam

(b) Cantilever beam

(c) Propped cantilever

(d) Fixed-fixed beam

(e) Overhanging beam

(f) Continuous beam

Figure 8.1

beyond the supports are called *overhanging beams* [Figure 8.1(e)]. If a beam is supported at more than two points [Figure 8.1(f)], it is referred to as a *continuous beam*. We shall use this nomenclature throughout.

Beams are also classified according to the shapes of their cross sections. For example, an I-beam has a cross section shaped like the letter I and a T-beam has a cross section shaped like the letter T.

8.2 SHEAR AND BENDING MOMENT DIAGRAMS

The loading on most beams is such that the stress resultant on planes perpendicular to the axis consists of a shear force, V, and a bending moment, M. In determining beam responses, it is highly convenient, if not essential, to first determine the *shear* and *bending moment diagrams* showing the values of these quantities along the entire length of the member. Because of the nature of the loading on beams, these diagrams are usually more complicated than the normal force and torque diagrams considered previously. Thus, we shall consider them in some detail.

The basic procedure for determining the shear and bending moment diagrams is to determine the values of V and M at various locations along the member by using the method of sections, as discussed in Section 5.2, and then plot the results. As an illustration, let us consider the beam shown in Figure 8.2(a). Note that the reactions have already been determined. Ordinarily, this would be the first step taken.

We choose coordinates with the x axis to the right along the beam and the y axis vertically upward. The exact location of the x axis within the cross section will be left unspecified for the present. Sectioning the beam perpendicular to its axis at an arbitrary location between points A and B, we obtain the free-body diagram shown in Figure 8.2(b). The shear force will be considered positive if it is directed downward on the right-hand face of the beam segment, and the bending moment

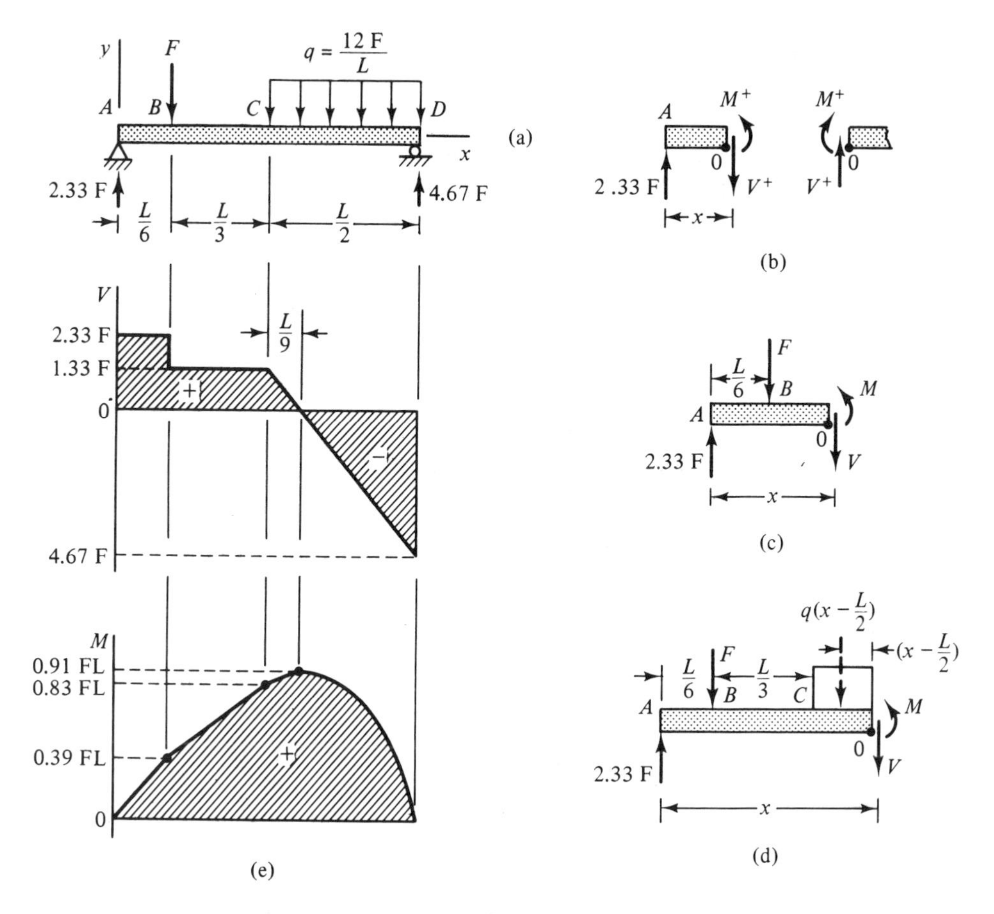

Figure 8.2

on this face will be considered positive if it is counterclockwise. Note that positive V and M have the opposite sense on the left-hand face of the adjoining segment, in accordance with Newton's third law. It is convenient to place V and M on the free-body diagram in their positive directions because this simplifies the interpretation of the signs on the answers.

The equilibrium equations for this segment of the beam are

$$\Sigma F_y = 2.33F - V = 0$$

$$\overset{+}{\curvearrowleft}\ \Sigma M_o = M - (2.33F)x = 0$$

from which we obtain

$$V = 2.33F \qquad M = 2.33Fx \qquad \left(0 \leqslant x \leqslant \frac{L}{6}\right)$$

These equations hold only over the left one-sixth of the beam because the free-body diagram is valid only over this range.

If the beam is sectioned between points B and C, the force F comes into play, and the free-body diagram is as shown in Figure 8.2(c). This diagram is valid over the central one-third of the beam. We have

$$\Sigma F_y = 2.33F - F - V = 0$$

$$\overset{+}{\Sigma M_o} = M - (2.33F)x + F\left(x - \frac{L}{6}\right) = 0$$

or

$$V = 1.33F \qquad M = \frac{FL}{6} + 1.33Fx \qquad \left(\frac{L}{6} \leqslant x \leqslant \frac{L}{2}\right)$$

Finally, if the beam is sectioned within portion CD, we obtain the free-body diagram shown in Figure 8.2(d). The equilibrium equations for this segment of the beam are

$$\Sigma F_y = 2.33F - F - q\left(x - \frac{L}{2}\right) - V = 0$$

$$\overset{+}{\Sigma M_o} = M - (2.33F)x + F\left(x - \frac{L}{6}\right) + \frac{q}{2}\left(x - \frac{L}{2}\right)\left(x - \frac{L}{2}\right) = 0$$

or

$$V = 7.33F - 12F\frac{x}{L}$$

$$M = -1.33FL + 7.33Fx - 6F\frac{x^2}{L} \qquad \left(\frac{L}{2} \leqslant x \leqslant L\right)$$

Notice that there is one set of equations for V and M for each portion of the beam and that each set applies until a change in loading is encountered. Plotting these equations, we obtain the shear and bending moment diagrams shown in Figure 8.2(e). It is convenient to plot the diagrams directly below the free-body diagram of the beam.

From Figure 8.2(e), we see that upward concentrated forces (A_y and D_y in this example) cause upward jumps in the shear diagram and downward concentrated forces (F in this example) cause downward jumps. The magnitudes of the jumps are equal to the magnitudes of the corresponding forces. Applied couples cause similar jumps in the moment diagram. Counterclockwise couples cause upward jumps and clockwise couples cause downward jumps (see Example 8.1).

The weight of a beam is often small compared to the applied loads, in which case the weight can be neglected with little error. It is a simple matter to include it, however. If the beam is uniform, its weight merely contributes an additional uniformly distributed loading. The weight per unit length of some common structural shapes is given in Tables A.3 through A.12 of the Appendix, and values for other members are available in handbooks.

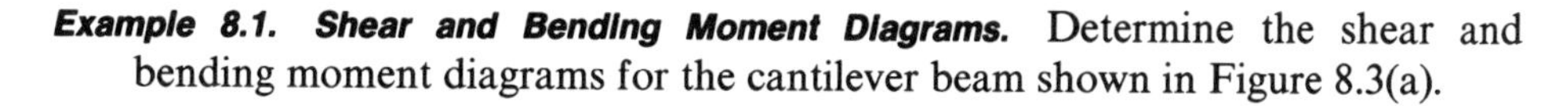

Example 8.1. Shear and Bending Moment Diagrams. Determine the shear and bending moment diagrams for the cantilever beam shown in Figure 8.3(a).

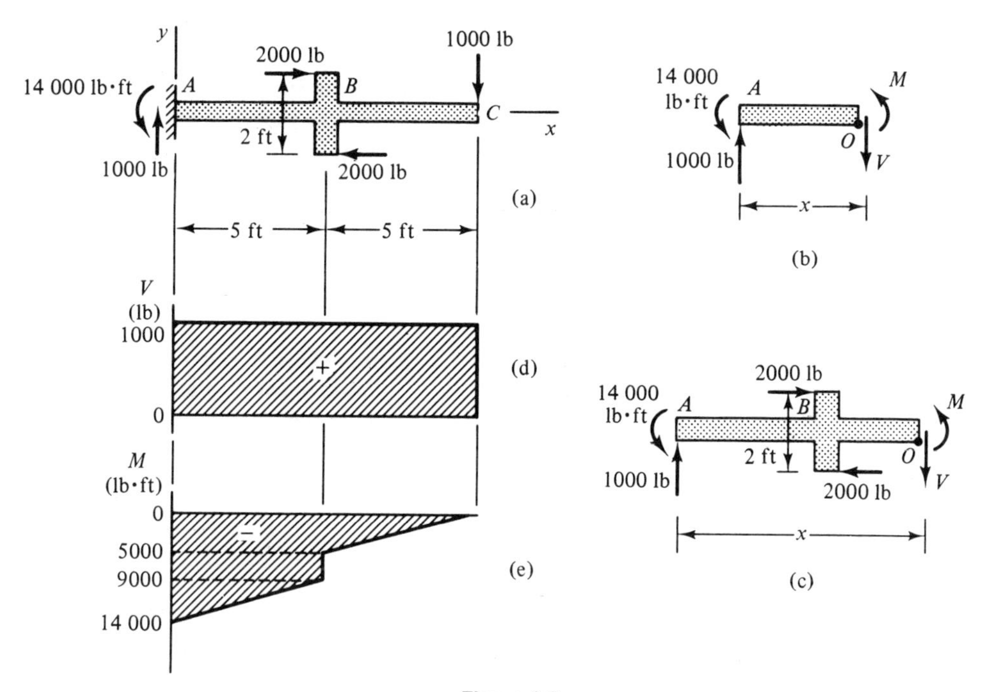

Figure 8.3

Solution. The first step is to determine the reactions. From $\Sigma F_y = 0$ and $\Sigma M_A = 0$, the reactions at the wall are found to be a 1000-lb upward force and a 14 000-lb·ft counterclockwise couple.

Sectioning the beam between A and B, we obtain the FBD shown in Figure 8.3(b). The equilibrium equations are

$$\Sigma F_y = 1000 \text{ lb} - V = 0$$

$$\overset{+}{\curvearrowleft} \Sigma M_o = M + 14\ 000 \text{ lb·ft} - (1000 \text{ lb})x = 0$$

from which we obtain

$$V = 1000 \text{ lb} \qquad M = -14\ 000 + 1000x \text{ lb·ft} \qquad (0 \leqslant x \leqslant 5)$$

The FBD obtained by sectioning the beam between B and C is shown in Figure 8.3(c). We have

$$\Sigma F_y = 1000 \text{ lb} - V = 0$$

$$\overset{+}{\curvearrowleft} \Sigma M_o = M + 14\ 000 \text{ lb·ft} - (1000 \text{ lb})x - 2000 \text{ lb}(2 \text{ ft}) = 0$$

or

$$V = 1000 \text{ lb} \qquad M = -10\ 000 + 1000x \text{ lb·ft} \qquad (5 \leqslant x \leqslant 10)$$

The corresponding shear and moment diagrams are shown in Figures 8.3(d) and 8.3(e), respectively.

Notice that the concentrated couples cause jumps in the moment diagram. The counterclockwise couple at the wall causes a downward jump, and the clockwise couple at the center of the beam causes an upward jump. The jumps are of the same magnitude as the corresponding couples.

PROBLEMS

8.1 to 8.8 For the beams shown, determine the equations for the shear force V and the bending moment M as a function of the distance x from the left end and construct the shear and moment diagrams. Neglect the weights of the members.

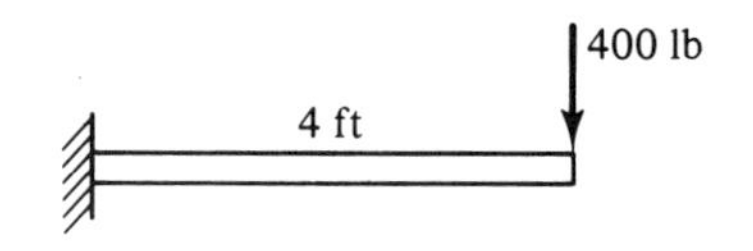

Problem 8.1

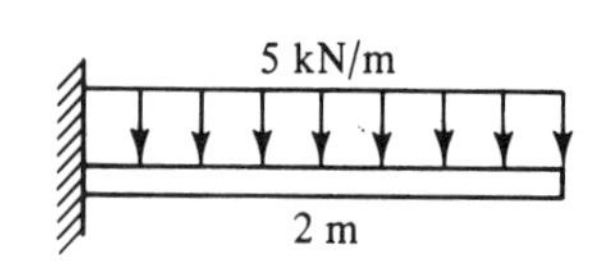

Problem 8.2

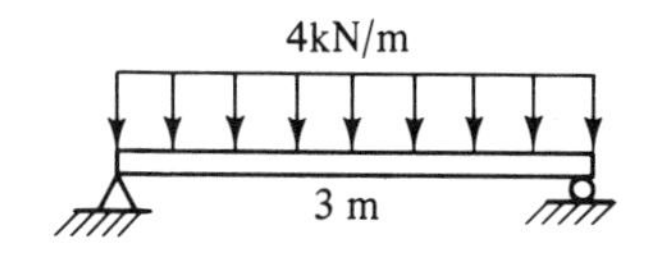

Problem 8.3

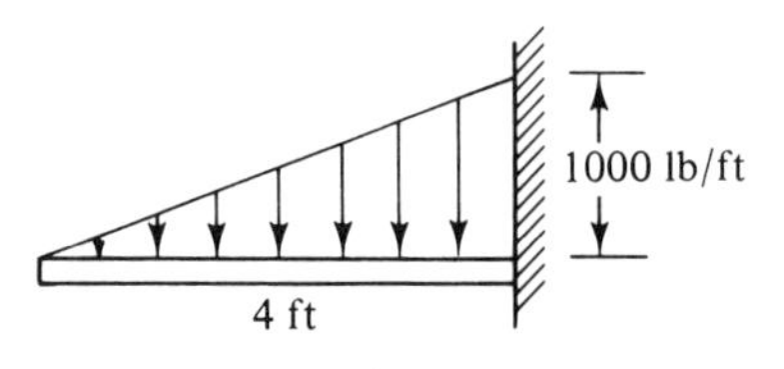

Problem 8.4

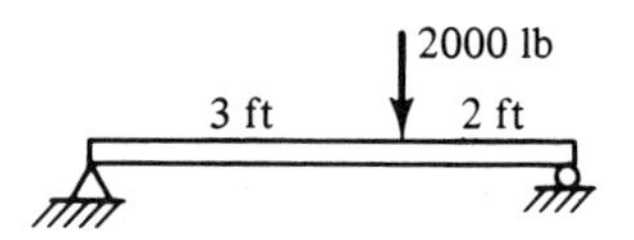

Problem 8.5

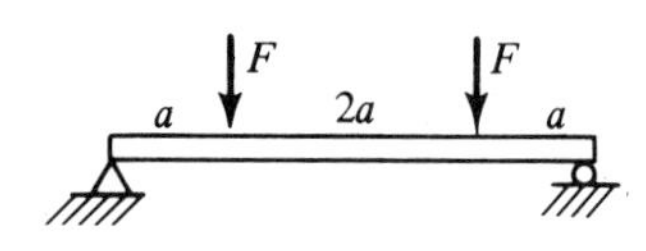

Problem 8.6

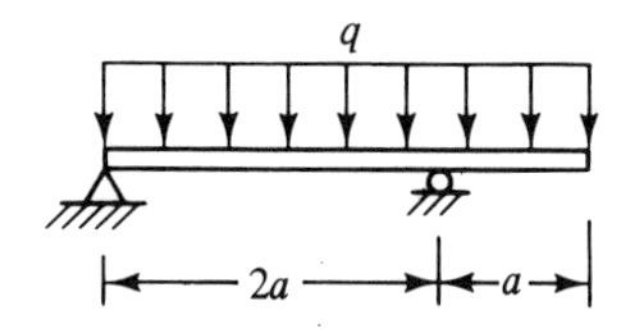

Problem 8.7

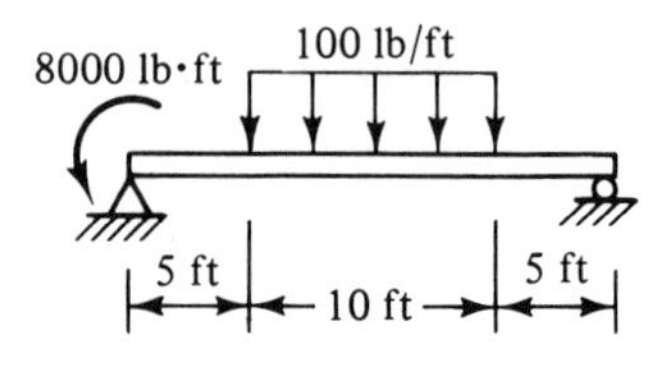

Problem 8.8

8.3 LOAD, SHEAR, AND MOMENT RELATIONS

Although the method of determining the shear and bending moment diagrams presented in the preceding section is straightforward, it is tedious to apply because it requires that the relations between the shear force, bending moment, and applied loads be derived for each different beam considered. An alternate approach is to derive relations between these quantities which apply to beams in general. As we shall see, this approach greatly facilitates the construction of the shear and bending moment diagrams, especially for simple loadings.

Figure 8.4(a) shows a beam subjected to a distributed load, $q(x)$, which will be considered positive if it acts in the positive y direction. We now consider the equilibrium of a short segment of the beam with length Δx located an arbitrary distance x from the end. The free-body diagram is as shown in Figure 8.4(b). Since the shear force and bending moment generally vary along the beam, their values on the left- and right-hand faces of the segment are shown to differ by a small amount ΔV and ΔM, respectively. Since the segment is short, the loading can be considered to be uniform over its length.

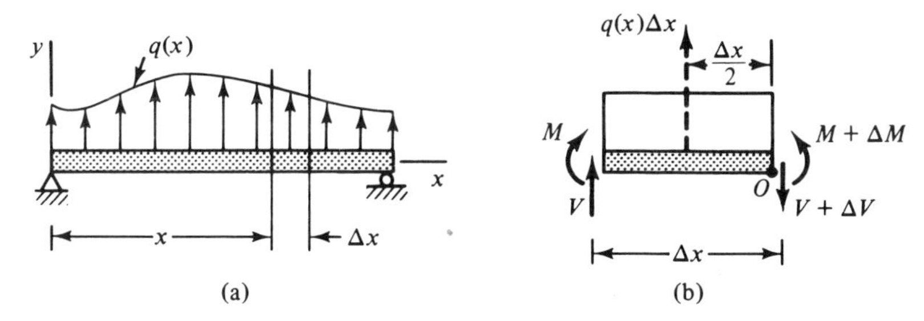

Figure 8.4

From the equilibrium equations

$$\Sigma F_y = V + q\Delta x - (V + \Delta V) = 0$$

$$\overset{+}{\curvearrowleft}\Sigma M_o = (M + \Delta M) - q\Delta x\left(\frac{\Delta x}{2}\right) - M - V\Delta x = 0$$

we have

$$\Delta V = q\Delta x$$

$$\Delta M = V\Delta x + q\frac{(\Delta x)^2}{2}$$

Dividing these expressions by Δx and taking the limit as Δx tends to zero, we obtain

$$\lim_{\Delta x \to 0} \frac{\Delta V}{\Delta x} = \frac{dV}{dx} = q \tag{8.1}$$

$$\lim_{\Delta x \to 0} \frac{\Delta M}{\Delta x} = \frac{dM}{dx} = V \tag{8.2}$$

These equations define the relationship between the shear force, bending moment, and distributed load. Any concentrated forces or couples that may be acting are not accounted for in these relations. Consequently, they apply only over the regions between concentrated loads.

Equations (8.1) and (8.2) can be expressed in an alternate form by multiplying them by dx and integrating from a location $x = x_1$ where the shear and bending moment have values V_1 and M_1 to another location $x = x_2$ where they have values V_2 and M_2:

$$V_2 - V_1 = \int_{x_1}^{x_2} q dx \tag{8.3}$$

$$M_2 - M_1 = \int_{x_1}^{x_2} V dx \tag{8.4}$$

Interpreted geometrically, Eqs. (8.1) and (8.2) state that the slope of the shear diagram at any location is equal to the intensity of the distributed loading at that location, and that the slope of the moment diagram at any point is equal to the value of the shear force at that point. It is also evident from Eq. (8.2) that the relative maxima and minima on the moment diagram, which correspond to points of zero slope, occur where $V = 0$.

The integrals in Eqs. (8.3) and (8.4) represent areas under the loading and shear diagrams, respectively. Consequently, Eq. (8.3) indicates that the change in shear between any two locations along the beam is equal to the area under the loading diagram between these locations. Similarly, Eq. (8.4) indicates that the change in bending moment between any two points is equal to the area under the shear diagram between these points. These areas are considered to be positive if q and V are positive, respectively, and negative if they are negative.

For simple loadings, the areas under the loading and shear diagrams can be easily determined by using only geometry, and the shear and moment diagrams can be sketched pretty much by inspection by using the geometrical interpretations of Eqs. (8.1) through (8.4). The jumps in the diagrams due to concentrated forces and couples must be added in separately, however, for they are not accounted for in these equations. For more complex loadings, the areas under the loading and shear diagrams are not so easily computed, and it is usually necessary to determine V and M by integration.

When sketching the shear and moment diagrams, it is important that their accuracy be checked. This can be done by noting whether or not they close (end up at zero value). Closure of the shear diagram indicates that the sum of the vertical forces acting upon the beam is zero, as it must be for equilibrium. Similarly, closure of the moment diagram implies that the sum of the moments for the entire beam is zero. If either of the diagrams fails to close, it indicates that there is an error in their construction or in the determination of the reactions.

Example 8.2. Shear and Moment Diagrams for a Simple Beam. Sketch the shear and bending moment diagrams for the simple beam shown in Figure 8.5(a).

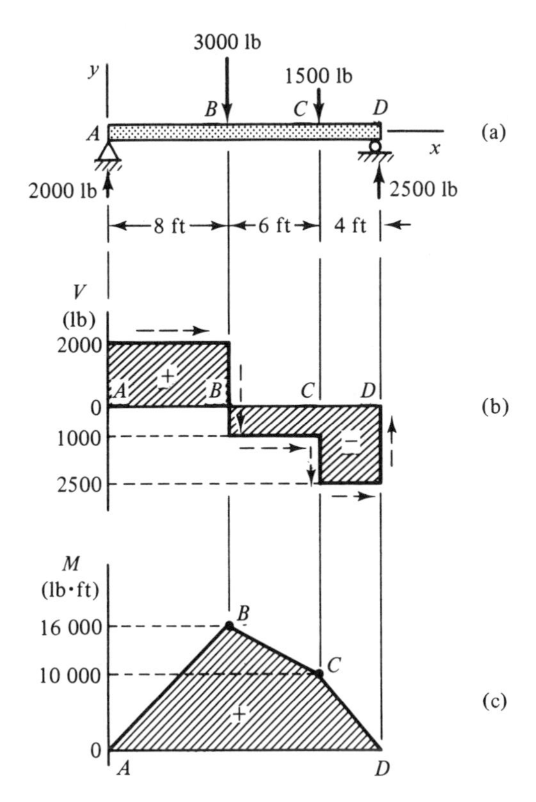

Figure 8.5

Solution. We first determine the reactions. The upward force at A is found to be 2000 lb and the upward force at D is 2500 lb.

We now sketch the shear diagram. From Eq. (8.1), we note that the diagram will consist of horizontal lines (zero slope) between the concentrated forces, since there is no distributed load ($q \equiv 0$). Working from left to right along the beam, we first encounter the upward concentrated force at A. Thus, the shear jumps up to a value of 2000 lb at this point [Figure 8.5(b)]. It then remains constant up to point B, where it suddenly decreases by an amount equal to the magnitude of the downward force acting there. Continuing across the beam in this fashion, we obtain the complete diagram shown. The order of progression is indicated by the dashed arrows.

The moment diagram is now obtained by referring to the shear diagram. There will be no jumps in the value of M because there are no couples acting. Consequently, $M = 0$ at the left end of the beam.

The shear is positive and constant from A to B; therefore, according to Eq. (8.2), the moment diagram has constant positive slope over this interval [Figure 8.5(c)]. Furthermore, the change in moment in going from A to B is equal to the area under the shear diagram between these points [Eq. (8.4)]:

$$M_B - M_A = (2000 \text{ lb})(8 \text{ ft}) = 16\ 000 \text{ lb·ft}$$

This establishes the value of the moment at B. From B to C, the diagram has constant negative slope because of the constant negative shear. The change in moment over this interval is

$$M_C - M_B = (-1000\ \text{lb})(6\ \text{ft}) = -6000\ \text{lb·ft}$$

for a total of 10 000 lb·ft at C. Continuing on across the beam in this fashion, we obtain the complete diagram shown.

Note that both diagrams close. This does not guarantee that they are correct, but it does provide a degree of confidence in them.

Example 8.3. Maximum Shear and Moment in a Beam. The overhanging beam shown in Figure 8.6(a) supports only its own weight. Determine the maximum shear force and bending moment in the beam and the locations at which they occur. The member is a 381-mm × 152-mm Universal Beam with an actual depth of 388.6 mm.

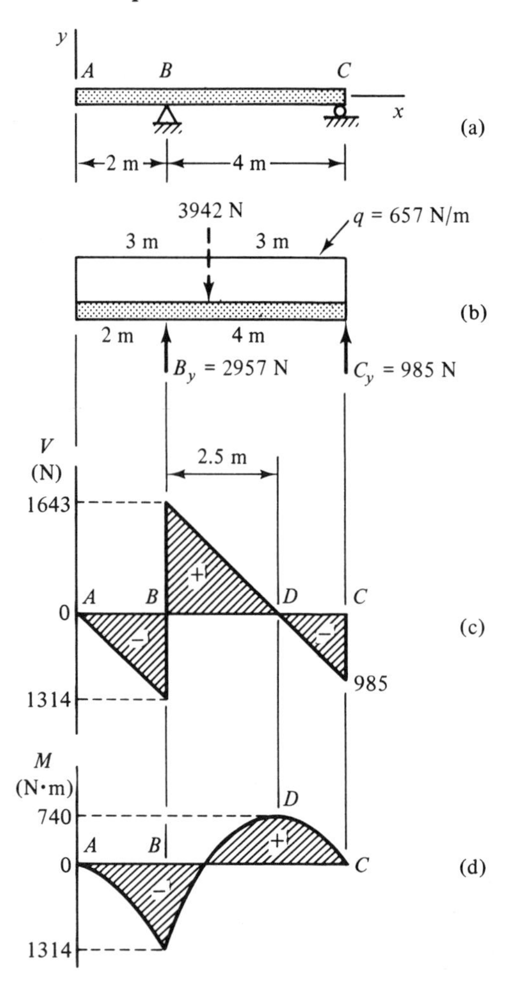

Figure 8.6

Solution. From Table A.8, we find that the beam has a mass of 67 kg/m. The weight per unit length is obtained by multiplying by the acceleration of gravity:

$$q = (67 \text{ kg/m})(9.81 \text{ m/s}^2) = 657 \text{ N/m}$$

From the FBD [Figure 8.6(b)], we have

$$\Sigma F_y = B_y + C_y - 3942 \text{ N} = 0$$

$$\overset{+}{\curvearrowleft} \Sigma M_C = -B_y(4 \text{ m}) + 3942 \text{ N}(3 \text{ m}) = 0$$

or

$$B_y = 2957 \text{ N} \qquad C_y = 985 \text{ N}$$

Because of the uniform downward (negative) distributed loading, each portion of the shear diagram will consist of straight lines with the same negative slope. The change in shear between A and B is equal to the area under the loading diagram between these points [Eq. (8.3)]:

$$V_B - V_A = (-657 \text{ N/m})(2 \text{ m}) = -1314 \text{ N}$$

This establishes the value of the shear just to the left of B [Figure 8.6(c)]. At B, the shear suddenly increases an amount equal to the magnitude of the upward force acting there. It then decreases linearly with distance from B to C. The change in shear in going from B to C is

$$V_C - V_B = (-657 \text{ N/m})(4 \text{ m}) = -2628 \text{ N}$$

for a total of -985 N just to the left of C. The upward force at C then brings the value of the shear back to zero.

The distance d from the left support to the point of zero shear can be determined from similar triangles. An equivalent procedure is to determine the distance required to reduce the shear from 1643 N to zero, given that it decreases at a rate of 657 N/m. Obviously, the required distance is $d = 1643 \text{ N}/(657 \text{ N/m}) = 2.5$ m. For more complex diagrams, it may be necessary to determine the points of zero shear by setting the equations for V equal to zero and solving for the corresponding values of x (see Example 8.4).

Since the shear is negative and increasing in magnitude from A to B, the moment diagram will have an increasing negative slope over this interval [Figure 8.6(d)]. The area under the shear diagram provides the change in moment:

$$M_B - M_A = \frac{1}{2}(-1314 \text{ N})(2 \text{ m}) = 1314 \text{ N·m}$$

This establishes the moment at B. Continuing on across the beam, we obtain the complete diagram shown. Note that the slope is positive and decreasing from B to D because of the positive decreasing shear. Similarly, the slope is negative and increasing from D to C. Also note that the diagram closes and that the relative maximum occurs at a point of zero shear.

From the shear and moment diagrams, we see that the maximum shear force is 1643 N and occurs just to the right of the support at B. The maximum bending moment occurs at this same support and has a magnitude of 1314 N·m.

Example 8.4. ***Shear and Moment Diagrams for a Cantilever Beam.*** A cantilever beam supports the loading shown in Figure 8.7(a). Obtain the shear and bending moment diagrams and determine the magnitude and location of the maximum bending moment. The reactions are as indicated.

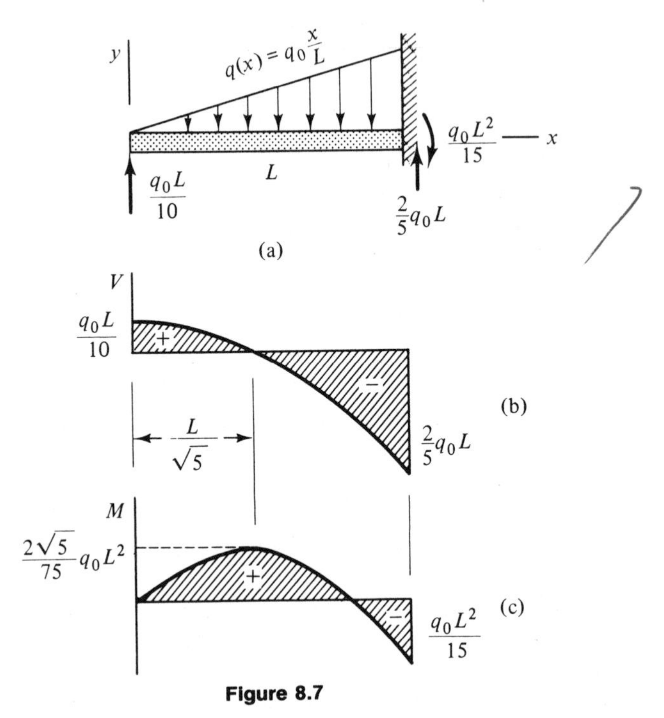

Figure 8.7

Solution. Since $q(x)$ varies linearly with x, V will be a function of x^2. Thus, the area under the shear diagram, which is a parabola, cannot be determined from simple geometry. V and M could be determined by using the procedure of Section 8.2, but we shall use integration in order to illustrate the steps involved.

At $x = 0$, the shear jumps up to a value $V(0) = q_o L/10$ due to the concentrated force acting there. According to Eq. (8.3), the change in shear between this location and any other arbitrary location x is

$$V(x) - V(0) = \int_0^x q\,dx = -\int_0^x \frac{q_o x}{L}\,dx$$

Thus,

$$V(x) = \frac{q_o L}{10} - \frac{q_o x^2}{2L}$$

A sketch of the shear diagram is shown in Figure 8.7(b). The point of zero shear is located by setting the equation for $V(x)$ equal to zero and solving for x. This gives $x = L/\sqrt{5}$.

The moment at the left end of the beam is zero because there is no couple acting there. From Eq. (8.4), the change in moment between this location and any other arbitrary location x is

$$M(x) - M(0) = \int_0^x V dx = \int_0^x \left(\frac{q_o L}{10} - \frac{q_o x^2}{2L} \right) dx$$

Thus,

$$M(x) = \frac{q_o L x}{10} - \frac{q_o x^3}{6L}$$

This curve is sketched in Figure 8.7(c). The relative maximum occurs at $x = L/\sqrt{5}$, where $V = 0$. The moment at this location is found to be $(2\sqrt{5}/75) q_o L^2$. This is not the maximum moment in the beam, however. It occurs just to the left of the wall where $M = -q_o L^2/15$. The clockwise couple acting at the wall then brings the diagram back to zero, as it should.

PROBLEMS

8.9 to 8.12 Sketch the shear and bending moment diagrams for the beams in Problems 8.5 through 8.8. Neglect the weights of the members.

8.13 to 8.21 Obtain the shear and bending moment diagrams for the beams shown and determine the magnitudes and locations of the maximum shear force and bending moment. Neglect the weights of the members.

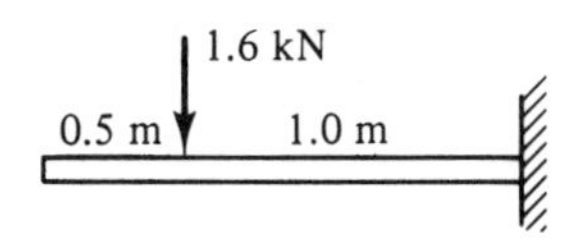

Problem 8.13

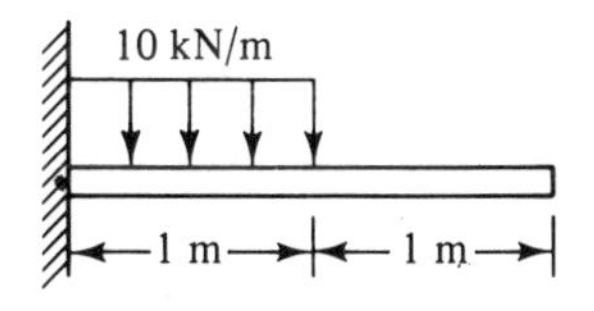

Problem 8.14

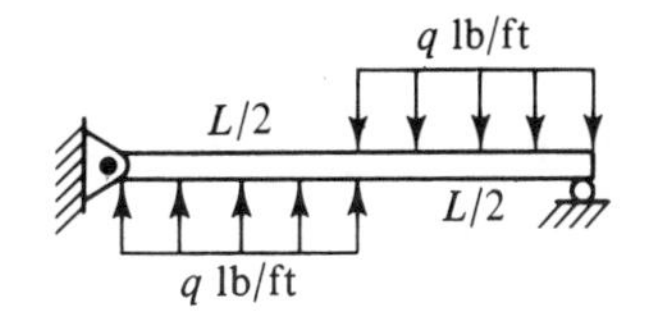

Problem 8.15

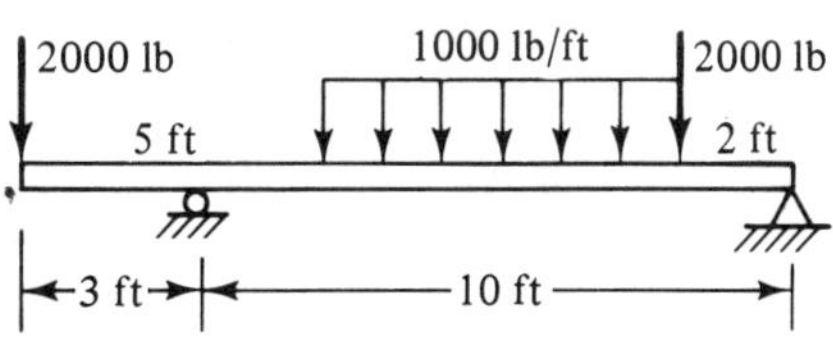

Problem 8.16

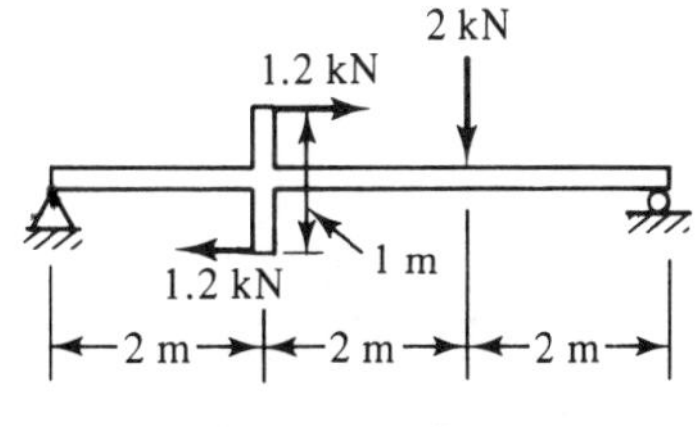

Problem 8.17

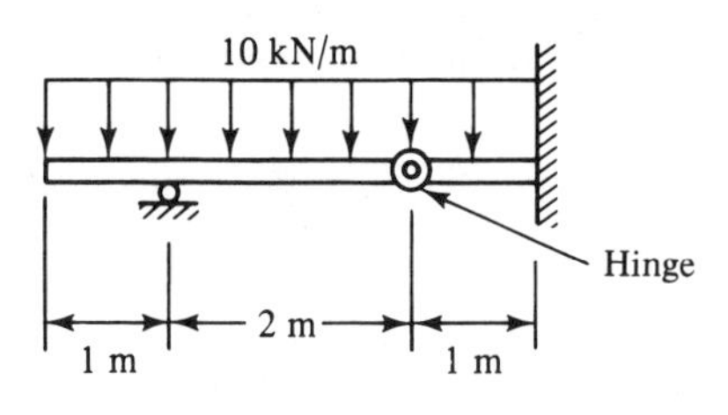

Problem 8.18

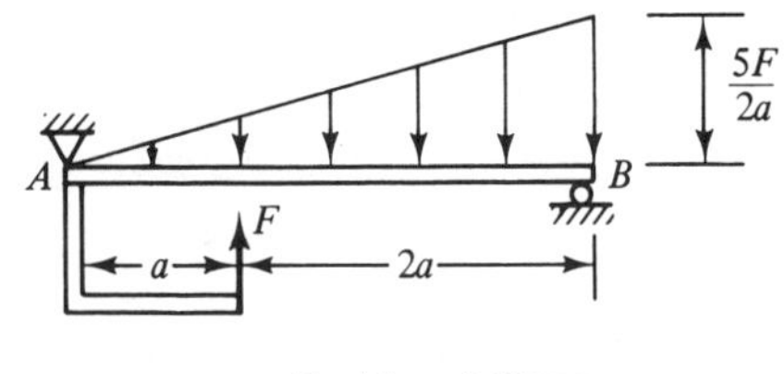

Problem 8.20

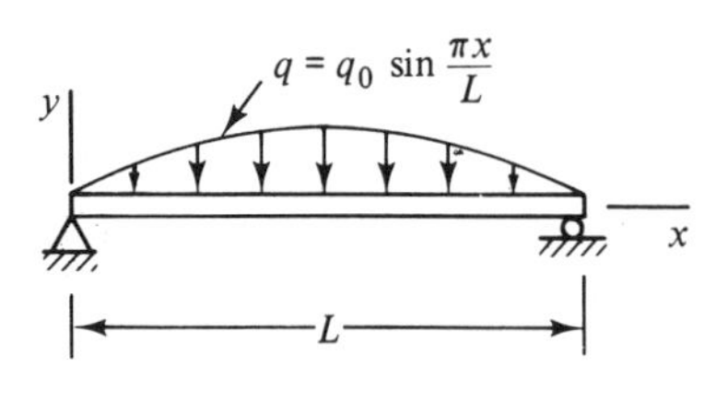

Problem 8.19

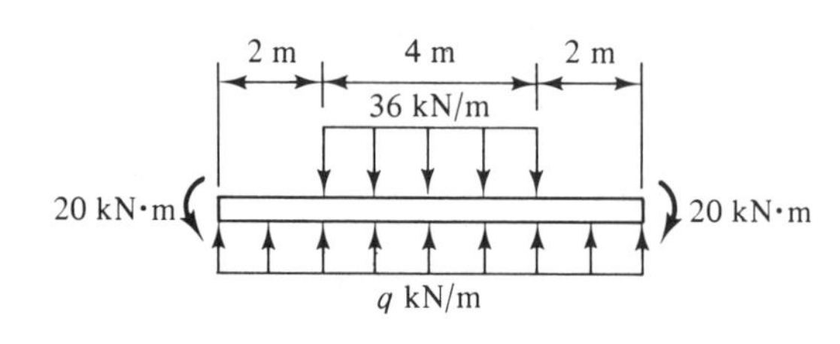

Problem 8.21

8.22 If the member in Problem 8.13 is a 254 mm × 146 mm Universal Beam with a mass of 43 kg/m, what is the percent error in the value of the maximum bending moment due to neglecting the weight of the member?

8.23 The beam in Problem 8.14 has a mass of 43 kg/m. What is the percent error in the value of the maximum bending moment if the weight of the member is neglected?

8.4 DEFORMATION PATTERN AND STRAINS

Now that procedures for determining the shear force and bending moment in a beam have been established, we shall turn our attention to the stresses and deformations associated with these quantities. We shall again follow the general procedure given in Section 5.10, the first step of which is to consider the deformation pattern. But, first, we must be more specific about the type of members and loadings considered.

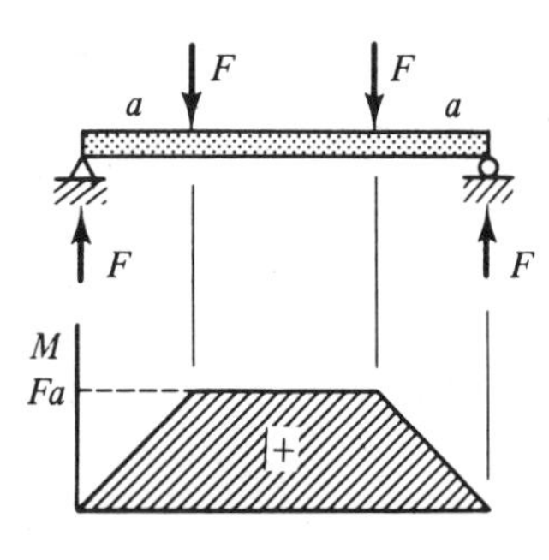

Figure 8.8

For the present, we assume that the loading is such that the stress resultant consists only of a bending moment and no shear force. In view of the relationship $dM/dx = V$ [Eq. (8.2)], this implies that the bending moment does not vary along the length of the member. This condition can be achieved by applying equal and opposite couples at the ends of the beam. It can also be achieved in other ways, at least over a portion of the length (see Figure 8.8).

We further assume that the beam has a longitudinal plane of symmetry, as most beams encountered in practice do, and that the bending moment lies within this plane. (The moment vector will be perpendicular to this plane.) If the material is homogeneous and isotropic, this restriction on the bending moment eliminates the possibility that the beam will twist as it deflects. Consequently, its primary response will be a bending within the longitudinal plane of symmetry. Under these conditions, the beam is said to be subjected to *pure bending*.

Experiments show that uniform beams made of homogeneous and isotropic materials and subjected to pure bending deform in such a way that planes perpendicular to the longitudinal axis before loading remain plane and perpendicular to the axis after loading. In other words, the beam cross sections rotate relative to one another when the beam deforms. This result, which also follows from symmetry, is illustrated in Figure 8.9.

(a) Before Loading

(b) After Loading.

Figure 8.9. Deformation of a Rubber Beam as Indicated by Grid Patterns

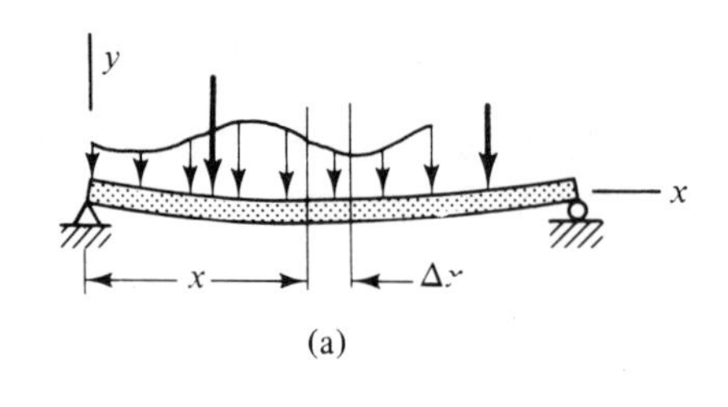

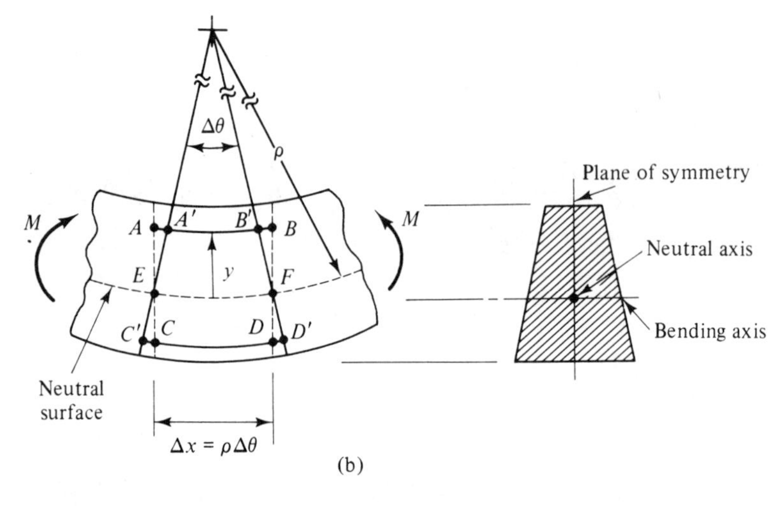

Figure 8.10

The deformations can be related to the strains by considering a small element of the beam of length Δx [Figure 8.10(a)]. In doing this, it will be helpful to think of the beam as consisting of a number of longitudinal fibers, or line elements, laid side by side.

If the bending moment is positive, as in Figure 8.10(b), a line element such as AB in the top portion of the beam will shorten while one in the bottom portion, such as CD, will elongate. At some point in between, there is a line element EF that does not change length. Actually, this line is the edge of an entire surface extending over the width and length of the beam, called the *neutral surface*. If the sense of the bending moment is reversed, the fibers at the top of the beam elongate while those at the bottom shorten. The deformations have been shown greatly exaggerated for clarity.

The intersection of the neutral surface with the longitudinal plane of symmetry is called the *neutral axis* of the beam, and its intersection with the beam cross section will be referred to as the *bending axis* [Figure 8.10(b)]. Notice that the beam cross sections rotate about the bending axis as the beam deforms. The exact location of the coordinate axes introduced in Section 8.2 can now be specified. The x axis is taken to coincide with the neutral axis in the undeformed beam, and the y axis is taken positive upward from it.

Now let us consider the strain of an arbitrary line element, such as AB, located a distance y from the neutral surface [Figure 8.10(b)]. From the geometry of the figure, the final length of the element is $A'B' = (\rho - y)\Delta\theta$, where ρ is the radius of curvature of the neutral axis in the deformed beam and $\Delta\theta$ is the included angle between adjacent cross sections. The initial length Δx of the element can be expressed in terms of $\Delta\theta$ by recognizing that line element EF does not change length; hence, $\Delta x = \rho\Delta\theta$. Thus, the normal strain in the longitudinal direction is

$$\varepsilon_{AB} = \frac{A'B' - AB}{AB} = \frac{(\rho - y)\Delta\theta - \rho\Delta\theta}{\rho\Delta\theta}$$

or

$$\varepsilon = -\frac{y}{\rho} \tag{8.5}$$

Equation (8.5) indicates that the normal strain varies linearly with the distance from the neutral surface and is largest at the top and bottom of the beam where the value of y is largest. The strain distribution is shown in Figure 8.11, where the height of the diagram denotes the relative magnitude of ε. Note that the strain does not vary across the width of the beam. Note also that the location of the neutral surface has not yet been established. This will be done in the following section.

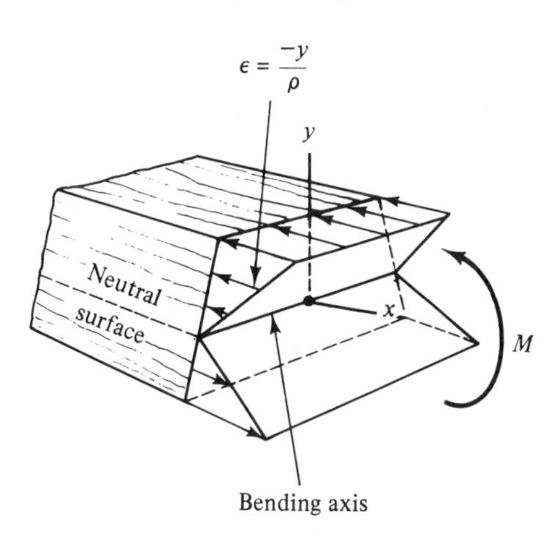

Figure 8.11

The question now arises as to what effect the shear forces, which are present in most beams, have on the deformation pattern and the strains. Experiments and more advanced analyses show that the shear forces have a negligible effect on these quantities, unless the beam is exceptionally short (of the same order of magnitude as the depth of the cross section). On the other hand, long beams with deep, narrow cross sections may buckle and undergo out-of-plane deformations, in which case Eq. (8.5) no longer applies. This behavior can be readily demonstrated by attempting to bend a strip of paper within the plane in which it lies. We shall assume that buckling does not occur.

In arriving at the strain distribution given in Eq. (8.5), we have invoked a number of assumptions and restrictions concerning the geometry and loading of the beam. There are also other conditions that must be met before Eq. (8.5) is valid. For convenience, all of these requirements are summarized below:

1. The beam must be straight and have constant cross-sectional area, or nearly so, for some distance on either side of the cross section of interest.

2. The beam must have a longitudinal plane of symmetry, and the bending moment must lie within this plane. Stated another way, the resultant applied loads must lie within the longitudinal plane of symmetry.

3. The member must not be so short that shearing effects are significant, nor so long and narrow that it buckles.

4. The section of interest must be some distance away from supports, connections, or points of load application.

5. The material must be homogeneous and isotropic.

Some of the preceding conditions can be relaxed somewhat, but they are sufficiently general for our purposes. The distortions in the strain distribution that occur near the supports, points of load application, and discontinuities in the cross section tend to be localized (St. Venant's principle) and generally die out over a distance approximately equal to the depth of the cross section. Note that the strain distribution depends only upon the geometry of the deformations and is completely independent of the stress-strain relations for the material.

8.5 NORMAL STRESSES—LINEARLY ELASTIC CASE

In this section we shall determine the distribution of normal stress and the moment-stress relation for a beam by combining the strain distribution with the stress-strain relation for the material and the conditions for static equivalence. The location of the neutral surface will also be determined in the process. Linearly elastic material behavior will be assumed for the present. The inelastic response of beams will be considered in Section 8.9.

Moment-stress relation. If the material is homogeneous and isotropic and the stresses are below the proportional limit, the stress-strain relation $\sigma = E\varepsilon$ applies throughout the beam. Combining this expression with the strain distribution of Eq. (8.5), we obtain for the stress distribution

$$\sigma = \frac{-Ey}{\rho} \tag{8.6}$$

Here, we have assumed that the modulus of elasticity, E, of the material is the same in tension and compression. As indicated in Chapter 5, this is a reasonable assumption for most common engineering materials. Equation (8.6) indicates that the normal stresses, like the normal strains, vary linearly with the distance y from the neutral surface [Figure 8.12(a)]. The normal stresses in beams are called *bending*, or *flexural*, *stresses* to distinguish them from those due to other types of loadings.

Now let us consider the moment-stress relation. Referring to Figure 8.12(b), we see that the force $dF = \sigma dA$ acting upon an element of area dA of the cross section has a moment about the bending axis (z axis) of magnitude $ydF = y\sigma dA$. The stresses are statically equivalent to the bending moment, M; therefore, we have

$$(\Sigma F_x)_I = (\Sigma F_x)_{II}: \quad \int \sigma dA = 0 \tag{8.7}$$

$$\left(\overset{+}{\curvearrowleft}\Sigma M_z\right)_I = \left(\overset{+}{\curvearrowleft}\Sigma M_z\right)_{II}: \quad -\int y\sigma dA = M \tag{8.8}$$

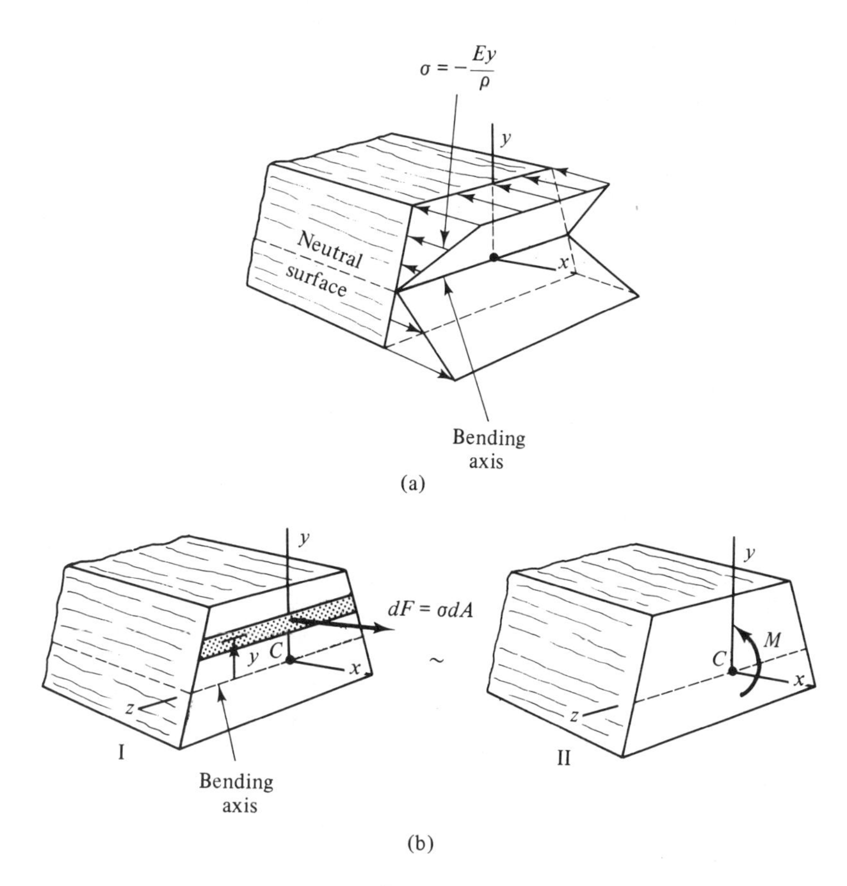

Figure 8.12

Equation (8.7) is a statement of the condition that the net normal force on the cross section must be zero.

Substituting Eq. (8.6) into Eq. (8.7) and recognizing that E and ρ do not vary over the cross section, we have

$$-\frac{E}{\rho}\int y dA = -\frac{E}{\rho}A\bar{y} = 0$$

where A is the cross-sectional area and $\bar{y}$ is the distance from the bending axis to its centroid. This equation can be satisfied only if $\bar{y} = 0$, which implies that the bending axis passes through the centroid C of the cross section. This establishes the location of the neutral surface.

Combining Eqs. (8.6) and (8.8), we obtain

$$M = \int y \frac{(Ey)}{\rho} dA = \frac{E}{\rho}\int y^2 dA$$

The latter integral is the moment of inertia, I, of the cross section about the bending axis; therefore, we have

$$M = \frac{EI}{\rho} \tag{8.9}$$

This equation relates the bending moment, M, to the curvature of the neutral axis, $1/\rho$, and is called the *moment-curvature relation*. We shall use this relation later to determine the deflection of beams.

The moment-stress relation is obtained by combining Eqs. (8.6) and (8.9):

$$\sigma = \frac{-My}{I} \qquad (\sigma \leqslant \sigma_{PL}) \tag{8.10}$$

Here, M is the bending moment acting upon the particular cross section of interest, I is the moment of inertia of that cross section about the bending axis, and y is the vertical distance from the bending axis to the particular point of interest within the cross section.

Note from Eq. (8.10) that a positive bending moment produces compression ($\sigma < 0$) in the top fibers of the beam where y is positive and tension ($\sigma > 0$) in the bottom fibers where y is negative. Conversely, a negative bending moment produces tension in the top fibers and compression in the bottom fibers. It is common practice to drop all signs in Eq. (8.10) and to use the expression only to compute the magnitude of the stress. The sense of the stress is then determined by inspection from the sense of the bending moment.

Equation (8.10) indicates that the larger the moment of inertia, I, of the cross section, the smaller the bending stresses. This explains why I-beams are so commonly used. With a cross section of this shape, it is possible to obtain a large moment of inertia with a minimum of material. Values of I for common structural shapes are given in handbooks; see Tables A.3 through A.12 of the Appendix, for example.

Design of beams. The process of selecting a beam of appropriate size to support a given loading is complicated by the fact that the maximum stress depends upon two cross-sectional parameters, y and I. This difficulty can be eliminated by combining these two parameters into a single parameter, $S = I/y_{max}$, called the *section modulus*. Thus, the magnitude of the maximum stress can be expressed as

$$\sigma_{max} = \frac{M}{(I/y_{max})} = \frac{M}{S} \tag{8.11}$$

where S has the units of length cubed. Values of S for common structural shapes are given in handbooks and in Tables A.3 through A.12 of the Appendix.

Once the maximum bending moment and allowable stress have been determined, the required section modulus can be computed from Eq. (8.11). It is then a simple matter to select the appropriate sized beam from tables of beam properties. There are usually several different sizes of beams with the required value of S. The one with the smallest weight per unit length involves the least amount of material, and, for economic reasons, it is the one that is usually used. Of course, the cross-sectional dimensions may also be a factor in the selection. For example, the depth of a floor beam must be compatible with the desired overall floor thickness. The procedure for "sizing" a beam will be illustrated in Example 8.7.

Validity of moment-stress relation. The moment-stress relation, Eq. (8.10), is based upon Hooke's law and applies only if the stresses are below the proportional limit. Furthermore, the modulus of elasticity of the material must be the same in tension and compression. For other types of material behavior, it is only necessary to use the appropriate stress-strain relation in Eq. (8.6) and then proceed as before. It is important to note that the location of the neutral surface may or may not coincide with the centroid of the cross section in such cases. As indicated by Eq. (8.7), this surface must always be located such that there is no net normal force on the cross section. Since the strain-curvature relation, Eq. (8.5), was used in the derivation, the conditions given in Section 8.4 concerning the geometry and loading of the beam must also be satisfied.

The effects of geometrical discontinuities, such as holes, fillets, and notches, usually can be accounted for by multiplying the maximum value of stress obtained from Eq. (8.10) or Eq. (8.11) by a stress concentration factor, K_σ:

$$\sigma_{\max} = K_\sigma\left(\frac{My_{\max}}{I}\right) = K_\sigma\left(\frac{M}{S}\right) \tag{8.12}$$

Values of K_σ for beams of rectangular cross section with fillets and semicircular grooves are given in Figure 8.13 for the linearly elastic case. These values apply to the stresses computed by using the dimensions of the minimum cross section. As has been mentioned several times before, geometrical discontinuities are most serious in members made of brittle materials.

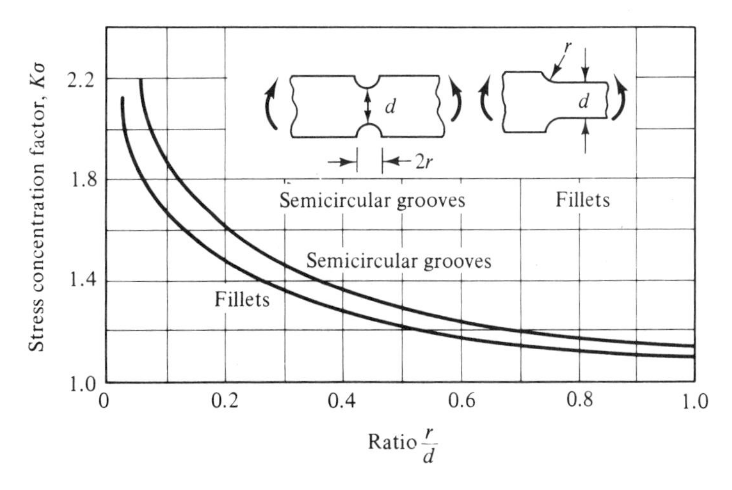

Figure 8.13. Stress-Concentration Factors for Beams of Rectangular Cross Section.

Example 8.5. Flexural Stresses in a Cantilever Beam. A steel cantilever beam with a triangular cross section is loaded as shown in Figure 8.14(a). Determine at the center of the beam (a) the normal stress and strain at a point 50 mm from the

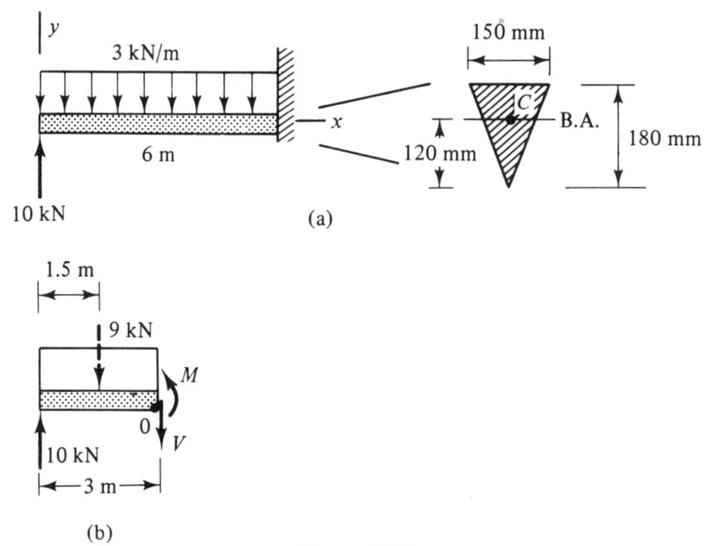

Figure 8.14

bottom of the cross section, (b) the maximum tensile bending stress, (c) the maximum compressive bending stress, and (d) the radius of curvature of the neutral axis. Assume linearly elastic material behavior.

Solution. From Table A.1, the moment of inertia about the bending axis is

$$I = \frac{bh^3}{36} = \frac{(0.15\text{ m})(0.18\text{ m})^3}{36} = 24.3 \times 10^{-6}\text{ m}^4$$

Sectioning the beam at the center and considering the piece to the left [Figure 8.14(b)], we have

$$\overset{+}{\curvearrowleft}\Sigma M_o = M + 9\text{ kN}(1.5\text{ m}) - 10\text{ kN}(3\text{ m}) = 0$$
$$M = 16.5\text{ kN·m}$$

Since the bending moment is positive, the stresses at this location will be compressive above the bending axis and tensile below it.

(a) A point 50 mm from the bottom of the cross section is 70 mm below the bending axis. From Eq. (8.10), the bending stress at this point is

$$\sigma = \frac{My}{I} = \frac{(16.5 \times 10^3\text{ N·m})(0.07\text{ m})}{24.3 \times 10^{-6}\text{ m}^4}$$
$$= 47.5 \times 10^6\text{ N/m}^2 \text{ or } 47.5\text{ MN/m}^2 \text{ (tension)} \qquad \textit{Answer}$$

Now, $E = 207\text{ GN/m}^2$ (Table A.2); therefore, the normal strain at this point is

$$\varepsilon = \frac{\sigma}{E} = \frac{47.5 \times 10^6\text{ N/m}^2}{207 \times 10^9\text{ N/m}^2} = 0.00023 \text{ (tension)} \qquad \textit{Answer}$$

(b) The maximum tensile bending stress occurs at the bottom of the cross section, and its magnitude is

$$(\sigma_t)_{\max} = \frac{My}{I} = \frac{(16.5 \times 10^3 \text{ N·m})(0.12 \text{ m})}{24.3 \times 10^{-6} \text{ m}^4}$$

$$= 81.5 \times 10^6 \text{ N/m}^2 \text{ or } 81.5 \text{ MN/m}^2 \qquad \textbf{\textit{Answer}}$$

(c) The maximum compressive bending stress occurs at the top of the cross section, and its magnitude is

$$(\sigma_c)_{\max} = \frac{My}{I} = \frac{(16.5 \times 10^3 \text{ N·m})(0.06 \text{ m})}{24.3 \times 10^{-6} \text{ m}^4}$$

$$= 40.7 \times 10^6 \text{ N/m}^2 \text{ or } 40.7 \text{ MN/m}^2 \qquad \textbf{\textit{Answer}}$$

(d) The radius of curvature, ρ, of the neutral axis can be determined either from the strain-curvature relation of Eq. (8.5) or the moment-curvature relation of Eq. (8.9). Using the former relation and the value of strain from part (a), we have

$$\rho = \left| \frac{y}{\varepsilon} \right| = \frac{0.07 \text{ m}}{0.23 \times 10^{-3}} = 304.3 \text{ m} \qquad \textbf{\textit{Answer}}$$

Note that ρ is many times larger than the length of the beam. This will always be the case for small deformations.

Example 8.6. Maximum Flexural Stresses in a Beam. The T-beam shown in Figure 8.15(a) is made of structural steel and supports a uniformly distributed load of 10 kips/ft. Determine the maximum tensile and compressive bending stresses and the locations at which they occur.

Solution. The first step is to determine the reactions. The upward force at A is found to be 42 kips and the upward force at B is 98 kips. We now sketch the moment diagram [Figure 8.15(b)]. The location of the centroid of the cross section and the moment of inertia about the bending axis are determined next. Breaking the area into two rectangles [Figure 8.15(c)] and applying Eq. (4.17), we find that the distance $\bar{y}$ from the bottom of the cross section to the centroid is

$$\bar{y} = \frac{\Sigma A_i \bar{y}_i}{\Sigma A_i} = \frac{(40 \text{ in.}^2)(5 \text{ in.}) + (48 \text{ in.}^2)(12 \text{ in.})}{40 \text{ in.}^2 + 48 \text{ in.}^2} = 8.82 \text{ in.}$$

Using the parallel axis theorem [Eq. (4.27)], the moment of inertia about the bending axis is found to be

$$I = \Sigma(\bar{I} + Ad^2) = \frac{1}{12}(4 \text{ in.})(10 \text{ in.})^3 + (40 \text{ in.}^2)(3.82 \text{ in.})^2$$

$$+ \frac{1}{12}(12 \text{ in.})(4 \text{ in.})^3 + (48 \text{ in.}^2)(3.18 \text{ in.})^2$$

$$= 1466 \text{ in.}^4$$

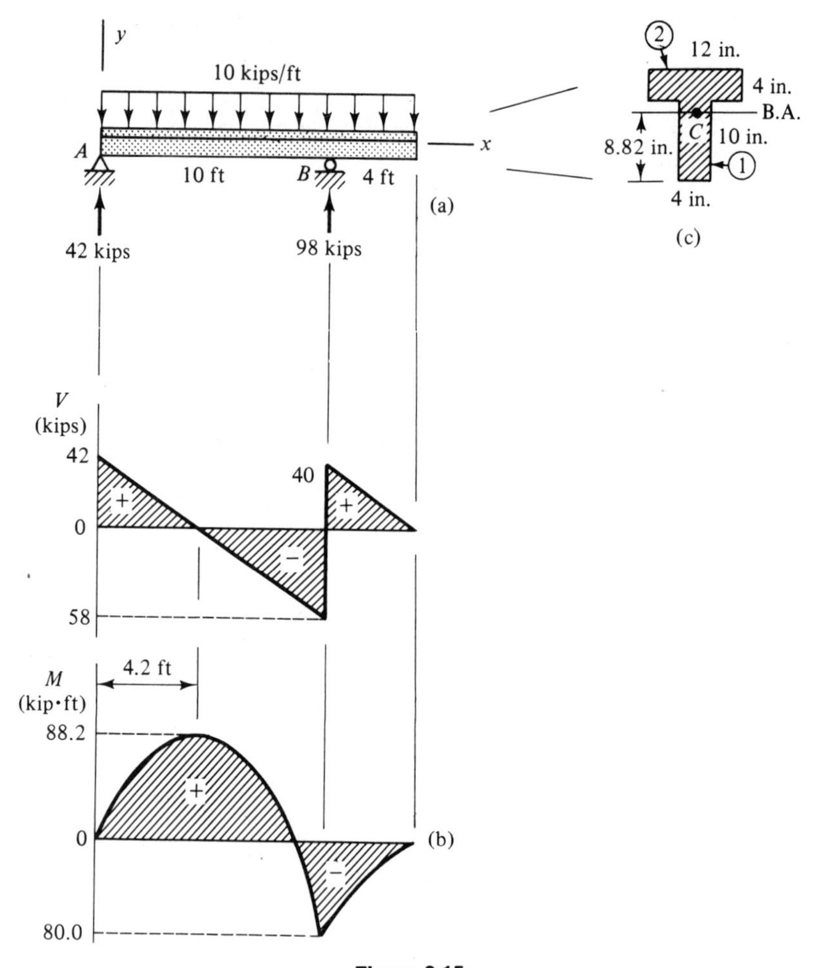

Figure 8.15

The maximum stresses will occur at the location of one or the other peaks on the moment diagram. We consider each location separately. At 4.2 ft from the left end, the top fibers of the beam are in compression and the bottom fibers are in tension because M is positive. Thus, the maximum stresses there are

$$(\sigma_t)_{\max} = \frac{(88\ 200\ \text{lb·ft})(12\text{in./ft})(8.82\ \text{in.})}{1466\ \text{in.}^4} = 6370\ \text{psi}$$

$$(\sigma_c)_{\max} = \frac{(88\ 200\ \text{lb·ft})(12\ \text{in./ft})(5.18\ \text{in.})}{1466\ \text{in.}^4} = 3740\ \text{psi}$$

The moment is negative at the right support; therefore, the maximum stresses there are

$$(\sigma_t)_{\max} = \frac{(80\ 000\ \text{lb·ft})(12\ \text{in./ft})(5.18\ \text{in.})}{1466\ \text{in.}^4} = 3390\ \text{psi}$$

$$(\sigma_c)_{\max} = \frac{(80\ 000\ \text{lb·ft})(12\ \text{in./ft})(8.82\ \text{in.})}{1466\ \text{in.}^4} = 5780\ \text{psi}$$

A comparison of the two sets of values of stresses shows that the largest tensile stress is 6370 psi and occurs at the bottom of the beam 4.2 ft from the left end. The largest compressive stress is 5780 psi and occurs at the bottom of the beam over the right support. Note that the largest compressive stress does not occur where the bending moment is largest. The larger value of y_{max} for compressive stresses at the right support more than compensates for the smaller bending moment there. Since the maximum stresses are well below the yield strength of structural steel (see Table A.2), our results are valid.

Example 8.7. Design of a Beam. A simple beam with a span of 10 ft is to support a 10 000-lb concentrated load at its center [Figure 8.16(a)]. Determine the size of W-Shape beam that will support this load and that would be most economical from a weight standpoint if (a) the weight of the beam is neglected and (b) the weight of the beam is included. The maximum allowable bending stress is 20 000 psi.

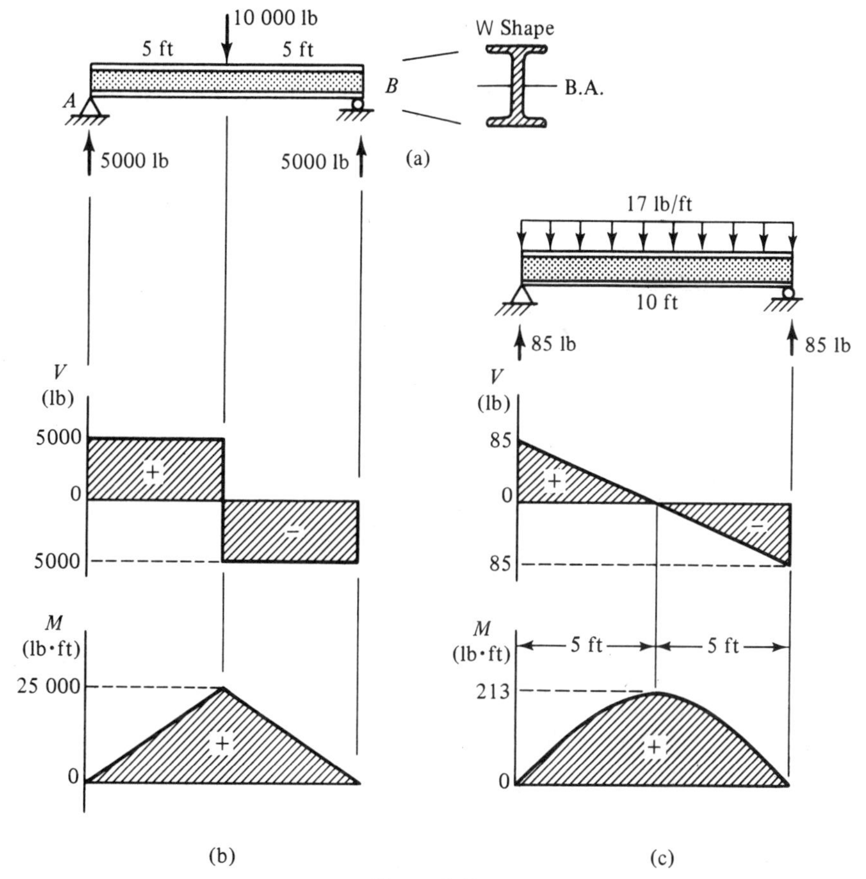

Figure 8.16

Solution. (a) We first determine the reactions and sketch the moment diagram [Figure 8.16(b)]. The maximum bending moment is found to be 25 000 lb·ft.

If we assume linearly elastic material behavior, we have from Eq. (8.11)

$$\sigma_{max} = \frac{M}{S} \leqslant \sigma_{allowable}$$

or

$$S \geqslant \frac{(25\ 000\ \text{lb·ft})(12\ \text{in./ft})}{20\ 000\ \text{lb/in.}^2} = 15.00\ \text{in.}^3$$

Turning to Table A.3 of the Appendix, we now look for those W-Shapes with a value of S (about axis x-x) of 15 in.^{3} or greater. We start with the smallest beams at the bottom of the page and work upward. The first shape encountered that meets the requirements is a W 6 × 25 shape with $S = 16.7$ in.^{3}. (The second number in the designation, 25 in this case, denotes the nominal weight of the member per foot of length.) We continue on up the page to see if there are any lighter weight beams that can be used. We next find a W 8 × 20 shape with an S of 17.0 in.^{3}. All other 8-in. shapes are heavier than this one. In the 10-in. shapes, we find a W 10 × 17 with $S = 16.2$ in.^{3}. This is the lightest beam with the required value of S.

(b) We now check to see if the W 10 × 17 shape is still satisfactory when the weight of the member is included. This can be done either by computing the maximum stress and comparing it with the allowable value or by computing the new required value of S and comparing it with the actual value for the shape. The latter approach will be used here.

The additional bending moment due to the weight of the beam has a maximum value of 213 lb·ft at the center [Figure 8.16(c)], for a total moment of 25 213 lb·ft at that location. The required value of S is now

$$S \geqslant \frac{M}{\sigma_{allowable}} = \frac{(25\ 213\ \text{lb·ft})(12\ \text{in./ft})}{20\ 000\ \text{lb/in.}^2} = 15.13\ \text{in.}^3$$

which is smaller than the actual value of 16.2 in.^{3}. Thus, the beam chosen by neglecting the weight is still satisfactory. This will often be the case, since the weight of the member is usually small compared to the applied loads.

PROBLEMS

8.24 Which can support the larger bending moment, a beam with a square cross section or a beam with a circular cross section of the same area? The maximum allowable stress is the same for each. Explain your answer.

8.25 A cast-iron beam with the cross section shown supports a positive bending mo-

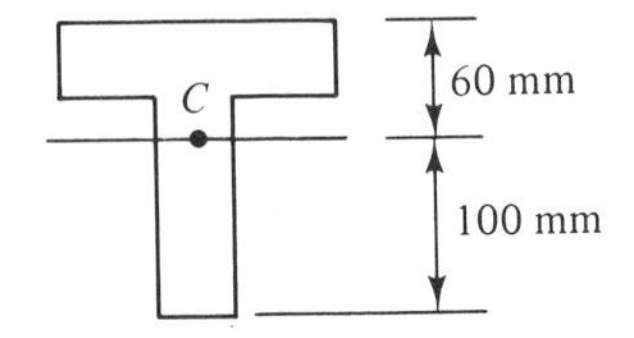

Problem 8.25

ment. Is the beam stronger with the flange placed up or down? Explain your answer. The ultimate strength of the material is 140 MN/m^2 in tension and 520 MN/m^2 in compression.

8.26 An elastic beam made of a 2-in. × 12-in. nominal plank is strengthened by nailing 2-in. × 4-in. nominal timbers to the bottom surface as shown. By what percentage is the moment-carrying capacity of the beam increased? The maximum allowable stress is the same in each case.

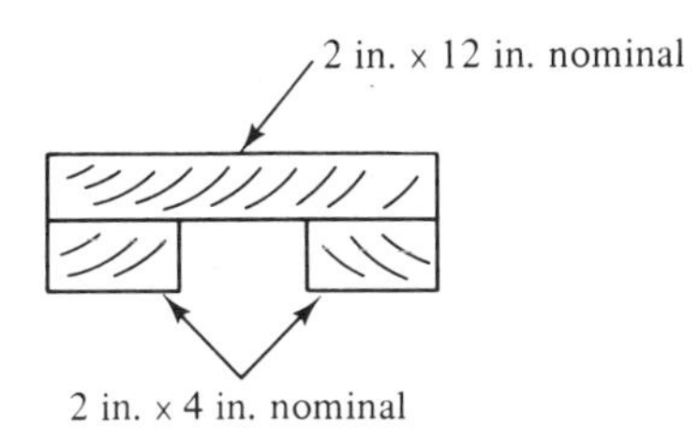

Problem 8.26

8.27 A 4 ft long steel (σ_{YS} = 36 ksi) cantilever beam supports a 400-lb downward load at the tip and is made of a C 6 × 13 American Standard Channel oriented with the flanges down. Determine the maximum tensile and compressive bending stresses in the beam and the radius of curvature of the neutral axis at the wall. What is the radius of curvature of the neutral axis at the free end? Neglect the weight of the member.

8.28 to 8.31 Determine the maximum tensile and compressive bending stresses in the beams shown and indicate where those stresses occur. Assume elastic behavior.

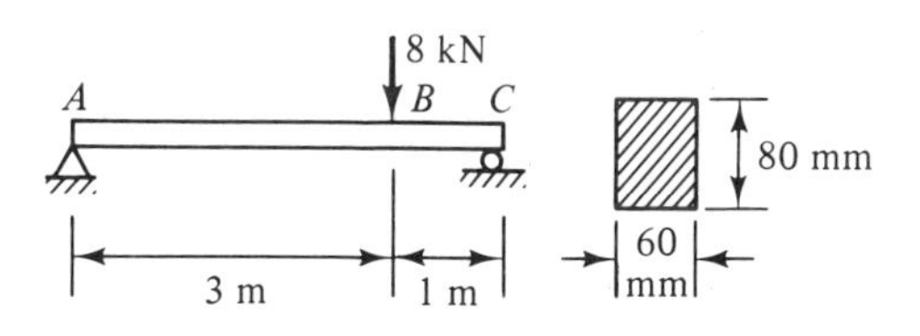

Problem 8.28

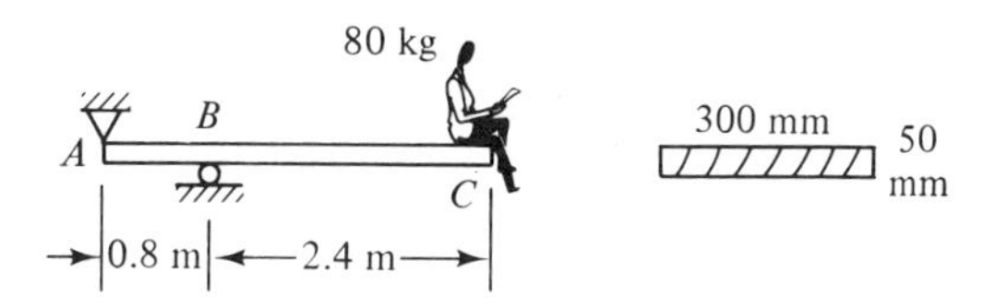

Problem 8.29

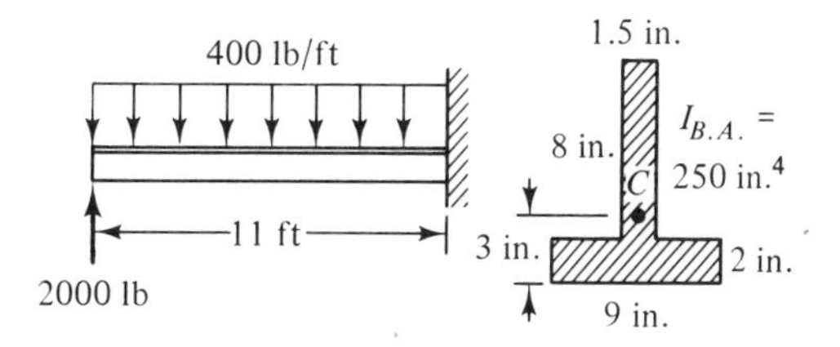

Problem 8.30

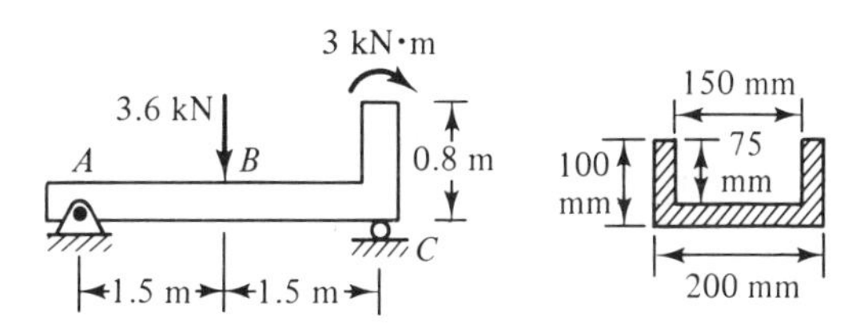

Problem 8.31

8.32 A skate board with the cross section shown is made of a molded plastic for which the maximum allowable tensile stress is 70 MN/m^2. If the rider has a mass of 58 kg, what safety factor does this design provide?

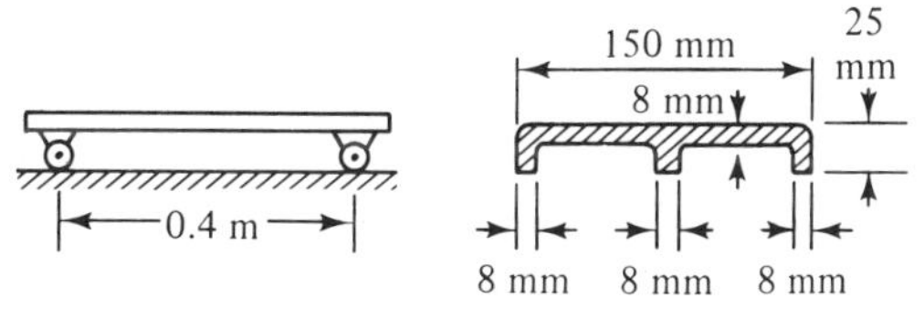

Problem 8.32

8.33 A simply supported 254-mm × 102-mm Universal Beam with a mass of 28 kg/m is to support a uniformly distributed load of

3 kN/m with a safety factor of 2.5 against failure by yielding. What is the maximum allowable span if (a) the weight of the beam is neglected and (b) the weight of the beam is included? Use $\sigma_{YS} = 250 \text{ MN/m}^2$.

8.34 What is the lightest weight steel channel section oriented with the flanges down that will support the loading in Problem 8.28 without failure by yielding if (a) the weight of the beam is neglected and (b) the weight of the beam is included. Use a safety factor of 2, with $\sigma_{YS} = 248 \text{ MN/m}^2$.

8.35 What is the lightest weight W-Shape beam that will support the loading in Problem 8.30 without failure by yielding? Use a safety factor of 4, with $\sigma_{YS} = 30$ ksi.

8.36 The beam shown supports a concentrated load F and a uniformly distributed load with total magnitude $4F$. What is the maximum allowable value of F if the tensile and compressive stresses are not to exceed 14 ksi and 9 ksi, respectively?

8.37 By what percentage do the semicircular grooves reduce the load-carrying capacity of the beam shown? Assume elastic behavior.

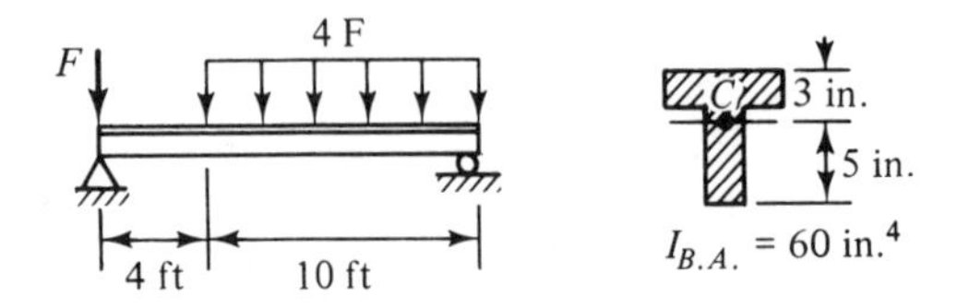

Problem 8.36

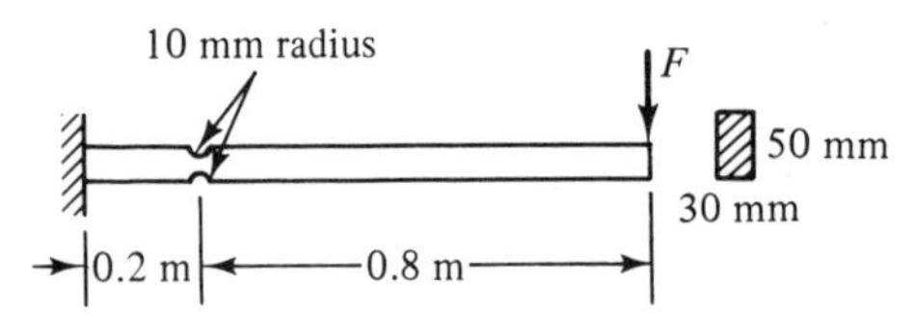

Problem 8.37

8.38 Determine the maximum flexural stress in the 2-in. thick member shown if (a) $d = 1$ ft and (b) $d = 2$ ft. Assume elastic behavior.

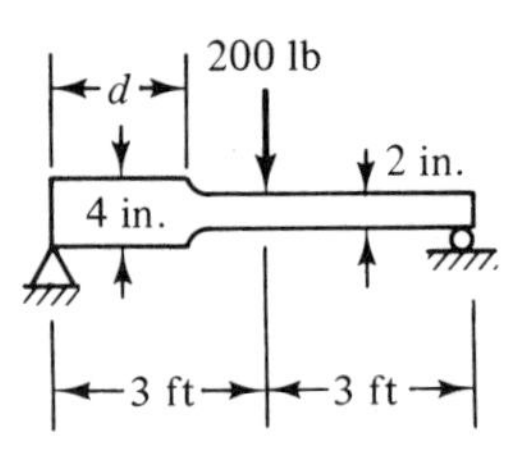

Problem 8.38

8.39 What is the minimum allowable dimension d for the $\frac{1}{2}$-in. thick beam shown if the maximum allowable stress is 10 ksi? Assume elastic behavior.

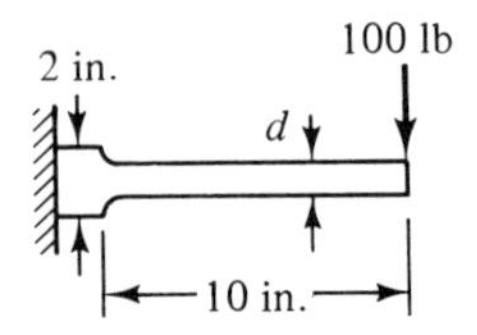

Problem 8.39

8.40 Derive the moment-stress relation for a beam made of a material whose stress-strain relation is $|\sigma| = k|\varepsilon|^{1/3}$, where k is a material constant. The cross section is rectangular with width b and depth h.

8.41 Same as Problem 8.40, except that the stress-strain relation is $|\sigma| = k|\varepsilon|^{1/2}$.

8.42 A steel band-saw blade is 8 mm wide and 0.5 mm thick. What is the minimum allowable pulley radius if the bending stress in the blade is not to exceed 500 MN/m^2? Use $E = 207 \text{ GN/m}^2$.

8.43 A $\frac{1}{2}$-in. diameter steel rod with a yield strength of 60 ksi is bent into a circular arc with a radius of 12 ft by couples applied to the ends. Determine the maximum bending stress in the rod and the magnitude C of the applied couples.

8.44 A 6-in.×6-in. wood beam is strengthened by attaching $\frac{1}{8}$-in. thick steel plates to the top and bottom surfaces. If the maximum allowable stress in the wood is 2500 psi, what is the percentage increase in the moment that can be supported? $E_{\text{steel}}/E_{\text{wood}} = 20$. *Hint*: The strains vary linearly with the distance from the bending axis regardless of the material properties.

8.45 The reinforced concrete beam shown supports a positive bending moment of 240 000 lb·ft. If the steel reinforcing rods have a combined cross-sectional area of 6 in.^{2}, what is the maximum compressive stress in the concrete and the tensile stress in the steel? Assume the concrete has zero tensile strength and a modulus of elasticity equal to one-twelfth that of steel. *Hint*: The strains vary linearly with the distance from the bending axis regardless of the material properties.

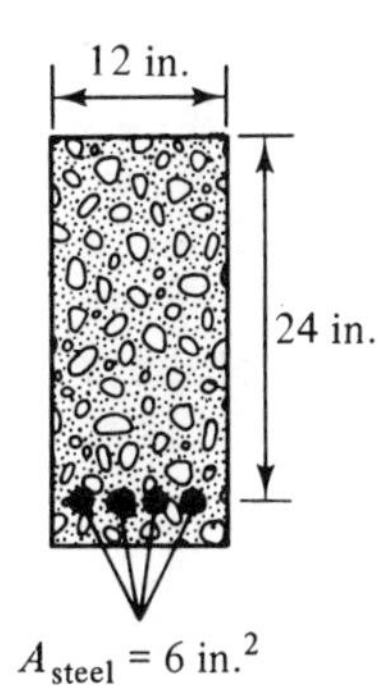

Problem 8.45

8.6 SHEAR STRESSES IN BEAMS

Since the stress resultant in a beam usually includes a shear force, it follows that shear stresses must also be present. These stresses will be considered in this section for the case of linearly elastic material behavior.

Consider two cross sections AB and CD of a beam located a small distance Δx apart [Figure 8.17(a)]. Let M denote the value of the bending moment at section

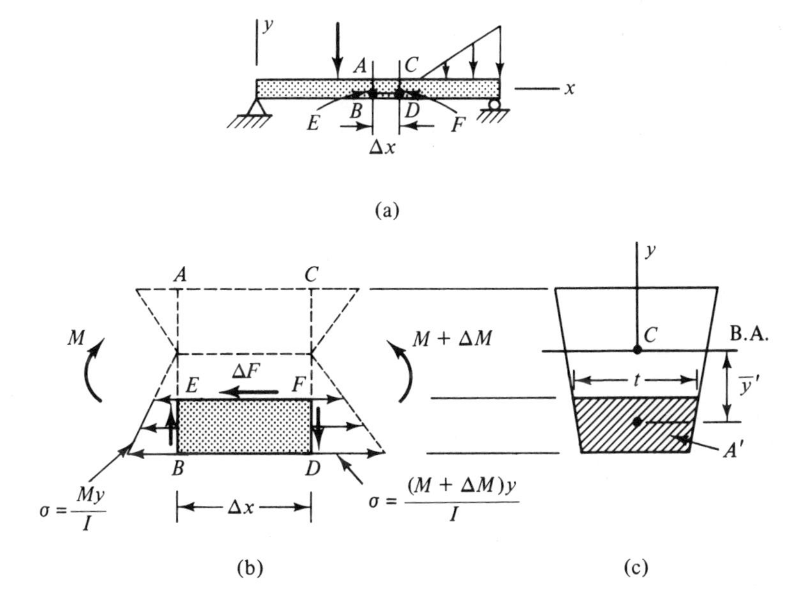

Figure 8.17

AB and $M + \Delta M$ the value at section CD. To determine the shear stress at a particular level in the beam, such as level EF, we section the beam horizontally along this level and between the two cross sections to obtain the segment $BEFD$ shown in Figure 8.17(b). Since the bending stresses acting upon the ends of the segment differ in magnitude, there must also be a horizontal shear force, ΔF, acting upon the top surface in order to maintain equilibrium. Shear forces also act over the ends of the segment.

Referring to Figure 8.17(b), we have the equilibrium condition

$$\Sigma F_x = \int (M + \Delta M) \frac{y}{I} dA - \Delta F - \int \frac{My}{I} dA = 0$$

or

$$\Delta F = \frac{\Delta M}{I} \int y dA$$

where the integration is to be carried out over the area A' of the end of the segment. This expression can be rewritten as

$$\Delta F = \frac{\Delta M}{I} A'\bar{y}' \tag{8.13}$$

where I is the moment of inertia of the entire beam cross section about the bending axis and $\bar{y}'$ is the distance from this axis to the centroid of the area A' [Figure 8.17(c)]. This same result is obtained if we consider a beam segment above the level of interest instead of the segment below it. Consequently, we can use either the area above or below the level of interest as the area A'.

The shear force per unit of length of the beam, denoted by q_s and called the *shear flow*, is

$$q_s = \lim_{\Delta x \to 0} \frac{\Delta F}{\Delta x}$$

Using the expression for ΔF given in Eq. (8.13), we have

$$q_s = \lim_{\Delta x \to 0} \frac{\Delta M}{\Delta x} \frac{A'\bar{y}'}{I} = \frac{dM}{dx} \frac{A'\bar{y}'}{I}$$

But $dM/dx = V$, where V is the shear force at the cross section of interest; therefore,

$$q_s = \frac{VA'\bar{y}'}{I} \tag{8.14}$$

Shear flow is an important concept in the study of the torsion and bending of thin-walled members and in the study of the connections in built-up beam sections.

Since the actual distribution of shear stress across the thickness of a beam is difficult to determine, a uniform distribution is usually assumed. From Eq. (8.14), the average shear stress across the thickness is

$$\tau = \frac{q_s}{t} = \frac{VA'\bar{y}'}{It} \tag{8.15}$$

where t is the thickness of the beam at the level of interest [Figure 8.17(c)]. The shear stresses associated with the shear force V are often called *direct shear stresses* to distinguish them from those due to other types of loadings, such as torsion.

Except for rectangular sections, Eq. (8.15) gives a meaningful estimate of the actual shear stresses only if t is small (thin-walled sections). It should also be noted that this equation and Eq. (8.14) for the shear flow are valid only when the bending stress formula is valid, since it was used in their derivation. In particular, Eqs. (8.14) and (8.15) apply only for linearly elastic material behavior.

Although we have considered only shear stresses on a plane parallel to the longitudinal axis, there are also vertical shear stresses acting over the cross section (Figure 8.18). The stresses on these two planes are equal in magnitude at any given point and, consequently, are both defined by Eq. (8.15). These results follow directly from the fact that if there is a shear stress acting upon one plane in a body, there must be a shear stress of equal magnitude acting upon a plane perpendicular to the first one, as shown in Section 7.4.

There are also horizontal shear stresses acting in the flanges of thin-walled members such as I-beams. Consideration of these stresses is beyond the scope of this text, but their existence can be easily demonstrated by considering the forces acting upon an element of the flange, as in Figure 8.19. Since the longitudinal forces due to the bending stresses differ by an amount ΔF when a shear force is present, it is clear that equilibrium can be maintained only if there are shear forces acting upon side $ABCD$ of the element and upon both ends. The latter forces give rise to the horizontal shear stresses.

It is evident from Eq. (8.15) that the maximum direct shear stress, τ_{max}, occurs on the cross section(s) of the beam where V is largest and at the level in the cross

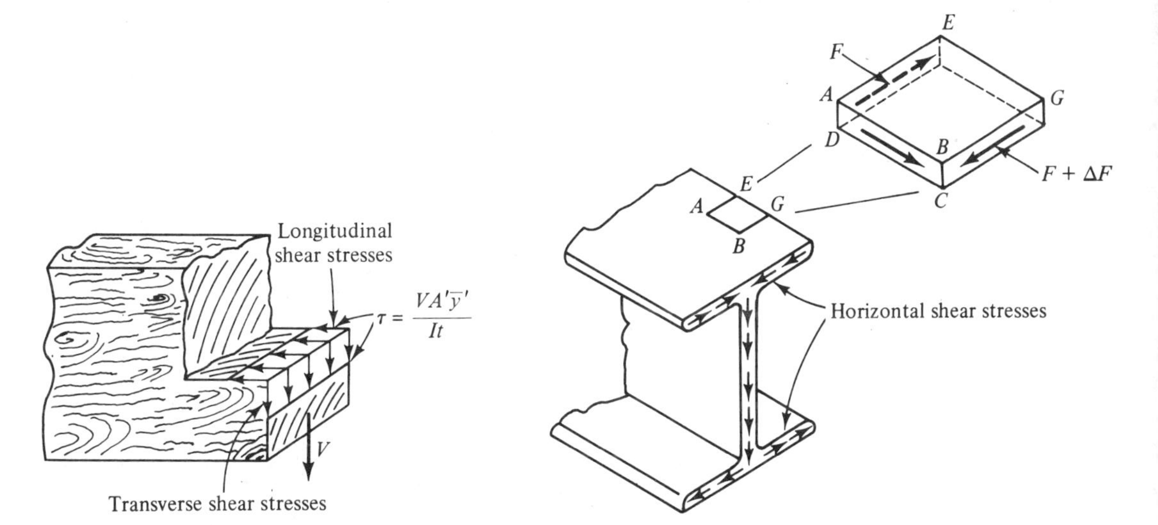

Figure 8.18

Figure 8.19

section where the quantity $A'\bar{y}'/t$ is a maximum. Little can be said in general about the location of τ_{max} for cross sections of arbitrary shape. For most common shapes, but not all, it occurs at the centroid of the section. We do know that the product $A'\bar{y}'$ is zero at the top and bottom of the cross section and increases as the level of interest moves toward the bending axis, at which point it attains its maximum value. This fact is usually sufficient to determine the location of τ_{max} for beams with cross sections consisting of rectangular segments. The shear flow is always largest at the bending axis.

The maximum direct shear stress in a beam is usually small compared to the maximum bending stresses, unless the member is exceptionally short. Nevertheless, the direct shear stresses are often of significance, particularly in wooden beams. Since wood is relatively weak in shear parallel to the grain, such members tend to fail by splitting longitudinally due to the shear stresses developed. In fact, the original derivation of Eq. (8.15) resulted from a study of the horizontal cracks in wood ties on railroad bridges conducted by the Russian engineer D. I. Jouravsky in 1855.

Example 8.8. Shear Stresses in a Rectangular Beam. Determine the distribution of shear stress in a beam with rectangular cross section and subjected to a shear force V. What is the maximum value of the shear stress and where does it occur?

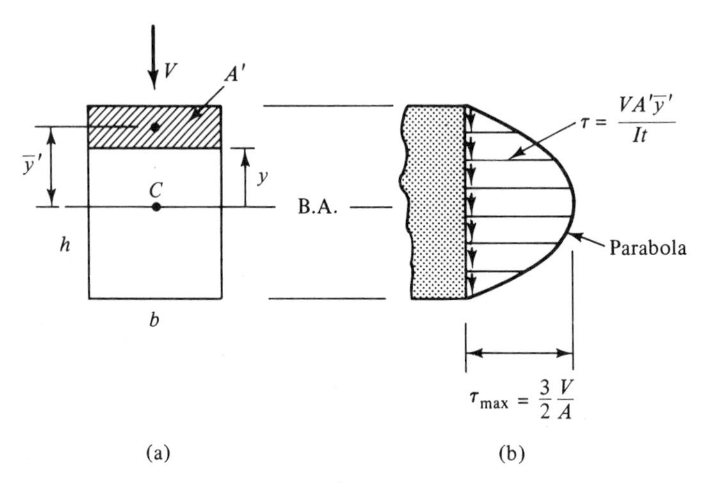

Figure 8.20

Solution. Selecting a point at an arbitrary distance y from the bending axis [Figure 8.20(a)], we find that the area above this point is $A' = b(h/2 - y)$ and the distance from the bending axis to its centroid is $\bar{y}' = (h/2 + y)/2$. Here, b is the width of the cross section and h is its height. At the point of interest, $t = b$. From Eq. (8.15), we obtain for the shear stress at this level

$$\tau = \frac{VA'\bar{y}'}{It} = \frac{Vb\left(\frac{h}{2} - y\right)\left(\frac{h}{2} + y\right)/2}{\left(\frac{1}{12}bh^3\right)b} = \frac{6V}{Ah^2}\left[\left(\frac{h}{2}\right)^2 - y^2\right] \qquad \textbf{\textit{Answer}}$$

where $A = bh$ is the total cross-sectional area.

The magnitude of the shear stress varies parabolically over the depth of the beam. This distribution is shown in Figure 8.20(b), where the height of the diagram represents the relative magnitude of τ. Note that $\tau = 0$ at the top and bottom of the beam. This will always be the case regardless of the shape of the cross section because the area A' above or below these respective levels is zero. The maximum shear stress occurs at the bending axis where $y = 0$, and is equal to

$$\tau_{\max} = \frac{3}{2}\frac{V}{A} \qquad \textbf{\textit{Answer}}$$

Example 8.9. Shear Stresses in an I-Beam. A 254-mm × 102-mm Universal Beam with a mass of 22 kg/m is subjected to a shear force of 5 kN. Determine the vertical shear stress at the three levels indicated in Figure 8.21(a) and sketch its distribution over the cross section.

Solution. The dimensions and geometric properties of the section obtained from Table A.8 are listed in Figure 8.21(a). Referring to Figure 8.21(b), we have at levels *a-a* and *b-b*

$$A'\bar{y}' = (101.6 \text{ mm})(5.8 \text{ mm})(127 \text{ mm} - 2.9 \text{ mm}) = 73.1 \times 10^3 \text{ mm}^3$$

At level *c-c* [Figure 8.21(c)], we have

$$A'\bar{y}' = A'\bar{y}' \text{ (flange)} + A'\bar{y}'\left(\frac{1}{2}\text{ web}\right)$$

$$= 73.1 \times 10^3 \text{ mm}^3 + \frac{(127 \text{ mm} - 5.8 \text{ mm})(6.8 \text{ mm})(127 \text{ mm} - 5.8 \text{ mm})}{2}$$

$$= 123.1 \times 10^3 \text{ mm}^3$$

From Eq. (8.15), the direct shear stresses at these levels are:

$$(a\text{-}a) \qquad \tau = \frac{VA'\bar{y}'}{It} = \frac{(5 \times 10^3 \text{ N})(73.1 \times 10^{-6} \text{ m}^3)}{(2863 \times 10^{-8} \text{ m}^4)(0.1016 \text{ m})}$$

$$= 0.13 \times 10^6 \text{ N/m}^2 \text{ or } 0.13 \text{ MN/m}^2 \qquad \textbf{\textit{Answer}}$$

$$(b\text{-}b) \qquad \tau = \frac{(5 \times 10^3 \text{ N})(73.1 \times 10^{-6} \text{ m}^3)}{(2863 \times 10^{-8} \text{ m}^4)(0.0068 \text{ m})}$$

$$= 1.88 \times 10^6 \text{ N/m}^2 \text{ or } 1.88 \text{ MN/m}^2 \qquad \textbf{\textit{Answer}}$$

$$(c\text{-}c) \qquad \tau = \frac{(5 \times 10^3 \text{ N})(123.1 \times 10^{-6} \text{ m}^3)}{(2863 \times 10^{-8} \text{ m}^4)(0.0068 \text{ m})}$$

$$= 3.16 \times 10^6 \text{ N/m}^2 \text{ or } 3.16 \text{ MN/m}^2 \qquad \textbf{\textit{Answer}}$$

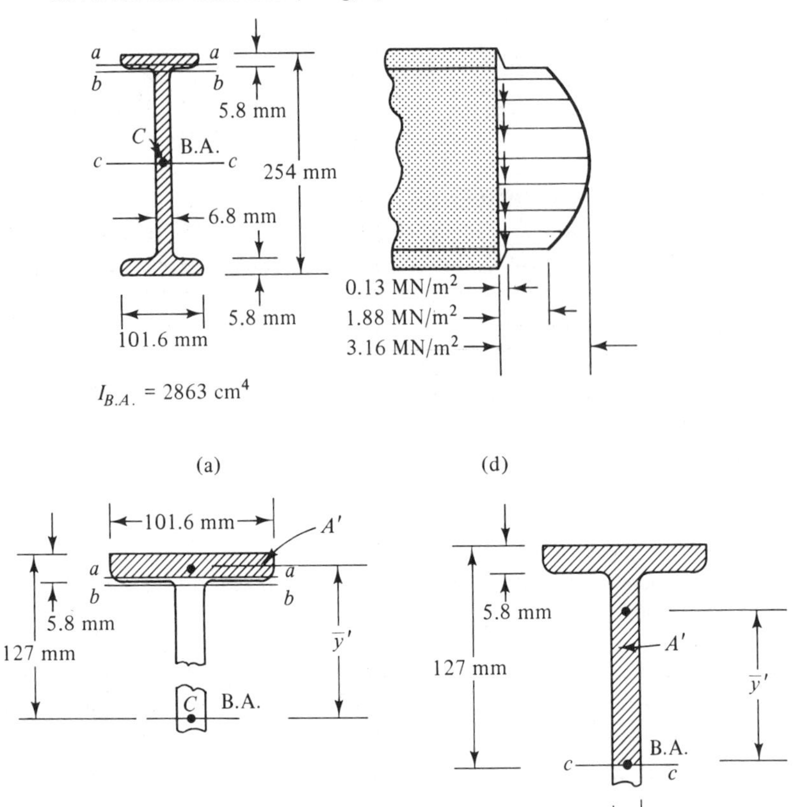

Figure 8.21

The stress distribution is sketched in Figure 8.21(d). The stress varies parabolically over the various portions of the cross section, which are very nearly rectangular, and is a maximum at the centroid. Note that the stresses in the web are much larger than those in the flanges. Thus, the web supports the majority of the shear load; any bending moment present is carried primarily by the flanges.

It is common practice to assume that the web of an I-Beam supports all the shear force and that the shear stress is uniformly distributed over the web area. In this case, the average stress in the web is

$$\tau = \frac{V}{A_{\text{web}}} = \frac{5 \times 10^3 \text{ N}}{(0.254 \text{ m} - 0.0116 \text{ m})(0.0068 \text{ m})}$$

$$= 3.03 \times 10^6 \text{ N/m}^2 \text{ or } 3.03 \text{ MN/m}^2$$

This approach usually gives a value of shear stress that is in close agreement with the maximum value obtained by using Eq. (8.15). For this particular example, the two values differ by only 4%.

Example 8.10. Shear Stresses in a T-Beam. Two timbers are joined together with lag screws to form the beam shown in 8.22(a). (a) What is the maximum uniformly distributed load, q, that can be supported if the allowable direct shear stress in the wood is 95 psi? (b) What is the maximum allowable spacing between the lag screws for this loading if each screw can support a total shear force of 370 lb?

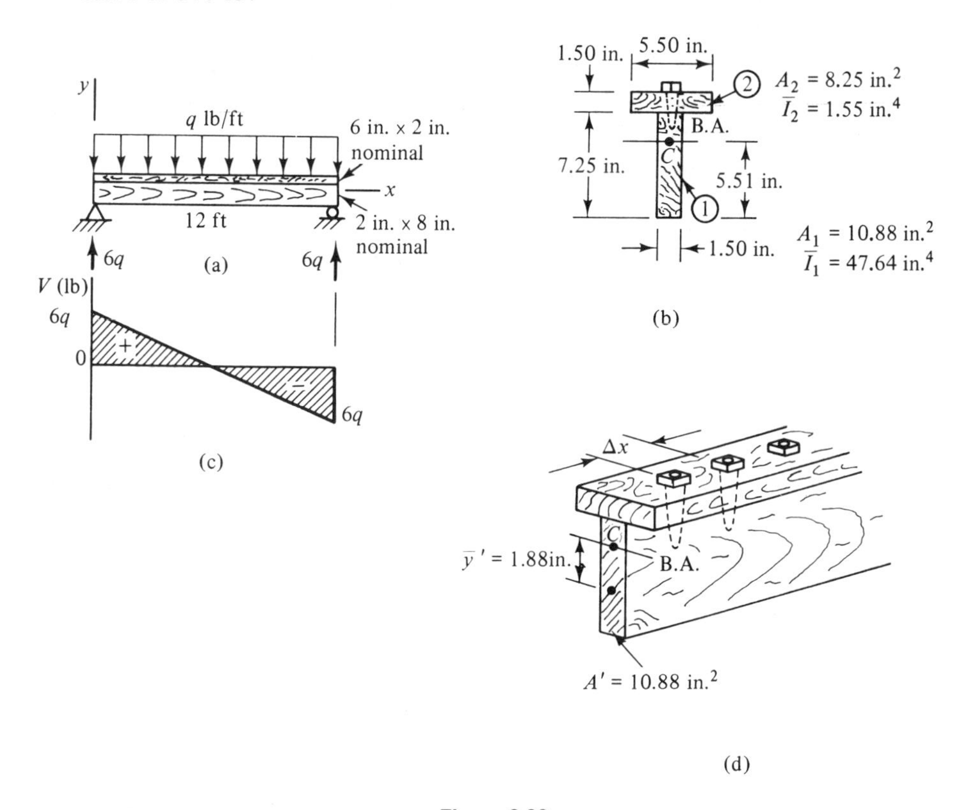

Figure 8.22

Solution. The actual dimensions and geometric properties of the timbers obtained from Table A.12 are listed in Fgiure 8.22(b). The centroid of the cross section is found to be located 5.51 in. above the bottom surface of the beam, and the moment of inertia of the cross section about the bending axis is 138.8 in.^{4}. From the shear diagram [Figure 8.22(c)], we see that the maximum shear force is $6q$ and occurs at the supports.

(a) The maximum direct shear stress occurs at the centroid of the section, since $A'\bar{y}'$ is largest there and t is smallest. Taking A' to be the area below this level, we obtain from Eq. (8.15)

$$\tau_{\max} = \frac{VA'\bar{y}'}{It} = \frac{(6q \text{ lb})(5.51 \text{ in.})(1.50 \text{ in.})(5.51 \text{ in.}/2)}{(138.8 \text{ in.}^4)(1.50 \text{ in.})} = 0.66q \text{ psi}$$

Thus, we have

$$0.66q \text{ psi} \leqslant 95 \text{ psi}$$

or

$$q \leqslant 144 \text{ lb/ft} \qquad \textbf{\textit{Answer}}$$

(b) We assume that the lag screws are equally spaced a distance Δx apart [Figure 8.22(d)]. Thus, each screw must support the longitudinal shear force developed over a distance Δx at the intersection between the timbers. Taking A' to be the area below this level and using the maximum value of the shear force ($6q = 864$ lb), we obtain for the shear flow

$$q_s = \frac{(864 \text{ lb})(10.88 \text{ in.}^2)(1.88 \text{ in.})}{138.8 \text{ in.}^4} = 127 \frac{\text{lb}}{\text{in.}}$$

The force supported by each screw is $q_s \Delta x$; therefore, we have

$$\left(127 \frac{\text{lb}}{\text{in.}}\right)(\Delta x \text{ in.}) \leqslant 370 \text{ lb}$$

$$\Delta x \leqslant 2.9 \text{ in.} \qquad \textbf{\textit{Answer}}$$

This spacing is based on the maximum shear force at the supports. A larger spacing could be used away from these locations where V is smaller. However, a uniform spacing is usually used for convenience and for the added margin of safety it provides.

PROBLEMS

8.46 and 8.47 Beams with the cross sections shown support a positive shear force $V = 20$ kN. Determine the direct shear stress at the levels indicated and sketch its distribution over the cross section.

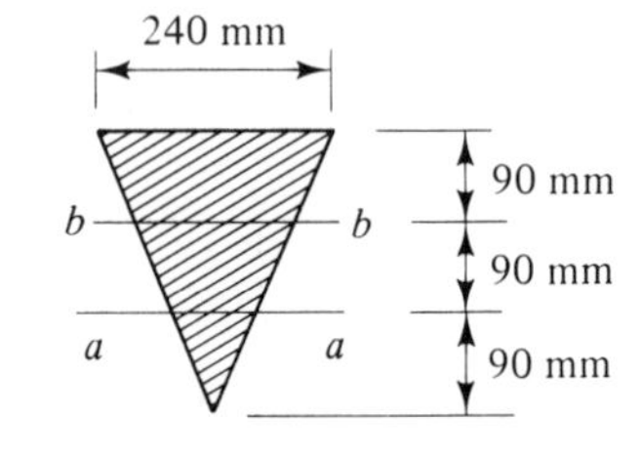

Problem 8.47

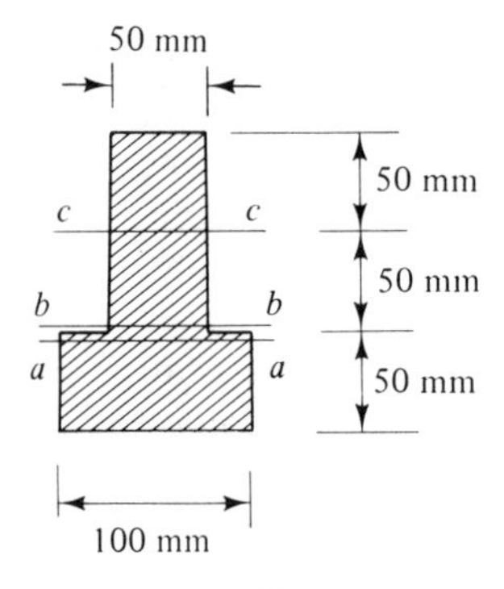

Problem 8.46

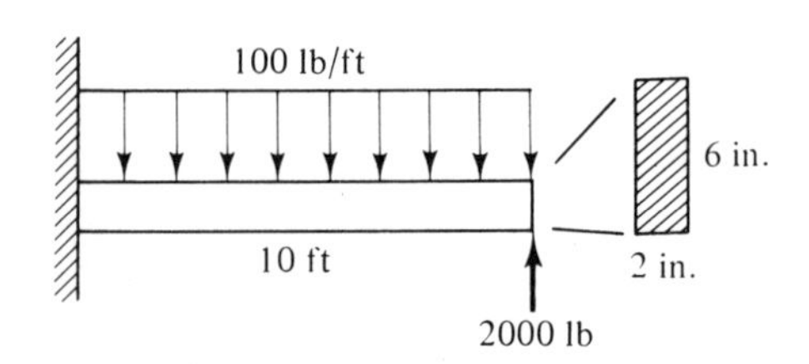

Problem 8.48

8.48 Determine the direct shear stress at a point 2 ft from the left end and 1 in. above the centerline of the beam shown.

8.49 What is the maximum direct shear stress in the beam in Problem 8.48 and where does it occur?

8.50 What is the maximum concentrated load F that can be supported at the center of a 3-m long simply supported beam made of a 100-mm × 150-mm timber oriented with the larger dimension vertical if the allowable shear stress in the wood is 0.7 MN/m^2?

8.51 Determine the ratio of the maximum bending stress to the maximum direct shear stress for the beam in Problem 8.30 (Section 8.5).

8.52 A beam is made of four 2-in. × 8-in. nominal planks glued together as shown. Determine the maximum direct shear stresses in the beam and in the glue and indicate where they occur.

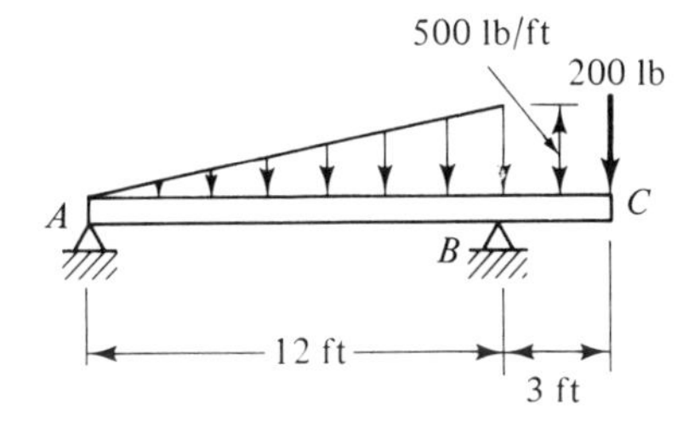

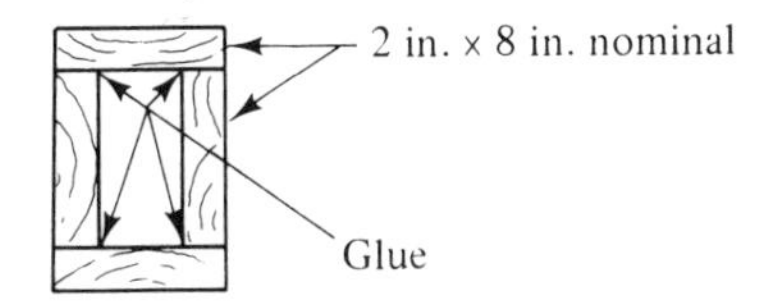

Problem 8.52

8.53 Determine the maximum allowable value of the load F for the beam shown if the allowable tensile and compressive stresses are 28 MN/m^2 and 70 MN/m^2, respectively, and the allowable direct shear stress is 2 MN/m^2.

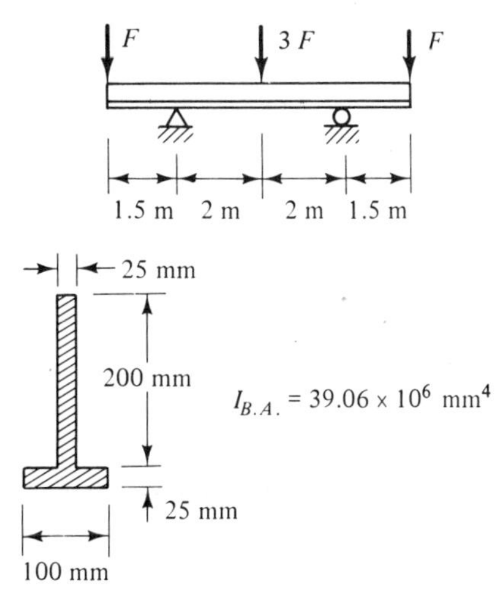

Problem 8.53

8.54 Two 50-mm × 150-mm steel plates are welded together to form a beam with the cross section shown. If each weld can withstand a shear force per unit length of 450 kN/m, what is the maximum shear force V that the beam can support.

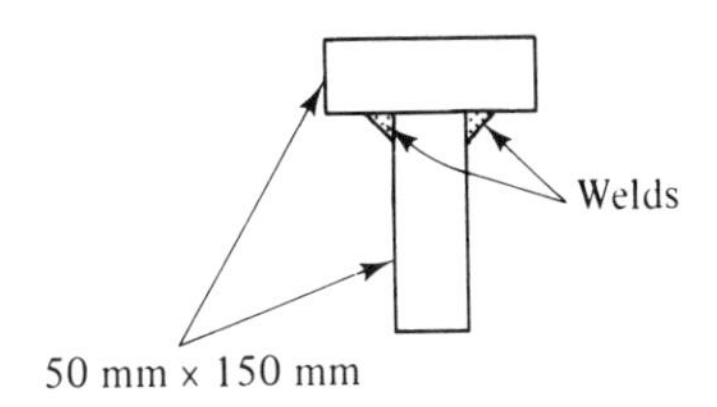

Problem 8.54

8.55 A laminated beam consists of three 50-mm × 100-mm boards glued together as shown. If the allowable shear stresses are 1.0 MN/m^2 in the glue and 0.7 MN/m^2 in the wood, what is the maximum allowable load F? What is the maximum bending stress in the member at this load?

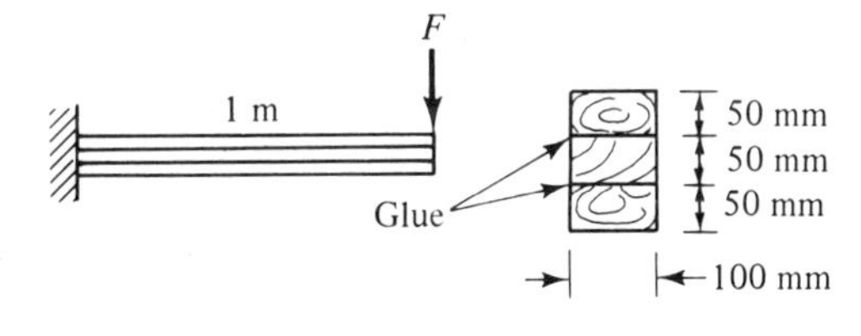

Problem 8.55

8.56 A simply supported Universal Beam 3 m long subjected to a 120-kN concentrated load at the center is found to be too weak in bending. It is strengthened by adding 125-mm wide × 10-mm thick steel plates to the top and bottom surfaces over the central one-third of the span where the bending stresses are largest. The plates are attached at each side with intermittent welds of equal length spaced 200 mm apart as shown. If each weld can support a shear force per unit length of 675 kN/m, what is the minimum required weld length L?

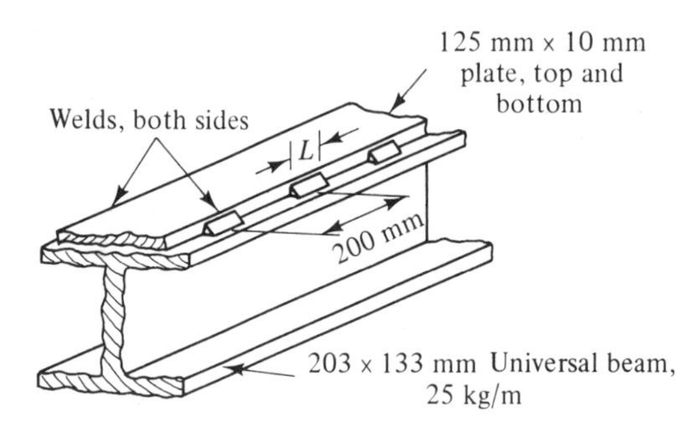

Problem 8.56

8.57 Two 2-in.×8-in. nominal planks oriented with the 8-in. dimension horizontal and connected with a single row of bolts are used as a simply supported beam with a span of 10 ft. What is the largest uniformly distributed load q that can be supported if the bending stress is not to exceed 2000 psi? What is the maximum allowable bolt spacing for this loading if the $\frac{1}{2}$-in. diameter bolts have an allowable shear stress of 15 ksi?

8.58 A beam is fabricated by attaching two 4-in.×4-in.× $\frac{1}{2}$-in. angle sections to a plate with rivets as shown. What is the minimum required rivet diameter if the beam must support a maximum shear force of 5000 lb? The rivets are spaced 6 in. apart and have an allowable shear stress of 15 ksi. *Hint*: The area A' for determining the shear stress in the rivets along the interface between the members is the area to one side of this interface, i.e., A' is the cross-sectional area of one of the angle sections.

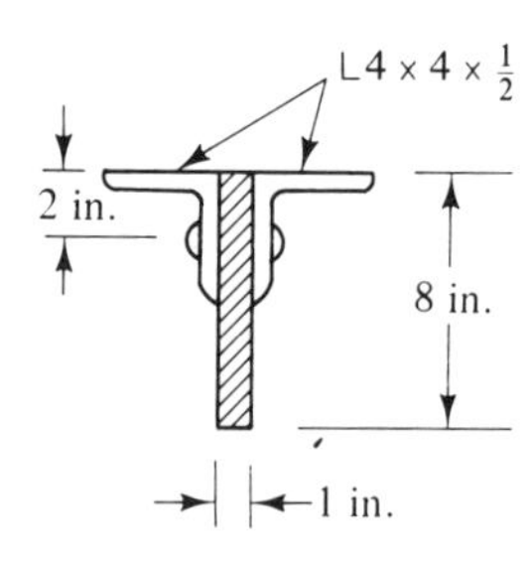

Problem 8.58

8.7 DEFLECTION OF BEAMS

When a beam is loaded, the neutral axis, which is originally straight, deforms into a curve (Figure 8.23). We shall denote this so-called *elastic curve* by $v(x)$. The value of v is measured from the original position of the neutral axis (x axis) and is called the *deflection* of the beam. It is considered positive if the beam deflects upward in the positive y direction and negative if it deflects in the negative y direction. The slope of the deformed beam at any location x is $dv(x)/dx$ and will be denoted by $\theta(x)$.

The deflections of a beam are important because they can result in a "failure" of the member if they are excessive. For example, a floor beam that deflects an excessive amount would be of little value, even though it may be in no danger of

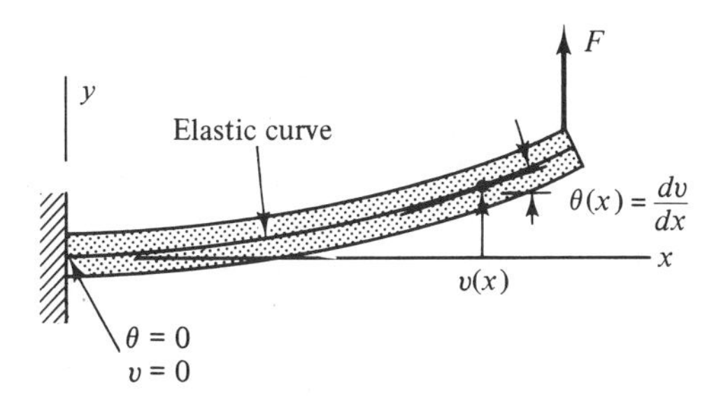

Figure 8.23

breaking. Consequently, limits are often placed upon the allowable deflections of a beam, as well as upon the stresses.

Our analysis of beam deflections will be based upon three fundamental assumptions. First, we assume that the material behavior is linearly elastic and that the beam geometry and loading conditions are such that the moment-curvature relation of Eq. (8.9) applies. Second, we assume that the deflections are small. And third, we assume that the contributions of the shear force, V, to the deflections are negligible. This last assumption is reasonable, except for exceptionally short beams.

The deflection, $v(x)$, of a beam is related to its curvature via the relation

$$\frac{1}{\rho} = \frac{\frac{d^2v}{dx^2}}{\left[1 + \left(\frac{dv}{dx}\right)^2\right]^{3/2}} \tag{8.16}$$

The derivation of this equation, which applies for any plane curve, is given in texts on analytic geometry. In applying this equation, it is important to note that x is the coordinate of a point on the elastic curve of the deformed beam. However, we shall be considering only small deflections, in which case x can be taken to be the coordinate along the undeformed neutral axis, as we have been doing. Small deflections also imply that the slope dv/dx will be small, in which case the term $(dv/dx)^2$ is negligible compared to unity. Thus, Eq. (8.16) reduces to

$$\frac{1}{\rho} \approx \frac{d^2v}{dx^2} \tag{8.17}$$

Combining Eq. (8.17) with the moment-curvature relation of Eq. (8.9), we obtain as the governing differential equation for the deflections

$$\frac{d^2v}{dx^2} = \frac{M}{EI} \tag{8.18}$$

In this equation, M is the bending moment, E is the modulus of elasticity of the material, and I is the cross-sectional moment of inertia about the bending axis.

Once the bending moment and the so-called *flexural rigidity* EI are expressed in terms of x, Eq. (8.18) can be integrated once to obtain the slope of the beam, dv/dx, and a second time to determine the deflection, $v(x)$. In the most problems, E and I are constants. For simple loadings, the moment equation can be readily determined by using the method of sections, as discussed in Section 8.2. For more complex loadings, the moment equation can be obtained by integrating the shear-load and moment-shear relations in Eqs. (8.1) and (8.2).

The integrated forms of Eq. (8.18) involve certain constants of integration, which are determined from the *boundary conditions* imposed upon the slope or deflection of the beam at the supports or other locations. For example, the supports of a simple beam prevent it from deflecting at these locations; therefore, the boundary conditions are that the deflection is zero at the ends $x = 0$ and $x = L$. Stated mathematically, we have $v(0) = v(L) = 0$. A fixed support prevents both rotation and deflection. Thus, the boundary conditions for the cantilever beam shown in Figure 8.23 are that the slope and deflection are zero at the support, or $v(0) = \theta(0) = 0$. A sketch of the expected shape of the elastic curve is often helpful in determining the appropriate boundary conditions.

If there is only one moment equation for the beam, the process of determining the slope and deflection is relatively straightforward (see Example 8.11). However, if there is more than one moment equation, as is often the case, the solution is more complex. In this case, there will be a different solution for $\theta(x)$ and $v(x)$ over each portion of the beam. These solutions must be pieced together by using the physical requirement that the neutral axis deform into a smooth curve without sudden jumps or sharp corners.

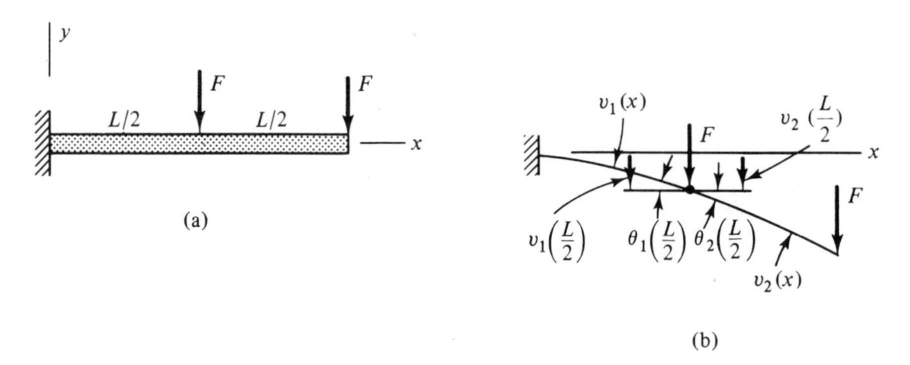

Figure 8.24

The procedure for matching the solutions is best illustrated by means of an example. Consider the beam shown in Figure 8.24(a). Since there is a different moment equation for each half of the member, there will be one set of solutions, $\theta_1(x)$ and $v_1(x)$, for the slope and deflection over the left half of the beam and another set, $\theta_2(x)$ and $v_2(x)$, for the corresponding quantities over the right half. Each set of solutions will involve two constants of integration, for a total of four.

Now, the elastic curve must be smooth, as indicated in Figure 8.24(b). From the figure, we see that this requires that both the slope and deflection have the same values at points immediately to the left and to the right of the center of the beam. Stated mathematically, we have for this example

$$\theta_1\left(\frac{L}{2}\right) = \theta_2\left(\frac{L}{2}\right)$$

$$v_1\left(\frac{L}{2}\right) = v_2\left(\frac{L}{2}\right)$$

These so-called *continuity conditions*, in addition to the two boundary conditions, are sufficient to determine the four constants of integration. This matching procedure will be further illustrated in Example 8.12.

Statically indeterminate problems can also be handled by using the basic procedures outlined in this section. In this case, there will be some reactions that cannot be determined from a force analysis and that will remain as unknowns in the moment equation. However, the additional support conditions that make the problem indeterminate also provide enough additional boundary conditions so that both the constants of integration and the unknown reactions can be determined (see Example 8.13).

Many different techniques for determining the slope and deflection of a beam have been developed over the years. Some of these methods are graphical or semigraphical, but all are based upon Eq. (8.18). These various techniques will not be considered in this text. The method of direct integration of Eq. (8.18) discussed in this section and the method of superposition to be discussed in the following section will be sufficient for our purposes.

Example 8.11. Deflection of a Simple Beam. For the beam shown in Figure 8.25(a), determine (a) the deflection equation, (b) the maximum deflection and the location at which it occurs, and (c) the slope at the left end. The flexural rigidity, EI, is constant.

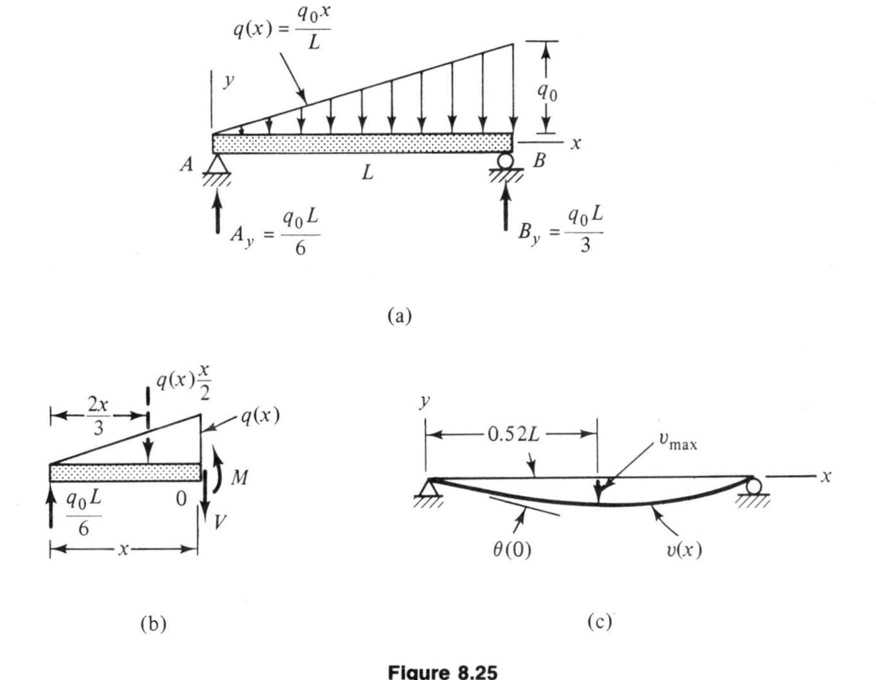

Figure 8.25

Solution. (a) The reactions at A and B are found to be

$$A_y = \frac{q_o L}{6} \qquad B_y = \frac{q_o L}{3}$$

Sectioning the beam at an arbitrary location and considering the equilibrium of the piece to the left [Figure 8.25(b)], we have

$$\overset{+}{\curvearrowleft}\Sigma M_o = M + q(x)\left(\frac{x}{2}\right)\left(\frac{x}{3}\right) - \frac{q_o L}{6} x = 0$$

From this equation and Eq. (8.18), we obtain

$$EI\frac{d^2 v}{dx^2} = M = q_o\left(\frac{Lx}{6} - \frac{x^3}{6L}\right)$$

Integrating both sides of this expression with respect to x, we have

$$EI\frac{dv}{dx} = EI\theta(x) = q_o\left(\frac{Lx^2}{12} - \frac{x^4}{24L}\right) + C_1$$

where C_1 is a constant of integration. A second integration yields

$$EIv(x) = q_o\left(\frac{Lx^3}{36} - \frac{x^5}{120L}\right) + C_1 x + C_2$$

The boundary conditions are $v(0) = 0$ and $v(L) = 0$. Applying the first condition to the preceding equation, we find that $C_2 = 0$. The second condition requires that

$$0 = q_o\left(\frac{L^4}{36} - \frac{L^4}{120}\right) + C_1 L$$

or

$$C_1 = -\frac{7}{360} q_o L^3$$

Thus, the deflection equation is

$$v(x) = \frac{q_o x L}{360EI}\left(10x^2 - 3\frac{x^4}{L^2} - 7L^2\right) \qquad \textit{Answer}$$

(b) The maximum deflection occurs at a point of zero slope. The location of this point is determined by setting the equation for $\theta(x)$ equal to zero and solving for x. We have

$$EI\theta(x) = q_o\left(\frac{Lx^2}{12} - \frac{x^4}{24L}\right) - \frac{7}{360} q_o L^3 = 0$$

or

$$x^4 - 2L^2x^2 + \frac{7}{15} L^4 = 0$$

This is a quadratic equation for x^2. From the quadratic formula, we obtain

$$x^2 = L^2 \pm \sqrt{L^4 - \frac{7}{15} L^4}$$
$$= L^2 \pm 0.73L^2 = 0.27L^2 \text{ or } 1.73L^2$$

The second solution is ruled out on the grounds that x cannot be greater than L; therefore,

$$x = \sqrt{0.27L^2} = 0.52L$$

Substituting this value of x into the deflection equation, we find that

$$v_{\max} = -0.0065 \frac{q_o L^4}{EI} \qquad \textbf{\textit{Answer}}$$

(c) The slope at the left end of the beam is

$$\theta(0) = \frac{C_1}{EI} = -\frac{7q_o L^3}{360EI} \qquad \textbf{\textit{Answer}}$$

The negative slope at this location is consistent with the downward deflection of the beam [Figure 8.25(c)].

Example 8.12. Deflection of a Simple Beam. A simple beam supports a single concentrated load F at a distance a from the left end [Figure 8.26(a)]. Determine the deflection equation if EI is a constant.

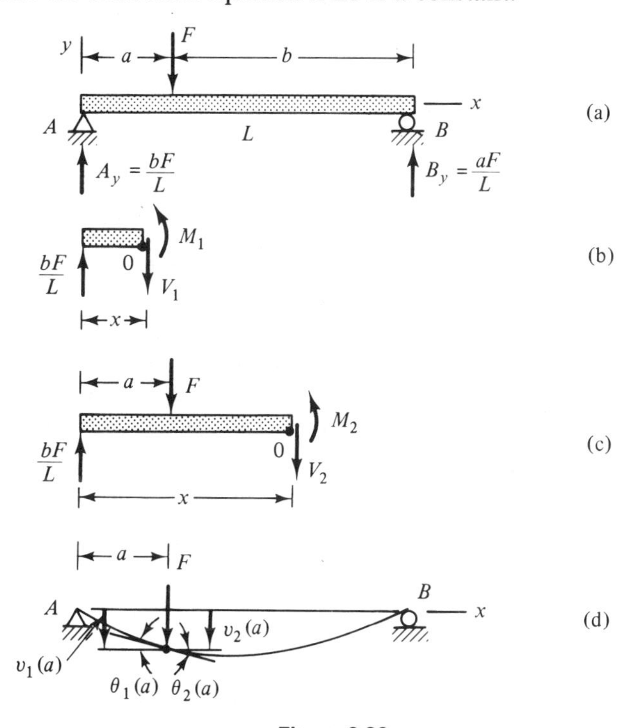

Figure 8.26

Solution. A force analysis of the entire beam reveals that the reactions are

$$A_y = \frac{bF}{L} \qquad B_y = \frac{aF}{L}$$

Sectioning the beam at a location between the left end and the applied load, we obtain the moment equation

$$M_1 = \frac{bFx}{L} \qquad (0 \leqslant x \leqslant a)$$

Similarly, the moment in the other portion of the beam [Figure 8.26(c)] is found to be

$$M_2 = \frac{bFx}{L} - F(x-a) \qquad (a \leqslant x \leqslant L)$$

Substituting the moment equations into the relation $EId^2v/dx^2 = M$ and integrating, we obtain for the slope and deflection in the first portion of the beam

$$EI\theta_1(x) = \frac{Fbx^2}{2L} + C_1$$

$$EIv_1(x) = \frac{Fbx^3}{6L} + C_1x + C_2$$

and for the corresponding quantities in the second portion

$$EI\theta_2(x) = \frac{Fbx^2}{2L} - \frac{F(x-a)^2}{2} + C_3$$

$$EIv_2(x) = \frac{Fbx^3}{6L} - \frac{F(x-a)^3}{6} + C_3x + C_4$$

Note that in determining θ_2 and v_2, the integral of terms of the form $(x-a)^n$ was writen as $\dfrac{(x-a)^{(n+1)}}{(n+1)}$. Ordinarily, the term would first be expanded and then integrated. However, the two results differ only by a constant, which can be combined with the constant of integration. The algebra involved in determining the constants of integration is greatly reduced by using this procedure.

A sketch of the elastic curve is shown in Figure 8.26(d). It is clear from this sketch that the boundary conditions are $v_1(0) = 0$ and $v_2(L) = 0$ and that the continuity conditions between the two portions of the beam are $\theta_1(a) = \theta_2(a)$ and $v_1(a) = v_2(a)$. Applying these conditions to our solutions, we have:

$$v_1(0) = 0: \quad 0 = C_2$$

$$v_2(L) = 0: \quad 0 = \frac{FbL^2}{6} - \frac{Fb^3}{6} + C_3L + C_4$$

$$\theta_1(a) = \theta_2(a): \quad \frac{Fba^2}{2L} + C_1 = \frac{Fba^2}{2L} + C_3$$

$$v_1(a) = v_2(a): \quad \frac{Fba^3}{6L} + C_1a = \frac{Fba^3}{6L} + C_3a + C_4$$

Solving these equations, we obtain

$$C_2 = C_4 = 0 \qquad C_1 = C_3 = \frac{Fb}{6L}(b^2 - L^2)$$

Thus, the deflection equations are, after rearrangement of terms,

$$v(x) = \frac{Fbx}{6EIL}(x^2 + b^2 - L^2) \qquad (x \leqslant a)$$

$$v(x) = \frac{Fb}{6EIL}\left[x^3 - \frac{L}{b}(x-a)^3 - x(L^2 - b^2)\right] \qquad (x \geqslant a)$$

Answer

Example 8.13. Analysis of an Indeterminate Beam. Determine the reactions and the deflection equation for the beam shown in Figure 8.27(a).

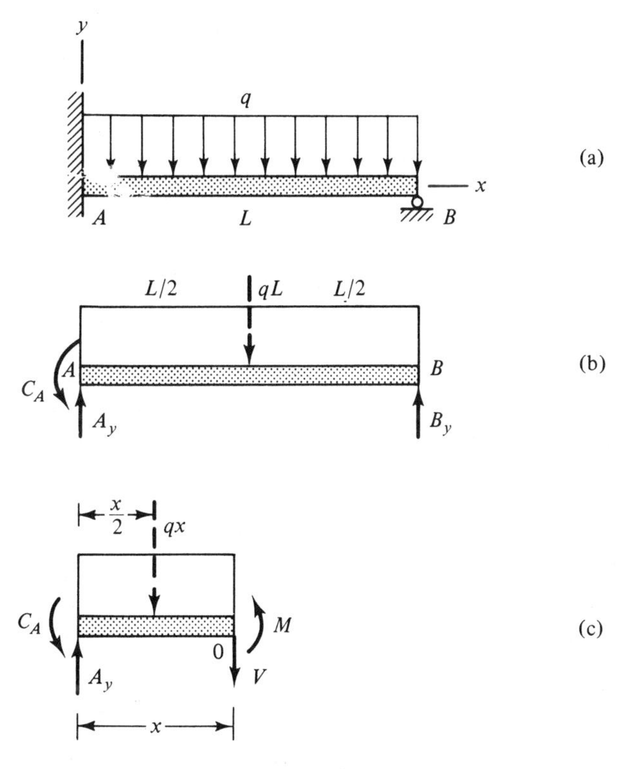

Figure 8.27

Solution. The FBD [Figure 8.27(b)] reveals that the problem is statically indeterminate because there are three unknown reactions and only two equations of equilibrium from which to determine them. These equations are

$$\Sigma F_y = A_y + B_y - qL = 0$$

$$\overset{+}{\curvearrowleft} \Sigma M_A = C_A + B_y L - \frac{qL^2}{2} = 0$$

To obtain a third equation, we must consider the deflections of the beam.

Sectioning the beam at an arbitrary distance x from the left end [Figure 8.27(c)]. we have

$$\overset{+}{\curvearrowleft} \Sigma M_o = M + qx\left(\frac{x}{2}\right) + C_A - A_y x = 0$$

From this equation and Eq. (8.18), we obtain

$$EI\frac{d^2v}{dx^2} = M = A_y x - \frac{qx^2}{2} - C_A$$

Two integrations yield

$$EI\theta(x) = \frac{A_y x^2}{2} - \frac{qx^3}{6} - C_A x + C_1$$

$$EIv(x) = \frac{A_y x^3}{6} - \frac{qx^4}{24} - \frac{C_A x^2}{2} + C_1 x + C_2$$

The boundary conditions $\theta(0) = 0$ and $v(0) = 0$ imply that $C_1 = 0$ and $C_2 = 0$, respectively. However, there is a third boundary condition in this problem; namely, $v(L) = 0$. This condition provides the necessary third equation for determining the reactions:

$$\frac{A_y L^3}{6} - \frac{qL^4}{24} - \frac{C_A L^2}{2} = 0$$

From this equation and the two equilibrium equations, we obtain

$$A_y = \frac{5}{8}qL \qquad B_y = \frac{3}{8}qL \qquad C_A = \frac{qL^2}{8} \qquad \textit{Answer}$$

Thus, the deflection equation is

$$v(x) = \frac{qx^2}{48EI}(5xL - 2x^2 - 3L^2) \qquad \textit{Answer}$$

PROBLEMS

8.59 to 8.61 Derive the deflection equations for the following uniform beams with flexural rigidity EI and length L and compare the results with those given in Table A.13 of the Appendix. Assume that cantilever beams are supported at the left end.

8.59 Cantilever beam with concentrated load F at tip.

8.60 Cantilever beam with uniformly distributed load q.

8.61 Simple beam with uniformly distributed load q.

8.62 Knowing that the water pressure varies linearly with depth, determine the deflection equation for the cantilevered dam shown by considering a strip one unit wide. What is the deflection and slope at the top of the dam?

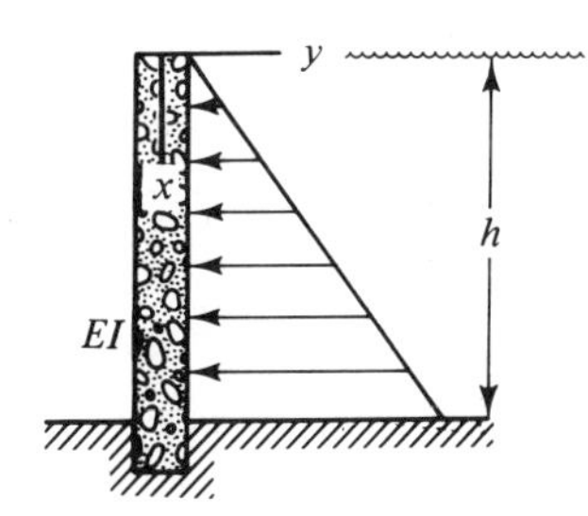

Problem 8.62

8.63 For the beam shown, derive the deflection equation and determine the magnitude and location of the maximum deflection. What is the slope at midspan?

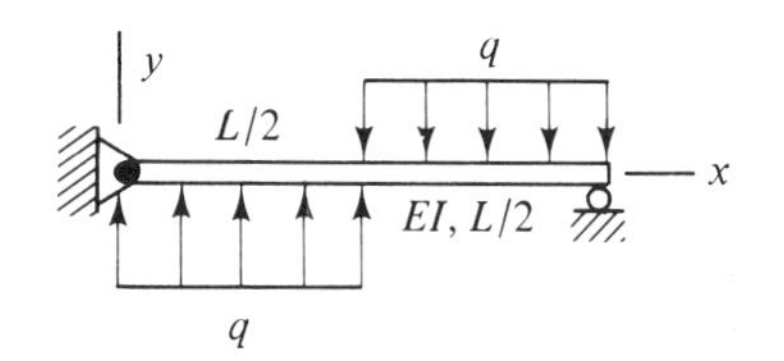

Problem 8.63

8.64 Derive the deflection equation for the beam shown and find the slope at the overhanging end. What is the maximum deflection and where does it occur?

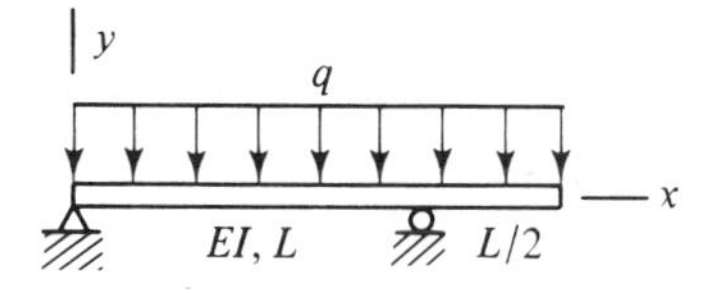

Problem 8.64

8.65 Determine the slope at midspan and the magnitude and location of the maximum deflection of the beam shown; $E = 24 \times 10^6$ psi and $I_{B.A.} = 144$ in.^{4}.

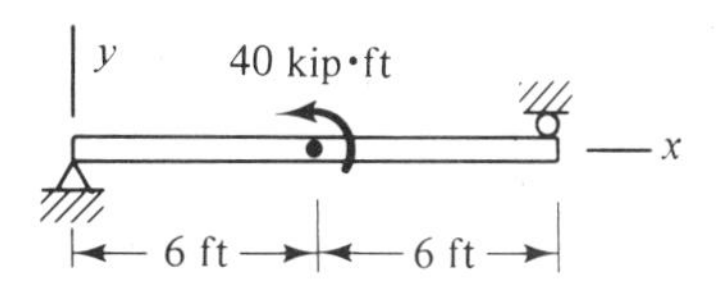

Problem 8.65

8.66 A 12-in. × 3-in. nominal plank is used as a diving board. What is the tip deflection and slope when the 180-lb diver stands at the end as shown? $E = 1.6 \times 10^6$ psi.

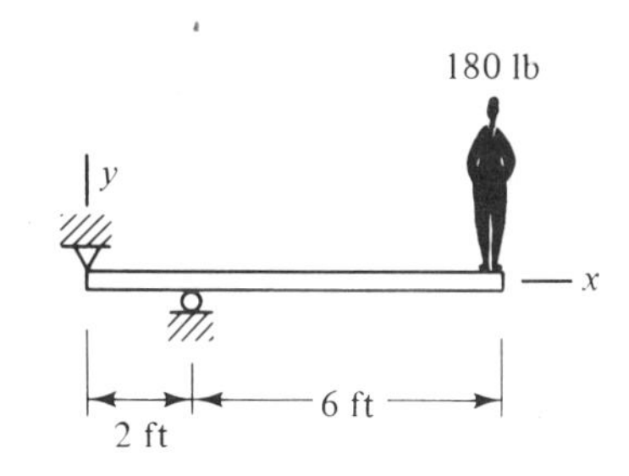

Problem 8.66

8.67 Using the results in Table A.13 of the Appendix, calculate the midspan deflection of the beam shown. $E = 207$ GN/m^2.

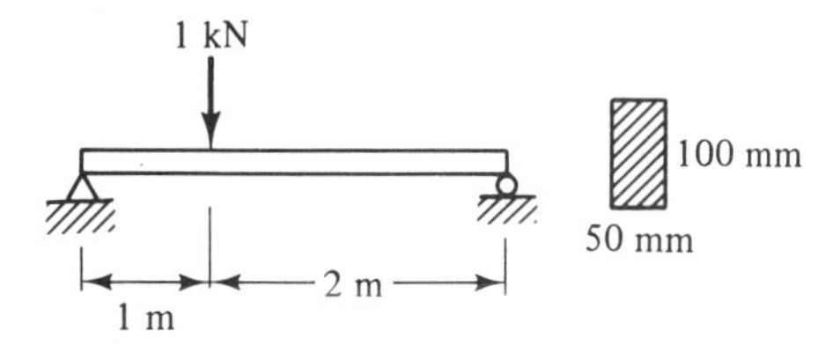

Problem 8.67

8.68 An aluminum strip 30 mm wide and 6 mm thick is to be bent between three fixed pins 4 mm in diameter as shown. What is the minimum allowable support spacing d if the strip is not to yield? Use Table A.13, with $\sigma_{YS} = 240$ MN/m^2 and $E = 70$ GN/m^2.

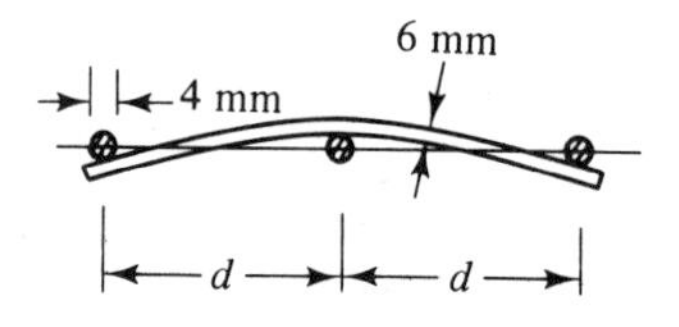

Problem 8.68

8.69 Calibrate the torque wrench shown by obtaining the relation between the angle θ between the pointer and the centerline of the arm at the scale and the couple C exerted by the wrench. Use Table A.13, and express the results in terms of the given dimensions and the flexural rigidity EI of the arm.

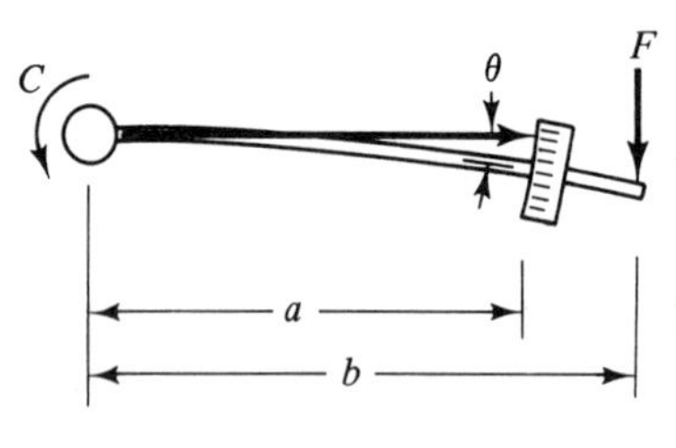

Problem 8.69

8.70 A 2-in.×6-in. nominal timber oriented with the larger dimension vertical is used as a simply supported beam with a span of 10 ft. What is the maximum uniformly distributed load q that can be supported if the deflection must not exceed $\frac{1}{360}$ of the span length and the allowable bending and direct shear stresses are 2000 psi and 120 psi, respectively; $E = 1.6 \times 10^6$ psi.

8.71 Determine the reactions at the supports and the deflection equation for the beam shown. Compare the midspan deflection with that of an identical simply supported beam carrying the same load.

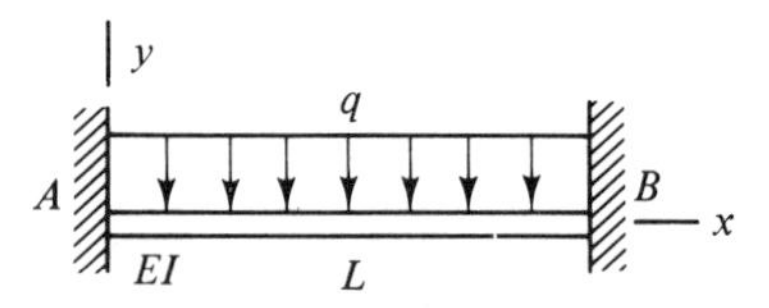

Problem 8.71

8.72 The uniform beam shown is fixed at each end. Because of improper soil preparation, support B settles a small distance Δ below support A. What is the maximum bending moment induced in the beam by the settlement and where does it occur?

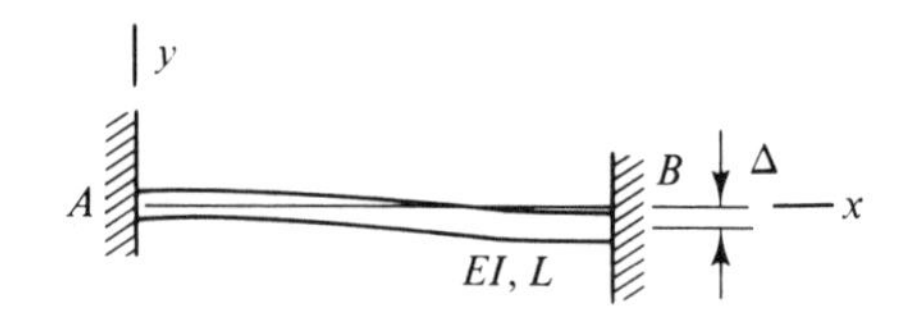

Problem 8.72

8.8 BEAM DEFLECTIONS BY SUPERPOSITION

It is an understatement to say that the integration method of determining beam deflections becomes lengthy when the loadings are complex. For example, consider the work involved in Example 8.12 in determining the deflection of a simple beam with a single concentrated load. Just imagine what the situation would be if there were, say, three concentrated loads. There would be four different moment equations and four sets of solutions for the slope and deflection involving a total of eight constants of integration! These four sets of solutions would have to be matched together at three different locations. Clearly, the amount of work involved would be prohibitive. Fortunately, there is an easier way.

The example problems considered in the preceding section show that the slope and deflection of a beam are linearly proportional to the applied loads. As indicated by Eq. (8.18), this will always be the case if the material is linearly elastic and the deflections are small. Consequently, the slope and deflection due to several loads are equal to the sum of those due to the individual loads. This result follows directly from the *principle of superposition*, which states that if an effect E_1 is linearly proportional to its cause C_1 and effect E_2 to its cause C_2, then C_1 and C_2 together will have an effect equivalent to E_1 and E_2 together. We have used this principle several times before, but without calling it by name.

The advantage of using superposition in beam deflection problems lies in the fact that solutions for the slope and deflection of common types of beams with simple loadings have been tabulated in engineering handbooks (see Table A.13 of the Appendix). Using superposition, we may combine these solutions to obtain the solution for more complicated loadings.

Superposition is also convenient to use for statically indeterminate problems. The procedure here is to remove enough supports to make the member statically determinate and to replace them by the reactions they exert. These reactions are then treated as part of the applied loading. From this point, the procedure is basically the same as that for statically determinate problems, except that the slopes and deflections due to the individual loadings must satisfy certain compatibility conditions so that the boundary conditions for the original problem are met. These compatibility conditions yield the additional equations necessary for determining the reactions.

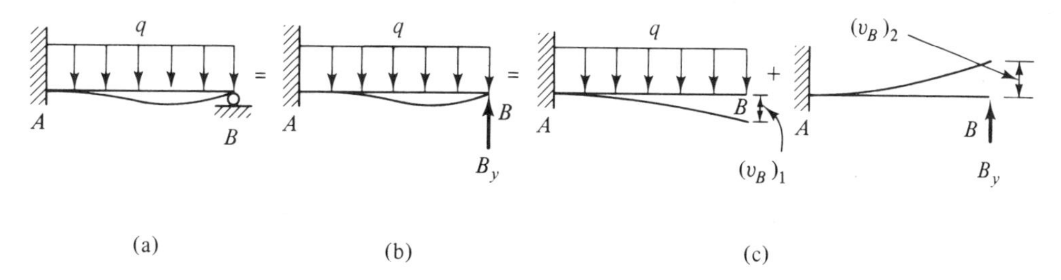

Figure 8.28

For example, if we remove the roller from under the beam shown in Figure 8.28(a) and replace it by the force that it exerts, we obtain the loading shown in Figure 8.28(b). Decomposing this loading, as in Figure 8.28(c), we have as the compatibility condition between the deflections at B

$$v_B = (v_B)_1 + (v_B)_2 = 0$$

Both $(v_B)_1$ and $(v_B)_2$ are given in Table A.13 (cases 1 and 3). Now $(v_B)_1$ involves the applied loading and $(v_B)_2$ involves the unknown reaction B_y. Thus, B_y can be determined from the compatability condition. The other reactions can then be determined from the equations of equilibrium. Once the reactions are known, any other quantities of interest, such as bending moments, stresses, or deflections, can be found.

Example 8.14. Deflection of a Simple Beam. Determine the midspan deflection of the beam shown in Figure 8.29(a).

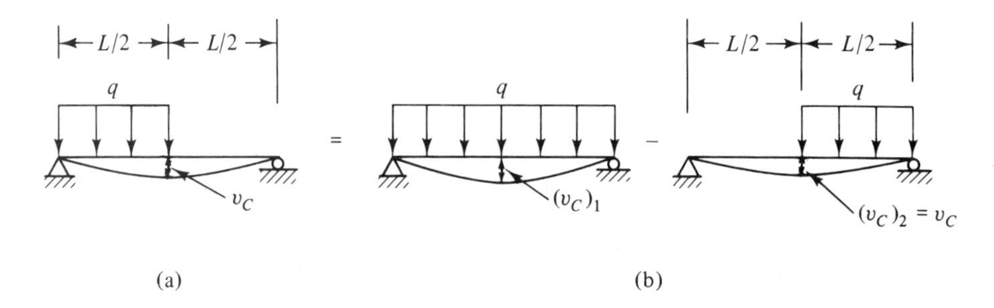

Figure 8.29

Solution. Decomposing the loading as shown in Figure 8.29(b), we have

$$v_C = (v_C)_1 - (v_C)_2$$

But $(v_C)_2 = v_C$, since the first and last beams are the same except that one is turned end for end with respect to the other. Thus,

$$v_C = \frac{1}{2}(v_C)_1 = -\frac{5qL^4}{768EI} \qquad \textit{Answer}$$

where the value of $(v_C)_1$ is obtained from Table A.13.

Example 8.15. Analysis of an Indeterminate Beam. Compare the maximum bending moment in the beam shown in Figure 8.30(a) with that in a simply supported beam with the same span and loading.

Solution. We remove the fixed support at the right end and replace it by the reactions it exerts [Figure 8.30(b)]. This loading is then broken down as shown in Figure 8.30(c). Referring to Table A.13 and using the conditions that the slope and deflection of the original beam must both be zero at the right end, we have

$$\theta_B = (\theta_B)_1 + (\theta_B)_2 + (\theta_B)_3 = \frac{-FL^2}{8EI} + \frac{B_y L^2}{2EI} - \frac{C_B L}{EI} = 0$$

or

$$-8C_B + L(4B_y - F) = 0$$

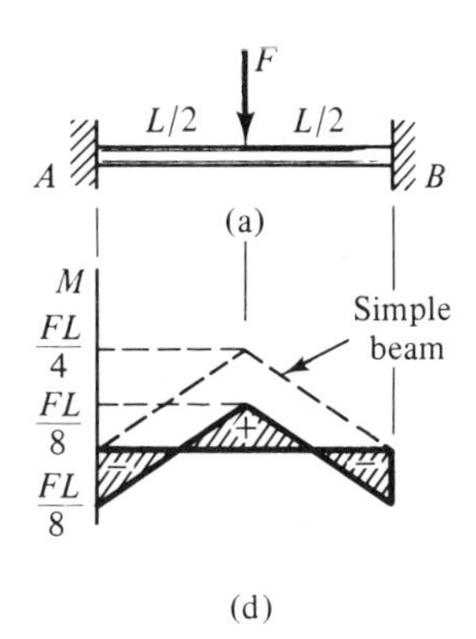

(b)

(c)

Figure 8.30

and

$$v_B = (v_B)_1 + (v_B)_2 + (v_B)_3 = \frac{-5FL^3}{48EI} + \frac{B_y L^3}{3EI} - \frac{C_B L^2}{2EI} = 0$$

or

$$-24C_B + L(16B_y - 5F) = 0$$

Solving these equations, we obtain

$$B_y = \frac{F}{2} \qquad C_B = \frac{FL}{8}$$

A force analysis of the beam shows that the reactions at the left end have the same values. Thus, the moment diagram is as shown in Figure 8.30(d). The diagram for the corresponding simply supported beam is indicated by the dashed lines. These diagrams show that the maximum bending moment in the indeterminate beam is only one-half that in the simple beam. Also, since the moments in the indeterminate beam are more nearly uniform over the length, the material in the member is utilized more effectively. It is also of interest to note that the maximum deflection of the indeterminate beam is only one-fourth that of the corresponding simple beam $[-FL^3/(192EI)$ compared to $-FL^3/(48EI)]$.

These results illustrate the reason for making a member indeterminate, which is to "stiffen" it and reduce the deformations and stresses. Stated another way, the load-carrying capacity of a member can be increased by making it statically indeterminate.

PROBLEMS

8.73 Determine the midspan deflection of the beam shown.

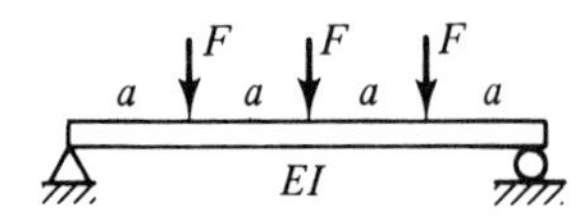

Problem 8.73

8.74 For the W 6 × 16 steel beam shown, what is the maximum allowable value of the distributed loading q if the midspan deflection is not to exceed $\frac{1}{360}$ of the span length?

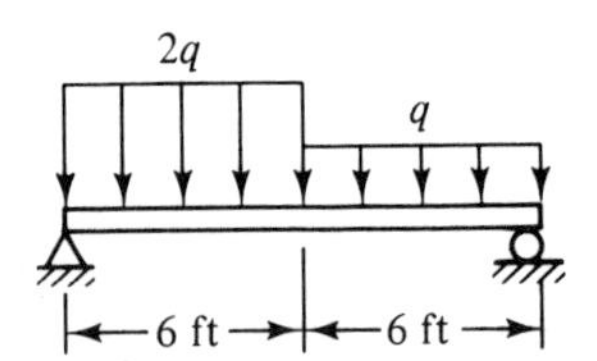

Problem 8.74

8.75 Use the results of Example 8.11 to determine the magnitude and location of the maximum deflection for the beam shown. $E = 10 \times 10^6$ psi and $I_{\text{B.A.}} = 60$ in.^{4}.

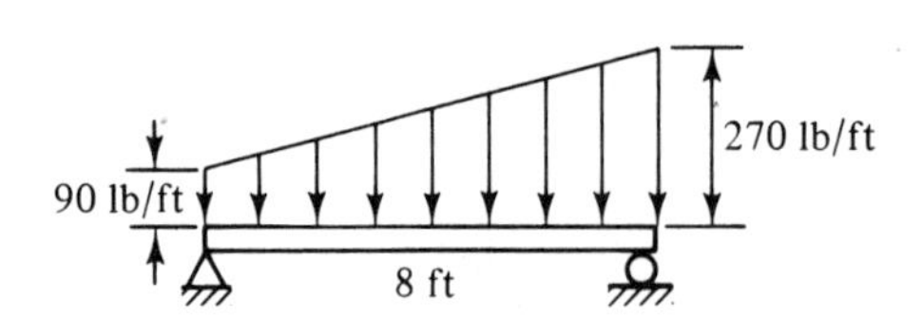

Problem 8.75

8.76 Find the magnitude C of the couple required to keep the pointer on the beam shown horizontal.

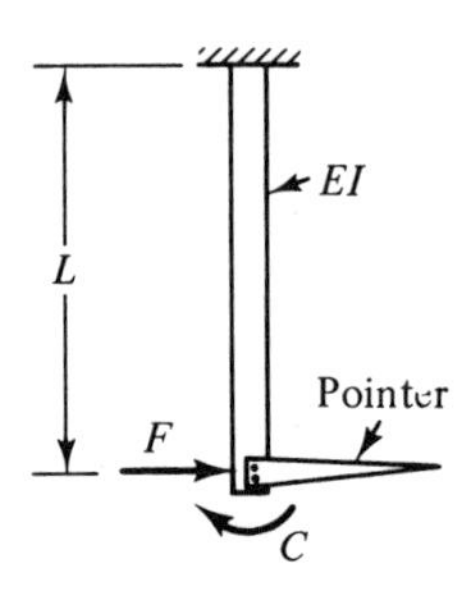

Problem 8.76

8.77 A simply supported floor beam in a house sags excessively. To correct the situation, a jack is used to raise the midpoint of the beam until it is level with the end supports. If the beam is 5 m long and carries a uniformly distributed load of 300 N/m over its entire length, what is the magnitude of the force exerted on the beam by the jack?

8.78 If a roller is placed under the free end of a cantilever beam subjected to a uniformly distributed load q over its entire length, by what percentages are the maximum bending stress and maximum deflection reduced?

8.79 Determine the midspan deflection of the beam shown.

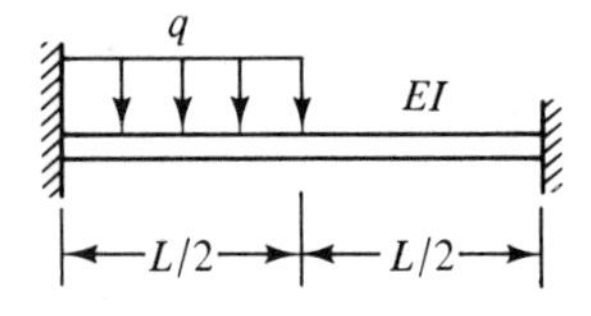

Problem 8.79

8.80 A steel wire is attached to the end of a round steel cantilever beam as shown. What is the largest couple C that can be applied to the beam if the maximum allowable tensile stress in the beam and in the wire is 20 ksi?

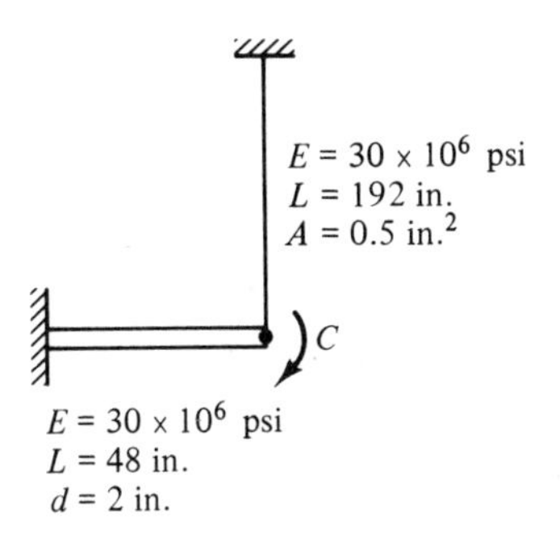

Problem 8.80

8.81 A 6-mm diameter steel wire is attached to the center of the simply supported steel beam shown. The wire is initially stress-free. If the temperature drops 40°C, what is the maximum tensile stress in the beam and in the wire? $E = 207$ GN/m^2; $\alpha = 11.7 \times 10^{-6}(°\text{C})^{-1}$.

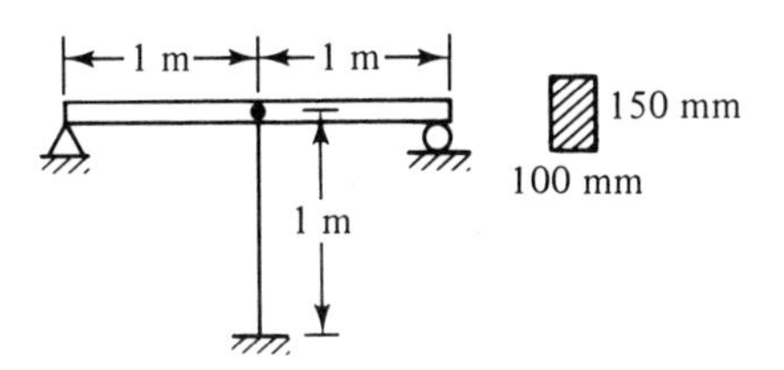

Problem 8.81

8.82 The right end of the 50-mm × 50-mm beam shown is initially 5 mm above the support. What is the reaction at the right end when the beam is loaded? $E = 70$ GN/m^2. What is the maximum bending stress in the beam?

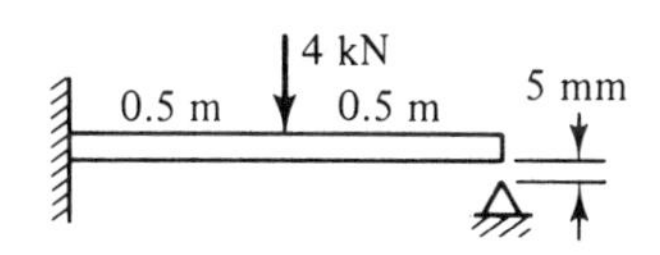

Problem 8.82

8.83 Two cantilever beams with the same flexural rigidity EI and length L are arranged as shown with a roller that just fits snugly between them. A uniformly distributed load q is then applied to the upper beam. Determine the force exerted on the beams by the roller.

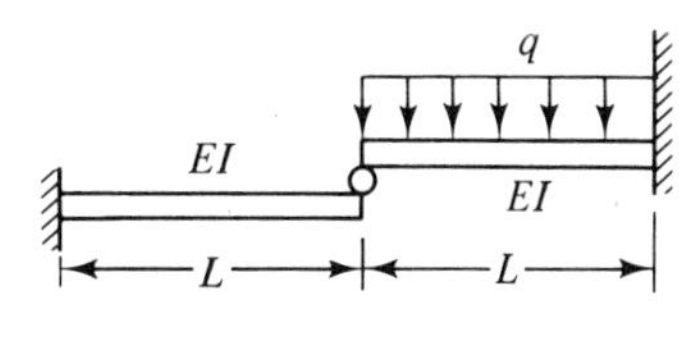

Problem 8.83

8.84 The two simply supported beams shown are made of 4-in.×6-in. nominal timbers oriented with the larger dimension vertical. What is the maximum load F that can be supported without exceeding the allowable bending stress of 5 ksi? What is the midspan deflection of the beams for this value of load? Use $E = 1.6 \times 10^6$ psi and assume that the beams just touch when unloaded.

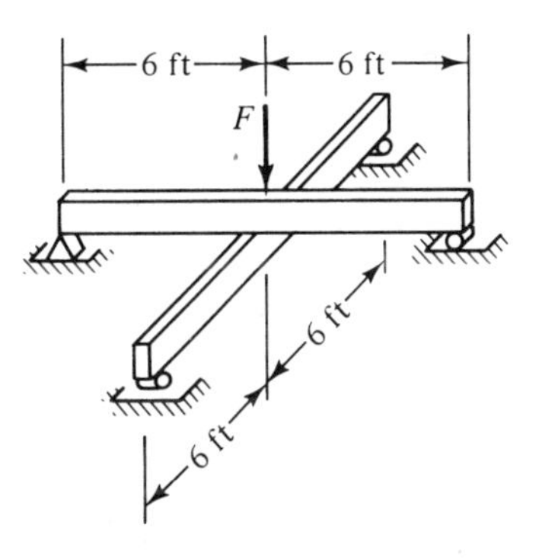

Problem 8.84

8.9 INELASTIC RESPONSE OF BEAMS

Now that we have considered the linearly elastic response of beams, we shall turn our attention to the inelastic response. All of the assumptions made in the preceding sections concerning the geometry and loading of the beam still apply, in addition to the assumption that the material behavior is essentially elastic-perfectly plastic with the same yield strength in tension and compression.

Stress distribution. As the beam undergoes inelastic deformations, the bending stress distribution tends to become uniform over the cross section, and the moment-stress and moment-curvature relations derived in the preceding sections no longer apply. The situation is illustrated in Figure 8.31, which

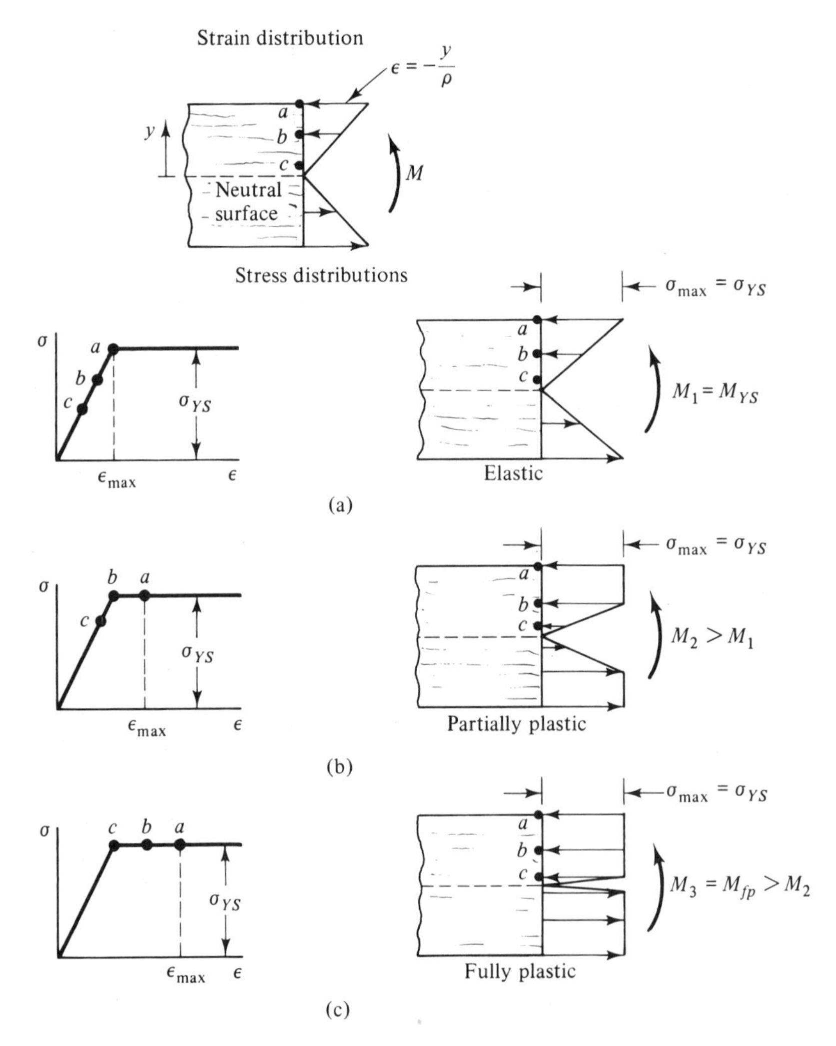

Figure 8.31

shows the bending stresses in a beam made of an elastic-plastic material for three different values of bending moment. The conditions at locations *a*, *b*, and *c* on the cross section are denoted by the corresponding points on the stress-strain diagram.

Figure 8.31(a) shows the situation when the moment is such that the maximum bending stress is just equal to the yield strength of the material. Up to this value of moment, the beam response is linearly elastic. If the moment is increased above this value, the curvature and strains increase, while the maximum bending stress remains equal to the yield strength. Since strains at points away from the top and bottom of the cross section are now at or above the proportional limit strain, we have the partially plastic stress distribution shown in Figure 8.31(b). Note that the stresses in the elastic central portion of the beam remain linearly distributed.

If the bending moment continues to increase, a situation is eventually reached where the strains are at or above the proportional limit strain everywhere over the cross section, except near the bending axis where the strain is always zero. The stresses are then very nearly uniform over the entire cross section, as shown in Figure 8.31(c). The bending moment at which this fully plastic condition is reached is called the *fully plastic moment* and will be denoted by M_{fp}.

The location of the bending axis in the partially plastic and fully plastic cases is determined from the requirement that there be no net normal force on the cross section. In other words, the net tensile force acting upon one part of the cross section must balance the net compressive force acting upon the other part of the cross section. For the fully plastic case in which the stresses are uniformly distributed and the tensile stress is equal to the compressive stress, this condition implies that the area in tension must equal the area in compression. In other words, the bending axis in the fully plastic case is located such that it divides the cross-sectional area in half. If the cross section has a horizontal axis of symmetry, as in the case of an I-beam or rectangular section, the bending axis will pass through the centroid. For unsymmetrical sections, such as a T-beam, it will not.

Fully plastic moment. Once the bending axis has been located, the fully plastic moment can be determined from the moment condition for static equivalence given in Eq. (8.8):

$$M = -\int y\sigma dA$$

However, it is simpler to determine it from first principles by simply computing the moment of the couple due to the resultant tensile and compressive forces acting upon each half of the cross section.

For example, consider a beam with a rectangular cross section (Figure 8.32). The magnitude of the net tensile and compressive forces acting is

$$F_T = F_C = \sigma_{YS}\left(\frac{bh}{2}\right)$$

and the distance between them is $h/2$. Thus,

$$M_{fp} = F_T\left(\frac{h}{2}\right) = \sigma_{YS}\left(\frac{bh^2}{4}\right) \tag{8.19}$$

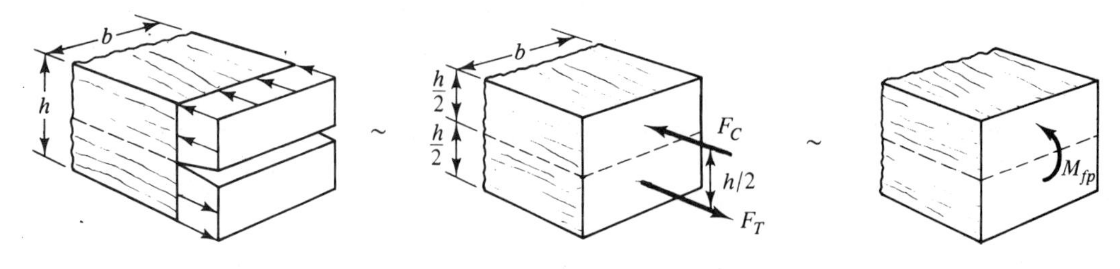

Figure 8.32

From Eq. (8.10), the moment M_{YS} at which yielding first occurs is

$$M_{YS} = \frac{\sigma_{YS} I}{(h/2)} = \sigma_{YS}\left(\frac{bh^2}{6}\right) \tag{8.20}$$

Comparing these equations, we find that for a rectangular section

$$M_{fp} = \frac{3}{2} M_{YS} \tag{8.21}$$

For cross sections with other shapes, the ratio between the two moments will be different.

These various bending moments and the different ranges of behavior associated with them are illustrated in Figure 8.33, which shows a portion of the

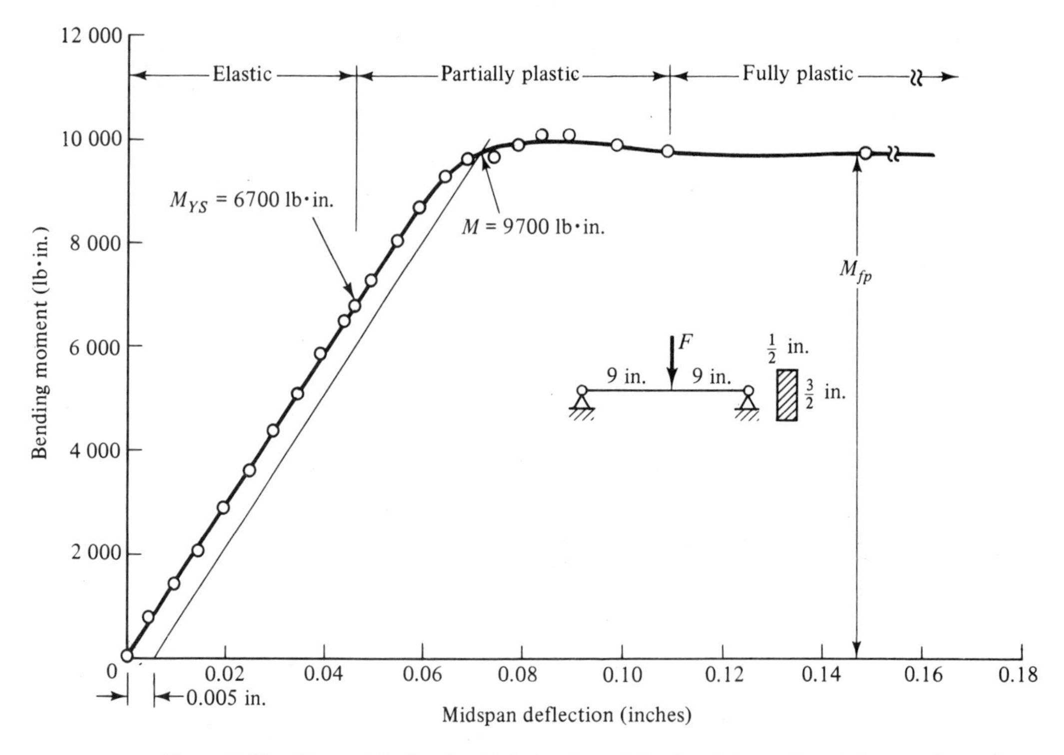

Figure 8.33. Moment-Deflection Relation for a Mild Steel Beam (Initial Portion Only).

experimentally determined moment-deflection curve for a mild steel beam. The point at which the curve first starts to deviate from a straight line corresponds to the onset of yielding. This point is difficult to locate accurately because the deviation from linearity is initially very slight. The horizontal portion of the curve corresponds to the fully plastic condition.

The test specimen used to obtain this curve is shown in Figure 8.34(a). The flaking of the mill scale evident at its center indicates the regions over which the material has yielded. Note that even though the member was severely deformed, it did not break and could still support a load. Compare this with the behavior of a cast-iron beam [Figure 8.34(b)], that suddenly breaks when the maximum tensile stress reaches the ultimate tensile strength of the material.

It is advantageous to allow some permanent deformation whenever possible, as can be seen from Figure 8.33. If no permanent deflection is allowed, the maximum bending moment that can be supported is M_{YS}, which for the particular beam tested is 6700 lb·in. If, however, a permanent midspan deflection of 0.005 in. is allowed, a moment of 9700 lb·in. can be supported. This is an increase of 45%.

(a)

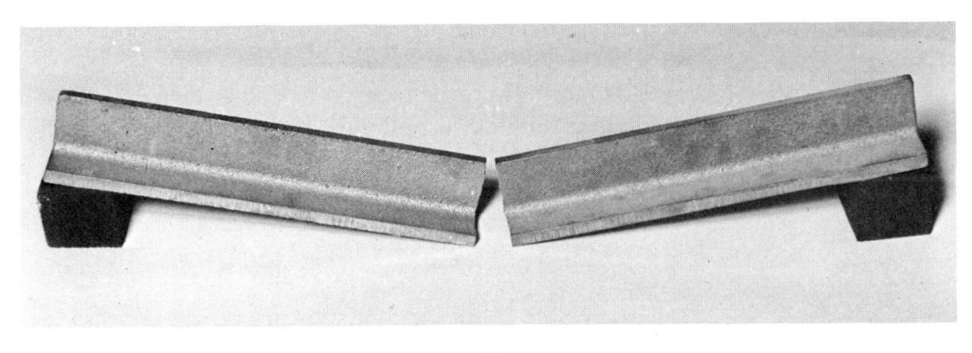

(b)

Figure 8.34. Response of (a) Mild Steel and (b) Cast Iron Beams when Overloaded.

The procedure for determining the latter value of moment is the same as that for finding the yield strength of a material.

There is, of course, a limit to the increase in load-carrying capacity that can be achieved by allowing permanent deformations. Once the fully plastic condition is reached, there is little or no additional resistance to bending at that section, and the beam behaves much as if it were hinged there. Consequently, the beam is said to have developed a *plastic hinge* at the location of the fully plastic moment. The formation of a plastic hinge can lead to large deflections with little or no increase in load, as indicated by the horizontal portion of the moment-deflection curve in Figure 8.33. Thus, a beam made of a ductile material usually fails by general yielding once the fully plastic condition is reached, unless the deflection is somehow restricted.

Example 8.16. Fully Plastic Moment for a T-Beam. Determine the fully plastic moment for a beam made of 2024-T3 aluminum with the cross section shown in Figure 8.35(a).

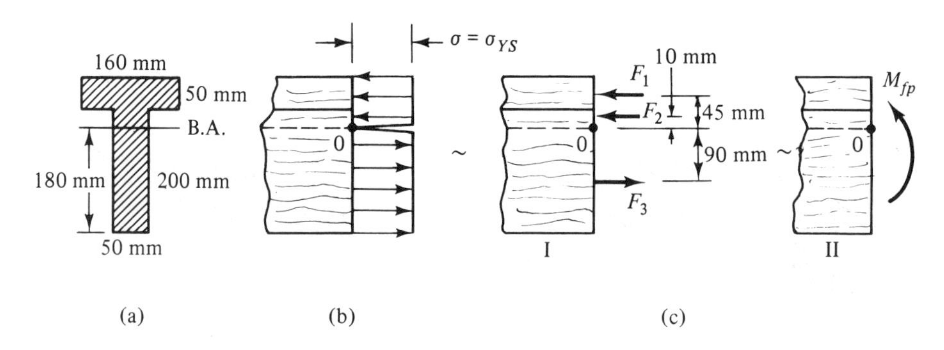

Figure 8.35

Solution. We first locate the bending axis, which divides the cross-sectional area in half in the fully plastic case. The total area is

$$A = (50\text{ mm})(160\text{ mm}) + (50\text{ mm})(200\text{ mm}) = 18\ 000\text{ mm}^2$$

Therefore, $A/2 = 9000\text{ mm}^2$. Thus, the bending axis is located 180 mm above the bottom of the section. The stress distribution is as shown in Figure 8.35(b). From Table A.2, the yield strength of the material is found to be 345 MN/m^2.

We now replace the stresses acting over each portion of the cross section by their resultants. This gives the three forces shown in Figure 8.35(c), where

$$F_1 = \sigma_{YS}A_1 = (345 \times 10^6\text{ N/m}^2)(0.008\text{ m}^2) = 2.76 \times 10^6\text{ N}$$

$$F_2 = \sigma_{YS}A_2 = (345 \times 10^6\text{ N/m}^2)(0.001\text{ m}^2) = 0.35 \times 10^6\text{ N}$$

$$F_3 = \sigma_{YS}A_3 = (345 \times 10^6\text{ N/m}^2)(0.009\text{ m}^2) = 3.11 \times 10^6\text{ N}$$

These three forces are statically equivalent to the fully plastic moment. Taking moments about the bending axis (point o), we have from the condition $(\Sigma M_o)_I = (\Sigma M_o)_{II}$

$$M_{fp} = (2.76 \times 10^6 \text{ N})(0.045 \text{ m}) + (0.35 \times 10^6 \text{ N})(0.010 \text{ m}) + (3.11 \times 10^6 \text{ N})(0.090 \text{ m})$$

$$= 408 \times 10^3 \text{N·m or } 408 \text{ kN·m} \qquad \textit{Answer}$$

Example 8.17. Failure Load for a Cantilever Beam. The cantilever beam shown in Figure 8.36(a) is made of structural steel with a yield strength of 36 ksi. What is the maximum load F that can be supported with a safety factor of 2 against failure by general yielding?

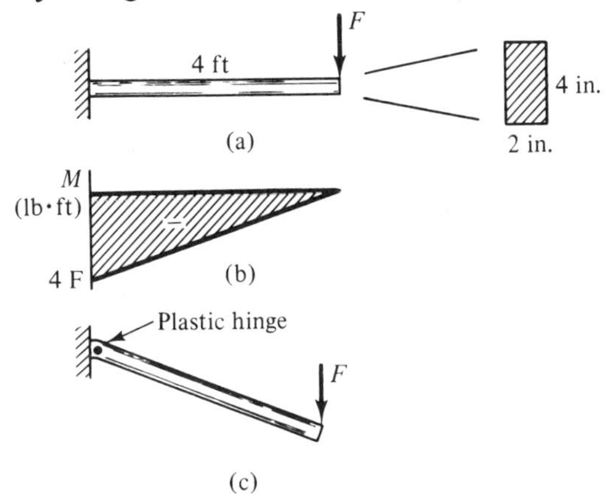

Figure 8.36

Solution. The beam will fail by general yielding once a fully plastic condition is reached somewhere along its length. From the moment diagram [Figure 8.36(b)], we see that the bending moment is largest at the wall. Thus, a plastic hinge will form there when the bending moment becomes equal to the fully plastic moment, allowing the beam to deflect approximately as indicated in Figure 8.36(c).

The fully plastic moment is computed from Eq. (8.19):

$$M_{fp} = \sigma_{YS}\left(\frac{bh^2}{4}\right) = \frac{(36 \times 10^3 \text{ lb/in.}^2)(2 \text{ in.})(4 \text{ in.})^3}{4} = 288 \times 10^3 \text{ lb·in.}$$

Setting this moment equal to the bending moment at the wall, we obtain for the failure load

$$48F \text{ lb·in.} = M_{fp} = 288 \times 10^3 \text{ lb·in.}$$

$$F = 6000 \text{ lb}$$

The working load is the failure load divided by the safety factor. Thus,

$$F_{\text{working}} = 3000 \text{ lb} \qquad \textit{Answer}$$

PROBLEMS

8.85 to 8.87 Determine the ratio of the fully plastic moment M_{fp} to the moment M_{YS} at which yielding first occurs for beams with the cross sections shown.

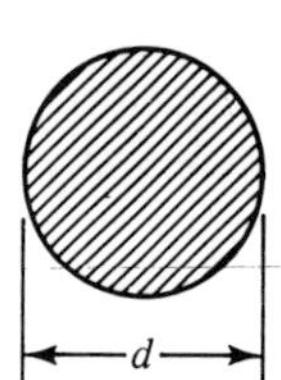

Problem 8.85

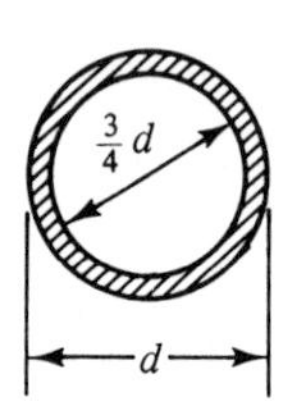

Problem 8.86

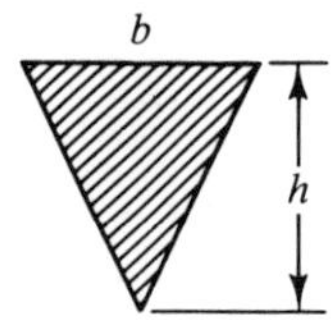

Problem 8.87

8.88 to 8.90 For the beams shown, determine the largest load that can be applied without causing collapse. The yield strength of the material and the required safety factor, if any, are as indicated in the figures.

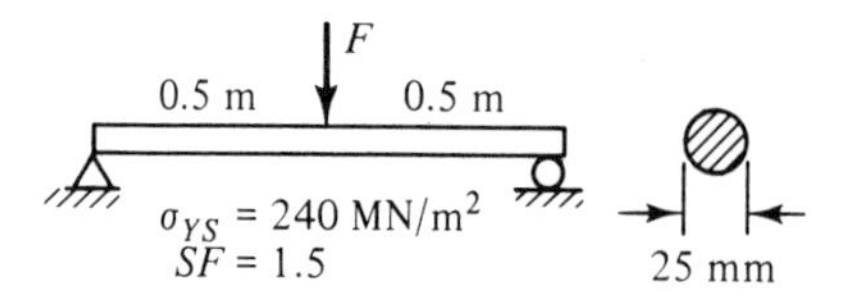

Problem 8.88

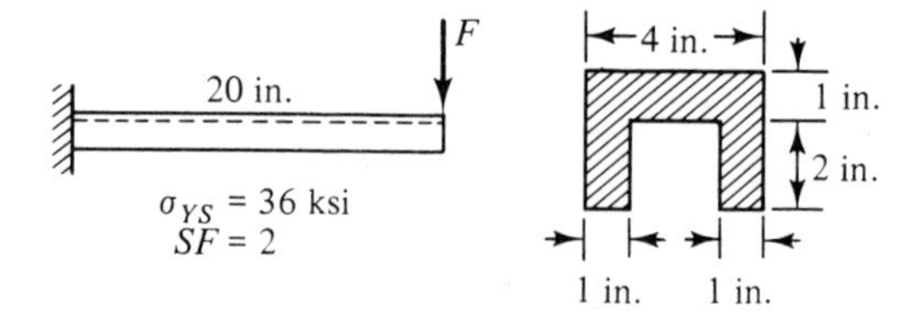

Problem 8.89

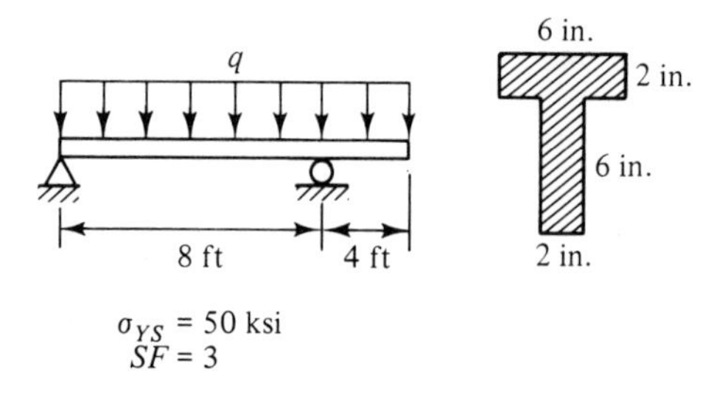

Problem 8.90

8.91 The beam shown is 30 mm thick and is made of steel with yield strength 250 MN/m². What load F will cause the beam to collapse if (a) $d = 0.5$ m and (b) $d = 0.3$ m?

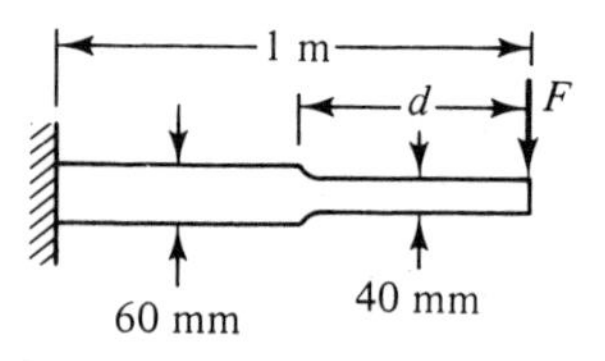

Problem 8.91

8.92 The beam shown is 40 mm wide by 120 mm deep. By what percentage does the presence of the semicircular grooves reduce the collapse load if (a) $d = L/4$ and (b) $d = 0.4L$? $\sigma_{YS} = 350$ MN/m².

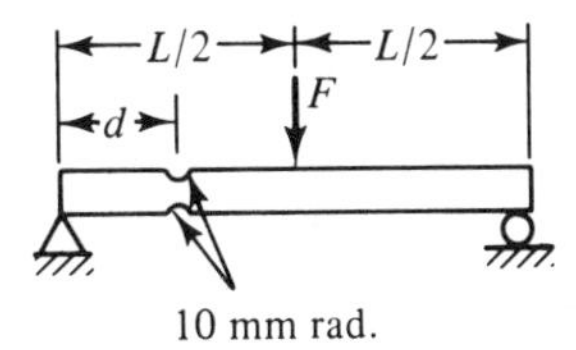

Problem 8.92

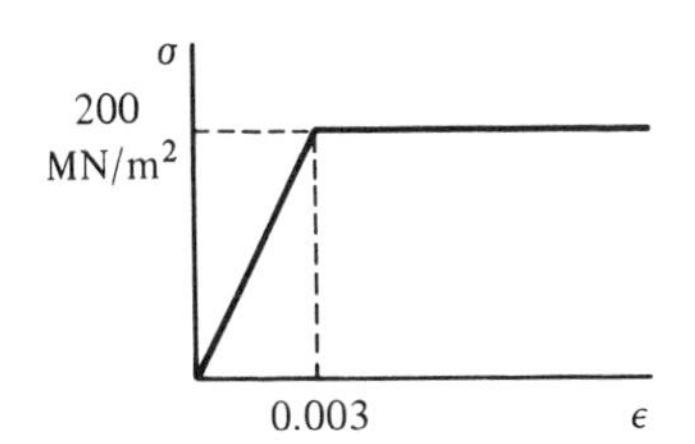

Problem 8.93

8.93 A 40-mm wide by 16-mm thick bar is bent by couples applied at the ends so that the radius of curvature of the neutral axis is 1 m. The stress-strain curve for the material (same in tension and compression) is as shown. Determine the magnitude C of the applied couples.

8.94 A strain gage mounted on the beam shown indicates that the longitudinal strain at midspan 1 in. above the bottom surface is $\varepsilon = 1600\mu$ in./in. If the material is mild steel with a yield point of 30 ksi, what is applied loading q?

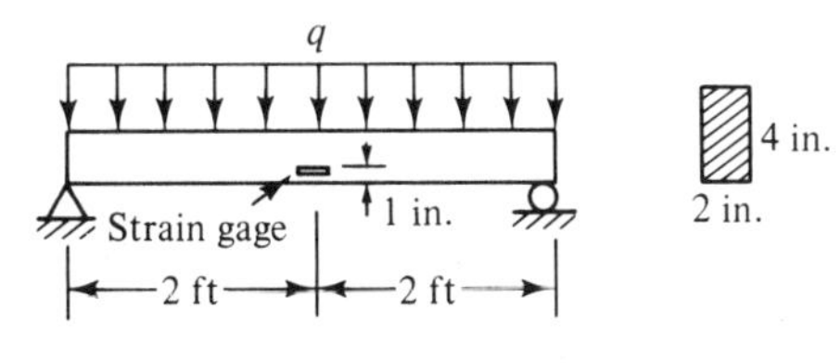

Problem 8.94

8.10 LIMIT LOADS

As indicated in the preceding section, a beam made of a ductile material usually fails by general yielding and the associated large deflections when a fully plastic condition is reached. If the member is statically indeterminate, however, the deflections are restricted by the additional support conditions, and further loading is possible.

To illustrate this, let us consider the example of a propped cantilever that supports a concentrated force F at midspan [Figure 8.37(a)]. The material behavior is assumed to be elastic-perfectly plastic. Now, plastic hinges form at the locations of maximum bending moment, which for a beam subjected to concentrated loads are always either at the loads or at the supports. Thus, plastic hinges can form at the wall and under the load.

An elastic analysis yields the moment diagram shown in Figure 8.37(b), which indicates that the largest bending moment occurs at the wall. Thus, the fully plastic condition will first be reached at this section. However, the formation of a plastic hinge at the wall does not lead to a collapse of the member because its deflection is restrained by the roller under the right end. Any further increases in load must be supported by sections away from the wall because the bending moment at the wall will remain equal to the fully plastic moment. Eventually, a second plastic hinge will form under the load, allowing the member to undergo large deflections. The

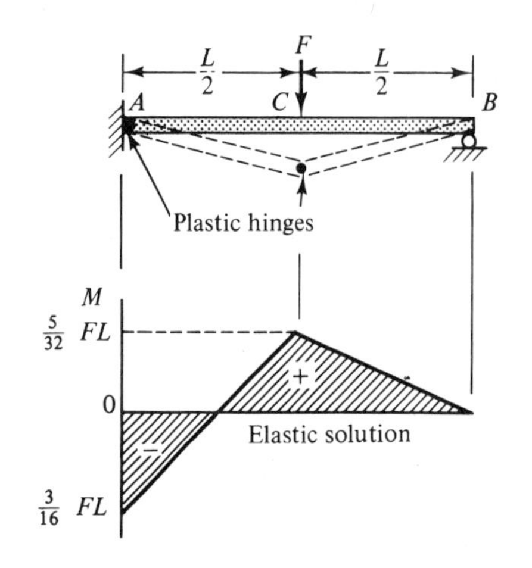

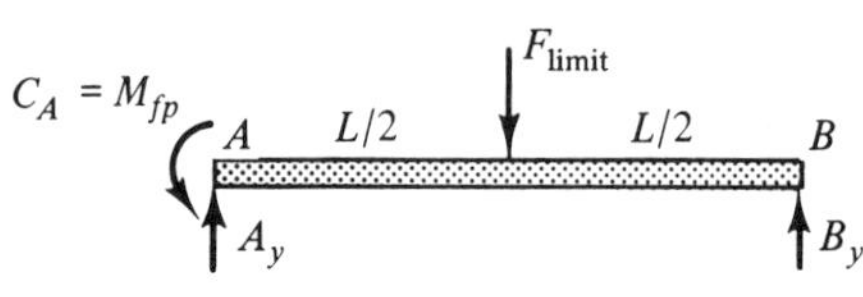

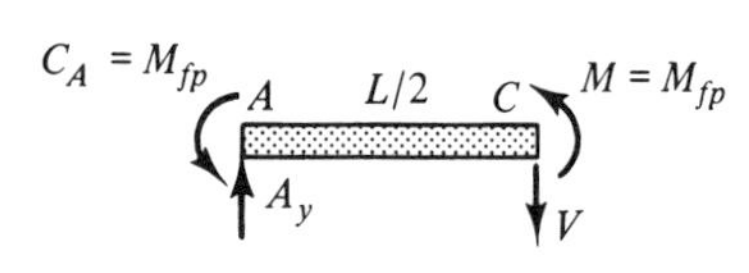

Figure 8.37

resulting shape of the beam is indicated by the dashed lines in Figure 8.37(a) and is called a *collapse mechanism*.

The limit load at which collapse occurs can be determined solely from equilibrium considerations because the bending moment at the plastic hinges is known to be equal to the fully plastic moment. This is evident from the free-body diagrams shown in Figure 8.37(c). The first diagram is for the entire beam and the second is for the portion obtained by sectioning the member just to the left of the applied load. There is one very important thing to note here—the moments at the plastic hinges are known quantities and, therefore, must be shown with the proper sense. The sense of these moments can usually be determined by inspection.

From the second diagram in Figure 8.37(c), we have

$$\stackrel{+}{\curvearrowleft}\Sigma M_C = M_{fp} + M_{fp} - A_y\left(\frac{L}{2}\right) = 0$$

and from the first diagram, we have

$$\stackrel{+}{\curvearrowleft}\Sigma M_B = F_{\text{limit}}\left(\frac{L}{2}\right) - A_y L + M_{fp} = 0$$

Solving these equations, we find that

$$F_{\text{limit}} = \frac{6}{L} M_{fp}$$

where M_{fp} is computed as explained in the preceding section.

This example points out the two steps involved in determining the limit load for beams and beam structures. One is an investigation of the geometry of the problem to determine the possible collapse mechanisms; the other is a force analysis of the beam and beam segments. If there is more than one possible collapse mechanism, the one that actually occurs is the one with the smallest limit load.

For distributed loads, the locations of the plastic hinges away from the supports are unknown. Furthermore, they cannot be determined from a moment diagram obtained from an elastic analysis of the deflections because the results become invalid once yielding occurs. However, we do know that the hinges form at points of maximum bending moment, which are also points of zero shear force for distributed loadings. (Recall from Section 8.3 that $dM/dx = V$.) This fact provides the additional information necessary to locate the plastic hinges. The procedure is illustrated in the following example.

Example 8.18. Limit Load for a Propped Cantilever Beam. Determine the limit load, q_{limit}, for the beam shown in Figure 8.38(a).

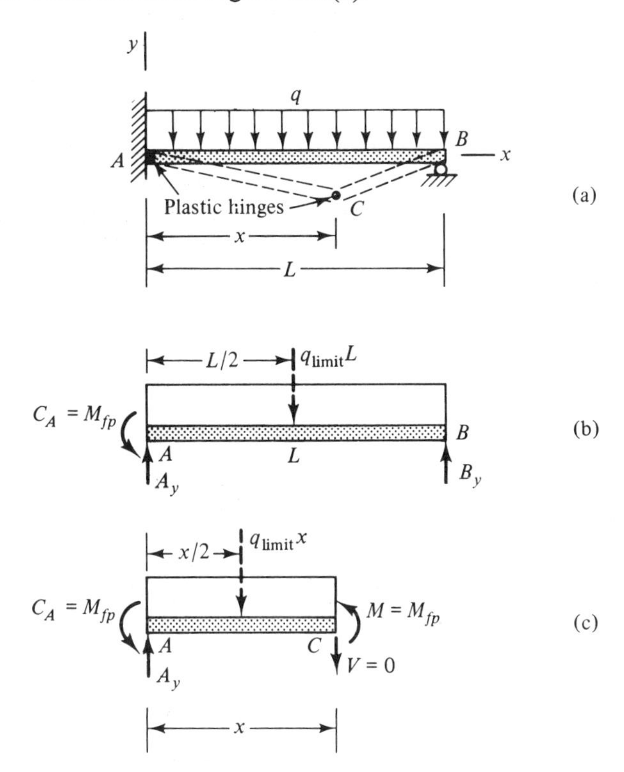

Figure 8.38

Solution. This problem is the same as that for the beam shown in Figure 8.37(a), except that the applied load is distributed instead of concentrated. The collapse mechanism will also be the same, but the location of the plastic hinge between the supports is unknown in this case.

Figure 8.38(b) shows the FBD for the entire member. Assuming that the second plastic hinge occurs at a distance x from the wall, and recalling that $V = 0$ and $M = M_{fp}$ at the hinge, we obtain the FBD shown in Figure 8.38(c) for the beam segment to the left of this section. From the first diagram, we have

$$\overset{+}{\curvearrowleft} \Sigma M_B = q_{\text{limit}} L\left(\frac{L}{2}\right) - A_y L + M_{fp} = 0$$

and from the second diagram, we have

$$\Sigma F_y = A_y - q_{\text{limit}} x = 0$$

$$\overset{+}{\curvearrowleft} \Sigma M_C = M_{fp} + q_{\text{limit}} x\left(\frac{x}{2}\right) - A_y x + M_{fp} = 0$$

Solving these equations, we get

$$x = 0.59L \qquad q_{\text{limit}} = 11.65 \frac{M_{fp}}{L^2} \qquad \textbf{\textit{Answer}}$$

where M_{fp} is the value of the fully plastic moment for this particular beam.

Example 8.19. Limit Load for a Beam Structure. The beam shown in Figure 8.39(a) is hinged at the left end and supported at the right by a vertical rod. Determine the limit load for this structure. Both members are made of the same material and have a yield strength of 30 ksi.

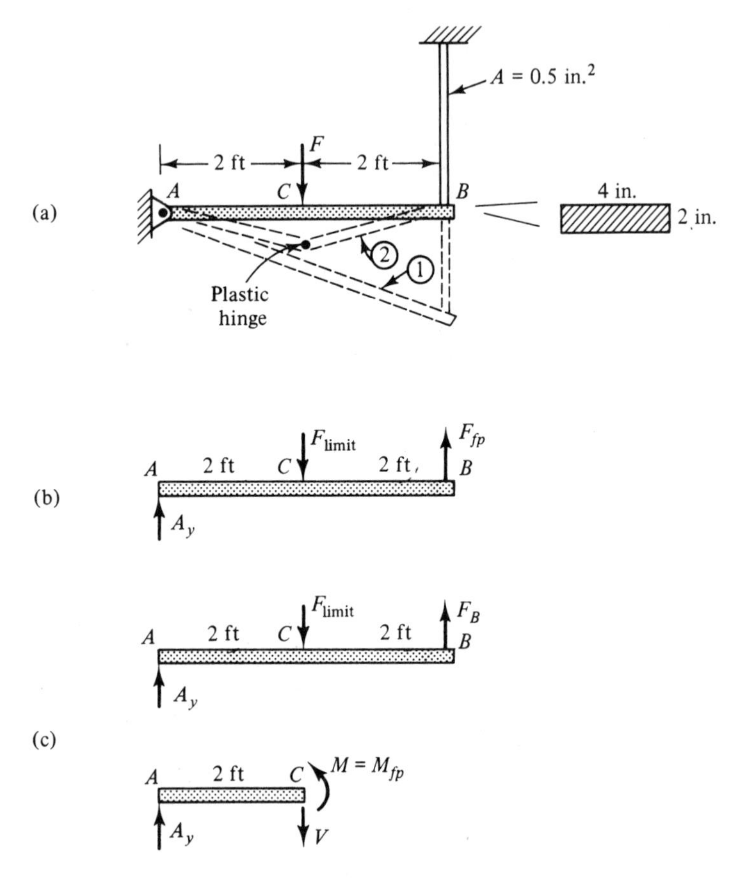

Figure 8.39

Solution. There are two possible collapse mechanisms, as indicated in Figure 8.39(a). The rod can yield, allowing the beam to rotate downward, or the rod can remain elastic with limited deformations and a plastic hinge can form under the load.

If the rod yields, the force on it will be

$$F_{fp} = \sigma_{YS} A_{\text{rod}} = (30\ 000\ \text{lb/in.}^2)(0.5\ \text{in.}^2) = 15\ 000\ \text{lb}$$

The FBD of the beam for this case is shown in Figure 8.39(b). We have

$$\overset{+}{\curvearrowleft}\Sigma M_A = -F_{\text{limit}}(2\ \text{ft}) + F_{fp}(4\ \text{ft}) = 0$$

$$F_{\text{limit}} = 2F_{fp} = 30\ 000\ \text{lb}$$

We now consider the second collapse mechanism. The FBDs of the beam and beam segment to the left of the load are shown in Figure 8.39(c). We have

$$\overset{+}{\curvearrowleft}\Sigma M_B = F_{\text{limit}}(2\ \text{ft}) - A_y(4\ \text{ft}) = 0$$

$$\overset{+}{\curvearrowleft}\Sigma M_C = M_{fp} - A_y(2\ \text{ft}) = 0$$

The fully plastic moment is determined from Eq. (8.19):

$$M_{fp} = \sigma_{YS}\frac{bh^2}{4} = \frac{(30\ 000\ \text{lb/in.}^2)(4\ \text{in.})(2\ \text{in.})^2}{4} = 120\ 000\ \text{lb·in.}$$

Solving these equations, we get

$$F_{\text{limit}} = 10\ 000\ \text{lb} \qquad \textbf{\textit{Answer}}$$

This is the actual limit load because it is the smaller of the two values. Thus, failure occurs due to formation of a plastic hinge under the load.

PROBLEMS

8.95 to 8.98 Identify the collapse mechanisms and determine the limit load for the beams or beam structures shown. The cross-sectional dimensions and yield strength of the material are as indicated; otherwise, express the results in terms of the fully plastic moment M_{fp} for the members.

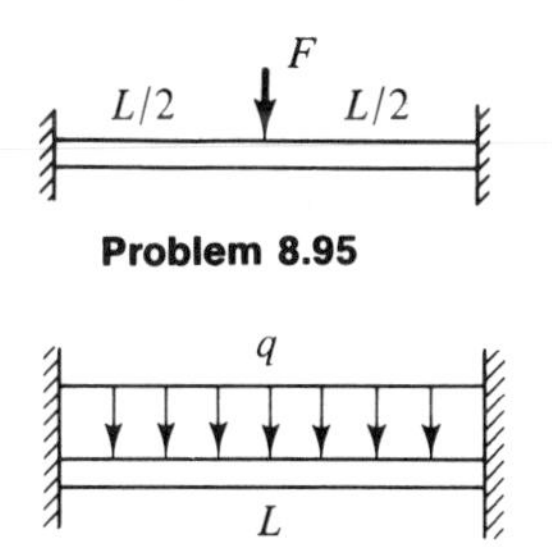

Problem 8.95

Problem 8.96

F F

4 in.

2 in.

2 ft 2 ft 2 ft 2 ft

σ_{YS} = 36 ksi

Problem 8.97

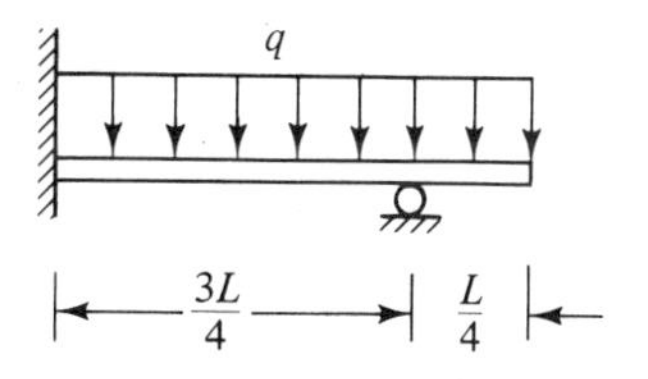

Problem 8.98

8.99 to 8.101 Identify the collapse mechanisms and determine the limit loads for the structures in Problems 8.79, 8.83, and 8.84 (Section 8.8). Express the results in terms of the fully plastic moment M_{fp} for the members.

8.102 Determine the couple C that will cause collapse of the structure in Problem 8.80 (Section 8.8). Assume that the beam and the rod have the same yield strength, σ_{YS}.

8.11 CLOSURE

In this chapter we considered the elastic and inelastic responses of beams, which are probably the most common type of load-carrying member. The analysis was more complicated than for axially loaded members or shafts, but for a very good reason. A review of the chapter shows that the loadings on beams are generally more complex and of greater variety than those for other members. Also, many different cross-sectional shapes were encountered. This is in contrast to shafts, where only circular members were considered.

Although we assumed that the beams are straight before loading, the results obtained also apply to members that are slightly curved. This is fortunate because there is no such thing as a perfectly straight beam. Under certain conditions, the results also apply to members with unsymmetrical cross sections. Furthermore, the basic concepts presented hold for composite members made of several different materials, the most common example of which is steel-reinforced concrete beams. However, a general treatment of composite members, or of members with significant initial curvature or unsymmetrical cross sections, is beyond the scope of this text. These topics are considered in more advanced works.

chapter nine COMBINED LOADINGS

9.1 INTRODUCTION

To this point, we have considered the response of members subjected to the separate effects of axial loadings, torsion, bending, and uniform pressure loadings. Load-stress relations were derived that provide the stresses acting upon certain planes, usually parallel and perpendicular to the longitudinal axis of the member, and equations for computing the deformations were developed.

Although these results enabled us to solve many meaningful problems, several important questions remain unanswered. For example, how does a member respond when several types of loading, say torsion and bending, act simultaneously, as they often do in practice? How does one determine the maximum stresses in the member under such conditions, or how does one determine the stresses on planes other than those for which the basic equations apply? These and other related questions will be considered in this chapter.

Primary emphasis will be placed upon the stresses, as opposed to the deformations, since they are usually of the most concern in combined loading problems. For the most part, we shall consider problems involving only linearly elastic material behavior. An analysis of the inelastic response of members subjected to combined loadings is considerably more complicated than that for a single type of loading and is beyond the scope of an introductory text such as this. Nevertheless, some of the basic results to be presented are independent of the material properties, and, therefore, apply in both the elastic and inelastic cases. These will be pointed out where appropriate.

9.2 STRESSES DUE TO COMBINED LOADINGS

As long as the deformations are small and the material behavior is linearly elastic, the stresses in a member will be proportional to the applied loads. In this case, the stresses due to several different loadings acting simultaneously can be determined by using the principle of superposition (see Section 8.8). The stresses due to the combined loading is simply the combination of those due to each of the individual loadings. Note that this implies that the presence of one loading does not affect the stresses due to another.

It will be convenient to show the various stresses present at a given point in a body as acting upon an infinitesimal element of material surrounding the point, as we have done on several previous occasions. Recall from Section 6.8 that this procedure merely provides a means of displaying the stresses acting upon mutually perpendicular planes through the particular point of interest.

To illustrate the preceding ideas, let us consider a pressurized thin-walled cylinder that is also twisted and subjected to an axial load [Figure 9.1(a)]. The stresses on planes parallel and perpendicular to the longitudinal axis due to each of the loadings can be computed by using the load-stress relations derived in preceding chapters. These stresses are the longitudinal and circumferential normal stresses due to the internal pressure, the longitudinal normal stress due to the axial load, and the shear stresses due to the applied couple. Figure 9.1(b) shows these stresses acting upon an element of material taken from the outside of the cylinder wall. The

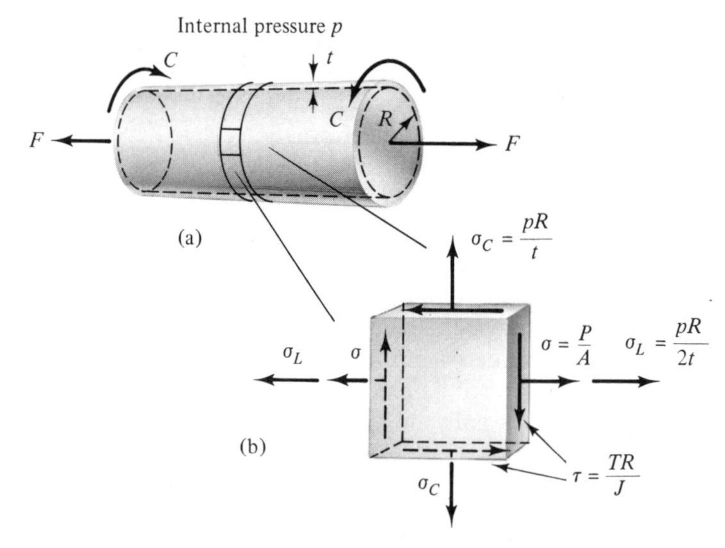

Figure 9.1

material behavior is assumed to be linearly elastic so that the torque-stress relation $\tau = T\rho/J$ applies.

The stresses acting upon a material element define what is known as the *state of stress* at that point in the body. The situation shown in Figure 9.1(b) is referred to as a *biaxial* state of stress because there are stresses acting in two directions. If there are stresses acting in one direction only, as in an axially loaded rod, the state of stress is said to be *uniaxial*. Both of these cases fall into a more general category known as a state of *plane stress*, wherein all the stresses acting lie in one plane. Most of the problems we shall encounter fall into the plane stress category.

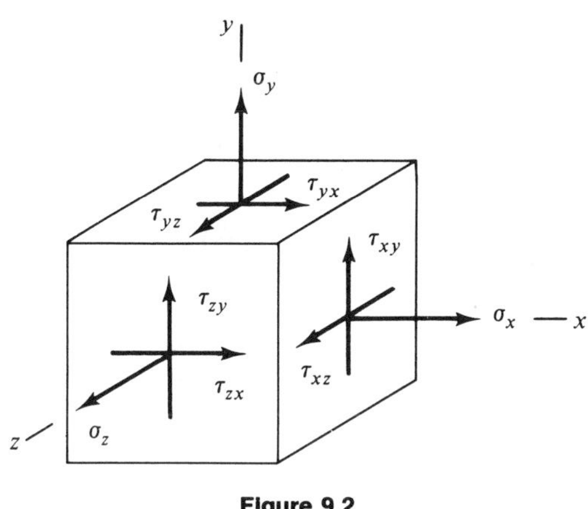

Figure 9.2

The most general situation possible involves a *triaxial* state of stress, wherein there are stresses on the element acting in all three directions (Figure 9.2). Subscripts are used to distinguish between the various stresses. For example, σ_x, σ_y, and σ_z denote the normal stresses on planes perpendicular to the x, y, and z axes, respectively. For brevity, we shall refer to these planes as the x, y, and z faces of the element. The first subscript on the shear stress denotes the face of the element upon which it acts and the second subscript indicates the direction of the stress. Thus, τ_{xy} and τ_{xz} are the shear stresses acting upon the x face in the y and z directions, and so forth. As in the preceding chapters, normal stresses are considered positive if they are tensile and negative if they are compressive. The positive directions for the shear stresses are as indicated in Figure 9.2.

In Figure 9.2 the stress components are shown on only three of the six faces of the element. It is assumed that components of stress of equal magnitude, but opposite sense, exist on the other three faces. Actually, the magnitudes of the stresses may, and usually do, vary slightly from one face of the element to another. However, these slight variations can be neglected for the types of problems considered in this text.

Since the shear stresses acting around an element in any one given plane must be equal in magnitude (see Section 7.4), we have from Figure 9.2

$$\tau_{yx} = \tau_{xy} \qquad \tau_{zx} = \tau_{xz} \qquad \tau_{zy} = \tau_{yz}$$

Thus, six components of stress (σ_x, σ_y, σ_z, τ_{xy}, τ_{xz}, τ_{yz}) are required to completely define a general state of stress. For plane stress, say in the x-y plane, only three components are required (σ_x, σ_y, τ_{xy}). In this case, a two-dimensional view of the element is sufficient to display the stresses (Figure 9.3).

The question now arises as to how the various stresses acting upon a material element combine. In general, special procedures are required because the stresses not only have magnitude and direction, but they also depend upon the orientation

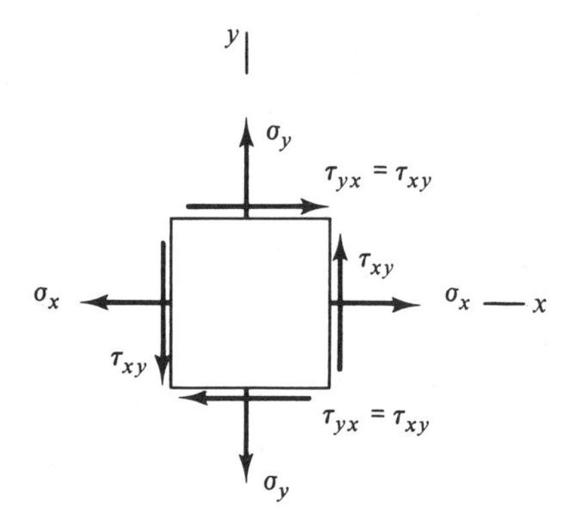

Figure 9.3

of the plane upon which they act. These procedures will be considered in the following sections. The one exception, which we shall consider in this section, involves like kinds of stresses (normal or shear) acting upon the same plane and in the same direction. In this case, the stresses can be combined algebraically. For example, the two normal stresses acting in the longitudinal direction in Figure 9.1(b) can be added to obtain the total normal stress in that direction.

Problems involving combined bending and axial loading or bending about two axes are probably the most common examples of situations in which the stresses can be added directly. Another common example involves the torsional and direct shear stresses in members subjected to simultaneous bending and torsion. The determination of the stresses in cases such as these involves no new principles, as can be seen from the following examples.

Example 9.1. Stresses in a Linkage. Determine the distribution of stress over the cross section in the central portion of the machine linkage shown in Figure 9.4(a). The member is 20 mm thick and is made of an alloy steel with a yield strength of 690 MN/m^2.

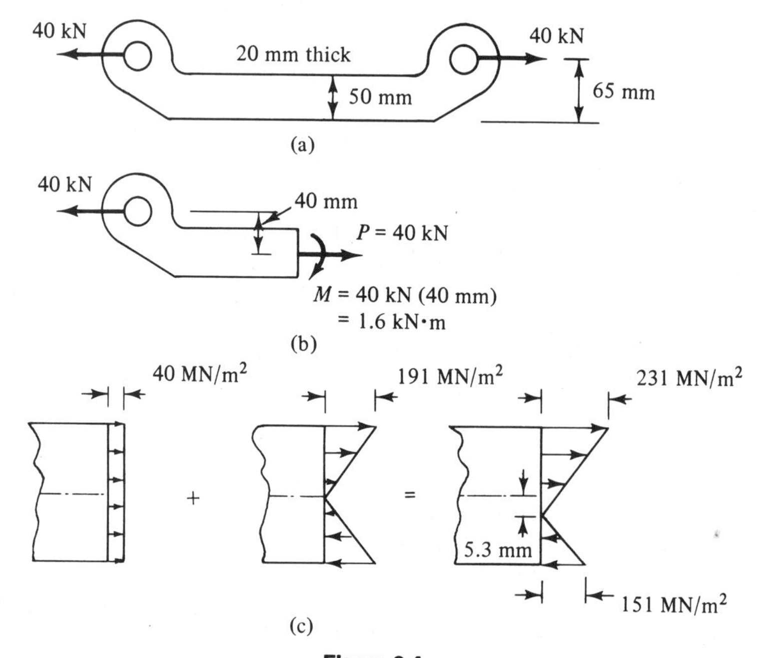

Figure 9.4

Solution. We first section the member and determine the stress resultants, as indicated in Figure 9.4(b). Note that the member is subjected to both axial loading and bending. The bending arises because of the offset between the line of action of the applied loads and the centroidal axis of the member. The bending moment and normal force are constant throughout the central portion of the linkage.

The stress due to the axial loading is

$$\sigma = \frac{P}{A} = \frac{40 \times 10^3 \text{ N}}{(0.05 \text{ m})(0.02 \text{ m})} = 40 \times 10^6 \text{ N/m}^2 \text{ or } 40 \text{ MN/m}^2$$

The moment of inertia about the bending axis is

$$I = \frac{1}{12} bh^3 = \frac{1}{12} (0.02 \text{ m})(0.05 \text{ m})^3 = 2.1 \times 10^{-7} \text{ m}^4$$

therefore, the maximum bending stress is

$$\sigma = \frac{My_{\text{max}}}{I} = \frac{(1600 \text{ N·m})(0.025 \text{ m})}{2.1 \times 10^{-7} \text{ m}^4} = 191 \times 10^6 \text{ N/m}^2 \text{ or } 191 \text{ MN/m}^2$$

Since both stresses act upon the same plane and in the same direction, they can be added directly [Figure 9.4(c)]. The maximum stress is below the yield strength of the material; therefore, all equations used are valid.

Note that the stresses still vary linearly over the depth of the cross section. The location of the level of zero stress can be obtained by setting the expression for the total stress equal to zero and solving for the corresponding value of y:

$$\sigma_{\text{total}} = \frac{P}{A} + \frac{My}{I} = 0$$

$$y = -\frac{PI}{MA} = \frac{-(40 \times 10^3 \text{ N})(2.1 \times 10^{-7} \text{ m}^4)}{(1600 \text{ N·m})(10 \times 10^{-4} \text{ m}^4)} = -5.3 \times 10^{-3} \text{ m or } -5.3 \text{ mm}$$

Thus, the level of zero stress lies 5.3 mm below the centroid of the cross section. Its location can also be determined from Figure 9.4(c) by using similar triangles. The shift of the level of zero stress away from the centroid of the cross section is due to the presence of the normal force.

Example 9.2. Bending About Two Axes. A steel cantilever beam made of a W 8 × 31 shape is loaded as shown in Figure 9.5(a). Determine the maximum tensile and compressive stresses in the member.

Solution. Resolving the applied load into components [Figure 9.5(b)], we see that the vertical component will produce bending with xx as the bending axis and the horizontal component will produce bending with yy as the bending axis. Thus, the member is subjected to bending about two axes. The maximum bending moments occur at the wall and have the values indicated in Figure 9.5(c). The geometric properties of the cross section obtained from Table A.3 of the Appendix are listed in Figure 9.5(b).

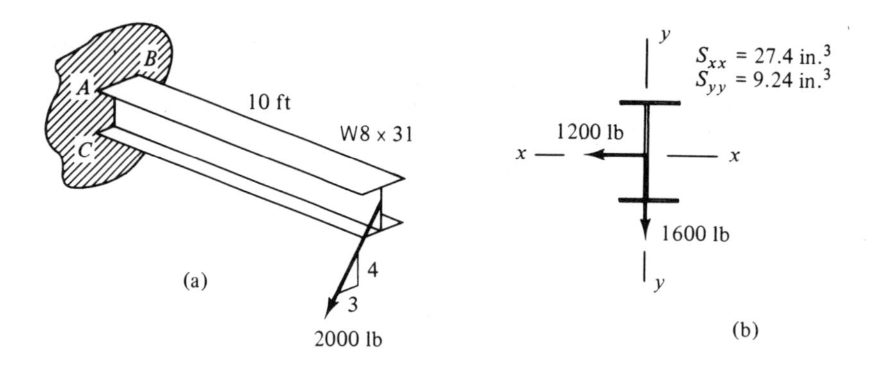

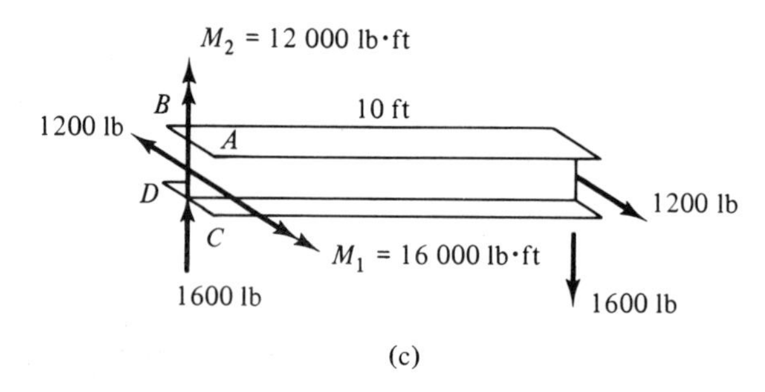

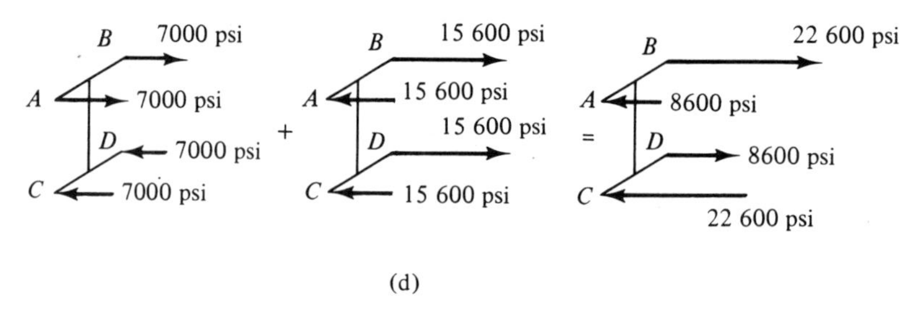

Figure 9.5

Since the maximum bending stresses occur at the extremities of the cross section, the maximum tensile and compressive stresses will occur at one of the four corners of the cross section at the wall. Thus, we need consider only these four points. The maximum stress due to the bending moment M_1 is

$$\sigma = \frac{M_1}{S_{xx}} = \frac{(16\ 000 \text{ lb}\cdot\text{ft})(12 \text{ in./ft})}{27.4 \text{ in.}^3} = 7000 \text{ psi}$$

and is tensile at the top of the cross section and compressive at the bottom. The moment M_2 produces a tensile stress at points B and D and a compressive stress at points A and C of magnitude

$$\sigma = \frac{M_2}{S_{yy}} = \frac{(12\ 000 \text{ lb}\cdot\text{ft})(12 \text{ in./ft})}{9.24 \text{ in.}^3} = 15\ 600 \text{ psi}$$

Since both sets of stresses act upon the same plane and in the same direction, they can be added directly [Figure 9.5(d)]. The maximum tensile and compressive stresses are equal in magnitude and occur at points B and C, respectively:

$$(\sigma_t)_{\max} = (\sigma_c)_{\max} = 22\ 600 \text{ psi} \qquad \textbf{\textit{Answer}}$$

Since the maximum stresses are below the yield strength of steel (see Table A.2), our results are valid.

Example 9.3. State of Stress In a Bar. Determine the state of stress at points A, B, C, and D in the member shown in Figure 9.6(a). Assume linearly elastic material behavior.

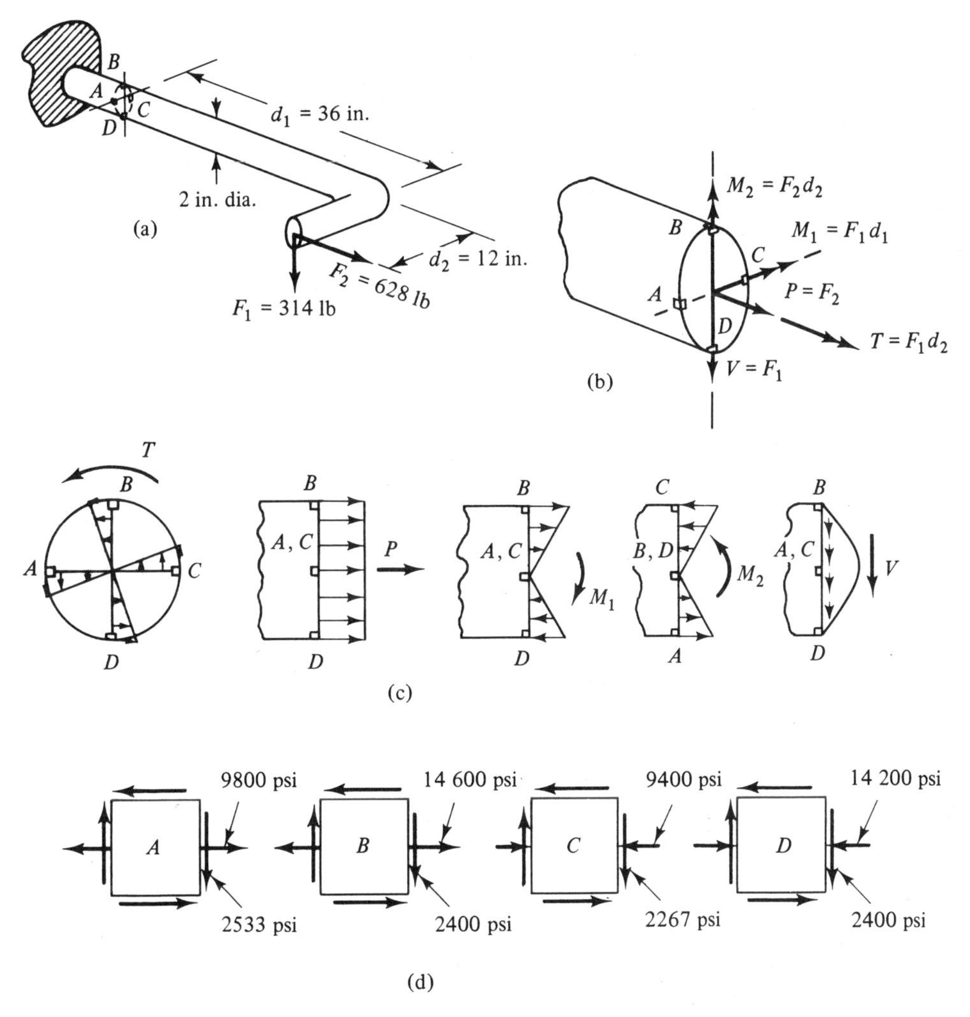

Figure 9.6

Solution. We first determine the stress resultants at the cross section of interest. The details of these computations will not be given here; the results are shown in Figure 9.6(b). We next consider the stresses due to the various loadings. These stresses are shown in Figure 9.6(c).

The area, moment of inertia about the bending axes, polar moment of inertia, and section modulus of the cross section are, respectively,

$$A = \frac{\pi D^2}{4} = \pi \text{ in.}^2 \qquad J = \frac{\pi D^4}{32} = \frac{\pi}{2} \text{ in.}^4$$

$$I = \frac{\pi D^4}{64} = \frac{\pi}{4} \text{ in.}^4 \quad S = \frac{I}{y_{\text{max}}} = \frac{\pi}{4} \text{ in.}^3$$

The torsional shear stress at the outside of the member is

$$\tau = \frac{TR}{J} = \frac{(12 \text{ in.})(314 \text{ lb})(1 \text{ in.})}{\pi/2 \text{ in.}^3} = 2400 \text{ psi}$$

and the maximum bending stresses due to the moments M_1 and M_2 are

$$\sigma = \frac{M_1}{S} = \frac{(36 \text{ in.})(314 \text{ lb})}{\pi/4 \text{ in.}^3} = 14\ 400 \text{ psi}$$

and

$$\sigma = \frac{M_2}{S} = \frac{(12 \text{ in.})(623 \text{ lb})}{\pi/4 \text{ in.}^3} = 9600 \text{ psi}$$

The normal stress due to the axial load is

$$\sigma = \frac{P}{A} = \frac{628 \text{ lb}}{\pi \text{ in.}^2} = 200 \text{ psi}$$

The direct shear stress is zero at points B and D. At points A and C it is

$$\tau = \frac{VA'\bar{y}'}{It} = \frac{(314 \text{ lb})(\pi/2 \text{ in.}^2)(4/3\pi \text{ in.})}{(\pi/4 \text{ in.}^4)(2 \text{ in.})} = 133 \text{ psi}$$

Referring to Figure 9.6(c) and noting the stresses acting at the various locations, we obtain the states of stress shown in Figure 9.6(d). The stresses at point C are as seen from the back side of the member, and those at D are as seen from below.

PROBLEMS

9.1 Determine the maximum tensile and compressive stresses in the link shown if $w = 60$ mm. The link is made of steel and is 30 mm thick.

9.2 A 3-in. diameter elastic hook supports a 6-kip load as shown. Determine the maximum tensile and compressive stresses in the hook and indicate where they occur.

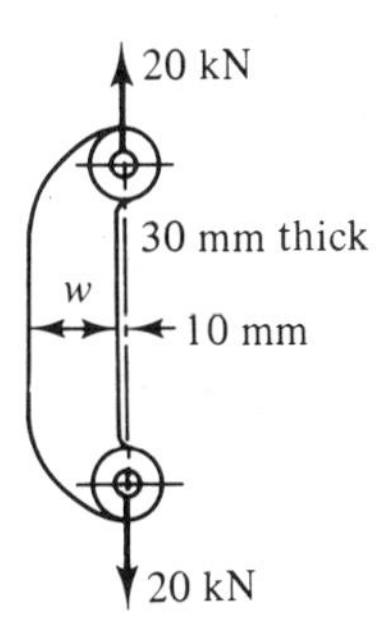

Problem 9.1

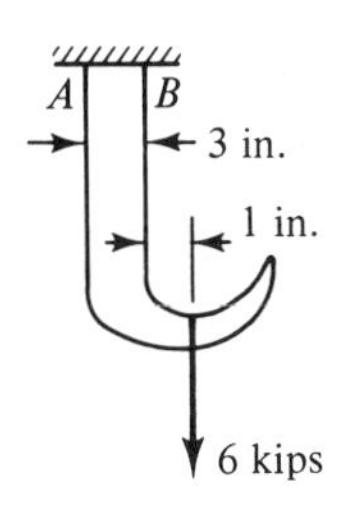

Problem 9.2

9.3 Determine the distribution of stress over the cross section at section a-a of the $\frac{1}{2}$-in. thick bar shown if $d = 3$ in.

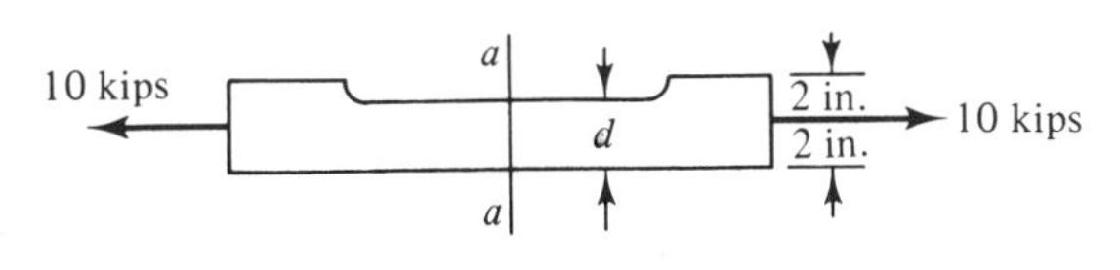

Problem 9.3

9.4 If the link in Problem 9.1 is made of cast iron with an ultimate strength of 140 MN/m^2 in tension and 520 MN/m^2 in compression, what is the required width w? Use a safety factor of 2.

9.5 If the bar in Problem 9.3 is made of 6061-T6 aluminum, what is the minimum allowable dimension d if the member is not to yield under the applied loading? Neglect stress concentrations.

9.6 What is the largest force F the 254-mm × 102-mm Universal Beam shown can support without yielding if it is made of steel with a yield point of 690 MN/m^2? The mass of the beam, which is 22 kg/m, can be neglected.

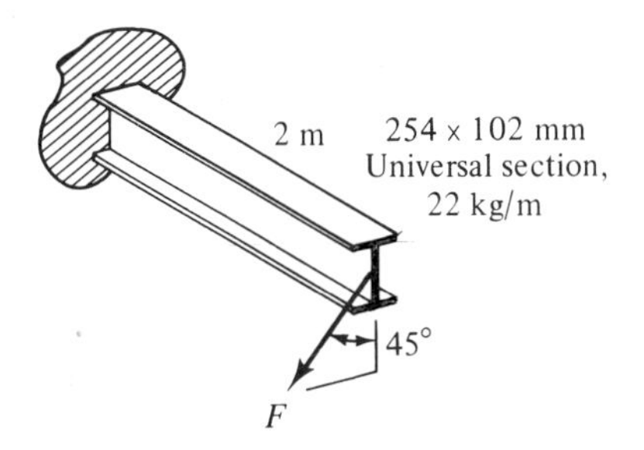

Problem 9.6

9.7 If the beam in Problem 9.6 is replaced with a 3 ft long 2-in.×6-in. nominal timber oriented with the larger dimension vertical, determine the magnitude and location of the maximum tensile stress for $F = 283$ lb.

9.8 What is the largest force F that can be supported by the block shown if the maximum allowable tensile stress is 10 000 psi?

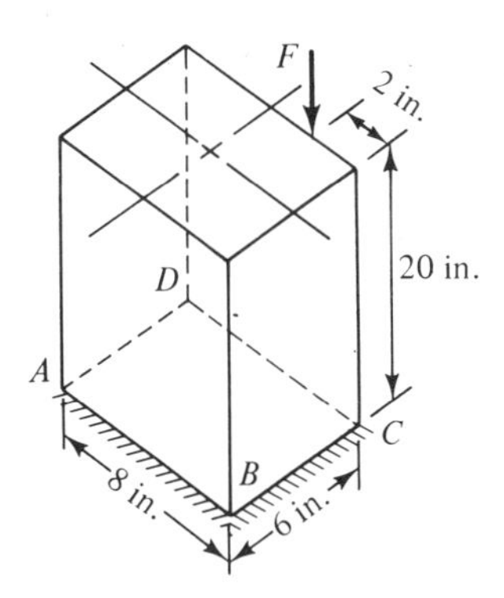

Problem 9.8

9.9 Determine the normal stresses at corners A, B, C, and D of the elastic block shown in Problem 9.8 if $F = 9600$ lb.

9.10 A rectangular brick pier, although strong in compression, has negligible strength in tension. Over what region R about the center of the pier can a downward vertical force F act without producing tensile stresses in the member? Neglect the weight of the pier; the cross-sectional dimensions are a and b.

9.11 Member ABC of the structure shown is made of a 76-mm × 38-mm steel channel section with a yield strength of 690 MN/m²? What is the safety factor against failure by the onset of yielding?

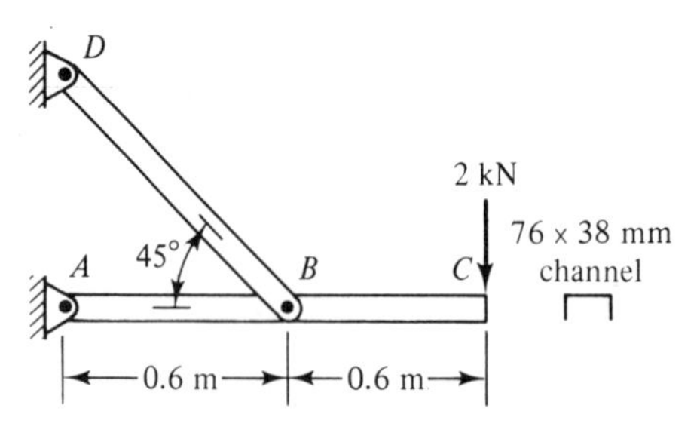

Problem 9.11

9.12 What is the largest force F that can be supported by the assembly shown if the beam is made of grey cast iron? Use a safety factor of 3. The pulley is frictionless, and all members have negligible weight.

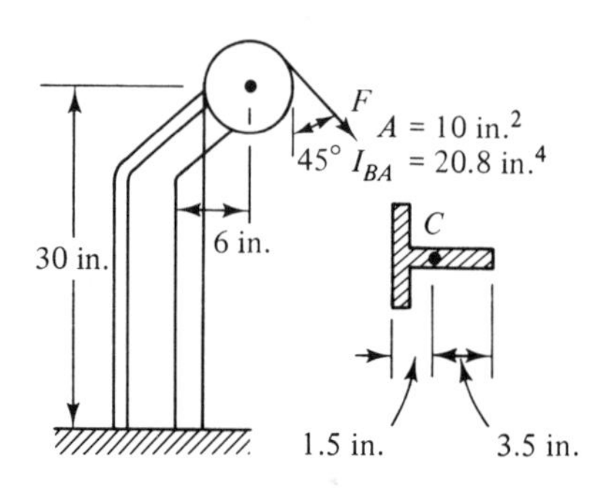

Problem 9.12

9.13 Experiments show that the thin-walled cylinder shown will buckle if the longitudinal compressive stress exceeds 5000 psi. If the internal pressure is 300 psi, what is the largest force F that can be applied as shown without buckling the cylinder?

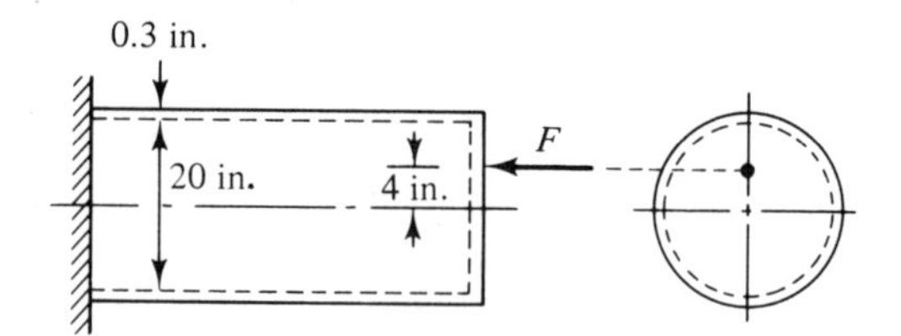

Problem 9.13

9.14 to 9.16 For the members shown, determine the state of stress at the point(s) indicated and show the results on small material elements. Neglect the weights of the members.

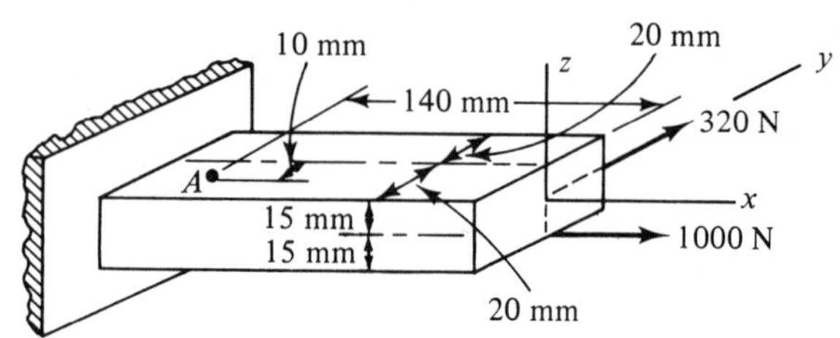

Problem 9.14

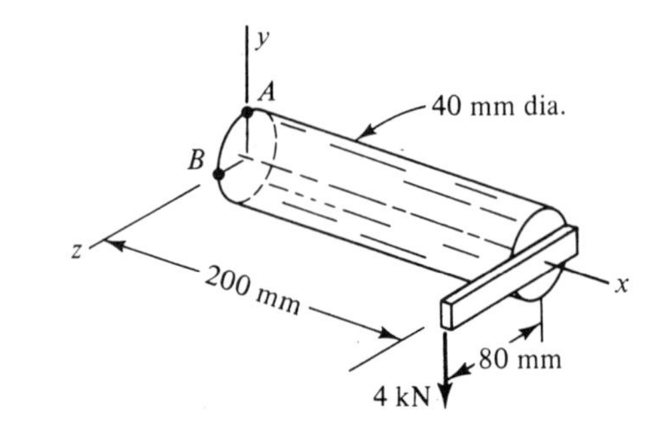

Problem 9.15

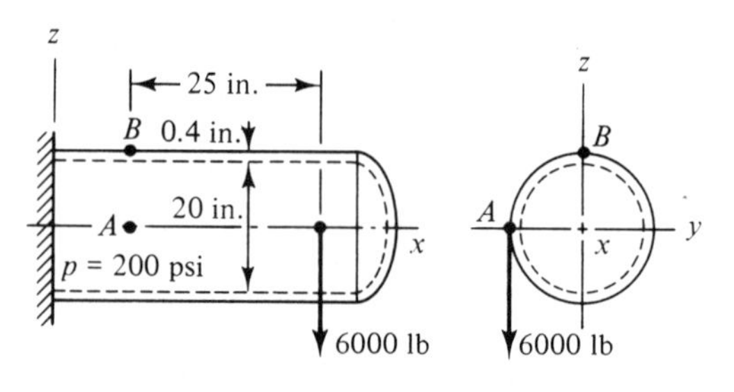

Problem 9.16

9.3 STRESSES ON AN INCLINED PLANE

Once the state of stress at a point in a body is known, the stresses on any other plane through that point can be determined. The procedure for doing this for the case of plane stress will be discussed in this section.

Stress transformation equations. Figure 9.7(a) shows the stress components with respect to xy axes acting at a point in a body subjected to a general state of plane stress. To determine the stresses on some other plane inclined to the coordinate axes, such as plane PQ, we section the element along the plane of interest and consider the equilibrium of one of the resulting pieces. This is the same procedure used in Section 7.4 to determine the stresses on planes inclined to the axis of a shaft. Once again, we note that we are dealing with infinitesimal material elements, so that any plane through the element actually

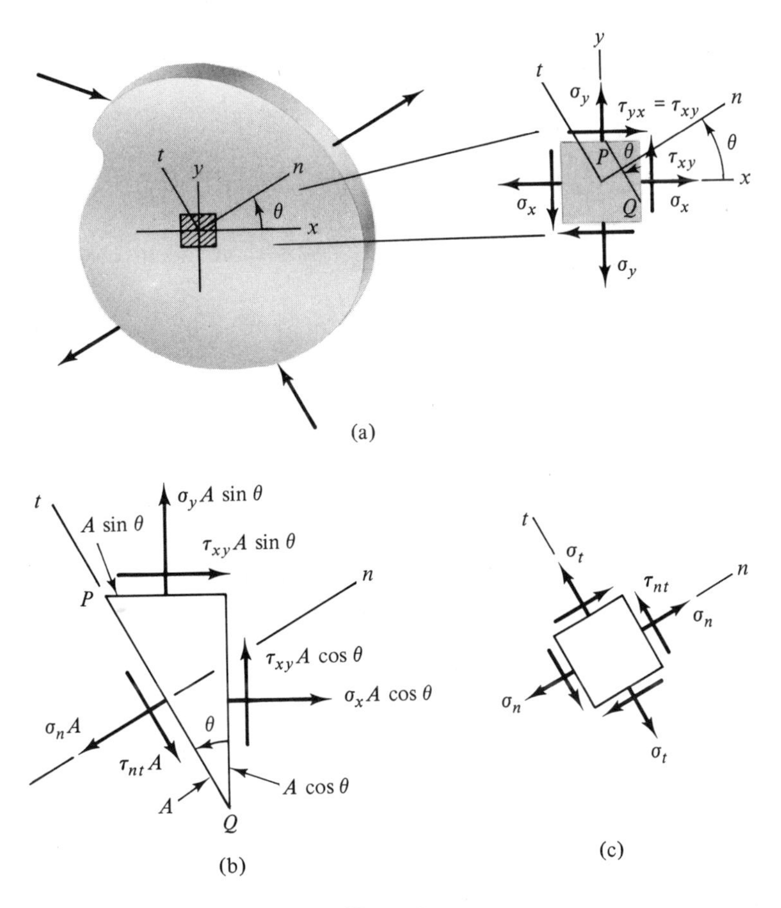

Figure 9.7

passes through the corresponding point in the body; the elements are shown enlarged for clarity only.

Sectioning the element shown in Figure 9.7(a) along plane PQ and converting the stresses to forces by multiplying by the respective areas over which they act, we obtain the free-body diagram shown in Figure 9.7(b). Here, n and t are axes perpendicular and parallel to the inclined plane, σ_n and τ_{nt} are the normal and shear stresses acting upon this plane, A is the area over which these stresses act, and θ is the angle through which the nt axes are rotated with respect to the xy axes, measured counterclockwise [Figure 9.7(a)].

The equations of equilibrium are

$$\begin{aligned}\Sigma F_n &= -\sigma_n A + (\sigma_x A \cos\theta)\cos\theta + (\tau_{xy} A \cos\theta)\sin\theta \\ &\quad + (\tau_{xy} A \sin\theta)\cos\theta + (\sigma_y A \sin\theta)\sin\theta = 0 \\ \Sigma F_t &= -\tau_{nt} A - (\sigma_x A \cos\theta)\sin\theta + (\tau_{xy} A \cos\theta)\cos\theta \\ &\quad - (\tau_{xy} A \sin\theta)\sin\theta + (\sigma_y A \sin\theta)\cos\theta = 0\end{aligned}$$

from which we obtain

$$\sigma_n = \sigma_x \cos^2\theta + \sigma_y \sin^2\theta + 2\tau_{xy}\sin\theta\cos\theta$$

and

$$\tau_{nt} = -(\sigma_x - \sigma_y)\sin\theta\cos\theta + \tau_{xy}(\cos^2\theta - \sin^2\theta)$$

These expressions can be put into a more convenient form by introducing the trigonometric identities

$$\begin{aligned}\cos^2\theta &= \frac{1}{2} + \frac{1}{2}\cos 2\theta \\ \sin^2\theta &= \frac{1}{2} - \frac{1}{2}\cos 2\theta \\ \sin\theta\cos\theta &= \frac{1}{2}\sin 2\theta\end{aligned}$$

We have

$$\sigma_n = \frac{\sigma_x + \sigma_y}{2} + \frac{\sigma_x - \sigma_y}{2}\cos 2\theta + \tau_{xy}\sin 2\theta \qquad (9.1)$$

$$\tau_{nt} = -\frac{(\sigma_x - \sigma_y)}{2}\sin 2\theta + \tau_{xy}\cos 2\theta \qquad (9.2)$$

These equations are known as the *stress transformation equations*. Given the values of the stress components with respect to xy axes, they determine the values with respect to the rotated axes nt. In other words, the transformation equations determine the stresses acting on a rotated element with sides aligned with the n and t axes [Figure 9.7(c)]. The normal stress, σ_t, on the t face of the rotated element is determined from Eq. (9.1) by replacing the value of θ by $\theta + \pi/2$.

It is of interest to note that Eqs. (5.3) and (5.4) for the normal and shear stresses on a plane inclined to the axis of a uniaxially loaded member are special cases of Eqs. (9.1) and (9.2) with $\sigma_y = \tau_{xy} = 0$. Similarly, Eqs. (7.11) and (7.12) for the stresses on planes inclined to the axis of a shaft are special cases of these equations with $\sigma_x = \sigma_y = 0$.

Considerable care must be taken to avoid computational errors when using the stress transformation equations. The signs on all quantities and the manner in which the angle θ is measured must be consistent with the conventions used in the derivations. However, these difficulties can be largely avoided by using a geometrical interpretation of these equations, which we shall consider in the following.

Mohr's circle of stress. If we transpose the first term on the right-hand side of Eq. (9.1) to the left side and then square both it and Eq. (9.2), we obtain

$$\left[\sigma_n - \left(\frac{\sigma_x + \sigma_y}{2}\right)\right]^2 = \left[\left(\frac{\sigma_x - \sigma_y}{2}\right)\cos 2\theta + \tau_{xy}\sin 2\theta\right]^2$$

$$\tau_{nt}^2 = \left[-\left(\frac{\sigma_x - \sigma_y}{2}\right)\sin 2\theta + \tau_{xy}\cos 2\theta\right]^2$$

Expanding the right-hand sides of these expressions and then adding them, we have

$$\left[\sigma_n - \left(\frac{\sigma_x + \sigma_y}{2}\right)\right]^2 + \tau_{nt}^2 = \left(\frac{\sigma_x - \sigma_y}{2}\right)^2 + \tau_{xy}^2 \tag{9.3}$$

This equation defines a circle in a plane with σ_n and τ_{nt} as coordinates, as can be seen by comparison with the general equation of a circle in rectangular coordinates

$$(x - a)^2 + (y - b)^2 = R^2 \tag{9.4}$$

This geometrical interpretation of the stress transformation equations is attributed to the German professor of structural mechanics, Otto Mohr (1835–1918), and the resulting circle is called *Mohr's circle of stress*.

The significance of Mohr's circle lies in the fact that every point on it represents the stresses on some plane through the corresponding element. Thus, to determine the stresses on any given plane through the element, or with respect to any rotated set of axes, we need only find the coordinates of the corresponding points on the circle. As we shall see, this can be done by using only simple geometry.

Given a state of plane stress with respect to xy axes [Figure 9.8(a)], the step-by-step procedure for constructing Mohr's circle and for determining the stresses with respect to a rotated set of axes, or on any plane inclined to the xy axes, is as follows. These steps are illustrated in Figure 9.8(b):

1. Set up a coordinate system with normal stress as the horizontal coordinate and shear stress as the vertical coordinate.

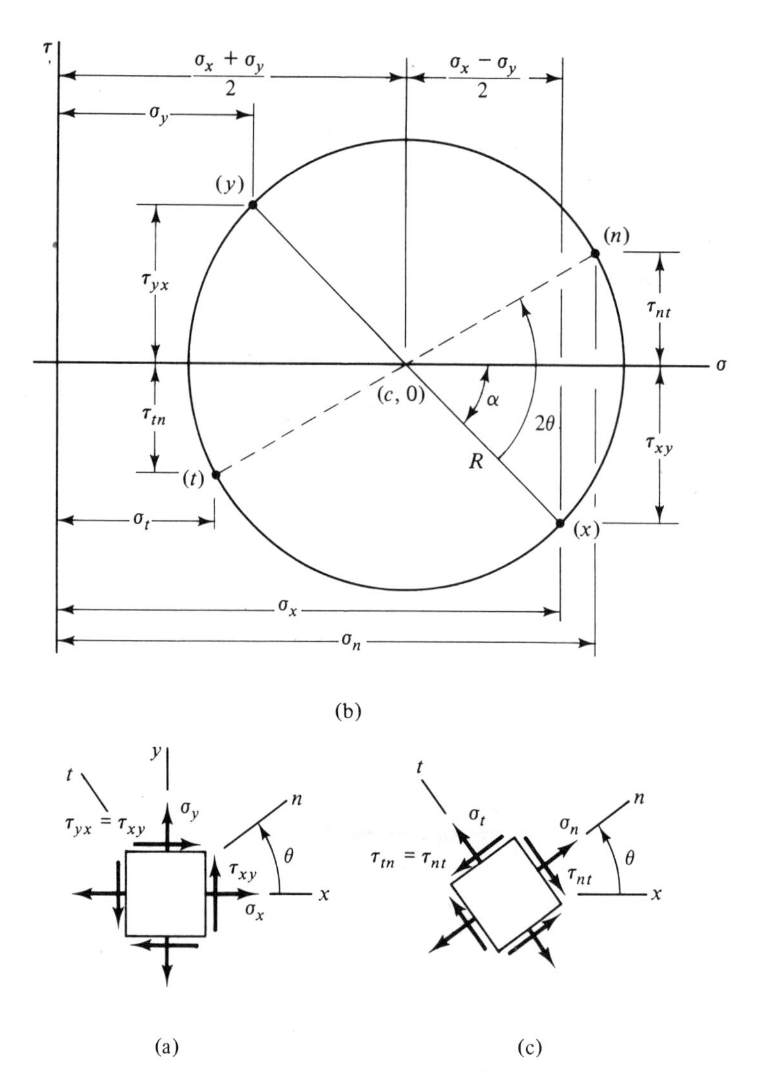

Figure 9.8

2. Plot the stresses on the x and y faces of the element as points in this coordinate system. These points are denoted by (x) and (y), respectively, in Figure 9.8(b). Normal stresses are considered positive if tensile and negative if compressive. We have shown each of the normal stresses as being positive and with $\sigma_x > \sigma_y$, but this need not be the case. Shear stresses are considered positive if they produce a clockwise moment about the center of the element and negative if they produce a counterclockwise moment. Thus, the shear stress on the x face of the element in Figure 9.8(a) is plotted as a negative quantity in Figure 9.8(b) and the shear stress on the y face is plotted as a positive quantity. It is important to note that *this is a special sign convention used only with Mohr's circle*. Otherwise, positive shear stresses are as defined in Figure 9.2.

3. Connect the points (x) and (y). This establishes a diameter of the circle. The complete circle can then be constructed or sketched in. From the geometry of Figure 9.8(b),

we see that the center of the circle will always lie on the horizontal axis. Furthermore, the coordinates $(c, 0)$ of the center of the circle, its radius R, and the angle α between the diameter (x)-(y) and the horizontal axis can be determined from the known coordinates of points (x) and (y) by using only simple geometry. We have

$$c = \frac{\sigma_x + \sigma_y}{2} \qquad R = \sqrt{\left(\frac{\sigma_x - \sigma_y}{2}\right)^2 + \tau_{xy}^2} \qquad \alpha = \tan^{-1}\frac{\tau_{xy}}{\left(\frac{\sigma_x - \sigma_y}{2}\right)} \tag{9.5}$$

The first two of these relations can also be obtained from a direct comparison of Eqs. (9.3) and (9.4).

4. To locate the points (n) and (t) on the circle associated with the rotated nt axes, we rotate the diameter (x)-(y) of the circle in the *same* direction as the axes are rotated, but through *twice* the angle. All angles on the circle are double those on the element because the stress transformation equations are functions of 2θ. Once points (n) and (t) have been located, their coordinates can be determined from the geometry of the circle. These coordinates provide the stresses on the n and t faces of the rotated element [Figure 9.8(c)]. The sense of these stresses is in accordance with the sign conventions given in step 2. This same procedure applies for determining the stresses on any given plane inclined to the xy axes. However, only the one point on the circle corresponding to the stresses on this plane need be considered in this case.

The procedures outlined here are further illustrated in the following examples. If Mohr's circle is constructed to scale, the stresses on any given plane can be determined graphically. However, the circle is usually used as a mental aid for computing these stresses analytically.

In concluding this section, we note that the stress transformation equations and Mohr's circle are based only upon equilibrium considerations. Consequently, they apply regardless of the material behavior.

Example 9.4. Stress Components with Respect to Rotated Axes. Determine the stress components with respect to the nt axes for the state of plane stress shown in Figure 9.9(a).

Solution. We first construct Mohr's circle, as shown in Figure 9.9(b). According to our sign convention, a positive shear stress produces a clockwise moment about the center of the element. Thus, the shear stress on the x face of the element is negative, while that on the y face is positive. Furthermore, the normal stress on the y face is considered negative because it is compressive.

From Figure 9.9(b), we see that the center of the circle is located a distance

$$c = \frac{10 + (-6)}{2} = 2 \text{ MN/m}^2$$

from the origin. Furthermore, the radius of the circle is

$$R = \sqrt{8^2 + 4^2} = 8.94 \text{ MN/m}^2$$

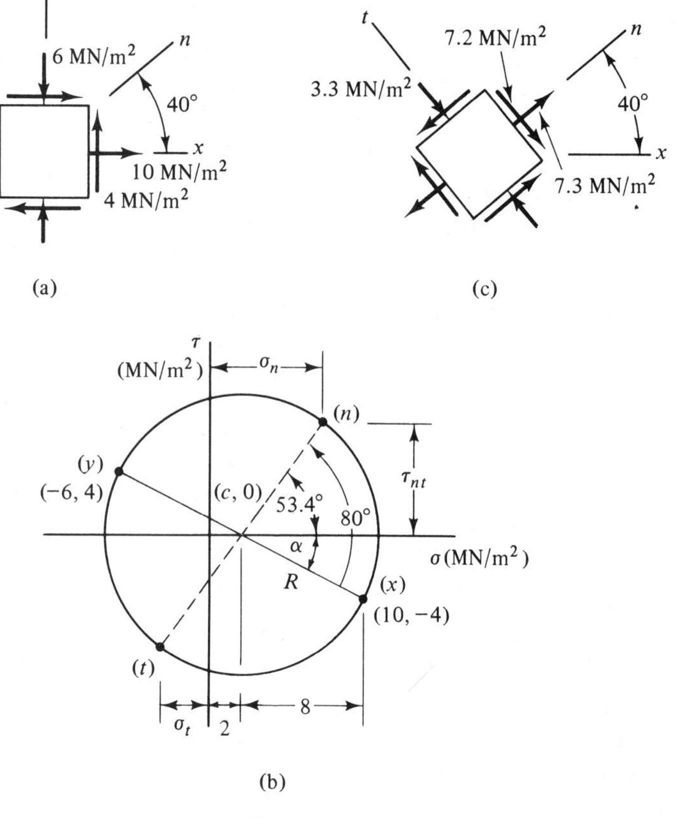

Figure 9.9

and

$$\tan \alpha = \frac{4}{8} \qquad \alpha = 26.6°$$

Since the *nt* axes are rotated 40° counterclockwise with respect to the *xy* axes, we rotate the diameter (*x*)-(*y*) of the circle 80° counterclockwise to determine the points (*n*) and (*t*) corresponding to the stresses associated with these axes. The angle between the diameter (*n*)-(*t*) and the horizontal axis is $80° - \alpha = 53.4°$. From the geometry of the circle, we then have

$$\sigma_n = c + R \cos 53.4° = 2 + 8.94 \cos 53.4°$$

$$= 2 + 5.3 = 7.3 \text{ MN/m}^2$$

$$\sigma_t = c - R \cos 53.4° = 2 - 5.3 = -3.3 \text{ MN/m}^2$$

$$\tau_{nt} = R \sin 53.4° = 8.94 \sin 53.4° = 7.2 \text{ MN/m}^2$$

Figure 9.9(c) shows these stresses acting upon the rotated element. Note that we need determine the shear stress on only one face of the element because the values on the other faces are the same. The sense of the stresses is in accordance with the sign conventions used for the circle.

It is not necessary to use a sketch of the rotated element to define the sense of the stresses, as we have done here. The sign convention given in Section 9.2 can be used. According to this convention, the value of the shear stress is $\tau_{nt} = -7.2\ MN/m^2$.

Example 9.5. Allowable Pressure in a Tank. A cylindrical tank with a radius of 10 in. and a wall thickness of 0.1 in. is welded along a helical seam that makes an angle of 55° with the longitudinal axis [Figure 9.10(a)]. What internal pressure, p, can the tank safely withstand if the tensile stress in the weld cannot exceed 30 ksi? Use a safety factor of 3. What is the shear stress in the weld at this pressure?

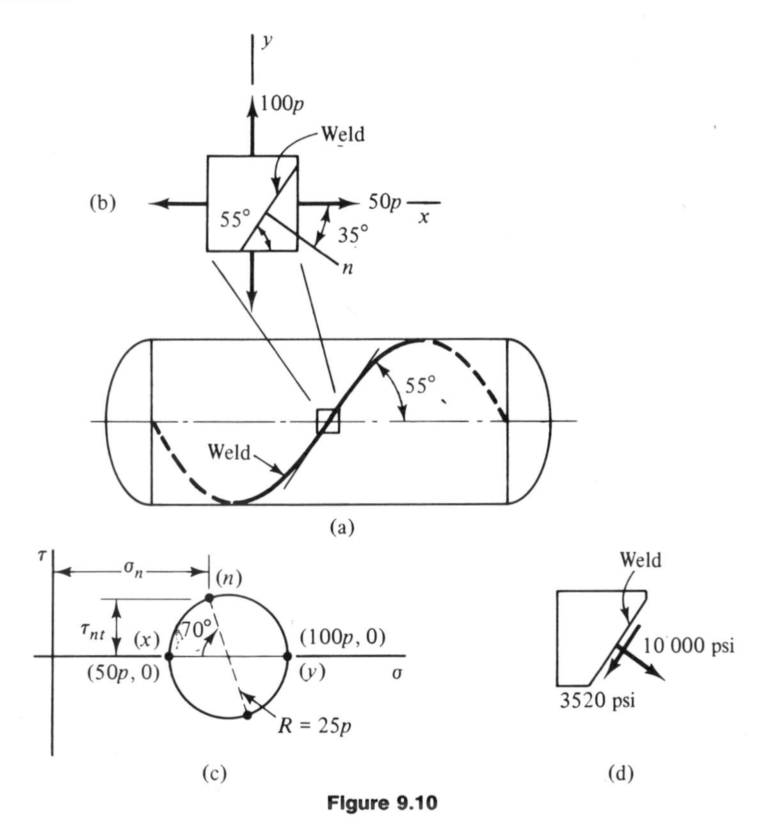

Figure 9.10

Solution. The longitudinal and circumferential stresses in the tank are (see Section 6.8)

$$\sigma_L = \frac{pR}{2t} = \frac{p(10\text{ in.})}{2(0.1\text{ in.})} = 50\text{ p}$$

$$\sigma_C = \frac{pR}{t} = 2\sigma_L = 100\text{ p}$$

Figure 9.10(b) shows these stresses acting upon a material element with sides oriented parallel and perpendicular to the longitudinal axis of the tank and containing a portion of the weld. Mohr's circle for this state of stress is shown in Figure 9.10(c).

We take the n axis perpendicular to the plane of interest. This axis is rotated 35° clockwise with respect to the x axis. Thus, the point (n) on the circle corresponding to the stresses on this plane lies 70° clockwise from point (x). From the geometry of the circle, we have for the tensile stress in the weld

$$\sigma_n = c - R\cos 70° = (75\text{ p}) - (25\text{ p})\cos 70° = 66.4\text{ p} \leqslant 30\ 000\text{ psi}$$

$$p \leqslant 450\text{ psi}$$

The working pressure is obtained by dividing by the safety factor:

$$p_{\text{working}} = \frac{450\text{ psi}}{3} = 150\text{ psi} \qquad \textbf{\textit{Answer}}$$

The shear stress in the weld at this pressure is

$$\tau_{nt} = R\sin 70° = (25\text{ p})\sin 70° = 25(150\text{ lb/in.}^2)\sin 70° = 3520\text{ psi} \qquad \textbf{\textit{Answer}}$$

Figure 9.10(d) shows the stresses acting upon the plane of the weld at the working pressure. The sense of these stresses is obtained from the sign conventions used for the circle.

It should be noted that our analysis accounts only for the stresses due to the internal pressure. Any stresses that may be present prior to the loading are not accounted for. Such initial stresses are often introduced during the fabrication process and are another reason why an adequate safety factor is required. Of course, initial stresses are a factor in most problems, not just pressurized cylinders.

PROBLEMS

9.17 to 9.22 Construct Mohr's circle for the states of stress shown. Determine the stress components with respect to the nt axes and show the results on a rotated element or determine the normal and shear stresses on the plane indicated and show the results on a sketch, as required.

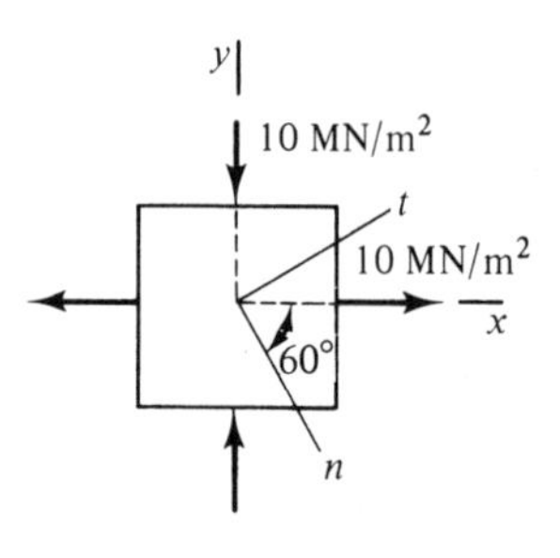

Problem 9.17

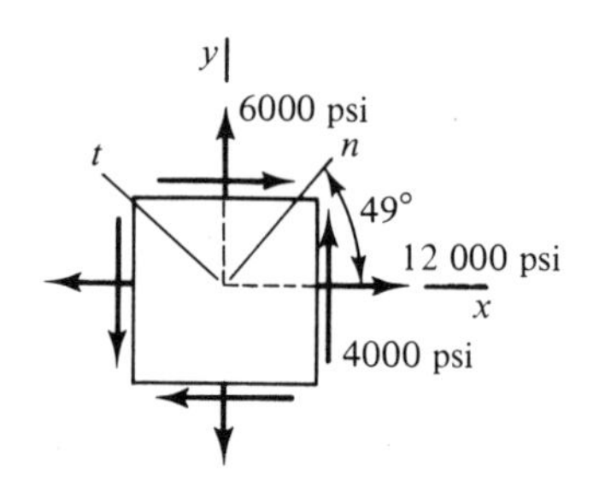

Problem 9.18

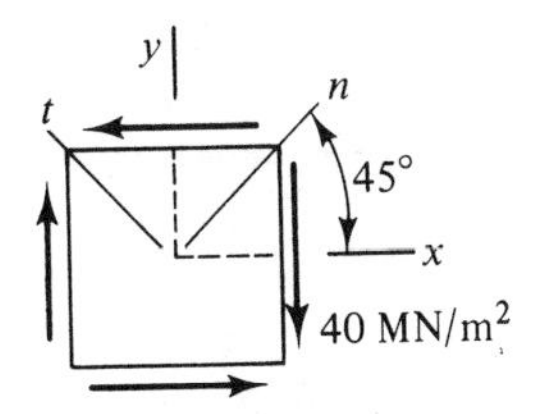

Problem 9.19

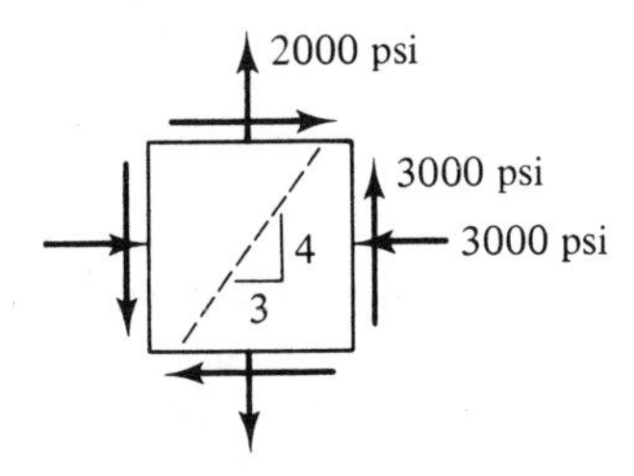

Problem 9.20

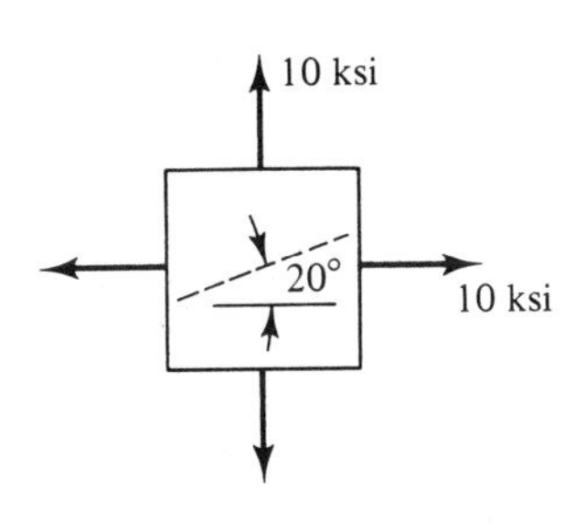

Problem 9.21

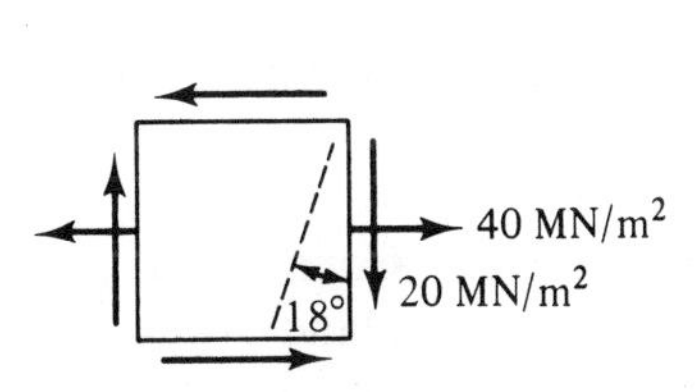

Problem 9.22

9.23 Two blocks are glued together as shown. If the normal stress on the joint is 15 MN/m², what is the value of σ_y?

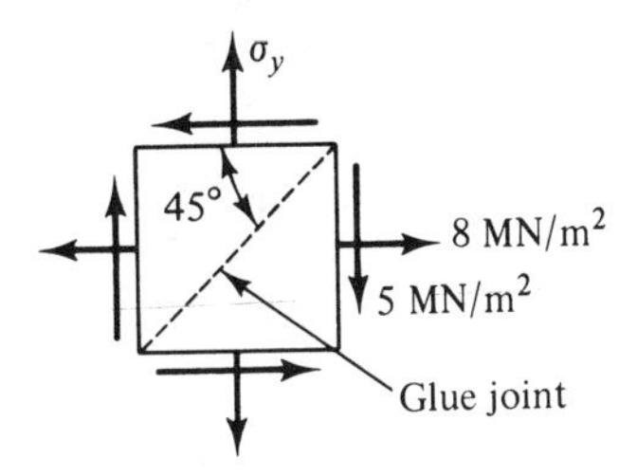

Problem 9.23

9.24 For the state of stress shown, it is known that $\sigma_n = -7.5$ ksi. Determine the angle θ and the stresses σ_t and τ_{nt}. Show the results on a properly oriented element.

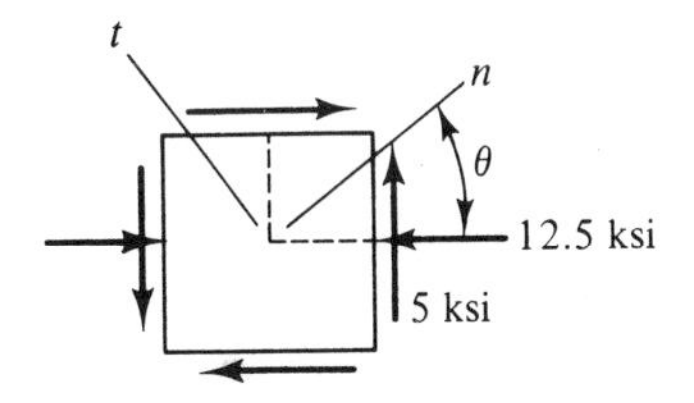

Problem 9.24

9.25 What is the maximum tensile stress σ_x to which the welded plate shown can be subjected if the shear stress in the weld is not to exceed 15 ksi?

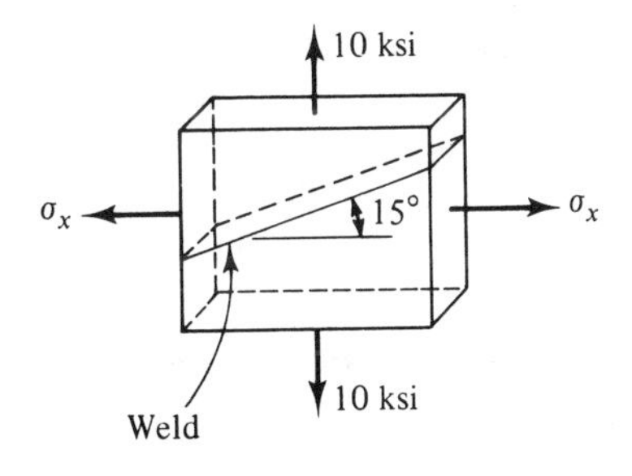

Problem 9.25

9.26 The stresses in the wall of a rocket motor casing are as shown, where p is the internal pressure. If the seam can withstand a tensile stress of 14 ksi, what pressure p can the casing withstand?

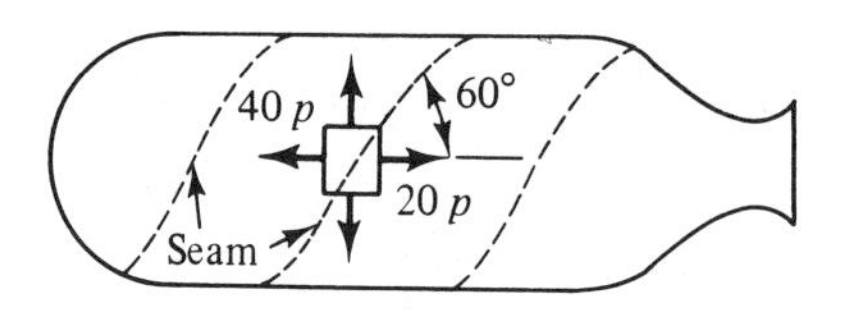

Problem 9.26

9.27 The block of wood shown will crack if the shear stress along the grain exceeds 8 MN/m^2. If $\sigma_x = 5$ MN/m^2 (tension), what range of values of σ_y can be applied without cracking the block?

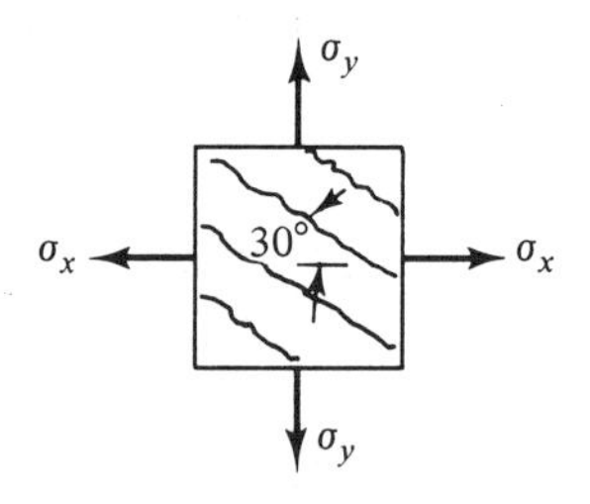

Problem 9.27

9.28 A steel plate $\frac{1}{2}$ in. thick × 5 in. wide × 10 in. long is subjected to uniformly distributed forces (not stresses) acting along its edges as shown. Use Mohr's circle to determine the normal and shear stresses acting along plane a-a.

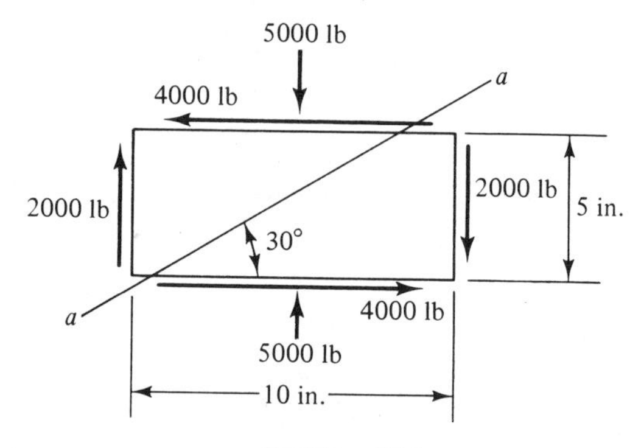

Problem 9.28

9.29 A skewed plate with unit thickness is subjected to uniformly distributed stresses along its sides as shown. Determine σ_x, σ_y, and τ_{xy}.

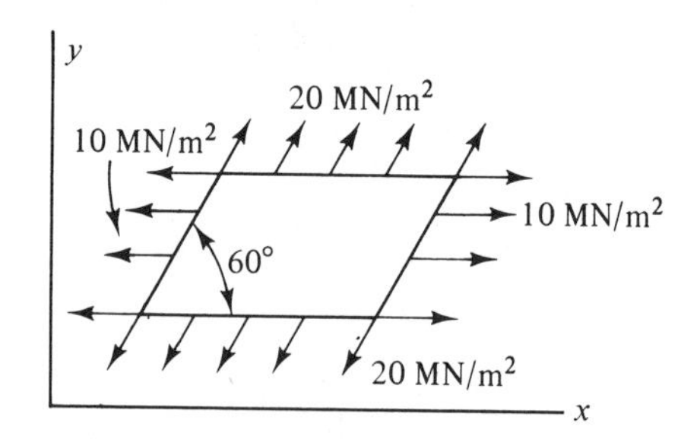

Problem 9.29

9.4 MAXIMUM STRESSES

In many problems it is the maximum normal and shear stresses that are of interest, rather than the stresses on any one particular plane. As we shall show in the following, these maximum stresses and the planes on which they occur can be readily determined by using Mohr's circle.

Principal stresses and maximum in-plane shear stress. Consider a general state of plane stress at a point in a body [Figure 9.11(a)] and the corresponding Mohr's circle [Figure 9.11(b)]. It is clear from the latter figure that the algebraic largest and smallest normal stresses

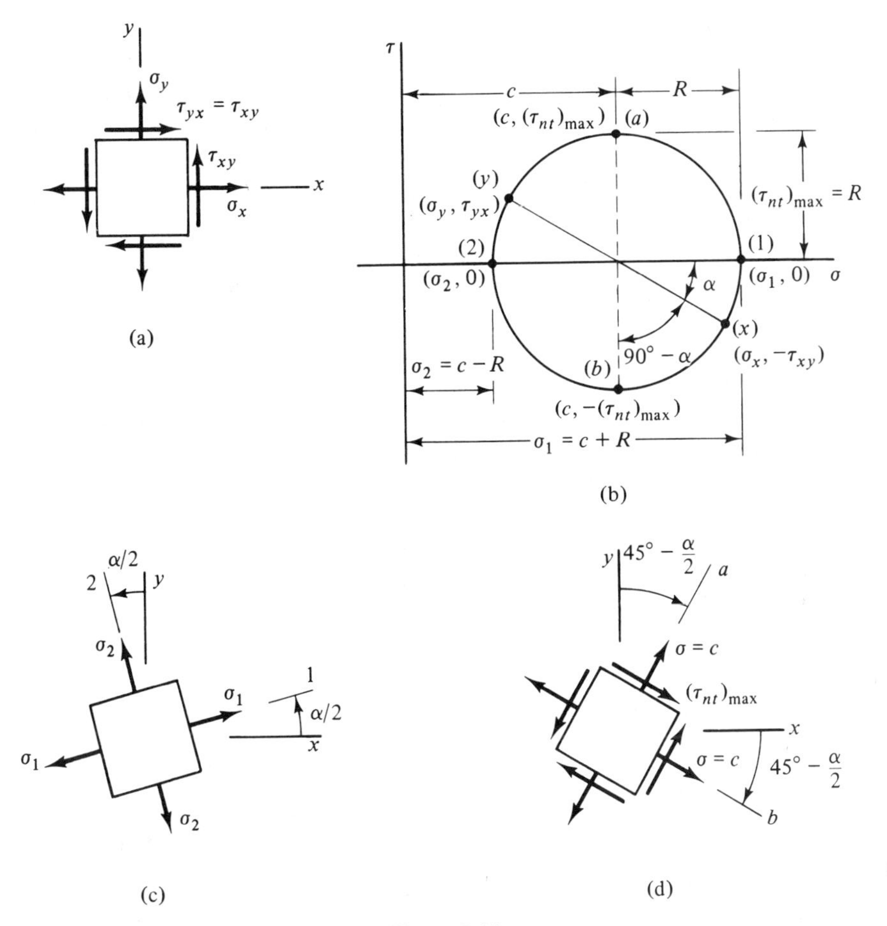

Figure 9.11

existing at this point correspond to the horizontal extremities of the circle. These stresses will be denoted by σ_1 and σ_2, respectively, and are called the *principal stresses*. The axes 1 and 2 associated with them are called the *principal axes of stress*. The points (a) and (b) at the extreme top and bottom of the circle correspond to the maximum in-plane shear stresses existing at this point in the body. These stresses will be denoted by $(\tau_{nt})_{max}$. There is no special name given the *ab* axes associated with them.

The principal stresses and the maximum in-plane shear stresses are easily determined. From the geometry of Mohr's circle [Figure 9.11(b)], we see that

$$\sigma_1 = c + R \qquad \sigma_2 = c - R \tag{9.6}$$

and

$$(\tau_{nt})_{max} = \pm R = \pm \frac{(\sigma_1 - \sigma_2)}{2} \tag{9.7}$$

where c is the horizontal coordinate of the center of the circle and R is its radius. Both of these quantities can be determined from the known coordinates of points

(x) and (y) on the circle, as was shown in the preceding section. The $\pm$ sign in Eq. (9.7) arises from the fact that the shear stresses are positive on one face of an element and negative on the adjoining face, according to the sign convention used for Mohr's circle. The principal stresses are shown in Figure 9.11(b) as being positive, but one or both may be negative.

The orientation of the principal axes is determined as follows. We rotate the diameter (x)-(y) of the circle in either direction until it is aligned with the diameter (1)-(2), which lies along the horizontal axis. The xy axes for the element are then rotated in the *same* direction, but through *one-half* the angle. This establishes the orientation of axes 1 and 2.

There is often some confusion as to which of the principal axes is axis 1 and which is axis 2. This question can be easily resolved by remembering that points on the circle correspond to axes on the element. Thus, if points (x) and (y) coincide with points (1) and (2), respectively, after rotation of the diameter of the circle, the rotated x axis is axis 1 and the rotated y axis is axis 2. This is the case shown in Figure 9.11(c). However, if point (x) coincides with point (2) after rotation, the rotated x axis is axis 2 and the rotated y axis is axis 1. As we shall show in Example 9.6, the direction in which the axes are rotated is immaterial. The results are the same in either case.

Once the principal axes have been determined, the principal stresses are shown on a rotated element with sides aligned with these axes, as in Figure 9.11(c). The faces of this element correspond to the planes upon which the principal stresses act. Note that there are no shear stresses on these planes. Thus, if a given state of stress involves only normal stresses, we can immediately say that they are the principal stresses.

This same procedure applies for determining the axes ab associated with the maximum in-plane shear stresses. In this case, the diameter (x)-(y) of Mohr's circle is rotated until it coincides with diameter (a)-(b), which is aligned vertically [Figure 9.11(b)]. Again, the direction of rotation is immaterial. Once the ab axes have been determined, the stresses associated with them are shown on a rotated element, as in Figure 9.11(d). In addition to the shear stresses, there are normal stresses acting upon all four faces of the element with magnitudes equal to the horizontal coordinate, c, of the center of the circle. The sense of all the stresses is determined from the sign convention used for Mohr's circle. Note that the ab axes will always be 45° away from the principal axes because the corresponding diameters on Mohr's circle are 90° apart.

Absolute maximum shear stress. In determining the maximum shear stress at a point in a body, we must not lose sight of the fact that the material element is three-dimensional and that we have been considering only a two-dimensional view of it. It is entirely possible that a different view of the element will yield a value of shear stress larger than $(\tau_{nt})_{max}$.

To illustrate this, let us consider a material element aligned with the principal axes 1 and 2 [Figure 9.12(a)]. Let us also suppose that there is a principal stress, σ_3,

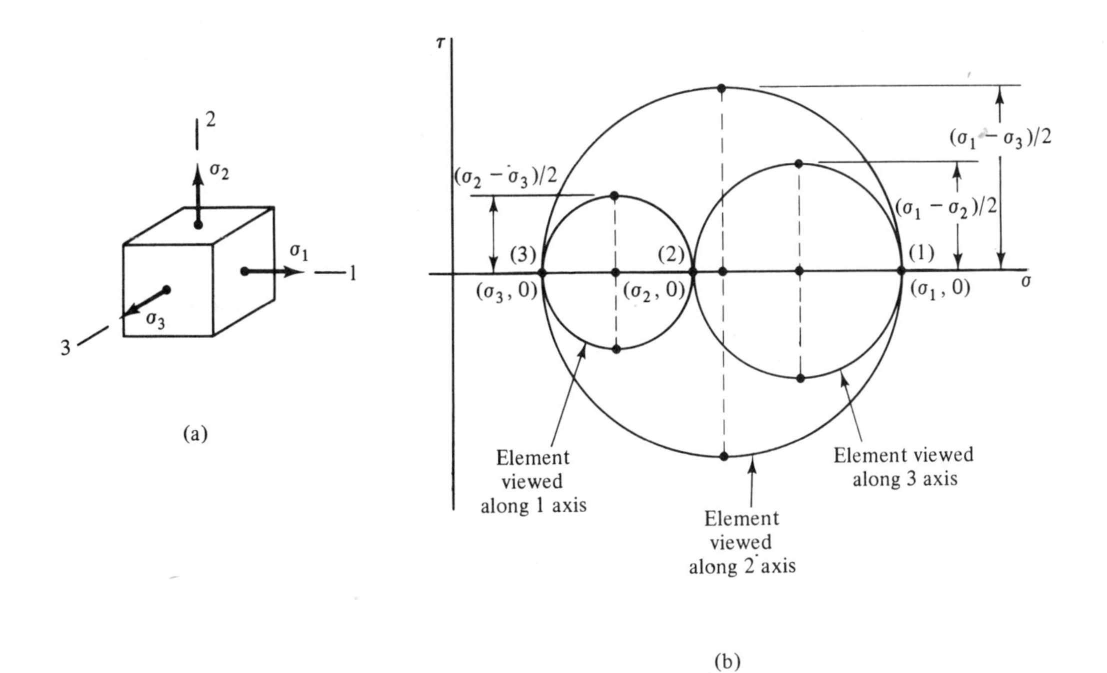

Figure 9.12

acting in the third direction along axis 3 and that $\sigma_1 > \sigma_2 > \sigma_3 > 0$. The results obtained will also hold if one or more of the principal stresses is negative.

If the element is viewed from along the 3 axis, as we have done previously, the corresponding Mohr's circle is as shown in Figure 9.12(b). The largest shear stress observed is $(\tau_{nt})_{\max}$, with magnitude $|\sigma_1 - \sigma_2|/2$. The Mohr's circles obtained by viewing the element from along the other two axes are also shown. Here, the largest shear stresses observed have magnitudes $|\sigma_1 - \sigma_3|/2$ and $|\sigma_2 - \sigma_3|/2$. The magnitude of the maximum shear stress existing at the point in question is the largest of the three values

$$\frac{|\sigma_1 - \sigma_2|}{2} \qquad \frac{|\sigma_1 - \sigma_3|}{2} \qquad \frac{|\sigma_2 - \sigma_3|}{2} \tag{9.8}$$

We shall refer to this stress as the *absolute maximum shear stress*, $\tau_{\max}$.

In many problems the principal stress σ_3 in the third direction is either zero or negligible compared to those in the other two directions. This does not necessarily mean that the maximum in-plane shear stress will be the largest shear stress in the body, however. All three values in Eq. (9.8) must still be checked.

Example 9.6. Determination of Principal Stresses and Maximum Shear Stresses. For the state of stress shown in Figure 9.13(a), determine the principal stresses, the maximum in-plane shear stress, and the orientation of the corresponding axes. Show the results on properly oriented elements. Also determine the absolute maximum shear stress.

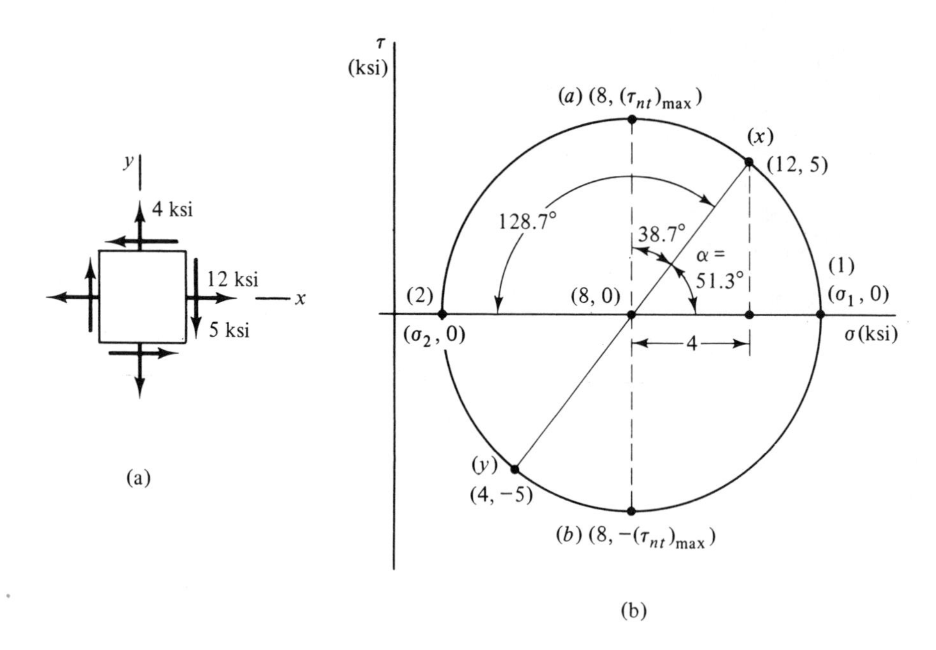

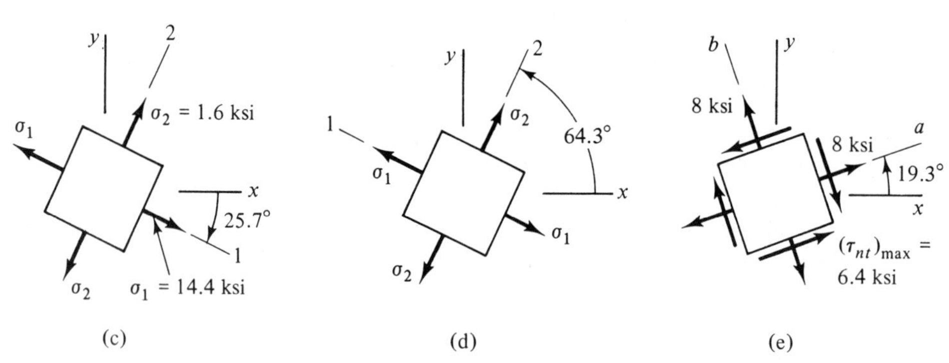

Figure 9.13

Solution. We first construct the Mohr's circle [Figure 9.13(b)]. The center of the circle has coordinates (8 ksi, 0) and its radius is

$$R = \sqrt{4^2 + 5^2} = 6.4 \text{ ksi}$$

Also,

$$\tan \alpha = \frac{5}{4} \qquad \alpha = 51.3°$$

From the geometry of the circle, we have

$$\sigma_1 = c + R = 8 + 6.4 = 14.4 \text{ ksi}$$

$$\sigma_2 = c - R = 8 - 6.4 = 1.6 \text{ ksi} \qquad \textbf{\textit{Answer}}$$

$$|(\tau_{nt})_{max}| = R = 6.4 \text{ ksi}$$

To locate the principal axes, we rotate the diameter (x)-(y) of the circle 51.3° clockwise until it is aligned with the horizontal diameter (1)-(2). The principal axes are then obtained by rotating the xy axes in the same direction, but through one-half the angle. Since points (x) and (y) on the circle coincide with points (1) and (2), respectively, after rotation of the diameter, the rotated x axis is axis 1 and the rotated y axis is axis 2. Figure 9.13(c) shows the principal stresses acting upon the rotated element.

The principal axes could also have been determined by rotating diameter (x)-(y) of the circle counterclockwise through an angle of 128.7°. In this case, the rotated x axis becomes axis 2 and the rotated y axis becomes axis 1. The resulting element, shown in Figure 9.13(d), is seen to be identical to the element in Figure 9.13(c). This confirms our earlier statement that the axes can be rotated in either direction.

The axes associated with the maximum in-plane shear stress are obtained by rotating diameter (x)-(y) of the circle 38.7° counterclockwise until it is aligned with diameter (a)-(b). The a and b axes then correspond to the rotated x and y axes, respectively (Figure 9.13(e)]. The stresses acting upon the rotated element are as indicated. It is left as an exercise to show that the same results are obtained if the diameter of the circle is rotated clockwise.

The absolute maximum shear stress is the largest of the three values given in Eq. (9.8). Since $\sigma_3 = 0$ in this problem, we have

$$\left|\frac{\sigma_1 - \sigma_2}{2}\right| = |(\tau_{nt})_{\max}| = 6.4 \text{ ksi}$$

$$\left|\frac{\sigma_1 - \sigma_3}{2}\right| = \frac{14.4 - 0}{2} = 7.2 \text{ ksi}$$

$$\left|\frac{\sigma_2 - \sigma_3}{2}\right| = \frac{1.6 - 0}{2} = 0.8 \text{ ksi}$$

Thus,

$$\tau_{\max} = 7.2 \text{ ksi} \qquad \textbf{\textit{Answer}}$$

PROBLEMS

9.30 to 9.32 For the states of stress shown, determine the principal stresses, the maximum in-plane shear stress, and the orientation of the

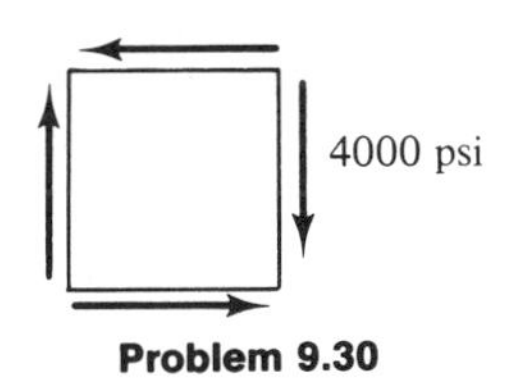

Problem 9.30

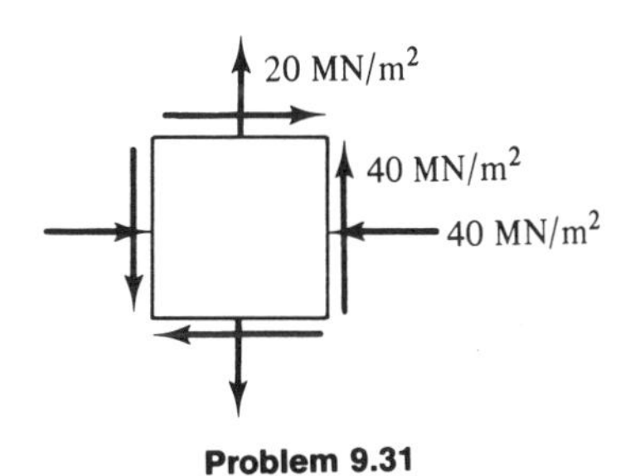

Problem 9.31

corresponding axes. Show the results on properly oriented elements. Also determine the absolute maximum shear stress.

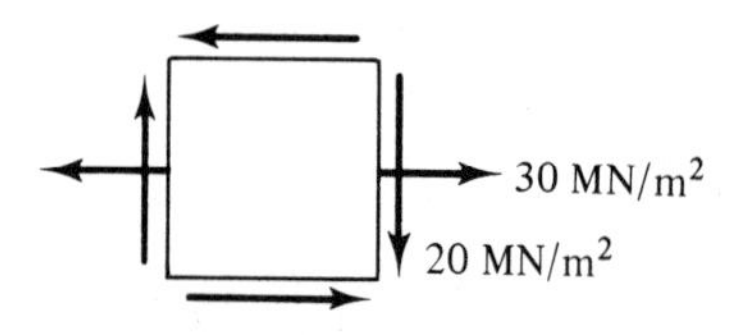

Problem 9.32

9.33 Under normal operating conditions, the state of stress on the outer surface of a drive shaft is found to be as shown. If the allowable stresses are $\sigma_{max} \leqslant 2000$ psi and $\tau_{max} \leqslant 800$ psi, is the shaft operating within specifications?

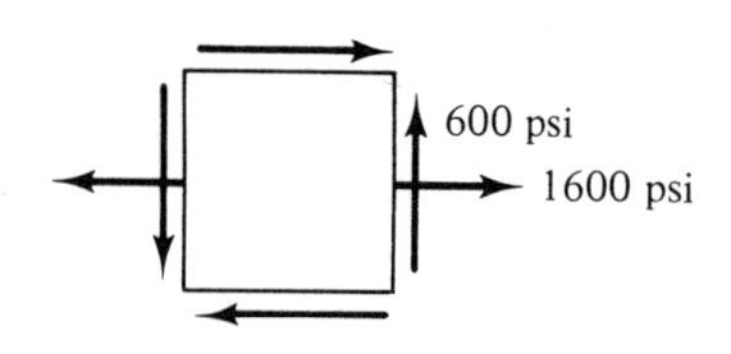

Problem 9.33

9.34 If the tensile principal stress is $\sigma_1 = 1200$ psi for the state of stress shown, what is the value of τ_{xy}?

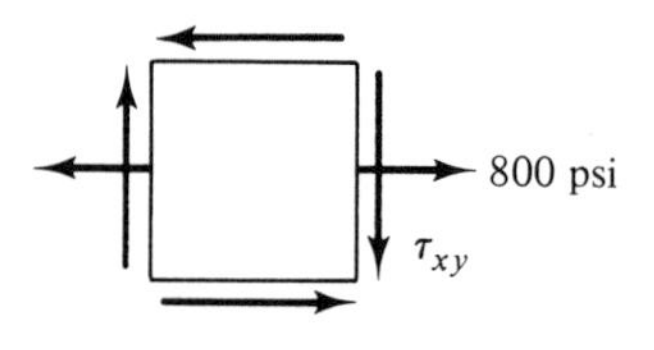

Problem 9.34

9.35 The uniform triangular plate ABC is acted upon by a compressive stress σ on face AB and a tensile stress 3σ on face BC. Find (a) the principal stresses and (b) the stresses σ_x, σ_y, and τ_{xy}. Show all results on properly oriented elements.

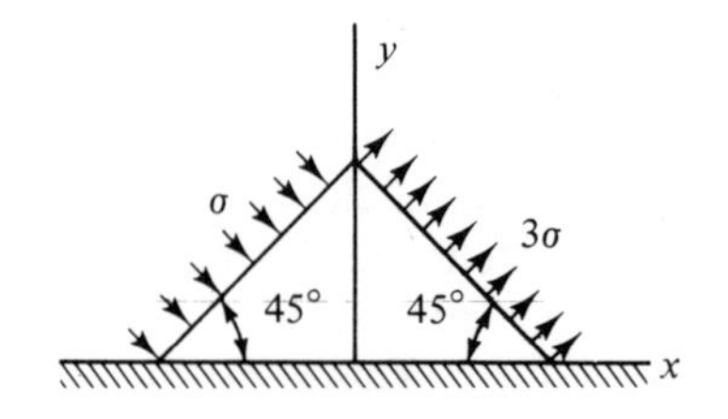

Problem 9.35

9.36 For the state of stress shown, the value of σ_n ranges between 10 MN/m² tension and 10 MN/m² compression. What are the corresponding ranges of values of σ_x and τ_{nt}, either of which may be positive or negative?

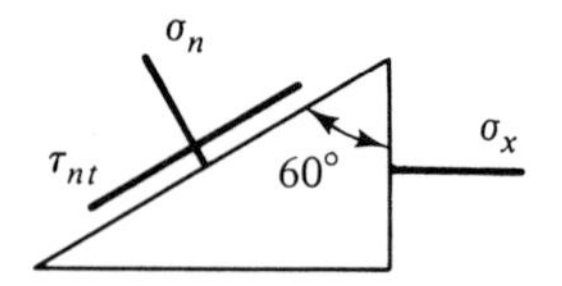

Problem 9.36

9.37 A 2-in. diameter shaft is subjected to combined torsion and bending. At a point on the top surface the bending stress is 18 ksi tension and the maximum principal stress is $\sigma_1 = 24$ ksi tension. Determine the magnitudes of the bending moment M and torque T in the shaft.

9.38 to 9.40 In Problems 9.14 through 9.16 (Section 9.2) determine the principal stresses and maximum in-plane shear stress at point A and show them on properly oriented elements.

9.5 FAILURE DUE TO COMBINED LOADINGS

Now that we have considered the states of stress produced by combined loadings, we need to determine how they relate to the failure of a member. In particular, we shall need to be able to predict the conditions under which members made of brittle materials will fracture and those made of ductile materials will undergo inelastic deformations. The latter case is particularly significant because some of our basic load-stress relations do not apply in the inelastic range. The following example illustrates the basic problem at hand.

Consider a circular shaft made of a ductile material and subjected to an axial load, F, and a twisting couple, C. The loads at which inelastic action will occur if F and C act separately can be readily determined. We need only set the maximum normal stress due to F equal to the tensile yield strength of the material or the maximum shear stress due to C equal to the yield strength in shear. But what criterion do we use to determine the combination of F and C that produces yielding of the material?

Similarly, the individual values of F and C at which the shaft will fracture if made of a brittle material can be determined by setting the maximum tensile stresses developed equal to the tensile ultimate strength of the material. Furthermore, the plane of fracture can be predicted. The obvious question now is, what combination of F and C will cause fracture and on what plane will it likely occur?

Numerous criteria, or *theories of failure*, have been developed for predicting the conditions under which brittle materials will fracture or ductile materials will undergo inelastic deformations when subjected to general states of stress. Only two of these theories will be discussed here—one for brittle materials and one for ductile materials. Other theories of failure are discussed in more advanced texts.

The *maximum tensile stress theory* states that a brittle material will fracture when the maximum tensile stress reaches the ultimate strength of the material in tension. Fracture usually occurs along the plane on which the maximum tensile stress acts. The *maximum shear stress theory* states that a ductile material will yield and undergo inelastic deformations when the absolute maximum shear stress reaches the material yield strength in shear. Experiments show that these theories give reliable results, and they are widely used in the design and analysis of members subjected to combined loadings. The application of these theories is illustrated in the following example.

Example 9.7. Allowable Loads on a Shaft. The shaft shown in Figure 9.14(a) is subjected to combined axial loading and torsion. What is the largest tensile force F that can be applied without causing yielding or fracture if (a) the shaft is made of a ductile material with shear yield strength $\tau_{YS} = 145\ \text{MN/m}^2$ and (b) it is made of a brittle material with ultimate tensile strength $\sigma_U = 140\ \text{MN/m}^2$.

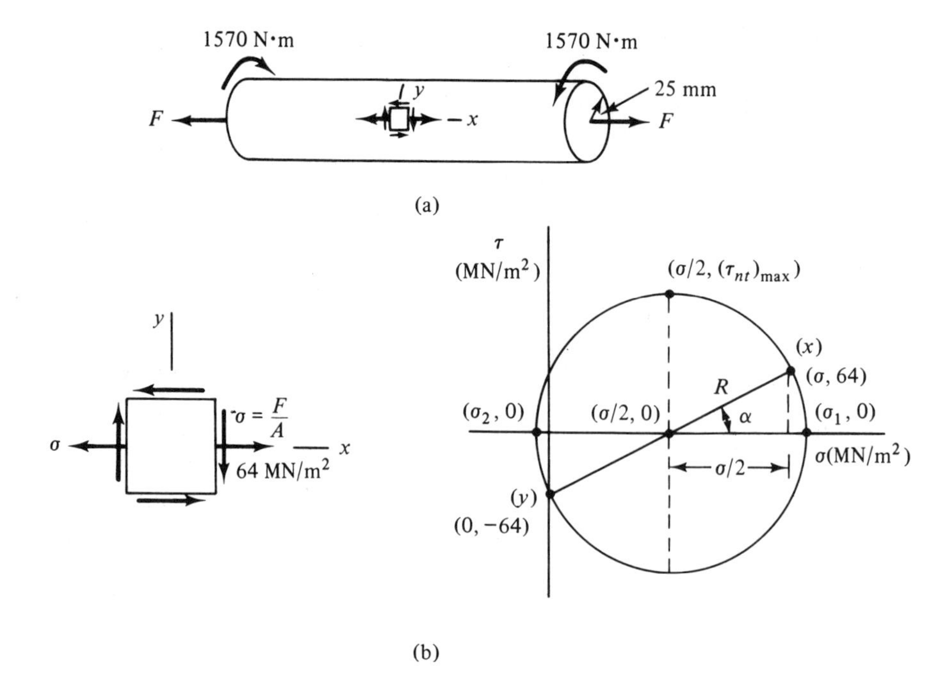

Figure 9.14

Solution. (a) The stresses are largest at the outside of the shaft, where the torsional shear stress is

$$\tau = \frac{TR}{J} = \frac{2T}{\pi R^3} = \frac{2(1570 \text{ N·m})}{\pi(0.025 \text{ m})^3} = 64 \times 10^6 \text{ N/m}^2 \text{ or } 64 \text{ MN/m}^2$$

Thus, the state of stress in the shaft and the corresponding Mohr's circle are as shown in Figure 9.14(b), where $\sigma = F/A$ is the tensile stress due to the axial load. The cross-sectional area of the member is

$$A = \pi R^2 = \pi(0.025 \text{ m})^2 = 2 \times 10^{-3} \text{ m}^2$$

The center of the Mohr's circle is located at $(\sigma/2, 0)$ and its radius is

$$R = \sqrt{\left(\frac{\sigma}{2}\right)^2 + (64^2)} \text{ MN/m}^2$$

Since σ_1 is positive, σ_2 is negative, and $\sigma_3 = 0$, the absolute maximum shear stress will be equal to the maximum in-plane shear stress [see Eqs. (9.8)]. Thus,

$$\tau_{\max} = |(\tau_{nt})_{\max}| = R \leqslant 145 \text{ MN/m}^2$$

or

$$R^2 = (\sigma/2)^2 + (64)^2 \leqslant (145)^2$$

Solving for σ, we get

$$\sigma \leqslant 260 \text{ MN/m}^2$$

therefore,

$$F \leqslant (260 \times 10^6 \text{ N/m}^2)(2 \times 10^{-3} \text{ m}^2)$$

$$F \leqslant 520 \text{ kN} \qquad \textbf{\textit{Answer}}$$

This is the maximum force that can be applied without causing inelastic deformations.

(b) The maximum tensile stress in the shaft is

$$\sigma_1 = \frac{\sigma}{2} + R \leqslant 140 \text{ MN/m}^2$$

so we have

$$R^2 = \left(\frac{\sigma}{2}\right)^2 + (64)^2 \leqslant \left(\frac{140 - \sigma}{2}\right)^2$$

or

$$\sigma \leqslant 110 \text{ MN/m}^2$$

Thus,

$$F \leqslant (110 \times 10^6 \text{ N/m}^2)(2 \times 10^{-3} \text{ m}^2)$$

$$F \leqslant 220 \text{ kN} \qquad \textbf{\textit{Answer}}$$

This is the largest force that can be applied without causing fracture.

PROBLEMS

9.41 A solid circular shaft made of a ductile material with shear yield strength $\tau_{YS} = 145 \text{ MN/m}^2$ can transmit a maximum torque of 3560 N·m without yielding. What torque can be transmitted without yielding if the shaft is also subjected to a tensile force $F = 200$ kN?

9.42 By what percentage is the maximum torque that can be transmitted without yielding in a 2-in. diameter shaft made of steel with shear yield strength $\tau_{YS} = 20$ ksi reduced if a 1000-lb flywheel is attached at the center of the shaft as shown? Assume the bearings act like simple supports with regard to bending.

9.43 Same as Problem 9.42, except that the shaft is hollow with 2 in. OD and 1 in. ID.

9.44 What is the minimum required radius r of the machine part shown if (a) it is made of 2024-T3 aluminum and (b) it is made of

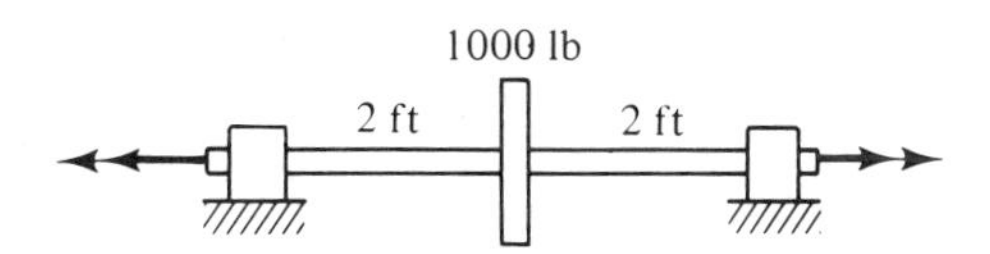

Problem 9.42

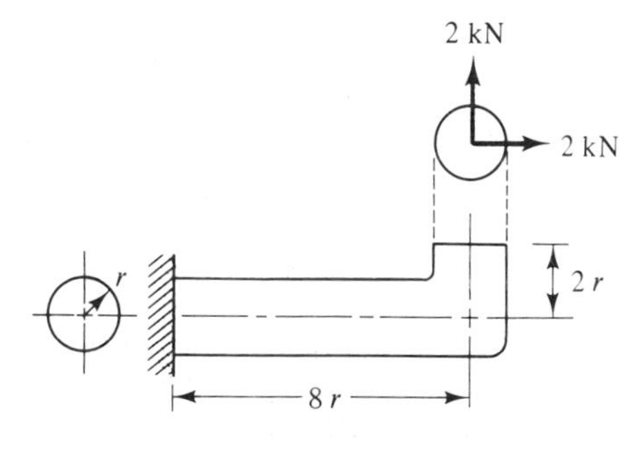

Problem 9.44

grey cast iron? For satisfactory performance, the part must not yield or fracture.

9.45 The design specifications for the 2-in. diameter shaft shown require that the maximum tensile stress not exceed 30 ksi and that the absolute maximum shear stress not exceed 20 ksi. What is the maximum allowable load F?

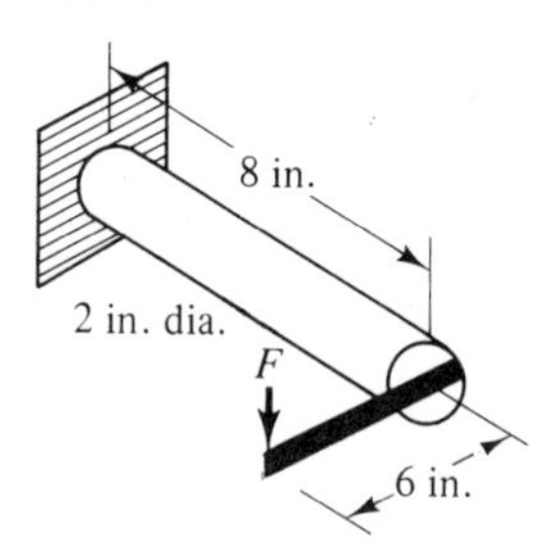

Problem 9.45

9.46 The design specifications for the 40-mm diameter steel shaft shown require that it transmit a torque of 800 N·m with a minimum safety factor of 2 against failure by yielding. Because of faulty installation, one end of the shaft is 6 mm lower than the other end. Will the shaft still operate within specifications? Assume the bearings act like fixed supports with regard to bending; $\tau_{YS} = 145\ \text{MN/m}^2$ and $E = 207\ \text{GN/m}^2$.

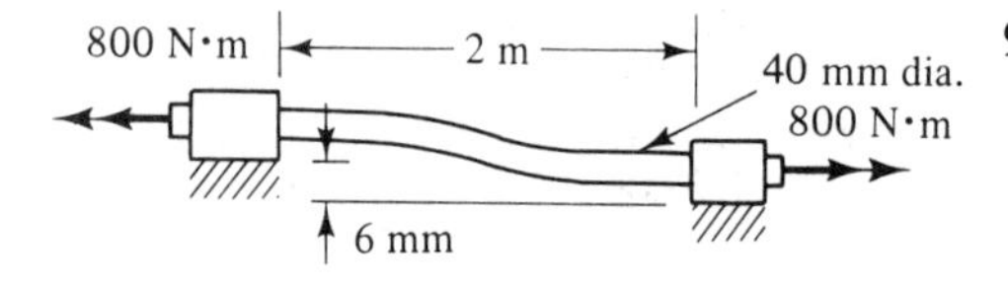

Problem 9.46

9.47 Same as Problem 9.46, except that one end of the shaft is 15 mm lower than the other end.

9.48 Derive an expression relating the torque T and bending moment M that a ductile circular shaft with shear yield strength τ_{YS} can support without yielding. Express the results in terms of T_{YS}, the torque required to produce yielding if acting alone, and M_{YS}, the moment required to produce yielding if acting alone.

9.49 Derive an expression relating the torque T and axial force P that a brittle circular shaft with ultimate tensile strength σ_U can support without breaking. Express the results in terms of the axial force P_U which would produce fracture if acting alone and the torque T_U which would produce fracture if acting alone.

9.50 The sign shown weighs 200 lb and is supported by a $3\frac{1}{2}$ in. nominal diameter steel pipe with shear yield strength $\tau_{YS} = 21$ ksi. If the wind exerts a maximum force of 100 lb on the sign, what is the safety factor against failure by yielding of the pipe? Include the effect of the pipe's weight.

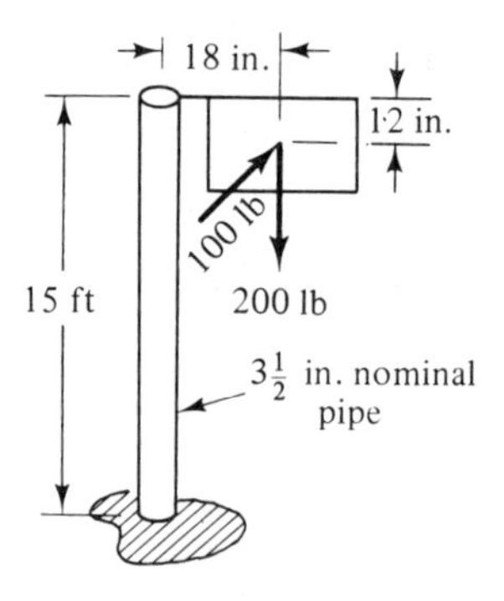

Problem 9.50

9.51 A 20-mm diameter acrylic rod with shear yield strength $\tau_{YS} = 20\ \text{MN/m}^2$ and ultimate tensile strength $\sigma_U = 35\ \text{MN/m}^2$ is loaded as shown. Will the member first yield or fracture as F is increased?

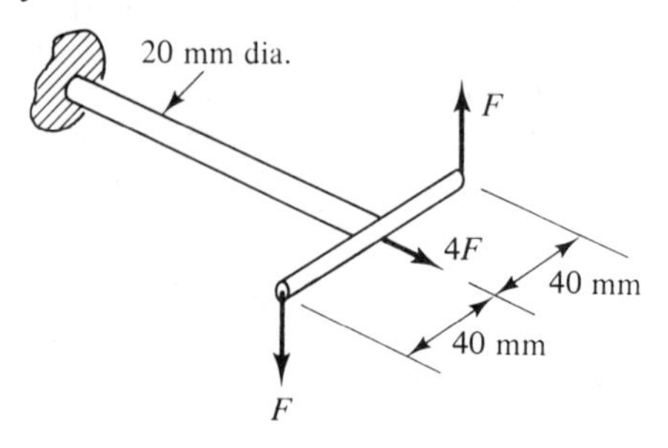

Problem 9.51

9.52 A closed-end brass tube with an inside radius of 30 mm and a 3-mm wall thickness supports an axial tensile force $F = 10$ kN and an internal pressure p. By considering all the stresses acting upon an element of material at the inside surface of the tube, determine the maximum pressure it can withstand without yielding; $\tau_{YS} = 165\ \text{MN/m}^2$. What is the maximum allowable pressure if the radial compressive stress is neglected?

9.53 Same as Problem 9.52, except that the tube is made of alloy steel with shear yield strength $\tau_{YS} = 420\ \text{MN/m}^2$.

9.6 GENERALIZED HOOKE'S LAW

The stresses acting upon a material element at a point in a body for a general triaxial state of stress were shown in Figure 9.2. It is clear that the normal stresses acting upon the element will tend to stretch or compress it and the shear stresses will tend to distort it. Our goal in this section is to determine the relationships between the various stresses and strains. We shall consider only the case of small deformations of linearly elastic isotropic materials. As a result, the strains will be proportional to the stresses, and the principle of superposition will apply. That is, the total strain due to a combination of stresses is the sum of those due to each of the individual stresses. We shall also make use of the fact that, for isotropic materials, a normal stress produces only normal strains and a particular component of shear stress produces only the corresponding component of shear strain. This important result follows from symmetry arguments, but a proof is beyond the scope of this text.

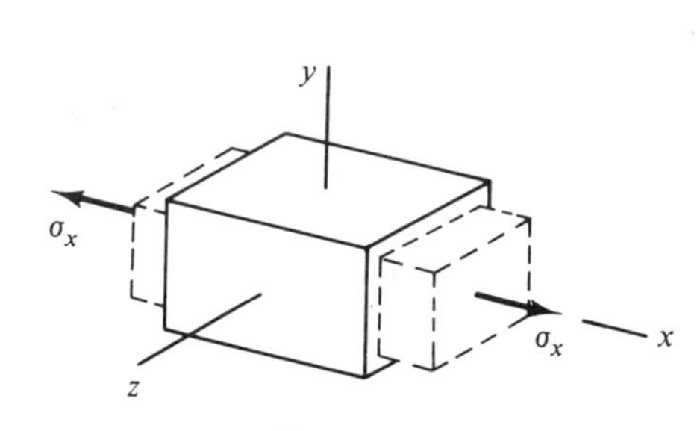

Figure 9.15

Figure 9.15 shows the resulting deformations of a material element due to a normal stress σ_x. The original shape of the element is indicated by the solid lines and the deformed shape by the dashed lines. The normal strain in the direction of the stress is determined by Hooke's law [Eq. (5.19)], and the strains in the other two directions are given by the definition of Poisson's ratio [Eq. (5.17)]. Thus, the stress σ_x produces a normal strain in the x direction of

$$\varepsilon_x^{(1)} = \frac{\sigma_x}{E}$$

and normal strains in the y and z directions of

$$\varepsilon_y^{(1)} = -\nu\varepsilon_x^{(1)} = -\frac{\nu\sigma_x}{E}$$

and

$$\varepsilon_z^{(1)} = -\nu\varepsilon_x^{(1)} = -\frac{\nu\sigma_x}{E}$$

Here E and ν are the modulus of elasticity and Poisson's ratio of the material, respectively. Similarly, stresses σ_y and σ_z produce strains of

$$\varepsilon_y^{(2)} = \frac{\sigma_y}{E} \qquad \varepsilon_x^{(2)} = \varepsilon_z^{(2)} = -\nu\varepsilon_y^{(2)} = -\frac{\nu\sigma_y}{E}$$

and

$$\varepsilon_z^{(3)} = \frac{\sigma_z}{E} \qquad \varepsilon_x^{(3)} = \varepsilon_y^{(3)} = -\nu\varepsilon_z^{(3)} = -\frac{\nu\sigma_z}{E}$$

The total normal strains in the three coordinate directions are

$$\begin{aligned}
\varepsilon_x &= \varepsilon_x^{(1)} + \varepsilon_x^{(2)} + \varepsilon_x^{(3)} = \frac{\sigma_x - \nu(\sigma_y + \sigma_z)}{E} \\
\varepsilon_y &= \varepsilon_y^{(1)} + \varepsilon_y^{(2)} + \varepsilon_y^{(3)} = \frac{\sigma_y - \nu(\sigma_x + \sigma_z)}{E} \\
\varepsilon_z &= \varepsilon_z^{(1)} + \varepsilon_z^{(3)} + \varepsilon_z^{(3)} = \frac{\sigma_z - \nu(\sigma_x + \sigma_y)}{E}
\end{aligned} \tag{9.9}$$

Since a component of shear stress produces only the corresponding component of shear strain, the relationships between the shear stresses and shear strains are

$$\gamma_{xy} = \frac{\tau_{xy}}{G} \qquad \gamma_{xz} = \frac{\tau_{xz}}{G} \qquad \gamma_{yz} = \frac{\tau_{yz}}{G} \tag{9.10}$$

where G is the shear modulus of the material.

Equations (9.9) and (9.10) are called the *generalized Hooke's law*. They define the stress-strain relations for a linearly elastic isotropic material subjected to a general triaxial state of stress. These relations apply for any set of mutually perpendicular axes, not just xyz.

It is sometimes convenient to invert Eqs. (9.9) so that the stresses are expressed in terms of the strains. Carrying out the necessary algebra, we obtain

$$\begin{aligned}
\sigma_x &= \frac{E}{1+\nu}\left[\varepsilon_x + \frac{\nu}{1-2\nu}(\varepsilon_x + \varepsilon_y + \varepsilon_z)\right] \\
\sigma_y &= \frac{E}{1+\nu}\left[\varepsilon_y + \frac{\nu}{1-2\nu}(\varepsilon_x + \varepsilon_y + \varepsilon_z)\right] \\
\sigma_z &= \frac{E}{1+\nu}\left[\varepsilon_z + \frac{\nu}{1-2\nu}(\varepsilon_x + \varepsilon_y + \varepsilon_z)\right]
\end{aligned} \tag{9.11}$$

For plane stress in the xy plane, $\sigma_z = \tau_{xz} = \tau_{yz} = 0$, and Eqs. (9.10) and (9.11) reduce to

$$\begin{aligned}
\sigma_x &= \frac{E}{(1-\nu^2)}(\varepsilon_x + \nu\varepsilon_y) \\
\sigma_y &= \frac{E}{(1-\nu^2)}(\varepsilon_y + \nu\varepsilon_x) \\
\tau_{xy} &= G\gamma_{xy}
\end{aligned} \tag{9.12}$$

The six components of strain (ε_x, ε_y, ε_z, γ_{xy}, γ_{xz}, γ_{yz}) define the *state of strain* at a point in a body, just as the six stress components define the state of stress. This is true regardless of the material behavior. If all the deformations occur in one plane, the strains associated with the third direction will be zero. This case is referred to as *plane strain*. For example, $\varepsilon_z = \gamma_{xz} = \gamma_{yz} = 0$ for plane strain in the xy plane, and the state of strain is completely defined by the three strain components (ε_x, ε_y, γ_{xy}).

The state of strain at a point in a body can be displayed by showing the associated deformations of an infinitesimal element of material located at that point. However, it is more convenient to represent the strains by arrows. Tensile and compressive normal strains are represented by outward and inward directed arrows, respectively, as shown in Figure 9.16(a). Figure 9.16(b) shows the representation for positive and negative shear strains. Note that the arrows depicting the shear strains are in the same directions as the accompanying shear stresses. However, it is emphasized that these arrows do not stand for stresses; they are only a shorthand notation for the strains.

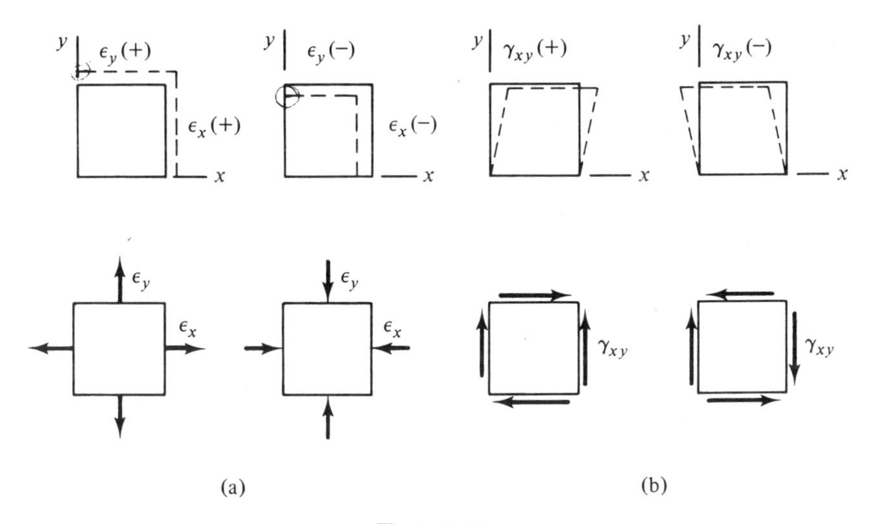

Figure 9.16

Example 9.8. Deformations of a Plate. A structural steel plate with $E = 210$ GN/m^2 and $\nu = 0.3$ has the dimensions shown in Figure 9.17 before loading. The plate is then subjected to a state of plane stress in the xy plane with $\sigma_x = 150$ MN/m^2. For what value of the stress σ_y will the dimension Y of the plate remain unchanged? What are the final dimensions of the plate in the other two directions?

Solution. For plane stress, $\sigma_z = 0$. Also, $\varepsilon_y = 0$, since the dimension Y does not change. Thus, we have from the second of Eqs. (9.9)

$$\varepsilon_y = \frac{\sigma_y - \nu\sigma_x}{E} = 0$$

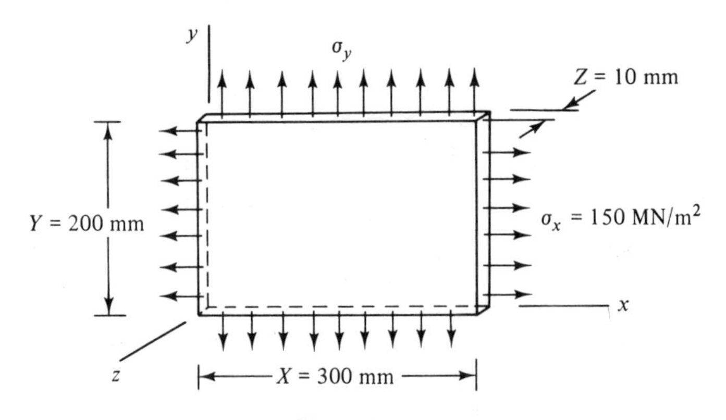

Figure 9.17

or

$$\sigma_y = \nu\sigma_x = 0.3(150\ \text{MN/m}^2) = 45\ \text{MN/m}^2 \qquad \textbf{\textit{Answer}}$$

From the first and third of Eqs. (9.9), we now obtain

$$\varepsilon_x = \frac{\sigma_x - \nu\sigma_y}{E} = \frac{(150 \times 10^6\ \text{N/m}^2) - 0.3(45 \times 10^6\ \text{N/m}^2)}{210 \times 10^9\ \text{N/m}^2} = 0.00065$$

$$\varepsilon_z = \frac{-\nu(\sigma_x + \sigma_y)}{E} = \frac{-0.3(195 \times 10^6\ \text{N/m}^2)}{210 \times 10^9\ \text{N/m}^2} = -0.00028$$

The final dimensions X' and Z' of the plate are

$$X' = X(1 + \varepsilon_x) = 300\ \text{mm}\ (1.00065) = 300.195\ \text{mm}$$
$$Z' = Z(1 + \varepsilon_z) = 10\ \text{mm}\ (0.99972) = 9.997\ \text{mm} \qquad \textbf{\textit{Answer}}$$

Example 9.9. Change in Radius of a Pressurized Cylinder. Determine the change in radius of a thin-walled cylinder with radius R and wall thickness t due to an internal pressure p.

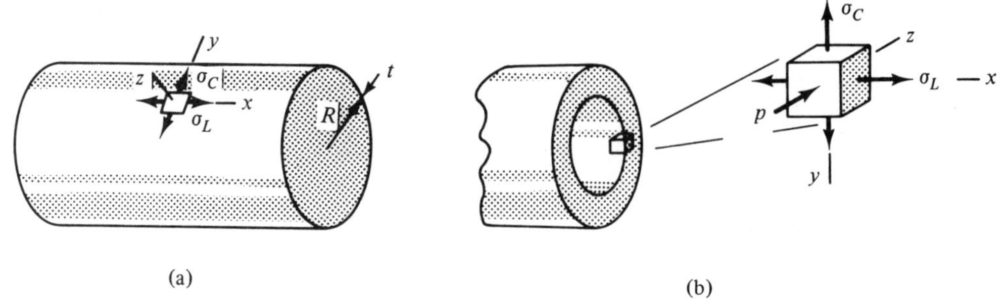

Figure 9.18

Solution. We take axes xyz with x in the longitudinal direction, y in the circumferential direction, and z in the radial direction [Figure 9.18(a)]. From the pressure-stress relations, Eqs. (6.7) and (6.8), we have

$$\sigma_x = \sigma_L = \frac{pR}{2t} \qquad \sigma_y = \sigma_C = \frac{pR}{t} \qquad \sigma_z = 0$$

The result $\sigma_z = 0$ is an approximation. At the inside of the cylinder, there is a compressive stress in the radial direction with magnitude equal to the internal pressure [Figure 9.18(b)]. This stress decreases through the wall thickness and is zero at the outside surface. For thin-walled pressure vessels, $R/t \gg 1$; therefore, $\sigma_C = 2\sigma_L \gg p$. Thus, the radial stress is much smaller than the longitudinal and circumferential stresses, and it is usually neglected.

Substituting the preceding values of stress into the second of Eqs. (9.9), we get for the circumferential strain

$$\varepsilon_y = \frac{\sigma_y - \nu\sigma_x}{E} = \frac{(pR/t) - \nu(pR/2t)}{E} = \frac{pR}{2Et}(2 - \nu)$$

This strain is also equal to the change in circumference of the cylinder divided by the original circumference. Letting ΔR denote the change in radius, we have

$$\varepsilon_y = \frac{2\pi(R + \Delta R) - 2\pi R}{2\pi R} = \frac{\Delta R}{R}$$

Equating the two expressions for ε_y, we obtain

$$\Delta R = \frac{pR^2}{2Et}(2 - \nu) \qquad \textit{Answer}$$

PROBLEMS

9.54 and 9.55 For the states of plane stress shown, determine the corresponding strains ε_x, ε_y, and γ_{xy} and sketch the deformed shape of the element. The material properties are as indicated.

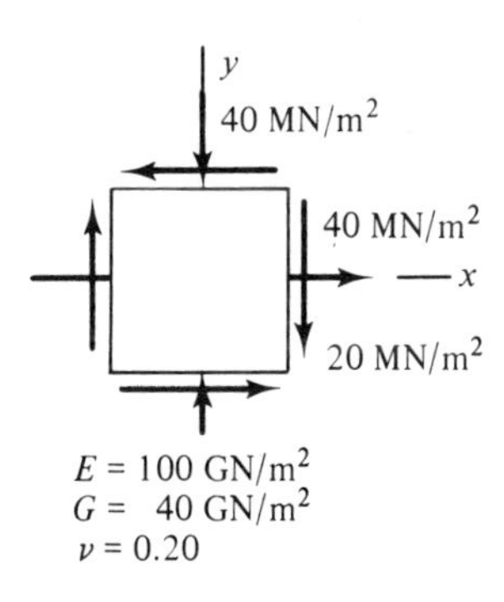

Problem 9.54

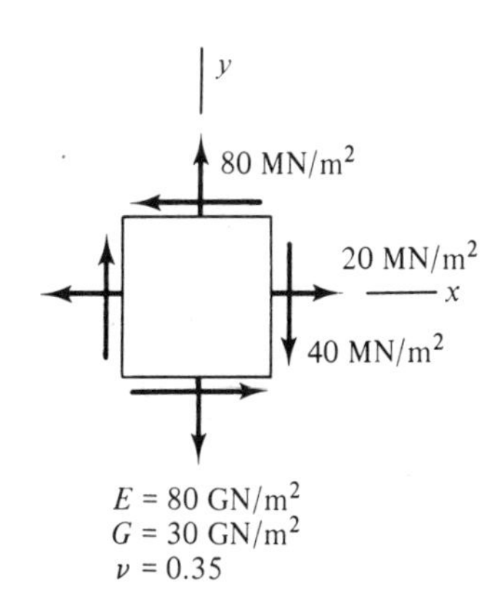

Problem 9.55

9.56 A steel block is subjected to uniformly distributed forces with the magnitudes shown. Determine the resulting change in each dimension of the block.

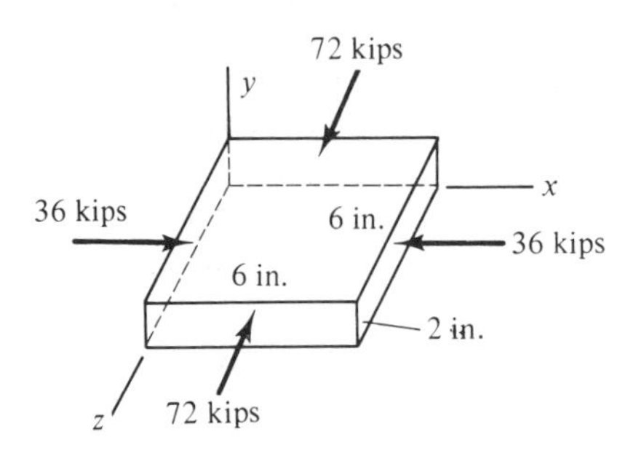

Problem 9.56

9.57 What is the change in volume of a brass cube 40 mm on a side when it is lowered into the ocean to a depth where the pressure is 10 MN/m²?

9.58 A thin plate 15 in. long in the x direction and 12 in. long in the y direction is subjected to uniformly distributed stresses σ_x and σ_y ($\sigma_z = 0$). If the 12-in. length becomes 12.006 in. when $\sigma_x = 30$ ksi, determine σ_y and the change in length in the x direction; $E = 30 \times 10^6$ psi and $\nu = \frac{1}{3}$.

9.59 A rectangular block is subjected to uniformly distributed forces with the magnitudes shown. For what value of the force F will there be no change in length in the y direction? What are the changes in the other dimensions for this value of F? $E = 10 \times 10^6$ psi and $\nu = \frac{1}{4}$.

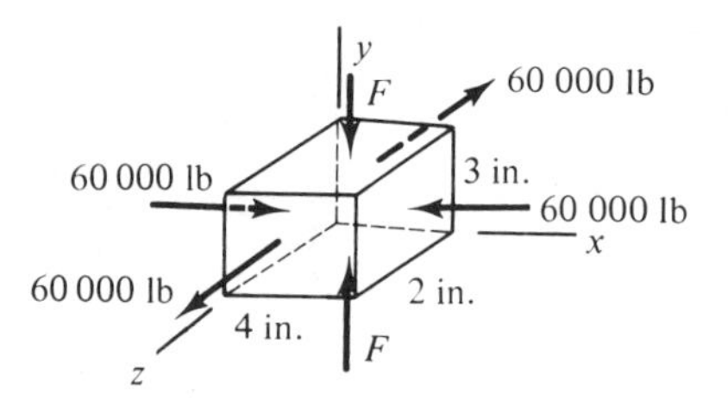

Problem 9.59

9.60 In Problem 9.59, what single loading acting in the z direction would produce the same change in the x dimension if $F =$ 40 000 lb?

9.61 A 2-in.×3-in.×6-in. long rubber block initially fits stress-free into a 6-in. deep cavity in a large steel block. When the block is subjected to a uniformly distributed compressive load F, it shortens 0.30 in. Determine F if $E = 52\ 000$ psi and $\nu = \frac{3}{7}$.

9.62 Calculate the change in the 8-in. dimension of the block shown if $E = 10^7$ psi and $\nu = 0.3$. What are the resultant forces exerted on the sides of the block by the smooth rigid walls?

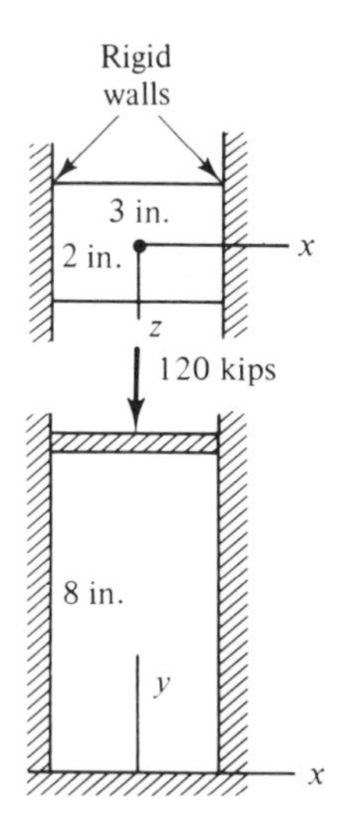

Problem 9.62

9.63 A thin-walled cylindrical tank with closed ends has radius R, wall thickness t, and length L. The tank just fits between rigid end walls when the internal pressure is zero. Calculate the forces exerted on the walls by the tank for an internal pressure p. Neglect the radial stresses in the tank.

9.64 A thin-walled tube with radius R, wall thickness t, and length L fits snugly into a smooth hole drilled into a rigid block. If the tube is subjected to a longitudinal compressive stress σ, what is the change in its length and what is the contact pressure p between it and the block? Neglect the radial stresses in the tube.

9.65 How is the generalized Hooke's law modified for a temperature change ΔT if the material has a coefficient of thermal expansion α?

9.66 A rectangular plate with thickness h is placed between rigid, frictionless walls as shown and subjected to a uniform temperature increase ΔT. Determine the stresses σ_x and σ_y in the plate and the change in its thickness. See Problem 9.65.

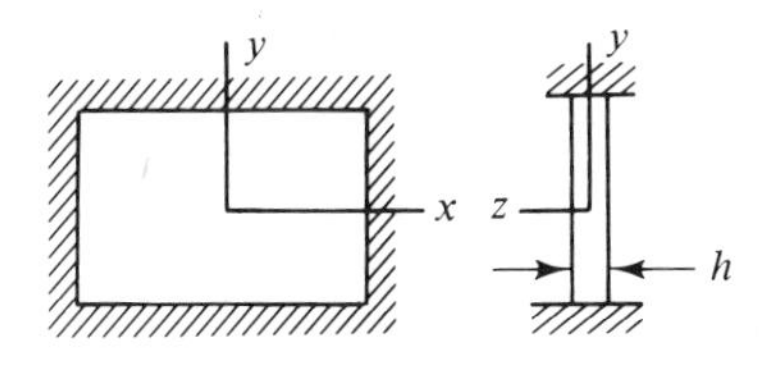

Problem 9.66

9.67 A 1-in. thick plate is loaded as shown ($\sigma_z = 0$) and is also subjected to a uniform temperature increase of 250°F. Determine the total change in each of the dimensions if $E = 25 \times 10^6$ psi, $\nu = \frac{1}{4}$, and $\alpha = 6 \times 10^{-6}\ (°\text{F})^{-1}$. See Problem 9.65.

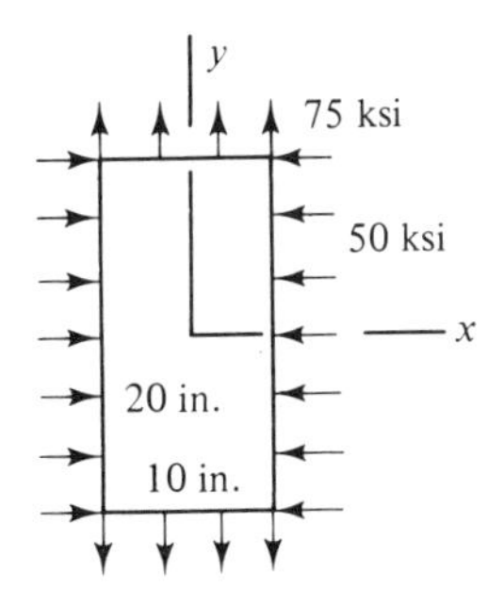

Problem 9.67

9.7 TRANSFORMATION OF STRAINS

Once the strain components with respect to one set of axes are known at a point in a body, the strain components with respect to any other rotated set of axes can be determined. In particular, the maximum and minimum normal strains and the maximum shear strain existing at the point in question can be found, as well as the orientation of the corresponding axes. The necessary computational procedures are discussed in this section for the case of plane strain.

Strain transformation equations. Consider axes xy and nt at a point P in an undeformed body, with nt rotated with respect to xy through an angle θ, measured counterclockwise [Figure 9.19(a)]. Let PQ and PR be short line segments in the n and t directions, respectively. When the body deforms, points P, Q, and R will move to new positions P', Q', and R' [Figure 9.19(b)]. In general, the lengths of the line segments and the angle between them will change. The problem is to express the resulting strain components (ε_n, ε_t, γ_{nt}) in terms of the strain components (ε_x, ε_y, γ_{xy}) and the angle θ.

To determine ε_n, we consider the rectangle $PAQB$ in Figure 9.19(a) with dimensions $\Delta x = \Delta L_n \cos\theta$ and $\Delta y = \Delta L_n \sin\theta$, where ΔL_n is the length of the diagonal PQ. Figure 9.19(c) shows the displacement of point Q relative to point P when this rectangle is subjected to each of the strains ε_x, ε_y, and γ_{xy}. Since the strains and the associated deformations are small, these displacements can be directly superimposed, as in Figure 9.19(d). The deformations are shown greatly exaggerated for clarity.

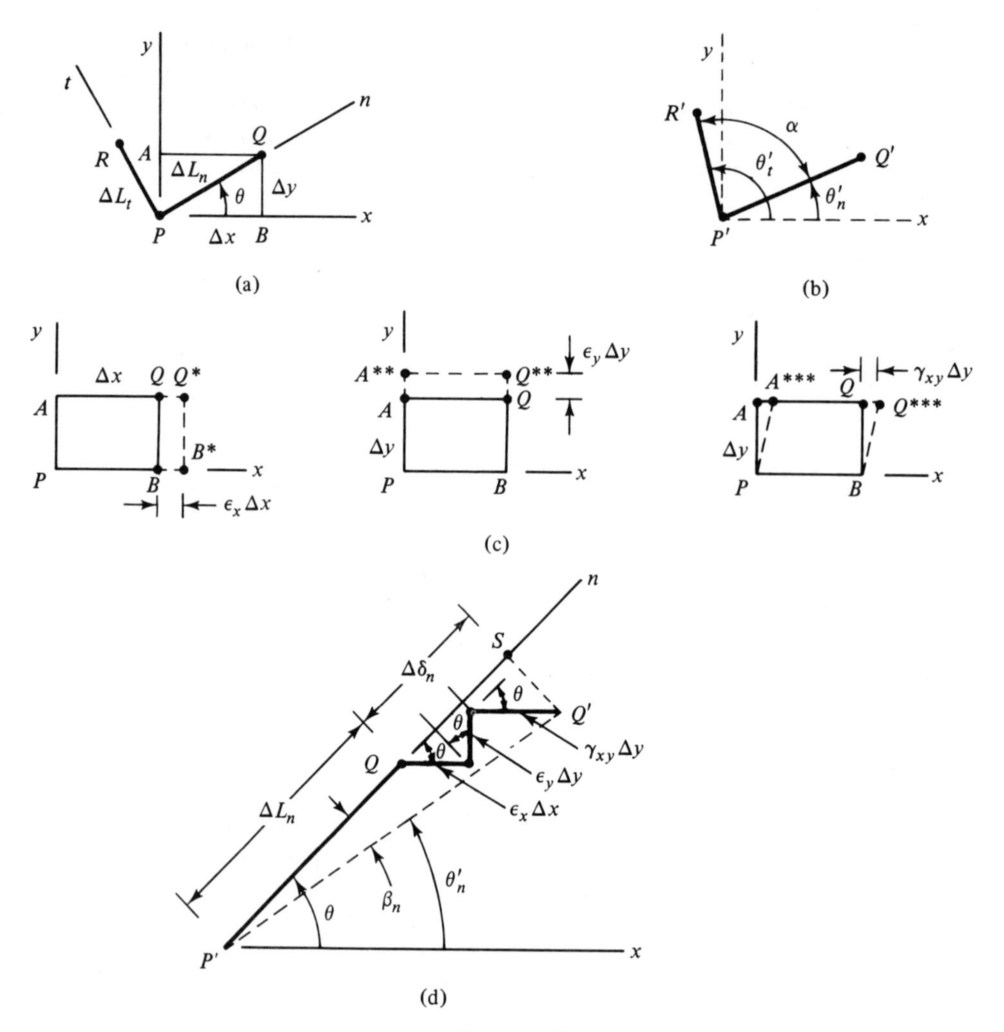

Figure 9.19

From the geometry of Figure 9.19(d), we find that the change in length of line segment PQ is

$$\Delta\delta_n = P'Q' - PQ = \varepsilon_x \Delta_x \cos\theta + \varepsilon_y \Delta_y \sin\theta + \gamma_{xy} \Delta_y \cos\theta$$

or, since $\Delta x = \Delta L_n \cos\theta$ and $\Delta y = \Delta L_n \sin\theta$,

$$\Delta\delta_n = \Delta L_n \left(\varepsilon_x \cos^2\theta + \varepsilon_y \sin^2\theta + \gamma_{xy} \sin\theta \cos\theta \right)$$

Substituting this expression into the definition of normal strain, Eq. (5.12), we obtain

$$\varepsilon_n = \lim_{\Delta L_n \to 0} \frac{\Delta\delta_n}{\Delta L_n} = \varepsilon_x \cos^2\theta + \varepsilon_y \sin^2\theta + \gamma_{xy} \sin\theta \cos\theta \qquad (9.13)$$

Since the angle θ is arbitrary, this equation gives the normal strain in any direction at the point P. In particular, the normal strain ε_t can be obtained by replacing the angle θ by $\theta + \pi/2$.

Now let us consider the shear strain γ_{nt}. When the body deforms, line segments PQ and PR will rotate slightly and assume final orientations defined by the angles θ_n' and θ_t' in Figure 9.19(b). Let β_n and β_t be the respective angles of rotation. From Figure 9.19(d), we see that $\theta_n' = \theta - \beta_n$. Similarly, $\theta_t' = \theta + \pi/2 - \beta_t$. Thus, the final angle between the line segments is

$$\alpha = \theta_t' - \theta_n' = \frac{\pi}{2} - (\beta_t - \beta_n)$$

From the definition of shear strain, Eq. (5.14), we have

$$\gamma_{nt} = \frac{\pi}{2} - \lim_{\substack{\Delta L_n \to 0 \\ \Delta L_t \to 0}} \alpha = \lim_{\substack{\Delta L_n \to 0 \\ \Delta L_t \to 0}} (\beta_t - \beta_n) \tag{9.14}$$

The angle β_n can be obtained from the geometry of Figure 9.19(d). Since β_n is small, the arc length $\beta_n(\Delta L_n + \Delta\delta_n)$ will be approximately equal to the chord length $Q'S$. Thus, we have

$$\beta_n(\Delta L_n + \Delta\delta_n) = Q'S = \varepsilon_x \Delta x \sin\theta - \varepsilon_y \Delta_y \cos\theta + \gamma_{xy}\Delta_y \sin\theta$$

or, since $\Delta x = \Delta L_n \cos\theta$ and $\Delta y = \Delta L_n \sin\theta$,

$$\beta_n = \frac{1}{(1 + \Delta\delta_n/\Delta L_n)}\left[(\varepsilon_x - \varepsilon_y)\sin\theta\cos\theta + \gamma_{xy}\sin^2\theta\right]$$

Taking the limit as $\Delta L_n \to 0$ and noting that the quantity $\lim \Delta\delta_n/\Delta L_{\mathrm{n}} = \varepsilon_n$ is very small compared to unity, we obtain

$$\lim_{\Delta L_n \to 0} \beta_n = (\varepsilon_x - \varepsilon_y)\sin\theta\cos\theta + \gamma_{xy}\sin^2\theta$$

The angle β_t can also be obtained from this expression by replacing θ by $\theta + \pi/2$. Making this substitution and recognizing that $\cos(\theta + \pi/2) = -\sin\theta$ and $\sin(\theta + \pi/2) = \cos\theta$, we have

$$\lim_{\Delta L_t \to 0} \beta_t = -(\varepsilon_x - \varepsilon_y)\sin\theta\cos\theta + \gamma_{xy}\cos^2\theta$$

The shear strain is obtained by substituting these results into Eq. (9.14):

$$\gamma_{nt} = -2(\varepsilon_x - \varepsilon_y)\sin\theta\cos\theta + \gamma_{xy}(\cos^2\theta - \sin^2\theta) \tag{9.15}$$

It is convenient to express Eqs. (9.13) and (9.15) in terms of 2θ by using the double-angle trigonometric identities listed in Section 9.3. We have

$$\varepsilon_n = \frac{\varepsilon_x + \varepsilon_y}{2} + \frac{\varepsilon_x - \varepsilon_y}{2}\cos 2\theta + \frac{\gamma_{xy}}{2}\sin 2\theta \tag{9.16}$$

$$\frac{\gamma_{nt}}{2} = -\frac{(\varepsilon_x - \varepsilon_y)}{2}\sin 2\theta + \frac{\gamma_{xy}}{2}\cos 2\theta \tag{9.17}$$

These equations are known as the *strain transformation equations*. Given the values of the strain components with respect to xy axes, they determine the values with respect to the rotated axes nt. These relations are based only upon geometry and, therefore, apply regardless of the material behavior.

Mohr's circle of strain. Equations (9.16) and (9.17) are identical in form to the stress transformation equations, Eqs. (9.1) and (9.2). The only differences are that normal stresses are replaced by normal strains and the shear stresses are replaced by one-half the shear strains. Consequently, the strain transformation equations can be represented by a *Mohr's circle of strain* in a plane with coordinates $(\varepsilon, \gamma/2)$.

Mohr's circle of strain is laid out in exactly the same way as the circle of stress, with clockwise shear strains considered positive (Figure 9.20). Again, *this is a special sign convention used only with Mohr's circle*. Otherwise, positive shear strains are as defined in Figure 9.16.

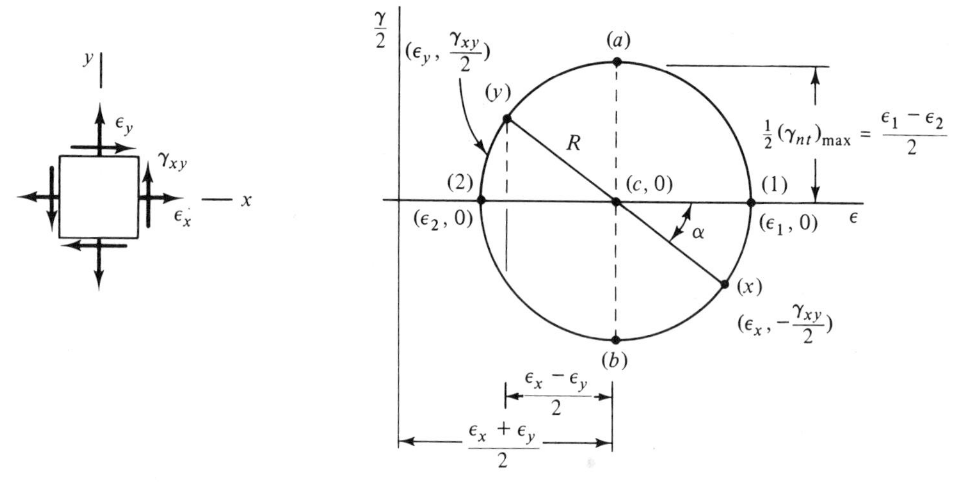

Figure 9.20

The algebraic maximum and minimum normal strains, denoted by ε_1 and ε_2, are called the *principal strains*. The associated axes 1 and 2 are called the *principal axes of strain*. Note that there are no shear strains associated with the principal axes. The extreme top and bottom of the circle correspond to one-half of the maximum shear strain, $(\gamma_{nt})_{max}$, which is accompanied by normal strains with magnitudes equal to the horizontal coordinate of the center of the circle. There is no special name for the axes ab corresponding to $(\gamma_{nt})_{max}$. For isotropic materials, the principal axes of strain coincide with the principal axes of stress. Furthermore, the axes associated with $(\gamma_{nt})_{max}$ coincide with those for the maximum in-plane shear stress, $(\tau_{nt})_{max}$.

The procedures for determining the strain components with respect to a given set of axes are the same as for stresses. This is also true of the procedures for finding the principal strains, the maximum shear strain, and the associated axes, as will be demonstrated in Example 9.10.

Relationship between E, G, and ν. We are now in a position to derive the relationship between the elastic constants E, G, and ν given in Eq. (5.18) of Section 5.8.

Consider an element of linearly elastic isotropic material subjected to principal stresses σ_1 and σ_2, with $\sigma_3 = 0$. From the generalized Hooke's law, Eqs. (9.9), we have for the principal strains

$$\varepsilon_1 = \frac{\sigma_1 - \nu\sigma_2}{E}$$

$$\varepsilon_2 = \frac{\sigma_2 - \nu\sigma_1}{E}$$

Subtracting the second of these equations from the first, we obtain

$$\varepsilon_1 - \varepsilon_2 = (\sigma_1 - \sigma_2)\frac{(1 + \nu)}{E}$$

Now $\sigma_1 - \sigma_2 = 2(\tau_{nt})_{max}$ and $\varepsilon_1 - \varepsilon_2 = (\gamma_{nt})_{max}$ from the corresponding Mohr's circles of stress and strain; therefore, the preceding expression can be written as

$$(\gamma_{nt})_{max} = (\tau_{nt})_{max}\frac{2(1 + \nu)}{E}$$

Introducing the stress-strain relation $(\tau_{nt})_{max} = G(\gamma_{nt})_{max}$, we obtain

$$(\gamma_{nt})_{max} = G(\gamma_{nt})_{max}\frac{2(1 + \nu)}{E}$$

or

$$G = \frac{E}{2(1 + \nu)}$$

This is the relationship given in Eq. (5.18).

Example 9.10. Determination of Principal Strains and Maximum Shear Strain. The strain components at a point in a body subjected to plane strain are $\varepsilon_x = 750\ \mu$ in./in., $\varepsilon_y = -250\ \mu$ in./in., and $\gamma_{xy} = -600\ \mu$ in./in. (Recall that 1 μ in./in. $= 10^{-6}$ in./in.) Determine the principal strains, the maximum shear strain and the accompanying normal strains, and the orientation of the corresponding axes. Sketch the original and deformed shapes of elements with sides aligned with these axes.

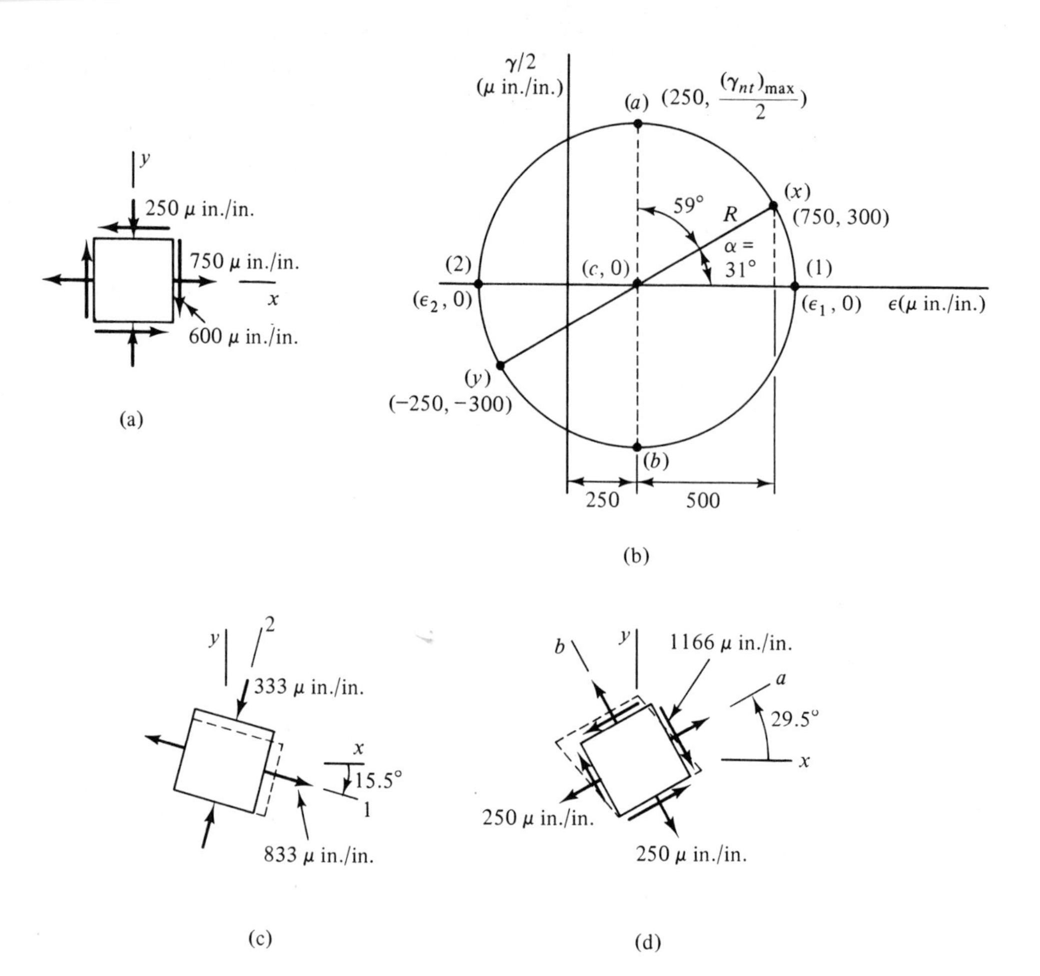

Figure 9.21

Solution. We first represent the strains by arrows acting upon a material element [Figure 9.21(a)] and then construct the corresponding Mohr's circle [Figure 9.21(b)]. The center of the circle has coordinates (250, 0) μ in./in. and its radius is

$$R = \sqrt{300^2 + 500^2} = 583\ \mu\ \text{in./in.}$$

For the angle α, we have

$$\tan \alpha = \frac{300}{500} \qquad \alpha = 31°$$

The principal strains are

$$\varepsilon_1 = c + R = 250 + 583 = 833\ \mu\ \text{in./in.}$$

$$\varepsilon_2 = c - R = 250 - 583 = -333\ \mu\ \text{in./in.}$$

and the magnitude of the maximum shear strain is

$$|(\gamma_{nt})_{max}| = 2R = 1166\ \mu\ \text{in./in.}$$

The normal strains associated with $(\gamma_{nt})_{max}$ are

$$\varepsilon_a = \varepsilon_b = c = 250\ \mu\ \text{in./in.}$$

Since diameter (1)-(2) of the circle lies 31° clockwise from diameter (x)-(y), the principal axes are oriented 15.5° clockwise from the xy axes [Figure 9.21(c)]. The corresponding principal strains are as shown. These strains represent a lengthening of the element in the 1 direction and a shortening in the 2 direction, as indicated by the dashed lines.

The axes ab associated with $(\gamma_{nt})_{max}$ lie 29.5° counterclockwise from xy, since the corresponding diameters on the circle are 59° apart. The strains associated with these axes are shown in Figure 9.21(d). They represent a lengthening of the element in the a and b directions and an increase in the angle between the a and b axes, as indicated by the dashed lines.

PROBLEMS

9.68 to 9.70 For the following states of strain, determine the principal strains, the maximum shear strain, and the orientation of the corresponding axes. Show the results on properly oriented elements and sketch the deformed shapes of these elements (recall that $\mu = 10^{-6}$).

9.68. $\varepsilon_x = -800\ \mu \quad \varepsilon_y = -200\ \mu \quad \gamma_{xy} = -600\ \mu$

9.69. $\varepsilon_x = 500\ \mu \quad \varepsilon_y = 100\ \mu \quad \gamma_{xy} = 600\ \mu$

9.70. $\varepsilon_x = 200\ \mu \quad \varepsilon_y = -600\ \mu \quad \gamma_{xy} = 800\ \mu$

9.71 A thin plate loaded in the xy plane has principal strains $\varepsilon_1 = 3.6 \times 10^{-4}$ and $\varepsilon_2 = -6.0 \times 10^{-4}$. If $E = 10^7$ psi and $\nu = \frac{1}{3}$, determine the stresses σ_x, σ_y, and τ_{xy}. The axes are oriented as shown.

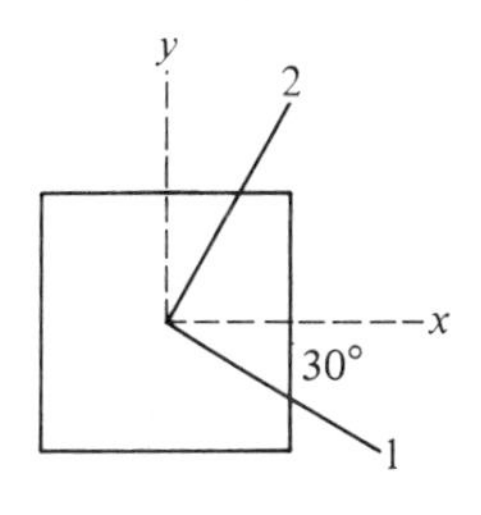

Problem 9.71

9.72 Same as Problem 9.71, except that $\varepsilon_1 = 5.4 \times 10^{-4}$ and $\varepsilon_2 = 3.2 \times 10^{-4}$.

9.73 Circles 10 mm in diameter are scribed on the surface of steel billets before cold rolling. After rolling, the circles become ellipses with major axes of 10.096 mm and minor axes of 9.944 mm. Determine the principal strains and the maximum shear strain.

9.74 The deformed shape of rectangle $OABC$ shown is indicated by the dashed lines. It is known that the principal strains are $\varepsilon_1 = 6 \times 10^{-4}$ and $\varepsilon_2 = -4 \times 10^{-4}$. Determine the displacements a and b and the orientation of the principal axes.

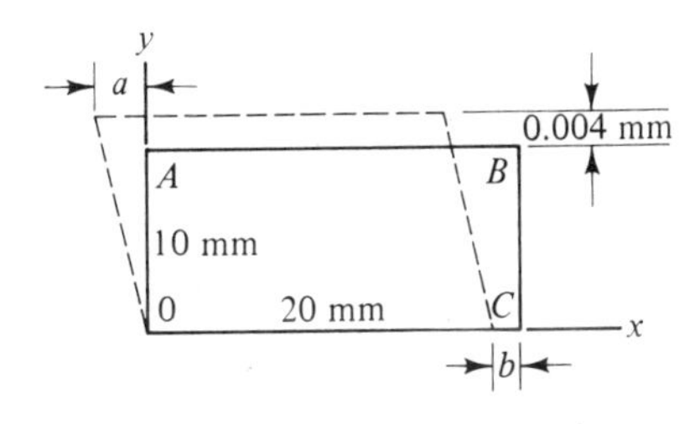

Problem 9.74

9.75 When deformed, the rectangle shown distorts into a parallelogram. Sides AB and CD elongate 0.02 mm and rotate 0.002 radian clockwise; sides AD and BC elongate 0.02 mm and rotate 0.003 radian clockwise. Find the principal strains and the orientations of the principal axes.

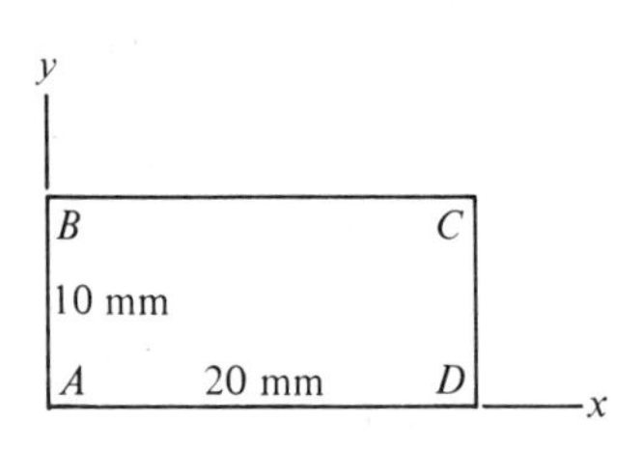

Problem 9.75

9.76 The strains at a point on the surface of a loaded body are found to be $\varepsilon_x = 100\ \mu$, $\varepsilon_y = -700\ \mu$, and $\gamma_{xy} = 600\ \mu$. Determine the principal stresses and the orientation of the principal axes. The material is steel with $E = 207\ \text{GN/m}^2$, $G = 83\ \text{GN/m}^2$, and $\nu = 0.3$.

9.77 Same as Problem 9.76, except that the strain components are $\varepsilon_x = 750\ \mu$, $\varepsilon_x = -250\ \mu$, and $\gamma_{xy} = -600\ \mu$.

9.78 If $\varepsilon_x = 600\ \mu, \varepsilon_y = -400\ \mu$, and $\gamma_{xy} = 250\mu$, determine the normal strain along a line oriented 30° counterclockwise from the x axis.

9.79 Calculate the normal strain in direction A-A for the state of stress shown if $E = 10^7$ psi and $\nu = \frac{1}{4}$. What is the magnitude of the maximum shear strain?

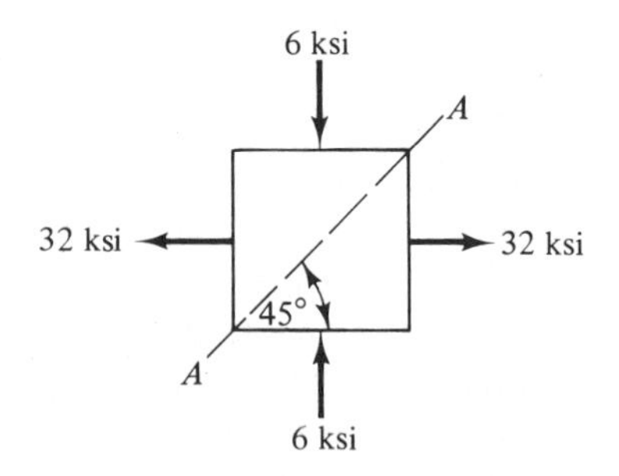

Problem 9.79

9.80 A thin-walled copper pipe with radius to thickness ratio $R/t = 10$ carries a gas under a pressure of 600 psi. If a gage for measuring normal strains is attached to the outer surface of the pipe, what strain will the gage indicate if it is oriented (a) parallel to the pipe axis, (b) perpendicular to the pipe axis, and (c) at an angle of 25° from the pipe axis? Use $E = 12 \times 10^6$ psi and $\nu = \frac{1}{3}$. Treat the pipe as an open-ended pressure vessel, i.e., $\sigma_L = 0$.

9.81 and 9.82 For the states of strain given in Problems 9.69 and 9.70, determine the strain components with respect to nt axes rotated 30° counterclockwise from the xy axes.

9.8 STRAIN GAGE ROSETTES

Although it is possible to derive equations for computing the stresses in a variety of members, there are many problems in which the loading or geometry is so complicated that the only practical approach is to determine the stresses experimentally. It is usually the maximum stresses that are of interest, and, with few exceptions, these occur at the surface of a member. Consequently, the stresses can be determined by measuring the strains on the surface and then calculating the stresses from the stress-strain relations. Since a state of plane stress exists at a free surface, it is only necessary to know the stress components with respect to two perpendicular axes in the plane of the surface, say σ_x, σ_y, and τ_{xy}. From the generalized Hooke's law for plane stress, Eqs. (9.12), we see that these stresses can be determined if the two normal strains ε_x and ε_y and the shear strain γ_{xy} are known.

Normal strains are relatively easy to measure, and electrical resistance strain gages (discussed in Chapter 11) are widely used for this purpose. However, shear strains are difficult to measure accurately because they involve very small changes in angles. Fortunately, they can be determined indirectly by measuring the normal strain in a third direction at the point of interest. Thus, the state of strain at a point on the surface of a member can be determined by measuring the normal strains in three different directions at that point. Special assemblages of strain gages called *strain gage rosettes* are available for making such measurements. Two common configurations are the rectangular rosette [Figure 9.22(a)], in which the three gages are 45° apart, and the delta rosette [Figure 9.22(b)] in which they are 60° apart.

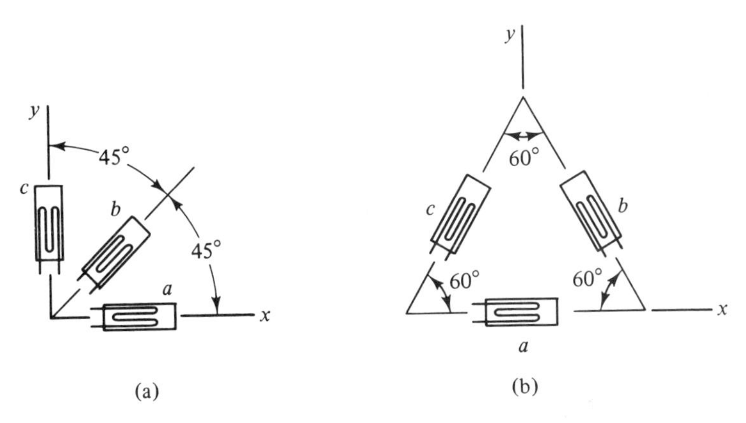

Figure 9.22

Each of the gages in a rosette detects the normal strain in the direction along which the gage is aligned. These three pieces of data are sufficient to determine the strain components ε_x, ε_y, and γ_{xy}.

For example, consider the rectangular rosette [Figure 9.22(a)]. Let ε_a, ε_b, and ε_c be the respective strain readings from the three gages. Since the choice of axes is arbitrary, we take axes xy as shown in the figure. Thus, the angles between the gages and the x axis are, respectively, $\theta_a = 0°$, $\theta_b = 45°$, and $\theta_c = 90°$. Applying the strain transformation equation, Eq. (9.13), in each of the directions a, b, and c, we obtain

$$\begin{aligned}
\varepsilon_a &= \varepsilon_x \cos^2 0° + \varepsilon_y \sin^2 0° + \gamma_{xy} \sin 0° \cos 0° = \varepsilon_x \\
\varepsilon_b &= \varepsilon_x \cos^2 45° + \varepsilon_y \sin^2 45° + \gamma_{xy} \sin 45° \cos 45° = \tfrac{1}{2}(\varepsilon_x + \varepsilon_y + \gamma_{xy}) \qquad (9.18) \\
\varepsilon_c &= \varepsilon_x \cos^2 90° + \varepsilon_y \sin^2 90° + \gamma_{xy} \sin 90° \cos 90° = \varepsilon_y
\end{aligned}$$

Thus, we have three equations from which to determine the three unknowns ε_x, ε_y, and γ_{xy}. Once these strains are known, the stress components can be determined from the generalized Hooke's law, provided the material is isotropic and linearly elastic.

The experimental procedure discussed here is widely used for determining the strains and stresses. However, a number of other procedures are available. Some of these are described in Chapter 11, and others are discussed in texts on experimental stress analysis.

Example 9.11. Experimental Determination of Principal Stresses. A delta rosette [Figure 9.23(a)] applied to the surface of a member made of 2024-T3 aluminum alloy gives the strain readings $\varepsilon_a = 1224\ \mu\text{m/m}$, $\varepsilon_b = -66\ \mu\text{m/m}$, and $\varepsilon_c = 442\ \mu\text{m/m}$. Determine the principal stresses at this location and the orientation of the principal axes.

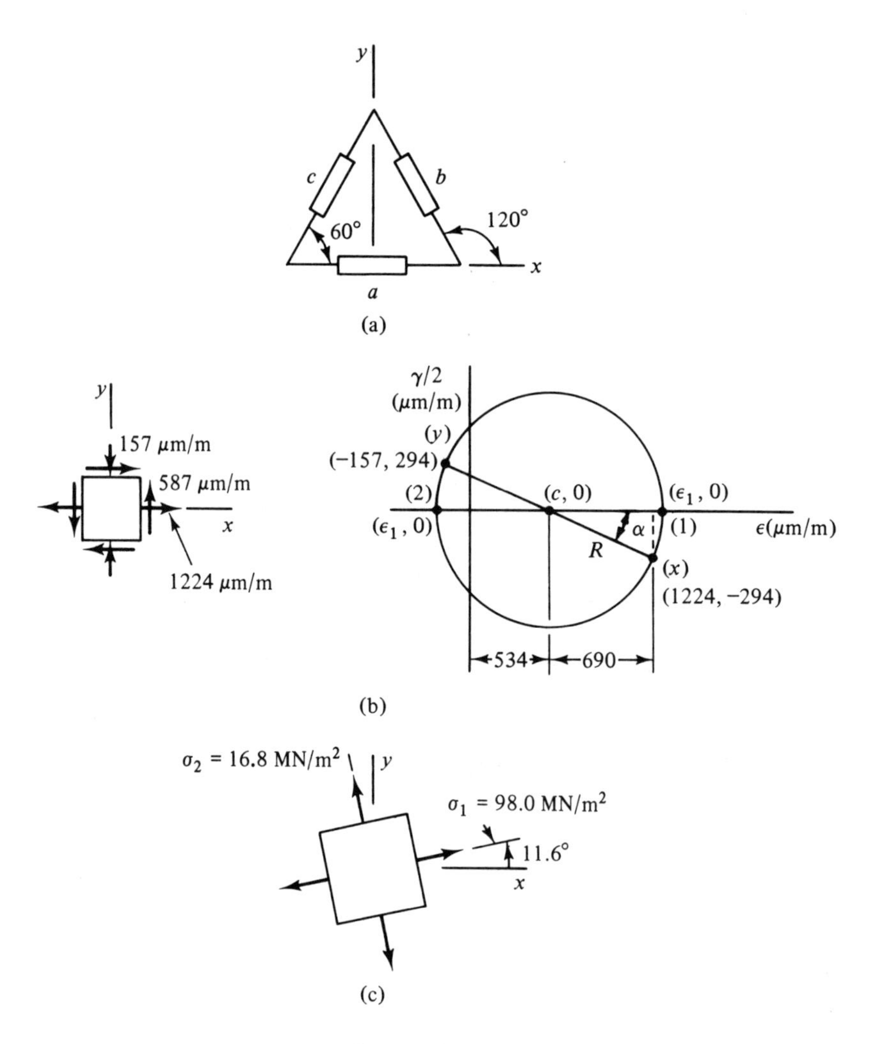

Figure 9.23

Solution. We choose axes xy as shown in Figure 9.23(a). Thus, the angles between the gages and the x axis are, respectively, $\theta_a = 0°$, $\theta_b = 120°$, and $\theta_c = 60°$. Applying the strain transformation equation, Eq. (9.13), in each of the three directions a, b, and c, we have

$$\varepsilon_a = \varepsilon_x \cos^2 0° + \varepsilon_y \sin^2 0° + \gamma_{xy} \sin 0° \cos 0° = \varepsilon_x$$

$$\varepsilon_b = \varepsilon_x \cos^2 120° + \varepsilon_y \sin^2 120° + \gamma_{xy} \sin 120° \cos 120° = \frac{\varepsilon_x}{4} + \frac{3\varepsilon_y}{4} - \frac{\sqrt{3}}{4}\gamma_{xy}$$

$$\varepsilon_c = \varepsilon_x \cos^2 60° + \varepsilon_y \sin^2 60° + \gamma_{xy} \sin 60° \cos 60° = \frac{\varepsilon_x}{4} + \frac{3\varepsilon_y}{4} + \frac{\sqrt{3}}{4}\gamma_{xy}$$

Substituting the given values of ε_a, ε_b, and ε_c into these equations and solving, we obtain

$$\varepsilon_x = 1224\ \mu\text{m/m} \qquad \varepsilon_y = -157\ \mu\text{m/m} \qquad \gamma_{xy} = 587\ \mu\text{m/m}$$

There are two ways to proceed from here. We can either determine σ_x, σ_y, and τ_{xy} from the generalized Hooke's law, Eqs. (9.12), and then use Mohr's circle to find the principal stresses or we can determine the principal strains ε_1 and ε_2 and then compute σ_1 and σ_2 from Eqs. (9.12). We shall use the latter approach here.

The Mohr's circle for strain is shown in Figure 9.23(b). The center of the circle is located at (534, 0) μm/m and its radius is

$$R = \sqrt{690^2 + 294^2} = 750\ \mu\text{m/m}$$

Thus,

$$\varepsilon_1 = c + R = 534 + 750 = 1284\ \mu\text{m/m}$$

$$\varepsilon_2 = c - R = 534 - 750 = -216\ \mu\text{m/m}$$

$$\tan\alpha = \frac{294}{690} \qquad \alpha = 23.1°$$

The principal stresses are computed from Eqs. (9.12) by using the values $E = 72\ \text{GN/m}^2$ and $\nu = 0.33$ obtained from Table A.2:

$$\sigma_1 = \frac{E}{(1-\nu^2)}(\varepsilon_1 + \nu\varepsilon_2) = \frac{72 \times 10^9\ \text{N/m}^2}{(1 - 0.33^2)}\left[1284 + 0.33(-216)\right]10^{-6}$$

$$= 98.0 \times 10^6\ \text{N/m}^2 \text{ or } 98.0\ \text{MN/m}^2 \qquad \textit{Answer}$$

$$\sigma_2 = \frac{E}{(1-\nu^2)}(\varepsilon_2 + \nu\varepsilon_1) = \frac{72 \times 10^9\ \text{N/m}^2}{(1 - 0.33^2)}\left[-216 + 0.33(1284)\right]10^{-6}$$

$$= 16.8 \times 10^6\ \text{N/m}^2 \text{ or } 16.8\ \text{MN/m}^2 \qquad \textit{Answer}$$

For an isotropic material, the principal axes of stress coincide with the principal axes of strain, which are oriented 11.6° counterclockwise from the xy axes [Figure 9.23(c)].

PROBLEMS

9.83 and 9.84 The strains on the surface of a member were measured by using a delta rosette oriented as shown in Figure 9.22(b). Determine the strain components ε_x, ε_y, and γ_{xy} for the gage readings indicated below. (Recall that $\mu = 10^{-6}$.)

9.83. $\varepsilon_a = 1360\ \mu$ $\varepsilon_b = -2450\ \mu$ $\varepsilon_c = -1310\ \mu$

9.84. $\varepsilon_a = 294\ \mu$ $\varepsilon_b = -427\ \mu$ $\varepsilon_c = 67\ \mu$

9.85 and 9.86 Same as Problems 9.83 and 9.84, except that the strains were measured by using a rectangular rosette oriented as shown in Figure 9.22(a).

9.87 Strains on the surface of a bulkhead were measured in the three directions shown. The strain readings are $\varepsilon_a = 1000\ \mu$, $\varepsilon_b = -700\ \mu$, and $\varepsilon_c = -400\ \mu$. Determine the principal strains, the maximum shear strain, and the orientation of the corresponding axes.

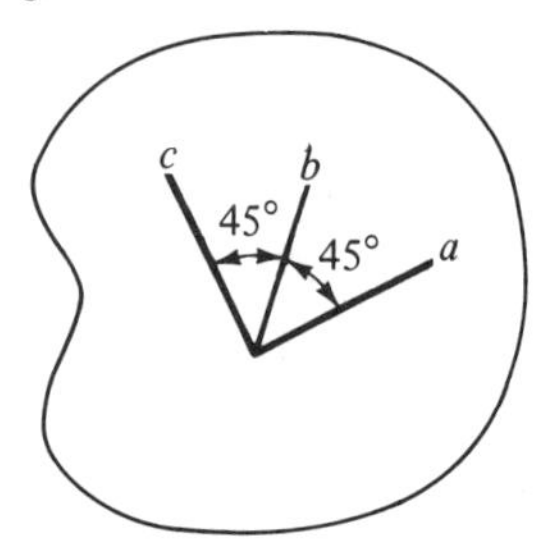

Problem 9.87

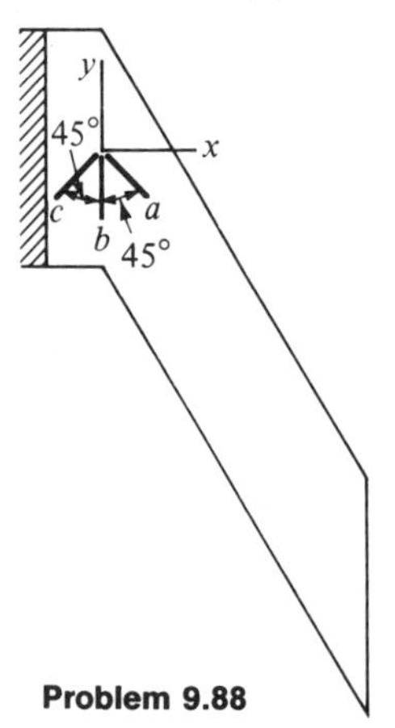

Problem 9.88

9.88 The response of a swept-back aircraft wing is simulated by using a thin steel plate with the shape shown loaded perpendicular to the plane of the figure. The readings from the strain rosette are $\varepsilon_a = 271\ \mu$, $\varepsilon_b = 66\ \mu$, and $\varepsilon_c = -50\ \mu$. Determine the principal strains, the maximum shear strain, and the orientation of the corresponding axes.

9.89 to 9.91 The strains on a free surface of a member were measured by using three strain gages oriented as shown. The strain readings and material properties are as indicated. Determine the principal stresses and maximum in-plane shear stress and show them

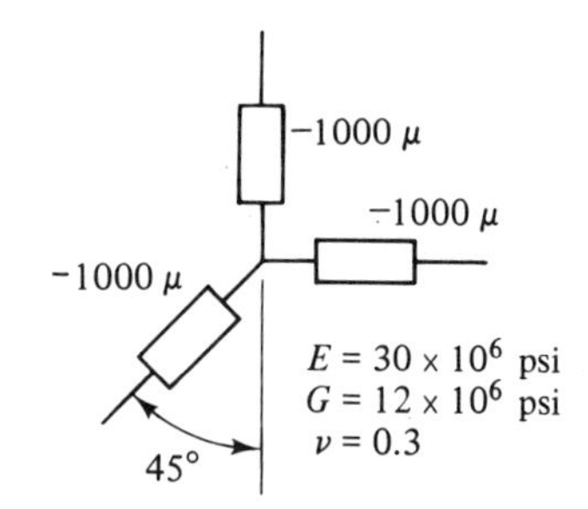

Problem 9.89

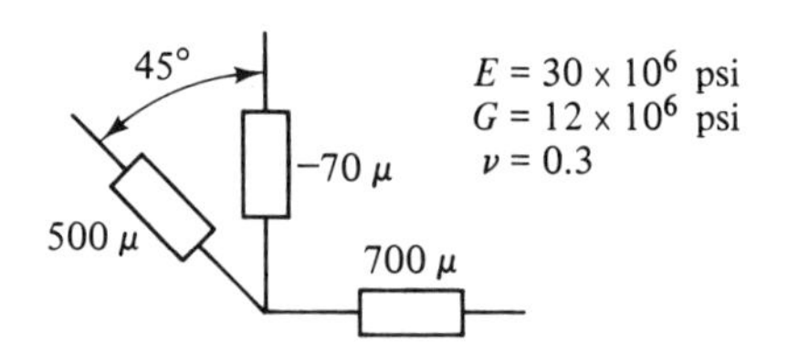

Problem 9.90

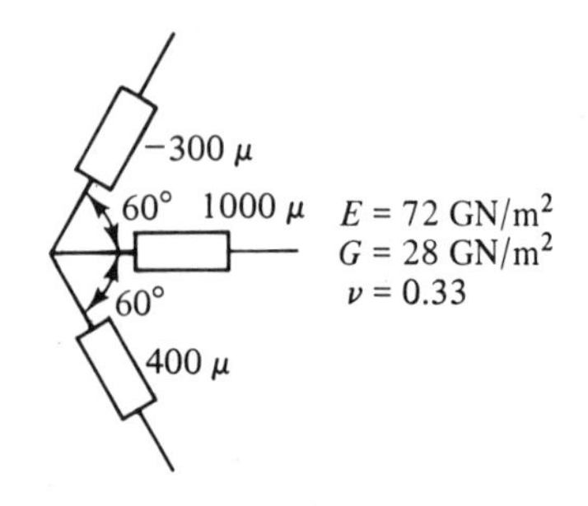

Problem 9.91

on properly oriented elements. Also determine the absolute maximum shear stress.

9.92 The readings from three strain gages mounted as shown on a 1-in. diameter shaft subjected to combined axial loading and torsion are $\varepsilon_a = -485\ \mu$, $\varepsilon_b = 191\ \mu$, and $\varepsilon_c = 618\ \mu$. Determine the axial load F and the magnitude C of the applied couples; $E = 30 \times 10^6$ psi, $G = 12 \times 10^6$ psi, and $\nu = 0.3$.

9.93 Same as Problem 9.92, except that the strain readings are $\varepsilon_a = -370\ \mu$, $\varepsilon_b = 255\ \mu$, and $\varepsilon_c = 549\ \mu$.

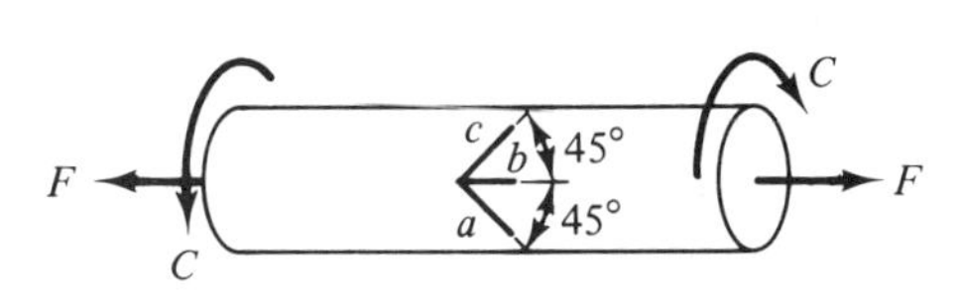

Problem 9.92

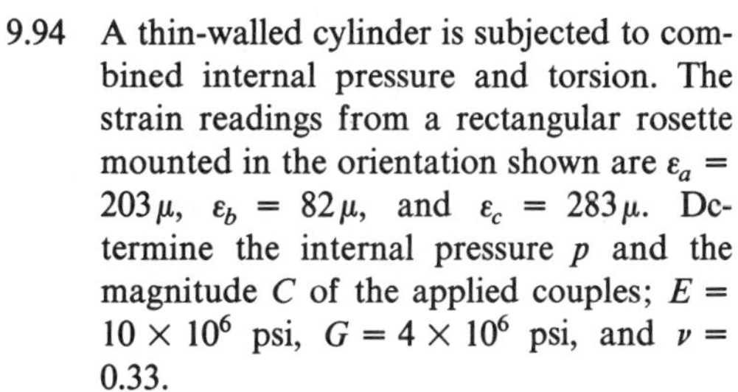

9.94 A thin-walled cylinder is subjected to combined internal pressure and torsion. The strain readings from a rectangular rosette mounted in the orientation shown are $\varepsilon_a = 203\,\mu$, $\varepsilon_b = 82\,\mu$, and $\varepsilon_c = 283\,\mu$. Determine the internal pressure p and the magnitude C of the applied couples; $E = 10 \times 10^6$ psi, $G = 4 \times 10^6$ psi, and $\nu = 0.33$.

9.95 Same as Problem 9.94, except that the strain readings are $\varepsilon_a = 338\ \mu$, $\varepsilon_b = 128\ \mu$, and $\varepsilon_c = 418\ \mu$.

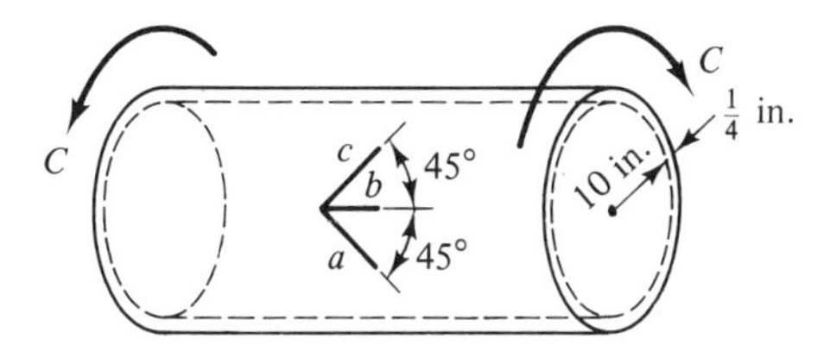

Problem 9.94

9.9 CLOSURE

In this chapter we were concerned with procedures for determining the states of stress and strain in a body subjected to various types of combined loadings. Particular emphasis was placed upon the determination of the maximum stresses because they usually determine whether or not a member will fail. There are two steps involved in determining the maximum stresses. First, we must use our knowledge of the stress distribution in members obtained from the basic load-stress relations to determine the points in the body where the maximum stresses will likely occur. Second, Mohr's circle is used to determine the maximum stresses at these points and the planes on which they act. It is sometimes necessary to check more than one point before the actual maximum stresses can be determined.

The basic tool for transferring stress and strain components from one set of axes to another is Mohr's circle. Both stresses and strains transfer in exactly the same way because they are both what is known in mathematics as *tensor* quantities, which transform in a specific way under a rotation of axes. Area moments of inertia are also tensor quantities.

Mohr's circle for stress and strain and the underlying transformation equations are based solely on equilibrium and geometric considerations. Consequently, they apply regardless of the material properties. However, the generalized Hooke's law used to relate the stresses and the strains holds only for linearly elastic isotropic materials.

chapter ten BUCKLING

10.1 INTRODUCTION

A straight slender member subjected to an axial compressive load is called a *column*. Such members are commonly encountered in trusses and in the framework of buildings where they serve primarily to carry the weight of the structure and the applied loads to the foundation. They are also found as machine linkages, flagpoles, signposts, supports for highway overpasses, and numerous other structural and machine elements.

If a compression member is relatively short, it will remain straight when loaded, and the load-stress and load-deformation relations presented in Chapters 5 and 6 apply. However, a different type of behavior is observed for longer members. When the compressive load reaches a certain critical value, the column undergoes a bending action in which the lateral deflection becomes very large with little increase in load. This response is called *buckling* and usually leads to collapse of the member. Failures due to buckling are frequently catastrophic because they occur suddenly with little warning when the critical load is reached.

Our goal in this chapter is to determine the critical loads at which various types of columns will buckle. The analysis is much the same as for other load-carrying members, but there are some fundamental differences. The most significant of these is that the equilibrium equations must be based on the deformed geometry of the member instead of the undeformed geometry used in all previous cases. In other words, the rigid-body approximation cannot be used in deriving the equations governing the buckling of a column.

The concept of buckling is closely allied to the idea of stability of an equilibrium position introduced in Section 4.2. The latter concept is, in turn, related to the behavior of a system in equilibrium when subjected to the bumps, jolts, vibration, extraneous forces, and other disturbances, irregularities, and imperfections that are always present. Such factors have been ignored until now, for they had no significant influence on the response of the members considered. This is not the case for columns and other members subject to buckling. The relationship between buckling and the stability of an equilibrium position will be discussed in this chapter.

Buckling can occur with most types of members under the proper conditions. For example, we have already mentioned in Chapter 7 that a thin-walled tube tends to wrinkle (buckle) when twisted. Similarly, a thin-walled cylinder will buckle when subjected to an excessive external pressure, and a beam with a deep, narrow cross section can buckle by twisting and deflecting sideways if the span is too long. Problems like these are not as common as those involving columns, however, and will not be considered in this text.

10.2 BASIC CONCEPTS

Buckling of perfect systems. The concepts of stability of an equilibrium position and buckling can probably best be described by means of an example. To this end, let us consider the system shown in Figure 10.1(a). The system consists of a rigid bar supported at the bottom by a hinge and torsional spring and subjected to a force that always remains vertical. This system is a crude model of an elastic column, with the bending resistance of the member represented by the torsional spring.

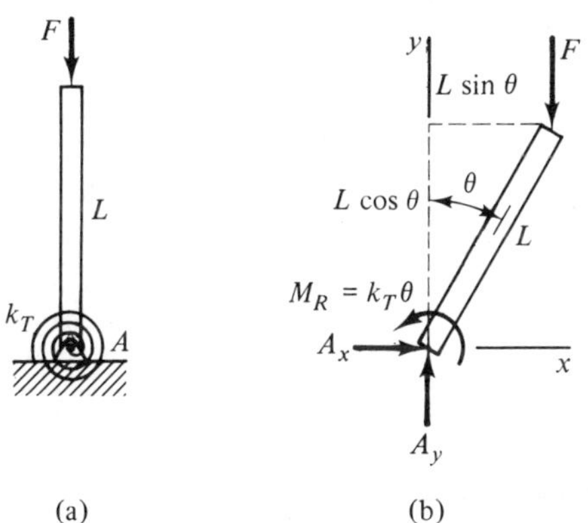

Figure 10.1

It is clear that the applied load will have no effect upon the system as long as the bar remains vertical and the load is aligned with its axis. However, if the system is disturbed so that the bar rotates through some angle θ [Figure 10.1(b)], there will be a moment of magnitude $FL \sin \theta$ tending to rotate the bar further in the same direction. This rotation is resisted by the moment exerted by the spring, which has magnitude $k_T\theta$. This moment tends to return the bar to its original position. We shall refer to these two moments as the *disturbing moment*, M_D, and the *restoring moment*, M_R, respectively.

Equilibrium of the bar requires that $M_D = M_R$, or

$$FL \sin \theta = k_T \theta \tag{10.1}$$

We wish to solve this equation for the angle θ through which the bar will rotate for a given load F. To do this, we rewrite the equation as

$$\sin \theta = \frac{\theta}{\beta} \tag{10.2}$$

where

$$\beta = \frac{FL}{k_T} \tag{10.3}$$

Equation (10.2) can be solved by choosing values of θ and determining the corresponding values of the load parameter β. However, it is more informative to solve it by plotting both sides of the equation and determining the values of θ at the intersection points [Figure 10.2(a)]. Note that the sine curve in Figure 10.2(a) represents the disturbing moment and the straight lines represent the restoring

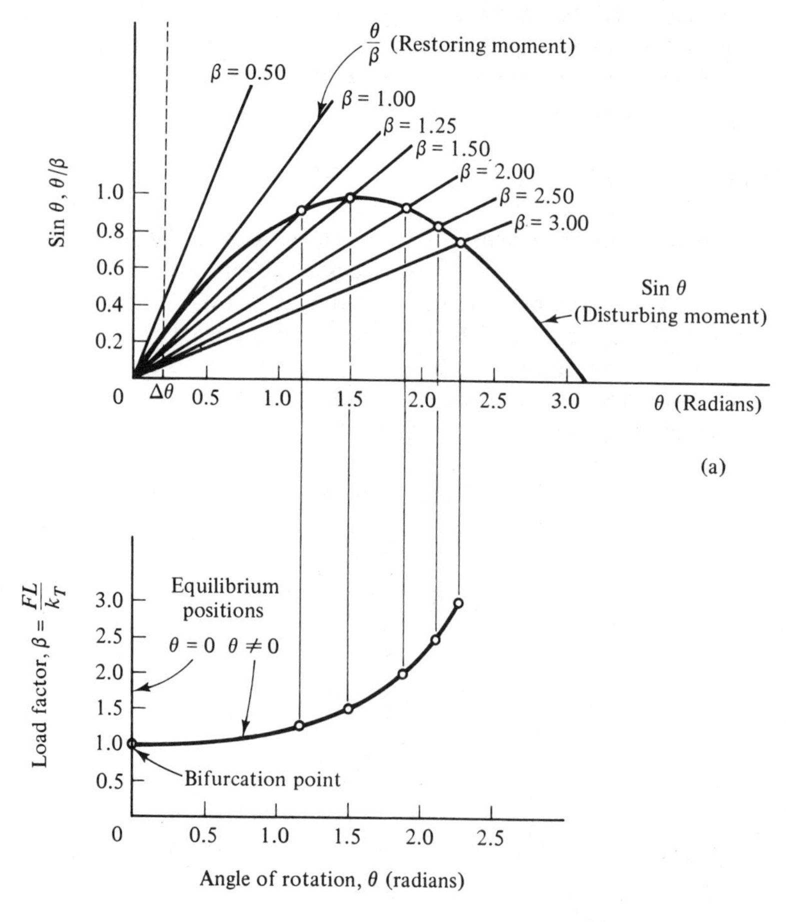

Figure 10.2

moment for various values of β. This procedure yields the solutions shown in Figure 10.2(b). There are other solutions corresponding to negative values of θ which are not shown.

The values of θ obtained from Eq. (10.2) define the equilibrium positions of the bar. Note that $\theta = 0$ is always a solution of this equation and is, therefore, always an equilibrium position. Furthermore, this is the only solution if $\beta < 1$ ($F < k_T/L$), as can be seen from Figure 10.2(a). For $\beta > 1$ ($F > k_T/L$), there is a second set of solutions for which θ is nonzero. The point at which the two solution curves cross is called the *bifurcation point* [Figure 10.2(b)].

Now let us interpret these solutions in terms of the mechanics of the problem. Suppose that the bar is disturbed some small amount $\Delta\theta$ from the equilibrium position $\theta = 0$. It is clear from Figure 10.2(a) that the restoring moment at this new position of the bar is larger than the disturbing moment for $\beta < 1$. Thus, the bar will return to its original position after the disturbance, in which case the equilibrium position $\theta = 0$ is said to be *stable*. There may be some oscillations, but they will eventually die out. In contrast, the disturbing moment is seen to be larger than the restoring moment for $\beta > 1$. The equilibrium position $\theta = 0$ is *unstable* in this case, and the bar will move away from this position if disturbed.

The dividing line between these two types of behavior corresponds to

$$\beta_{\text{cr}} = 1 \tag{10.4}$$

or

$$F_{\text{cr}} = \frac{k_T}{L} \tag{10.5}$$

This expression defines the so-called *critical load*, F_{cr}, for the system.

Using this same procedure, it can be shown that the nonzero equilibrium positions, which exist only for $\beta > 1$, are stable. Thus, the stable and unstable equilibrium positions of the bar are as indicated in Figure 10.3.

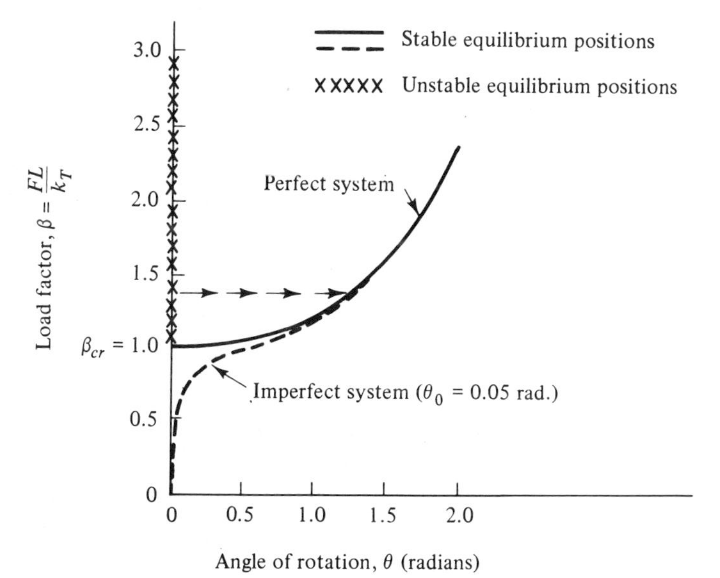

Figure 10.3

Once the critical load is exceeded, any disturbance will cause the bar to undergo a relatively large rotation to the nonzero equilibrium position, which is now the stable one. It is this type of response, indicated by the arrows in Figure 10.3, that was referred to earlier as buckling. Notice that the angle of rotation of the bar increases very rapidly as the applied load is increased above the critical value.

Systems with Imperfections. To this point, we have assumed a perfect system. This is not entirely realistic because there will always be some imperfections present. For example, the bar may not be exactly vertical, or the force may not be in perfect alignment with the axis of the bar. The question is, what effect do such imperfections have on the response of the system?

This question can be answered by considering an imperfection in the form of a small initial angle θ_0 between the bar and the vertical. The equilibrium equation in this case is

$$FL \sin(\theta + \theta_0) = k_T\theta \tag{10.6}$$

and its solution for $\theta_0 = 0.05$ radian (2.9°) is indicated by the dashed curve in Figure 10.3. Notice that, with the imperfection, there is some rotation of the bar for all nonzero values of load. Furthermore, all equilibrium positions are now stable, and there is no well-defined critical load. However, the angle of rotation increases rapidly and becomes very large in the vicinity of the critical load for the perfect system, so that load can still be taken as the buckling load. In other words, the buckling load for the system is not affected by small imperfections.

Determination of the critical load. If only the critical load is of interest, as is the case in most problems, it is not necessary to go through so complete an analysis. The desired information can be determined from a linear analysis, as we shall now show.

For small values of θ, $\sin\theta \cong \theta$, and Eq. (10.2) becomes

$$\theta = \frac{\theta}{\beta} \tag{10.7}$$

This equation has a nonzero solution only if $\beta = 1$, which corresponds to the critical load for the system. Similarly, Eq. (10.6) for the imperfect system reduces to

$$\theta + \theta_0 = \frac{\theta}{\beta} \tag{10.8}$$

which has the solution

$$\theta = \frac{\beta\theta_0}{1 - \beta} \tag{10.9}$$

This equation predicts an infinite angle of rotation for $\beta = 1$, which again corresponds to the critical load.

These results indicate that, for a perfect system, the critical load can be obtained from the condition that the linear equilibrium equation has a nonzero solution. In other words, we seek the load for which the system will remain in equilibrium in a slightly deflected position. For an imperfect system, the critical load is obtained from the condition that the displacements become infinite.

The results presented in this section were obtained for a specific example, but they also apply to the buckling of columns and many other common types of structural elements.

Example 10.1. Critical Load for a Bar-Spring System. Another simple model of a column consists of a rigid bar hinged at the bottom and supported at the top by a horizontal spring with spring constant k [Figure 10.4(a)]. The spring exerts zero force when the bar is vertical, and the applied force F always acts vertically downward. Determine the critical load for this system if (a) it is perfect and (b) there is an imperfection in the form of a small horizontal force F_H acting at the tip of the bar.

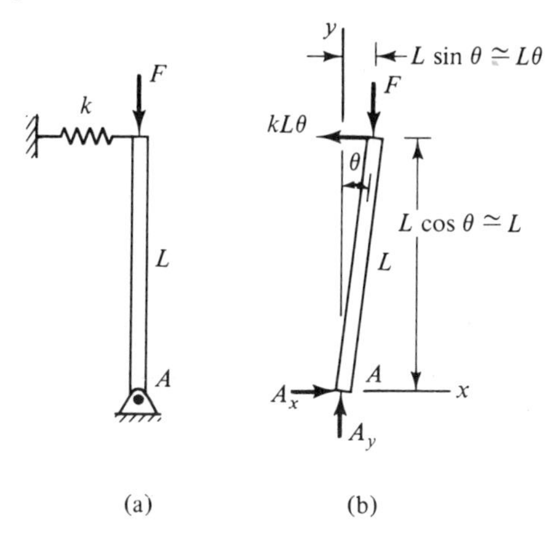

Figure 10.4

Solution. (a) Figure 10.4(b) shows the FBD of the bar when it is in a slightly deflected position. The spring exerts a force proportional to its elongation, which, for small values of θ, is $L\theta$. Summing moments about point A, we obtain

$$\overset{+}{\curvearrowleft}\Sigma M_A = kL^2\theta - FL\theta = 0$$

or

$$kL^2\theta = FL\theta$$

This equation yields a nonzero value of θ only if

$$F = F_{cr} = kL \qquad \textbf{\textit{Answer}}$$

(b) Assuming that the horizontal force F_H acts toward the right, we have

$$\overset{+}{\curvearrowleft}\Sigma M_A = kL^2\theta - FL\theta - F_H L = 0$$

or

$$\theta = \frac{F_H}{kL - F}$$

The critical load is the load at which θ becomes infinite:

$$F_{cr} = kL \qquad \textbf{\textit{Answer}}$$

This is the same result obtained in (a), which means that the presence of the horizontal force does not affect the buckling load for the system.

PROBLEMS

10.1 to 10.4 Determine the critical load for the systems shown. The rigid bars have negligible weight, and all connections and supports are frictionless. The torsional springs exert a moment $M = k_T\theta$, where kT is the torsional spring constant and θ is the angle through which the ends of the springs rotate.

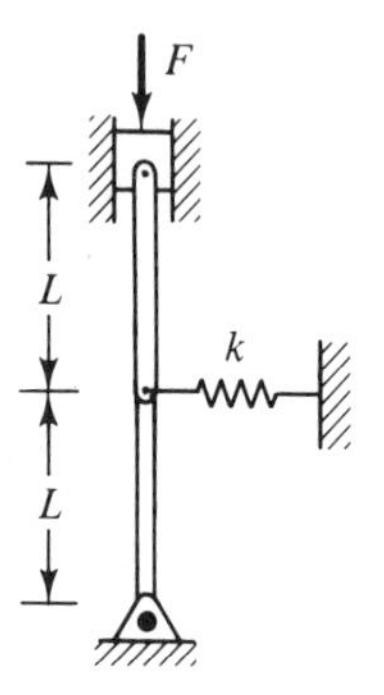

Problem 10.1

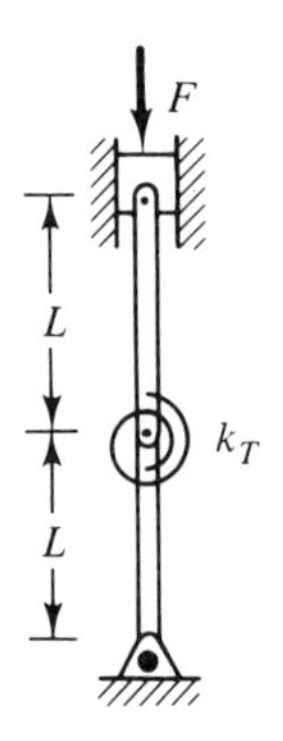

Problem 10.2

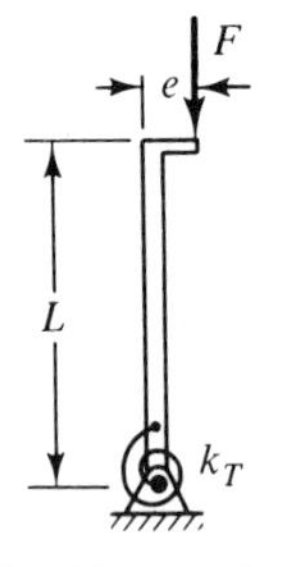

Problem 10.3

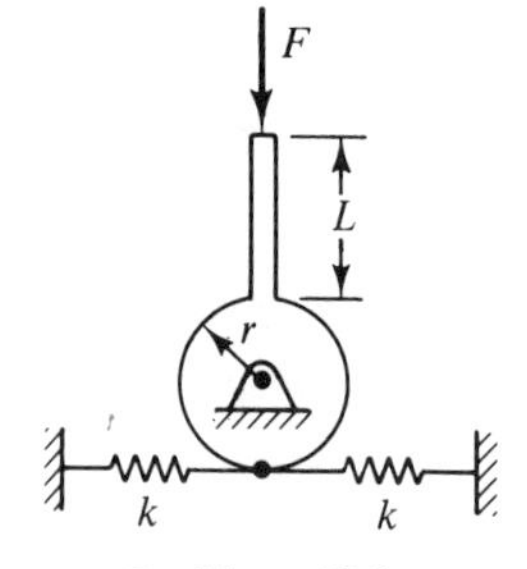

Problem 10.4

10.5 Same as Problem 10.1, except that each of the uniform bars has weight W.

10.6 What is the minimum value of the mass m_2 of the block for which the system will be

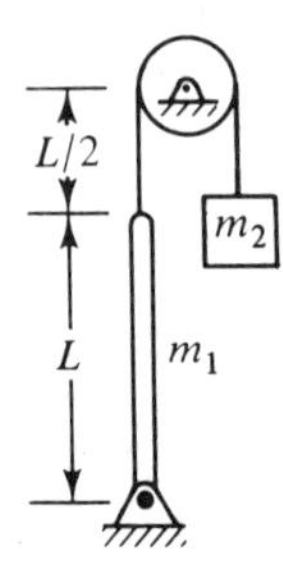

Problem 10.6

stable in the position shown? The bar is uniform and has mass m_1; the pulley is frictionless.

10.7 A rigid bar welded to the end of a cantilever beam is subjected to equal and opposite forces that always remain vertical. Determine the critical load for in-plane buckling of the system by considering the equilibrium of the bar when it is rotated through a small angle θ, as shown. Express the answer in terms of the dimension L and the flexural rigidity EI of the beam.

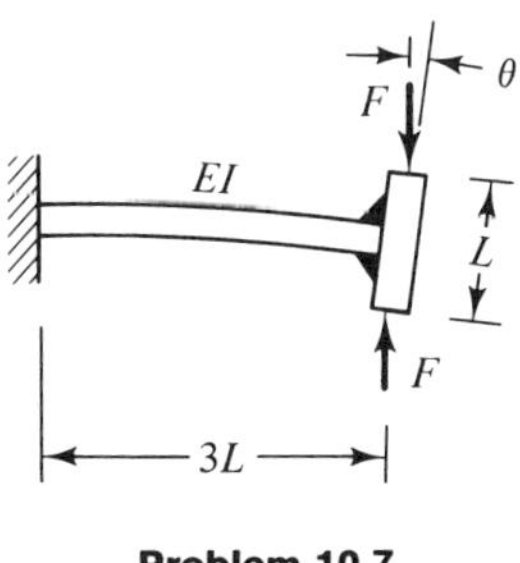

Problem 10.7

10.3 BUCKLING OF ELASTIC COLUMNS

Consider a hinge-ended column supported by smooth pins, or the equivalent, at each end and subjected to an axial compressive load F that always retains its original orientation [Figure 10.5(a)]. We assume that one end of the column is free to move longitudinally with respect to the other, so that the member must support the entire compressive load. It is further assumed that the material behavior is linearly elastic and that there are no imperfections, such as an initial deviation from straightness of the member or an eccentricity of the applied load. The problem is to determine the load at which the column will buckle.

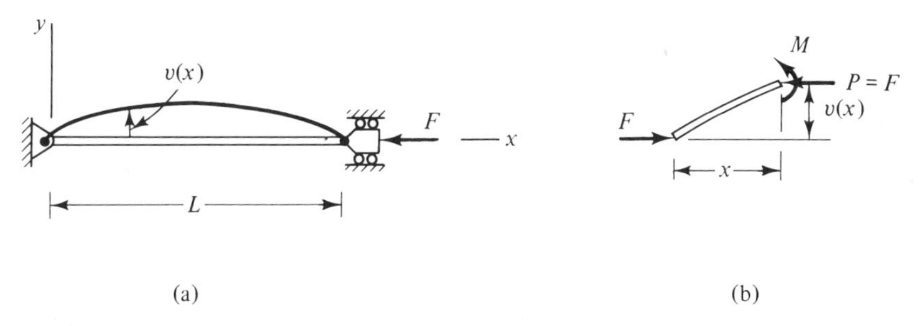

Figure 10.5

According to the procedure given in Section 10.2, the buckling load can be obtained by deflecting the column slightly and determining the value of the compressive load for which it will stay in this position. Figure 10.5(b) shows the free-body diagram of a portion of the column when it is deflected an amount $v(x)$. The bending moment is seen to be

$$M = -Fv$$

Substituting this expression into Eq. (8.18) for the small deflection of a linearly elastic beam, we obtain as the governing differential equation for the bending of the column

$$EI\frac{d^2v}{dx^2} = -Fv \tag{10.10}$$

or

$$\frac{d^2v}{dx^2} + k^2v = 0 \tag{10.11}$$

where

$$k^2 = \frac{F}{EI} \tag{10.12}$$

The term on the right-hand side of Eq. (10.10) represents the disturbing moment, which tends to deflect the column, and the term on the left-hand side represents the restoring moment, which tends to bring it back to its undeflected equilibrium position $v(x) = 0$.

The general solution to Eq. (10.11) is

$$v(x) = A\cos kx + B\sin kx \tag{10.13}$$

where A and B are constants. This can be readily confirmed by direct substitution. Applying the boundary conditions $v(0) = 0$ and $v(L) = 0$, we have

$$v(0) = 0: \quad 0 = A$$

$$v(L) = 0: \quad 0 = B\sin kL$$

It is clear from the second of these relations that $\sin kL$ must be zero in order to have a nonzero solution ($B \neq 0$). This implies that $kL = n\pi$, where n is an integer. Substituting this result into Eq.(10.12), we obtain

$$F_{cr} = \frac{n^2\pi^2EI}{L^2} \qquad n = 1, 2, 3, \ldots \tag{10.14}$$

Equation (10.14) has physical significance only for $n = 1$ because the undeflected equilibrium position $v(x) = 0$ is unstable at loads above the lowest critical load. The values of F_{cr} corresponding to $n > 1$ cannot be attained physically. Thus, the buckling load for the column is

$$F_{cr} = \frac{\pi^2EI}{L^2} \tag{10.15}$$

This result is attributed to the Swiss mathematican Leonhard Euler (1707–1783). Accordingly, Eq. (10.15) is called *Euler's formula*, and the corresponding load is called the *Euler critical load*.

In Eq. (10.15), E is the modulus of elasticity of the material, L is the length of the column, and I is the moment of inertia of the cross section about the bending axis. If the column is free to deflect in any direction, as in the case of ball and socket supports, it will tend to bend about the axis corresponding to the smallest value of I. This axis offers the least resistance to bending, and, as can be seen from

Eq. (10.15), gives the smallest value of the critical load. Of course, if the column is constrained to bend about a certain axis, the value of I about that axis must be used.

According to Eqs. (10.13) and (10.15), the deflected shape of the column is

$$v(x) = B \sin kx = B \sin\left(\frac{\pi x}{L}\right) \tag{10.16}$$

That is, the column buckles into a sine wave. Note that the constant B, which defines the amplitude of the deflection, remains undetermined. It can be obtained only from a nonlinear analysis. This is of little concern because we are usually interested only in the critical load.

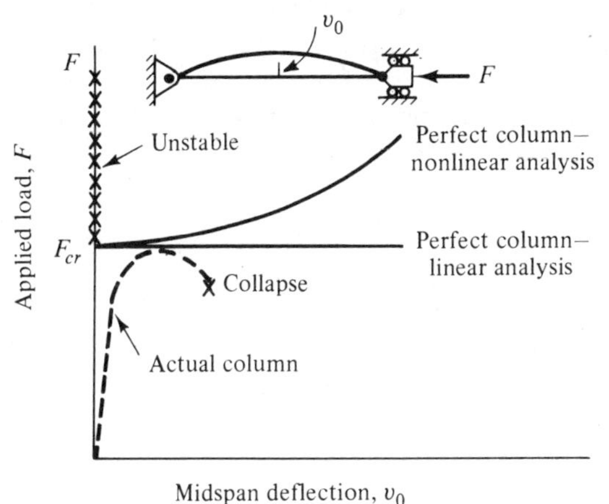

Figure 10.6

The load-deflection curve corresponding to Eq. (10.16) is shown in Figure 10.6, which is a plot of the applied load, F, versus the midspan deflection, v_0. For comparison purposes, the curve obtained from the exact equilibrium equations for large deflections is also shown. Both of these curves are based on the assumption of linearly elastic material behavior.

In actuality, the bending moment and stresses become large as the deflection increases, and the material will yield if it is ductile. Eventually, a plastic hinge will form at the most highly stressed section, and the member will collapse. Thus, the actual load-deflection behavior of a typical column made of a ductile material is as indicated by the dashed curve in Figure 10.6. Because of imperfections, the maximum load that can be supported is somewhat less than the Euler critical load. Consequently, some safety factor should always be used with Eq. (10.15). Imperfections are also responsible for there being some deflection at all values of load.

Euler's formula for the critical load is based upon the assumption of linearly elastic material behavior. Since many of the columns encountered in practice are such that this condition is not met, the validity of Eq. (10.15) should always be checked.

Since most columns remain very nearly straight prior to buckling, the average compressive stress at the critical load is $\sigma_{cr} = F_{cr}/A$. Writing the moment of inertia as $I = Ar^2$, where r is the radius of gyration, we have from Eq. (10.15)

$$\sigma_{cr} = \frac{F_{cr}}{A} = \frac{\pi^2 E}{(L/r)^2} \tag{10.17}$$

Thus, the condition for the validity of Euler's formula can be expressed as

$$\sigma_{cr} = \frac{\pi^2 E}{(L/r)^2} \leqslant \sigma_{PL} \tag{10.18}$$

or

$$(L/r)^2 \geqslant \frac{\pi^2 E}{\sigma_{PL}} \tag{10.19}$$

where σ_{PL} is the proportional limit of the material. The quantity L/r in these equations is called the *slenderness ratio* of the column.

The various ranges of column behavior are conveniently displayed on a plot of the critical stress versus the slenderness ratio (Figure 10.7). Columns that satisfy the condition given in Eq. (10.18) or Eq. (10.19), and for which Euler's formula applies, are designated *long columns*. For steel with $E = 30 \times 10^6$ psi and $\sigma_{PL} = 30$ ksi, a long column is one for which $L/r \geqslant 100$. Columns with $L/r > 200$ are seldom used because they can support very little load.

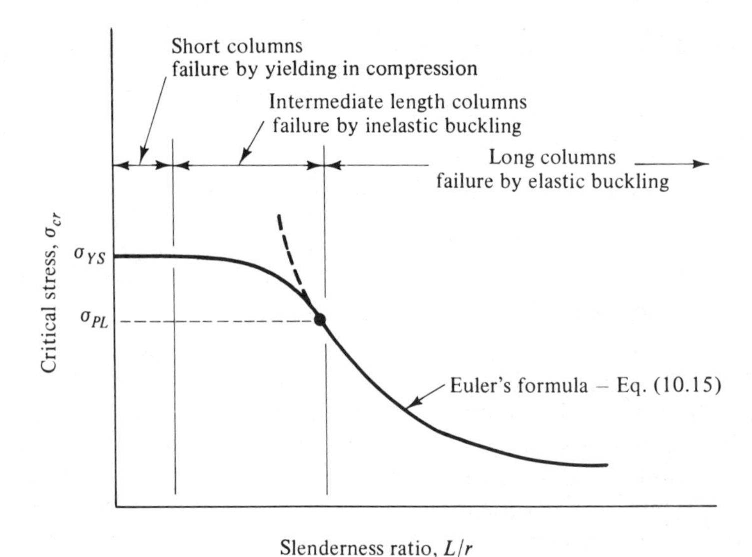

Figure 10.7

Compression members with a slenderness ratio of approximately 10 or less do not usually buckle. They fail due to general yielding when the compressive stress reaches the yield strength of the material. These so-called *short columns*, or *compression blocks*, were considered in Chapters 5 and 6.

Columns that fall in between these two extremes are said to be of *intermediate length*. They fail by buckling, but Euler's formula does not apply because the stresses exceed the proportional limit. The inelastic buckling of columns will be considered in Section 10.5.

Euler's formula for the critical load for an elastic column is remarkable in that it contains no measure of the strength of the material. The failure load depends only upon the material stiffness and the geometry of the member. This is in contrast to the members considered in earlier chapters, for which the failure load was usually found to depend upon the ultimate or yield strength of the material.

Example 10.2. Adequacy of a Column. It is proposed that 4-in.×4-in. nominal timber columns with $E = 1.6 \times 10^6$ psi and $\sigma_{PL} = 5400$ psi be used as intermediate supports under a floor beam in a building. Each column is 8 ft long and must support an estimated load of 10 000 lb. Are the proposed columns adequate, and, if so, what safety factor do they provide? What is the slenderness ratio of the columns? Assume hinged ends.

Solution. The geometric properties of the members are obtained from Table A.12: $I = 12.51$ in.^{4} and $A = 12.25$ in.^{2}. Assuming linearly elastic material behavior, we obtain from Euler's formula [Eq. (10.15)]

$$F_{cr} = \frac{\pi^2 EI}{L^2} = \pi^2 \frac{(1.6 \times 10^6 \text{ lb/in.}^2)(12.51 \text{ in.}^4)}{(96 \text{ in.})^2} = 21\ 400 \text{ lb}$$

The corresponding critical stress is

$$\sigma_{cr} = \frac{F_{cr}}{A} = \frac{21\ 400 \text{ lb}}{12.25 \text{ in.}^2} = 1750 \text{ psi}$$

which is well below the proportional limit of 5400 psi. Thus, the assumption of linearly elastic material behavior is correct, and Euler's formula is valid.

The columns are adequate because the applied load is less than the buckling load. The safety factor is

$$SF = \frac{\text{failure load}}{\text{working load}} = \frac{21\ 400 \text{ lb}}{10\ 000 \text{ lb}} = 2.14 \qquad \textit{Answer}$$

The radius of gyration and slenderness ratio are, respectively,

$$r = \sqrt{\frac{I}{A}} = \sqrt{\frac{12.51 \text{ in.}^4}{12.25 \text{ in.}^2}} = 1.01 \text{ in.}$$

and

$$\frac{L}{r} = \frac{96 \text{ in.}}{1.01 \text{ in.}} = 95 \qquad \textit{Answer}$$

Example 10.3. Sizing of a Column. What is the most economical (from a weight standpoint) unequal angle section to use for member BC of the truss shown in Figure 10.8(a)? The member is to be made of steel with $E = 207$ GN/m^2 and $\sigma_{YS} = 210$ MN/m^2. Use a safety factor of 2 and assume that the member is free to buckle in any direction.

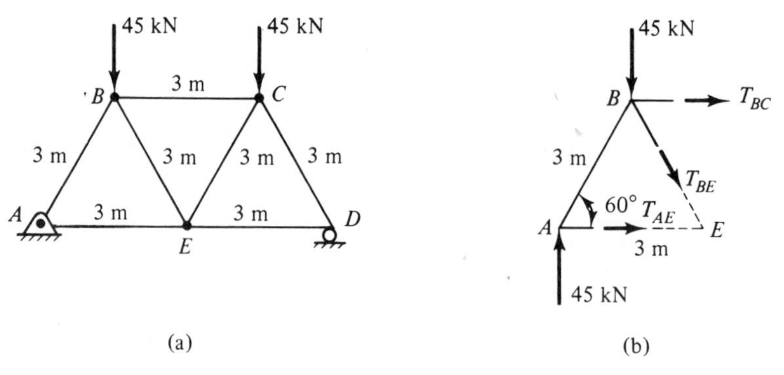

Figure 10.8

Solution. Using the method of sections [Figure 10.8(b)], we have

$$\overset{+}{\curvearrowleft}\ \Sigma M_E = -T_{BC}(3 \sin 60° \text{ m}) - 45 \text{ kN } (3 \text{ m}) + 45 \text{ kN } (3 \cos 60° \text{ m}) = 0$$

$$T_{BC} = -26 \text{ kN} \quad \text{or} \quad 26 \text{ kN (compression)}$$

The buckling load must be at least as large as the design load, which is the safety factor times the actual load, or 52 kN:

$$F_{cr} = \frac{\pi^2 EI}{L^2} \geqslant 52 \text{ kN}$$

The member will tend to buckle about the axis of least moment of inertia; therefore,

$$I_{min} \geqslant \frac{(52 \times 10^3 \text{ N})(3 \text{ m})^2}{\pi^2(207 \times 10^9 \text{ N/m}^2)} = 22.91 \times 10^{-8} \text{ m}^4 \text{ or } 22.91 \text{ cm}^4$$

Turning to Table A.11, we seek the angle section with the smallest mass per unit length that has a minimum moment of inertia of 22.91 cm^4, or larger. (Note that I is minimum about axis VV.) Starting at the bottom of the table and working upward, we find that there are several possibilities:

Nominal Size (mm)	I_{VV}*(cm^4)*	*Mass/Length (kg/m)*
76 × 64	24.4	11.17
89 × 64	24.6	10.57
89 × 76	25.5	7.89
102 × 64	23.2	9.69

There are other sections in each group in the table that have the required value of I, but they have a larger mass than those listed. The 89-mm × 76-mm section with a mass per unit length of 7.89 kg/m is seen to be the most economical.

Since Euler's formula was used in the computations, we must check to see if it is valid. From Table A.11, the area of the section selected is found to be 10.05 cm.^{2} Thus,

$$\sigma_{cr} = \frac{F_{cr}}{A} = \frac{52 \times 10^3 \text{ N}}{10.05 \times 10^{-4} \text{ m}^2} = 5.17 \times 10^7 \text{ N/m}^2 \text{ or } 51.7 \text{ MN/m}^2$$

which is well below the yield strength.

PROBLEMS

10.8 Determine the slenderness ratio and critical stress for a 50-mm × 50-mm steel column 2 m long. Assume hinged ends and elastic material behavior.

10.9 What is the slenderness ratio and critical stress for a 12-ft hinge-ended column made of a 4-in. nominal diameter standard steel pipe with a yield strength of 36 ksi?

10.10 What is the maximum allowable length for a 2-in.×4-in. nominal hinge-ended timber column that must support a compressive load of 1000 lb with a safety factor of 3? $E = 1.6 \times 10^6$ psi and $\sigma_{PL} = 5400$ psi. Assume the member is free to buckle in any direction.

10.11 What compressive load F can a 10-ft hinge-ended column made of a W8 × 17 section safely support? Use a safety factor of 4, and assume the member is free to buckle in any direction. The material is steel with a yield strength of 50 ksi.

10.12 Brace BD of the structure shown is made of a 76-mm × 38-mm steel channel section with a yield point of 250 MN/m^2. Is the brace adequate to support the applied load of 10 kN? If so, what is the safety factor against failure by buckling?

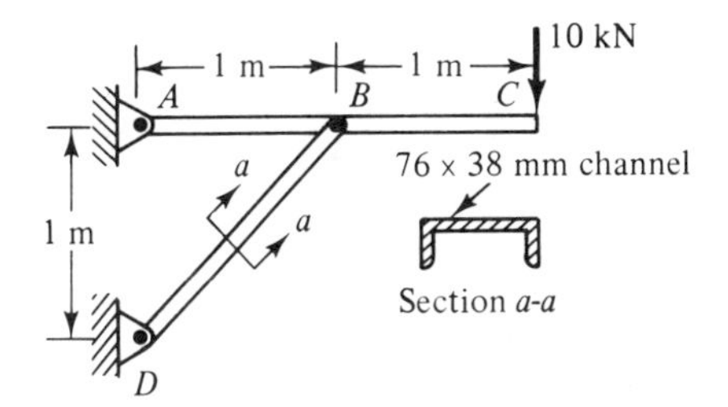

Problem 10.12

10.13 Brace BD of the structure shown is a 2024-T3 aluminum rod 30 mm in diameter. The design specifications require that the brace have a safety factor of 2 for an applied load $F = 4$ kN. Are the specifications met? If not, what diameter rod should be used?

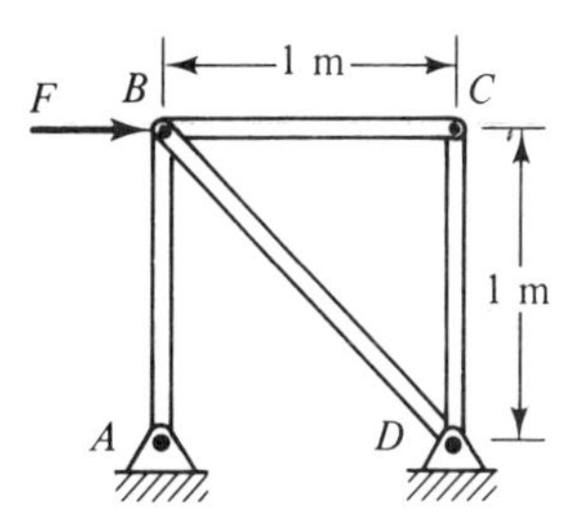

Problem 10.13

10.14 Same as Problem 10.13, except that the applied load is $F = 8$ kN.

10.15 The ceiling in a building is supported by 4-in.×4-in. nominal timber columns with $E = 1.6 \times 10^6$ psi and $\sigma_{PL} = 5400$ psi. Each column is 8 ft long and supports a compressive load of 10 000 lb. If during a fire the columns burn away on all sides at a rate of $\frac{1}{4}$ in. every 15 minutes, approximately how long will it be before the ceiling collapses?

10.16 A cover over a sidewalk at a shopping center is supported on each side by 10-ft columns made of 2-in. nominal square steel tubes with cross-sectional area $A = 1.27\ \text{in.}^2$ and radius of gyration $r = 0.726$ in. If the cover weighs 300 lb/ft of length, including estimated snow loads, how far apart can the columns be spaced? Assume hinged ends and use a safety factor of 6. The material has a yield strength of 36 ksi.

10.17 A $\frac{1}{4}$-in.×$\frac{1}{4}$-in. steel bar 18 in. long is placed between immovable walls. What increase in termperature is required to buckle the bar if $\alpha = 6.5 \times 10^{-6}(°\text{F})^{-1}$?

10.18 Select the most economical equal angle section to use as a 3-m long hinge-ended column that must support a compressive load of 10.9 kN with a safety factor of

2.5. The section is made of steel with $E = 207\ \mathrm{GN/m^2}$ and $\sigma_{YS} = 210\ \mathrm{MN/m^2}$.

10.19 Select the most economical American Standard Channel section that will support a compressive load of 8000 lb with a safety factor of 4. The member is 8 ft long and is made of steel with a yield strength of 36 ksi.

10.20 If the columns in Problem 10.16 are to be made of standard steel pipes spaced 12 ft apart, what size pipe is required?

10.4 EFFECT OF SUPPORT CONDITIONS

The manner in which a column is supported has a significant influence upon its buckling load. This will be demonstrated in this section by determining the critical load for elastic columns with various combinations of support conditions. The basic procedure will be the same as that used in the preceding section for hinge-ended columns.

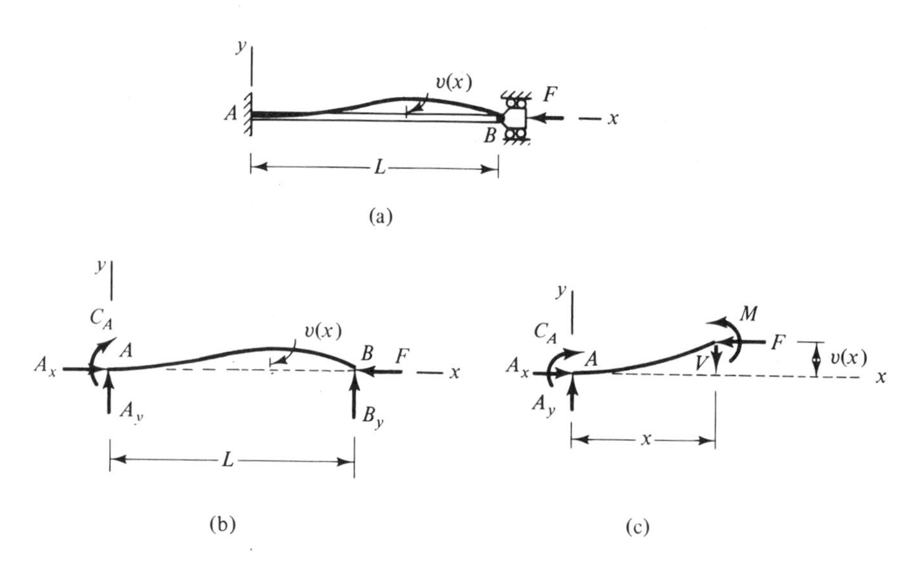

Figure 10.9

Consider a column fixed at one end and hinged at the other and subjected to an axial compressive load F [Figure 10.9(a)]. The free-body diagram of the entire member when it is in a slightly deflected position is shown in Figure 10.9(b). Note that the problem is statically indeterminate because the reactions cannot all be determined from the equations of equilibrium. However, these equations do yield the relationships

$$A_x = F \qquad C_A = -A_y L \qquad A_y = -B_y$$

From the free-body diagram of Figure 10.9(c), the bending moment is found to be

$$M = A_y x + C_A - A_x v = A_y x - A_y L - Fv$$

Substituting this expression into Eq. (8.18) for the small deflection of linearly elastic beams, we obtain as the governing differential equation for the bending of the column

$$EI \frac{d^2v}{dx^2} = A_y x - A_y L - Fv$$

It is convenient to rewrite this equation as

$$\frac{d^2v}{dx^2} + k^2 v = \frac{A_y}{EI}(x - L) \quad ; \quad k^2 = \frac{F}{EI} \tag{10.20}$$

The general solution to Eq. (10.20) consists of the complementary solution, v_C, plus a particular solution, v_P. The complementary solution is obtained by setting the right-hand side of the equation equal to zero, in which case it reduces to Eq. (10.11). Thus, v_C is given by Eq. (10.13):

$$v_C = A \cos kx + B \sin kx$$

The particular solution is assumed to be of the same form as the terms on the right-hand side of the equation, or

$$v_P = Cx + D$$

where C and D are constants. Substituting this expression into Eq. (10.20), we find that it does satisfy the equation for $C = A_y/F$ and $D = -A_y L/F$. Thus, the complete solution is

$$v = v_C + v_P = A \cos kx + B \cos kx + \frac{A_y}{F}(x - L) \tag{10.21}$$

as can be readily confirmed by direct substitution.

Applying the boundary conditions $v(0) = 0$, $\theta(0) = 0$, and $v(L) = 0$, we have

$$\begin{aligned} v(0) = 0&: \quad 0 = A - \frac{A_y L}{F} \\ \theta(0) = 0&: \quad 0 = Bk + \frac{A_y}{F} \\ v(L) = 0&: \quad 0 = A \cos kL + B \sin kL \end{aligned} \tag{10.22}$$

Elimination of A_y between the first two of these equations gives $A = -BkL$. Thus, the third condition becomes

$$B(\sin kL - kL \cos kL) = 0$$

One possibility is $B = 0$. However, this corresponds to the null solution $v(x) \equiv 0$, as can be seen from Eqs. (10.22) and (10.21). The nonzero solution is obtained when the terms in parentheses are zero. Thus, the critical load is defined by the condition

$$\tan kL = kL \tag{10.23}$$

The roots of Eq. (10.23) can be obtained by plotting both sides of the equation and determining the values of kL at the points of intersection. They can also be obtained by trial and error or other numerical schemes. Suffice it to say that the smallest nonzero value of kL that satisfies Eq. (10.23) is $kL = 1.43\pi$. This corresponds to a critical load of

$$F_{cr} = k^2EI = \frac{2\pi^2EI}{L^2} \tag{10.24}$$

where E, I, and L are as defined previously.

Although we shall not go through the details here, this same procedure applies for other columns with various combinations of fixed, free, or hinged ends. In every case, the critical load is found to differ from that for a hinge-ended column only by a constant factor. Thus, Euler's formula can be expressed in the general form

$$F_{cr} = \frac{C\pi^2EI}{L^2} \tag{10.25}$$

where C is a constant, called the *end-fixity factor*, which depends upon the end conditions. The values of C for several common cases are given in Figure 10.10. Notice that the critical load increases as the degree of constraint increases. It is smallest for a column fixed at one end and free at the other end and it is largest for one fixed at both ends.

All of the results presented in the preceding section for hinge-ended columns also apply for columns with other end conditions if the general Euler's formula [Eq. (10.25)] is used in place of Eq. (10.15). Thus, the critical stress becomes

$$\sigma_{cr} = \frac{F_{cr}}{A} = \frac{C\pi^2E}{(L/r)^2} \tag{10.26}$$

and the conditions for the validity of Euler's formula are

$$\sigma_{cr} = \frac{C\pi^2E}{(L/r)^2} \leqslant \sigma_{PL} \tag{10.27}$$

or

$$(L/r)^2 \geqslant \frac{C\pi^2E}{\sigma_{PL}} \tag{10.28}$$

In these equations, L/r is the slenderness ratio of the column and σ_{PL} is the proportional limit of the material.

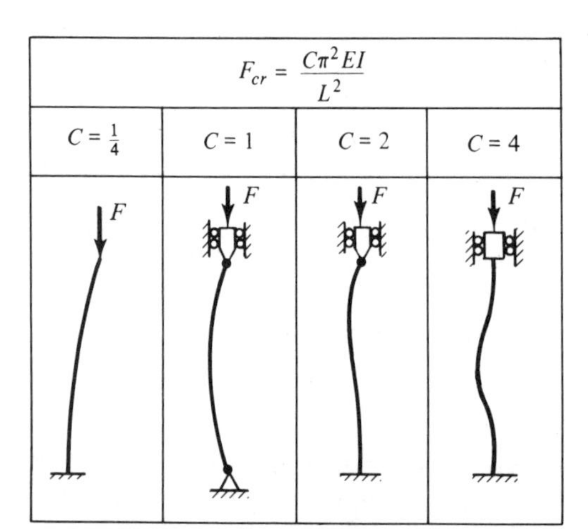

Figure 10.10

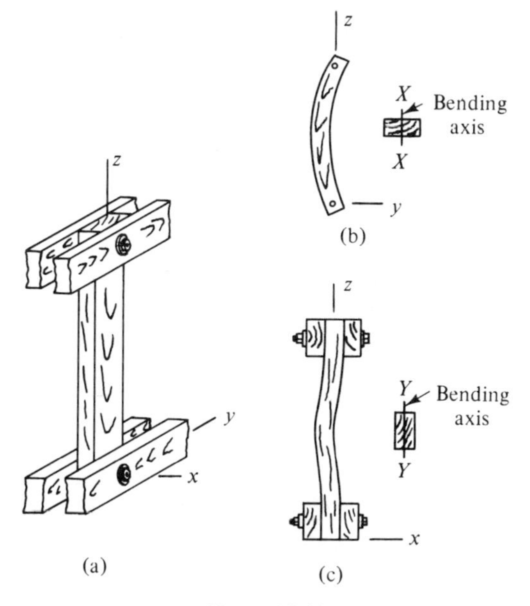

Figure 10.11

Equation (10.28) indicates that the range of slenderness ratios for which Euler's formula applies, and for which a column can be considered to be long, depends upon the value of C. For example, we mentioned in the preceding section that a hinge-ended steel column ($C = 1$) with $E = 30 \times 10^6$ psi and $\sigma_{PL} = 30$ ksi can be considered to be long if $L/r \geqslant 100$. However, if one end is fixed and the other is free ($C = \frac{1}{4}$), it can be considered to be long if $L/r \geqslant 50$.

The end conditions for a column often depend upon the direction in which it buckles. For example, consider the column shown in Figure 10.11(a), which is fastened between massive cross members at each end by a single bolt. If the column buckles in the yz plane, as in Figure 10.11(b), the bolts act as hinges. However, if it buckles in the xz plane [Figure 10.11(c)], the supports act as fixed ends. The manner in which the column actually buckles can be determined only by computing the critical load for each case and seeing which is smallest (see Example 10.4). Note that the value of I is different in each case because the column bends about different axes.

Actually, the support conditions shown in Figure 10.10 are idealizations, as we have already discussed in Chapter 3. Few, if any, real columns duplicate these conditions exactly. In the preceding example, for instance, there will likely be some friction and resistance to rotation about the bolts. Thus, the support does not act exactly like a hinged end, but it certainly isn't fixed or free either. The hinged end is probably the closest approximation to reality for buckling in the xy plane.

Engineering judgment is usually required in the modeling of the support conditions for columns, just as it is for any other type of member. Since the buckling load decreases as the degree of constraint decreases, the computed critical load will be on the conservative side if we assume less support than actually exists. The wide range in the values of the constant C in the general Euler's formula ($\frac{1}{4} \leqslant C \leqslant 4$ for the cases shown in Figure 10.10) and the degree of uncertainty in determining its actual value are additional reasons why substantial safety factors are required in the design of columns.

The results presented in this and the preceding sections also apply to some columns with intermediate supports. The requirement is that the intermediate supports divide the member into identical sections. For example, each portion of the column shown in Figure 10.12(a) looks and behaves like a hinge-ended column with length $L/3$; therefore, the buckling load is equal to that for one of the segments: $F_{cr} = \pi^2 EI/(L/3)^2$. A second example to which our results apply is shown in Figure 10.12(b). Here, each segment acts like a fixed-hinged column with

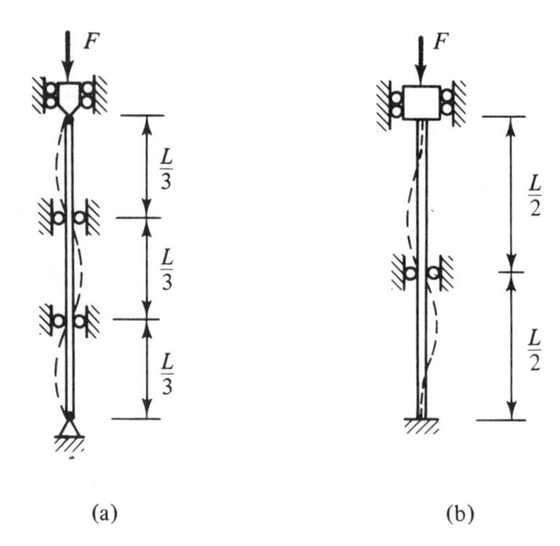

Figure 10.12

length $L/2$; therefore, the buckling load is $F_{cr} = 2\pi^2 EI/(L/2)^2$. Other problems involving intermediate supports are treated in texts on elastic stability.

Example 10.4. Buckling Load of a Column. Determine the buckling load for the column shown in Figure 10.11(a) if it is 10 ft long and made of a 2-in.×4-in. nominal timber with $E = 1.6 \times 10^6$ psi and $\sigma_{PL} = 5400$ psi.

Solution. The geometric properties of the member obtained from Table A.12 are $A = 5.25$ in.,^{2}. $I_{XX} = 5.359$ in.^{4}, and $I_{YY} = 0.984$ in.^{4} Since the direction in which the column will buckle is not known beforehand, both possibilities must be checked. If it buckles in the yz plane, as in Figure 10.11(b), axis XX is the bending axis, and the supports act like hinged ends. The critical load for this case is

$$F_{cr} = \frac{\pi^2 EI}{L^2} = \pi^2 \frac{(1.6 \times 10^6 \text{ lb/in.}^2)(5.359 \text{ in.}^4)}{(120 \text{ in.})^2} = 5880 \text{ lb}$$

If the member buckles in the xz plane [Figure 10.11(c)], the supports act like fixed ends, and YY is the bending axis. For this case, we have

$$F_{cr} = \frac{4\pi^2 EI}{L^2} = 4\pi^2 \frac{(1.6 \times 10^6 \text{ lb/in.}^2)(0.984 \text{ in.}^2)}{(120 \text{ in.})^2} = 4320 \text{ lb}$$

Since the critical load is smallest in the second case, that is the manner in which the column will buckle. The buckling load is

$$F_{cr} = 4320 \text{ lb} \qquad \textbf{\textit{Answer}}$$

The critical stress is

$$\sigma_{cr} = \frac{F_{cr}}{A} = \frac{4320 \text{ lb}}{5.25 \text{ in.}^2} = 823 \text{ psi}$$

which is well below the proportional limit.

PROBLEMS

10.21 What is the maximum allowable length for a $\frac{1}{4}$-in. diameter brass rod that must support a compressive load of 100 lb with a safety factor of 2 if the rod is (a) fixed at each end and (b) fixed at one end and free at the other end?

10.22 A homeowner intends to use a $1\frac{1}{2}$-in. nominal diameter standard steel pipe with a yield strength of 36 ksi to support a 50-lb martin house 10 ft off the ground. The bottom end of the pipe fits snugly into another pipe buried in concrete. Is the pipe large enough to prevent failure by buckling?

10.23 The actuating arm of the hydraulic cylinder in the trash compactor shown in 30 mm in diameter and 1 m long. It is made of steel with a yield strength of 690 MN/m^2. If the cylinder exerts a maximum force of 50 kN, what is the safety factor against buckling when the arm is fully extended?

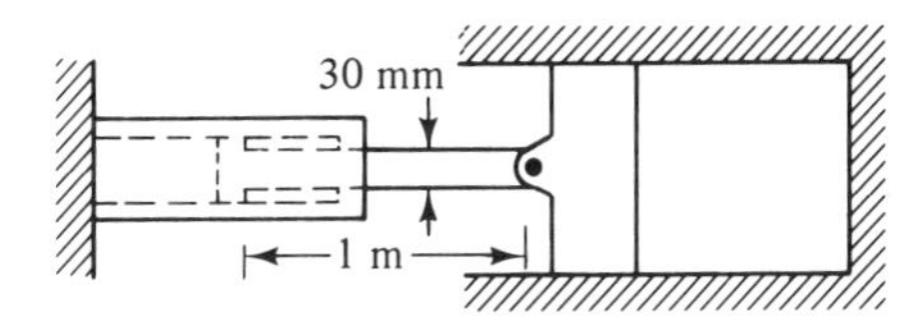

Problem 10.23

10.24 A square stool with 10-in. long legs rigidly attached at each corner rests on a smooth, horizontal surface. The legs are 1 in. × 1 in. with rounded ends and are made of a material with $E = 12 \times 10^6$ psi and $\sigma_{YS} = 28$ ksi. What is the largest vertical load Q that can be safely applied at the center of the stool?

10.25 A thin walled tube that tends to buckle when used as a column is strengthened by placing six other identical tubes around it as shown and strapping them together so that they act as a unit. Since the assembly contains seven tubes, the designer claims that it is seven times stronger than the original tube. Is he right? Each tube has an outside radius R and a wall thickness equal to one-tenth the radius.

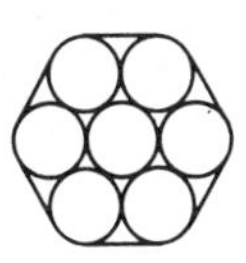

Problem 10.25

10.26 A 6061-T6 aluminum bar is pinned to a rigid support at each end as shown. If the bar is stress-free at 20°C, at what temperature will it buckle? $E = 68$ GN/m^2, $\sigma_{YS} = 240$ MN/m^2 and, $\alpha = 23.6 \times 10^6 (°C)^{-1}$.

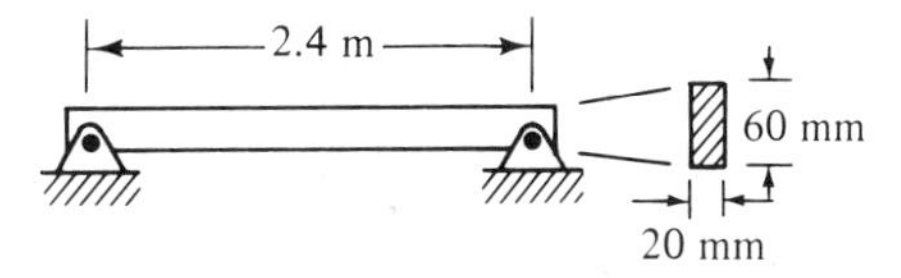

Problem 10.26

10.27 The modernistic lamp shown is made of 36 equally spaced acrylic rods fitted into holes drilled into the top and bottom

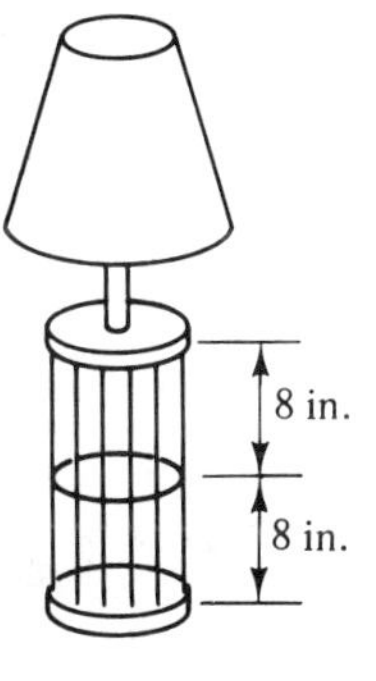

Problem 10.27

pieces and restrained by a wire hoop attached at the center. The hoop prevents deflection, but not rotation, of the rods. What is the minimum required diameter of the rods if they are to support the 4-lb upper portion of the lamp with a safety factor of 2? Use $E = 0.4 \times 10^6$ psi and $\sigma_{YS} = 6000$ psi.

10.28 The boom shown is made of a 2-in.×4-in. nominal timber oriented with the largest dimension vertical. Determine the weight of the largest object that can be supported without causing the boom to buckle. Use a safety factor of 4; $E = 1.8 \times 10^6$ psi and $\sigma_{YS} = 5.5$ ksi.

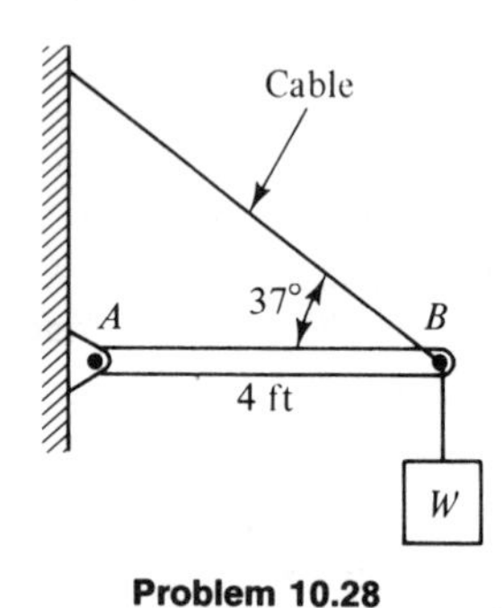

Problem 10.28

10.29 A machine linkage with the cross section shown is hinged at each end and passes through a slot cut in a plate at its middle. Determine the buckling load for the linkage if it is made of aluminum with $E = 70\ \text{GN/m}^2$ and $\sigma_{YS} = 240\ \text{MN/m}^2$.

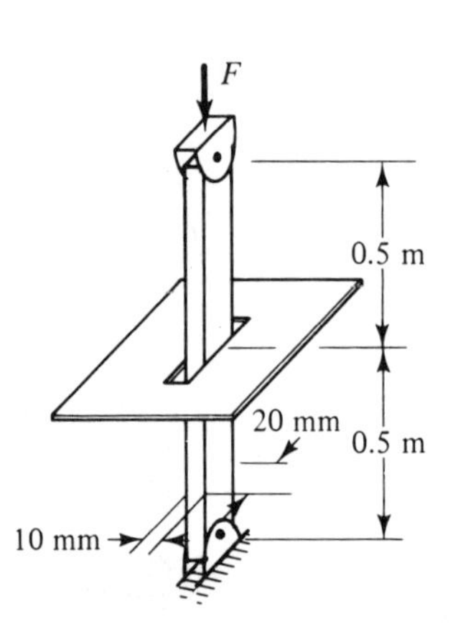

Problem 10.29

10.30 Same as Problem 10.29, except that the member is made of a titanium alloy with $E = 110\ \text{GN/m}^2$ and $\sigma_{YS} = 800\ \text{MN/m}^2$.

10.31 A gardener finds that his climbing vine will buckle unless it is tied to the supporting stake every 8 in. as shown. If the vine has an average diameter of $\frac{1}{8}$ in. and the bloom at the top has an estimated weight of 2 oz, what is the approximate modulus of elasticity of the vine?

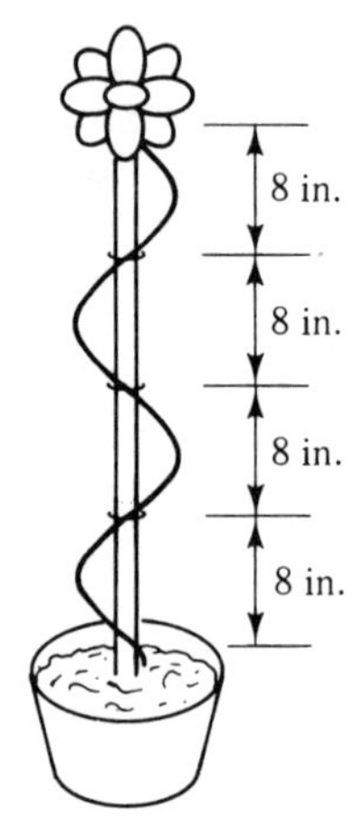

Problem 10.31

10.32 The 50-foot tower shown supports a 50 000-lb antenna and consists of a lat-

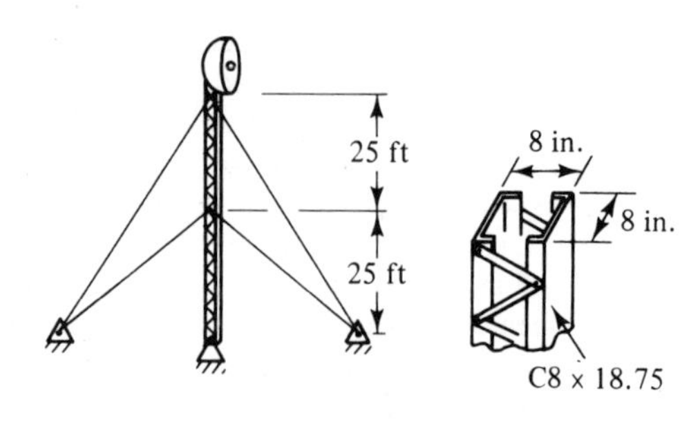

Problem 10.32

ticed column made of two C8 × 18.75 steel channel sections with yield strength $\sigma_{YS} = 36$ ksi. The tower can be considered to be pinned at the bottom, and it is supported at the top and midpoint with guy wires. It is proposed that a second identical antenna be placed atop the tower. Can the additional weight be supported without the safety factor for failure by buckling dropping below the required value of 4? If not, how can the tower be strengthened?

10.5 INELASTIC BEHAVIOR

As indicated in Section 10.3, intermediate length columns buckle at stresses that exceed the proportional limit of the material. Consequently, Euler's formula does not apply, and a separate analysis is required. Since the stiffness of most materials decreases significantly above the proportional limit, inelastic buckling can occur at loads well below those predicted by Euler's formula. The procedure for determining the inelastic buckling load of columns will be considered in this section.

Consider a column of intermediate length subjected to an axial compressive load F that is just slightly less than the buckling load, whatever that may be [Figure 10.13(a)]. The stress-strain diagram for the material is shown in Figure 10.13(b). Since we assume that the column remains straight prior to buckling, the stress is $\sigma = F/A$.

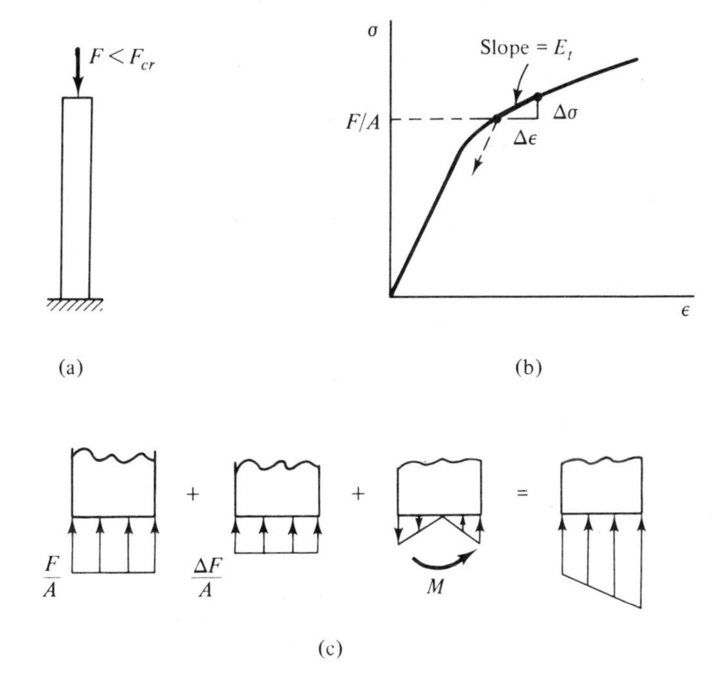

Figure 10.13

Now let us suppose that the applied load is increased by a small amount ΔF and that the column simultaneously starts to buckle. This increment in load produces an additional compressive stress of magnitude $\Delta F/A$ and a flexural stress due to the bending. The final stress distribution is as shown in Figure 10.13(c). It is assumed that the increment in compressive stress is larger than the bending stress, in which case the total stress increases at all points over the cross section and there is no unloading of any of the fibers.

As a result of the preceding assumptions, the increment of stress $\Delta\sigma$ on a typical fiber and the corresponding increment of strain $\Delta\varepsilon$, both of which are small for a small increment in load, must be as shown in Figure 10.13(b). The relationship between them is

$$\Delta\sigma = E_t \Delta\varepsilon \tag{10.29}$$

where E_t is the tangent modulus of the material at the stress $\sigma = F/A$. Equation (10.29) holds at all points over the cross section. This would not be true if the stress on some of the fibers decreased. In this case, the stresses would follow the path indicated by the dashed line in Figure 10.13(b), and the stress-strain relation for the fibers that experience unloading would be $\Delta\sigma = E\Delta\varepsilon$.

Now let us focus our attention upon the bending response, since it is directly associated with the buckling. The only difference between the bending encountered here and the linearly elastic bending considered in Chapter 8 is that the stress-strain relation is given by Eq. (10.29) instead of Hooke's law. Thus, all of our results for linearly elastic beams still apply if the elastic modulus, E, is replaced by the tangent modulus, E_t. In particular, Eq. (8.18) for the small deflections of beams still holds. Thus, the inelastic bending of the column is governed by the equation

$$E_t I \frac{d^2 v}{dx^2} = M \tag{10.30}$$

The expressions for the bending moment, M, and the boundary conditions are not affected by the material properties; therefore, the solutions of Eq. (10.30) will be the same as for the linearly elastic case, except that E is replaced by E_t. Consequently, we conclude that the buckling load and critical stress for inelastic columns are, respectively,

$$F_{\text{cr}} = \frac{C\pi^2 E_t I}{L^2} \tag{10.31}$$

and

$$\sigma_{\text{cr}} = \frac{F_{\text{cr}}}{A} = \frac{C\pi^2 E_t}{(L/r)^2} \tag{10.32}$$

In these equations, E_t is the tangent modulus of the material at the critical stress. All other quantities are as defined previously, including the end-fixity factor C, which has the same values as for elastic columns.

Equation (10.31) is called the *tangent modulus formula*. Although one may question the validity of the assumptions upon which this formula is based, it gives

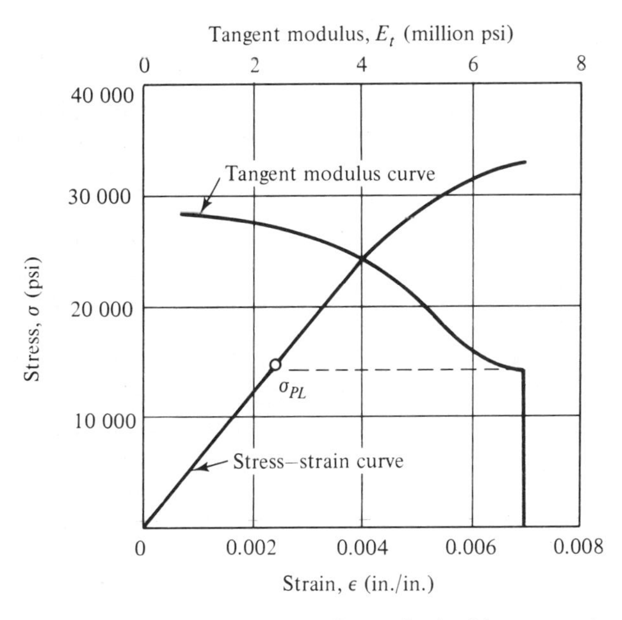

Figure 10.14. Compression Stress-Strain Diagram and Tangent Modulus Curve for AZ 31B-H24 Magnesium Alloy.

values of the buckling load that are in satisfactory agreement with experimental results.

Although the tangent modulus formula is identical in form to Euler's formula, it is more difficult to apply because the tangent modulus, E_t, is not a constant, but rather depends upon the magnitude of the stress. The procedure to be followed depends upon what is known and what is to be determined, as we shall see in the following.

The stress-strain diagram for the material`is usually available in the form of a graph. In this case, it is convenient to first determine the values of E_t at several points and plot its variation with the stress level, as in Figure 10.14. Plots of E_t for many common engineering materials are available in handbooks.

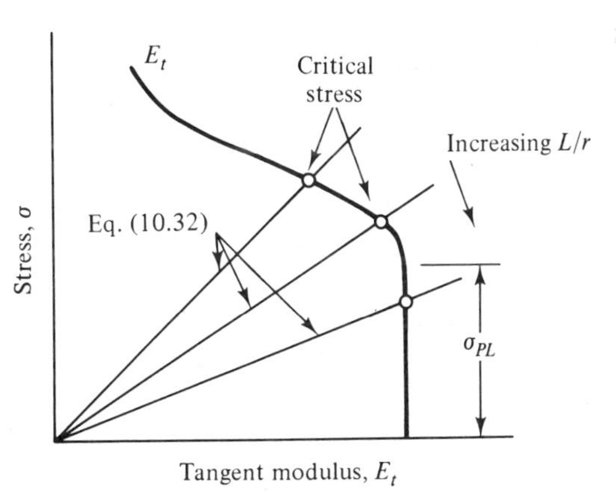

Figure 10.15

The simplest way to determine the buckling load for a given column is to plot Eq. (10.32) on the same graph as E_t (Figure 10.15). As indicated in the figure, this equation plots as a straight line whose slope depends upon the slenderness ratio. The point of intersection between this line and the curve of E_t determines the critical stress, σ_{cr}. Then, $F_{cr} = \sigma_{cr}A$, where A is the cross-sectional area of the member. Note that, for large values of the slenderness ratio, the value of the stress at the point of intersection is below the proportional limit. In this case, $E_t = E$, and the tangent modulus formula reduces to Euler's formula.

An alternate approach is to construct the curve showing the values of σ_{cr} at different values of the slenderness ratio for the material of interest. This is easily done by choosing several values of σ_{cr}, determining the corresponding values of E_t from the curve of E_t, and solving for L/r from Eq. (10.32). The central portion of the curve in Figure 10.7 is an example of such a curve. Once these curves are available, the critical stress for a given column can be read directly from the plot.

This plot is also convenient for showing the effect of the end conditions in the inelastic range. Figure 10.16 shows the variation of the critical stress with the slenderness ratio for various values of C. Notice that the end conditions have less effect upon the critical stress for columns of intermediate length than they do for long columns. In particular, the critical stress, and hence the critical load, is not proportional to C, as in the elastic case. For example, the inelastic buckling load of a column with fixed ends is not four times that for one with hinged ends. The buckling load must be determined separately for each different set of end conditions considered.

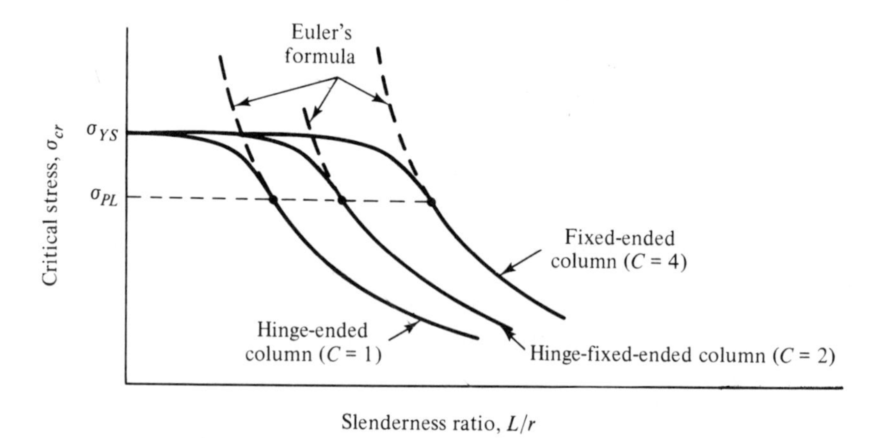

Figure 10.16

A trial-and-error procedure (or other numerical or graphical scheme), is usually required to determine the size of a column needed to support a given load. Several possible approaches are illustrated in Example 10.5.

If the stress-strain curve for the material is known in equation form, the relation between the tangent modulus and the stress can be determined directly from the definition of E_t:

$$E_t = \frac{d\sigma}{d\varepsilon} \tag{10.33}$$

In this case, the critical stress can be expressed directly in terms of the slenderness ratio, and the design or analysis of the column can be carried out analytically. For example, if the equation for the stress-strain curve is

$$\sigma = k\sqrt{\varepsilon}$$

where k is a constant, we have

$$E_t = \frac{d\sigma}{d\varepsilon} = \frac{1}{2}\frac{k}{\sqrt{\varepsilon}} = \frac{k^2}{2\sigma}$$

Evaluating E_t at the critical stress and substituting it into Eq. (10.32), we get the desired relationship

$$\sigma_{cr}^{\ 2} = \frac{C\pi^2 k^2}{2(L/r)^2}$$

In addition to the tangent modulus formula, there are a number of empirical formulas available for predicting the critical stress for intermediate length columns. These formulas are obtained by fitting simple equations to plots of the experimentally determined values of the critical stress for members with different slenderness ratios. These equations are widely used in column design because they are often simpler to apply than the tangent modulus formula. However, since many of these equations hold only for specific materials, we will not consider them here. They are discussed in texts on structural design and stability.

Example 10.5. Determination of Inelastic Buckling Load. A column with a 2-in. × 2-in. square cross section is made of 6061-T6 aluminum alloy with the properties shown in Figure 10.17. The member is $3\frac{1}{2}$ ft long and is fixed at one end and hinged at the other. (a) What is the buckling load and (b) what size square cross section would be needed to support this same value of load if both ends were fixed?

Solution. (a) The moment of inertia of the cross section is

$$I = \frac{1}{12}bh^3 = \frac{1}{12}(2\text{ in.})^4 = 1.33\text{ in.}^4$$

From Eq. (10.31), we have

$$F_{cr} = \frac{C\pi^2 E_t I}{L^2} = \frac{2\pi^2(1.33\text{ in.}^4)}{(42\text{ in.})^2}E_t = (14.88 \times 10^{-3})E_t$$

therefore,

$$\sigma_{cr} = \frac{F_{cr}}{A} = \frac{(14.88 \times 10^{-3})}{4\text{ in.}^2}E_t = (3.72 \times 10^{-3})E_t$$

This expression for σ_{cr} is now plotted on the same graph as E_t (Figure 10.17). The two curves intersect at a stress of approximately 31 300 psi; therefore,

$$F_{cr} = \sigma_{cr}A = (31\ 300\text{ lb/in.}^2)(4\text{ in.}^2) = 125\ 200\text{ lb} \qquad \textbf{\textit{Answer}}$$

(b) Let a denote the length of each side of the cross section. Then

$$F_{cr} = \frac{4\pi^2 E_t(a^4/12)}{L^2} = 125\ 200\text{ lb}$$

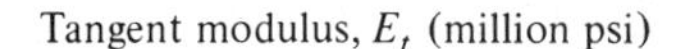

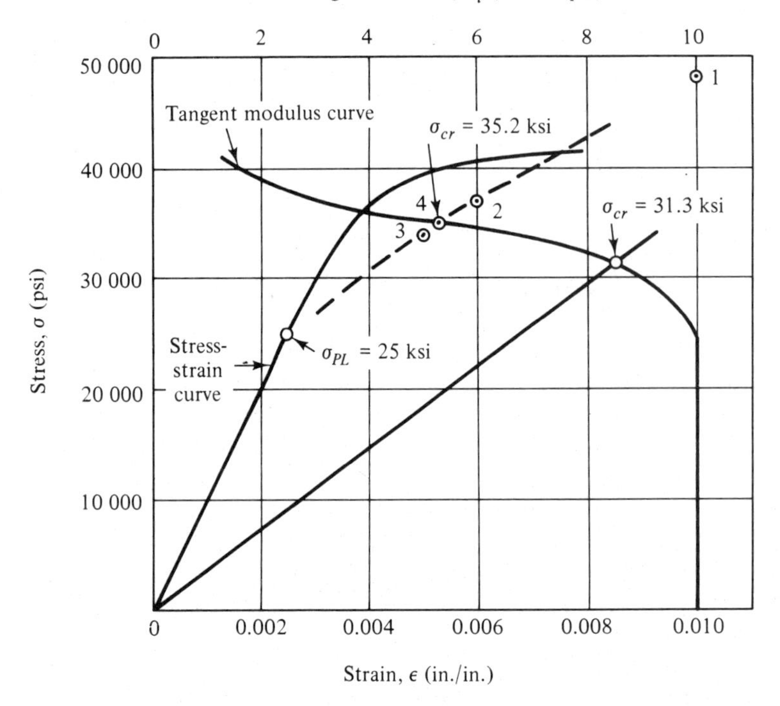

Figure 10.17. Compression Stress-Strain Diagram and Tangent Modulus Curve for 6061-T6 Aluminum Alloy.

and

$$\sigma_{cr} = \frac{F_{cr}}{A} = \frac{125\ 200}{a^2} \text{ psi}$$

The problem is to determine a value of E_t and σ_{cr} from the tangent modulus curve such that both of these relations are satisfied. There are several ways to do this. One is trial and error.

We pick a value of E_t, solve for a from the expression for F_{cr}, and then compute σ_{cr}. If this value of σ_{cr} lies on the tangent modulus curve, it is the critical stress for the column, and the problem is solved. If not, the value of E_t is adjusted until this condition is met. Since there is always a possibility that Euler's equation will apply, we take $E_t = E$ as our first guess. The computations are summarized below:

1. $E_t = 10 \times 10^6$ psi $\sigma_{cr} = 48\ 300$ psi (too high)
2. $E_t = 6 \times 10^6$ psi $\sigma_{cr} = 37\ 400$ psi (too high)
3. $E_t = 5 \times 10^6$ psi $\sigma_{cr} = 34\ 200$ psi (too low)
4. $E_t = 5.3 \times 10^6$ psi $\sigma_{cr} = 35\ 200$ psi (close)

$$a^2 = \frac{F_{cr}}{\sigma_{cr}} = \frac{125\ 200 \text{ lb}}{35\ 200 \text{ lb/in.}^2} = 3.56 \text{ in.}^2$$

$$a = 1.9 \text{ in.} \qquad \textbf{\textit{Answer}}$$

Notice that the required cross-sectional dimensions are very nearly the same as those in part (a). This is in keeping with our previous finding that the end conditions have little effect upon the critical stress for inelastic buckling.

An alternate procedure is to plot the stresses obtained from the various trials on the same graph as E_t. These points are labeled 1, 2, 3, and 4, respectively, in Figure 10.17. The intersection of the curve through these points and the E_t curve defines the critical stress for the column. The advantage of this procedure is that we need consider only enough points to establish the curve through them.

Still another possibility is to obtain the equation relating σ_{cr} and E_t for the column by eliminating a between the expressions for F_{cr} and σ_{cr}. We have

$$\sigma_{cr}^2 = \frac{\pi^2 F_{cr} E_t}{3L^2} = \frac{\pi^2(125\ 200\ \text{lb})}{3(42\ \text{in.})^2} E_t = 233.5E_t$$

The plot of this equation is indicated by the dashed curve in Figure 10.17. The intersection of this curve with the curve of E_t defines the critical stress for the column. This approach is the same as the preceding one, except that we actually obtain the equation relating σ_{cr} and E_t. Note that this is possible only if the cross-sectional area and moment of inertia can be expressed in terms of a single dimensional parameter. In contrast, the trial-and-error and first graphical procedure work for any problem.

PROBLEMS

10.33 Determine the buckling load for a 1-in. × 1-in. hinge-ended stainless steel column 2 ft long with the compressive tangent modulus shown.

10.34 Same as Problem 10.33, except that the column is fixed at one end and hinged at the other end.

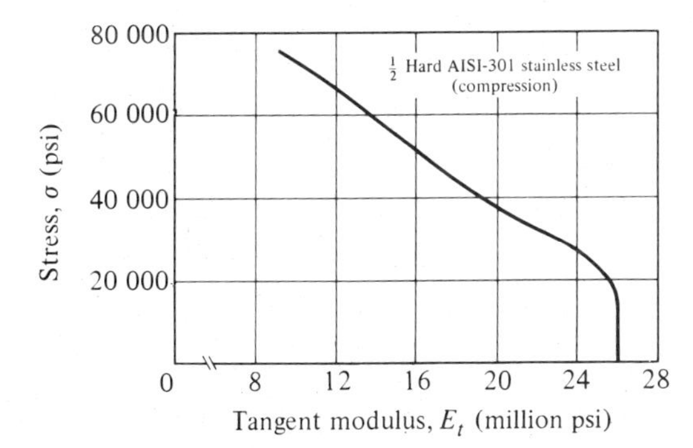

Problem 10.33

10.35 A 1-in. diameter column made of 6061-T6 aluminum with the properties shown in Figure 10.17 is fixed at one end and free at the other end. What is the maximum allowable length of the column if it must support a compressive load of 13 kips with a safety factor of 2?

10.36 A square magnesium column with the properties given in Figure 10.14 has a cross-sectional area of 2 in.^{2} and is 16.4 in. long. The column is hinged at each end and supports a compressive load of 20 kips. What is the safety factor against failure by buckling.

10.37 A 4-ft stainless steel tube with an outside diameter of 2 in. and a $\frac{1}{4}$-in. wall thickness is hinged at each end. By what percentage can the buckling load be increased by fixing one end? Use the material properties given in Problem 10.33.

10.38 Construct a curve showing the critical stress σ_{cr} as a function of the slenderness ratio L/r for hinge-ended columns made of 6061-T6 aluminum with the properties shown in Figure 10.17. Show that this curve can be used for columns with other end conditions by introducing an equivalent slenderness ratio $L/(\sqrt{C}\, r)$, where C is the end-fixity factor. Use this curve to obtain the buckling load for a 1-in. diameter column 14 in. long that is (a) hinged at each end and (b) fixed at one end and free at the other end.

10.39 Determine the minimum required diameter of an 18-in. stainless steel column fixed at one end and free at the other end that must support a compressive load of 11.5 kips with a safety factor of 4. The material properties are as given in Problem 10.33.

10.40 A 1-in. diameter magnesium column 20 in. long is fixed at one end and hinged at the other end. What is the buckling load for this column? What diameter column is required to support this same load if both ends are fixed? The material properties are as shown in Figure 10.14.

10.41 A hinge-ended column with slenderness ratio $L/r = 20$ is made of a material whose compressive stress-strain diagram can be described by the equation $\sigma = 10^5\sqrt{\varepsilon}$ over the stress range of interest. What is the critical stress for the column?

10.42 Same as Problem 10.41, except that the stress-strain equation is $\sigma = 10^5\varepsilon^{1/3}$.

10.6 CLOSURE

In this chapter we were primarily concerned with the elastic and inelastic buckling of columns. There are several factors that distinguish buckling from other types of failures. First, the failure load does not depend upon the strength of the material. It depends only upon the material stiffness and the geometry of the member. Consequently, failure by buckling can occur at very small loads and stresses.

Second, since the deformations remain small until just prior to buckling, there is little warning of the impending failure. This is true even for members made of ductile materials. As a result, buckling failures tend to be catastrophic. In contrast, the other types of members considered in this text, when made of ductile materials, generally undergo large deformations prior to failure. In these cases, the deformations may provide a visual indication that a member is overloaded, thus making it possible to take corrective action.

Finally, the physical nature of the stability of an equilibrium position and buckling and the manner in which they are defined mathematically make it necessary to use the deformed geometry of the member in any force analysis. This is true even for small deformations. This is in contrast to the other types of members considered, for which the rigid-body assumption can be used when the deformations are small.

chapter eleven

EXPERIMENTAL STRAIN AND STRESS ANALYSIS

11.1 INTRODUCTION

In the preceding chapters we developed equations for determining the stresses and deformations in simple load-carrying members. More complicated members and loadings are considered in advanced texts, and, with the use of modern computerized methods of analysis, it is now possible to solve exceedingly complex problems. Nevertheless, there is a continuing need for experiments. Some problems are so complex that they still are not amenable to analysis, in which case the stresses and deformations can only be determined experimentally. Experimental results are also needed to assist in the development of new methods of analysis and to confirm the validity of existing ones.

Some of the common experimental techniques for determining stresses and strains will be considered in this chapter. Our discussions must necessarily be brief; for additional information, the reader is referred to texts on experimental stress analysis. Manufacturers of experimental equipment and supplies are also an excellent source of information. A list of references is given in Section 11.8.

Throughout our discussions it will be helpful to keep in mind that stresses and strains are mathematical quantities, and, as such, they cannot be measured directly. Instead they must be determined indirectly through their definitions and the measurement of physical quantities such as force and deformation.

11.2 ELECTRICAL RESISTANCE STRAIN GAGES

In Chapter 5 we briefly discussed instruments for determining the strains in materials tests. However, these devices are designed specifically for use on slender tension or compression specimens and would generally not be suitable for determining strains in members of more complicated geometry. One device commonly used to determine strains in such instances is the *electrical resistance strain gage*. This device operates on a physical principle discovered in 1854 by Lord Kelvin; namely, that the electrical resistance of a conductor changes when it is strained.

Electrical resistance strain gages typically consist of a grid attached to a piece of backing material and formed from a fine wire or a thin sheet of foil etched away to form the grid pattern (Figure 11.1). The entire unit is bonded to the surface of the test specimen, as illustrated in Figure 11.2, and a small electrical current is passed through the grid.

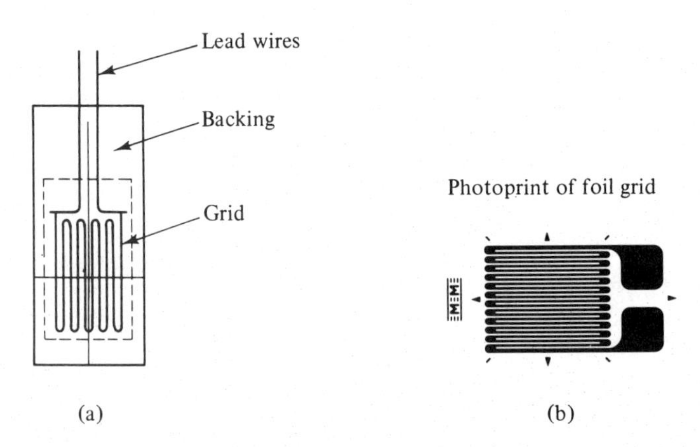

Figure 11.1 Wire Gage (a) and Etched-Foil Grid (*Courtesy* Measurements Group-Vishay Intertechnology, Inc.) (b).

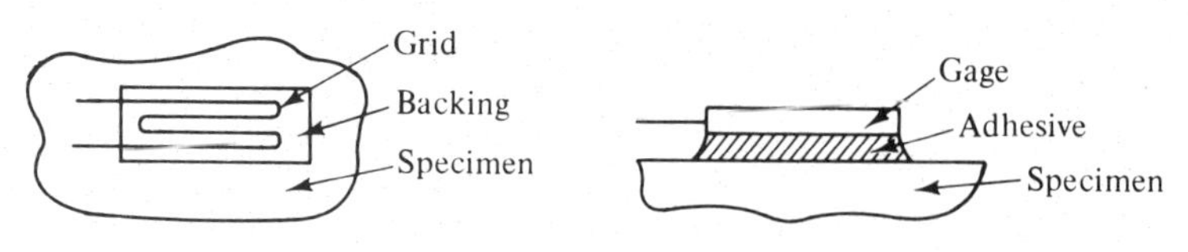

Figure 11.2

The adhesive transmits the deformations of the specimen to the grid, resulting in a change in its electrical resistance. This change in resistance is the physical quantity measured. When combined with the known strain-resistance characteristics of the gage, this information yields the normal strain in the direction along which the gage is aligned. The necessary conversion from change in resistance to strain is usually performed automatically by the instrumentation used. Both tensile and compressive strains can be measured because the gage is bonded over its entire length.

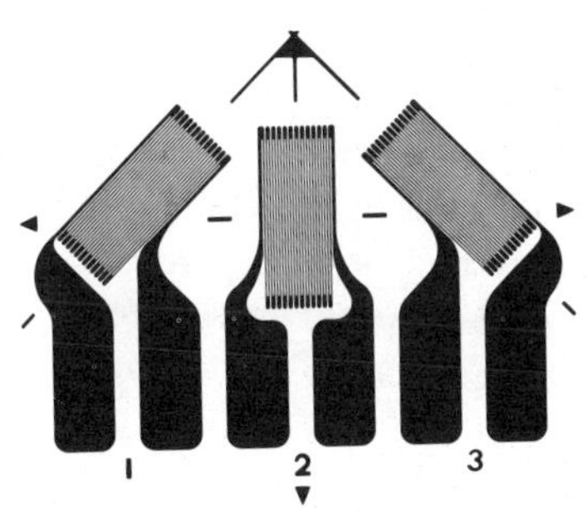

Figure 11.3. Strain Gage Rosette. (*Courtesy* Measurements Group-Vishay Intertechnology, Inc.)

Figure 11.3 shows a typical *strain gage rosette* used to measure the strains in three different directions at a point. As explained in Section 9.8, this information is needed to determine completely the states of strain and stress in general two-dimensional problems. The rosettes consist of three separate grids with different orientations attached to the same piece of backing material (not shown in the figure) and are mounted to the specimen in the same way as a single gage. Each grid responds as an individual gage and provides the normal strain in the direction along which it is aligned. Two common rosette configurations are the *rectangular rosette* shown in Figure 11.3, in which the grids are 45° apart, and the *delta rosette*, in which they are 60° apart (see also Figure 9.22).

Three individual gages can be used in place of a rosette, but the rosette is more convenient to mount. Also, the angles between the grids are known to a higher degree of accuracy with a rosette, and the three grids are located more nearly at the same point.

The important thing to note here is that the strains generally must be measured in three different directions at each point of interest. The only exceptions are for those problems in which some information about the strains is known beforehand. For example, if the directions of the principal strains are known, two gages, one in each of the principal directions, are sufficient to determine the state of strain (see Example 11.1). If only the strain in one principal direction is of interest, it can be determined by using a single gage aligned in that direction. Thus, a single gage is sufficient for determining the longitudinal or transverse strain in an axially loaded member. It is also sufficient for determining the shear strain in a shaft subjected to pure torsion (see Problem 11.5) and the bending strain in a beam subjected to pure bending.

The sensing element in electrical resistance strain gages is made of various metallic alloys and is formed into a grid pattern to provide a suitable length of conductor within a small overall distance. The backing material facilitates handling of the grid and provides the necessary electrical insulation between it and the specimen. Various materials are used for the backing, depending upon the environmental conditions to which the gage will be subjected. Paper and phenolic backings are common, but other materials, such as epoxies, polyesters, ceramics, and metals are also used.

Various kinds of adhesives are used to mount the gages, depending upon the loading and environmental conditions to be encountered and the type of backing material. Paper-backed gages are commonly mounted with cellulose nitrate cement (ordinary Duco household cement), and cyanoacrylate cement (Eastman 910) is used in a wide variety of cases. Phenolic and epoxy cements are also used. The manufacturers' recommendations and procedures should be followed closely to assure the integrity of the bond between the gage and the specimen. Obviously, a good bond is essential for proper performance of the gage.

All strain measuring devices give an average value of strain over their gage length, which, for an electrical resistance strain gage, is the length of the grid. Short gage lengths are required for accurate measurements at locations where the strains vary rapidly with position, such as near notches, holes, and other sources of strain concentration. Longer gage lengths are required whenever it is necessary to "average out" local variations in the material properties, such as occur in concrete and wood. Wire gages are available with gage lengths ranging from $\frac{1}{16}$ in. to 8 in.; for foil gages, the gage lengths range from $\frac{1}{64}$ in. to 1 in.

The sensitivity of a strain gage is expressed in terms of the *gage factor* G, which is defined as

$$G = \frac{\Delta R/R}{\varepsilon} \tag{11.1}$$

Here, ε is the normal strain in the direction of the gage axis, R is the gage resistance, and ΔR is the change in resistance due to the strain. The larger the gage factor, the more sensitive the gage, i.e., the larger the change in resistance for a given strain. Standard gages have a resistance of 120 ohms and a gage factor of approximately 2.

Values of the gage factor are provided by the manufacturer and are determined experimentally by subjecting samples from each lot of gages manufactured to a known strain and measuring the resulting change in resistance. The resistance of the undeformed gage is also measured, and the value of G is obtained from Eq. (11.1). Careful quality control in the manufacturing process assures that the value of G obtained for the samples is an accurate indication of the gage factor for the entire lot.

Electrical resistance strain gages have the following attractive features:

1. They are easy to use and are relatively inexpensive. Standard single gages start at several dollars each, with rosettes costing approximately ten times that amount. The necessary instrumentation may be expensive, however, as are special purpose gages.
2. Their use is not restricted to the laboratory. Measurements can be made "in the field," and the gage can be read either on location or remotely.
3. They have high sensitivity and provide very good accuracy when properly installed and used. They also have a good range of response. Strains of the order of 1 μ in./in. can be detected, and an overall measurement system accuracy of ± 5 μ in./in. can be achieved in some instances. The largest strains that can be measured range from 1% to 3%, which is well into the inelastic range for ductile metals (mild steel yields at a strain of approximately 0.1%). Special gages are available with ranges up to a strain of 10 %.
4. Their short gage lengths make it possible to obtain accurate results when the strains vary rapidly with position, such as at locations of strain concentration.
5. They are small and have no significant effect upon the response of the specimen, unless it is very thin or has a very low modulus of elasticity.

6. They can be used for both static and dynamic loadings.

7. They operate over a wide range of environmental conditions. Depending upon the type of backing material, the gages can be used at temperatures ranging from approximately -400 to $+1800°F$. With proper protection, they can be used underwater and in other adverse environments.

The gages also have some unattractive features. Since it is not feasible to place them over the entire surface of a member, the critical areas of high strain must be identified before the gages are mounted. This often requires using some other form of strain analysis, such as brittle coatings (see Section 11.4).

Changes in temperature during testing can also cause problems. Since the coefficient of thermal expansion of the gage metal is often different from that of the specimen, the two tend to expand or contract at different rates. However, since the two are bonded together, the gage will be stretched or compressed right along with the specimen as it expands or contracts. This leads to an additional change in resistance of the grid and a false indication of the strain due to the applied loads.

Fortunately, problems due to temperature changes can be circumvented by using the appropriate circuitry (see Section 11.3). There are also available *self-compensating gages* designed to expand or contract at the same rate as the metals and metallic alloys commonly used in design. However, these gages are expensive and provide temperature compensation only over a limited temperature range.

The fact that strain gages can be used only on the surface of a member is not a serious disadvantage because the maximum stresses and strains usually occur at the surface. Recall, for example, that the maximum stresses in a twisted shaft occur at the outside surface and the maximum bending stresses in a beam occur at the top and bottom of the section.

Example 11.1. Strain Measurement in Principal Directions. Show that two strain gages, aligned in the directions of the principal strains, are sufficient for determining the state of strain at a point.

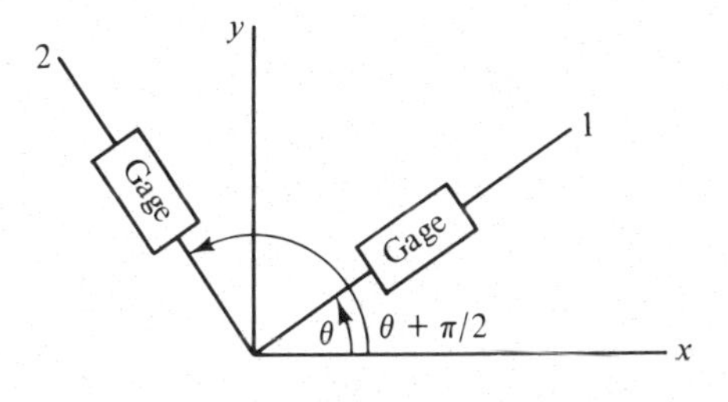

Figure 11.4

Solution. Let x and y be an arbitrary set of axes at the point and let 1 and 2 be the principal axes (Figure 11.4). Applying the strain transformation equations, Eqs. (9.13) and (9.15), in the principal directions and recognizing that the shear strain associated with the principal axes is zero, we obtain

$$\varepsilon_1 = \varepsilon_x \cos^2\theta + \varepsilon_y \sin^2\theta + \gamma_{xy} \sin\theta \cos\theta$$

$$\begin{aligned}\varepsilon_2 &= \varepsilon_x \cos^2(\theta + \pi/2) + \varepsilon_y \sin^2(\theta + \pi/2) + \gamma_{xy} \sin(\theta + \pi/2)\cos(\theta + \pi/2)\\ &= \varepsilon_x \sin^2\theta + \varepsilon_y \cos^2\theta - \gamma_{xy} \sin\theta \cos\theta\end{aligned}$$

$$\gamma_{12} = -2(\varepsilon_x - \varepsilon_y)\sin\theta \cos\theta + \gamma_{xy}(\cos^2\theta - \sin^2\theta) = 0$$

Thus, if the strains ε_1 and ε_2 are measured, we have three equations from which to determine the three unknowns ε_x, ε_y, and γ_{xy} for any choice of axes x and y. The angle θ is a known quantity because the principal axes are presumed known.

PROBLEMS

11.1 What is the change in resistance of a strain gage with gage factor $G = 2.09$ and resistance $R = 120$ ohms when subjected to a strain of 200 μ in./in.?

11.2 If the change in resistance of a strain gage with gage factor $G = 1.98$ and resistance $R = 120$ ohms is 0.2 ohm, what is the normal strain in the direction of the gage axis?

11.3 A strain gage is calibrated by mounting it on a 0.040-in. thick strip that is then bent around a cylindrical surface with a radius of 40 in. If the gage resistance is $R = 120$ ohms and the change in resistance is 0.12 ohm, what is the gage factor G?

11.4 A strain gage with resistance $R = 500$ ohms is mounted on the top surface of a beam that is loaded as shown. The mid-span deflection, measured with a micrometer, is 0.1 mm and the change in gage resistance is 0.31 ohm. What is the gage factor G?

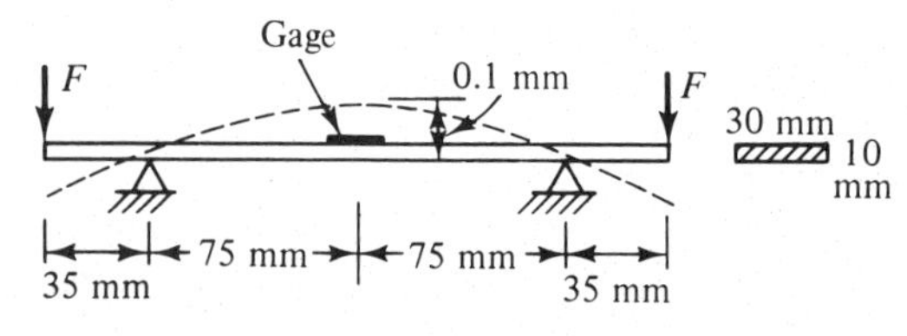

Problem 11.4

11.5 Show that a single strain gage oriented 45° from the longitudinal axis is sufficient to determine the state of strain in a shaft subjected to pure torsion.

11.6 If the shaft in Problem 11.5 is 4 ft. long and has a diameter of 1 in., what is the anticipated strain gage reading for an angle of twist of 5°? Does the gage reading depend upon the material properties?

11.7 Two strain gages are mounted on a 20-mm × 20-mm square steel bar subjected to a tensile load F. One gage is oriented in the longitudinal direction and the other in the transverse direction; $G = 2.1$ for the longitudinal gage and 1.9 for the transverse gage. If the longitudinal gage gives a reading of 600 μ m/m, determine the anticipated reading of the transverse gage and the applied load F.

11.8 Can a single strain gage be used to determine the state of strain on the surface of a thin-walled cylindrical pressure vessel at a point away from the ends? If so, how should the gage be aligned? Assume the material is linearly elastic with known properties.

11.9 A strain gage is mounted on a steel bar as shown. During initial checkout the gage reading is 520 μ m/m for a load of 40 kN. Is the gage functioning properly?

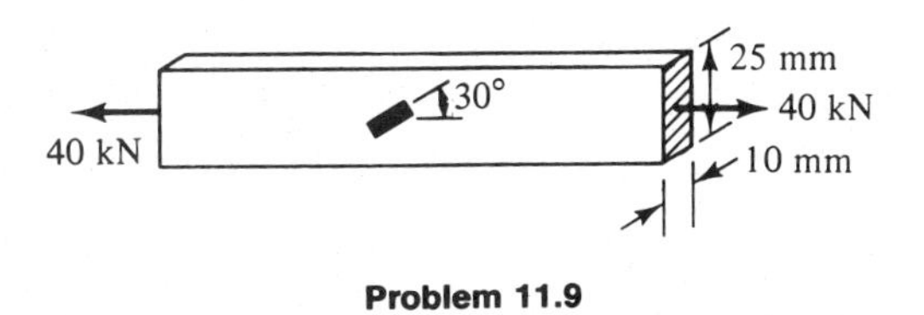

Problem 11.9

11.10 Same as Problem 11.9, except that the gage reading is 320 μ m/m.

11.11 A strain gage mounted on the outer surface of a thin-walled spherical tank made of a copper alloy with $E = 15 \times 10^6$ psi and $\nu = 0.35$ gives a reading of 125 μ in./in. If the ratio of radius to wall thickness is 40, what is the pressure in the tank?

11.12 Same as Problem 11.11, except that the tank is made of stainless steel with $E = 28 \times 10^6$ psi and $\nu = 0.3$.

11.3 STRAIN-GAGE CIRCUITRY AND TRANSDUCERS

According to Eq. (11.1), an electrical resistance strain gage with a resistance of 120 ohms and a gage factor of 2 undergoes a change in resistance of $\Delta R = 0.24$ ohm for a strain of 1000 μ in./in. (approximately the strain at yielding in mild steel). Smaller strains produce correspondingly smaller changes in resistance. Percentage-wise, these changes in resistance are very small and can be measured accurately only by using the appropriate instrumentation. Most of the instruments used with strain gages are based on some form of Wheatstone-bridge circuit.

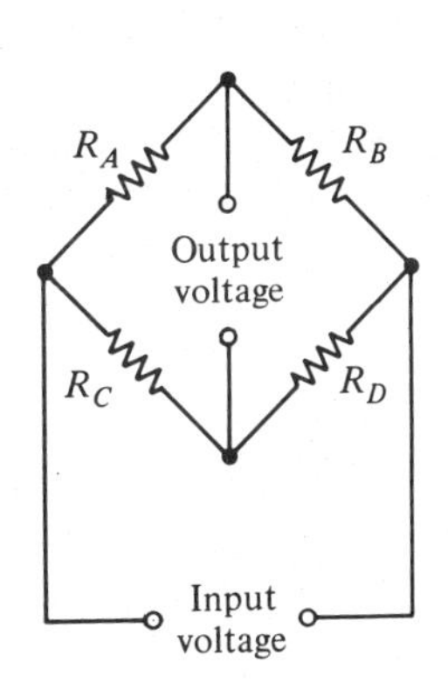

Figure 11.5

The basic Wheatstone-bridge circuit is shown in Figure 11.5, in which one or more of the arms is a strain gage and the others are resistors. When the resistances of the arms satisfy the condition

$$R_A R_D = R_B R_C \tag{11.2}$$

the bridge is balanced and there is zero output voltage.

A change in resistance of one or more of the arms of the bridge creates an unbalanced condition and an output voltage proportional to the unbalance. If each arm undergoes a small change in resistance ΔR, the amount of unbalance is

$$U = (R_A + \Delta R_A)(R_D + \Delta R_D) - (R_B + \Delta R_B)(R_C + \Delta R_C)$$

Expanding this expression, making use of the balance condition of Eq. (11.2), and neglecting second-order terms involving the square and products of the ΔR's, we obtain

$$U = (R_A \Delta R_D + R_D \Delta R_A) - (R_B \Delta R_C + R_C \Delta R_B) \tag{11.3}$$

If each arm has the same initial resistance R, as is often the case, this expression reduces to

$$U = R\big[(\Delta R_A + \Delta R_D) - (\Delta R_B + \Delta R_C)\big] \tag{11.4}$$

Equation (11.4) indicates that changes in resistance of opposite arms of the bridge add while those of adjacent arms subtract. This fact makes it possible to connect the gage, or gages, into the bridge circuit in such a way as to achieve certain desired effects, one of which is to provide compensation for changes in temperature.

Temperature compensation can be achieved by using the *active gage* mounted to the specimen as one arm of the bridge, say arm A, and a *dummy gage* as the opposite arm. The dummy gage is identical to the active gage and is mounted to a piece made of the same material as the test specimen. This piece is placed in the immediate vicinity of the active gage, but it is not loaded. The idea is to make the two gages experience the same changes in temperature. If they do, they will experience very nearly the same temperature-induced changes in resistance, ΔR_T. The active gage undergoes an additional change in resistance ΔR_L due to the applied loads. Since the resistances of the other two arms of the bridge don't change, we have from Eq. (11.4)

$$U = R[(\Delta R_T + \Delta R_L) - \Delta R_T] = R\Delta R_L$$

Thus, the temperature-induced change in resistance is canceled out by the circuitry, and the bridge output registers only the effect of the applied loads.

Other effects, such as increasing the bridge output for a given strain or canceling out strains due to certain types of loadings, can also be achieved by arranging the gages in an appropriate way. For example, suppose that we wish to determine only the bending strains in the member shown in Figure 11.6(a). This can be done by mounting one gage on top of the member and another on the

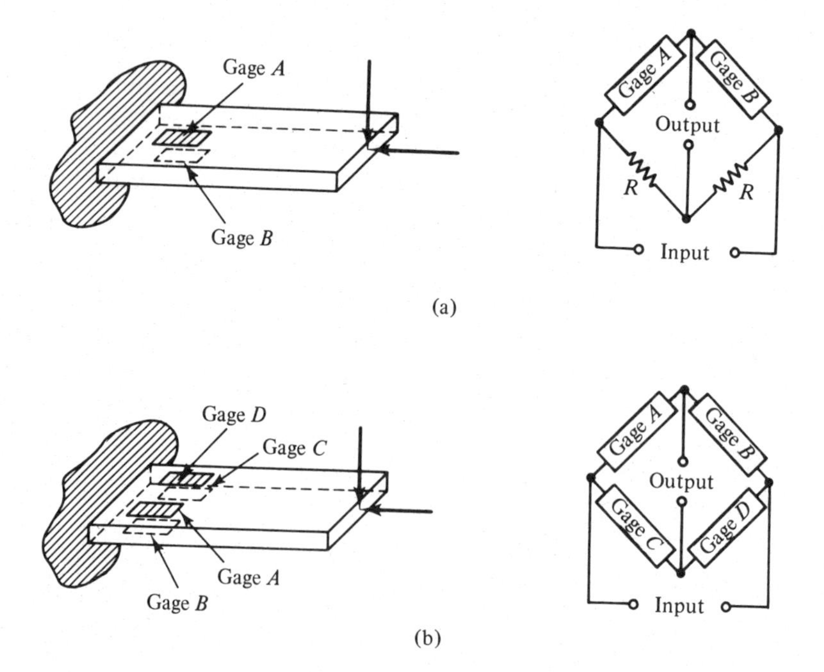

Figure 11.6

bottom and connecting them into adjacent arms of a bridge circuit, as shown. This arrangement of the gages causes the strains due to the compressive load to cancel out.

To verify this, we note that gage A will experience a change in resistance ΔR_b due to the tensile bending strain and gage B will experience an equal and opposite change, $-\Delta R_b$, due to the compressive bending strain. Both gages experience the same changes $-\Delta R_c$ due to the compressive load. Let us also suppose that the gages undergo a change in resistance ΔR_T due to a temperature change. From (11.4), we obtain

$$U = R\left[(\Delta R_b - \Delta R_c + \Delta R_T) - (-\Delta R_b - \Delta R_c + \Delta R_T)\right] = 2R\Delta R_b$$

Thus, the output of the bridge registers only the bending strain, as desired. This arrangement of the gages also provides temperature compensation and gives twice the output of a single gage. The output can be doubled again by using four active gages arranged as shown in Figure 11.6(b).

If the gages A and B on the top and bottom of the beam shown in Figure 11.6(a) are connected into opposite arms of the bridge, the effects of the bending strains will cancel and only the strains due to the compressive load will be measured. This arrangement also gives twice the output of a single gage. However, it does not provide temperature compensation, unless dummy gages are used as the other two arms of the bridge.

Strain gages mounted to various types of members can also be used as *transducers* for measuring force, displacement, torque, and other physical quantities. For example, the cantilever beam shown in Figure 11.6(a) or 11.6(b) can be used as a transducer for measuring either lateral force or displacement. The system can be calibrated to measure force by hanging various known weights from the end of the beam and measuring the corresponding output of the bridge. It can be calibrated to measure displacement by deflecting the tip of the beam various known amounts, say with a micrometer, and noting the corresponding bridge output.

Figure 11.7(a) shows a load cell for measuring tensile and compressive forces. It consists of a bar with strain gages mounted to the uniform central portion and arranged in a bridge circuit as shown. The cell can be calibrated by applying various known forces and measuring the corresponding bridge output. The arrangement shown in Figure 11.7(b) can be used to measure the torque in a shaft or its angle of twist. A transducer for measuring acceleration is shown in Figure 11.7(c). The acceleration of the mass results in bending of the beam element, which is detected by the strain gages.

The only basic requirement for a strain gage transducer is that the quantity to be measured produce a strain that can be detected by the gages. Thus, the number of possibilities is limited only by the imagination of the designer. Four active gages are used in most transducers in order to increase sensitivity. Whenever possible, the gages are arranged to provide temperature compensation. Most transducers can be designed so that they are linear (output proportional to input) over most, if not all, of their operating range.

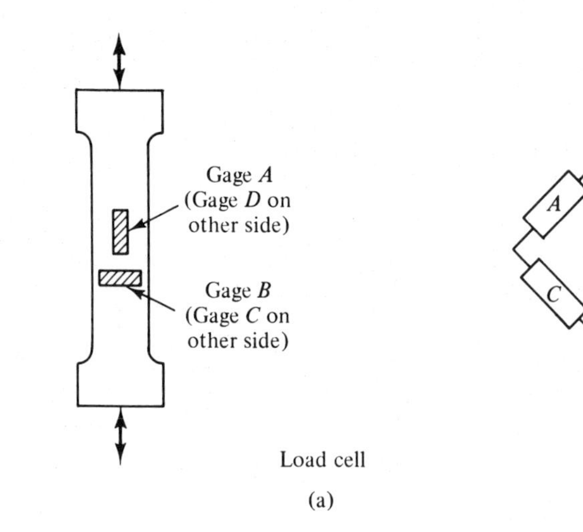

Load cell

(a)

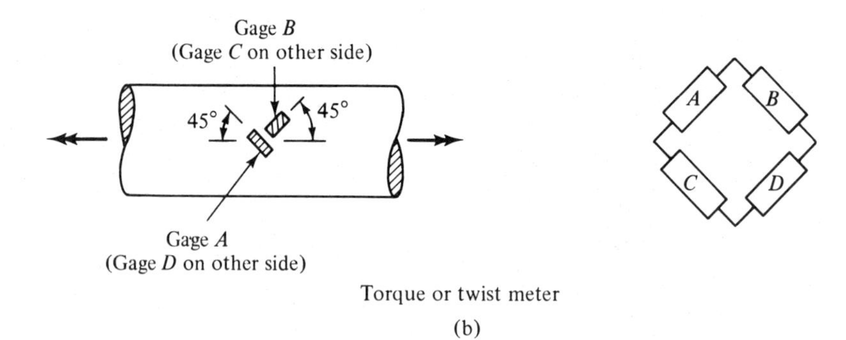

Torque or twist meter

(b)

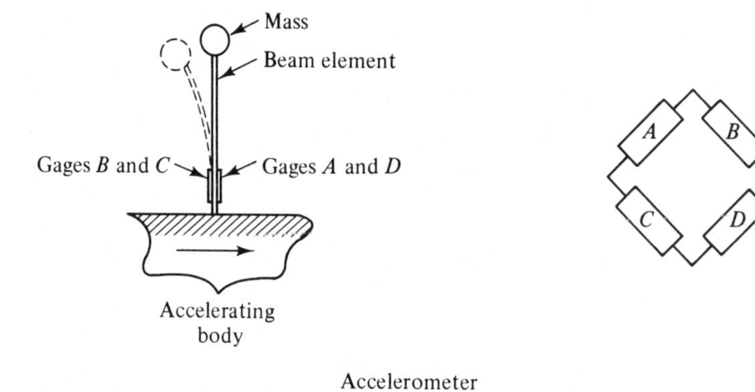

Accelerometer

(c)

Figure 11.7

Example 11.2. Measurement of Strains in a Shaft. The shear strain in a shaft is to be measured by using two electrical resistance strain gages aligned as shown in Figure 11.8(a). Should the gages be placed in adjacent or opposite arms of a bridge circuit? Does this arrangement provide temperature compensation?

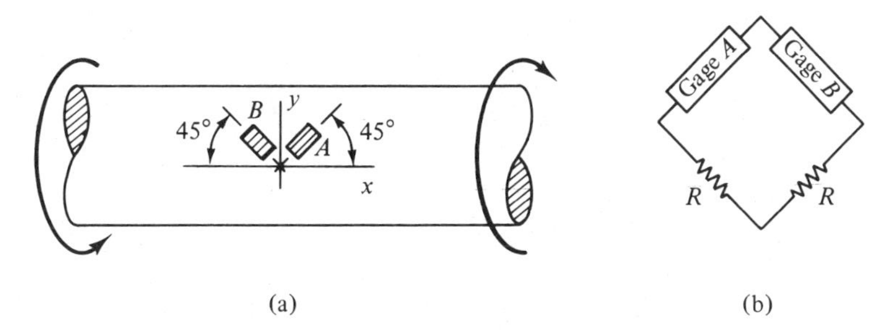

Figure 11.8

Solution. According to the strain transformation equation, Eq. (9.13), the respective strains experienced by gages A and B are

$$\varepsilon_A = \frac{\gamma_{xy}}{2} \qquad \varepsilon_B = -\frac{\gamma_{xy}}{2}$$

Thus, the changes in resistance of the gages will be equal in magnitude and opposite in sign. Accordingly, the gages should be placed in adjacent arms of the bridge [Figure 11.8(b)] since the changes in resistance of these arms subtract. If placed in opposite arms, the outputs from the two gages add and would cancel out. Since a temperature change produces the same change in resistance in each gage, this arrangement does give temperature compensation.

To verify our conclusions, let ΔR_L be the change in gage resistance due to the applied loads (couples) and ΔR_T the change due to temperature effects. From Eq. (11.4), we obtain

$$U = R\left[(\Delta R_L + \Delta R_T) - (-\Delta R_L + \Delta R_T)\right] = 2R\Delta R_L$$

PROBLEMS

11.13 Is the load cell shown in Figure 11.7(a) temperature compensated? What is the output of the cell compared to one with only a single longitudinal gage?

11.14 Is the torque meter shown in Figure 11.7(b) temperature compensated? What is the output of the meter compared to one with only a single strain gage?

11.15 If the twist meter shown in Figure 11.7(b) consists of a 1-in. diameter shaft 4 ft long, determine the strain reading from the meter per degree of angle of twist.

11.16 Determine the strain reading from the torque meter shown in Figure 11.7(b) per lb·in. of torque if the shaft is 4 ft long, has a diameter of 1 in., and is made of steel.

11.17 Determine the strain reading from the load cell shown in Figure 11.7(a) per kilonewton of applied force. The member is made of 6061-T6 aluminum alloy and has a cross-sectional area of 400 mm^2.

11.18 In the accelerometer shown in Figure 11.7(c), the acceleration $\boldsymbol{a}$ of the body produces an equivalent force on the mass m of magnitude ma acting in the direction opposite to $\boldsymbol{a}$. Obtain an expression for the strain reading from the accelerometer in terms of a, the distance L from the mass to the strain gage, and the flexural rigidity EI and thickness t of the beam element. Assume small deflections.

11.19 Two strain gages are mounted on opposite sides of a tensile specimen as shown. Show the gages arranged in a bridge circuit for measuring the longitudinal strain. Does this arrangement of the gages give temperature compensation? If not, how can it be achieved? Is the bridge output affected by any bending strains that may be present?

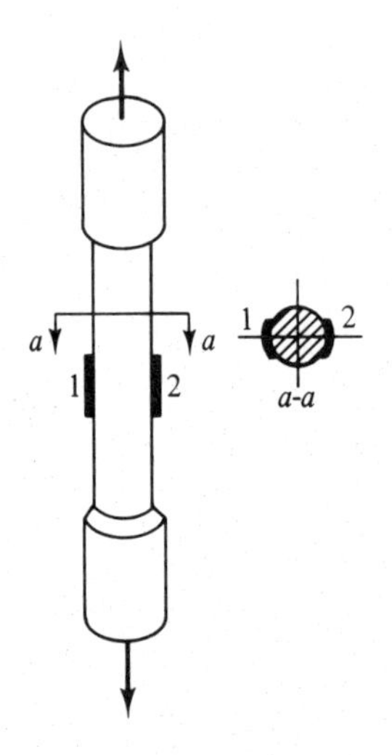

Problem 11.19

11.20 Two strain gages are mounted as shown on a shaft subjected to combined torsion and axial loading. Show the gages arranged in a temperature-compensated bridge circuit for measuring (a) the torque and (b) the axial load. Use dummy gages as required.

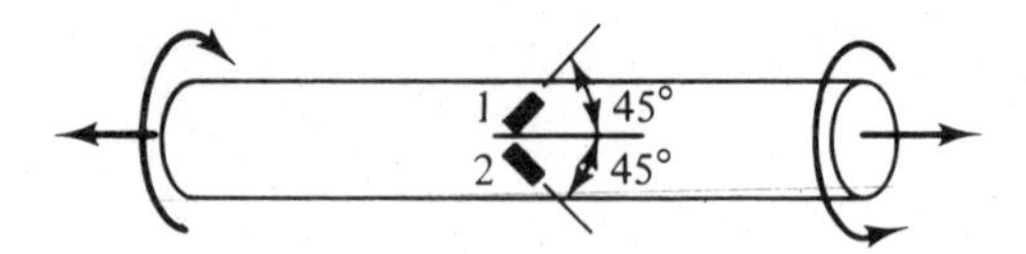

Problem 11.20

11.21 A pressure gage for a cylindrical air tank consists of two strain gages, one mounted in the longitudinal direction and the other in the circumferential direction. Show the gages arranged in a bridge circuit to give maximum output. Does this arrangement of the gages give temperature compensation? If not, how can it be obtained?

11.22 A load cell for measuring compressive loads consists of a rigid platen attached to a hollow tube with four strain gages

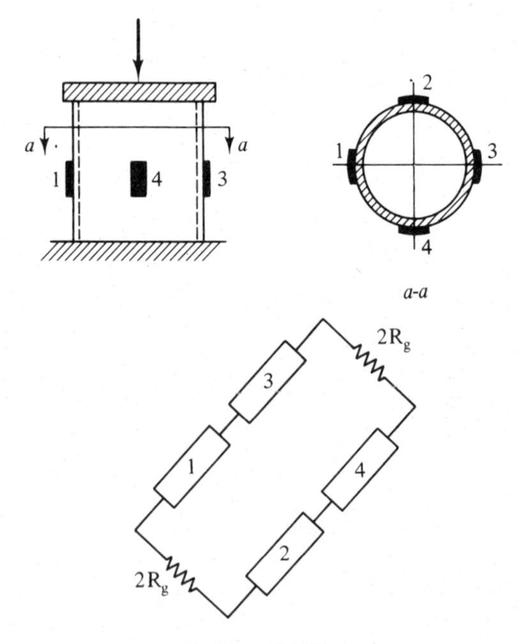

Problem 11.22

equally spaced around its circumference and arranged in a bridge circuit as shown. The manufacturer claims that the cell gives a true indication of the load no matter where it acts on the platen. Is this claim justified? Explain.

11.23 An extensometer for measuring the elongation or shortening of a specimen in tension or compression tests consists of two knife edges rigidly attached to a thin, flexible strip with a strain gage mounted on each side. Show the gages arranged in a bridge circuit and obtain an expression for the strain reading from the extensometer in terms of the dimensions L, a, and t, the flexural rigidity EI of the strip, and the change in length δ of the specimen. Assume small deformations and neglect the effect of the axial force on the strip.

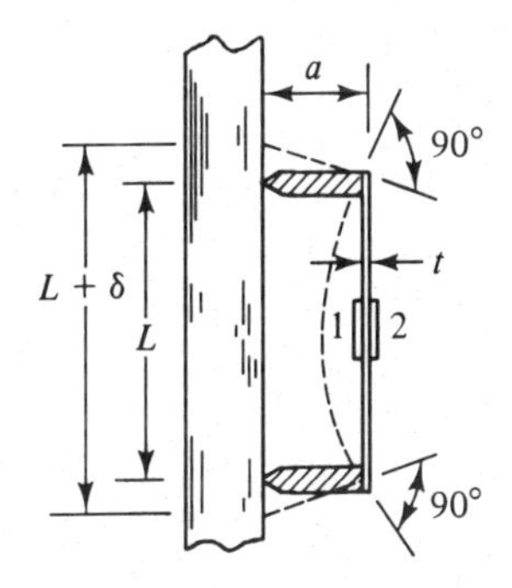

Problem 11.23

11.24 Strain measuring systems are often calibrated by using simulated strains produced by connecting a resistor in parallel with the strain gage as shown. Given that the equivalent resistance of the parallel combination is $R_{eq} = RR_g/(R + R_g)$, determine the resistance R necessary to simulate a strain of 1000 μ in./in. if R_g = 120 ohms and $G = 2$.

11.25 The cantilever beam shown is made of steel and is to be used as a scale. When the scale pan is empty, the meter reading is zero. If a 100 000-ohm resistor produces a meter reading of 60 units when connected in parallel with one of the gages, what is the weight of an object that produces a meter reading of 104 units when placed in the pan? (See Problem 11.24.) R_g = 120 ohms and $G = 2$.

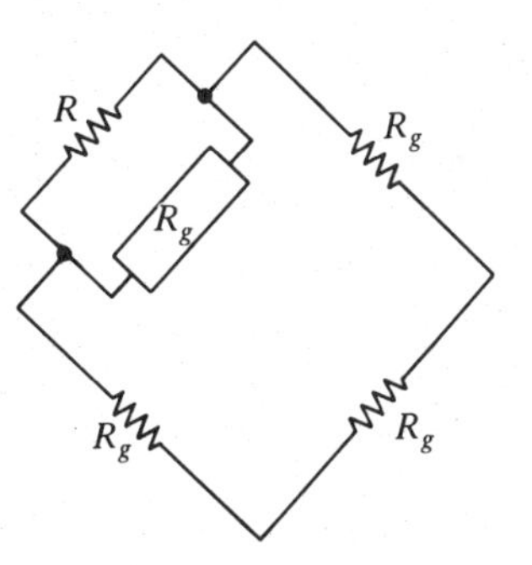

Problem 11.24

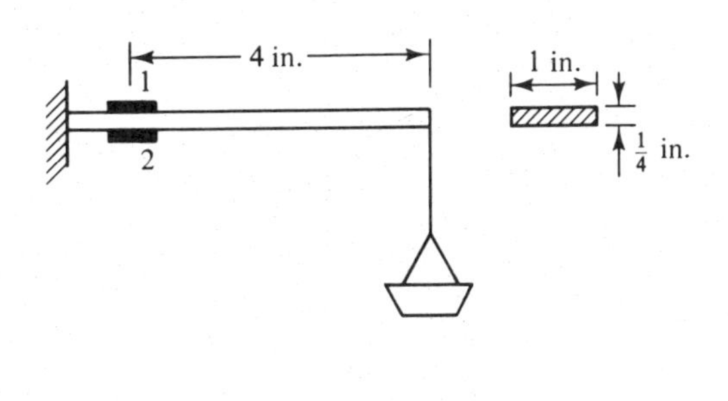

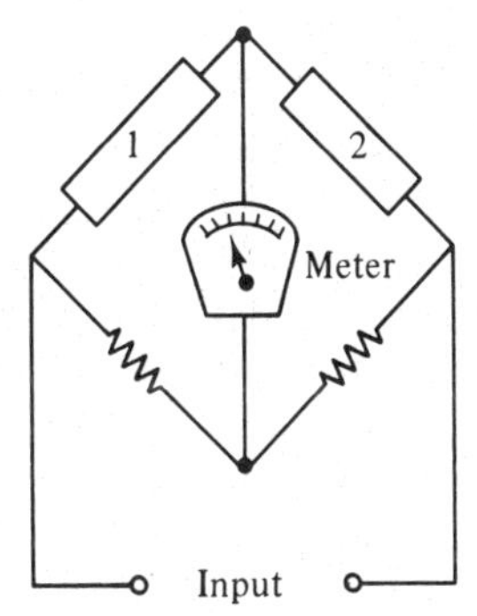

Problem 11.25

11.4 STRAIN-GAGE INSTRUMENTATION

A wide variety of instrumentation is available for detecting, displaying, and recording the output from electrical resistance strain gages. The choice of instruments used depends on such factors as whether the loading is static or dynamic, the total number of gages involved, and whether or not a permanent record of the data is required.

Instruments called null-balance *strain indicators* (a typical example is shown in Figure 11.9) are commonly used for static tests. These instruments operate on AC line current transformed to DC or an internal battery. Thus, they are completely portable and can be used in the field. They also have internal resistors that can be used in completing the strain-gage bridge circuit.

Strain indicators are very easy to operate. First, a knob is turned to adjust the instrument to the gage factor of the particular gages used. Next, the strain-gage bridge is connected to the instrument and the circuit is balanced by rotating a second knob until a zero meter reading is obtained. The test specimen is then loaded. The resulting changes in resistance of the active gages cause an unbalance in the bridge, which is indicated by a nonzero meter reading. When it is adjusted so that the meter reading is again zero, the indicator provides a direct reading of the strain in micro inches per inch. The particular instrument shown in Figure 11.9 displays the value of strain in digital form; others display it in the form of a dial reading.

The indicators are designed for use with a single strain-gage bridge with one, two, or four active gages. If readings are to be taken from more than one set of gages, it is convenient to use a switch and balance unit. The unit shown in Figure 11.10 has provisions for taking readings from ten different sets of gages. By turning a switch, each different set can, in turn, be connected with the strain indicator. There are also provisions for initial balancing of each of the ten circuits.

For dynamic loadings, it is necessary to use instruments that record the variation of the strain-gage signals with time. It is essential that the instrument have the proper frequency response, so that it doesn't distort the signals.

Oscillographs using galvanometers can be used to record dynamic strain signals involving components with frequencies in the range from 0 Hz to approximately 2000 Hz. The output from the strain gages causes the galvanometers to rotate slightly. A record of the signal is provided by a light beam reflected from a small mirror attached to the galvanometer and falling onto a light sensitive paper that passes through the instrument at a set, but changeable, speed. Oscillographs that record with a mechanical pen can be used for low-frequency signals (approximately 100 Hz and below). The frequency response of these instruments is limited by the mechanical inertia of the pen mechanism.

Cathode-ray oscilloscopes are commonly used for recording high-frequency strain signals, such as occur with impact loadings. A permanent record showing the variation of the strain with time can be obtained by photographing the signal trace on the screen of the scope with a still or moving-film camera. It is usually necessary

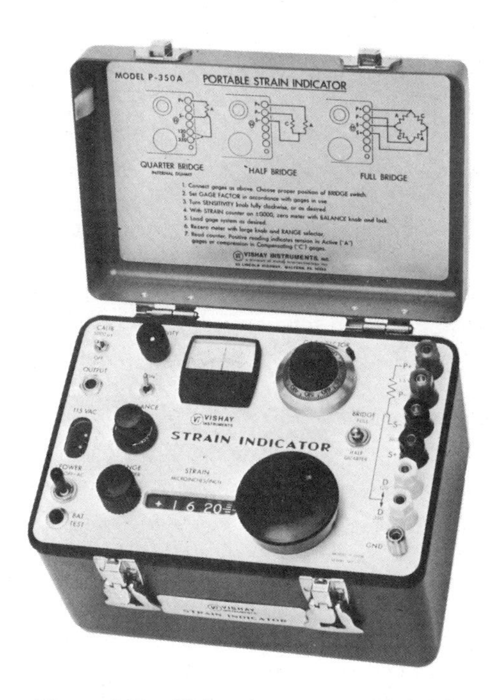

Figure 11.9. Vishay Instruments Model P-350A Portable Strain Indicator. (*Courtesy* Measurements Group-Vishay Intertechnology, Inc.)

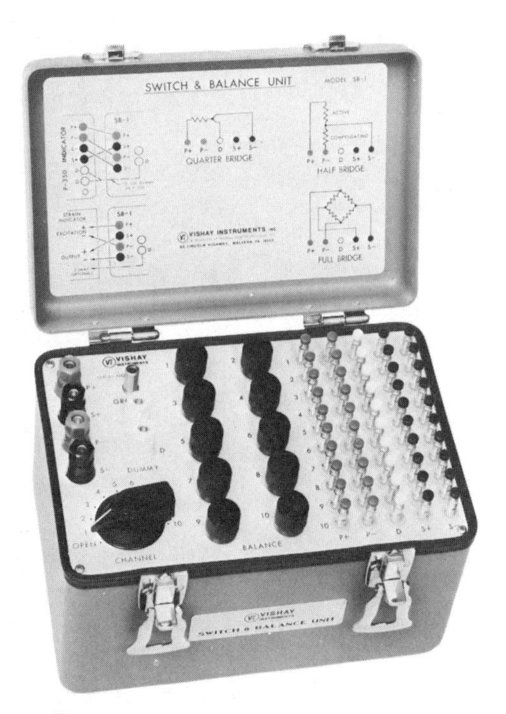

Figure 11.10. Vishay Instruments Model SB-1 Ten Channel Switch and Balance Unit. (*Courtesy* Measurements Group-Vishay Intertechnology, Inc.)

to amplify the strain-gage signals before they are fed into the scope. Some strain indicators can be used for this purpose, and numerous other types of amplifiers are commercially available.

11.5 BRITTLE COATINGS

There are several materials that form a thin, brittle coating when applied to the surface of a member and allowed to cure. When the member is loaded, the deformations in the plane of the surface are transmitted to the coating, which adheres tightly to the surface. Since the coating is brittle, it is relatively weak in tension and cracks when the maximum tensile strain reaches a certain critical value. By observing the resulting crack patterns in the coating, the maximum tensile surface strains within the member and their directions can be determined. Figure 11.11 shows the crack pattern in a coating applied to an automotive steering component.

The idea for this method of strain analysis was obtained by observing the cracking of glazes on old pottery and the flaking of mill scale from hot-rolled steel when it yields. Whitewash was one of the first coatings used, but it has low sensitivity and does not crack until the strains are very large. Most of the coatings

Figure 11.11. Crack Pattern in Brittle Coating on an Automotive Steering Component. (*Courtesy* Measurements Group-Vishay Intertechnology, Inc.)

in current use are a type of lacquer and are designed to crack at strains well below the yield strains for metals. They are available in different compositions for use at different temperatures and are sprayed onto the surface. Some skill is required to produce a uniform coating without runs and bubbles. The coatings must cure for approximately 24 hours at room temperature before they can be used. Curing at elevated temperatures is often desirable.

Before the specimen can be tested, the coating must be calibrated. This is done by using calibration strips in the form of rectangular bars that are coated at the same time as the specimen. These strips are loaded as cantilever beams subjected to a prescribed tip displacement by a cam arrangement shown schematically in Figure 11.12(a). For this loading, the longitudinal normal strain is largest at the support and decreases linearly to a value of zero at the tip of the beam. The coating cracks near the support where the strain is large and remains uncracked near the tip where it is small [Figure 11.12(b)]. The strain at the location where the cracks stop is the threshold strain at which the coating fails. A scale is available for determining the value of this strain without computation.

Once the coating has been calibrated, the specimen can be tested. By increasing the loads on the specimen in increments and observing how the cracks form and grow, it is possible to map out the variation of the tensile strains over the surface of the member. Compressive strains can be determined by applying the coating to a loaded specimen and then removing these loads so that the coating is placed in tension.

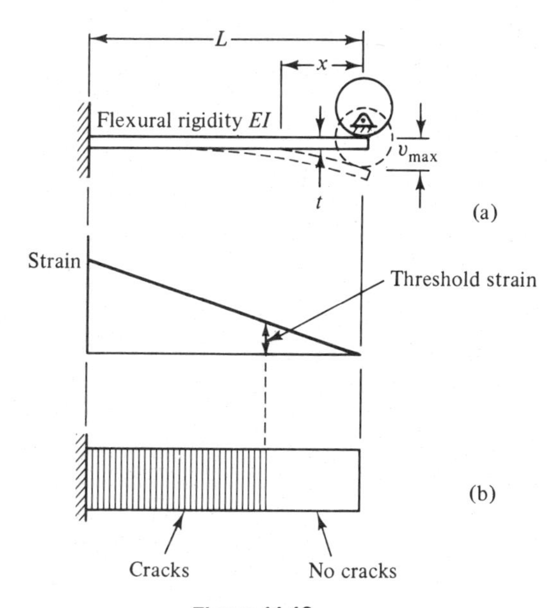

Figure 11.12

The data analysis is simple. When the coating first cracks at a point, the maximum strain there in the direction perpendicular to the crack is taken to be equal to the calibration value for the coating. The maximum stress is then computed by multiplying the strain by the modulus of elasticity of the specimen. Since this procedure neglects any effects due to the two-dimensional states of strain and stress that usually exist, the values obtained for the stresses and strains are not highly accurate. The accuracy is also affected by such factors as changes in temperature and humidity and variations in the thickness of the coating. An accuracy of $\pm$ 10% in the value of the strain is about the best that can be achieved under ideal conditions. However, this is sufficient for many applications.

The coatings do give an accurate indication of the location of maximum strains and the directions of the principal strains. Thus, they are widely used as an aid in determining where and how to locate strain gages. The regions of high strain are distinguished by closely spaced cracks; the closer the cracks are together, the larger the strain. Since the coating fails due to the maximum tensile strain, the principal strains are oriented perpendicular and parallel to the cracks. This information makes it possible to use two strain gages at each point instead of three (see Example 11.1). Of course, the coating must be removed before the gages are mounted.

Brittle coatings have the advantage that they give an indication of the stresses and strains over the entire surface. Their use requires little or no special equipment, and they can be easily removed after testing. The common coatings can be used at temperatures up to approximately 100°F, and special ceramic coatings for use at temperatures up to 700°F are available.

Example 11.3. Determination of Pressure in a Cylinder. A thin-walled cylindrical vessel covered with a brittle lacquer is pressurized until longitudinal cracks form in the coating. Testing of a calibration strip indicates that the threshold value of strain for the coating is 600 μ in./in. Estimate the pressure in the cylinder if it is made of aluminum with $E = 10 \times 10^6$ psi and has a radius to thickness ratio of $R/t = 20$.

Solution. Taking the circumferential strain (perpendicular to the cracks) to be equal to the calibration value for the coating and multiplying by the modulus of elasticity of the cylinder, we obtain for the circumferential stress

$$\sigma_C = E\varepsilon_C = (10 \times 10^6 \text{ lb/in.}^2)(600 \times 10^{-6} \text{ in./in.}) = 6000 \text{ psi}$$

From Eq. (6.8), we have

$$\sigma_C = \frac{pR}{t} = 6000 \text{ psi}$$

or

$$p = \frac{6000 \text{ psi}}{20} = 300 \text{ psi} \qquad \textbf{\textit{Answer}}$$

This value is only an estimate because the effects of the longitudinal stress are ignored in the analysis.

PROBLEMS

11.26 Obtain an expression for the strain in the brittle coating calibration strip shown in Figure 11.12 as a function of the distance x from the tip. Thus, verify that the linear variation of strain indicated in Figure 11.12(a) is correct. Does the strain at a given location depend upon the material properties of the strip?

11.27 A brittle coating is applied to a 1-in. wide × $\frac{1}{4}$-in. thick aluminum calibration strip with $E = 10 \times 10^6$ psi. The strip is 10 in. long and is subjected to a tip displacement of $\frac{1}{4}$ in. If the last visible crack is 4 in. from the tip, what is the threshold strain for the coating?

11.28 Same as Problem 11.27, except that the last visible crack is 5 in. from the tip.

11.29 A brittle coating with a threshold strain of 600 μ m/m is applied to a machine part made of grey cast iron with $E = 100$ GN/m^2. What is the maximum tensile stress in the part when the coating first cracks? If strain gages are to be used to determine more accurate results, how many gages are required and how should they be oriented?

11.30 Same as Problem 11.29, except that the specimen is an aircraft part made of a titanium alloy with $E = 110$ GN/m^2.

11.31 A 1-in. diameter steel shaft covered with a brittle coating is loaded in pure torsion until the coating cracks. The cracks form at 45° from the longitudinal axis. Calibration shows that the threshold strain for the coating is 400 μ in./in. What is the magnitude of the applied couples?

11.32 Same as Problem 11.31, except that the shaft is hollow and has an inside diameter equal to one-half the outside diameter.

11.33 Must a brittle coating calibration strip have the same modulus of elasticity as the test specimen? If not, is it necessary to make any corrections for the different values of E when determining stresses in the specimen? Explain.

11.34 Proof testing of a steel spherical tank with portholes requires that the tank be subjected to twice the normal operating pressure without showing evidence of yielding. Explain how a brittle coating could be used in this test. What threshold strain for the coating would be required?

11.6 MOIRÉ STRAIN ANALYSIS

When two grids consisting of equally spaced parallel lines on a transparent background are superposed and then displaced relative to one another, interference patterns are formed where the two sets of lines overlap. These patterns appear as

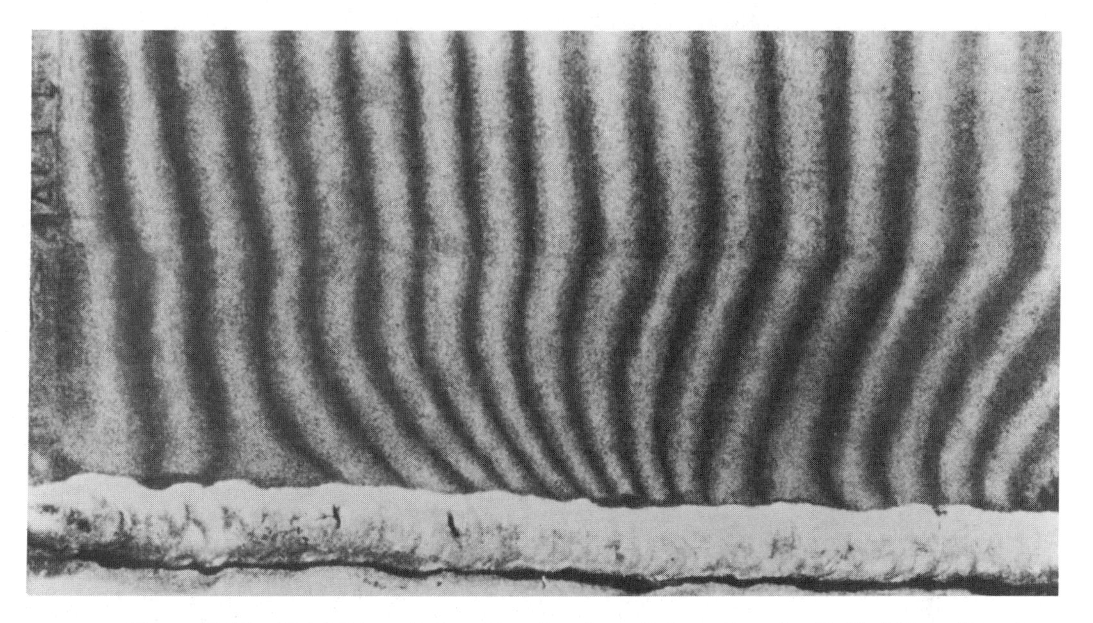

Figure 11.13. Moiré Fringe Pattern Resulting from Deformation of a Welded Plate. (*Courtesy* Measurements Group-Vishay Intertechnology, Inc.)

dark bands, or fringes. This phenomenon is called the *moiré effect* and can be easily demonstrated by using two pieces of screen wire.

If one of the grids is applied to the surface of a member and the other is held stationary over it, the pattern formed when the member deforms can be used to determine the surface displacements and strains. The grid can be applied to the member photographically or it can be formed separately and attached with an adhesive. Figure 11.13 shows the moiré fringe pattern resulting from the deformation of a flat plate that has been welded to another plate. The distortions in the pattern are caused by residual stresses set up during the welding process.

Each fringe in the moiré pattern is a loci of points that undergo the same displacement δ in the direction perpendicular to the stationary grid lines. The change $\Delta\delta$ in the value of displacement from one fringe to the next is equal to the pitch p of the grid [Figure 11.14(a)]. Referring to the figure, we see that the average strain of a line element, such as AB, between the two fringes is

$$\varepsilon_{\text{avg}} = \frac{\Delta\delta}{\Delta L} = \frac{p}{d} \tag{11.5}$$

where $d = AB$ is the fringe spacing.

The actual strain at a point, $\varepsilon = d\delta/dL$, can be obtained by plotting the variation of the displacement with distance and then determining the slope of the resulting curve [Figure 11.14(b)]. Since only the change in δ is important in determining the strain, its value can be set equal to an arbitrary constant at one of the fringes. This procedure gives the variations of the normal strain along a line perpendicular to the stationary grid lines. By repeating the process along other such lines, the variation of the strain over the entire surface can be determined. The strains in other directions can be found by changing the orientation of the grids.

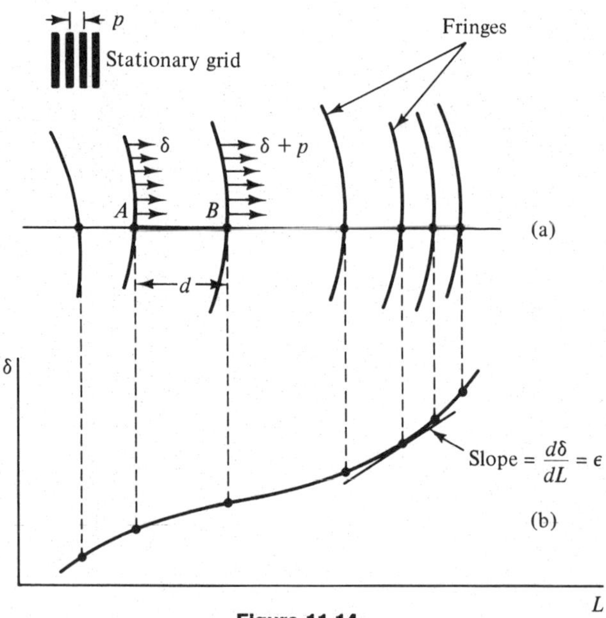

Figure 11.14

The complete state of strain at a point can be determined by measuring the normal strains in three different directions and applying the strain transformation equations [Eqs. (9.13) and (9.15)], as in the case of strain gages. It is also possible to determine the state of strain by using grids aligned in two perpendicular directions, but we shall not go into that here.

Moiré strain analysis is useful primarily for problems involving large strains. If the strains are small, the fringes are far apart and there is not enough of a pattern to permit accurate analysis. The resolution can be enhanced by making the pitch p of the grid smaller, but there is a limit to how close the lines can be spaced. The most sensitive grids currently available have approximately 2000 lines per inch ($p = 0.0005$ in.).

Like brittle coatings, moiré analysis gives an indication of the strains over the entire surface.

Example 11.4. Moiré Patterns for a Plate. A flat plate is loaded such that the longitudinal strain ε is uniform throughout. What would the resulting moiré fringe pattern look like if the grids used have 2000 lines/in. and are oriented perpendicular to the plate axis? What is the spacing between the fringes for $\varepsilon = 1000\ \mu$ in./in.?

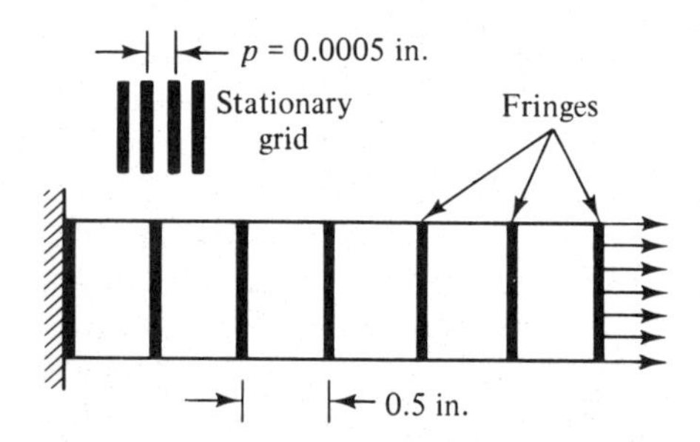

Figure 11.15

Solution. Since the strain does not vary along the length of the member, $\varepsilon = \varepsilon_{avg}$. From Eq. (11.5), we have

$$d = \frac{p}{\varepsilon} = \frac{0.0005 \text{ in.}}{0.001 \text{ in./in.}} = 0.5 \text{ in.}$$

Since d is a constant, the moiré pattern will consist of a series of equally spaced fringes 0.5 in. apart (Figure 11.15).

PROBLEMS

11.35 At a point on the surface of a rubber test specimen the spacing between adjacent moiré fringes is 0.5 in., measured in the direction perpendicular to the stationary grid lines. If the grids have 200 lines/in., what is the average strain at this point?

11.36 Same as Problem 11.35, except that the grids have 500 lines/in.

11.37 A rectangular epoxy plate with $E = 0.6 \times 10^6$ psi and $\nu = 0.4$ is subjected to uniformly distributed tensile stresses σ_x and σ_y as shown. Sketch the resulting moiré fringe pattern produced by the deformation of the plate and determine the fringe spacing if $\sigma_x = 1000$ psi and $\sigma_y = 500$ psi. The grids have 2000 lines/in. and are oriented with the grid lines parallel to the y axis.

11.38 When the epoxy plate in Problem 11.37 is subjected to stresses $\sigma_x = 1200$ psi and $\sigma_y = 0$, the resulting moiré fringes are found to be 0.25 in. apart. A stress σ_y is then applied, and the fringe spacing changes to 0.30 in. Determine σ_y.

11.39 Same as Problem 11.38, except that the fringe spacing changes to 0.20 in. after the stress σ_y is applied.

11.40 Same as Problem 11.37, except that the grid lines are parallel to the x axis.

11.41 A tracing of the moiré fringe pattern for the welded plate of Figure 11.13 is as shown. Determine the variation of the strain along line AB adjacent to the weld. The 2000 lines/in.-grids were oriented with the grid lines perpendicular to the weld.

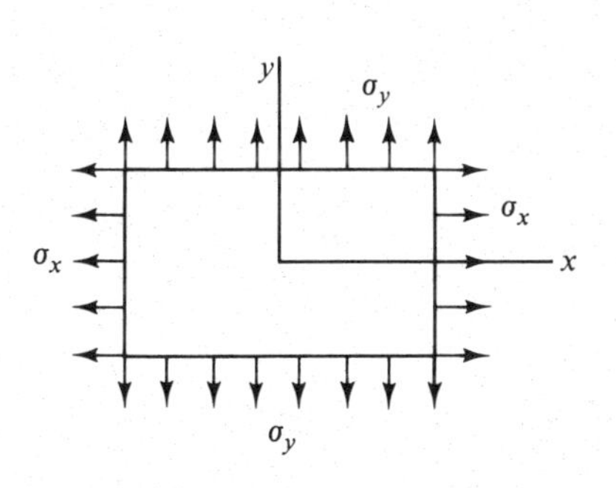

Problem 11.37

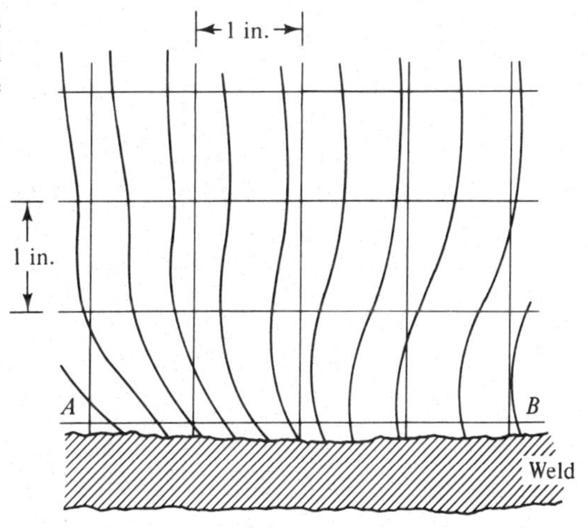

Problem 11.41

11.7 PHOTOELASTICITY

The photoelastic method of stress analysis is based on the optical fringe patterns formed when a transparent model of a member is loaded in the presence of polarized light. An analysis of these fringe patterns yields the stresses in the model, which are then related to the prototype by using the appropriate model theory.

Before attempting to describe the photoelastic method in more detail, let us briefly review some of the relevant properties of light. Light is an electromagnetic radiation, which, for our purposes, can be thought of as consisting of waves vibrating in planes perpendicular to the direction of propagation. Color is associated with the frequency of the radiation. Monochromatic light is a radiation with a single frequency; white light is a mixture of radiations of different frequencies.

Unpolarized light consists of waves vibrating in an infinite number of planes. If unpolarized light is passed through a polarizing element, only the waves vibrating in the plane of polarization are transmitted. Thus, polarized light consists of waves vibrating in parallel planes.

Doubly refracting materials transmit light waves vibrating in two perpendicular planes. When polarized light passes through such a material, the light wave is broken into two perpendicular components. These components travel through the material at different speeds, and, as a result, appear to be shifted in space relative to one another through a distance called the *phase shift*.

Most materials that are not doubly refracting in their natural state become doubly refracting when they are stressed. The planes of light transmission at any one given point in the material are aligned with the principal axes of stress, and the phase shift between the transmitted waves is proportional to the difference of the principal stresses at that point. This so-called *photoelastic effect* was discovered by Sir David Brewster in 1816, and it is the basis of the photoelastic method of stress analysis.

The first step in a photoelastic investigation is to make a model of the member of interest out of a suitable transparent material. This model is then placed in the optical field of an apparatus called a *polariscope*, which consists of a series of elements arranged as shown in Figure 11.16. In addition to the light source and collimating lens, there are two polarizing elements, the polarizer and analyzer, and two quarter-wave plates, which are discs of doubly refracting material whose thickness has been adjusted to give a phase shift equal to one-quarter of the

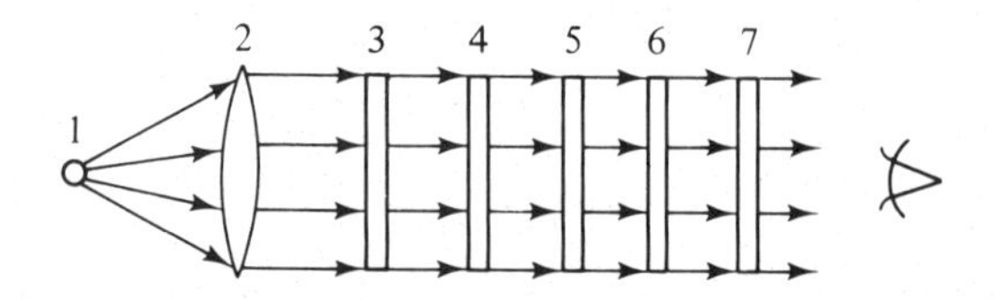

1. Light source
2. Collimating lens
3. Polarizer
4. Quarter-wave plate
5. Model
6. Quarter-wave plate
7. Analyzer

Figure 11.16

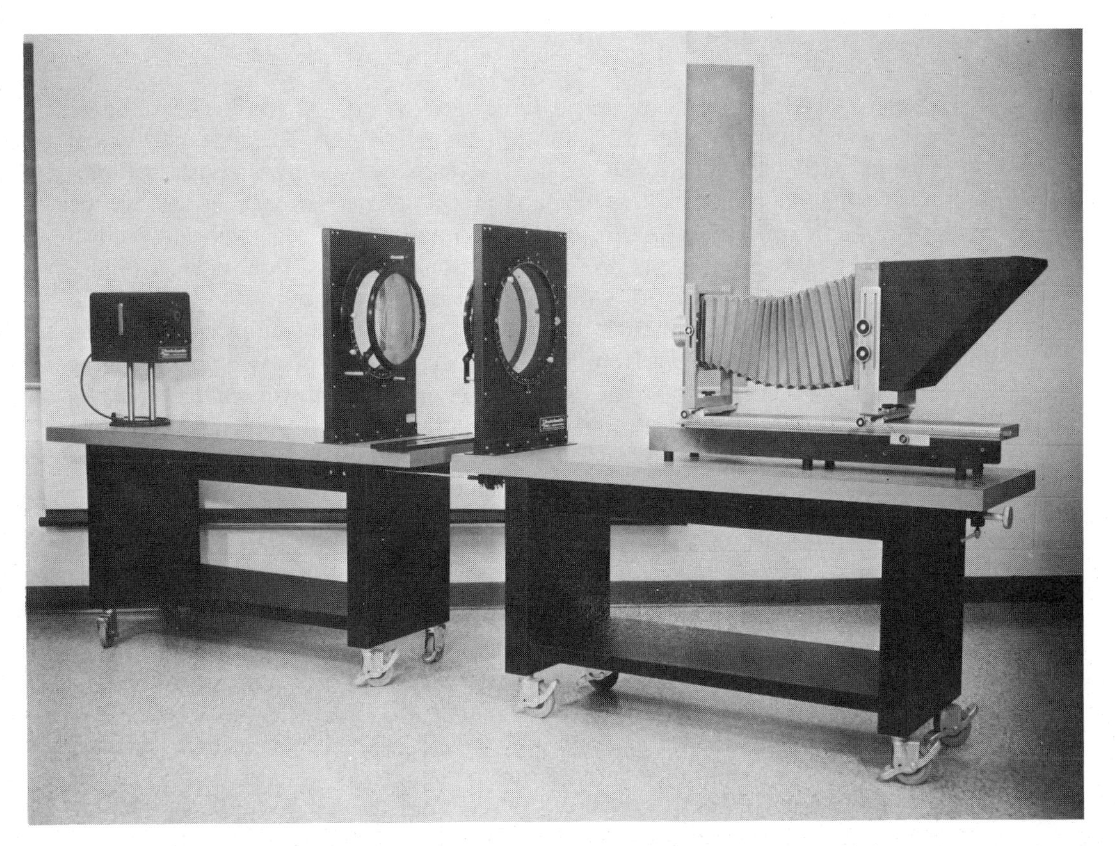

Figure 11.17. Photoelastic Inc. Model 261 Collimated Light Split Bench Polariscope. (*Courtesy Measurements Group-Vishay Intertechnology, Inc.*)

wavelength of the particular monochromatic light used. Figure 11.17 is a photograph of a commercially available polariscope. The loaded specimen is placed at the center, between the two benches, and the camera at the right is used to record the optical patterns in the model.

When the stresses at a point in the model are such that the resulting phase shift is an integral multiple of the wavelength of the monochromatic light source, the light transmitted through the model will be the same as the background light. This condition can be expressed as

$$\sigma_1 - \sigma_2 = \frac{Kn}{t} \tag{11.6}$$

where σ_1 and σ_2 are the principal stresses, K is the experimentally determined *photoelastic constant* for the material, t is the model thickness, and n is an integer ($n = 0, 1, 2, \ldots$).

The loci of points in the model at which the condition in Eq. (11.6) is satisfied appear as a series of bands, called *integer isochromatic fringes*. The loci of points at which the phase shift is one-half the wavelength of the light source, or any odd multiple thereof, also form a series of bands. These bands are called *half-order isochromatic fringes* and correspond to the values $n = \frac{1}{2}, \frac{3}{2}, \frac{5}{2}, \ldots,$ in Eq. (11.6).

If the polariscope is arranged to give a dark background, the whole-order fringes will be dark and the half-order fringes will be light. The reverse is true with a light background. Along any one given fringe, the quantity $\sigma_1 - \sigma_2$ is constant. Since $\sigma_1 - \sigma_2$ is equal to twice the maximum in-plane shear stress [see Eq. (9.7)], the fringes can be thought of as lines of constant shear stress.

Figure 11.18 is a photograph of the isochromatic fringe pattern in a photoelastic model of a knee-frame, with the polariscope arranged to give a light background. Note the numbering of the fringes. The fringe order n can be determined either by counting from a known point of zero stress or by watching the formation of the fringes as the model is loaded. The maximum stress in the frame occurs at the fillet where $n = 8$. Also note that the fringes are narrowest where the stresses are highest. This fact can be used to locate regions of high stress in a member.

At the boundaries of the model, and at any other points where one of the principal stresses is known to be zero, the remaining principal stress can be determined directly from Eq. (11.6) and the observed fringe pattern. Consequently, the photoelastic method is particularly convenient for determining the stress concentration factor at holes, notches, and other geometric discontinuities. Many of the stress concentration factors given in handbooks were obtained by this method.

If both principal stresses are nonzero, they cannot be determined from the fringe pattern alone; additional information is required. This information can be obtained by removing the quarter-wave plates from the polariscope and using a

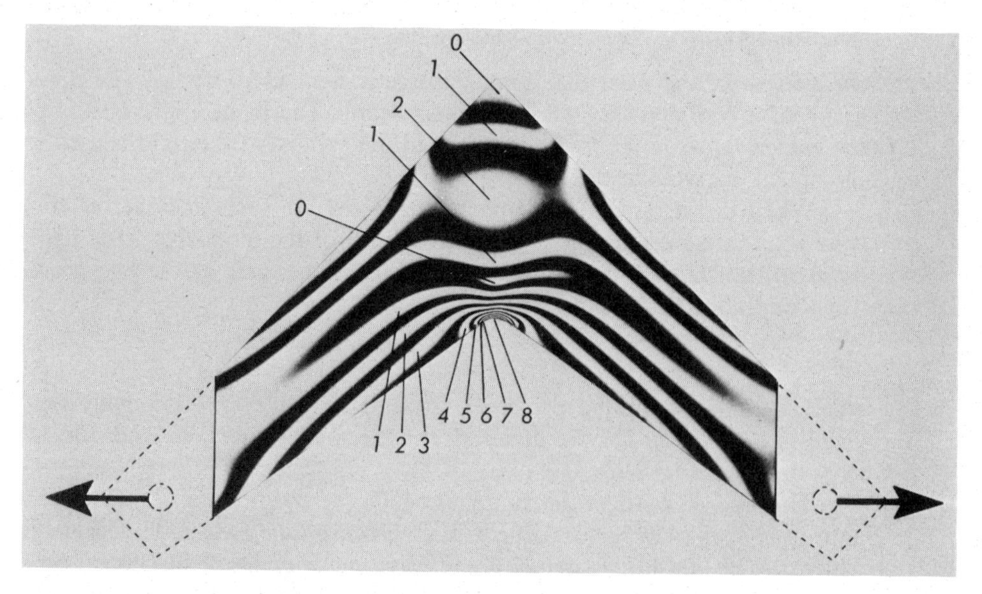

Figure 11.18. Isochromatic Fringe Pattern in a Photoelastic Model of a Knee-frame. (*Courtesy* Measurements Group-Vishay Intertechnology, Inc.)

white light source. With this setup, the isochromatic fringes appear as colored bands; hence the name isochromatics. Black bands, called *isoclinics*, will also be present. The isoclinics define the loci of points in the model at which the principal axes of stress are aligned parallel and perpendicular to the plane of polarization. By rotating the polarizer and analyzer together in increments, the directions of the principal stresses can be determined at all points in the model. The individual stresses σ_1 and σ_2 can then be obtained by combining this information with that obtained from the isochromatic fringe pattern. However, the procedure is tedious and involves many computations.

There are a number of different photoelastic materials that can be used for the model. Desirable properties are transparency, high optical sensitivity (low value of K), ease of machining, and linearly elastic response over a suitable range of stresses. The most commonly used materials are urethane rubbers, polyesters, epoxies, and types of phenolic.

The photoelastic constant K for the material is obtained from the fringe pattern in a specimen loaded in such a way that the stresses can be determined analytically. For example, if a strip of the material is loaded as a beam in pure bending, one principal stress will be equal to the bending stress and the other will be zero. The resulting fringe pattern is a series of equally spaced longitudinal bands, as illustrated in Figure 11.19. To determine K, we compute the stress at the location of each of the fringes from the bending stress formula and plot it versus the fringe order n. As can be seen from Eq. (11.6), the slope of the resulting straight line is equal to K/t (see also Example 11.5).

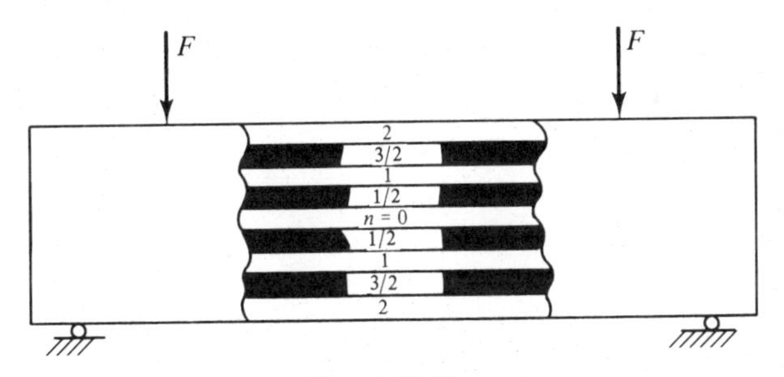

Figure 11.19

Photoelasticity gives an indication of the stresses throughout the entire member, and it can be used for both static and dynamic loadings. Three-dimensional problems can be analyzed by loading the model at an elevated temperature and letting it cool with the loads in place, thus causing the photoelastic patterns to be "frozen" into the material. The stresses over various cross sections can then be determined by examining thin slices cut from the model. Optical techniques for the direct analysis of three-dimensional problems have also been developed. Photoelasticity is particularly convenient for determining qualitative information about the stresses in a member. All that is required is a hand-loaded model held between two sheets of polaroid plastic.

The major disadvantage of the photoelastic method of stress analysis is that a model must be prepared and the stresses in the model related back to the prototype, a step that often leaves the final results open to question. This disadvantage has largely been overcome by the development of *photoelastic coatings* that can be bonded directly to the test specimen, in much the same way as a strain gage is mounted. The surface deformations of the specimen are transferred to the coating by the adhesive; the resulting photoelastic patterns give an indication of the surface stresses in the member without use of a model. A special type of polariscope is required because the fringe patterns observed are formed by light reflected off the back surface of the coating. Otherwise, the basic procedure is the same as that for photoelastic models.

Example 11.5. Determination of Photoelastic Constant. If the beam shown in Figure 11.19 has a rectangular cross section 1.8 in. deep × 0.5 in. wide and the bending moment is $M = 200$ lb·in., what is the photoelastic constant K for the material? Careful measurement shows that the fringes are of equal width.

Solution. Since the fringes are of equal width, their centers are 0.2 in. apart. Computing the bending stress at the centers of each of the fringes, we obtain

$$\begin{aligned} n &= 0 & \sigma &= 0 \\ n &= \tfrac{1}{2} & \sigma &= 166.7 \text{ psi} \\ n &= 1 & \sigma &= 333.3 \text{ psi} \\ n &= \tfrac{3}{2} & \sigma &= 500.0 \text{ psi} \\ n &= 2 & \sigma &= 666.7 \text{ psi} \end{aligned}$$

The stresses are plotted versus the fringe order n in Figure 11.20. The slope of the

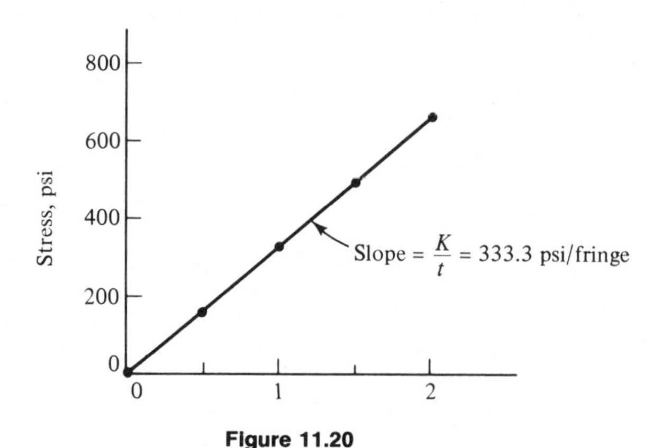

Figure 11.20

resulting straight line is K/t [see Eq. (11.6)]

$$\frac{K}{t} = 333.3 \frac{\text{lb/in.}^2}{\text{fringe}}$$

$$K = 167 \frac{\text{lb/in.}}{\text{fringe}} \qquad \textbf{\textit{Answer}}$$

If the fringe order is known to be proportional to the stress, as it is in this example, the value of K can be obtained directly from Eq. (11.6) by using the value of the stress at any one of the fringes. However, it is usually best to work with a plot of the stress versus the fringe order because it gives a direct indication of the range of stress over which the material exhibits a linear optical response.

PROBLEMS

11.42 When a tensile specimen of a photoelastic material is tested in a polariscope, the loads at which the various isochromatic fringes appear are as follows:

Load (newtons)	0	340	700	1060	1400	1780	2120
Fringe order	0	1	2	3	4	5	6

If the specimen is 25 mm wide and 8 mm thick, what is the photoelastic constant for the material?

11.43 Rework Example 11.5 with the bending moment $M = 150$ lb·in.

11.44 Determine the stress at the fillet of the knee-frame model shown in Figure 11.18. The photoelastic constant is $K = 80$ (lb/in.)/fringe and the model is $\frac{1}{4}$ in. thick.

11.45 Determine the stress at a point on the boundary of a photoelastic model where the fringe order is $n = 4$. The photoelastic constant for the material is $K = 15$ (kN/m)/fringe and the model is 10 mm thick.

11.46 A photoelastic model of a rectangular bar with a circular hole in the middle is loaded in tension and the fringe order at the side of the hole found to be $n = 4.5$. At some distance away from the hole, the fringe order is $n = 2$. What is the stress concentration factor due to the hole?

11.47 A photoelastic model of a rectangular bar with semicircular grooves is loaded in tension. The fringe order at the bottom of the grooves is $n = 6.5$, while at some distance away it is $n = 3.5$. What is the stress concentration factor due to the grooves?

11.8 CLOSURE

The experimental methods discussed in this chapter are the ones commonly used in the solution of day-to-day engineering problems. Further information concerning their application can be obtained from the following texts.

DALLY, J. W., and W. F. RILEY, *Experimental Stress Analysis*, McGraw-Hill Book Company, New York, 1965.

DOVE, R. C., and P. H. ADAMS, *Experimental Stress Analysis and Motion Measurement*, Charles E. Merrill Publishing Company, Columbus, Ohio, 1964.

DURELLI, A. J., E. A. PHILLIPS, and C. H. TSAO, *Introduction to the Theoretical and Experimental Analysis of Strain*, McGraw-Hill Book Company, New York, 1958.

DURELLI, A. J., and W. F. RILEY, *Introduction to Photomechanics*, Prentice-Hall, Inc., Englewood Cliffs, New Jersey, 1965.

HETENYI, M. (ed.), *Handbook of Experimental Stress Analysis*, John Wiley & Sons, Inc. 1950.

PERRY, C. C., and H. R. LISSNER, *The Strain Gage Primer*, 2nd ed., McGraw-Hill Book Company, New York, 1962.

Information on experimental equipment and supplies and their application can be obtained from the following manufacturers of experimental equipment and supplies (partial listing):

BLH Electronics, 42 Fourth Avenue, Waltham, Massachusetts 02154 (strain gages).

Magnaflux Corporation, 7300 West Lawrence Avenue, Chicago, Illinois 60656 (strain gages and brittle coatings).

Measurements Group–Vishay Intertechnology, Inc., 63 Lincoln Highway, Malvern, Pennsylvania 19355 (strain gages, brittle coatings, moiré, and photoelasticity).

APPENDIX A

Table A.1 Geometric Properties of Some Common Shapes

Shape		$\bar{x}$	$\bar{y}$	Area, Length, or Volume	Moments of Inertia
Triangle			$\frac{h}{3}$	$\frac{bh}{2}$	$\bar{I}_{x'} = \frac{1}{36} bh^3$ $I_x = \frac{1}{12} bh^3$
Circle		0	0	πr^2	$\bar{I}_x = \bar{I}_y = \frac{1}{4} \pi r^4$ $J_z = \frac{1}{2} \pi r^4$
Quarter circle		$\frac{4r}{3\pi}$	$\frac{4r}{3\pi}$	$\frac{\pi r^2}{4}$	$I_x = I_y = \frac{1}{16} \pi r^4$
Semi circle		0	$\frac{4r}{3\pi}$	$\frac{\pi r^2}{2}$	$I_x = I_y = \frac{1}{8} \pi r^4$
Rectangle		$\frac{b}{2}$	$\frac{h}{2}$	bh	$\bar{I}_{x'} = \frac{1}{12} bh^3$ $I_x = \frac{1}{3} bh^3$ $\bar{I}_{y'} = \frac{1}{12} hb^3$ $I_y = \frac{1}{3} hb^3$
Parabolic spandrel		$\frac{3b}{4}$	$\frac{3h}{10}$	$\frac{bh}{3}$	
Arc of circle		$\frac{r\ sm\ \alpha}{\alpha}$	0	$2\,\alpha r$	
Solid hemisphere		$\frac{3r}{8}$		$\frac{2}{3} \pi r^3$	
Solid cone		$\frac{h}{4}$		$\frac{1}{3} \pi r^2 h$	

Table A.2 Average Mechanical Properties of Some Common Engineering Materials
(Values in SI are given in parentheses following the corresponding values in US-British units)

Material	*Yield Strength*[a] *10^3 psi (MN/m^2)*		*Ultimate Strength* *10^3 psi (MN/m^2)*			*Elastic Moduli* *10^6 psi (GN/m^2)*				*Coeff. Ther. Expansion*	*Specific Weight*
Material	*Tension*[b]	*Shear*	*Tension*	*Comp.*	*Shear*	*Tension or Comp.*	*Shear*	*Poisson's Ratio*	*% Elong. in 2 in.*	*10^{-6} per °F (10^{-6} per °C)*	*lb/ft^3 (kN/m^3)*
Structural steel (A36)	36 (248)	21 (145)	65 (448)	(c)	45 (310)	30 (207)	12 (83)	0.29	30	6.5 (11.7)	490 (77.0)
Alloy steel (4130) (heat tr.)	100 (689)	60 (414)	125 (862)	(c)	80 (552)	30 (207)	12 (83)	0.30	23	6.5 (11.7)	490 (77.0)
Aluminum alloy 2024-T3	50 (345)	30 (207)	68 (469)	(c)	40 (276)	10.5 (72)	4 (28)	0.33	15	12.6 (22.7)	173 (27.2)
Aluminum alloy 6061-T6	35 (241)	21 (145)	42 (290)	(c)	27 (186)	9.9 (68)	3.8 (26)	0.33	10	13.1 (23.6)	169 (26.5)
Titanium alloy 6M-4V	120 (827)	72 (496)	130 (896)	(c)	76 (524)	16 (110)	6.2 (43)	0.34	10	4.6 (8.3)	276 (43.4)
Brass	40 (276)	24 (165)	60 (414)	(c)	—	12 (83)	4.5 (31)	0.35	30	10.5 (18.9)	529 (83.1)
Grey cast iron	—	—	20 (138)	75 (517)	30 (207)	15 (103)	6.3 (43)	0.20	1	6.7 (12.1)	449 (70.5)
Concrete (high strength)	—	—	—	5 (34)	—	3 (21)	—	0.15	—	6.0 (10.8)	150 (23.6)
Douglas fir (par. to grain)	—	—	—	6.8 (47)	0.9 (6.2)	1.6 (11)	—	—	—	3.0 (5.4)	28 (4.4)

Table A.3*

W SHAPES
Properties for designing

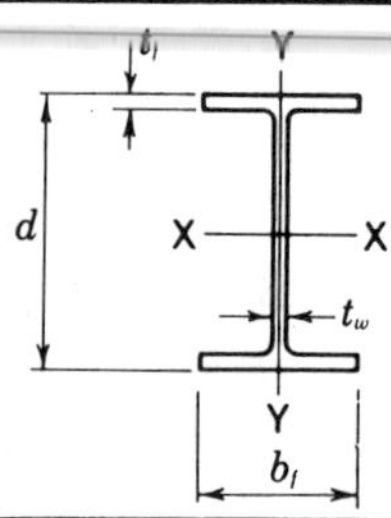

Designation	Area A	Depth d	Flange Width b_f	Flange Thickness t_f	Web Thickness t_w	Axis X-X I	Axis X-X S	Axis X-X r	Axis Y-Y I	Axis Y-Y S	Axis Y-Y r
	In.^{2}	In.	In.	In.	In.	In.^{4}	In.^{3}	In.	In.^{4}	In.^{3}	In.
W 10× 29	8.54	10.22	5.799	0.500	0.289	158	30.8	4.30	16.3	5.61	1.38
× 25	7.36	10.08	5.762	0.430	0.252	133	26.5	4.26	13.7	4.76	1.37
× 21	6.20	9.90	5.750	0.340	0.240	107	21.5	4.15	10.8	3.75	1.32
W 10× 19	5.61	10.25	4.020	0.394	0.250	96.3	18.8	4.14	4.28	2.13	0.874
× 17	4.99	10.12	4.010	0.329	0.240	81.9	16.2	4.05	3.55	1.77	0.844
× 15	4.41	10.00	4.000	0.269	0.230	68.9	13.8	3.95	2.88	1.44	0.809
× 11.5	3.39	9.87	3.950	0.204	0.180	52.0	10.5	3.92	2.10	1.06	0.787
W 8×67	19.7	9.00	8.287	0.933	0.575	272	60.4	3.71	88.6	21.4	2.12
×58	17.1	8.75	8.222	0.808	0.510	227	52.0	3.65	74.9	18.2	2.10
×48	14.1	8.50	8.117	0.683	0.405	184	43.2	3.61	60.9	15.0	2.08
×40	11.8	8.25	8.077	0.558	0.365	146	35.5	3.53	49.0	12.1	2.04
×35	10.3	8.12	8.027	0.493	0.315	126	31.1	3.50	42.5	10.6	2.03
×31	9.12	8.00	8.000	0.433	0.288	110	27.4	3.47	37.0	9.24	2.01
W 8×28	8.23	8.06	6.540	0.463	0.285	97.8	24.3	3.45	21.6	6.61	1.62
×24	7.06	7.93	6.500	0.398	0.245	82.5	20.8	3.42	18.2	5.61	1.61
W 8×20	5.89	8.14	5.268	0.378	0.248	69.4	17.0	3.43	9.22	3.50	1.25
×17	5.01	8.00	5.250	0.308	0.230	56.6	14.1	3.36	7.44	2.83	1.22
W 8×15	4.43	8.12	4.015	0.314	0.245	48.1	11.8	3.29	3.40	1.69	0.876
×13	3.83	8.00	4.000	0.254	0.230	39.6	9.90	3.21	2.72	1.36	0.842
×10	2.96	7.90	3.940	0.204	0.170	30.8	7.80	3.23	2.08	1.06	0.839
W 6×25	7.35	6.37	6.080	0.456	0.320	53.3	16.7	2.69	17.1	5.62	1.53
×20	5.88	6.20	6.018	0.367	0.258	41.5	13.4	2.66	13.3	4.43	1.51
×15.5	4.56	6.00	5.995	0.269	0.235	30.1	10.0	2.57	9.67	3.23	1.46
W 6×16	4.72	6.25	4.030	0.404	0.260	31.7	10.2	2.59	4.42	2.19	0.967
×12	3.54	6.00	4.000	0.279	0.230	21.7	7.25	2.48	2.98	1.49	0.918
× 8.5	2.51	5.83	3.940	0.194	0.170	14.8	5.08	2.43	1.98	1.01	0.889

*From *Manual of Steel Construction*, Seventh Edition, Courtesy of American Institute of Steel Construction, Inc.

Table A.4*

CHANNELS
AMERICAN STANDARD
Properties for designing

Designation	Area A	Depth d	Flange		Web thickness t_w	$\frac{d}{A_f}$	Axis X-X			Axis Y-Y			$\bar{x}$
			Width b_f	Average thickness t_f			I	S	r	I	S	r	
	In.^{2}	In.	In.	In.	In.		In.^{4}	In.^{3}	In.	In.^{4}	In.^{3}	In.	In.
C 9×20	5.88	9.00	2.648	0.413	0.448	8.22	60.9	13.5	3.22	2.42	1.17	0.642	0.583
×15	4.41	9.00	2.485	0.413	0.285	8.76	51.0	11.3	3.40	1.93	1.01	0.661	0.586
×13.4	3.94	9.00	2.433	0.413	0.233	8.95	47.9	10.6	3.48	1.76	0.962	0.668	0.601
C 8×18.75	5.51	8.00	2.527	0.390	0.487	8.12	44.0	11.0	2.82	1.98	1.01	0.599	0.565
×13.75	4.04	8.00	2.343	0.390	0.303	8.75	36.1	9.03	2.99	1.53	0.853	0.615	0.553
×11.5	3.38	8.00	2.260	0.390	0.220	9.08	32.6	8.14	3.11	1.32	0.781	0.625	0.571
C 7×14.75	4.33	7.00	2.299	0.366	0.419	8.31	27.2	7.78	2.51	1.38	0.779	0.564	0.532
×12.25	3.60	7.00	2.194	0.366	0.314	8.71	24.2	6.93	2.60	1.17	0.702	0.571	0.525
× 9.8	2.87	7.00	2.090	0.366	0.210	9.14	21.3	6.08	2.72	0.968	0.625	0.581	0.541
C 6×13	3.83	6.00	2.157	0.343	0.437	8.10	17.4	5.80	2.13	1.05	0.642	0.525	0.514
×10.5	3.09	6.00	2.034	0.343	0.314	8.59	15.2	5.06	2.22	0.865	0.564	0.529	0.500
× 8.2	2.40	6.00	1.920	0.343	0.200	9.10	13.1	4.38	2.34	0.692	0.492	0.537	0.512
C 5× 9	2.64	5.00	1.885	0.320	0.325	8.29	8.90	3.56	1.83	0.632	0.449	0.489	0.478
× 6.7	1.97	5.00	1.750	0.320	0.190	8.93	7.49	3.00	1.95	0.478	0.378	0.493	0.484
C 4× 7.25	2.13	4.00	1.721	0.296	0.321	7.84	4.59	2.29	1.47	0.432	0.343	0.450	0.459
× 5.4	1.59	4.00	1.584	0.296	0.184	8.52	3.85	1.93	1.56	0.319	0.283	0.449	0.458

*From *Manual of Steel Construction Seventh Edition*, Courtesy of American Institute of Steel Construction, Inc.

Table A.5*

ANGLES
Equal legs
Properties for designing

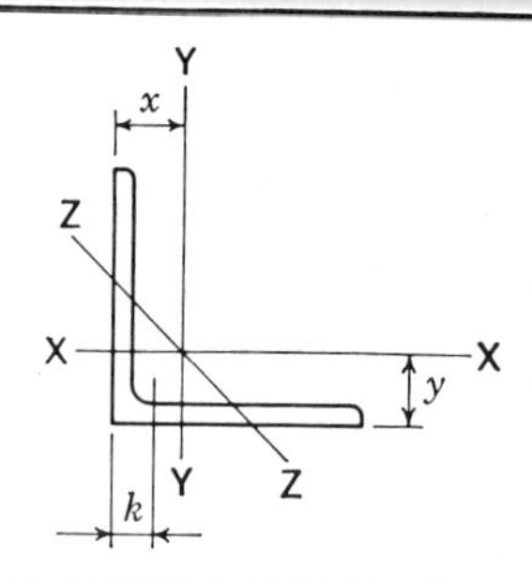

Size and Thickness	k	Weight per Foot	Area	AXIS X-X AND AXIS Y-Y				AXIS Z-Z
				I	S	r	x or y	r
In.	In.	Lb.	In.²	In.⁴	In.³	In.	In.	In.
L 8 × 8 × 1⅛	1¾	56.9	16.7	98.0	17.5	2.42	2.41	1.56
1	1⅝	51.0	15.0	89.0	15.8	2.44	2.37	1.56
⅞	1½	45.0	13.2	79.6	14.0	2.45	2.32	1.57
¾	1⅜	38.9	11.4	69.7	12.2	2.47	2.28	1.58
⅝	1¼	32.7	9.61	59.4	10.3	2.49	2.23	1.58
9/16	1 3/16	29.6	8.68	54.1	9.34	2.50	2.21	1.59
½	1⅛	26.4	7.75	48.6	8.36	2.50	2.19	1.59
L 6 × 6 × 1	1½	37.4	11.0	35.5	8.57	1.80	1.86	1.17
⅞	1⅜	33.1	9.73	31.9	7.63	1.81	1.82	1.17
¾	1¼	28.7	8.44	28.2	6.66	1.83	1.78	1.17
⅝	1⅛	24.2	7.11	24.2	5.66	1.84	1.73	1.18
9/16	1 1/16	21.9	6.43	22.1	5.14	1.85	1.71	1.18
½	1	19.6	5.75	19.9	4.61	1.86	1.68	1.18
7/16	15/16	17.2	5.06	17.7	4.08	1.87	1.66	1.19
⅜	⅞	14.9	4.36	15.4	3.53	1.88	1.64	1.19
5/16	13/16	12.4	3.65	13.0	2.97	1.89	1.62	1.20
L 5 × 5 × ⅞	1⅜	27.2	7.98	17.8	5.17	1.49	1.57	.973
¾	1¼	23.6	6.94	15.7	4.53	1.51	1.52	.975
⅝	1⅛	20.0	5.86	13.6	3.86	1.52	1.48	.978
½	1	16.2	4.75	11.3	3.16	1.54	1.43	.983
7/16	15/16	14.3	4.18	10.0	2.79	1.55	1.41	.986
⅜	⅞	12.3	3.61	8.74	2.42	1.56	1.39	.990
5/16	13/16	10.3	3.03	7.42	2.04	1.57	1.37	.994
L 4 × 4 × ¾	1⅛	18.5	5.44	7.67	2.81	1.19	1.27	.778
⅝	1	15.7	4.61	6.66	2.40	1.20	1.23	.779
½	⅞	12.8	3.75	5.56	1.97	1.22	1.18	.782
7/16	13/16	11.3	3.31	4.97	1.75	1.23	1.16	.785
⅜	¾	9.8	2.86	4.36	1.52	1.23	1.14	.788
5/16	11/16	8.2	2.40	3.71	1.29	1.24	1.12	.791
¼	⅝	6.6	1.94	3.04	1.05	1.25	1.09	.795

*From *Manual of Steel Construction*, Seventh Edition, Courtesy of American Institute of Steel Construction, Inc.

Table A.6*

L ANGLES
Unequal legs
Properties for designing

Size and Thickness	k	Weight per Foot	Area	AXIS X-X				AXIS Y-Y				AXIS Z-Z	
				I	S	r	y	I	S	r	x	r	Tan α
In.	In.	Lb.	In.^{2}	In.^{4}	In.^{3}	In.	In.	In.^{4}	In.^{3}	In.	In.	In.	
L 4 × 3 1/2 × 5/8	1 1/16	14.7	4.30	6.37	2.35	1.22	1.29	4.52	1.84	1.03	1.04	.719	.745
1/2	15/16	11.9	3.50	5.32	1.94	1.23	1.25	3.79	1.52	1.04	1.00	.722	.750
7/16	7/8	10.6	3.09	4.76	1.72	1.24	1.23	3.40	1.35	1.05	.978	.724	.753
3/8	13/16	9.1	2.67	4.18	1.49	1.25	1.21	2.95	1.17	1.06	.955	.727	.755
5/16	3/4	7.7	2.25	3.56	1.26	1.26	1.18	2.55	.994	1.07	.932	.730	.757
1/4	11/16	6.2	1.81	2.91	1.03	1.27	1.16	2.09	.808	1.07	.909	.734	.759
L 4 × 3 × 5/8	1 1/16	13.6	3.98	6.03	2.30	1.23	1.37	2.87	1.35	.849	.871	.637	.534
1/2	15/16	11.1	3.25	5.05	1.89	1.25	1.33	2.42	1.12	.864	.827	.639	.543
7/16	7/8	9.8	2.87	4.52	1.68	1.25	1.30	2.18	.992	.871	.804	.641	.547
3/8	13/16	8.5	2.48	3.96	1.46	1.26	1.28	1.92	.866	.879	.782	.644	.551
5/16	3/4	7.2	2.09	3.38	1.23	1.27	1.26	1.65	.734	.887	.759	.647	.554
1/4	11/16	5.8	1.69	2.77	1.00	1.28	1.24	1.36	.599	.896	.736	.651	.558
L 3 1/2 × 3 × 1/2	15/16	10.2	3.00	3.45	1.45	1.07	1.13	2.33	1.10	.881	.875	.621	.714
7/16	7/8	9.1	2.65	3.10	1.29	1.08	1.10	2.09	.975	.889	.853	.622	.718
3/8	13/16	7.9	2.30	2.72	1.13	1.09	1.08	1.85	.851	.897	.830	.625	.721
5/16	3/4	6.6	1.93	2.33	.954	1.10	1.06	1.58	.722	.905	.808	.627	.724
1/4	11/16	5.4	1.56	1.91	.776	1.11	1.04	1.30	.589	.914	.785	.631	.727
L 3 1/2 × 2 1/2 × 1/2	15/16	9.4	2.75	3.24	1.41	1.09	1.20	1.36	.760	.704	.705	.534	.486
7/16	7/8	8.3	2.43	2.91	1.26	1.09	1.18	1.23	.677	.711	.682	.535	.491
3/8	13/16	7.2	2.11	2.56	1.09	1.10	1.16	1.09	.592	.719	.660	.537	.496
5/16	3/4	6.1	1.78	2.19	.927	1.11	1.14	.939	.504	.727	.637	.540	.501
1/4	11/16	4.9	1.44	1.80	.755	1.12	1.11	.777	.412	.735	.614	.544	.506
L 3 × 2 1/2 × 1/2	7/8	8.5	2.50	2.08	1.04	.913	1.00	1.30	.744	.722	.750	.520	.667
7/16	13/16	7.6	2.21	1.88	.928	.920	.978	1.18	.664	.729	.728	.521	.672
3/8	3/4	6.6	1.92	1.66	.810	.928	.956	1.04	.581	.736	.706	.522	.676
5/16	11/16	5.6	1.62	1.42	.688	.937	.933	.898	.494	.744	.683	.525	.680
1/4	5/8	4.5	1.31	1.17	.561	.945	.911	.743	.404	.753	.661	.528	.684
3/16	9/16	3.39	.996	.907	.430	.954	.888	.577	.310	.761	.638	.533	.688

*From *Manual of Steel Construction*, Seventh Edition, Courtesy of American Institute of Steel Construction, Inc.

Table A.7*

PIPE
Dimensions and properties

Dimension				Weight per Foot Lbs. Plain Ends	Properties			
Nominal Diameter In.	Outside Diameter In.	Inside Diameter In.	Wall Thickness In.		A In.^{2}	I In.^{4}	S In.^{3}	r In.
Standard Weight								
½	.840	.622	.109	.85	.250	.017	.041	.261
¾	1.050	.824	.113	1.13	.333	.037	.071	.334
1	1.315	1.049	.133	1.68	.494	.087	.133	.421
1¼	1.660	1.380	.140	2.27	.669	.195	.235	.540
1½	1.900	1.610	.145	2.72	.799	.310	.326	.623
2	2.375	2.067	.154	3.65	1.07	.666	.561	.787
2½	2.875	2.469	.203	5.79	1.70	1.53	1.06	.947
3	3.500	3.068	.216	7.58	2.23	3.02	1.72	1.16
3½	4.000	3.548	.226	9.11	2.68	4.79	2.39	1.34
4	4.500	4.026	.237	10.79	3.17	7.23	3.21	1.51
5	5.563	5.047	.258	14.62	4.30	15.2	5.45	1.88
6	6.625	6.065	.280	18.97	5.58	28.1	8.50	2.25
8	8.625	7.981	.322	28.55	8.40	72.5	16.8	2.94
10	10.750	10.020	.365	40.48	11.9	161	29.9	3.67
12	12.750	12.000	.375	49.56	14.6	279	43.8	4.38

*From *Manual of Steel Construction*, Seventh Edition, Courtesy of American Institute of Steel Construction, Inc.

Table A.8*

UNIVERSAL BEAMS

DIMENSIONS AND PROPERTIES

Serial Size	Mass per metre	Depth of Section D	Width of Section B	Thickness		Area of Section	Moment of Inertia		Radius of Gyration		Elastic Modulus	
				Web t	Flange T		Axis x–x	Axis y–y	Axis x–x	Axis y–y	Axis x–x	Axis y–y
mm	kg	mm	mm	mm	mm	cm^2	cm^4	cm^4	cm	cm	cm^3	cm^3
381 × 152	67	388.6	154.3	9.7	16.3	85.4	21276	947	15.8	3.33	1095	122.7
	60	384.8	153.4	8.7	14.4	75.9	18632	814	15.7	3.27	968.4	106.2
	52	381.0	152.4	7.8	12.4	66.4	16046	685	15.5	3.21	842.3	89.96
356 × 171	67	364.0	173.2	9.1	15.7	85.3	19483	1278	15.1	3.87	1071	147.6
	57	358.6	172.1	8.0	13.0	72.1	16038	1026	14.9	3.77	894.3	119.2
	51	355.6	171.5	7.3	11.5	64.5	14118	885	14.8	3.71	794.0	103.3
	45	352.0	171.0	6.9	9.7	56.9	12052	730	14.6	3.58	684.7	85.39
356 × 127	39	352.8	126.0	6.5	10.7	49.3	10054	333	14.3	2.60	570.0	52.87
	33	348.5	125.4	5.9	8.5	41.7	8167	257	14.0	2.48	468.7	40.99
305 × 165	54	310.9	166.8	7.7	13.7	68.3	11686	988	13.1	3.80	751.8	118.5
	46	307.1	165.7	6.7	11.8	58.8	9924	825	13.0	3.74	646.4	99.54
	40	303.8	165.1	6.1	10.2	51.4	8500	691	12.9	3.67	559.6	83.71
305 × 127	48	310.4	125.2	8.9	14.0	60.8	9485	438	12.5	2.68	611.1	69.94
	42	306.6	124.3	8.0	12.1	53.1	8124	367	12.4	2.63	530.0	58.99
	37	303.8	123.5	7.2	10.7	47.4	7143	316	12.3	2.58	470.3	51.11
305 × 102	33	312.7	102.4	6.6	10.8	41.8	6482	189	12.5	2.13	414.6	37.00
	28	308.9	101.9	6.1	8.9	36.3	5415	153	12.2	2.05	350.7	30.01
	25	304.8	101.6	5.8	6.8	31.4	4381	116	11.8	1.92	287.5	22.85
254 × 146	43	259.6	147.3	7.3	12.7	55.0	6546	633	10.9	3.39	504.3	85.97
	37	256.0	146.4	6.4	10.9	47.4	5544	528	10.8	3.34	433.1	72.11
	31	251.5	146.1	6.1	8.6	39.9	4427	406	10.5	3.19	352.1	55.53
254 × 102	28	260.4	102.1	6.4	10.0	36.2	4004	174	10.5	2.19	307.6	34.13
	25	257.0	101.9	6.1	8.4	32.1	3404	144	10.3	2.11	264.9	28.23
	22	254.0	101.6	5.8	6.8	28.4	2863	116	10.0	2.02	225.4	22.84
203 × 133	30	206.8	133.8	6.3	9.6	38.0	2880	354	8.71	3.05	278.5	52.85
	25	203.2	133.4	5.8	7.8	32.3	2348	280	8.53	2.94	231.1	41.92

*From *Handbook on Structural Steelwork—Metric Properties and Safe Loads*, 1971, Courtesy of British Standards Institution, The Constructional Steel Research and Development Organization, and The British Constructional Steelwork Association, Ltd.

Table A.9*

CHANNELS

DIMENSIONS AND PROPERTIES

Nominal Size	Mass per metre	Depth of Section D	Width of Section B	Thickness		Area of Section	Dimension p	Moment of Inertia		Radius of Gyration		Elastic Modulus	
				Web t	Flange T			Axis x–x	Axis y–y	Axis x–x	Axis y–y	Axis x–x	Axis y–y
mm	in kg	mm	mm	mm	mm	cm^2	cm	cm^4	cm^4	cm	cm	cm^3	cm^3
432 × 102	65.54	431.8	101.6	12.2	16.8	83.49	2.32	21399	628.6	16.0	2.74	991.1	80.15
381 × 102	55.10	381.0	101.6	10.4	16.3	70.19	2.52	14894	579.8	14.6	2.87	781.8	75.87
305 × 102	46.18	304.8	101.6	10.2	14.8	58.83	2.66	8214	499.5	11.8	2.91	539.0	66.60
305 × 89	41.69	304.8	88.9	10.2	13.7	53.11	2.18	7061	325.4	11.5	2.48	463.3	48.49
254 × 89	35.74	254.0	88.9	9.1	13.6	45.52	2.42	4448	302.4	9.88	2.58	350.2	46.71
254 × 76	28.29	254.0	76.2	8.1	10.9	36.03	1.86	3367	162.6	9.67	2.12	265.1	28.22
229 × 89	32.76	228.6	88.9	8.6	13.3	41.73	2.53	3387	285.0	9.01	2.61	296.4	44.82
229 × 76	26.06	228.6	76.2	7.6	11.2	33.20	2.00	2610	158.7	8.87	2.19	228.3	28.22
203 × 89	29.78	203.2	88.9	8.1	12.9	37.94	2.65	2491	264.4	8.10	2.64	245.2	42.34
203 × 76	23.82	203.2	76.2	7.1	11.2	30.34	2.13	1950	151.4	8.02	2.23	192.0	27.59
178 × 89	26.81	177.8	88.9	7.6	12.3	34.15	2.76	1753	241.0	7.16	2.66	197.2	39.29
178 × 76	20.84	177.8	76.2	6.6	10.3	26.54	2.20	1337	134.0	7.10	2.25	150.4	24.73
152 × 89	23.84	152.4	88.9	7.1	11.6	30.36	2.86	1166	215.1	6.20	2.66	153.0	35.70
152 × 76	17.88	152.4	76.2	6.4	9.0	22.77	2.21	851.6	113.8	6.12	2.24	111.8	21.05
127 × 64	14.90	127.0	63.5	6.4	9.2	18.98	1.94	482.6	67.24	5.04	1.88	75.99	15.25
102 × 51	10.42	101.6	50.8	6.1	7.6	13.28	1.51	207.7	29.10	3.96	1.48	40.89	8.16
76 × 38	6.70	76.2	38.1	5.1	6.8	8.53	1.19	74.14	10.66	2.95	1.12	19.46	4.07

*From *Handbook on Structural Steelwork—Metric Properties and Safe Loads*, 1971, Courtesy of British Standards Institution, The Constructional Steel Research and Development Organization, and The British Constructional Steelwork Association, Ltd.

Table A.10*

EQUAL ANGLES

DIMENSIONS AND PROPERTIES

Nominal Size	Leg Lengths A×B	Actual Thickness	Mass per metre	Area of Section	Centre of Gravity		Moment of Inertia			Radius of Gyration		
					Cx	Cy	Axis x–x or y–y	Axis u–u Max.	Axis v–v Min.	Axis x–x or y–y	Axis u–u Max.	Axis v–v Min.
mm	mm	mm	kg	cm²	cm	cm	cm⁴	cm⁴	cm⁴	cm	cm	cm
89 × **89**	88.9 × 88.9	15.8	20.10	25.61	2.78	2.78	178	280	75.7	2.63	3.30	1.72
		14.2	18.31	23.32	2.72	2.72	164	259	69.1	2.65	3.33	1.72
		12.6	16.38	20.87	2.66	2.66	149	235	62.0	2.67	3.36	1.72
		11.0	14.44	18.40	2.60	2.60	133	211	55.0	2.69	3.38	1.73
		9.4	12.50	15.92	2.54	2.54	116	185	47.9	2.70	3.41	1.74
		7.9	10.58	13.47	2.48	2.48	99.8	159	41.0	2.72	3.43	1.74
		6.3	8.49	10.81	2.41	2.41	81.0	129	33.3	2.74	3.45	1.75
76 × **76**	76.2 × 76.2	14.3	15.50	19.74	2.41	2.41	99.6	157	42.7	2.25	2.82	1.47
		12.6	13.85	17.64	2.35	2.35	90.4	143	38.2	2.26	2.84	1.47
		11.0	12.20	15.55	2.29	2.29	80.9	128	33.8	2.28	2.87	1.47
		9.4	10.57	13.47	2.23	2.23	71.1	113	29.5	2.30	2.89	1.48
		7.8	8.93	11.37	2.16	2.16	60.9	96.8	25.1	2.31	2.92	1.49
		6.2	7.16	9.12	2.10	2.10	49.6	78.8	20.3	2.33	2.94	1.49
64 × **64**	63.5 × 63.5	12.5	11.31	14.41	2.03	2.03	50.4	78.9	21.8	1.87	2.34	1.23
		11.0	10.12	12.89	1.98	1.98	45.8	72.1	19.5	1.89	2.37	1.23
		9.4	8.78	11.18	1.92	1.92	40.5	64.0	17.0	1.90	2.39	1.23
		7.9	7.45	9.48	1.86	1.86	35.0	55.5	14.6	1.92	2.42	1.24
		6.2	5.96	7.59	1.80	1.80	28.6	45.4	11.8	1.94	2.45	1.25
57 × **57**	57.2 × 57.2	9.3	7.74	9.86	1.76	1.76	28.6	45.0	12.1	1.70	2.14	1.11
		7.8	6.55	8.35	1.70	1.70	24.7	39.1	10.3	1.72	2.16	1.11
		6.2	5.35	6.82	1.64	1.64	20.6	32.6	8.53	1.74	2.19	1.12
		4.6	4.01	5.11	1.57	1.57	15.8	25.0	6.51	1.76	2.21	1.13
51 × **51**	50.8 × 50.8	9.4	6.85	8.72	1.60	1.60	19.6	30.8	8.42	1.50	1.88	.98
		7.8	5.80	7.39	1.54	1.54	17.0	26.8	7.17	1.52	1.91	.98
		6.3	4.77	6.08	1.49	1.49	14.3	22.7	5.95	1.53	1.93	.99
		4.6	3.58	4.56	1.42	1.42	11.0	17.4	4.54	1.55	1.95	1.00

*From *Handbook on Structural Steelwork—Metric Properties and Safe Loads*, 1971, Courtesy of British Standards Institution, The Constructional Steel Research and Development Organization, and The British Constructional Steelwork Association, Ltd.

Table A.11*

UNEQUAL ANGLES

DIMENSIONS AND PROPERTIES

Nominal Size	Leg Lengths A×B	Actual Thickness	Mass per metre	Area of Section	Centre of Gravity		Moment of Inertia				Angle
					Cx	Cy	Axis x–x	Axis y–y	Axis u–u Max.	Axis v–v Min.	Axis x–x to Axis u–u
mm	mm	mm	kg	cm^2	cm	cm	cm^4	cm^4	cm^4	cm^4	tan α
102 × 64	101.6 × 63.5	11.0	13.40	17.07	3.51	1.62	174	51.9	194	31.4	.380
		9.5	11.61	14.79	3.45	1.56	152	45.8	171	27.4	.383
		7.8	9.69	12.35	3.38	1.49	129	38.9	145	23.2	.386
		6.3	7.89	10.05	3.31	1.43	106	32.2	119	19.1	.388
89 × 76	88.9 × 76.2	14.2	16.83	21.44	2.89	2.26	155	104	207	52.4	.710
		12.7	15.20	19.36	2.84	2.21	142	95.4	190	47.5	.713
		11.0	13.40	17.07	2.77	2.14	127	85.4	170	42.0	.715
		9.5	11.61	14.79	2.71	2.08	111	75.1	150	36.7	.718
		7.8	9.69	12.35	2.65	2.02	94.2	63.7	127	30.9	.720
		6.3	7.89	10.05	2.58	1.96	77.5	52.5	104	25.5	.721
89 × 64	88.9 × 63.5	11.0	12.20	15.55	2.97	1.71	118	49.8	140	28.2	.489
		9.4	10.57	13.47	2.91	1.65	104	43.9	123	24.6	.493
		7.8	8.93	11.37	2.85	1.59	88.8	37.7	106	21.0	.496
		6.2	7.16	9.12	2.78	1.53	72.1	30.7	85.8	17.0	.498
76 × 64	76.2 × 63.5	11.0	11.17	14.23	2.46	1.83	76.6	47.8	99.9	24.4	.669
		9.4	9.68	12.33	2.40	1.77	67.3	42.1	88.2	21.3	.673
		7.9	8.19	10.43	2.34	1.71	57.8	36.2	75.9	18.1	.676
		6.2	6.56	8.36	2.27	1.64	46.9	29.5	61.7	14.7	.678

*From *Handbook on Structural Steelwork—Metric Properties and Safe Loads*, 1971, Courtesy of British Standards Institution, The Constructional Steel Research and Development Organization, and The British Constructional Steelwork Association, Ltd.

Table A.12*

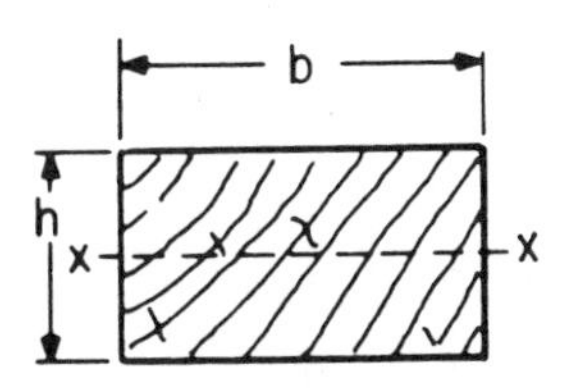

PROPERTIES OF STRUCTURAL LUMBER – Standard Dressed Sizes

NOMINAL SIZE b(inches)h	STANDARD DRESSED SIZE (S4S) b(inches)h	AREA OF SECTION A	MOMENT OF INERTIA I_{xx}	SECTION MODULUS S	Weight in lb/ft when γ equals: 25lb/ft^3	30lb/ft^3
2 x 4	1 1/2 x 3 1/2	5.250	5.359	3.063	0.911	1.094
2 x 6	1 1/2 x 5 1/2	8.250	20.797	7.563	1.432	1.719
2 x 8	1 1/2 x 7 1/4	10.875	47.635	13.141	1.888	2.266
2 x 10	1 1/2 x 9 1/4	13.875	98.932	21.391	2.409	2.891
2 x 12	1 1/2 x 11 1/4	16.875	177.979	31.641	2.930	3.516
4 x 2	3 1/2 x 1 1/2	5.250	0.984	1.313	0.911	1.094
4 x 3	3 1/2 x 2 1/2	8.750	4.557	3.646	1.519	1.823
4 x 4	3 1/2 x 3 1/2	12.250	12.505	7.146	2.127	2.552
4 x 6	3 1/2 x 5 1/2	19.250	48.526	17.646	3.342	4.010
4 x 8	3 1/2 x 7 1/4	25.375	111.148	30.661	4.405	5.286
4 x 10	3 1/2 x 9 1/4	32.375	230.840	49.911	5.621	6.745
4 x 12	3 1/2 x 11 1/4	39.375	415.283	73.828	6.836	8.203
6 x 2	5 1/2 x 1 1/2	8.250	1.547	2.063	1.432	1.719
6 x 3	5 1/2 x 2 1/2	13.750	7.161	5.729	2.387	2.865
6 x 4	5 1/2 x 3 1/2	19.250	19.651	11.229	3.342	4.010
6 x 6	5 1/2 x 5 1/2	30.250	76.255	27.729	5.252	6.302
6 x 8	5 1/2 x 7 1/2	41.250	193.359	51.563	7.161	8.594
6 x 10	5 1/2 x 9 1/2	52.250	392.963	82.729	9.071	10.885
6 x 12	5 1/2 x 11 1/2	63.250	697.068	121.229	10.981	13.177
8 x 2	7 1/4 x 1 1/2	10.875	2.039	2.719	1.888	2.266
8 x 3	7 1/4 x 2 1/2	18.125	9.440	7.552	3.147	3.776
8 x 4	7 1/4 x 3 1/2	25.375	25.904	14.802	4.405	5.286
8 x 6	7 1/2 x 5 1/2	41.250	103.984	37.813	7.161	8.594
8 x 8	7 1/2 x 7 1/2	56.250	263.672	70.313	9.766	11.719
8 x 10	7 1/2 x 9 1/2	71.250	535.859	112.813	12.370	14.844
8 x 12	7 1/2 x 11 1/2	86.250	950.547	165.313	14.974	17.969
10 x 2	9 1/4 x 1 1/2	13.875	2.602	3.469	2.409	2.891
10 x 3	9 1/4 x 2 1/2	23.125	12.044	9.635	4.015	4.818
10 x 4	9 1/4 x 3 1/2	32.375	33.049	18.885	5.621	6.745
10 x 6	9 1/2 x 5 1/2	52.250	131.714	47.896	9.071	10.885
10 x 8	9 1/2 x 7 1/2	71.250	333.984	89.063	12.370	14.844
10 x 10	9 1/2 x 9 1/2	90.250	678.755	142.896	15.668	18.802
10 x 12	9 1/2 x 11 1/2	109.250	1204.026	209.396	18.967	22.760
12 x 2	11 1/4 x 1 1/2	16.875	3.164	4.219	2.930	3.516
12 x 3	11 1/4 x 2 1/2	28.125	14.648	11.719	4.883	5.859
12 x 4	11 1/4 x 3 1/2	39.375	40.195	22.969	6.836	8.203
12 x 6	11 1/2 x 5 1/2	63.250	159.443	57.979	10.981	13.177
12 x 8	11 1/2 x 7 1/2	86.250	404.297	107.813	14.974	17.969

*From *National Design Specification for Stress-Grade Lumber and Its Fastenings*, 1973 Ed., Courtesy of National Forest Products Association.

Table A.13 Slopes and Deflections of Common Beams

Case	Deflection equation	Maximum deflection	Maximum slope
① Cantilever, end load F (reactions F, FL)	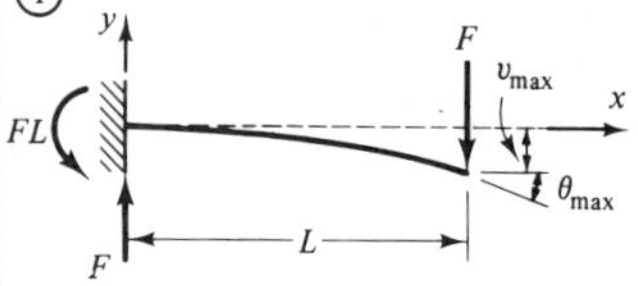 $v = \frac{Fx^2}{6EI}(x - 3L)$	$v_{max} = -\frac{FL^3}{3EI}$ at free end	$\theta_{max} = -\frac{FL^2}{2EI}$ at free end
② 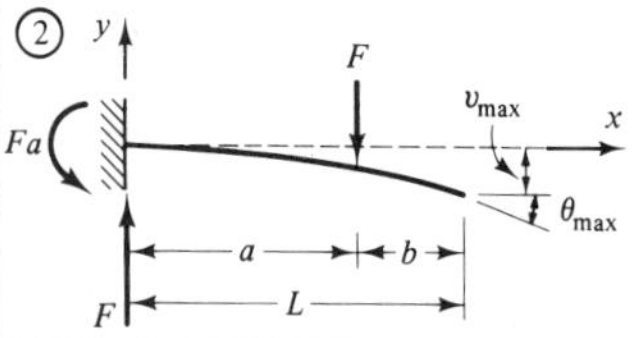 Cantilever, load F at a (reactions F, Fa)	$v = \frac{Fx^2}{6EI}(x - 3a) \quad (x \leqslant a)$; $= \frac{F}{6EJ}[x^3 - 3x^2a - (x-a)^3] \quad (x \geqslant a)$	$v_{max} = -\frac{Fa^2(3L - a)}{6EI}$ at free end	$\theta_{max} = -\frac{Fa^2}{2EI}$ at free end
③ 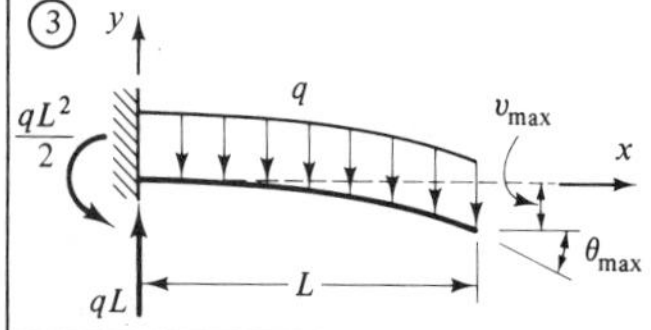 Cantilever, uniform load q (reactions qL, $\frac{qL^2}{2}$)	$v = \frac{qx^2}{24EI}(4Lx - x^2 - 6L^2)$	$v_{max} = -\frac{qL^3}{8EI}$ at free end	$\theta_{max} = -\frac{qL^3}{6EI}$ at free end
④ 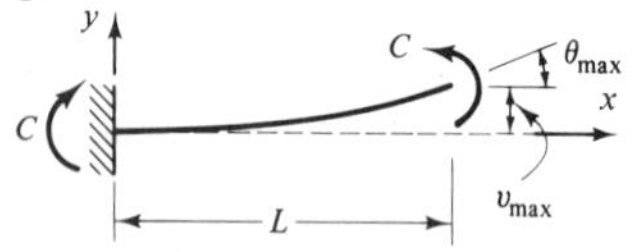 Cantilever, end moment C	$v = \frac{Cx^2}{2EI}$	$v_{max} = \frac{CL^2}{2EI}$ at free end	$\theta_{max} = \frac{CL}{EI}$ at free end
⑤ 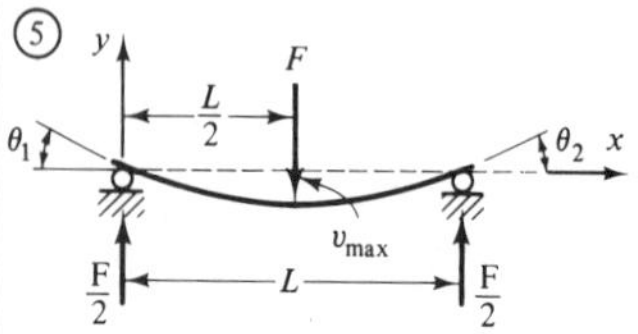 Simply supported, center load F (reactions $\frac{F}{2}$, $\frac{F}{2}$)	$v = \frac{Fx}{48EI}(4x^2 - 3L^2) \quad (x \leqslant \frac{L}{2})$	$v_{max} = -\frac{FL^3}{48EI}$ at center	$\theta_2 = -\theta_1 = \frac{FL^2}{16EI}$
⑥ 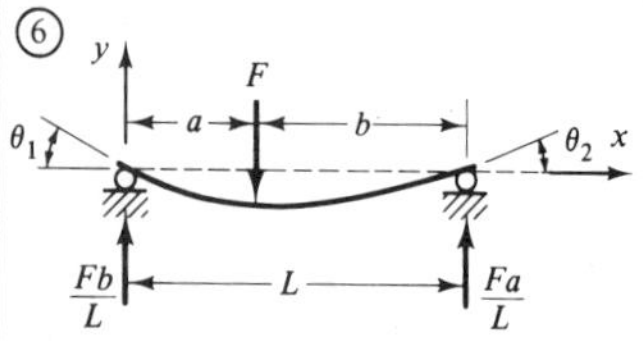 Simply supported, load F at a (reactions $\frac{Fb}{L}$, $\frac{Fa}{L}$)	$v = \frac{Fbx}{6EIL}[x^2 - (L^2 - b^2)] \quad (x \leqslant a)$; $= \frac{Fb}{6EIL}[x^3 - (L^2 - b^2)x - \frac{L}{b}(x-a)^3] \quad (x \geqslant a)$	$v_{max} = -Fb(L^2 - b^2)^{3/2}$ at $x = \sqrt{\frac{L^2 - b^2}{3}}$ for $a > b$	$\theta_1 = \frac{-Fb}{6EIL}(L^2 - b^2)$; $\theta_2 = \frac{Fb}{6EIL}(L - b)(2L - b)$
⑦ 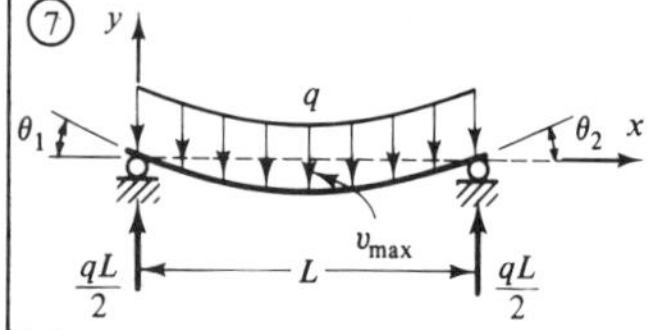 Simply supported, uniform load q (reactions $\frac{qL}{2}$, $\frac{qL}{2}$)	$v = -\frac{qx}{24EI}(2Lx^2 - x^3 - L^3)$	$v_{max} = -\frac{5qL^4}{384EI}$ at center	$\theta_2 = -\theta_1 = \frac{qL^3}{24EI}$

Table A.14 Conversion Factors: U. S.-British Units to SI

To convert from	*To*	*Multiply by**
(Length)		
foot (ft)	metre (m)	3.048×10^{-1}
inch (in.)	metre (m)	2.540×10^{-2}
(Area)		
foot^2 (ft^2)	metre^2 (m^2)	9.290×10^{-2}
inch^2 (in.^{2})	metre^2 (m^2)	6.452×10^{-4}
(Volume)		
foot^3 (ft^3)	metre^3 (m^3)	2.832×10^{-2}
inch^3 (in.^{3})	metre^3 (m^3)	1.639×10^{-5}
(Force)		
pound (lb)	newton (N)	4.448
(Moment)		
pound·foot (lb·ft)	newton·metre (N·m)	1.356
pound·inch (lb·in.)	newton·metre (N·m)	1.130×10^{-1}
(Pressure, Stress)		
pound/foot^2 (lb/ft^2)	newton/metre^2 (N/m^2)	4.788×10
pound/inch^2 (lb/in.^{2})	newton/metre^2 (N/m^2)	6.895×10^3
(Specific Weight)		
pound/foot^3 (lb/ft^3)	newton/metre^3 (N/m^3)	1.571×10^2
pound/inch^3 (lb/in.^{3})	newton/metre^3 (N/m^3)	2.714×10^5

*To convert from SI to U.S.-British units, divide by the factor given.

ANSWERS TO EVEN-NUMBERED PROBLEMS

Chapter 1

1.2 1640 miles
1.4 0.124 in./lb
1.6 no
1.8 7850 kg/m^3
1.10 37.4°F

Chapter 2

2.2 $\mathbf{T}$ = 100 N ∠ 30°, $\mathbf{W}$ = 883 N↓
2.4 $\mathbf{V}_A$ = 55 mph ∠45°, $\mathbf{V}_B$ = 30 mph →; length is doubled; sense is reversed
2.6 154.6 kN ∠ 31.9°
2.8 $\mathbf{A} - \mathbf{B}$ = 200 kips ↓ 6.4°, $\mathbf{B} - \mathbf{A}$ = 200 kips ↑ 6.4°
2.10 80.7 km ↖ 29.7°
2.12 $\mathbf{A}_H$ = 63.6 kN →, $\mathbf{A}_V$ = 63.6 kN↑, $\mathbf{B}_H$ = 67.6 kN →, $\mathbf{B}_V$ = 18.1 kN↑; $\mathbf{A}_V$ = 46.6 kN↑, $\mathbf{A}_B$ = 65.9 kN ∠ 15°
2.14 A = 100 lb, B = 200 lb
2.16 T_1 = 8.8 lb, T_2 = 6.2 lb; $\theta = 30°$, $(T_2)_{minimum}$ = 6 lb
2.18 6 m →, 2 m↓, 3 m↙
2.20 (a) −15.5 lb (b) −20.7 lb (c) −1242 lb^2 (d) 42.3 lb, 40.0 lb (assuming upward positive)
2.22 $\mathbf{A}$ = 10.0 ft ∠ 37°, $\mathbf{B}$ = 14.9 ft ↘ 23.7°
2.24 $\mathbf{F} = 39.0\mathbf{i} + 53.6\mathbf{j} + 22.5\mathbf{k}$ lb
2.26 $\mathbf{F} = -79.6\mathbf{i} + 34.2\mathbf{j} - 50.0\mathbf{k}$ N
2.28 $\theta_x = 31.0°$, $\theta_y = 73.4°$, $\theta_z = 115.4°$
2.30 $14\mathbf{i} + 7\mathbf{j} + 9\mathbf{k}$, $-10\mathbf{i} - 1\mathbf{j} + 3\mathbf{k}$, $30\mathbf{i} + 17\mathbf{j} + 24\mathbf{k}$, 54; $\theta = 53.6°$
2.32 56.6 lb ↘ 45°
2.34 (a) $-9.9\mathbf{i} + 9.0\mathbf{j} + 1.0\mathbf{k}$ m (b) 26.7 m^2 (c) 59.1°; S = 13.4 m, $\theta_x = 137.6°$, $\theta_y = 47.8°$, $\theta_z = 85.7°$

2.36 $T_1 = T_2 = T_3 = 1.92$ kips

2.38 (a) $\mathbf{r}_{OP} = -2\mathbf{i} + 1\mathbf{j} - 1\mathbf{k}$ ft, $\mathbf{r}_{OQ} = 4\mathbf{i} - 3\mathbf{j} + 2\mathbf{k}$ ft (b) $6\mathbf{i} - 4\mathbf{j} + 3\mathbf{k}$ ft (c) $0.77\mathbf{i} - 0.51\mathbf{j} + 0.38\mathbf{k}$

2.40 $\mathbf{F} = 0.32\mathbf{i} + 0.48\mathbf{j} - 0.40\mathbf{k}$ GN

2.42 -2.57 lb, $\mathbf{F}_{OB} = -2.20\mathbf{i} - 0.73\mathbf{j} + 1.10\mathbf{k}$ lb

2.44 $\mathbf{T} = 6\mathbf{i} - 2\mathbf{j} + 4\mathbf{k}$ kN, $\theta_x = 36.9°$, $\theta_y = 105.5°$, $\theta_z = 57.8°$

2.46 $15\mathbf{i} - 21\mathbf{j} - 33\mathbf{k}$

2.50 $\mathbf{e}_n = \pm[(\sqrt{3}/3)\mathbf{i} - (\sqrt{3}/3)\mathbf{j} + (\sqrt{3}/3)\mathbf{k}]$

2.52 $-30.0\mathbf{i} - 32.7\mathbf{j} - 2.7\mathbf{k}$ m^2

2.54 26 N·m ↓

2.56 (a) 12.2 N·m ↺ (b) 81.3 N → (c) 36.4 N ↘ 26.6°

2.58 $\mathbf{M}_{max} = 41.2$ N·m ↺ for $\theta = 104°$ or 41.2 N·m ↻ for $\theta = -76°$; $\mathbf{M}_{min} = \mathbf{0}$ for $\theta = 14°$ or $\theta = -166°$

2.60 24 MN·m ↻

2.62 (a) $23\mathbf{i} - 5\mathbf{j} + 19\mathbf{k}$ kN·m (b) 3.52 m

2.64 $\mathbf{M}_O = \mathbf{M}_C = \mathbf{O}$, $\mathbf{M}_A = 1600\mathbf{j} + 800\mathbf{k}$ N·m, $\mathbf{M}_B = -800\mathbf{i} + 800\mathbf{k}$ N·m

2.66 $\mathbf{M}_O = -626\mathbf{i} + 402\mathbf{j} + 280\mathbf{k}$ lb·in., d = 2.65 in.

2.68 -49

2.70 63.3 lb·ft ↑

2.72 (a) $\mathbf{O}$ (b) $800\mathbf{k}$ N·m (c) $320\mathbf{j} + 640\mathbf{k}$ N·m

2.74 0.121 N·m ↓

2.76 $-400\mathbf{i} - 400\mathbf{j} - 400\mathbf{k}$ lb·ft

2.80 no; $\mathbf{F} = 23.2$ lb ←, $\mathbf{C} = 400$ lb·ft ↺

2.82 10 lb·in. ↑

2.84 44.4 N at A and C or at B and D

2.86 1964 lb·in. ↺

2.88 $\mathbf{M}_A = 200$ lb·in. ↻, $\mathbf{M}_B = 200$ lb·in. ↻, $\mathbf{M}_C = 546$ lb·in. ↻

CHAPTER 3

3.2 On plant: upward force exerted by rope and downward force exerted by earth; $T = 19.6$ N

3.4 On pole: $\mathbf{F}_A$ exerted by rock, $\mathbf{F}_B$ exerted by fulcrum, and $\mathbf{F}_C$ exerted by hands On rock: $-\mathbf{F}_A$ exerted by pole, $\mathbf{W}$↓ due to earth's gravitational attraction, and $\mathbf{N}$↑ exerted by earth's surface

3.6 At A: $\mathbf{A}_x$ → and $\mathbf{A}_y$ ↑; at C: $\mathbf{F}_C$ (slope 2:1) (and applied load)

3.8 At A: $\mathbf{A}_x$ →, $\mathbf{A}_y$↑, and $\mathbf{C}_A$ ↻ (and applied load)

3.10 At O: $\mathbf{O}_x$ → and $\mathbf{O}_y$ ↑; on ropes: $\mathbf{T}_1 = mg$ ↓ and $\mathbf{T}_2$ (slope 1:2)

3.12 At A: $\mathbf{A}_x\mathbf{i} + A_y\mathbf{j} + A_z\mathbf{k}$ and $C_y\mathbf{j}$; at G: $-196\mathbf{j}$ N; at B: $T\mathbf{j}$ (and applied load)

3.14 On 1: 981 N ↓ (at G), $\mathbf{N}_C$ ∠ 60°, $\mathbf{N}_D$ ←; On 2: 981 N ↓ (at G), $\mathbf{N}_A$ ∠ 30°, $\mathbf{N}_B$ ↑, $\mathbf{N}_C$ ∠ 60°

3.16 On 1: $\mathbf{T}$ ↑, 491 N ↓, $\mathbf{N}_1$ ↑; On 2: $\mathbf{A}_x$ →, $\mathbf{A}_y$ ↑, $\mathbf{N}_1$ ↓ (at B), 98.1 N ↓ (at G), $\mathbf{T}$ ∠ 37° (at C)

3.18 579 N →, 1490 N ↑

3.20 $\mathbf{N}_A = 93.8$ lb ∠ 40°, $\mathbf{N}_B = 66.6$ lb ∠ 65°

3.22 (a) $\mathbf{A} = 14$ kN ↓, $\mathbf{B} = 17$ kN ↑ (b) 2.83 m

3.24 (a) $\mathbf{C} = 188$ lb · ft ↺ (b) $\mathbf{O}_x = 100$ lb ←, $\mathbf{O}_y = 75$ lb ↑

3.26 (a) 42.4 lb (b) 22 lb, 10.7 in. from right side

3.28 $\mathbf{B}_x = 208$ lb ←, $\mathbf{B}_y = 594$ lb ↑, $\mathbf{C}_B = 4170$ lb·ft ↻

3.30 34.8°

3.32 $T_A = 453$ N, $T_B = 401$ N

3.34 $T = 2.08$ kN, $\mathbf{A}_x = 2.60$ kN ←, $\mathbf{A}_y = 3.12$ kN ↓

3.36 $T_{AB} = 14.8$ kN(C), $T_{BD} = 15.9$ kN(T), $T_{CB} = 3.1$ kN(T)

3.38 11.3 lb

3.40 (a) 2m (b) $(W_1/3)\mathbf{j} - (2W_1/3)\mathbf{k}$

3.42 $F_A = F_B = F/6$, $F_C = 2F/3$

3.44 $1.18W$

3.46 (a) 400 lb·ft (b) $\mathbf{A} = -840\mathbf{j} - 280\mathbf{k}$ lb, $\mathbf{B} = -560\mathbf{j} + 1680\mathbf{k}$ lb

3.48 $\mathbf{A} = 150\mathbf{j} + 100\mathbf{k}$ N, $\mathbf{C}_A = -30\mathbf{i} - 22\mathbf{j} + 33\mathbf{k}$ N·m

3.50 $T_{ECF} = 725$ lb, $T_{FBG} = 2170$ lb

3.52 $T_{BG} = 405$ lb, $\mathbf{A} = 270\mathbf{i} + 135\mathbf{j} + 270\mathbf{k}$ lb, $\mathbf{C}_A = 2160\mathbf{j}$ lb·ft

3.54 Supports not adequate; could use link or cable at E instead of roller

3.56 **A** = 115 N →, **B** = 144 N ↖ 30°, **D** = 50 N ↗ 30°, **W** = 100 N ↓

3.60 $\mathbf{A}_x$ = 50 lb ←, $\mathbf{A}_y$ = 50 lb ↑, **D** = 50 lb ↓

3.62 $\mathbf{B}_x$ = 120 lb ←, $\mathbf{B}_y$ = 20 lb ↓, $\mathbf{C}_x$ = 90 lb →, $\mathbf{C}_y$ = 60 lb ↑

3.64 $\mathbf{A}_x$ = 0.87 kN ←, $\mathbf{A}_y$ = 0.47 kN ↓, $\mathbf{C}_x$ = 0.56 kN →, $\mathbf{C}_y$ = 0.65 kN ↑

3.66 $\mathbf{A}_x = F$ ←, $\mathbf{A}_y = F/4$ ↓, $\mathbf{B}_x = F$ →, $\mathbf{B}_y = F/4$ ↑; on *AED*: $\mathbf{E} = 3F/4$ →, $\mathbf{D} = \sqrt{2}\,F/4$ ↖ 45° (and $\mathbf{A}_x$ and $\mathbf{A}_y$)

3.68 On *ACD*: $\mathbf{A}_x$ = 0.5 kN →, $\mathbf{A}_y$ = 4 kN ↓, $\mathbf{C}_x$ = 1.5 kN →, $\mathbf{C}_y$ = 6 kN ↑, $\mathbf{D}_x$ = 2 kN ←, $\mathbf{D}_y$ = 2 kN ↓
On *BED*: $\mathbf{B}_x$ = 0.5 kN ←, $\mathbf{B}_y$ = 6 kN ↑, $\mathbf{C}_x$ = 1.5 kN ←, $\mathbf{C}_y$ = 6 kN ↓, **E** = 2 kN →

3.70 On *AB*: $\mathbf{A}_x$ = 0.6 kips ←, $\mathbf{A}_y$ = 0.3 kips ↓, $\mathbf{B}_x$ = 0.6 kips →, $\mathbf{B}_y$ = 0.3 kips ↑
On *BC*: $\mathbf{B}_x$ = 1.4 kips →, $\mathbf{B}_y$ = 2.3 kips ↓, $\mathbf{C}_x$ = 1.4 kips ←, $\mathbf{C}_y$ = 2.3 kips ↑

3.72 $\mathbf{F}_{DF}$ = 7.35 kips ↕ 10°, F_{CE} = 3.70 kips (C)

3.74 800 lb·in. ⤴

3.76 360 N·m ⟶

3.78 $F_{AB} = F_{CD}$ = 693 lb (C), $F_{AE} = F_{DE}$ = 347 lb (T), $F_{BE} = F_{CE}$ = 277 lb (T), F_{BC} = 486 lb (C)

3.80 F_{BC} = 1 kN (C), F_{BD} = 3 kN (C)

3.82 F_{DE} = 1380 lb (C), F_{EI} = 50 lb (T)

3.84 F_{CD} = 2380 lb (C), F_{DF} = 730 lb (T), F_{FG} = 2830 lb (T), F_{CG} = 1800 lb (T)

3.86 $F_{BD} = F_{DE} = F_{HJ} = 0$, F_{AB} = 2000 lb (T), F_{DF} = 4500 lb (T), F_{DG} = 2500 lb (T), F_{FG} = 6000 lb (C), F_{JK} = 1000 lb (C), F_{IK} = 1500 lb (C)

3.88 $F_{BG} = F_{BF} = 0$, F_{EF} = 43.5 kN (T), $F_{CE} = F_{CF}$ = 53.4 kN (C), $F_{DE} = F_{FG} = F_{GH}$ = 21.4 kN (T)

3.90 $F_{AB} = F_{AC}$ = 252 N (C), F_{AD} = 135 N (C)

3.92 $F \geqslant 120$ lb

3.94 39N $\leqslant F \leqslant$ 196 N

3.96 (a) 19.5° (b) 0.35

3.98 2650 lb

3.100 (a) 23° (b) 25° (c) 42°

3.102 not in equilibrium; $\mathbf{f}_A$ = 136 N (slope 21/20), $\mathbf{f}_B$ = 75 N (slope 7/24)

3.104 18.6 lb

3.106 7.5 N

3.108 7 lb

3.110 (a) $T \geqslant 7.3$ N (b) stronger spring required

CHAPTER 4

4.2 not equivalent; add **C** = − 1.2**i** + 1.2**j** kN·m

4.4 **F** = 860 N ⟵ 1.3°, **C** = 249 N·m ⤴

4.6 (a) **F** = 100**i** − 173**j** N, **C** = − 12.3**k** N·m
(b) $\mathbf{F}_A$ = − 310**i** − 357**j** N, $\mathbf{F}_B$ = 450 N ↗ 24.1°

4.8 **F** = 40 lb ↓, 16 in. to right of *AC*

4.10 **F** = − 2**k** kN; **C** = − 100**i** + 100**j** N·m; (a) − 100**i** N·m (b) 100**j** N·m; no

4.12 **R** = 2**i** + 5**j** − 1**k** lb, $\mathbf{C}^R$ = 19**i** − 12**j** − 62**k** lb·in; no wrong

4.14 **R** = 637 lb 58.5°, at *A*

4.16 **R** = 46.5 N ↓, 1.15 m to right of left side

4.18 **R** = 237**i** − 36.6**j** lb, through $x = 0$, y = 3.58 ft

4.20 **R** = 8 kN ↓, 1.75 m to right of left end

4.22 F_1 = 1.33 kN, F_2 = 1.60 kN, F_3 = 0.96 kN; counterclockwise

4.24 **R** = − 15**j** kips, through (10.0, 0, 10.8) ft

4.26 γ = 76.7 kN/m³, ρ = 7.82 Mg/m³

4.28 5.31 ft/s²

4.30 $\bar{x} = \bar{y} = 0$

4.32 along centerline, 0.75 in. above base

4.34 $a\sqrt{3}/6$ from each side

4.38 1190 lb, 4.57 ft; no

4.40 $\bar{x}$ = 4.8 ft, $\bar{y}$ = − 1.0 ft, $\bar{z} = 0$

4.42 $\bar{x} = \bar{y} = 4R/(3\pi)$

4.44 $\bar{x} = 3a/8$, $\bar{y} = 3a/5$

4.46 $\bar{x} = \bar{y} = 1.80$

4.48 $\bar{x} = 1.82$, $\bar{y} = 1.24$

4.50 along centerline, $R/2$ above base

4.52 along centerline, $h/4$ above base

4.54 $\bar{x} = \bar{z} = 0$, $\bar{y}$ = 0.33 ft

4.56 $\bar{x} = 4.0$ ft, $\bar{y} = 1.2$ ft, $\bar{z} = -4.5$ ft
4.58 $A = 4\pi R^2$, $V = (4/3)\pi R^3$
4.60 $A = 21.0$ m^2, $V = 4.71$ m^3
4.62 $\bar{x} = 9.58$ in., $\bar{y} = 3.22$ in.
4.64 $\bar{x} = \bar{y} = 0.78a$
4.66 $\bar{x} = -0.43$ in., $\bar{y} = 1.80$ in.
4.68 $\bar{x} = 0$, $\bar{y} = 1.36$ in.
4.70 $d = 5.64$ in.; $\bar{y} = 3.50$ in.
4.72 along centerline, 5.43 in. above base
4.74 $\bar{x} = 3.53$ in., $\bar{y} = 1.52$ in., $\bar{z} = -2.00$ in.
4.76 (a) along centerline, 6.29 in. from left end
(b) along centerline, 5.92 in. from left end
4.78 4.07 kg
4.80 total load = 294.3 kN
4.82 **A** = 10.8 kN ↑, **B** = 19.2 kN ↑
4.84 **A** = 216 lb ↑, **B** = 648 lb ↑
4.86 53 760 lb, 2.57 ft above bottom center
4.88 47.9 kN, 0.95 m above base
4.90 993 kg
4.92 20 560 lb, acting 16 ft vertically below surface
4.94 $R = \pi a^2 \gamma h$, $d = d_C + a^2/(4d_C)$
4.96 36.2 kN
4.98 90%
4.100 $I_x = I_y = \pi R^4/4$, $J_z = \pi R^4/2$
4.102 $I_x = \pi ab^3/16$, $I_y = \pi a^3 b/16$, $J_z = \pi ab(a^2 + b^2)/16$
4.104 324 in.^{4}
4.106 $I_x = 1600$ in.^{4}, $I_y = 419$ in.^{4}
4.108 0.0034 m^4
4.110 10.7 in.
4.112 $J = \frac{\pi}{2}(R_0^4 - R_i^4)$, $r = \left(\frac{R_0^2 + R_i^2}{2}\right)^{1/2}$

Chapter 5

5.2 On *A-a*: **P** = 10 kN ←, **V** = 5 kN ↓, **M** = 1.5 kN·m ⤴
On *B-b*: **P** = **O**, **V** = 10 kN →, **M** = 1 kN·m ↺
5.4 On *E-a*: **P** = 336 lb ↖ 41.6°, **V** = 299 lb ↗ 41.6°, **M** = 1200 lb·ft ↻
On *C-b*: **P** = 450 lb ↑, **V** = **O**, **M** = **O**
5.6 On *A-a*: **P** = **O**, **V** = 400 lb ↓, **M** = 1200 lb·ft ⤴
On *B-b*: **P** = **O**, **V** = 400 lb ↓, **M** = 3200 lb·ft ↺
5.8 On *A-a*: **P** = 4 kN →, **V** = 2 kN ↓, **M** = 1 kN·m ⤴
On *A-b*: **P** = 4 kN →, **V** = **O**, **M** = 2 kN·m ⤴
5.10 On *B-a*: $\mathbf{R} = -3\mathbf{i} + 4\mathbf{j}$ kN, $\mathbf{C}^R = -0.32\mathbf{i} - 0.24\mathbf{j} + 0.24\mathbf{k}$ N·m
5.12 $d \geqslant 28.9$ mm
5.14 (a) $\sigma = 1000$ psi (tension), $\tau = 0$ (b) $\sigma = 750$ psi (tension), $\tau = 433$ psi
5.16 2230 lb
5.18 $A_{BC} \geqslant 286$ mm^2, $A_{CI} \geqslant 673$ mm^2
5.20 $d \geqslant 6.6$ mm
5.22 14 500 lb
5.24 $\sigma_{max} = 12\,000$ psi (compression), $|\tau_{max}| = 6000$ psi
5.26 $r = r_0 \exp[\pi \gamma r_0^2 y/(2F)]$
5.28 56.6 kN
5.30 $d_{cable} \geqslant 26.6$ mm, $d_{pin} \geqslant 22.5$ mm
5.32 $\tau_{avg} = 17\,640$ psi, $\sigma_{bearing} = 63$ psi
5.34 (a) 15 280 psi (b) 4800 psi (c) 24 000 psi
5.36 3.39 kN
5.38 $h/d = 3/2$, $a/d = 1/2$
5.40 250 μm/m
5.42 0.50 in./in.
5.44 1.20 in.^{3}
5.46 -300 μm/m
5.48 0.0130 in./in.
5.50 (a) $2ay/h^2$ (b) 0 (c) a/h
5.52 $\varepsilon_x = \varepsilon_y = -499$ μin./in., $\gamma_{xy} = -3000$ μin./in.
5.54 $E = 275$ MN/m^2, $\nu = 0.25$
5.56 (a) 10.2×10^6 psi (b) 0.33 (c) 3.9×10^6 psi
5.58 (a) 10×10^6 psi (b) 43 000 psi (c) 47 500 psi
5.60 $\sigma_{YP} = 37\,000$ psi, $\sigma_U = 59\,000$ psi, % Elongation = 26%
5.62 % Reduction Area = 49.2%, % Elongation = 27.5%
5.64 $E = 2 \times 10^6$ psi, $\sigma_U = 7400$ psi

Chapter 6

6.2 0.29 m
6.4 $\delta_{AB} = -0.76$ mm, $\delta_{BD} = 0.38$ mm

6.6 4000 lb; $\delta_{BC} = 0.0096$ in., $\delta_{total} = 0.0400$ in.
6.8 82.6 kips, 2290 psi
6.10 W 8 × 15 or W 10 × 15
6.12 (a) 0.0040 in. (b) 0.0048 in.
6.14 (a) 12 000 lb (b) 0.060 in. (c) 20.040 in.
6.16 $\delta = \frac{n}{n+1}\left(\frac{\gamma}{k}\right)^{1/n} L^{(n+1/n}$
6.18 5670 lb
6.20 $F_{concrete} = 12.07$ kN, $F_{wood} = 7.24$ kN
6.22 294 kN; 68% larger
6.24 $\sigma_{aluminum} = 2500$ psi, $\sigma_{steel} = 17\ 500$ psi; $\delta_{steel} = 0.070$ in.
6.26 **A** = 120 kN ↑, **C** = 180 kN ↑
6.28 $\sigma_{AB} = 5500$ psi (T), $\sigma_{BC} = 4500$ psi (C)
6.30 $\sigma_{AB} = 15$ ksi (T), $\sigma_{BC} = 20$ ksi (C), $F =$ 60 kips
6.32 110 in.
6.34 7.62 mm, −2.07 μm
6.36 0.28 in.^{2}; 92°F
6.38 $\sigma_{AB} = 22$ ksi (T), $\sigma_{BC} = 32$ ksi (C)
6.40 $\sigma_{aluminum} = 45.1$ MN/m^2, $\sigma_{steel} = 240.7$ MN/m^2; $\theta = 0.023°$ counterclockwise
6.42 $\sigma_{steel} = 6040$ psi (T), $\sigma_{brass} = 3440$ psi (C), −52°F
6.44 (a) 1.60 (b) 1.60 (c) 1.26
6.46 (a) 8260 lb (b) 17 500 lb (c) 10 000 lb
6.48 0.31 in.
6.50 (a) 43% (b) 0%
6.52 1.10×10^6 lb
6.54 148 kN
6.56 1.73 $\sigma_{YS}A$
6.58 2 $\sigma_{YS}A$
6.60 1.66 MN
6.62 1.2 m
6.64 0.67 in.
6.66 3.92 MN/m^2
6.68 $\sigma_L = 32.0$ MN/m^2, $\sigma_C = 107.5$ MN/m^2
6.70 $\sigma_C = 66.7$ MN/m^2; $p = 860$ kN/m^2
6.72 $\sigma_{steel} = 3800$ psi (T), $\sigma_{aluminum} = 3800$ psi (C), p = 253 psi

CHAPTER 7

7.2 798 lb·in.
7.4 $d = 76.1$ mm, $m_{solid}/m_{hollow} = 2.8$
7.6 1.93 in.
7.8 47.8 MN/m^2
7.10 13 240 psi
7.12 1.70 kN·m
7.14 2.17 in.
7.16 247 N·m
7.18 $\sigma_\theta = 52$ MN/m^2, $\tau_\theta = 30$ MN/m^2
7.20 $L \geqslant 3.26$ m
7.22 1.51 kN·m
7.24 $ID = 2$ in., $OD = 4$ in.
7.26 0.086° at free end
7.28 0.8 m
7.30 $\theta_A = 32.8°$, $\tau_{AB} = 9550$ psi, $\tau_{DE} =$ 19 100 psi
7.32 $\mathbf{C}_A = C/3$ ←, $\mathbf{C}_B = 2C/3$ ←
7.34 $\tau_{steel} = 8910$ psi, $\tau_{bronze} = 6530$ psi
7.36 45.8 MN/m^2
7.38 $T = \frac{4\pi}{7}\tau R^3\left(\frac{R}{\rho}\right)^{\frac{1}{2}}$, $T = \frac{4\pi}{7} kR^3\left(\frac{R\theta}{L}\right)^{\frac{1}{2}}$
7.40 (a) 491 lb·in. (b) 2620 lb·in.
7.42 $T = 5.0$ kN·m
7.44 $T = 120$ kip·in.
7.46 72.2 kN·m
7.48 $\rho_E = 1.27$ in., $\theta = 11.6°$
7.50 $C_{limit} = 2T_{fp}/3$
7.52 $C_{limit} = 13.4$ kN·m
7.54 209 kip·in.

CHAPTER 8

8.2 $V = 10 - 5x$ kN, $M = -10 + 10x - 2.5x^2$ kN·m
8.4 $V = -125x^2$ lb, $M = -41.7x^3$ lb·ft
8.6 $V = F$, $M = Fx$ for $(0 \leqslant x \leqslant a)$; $V = 0$, $M = Fa$ for $(a \leqslant x \leqslant 3a)$; $V = -F$, $M = 4Fa - Fx$ for $(3a \leqslant x \leqslant 4a)$
8.8 $V = 900$ lb, $M = -8000 + 900x$ lb·ft for $(0 \leqslant x \leqslant 5$ ft); $V = 1400 - 100x$ lb, $M = -9250 + 1400x - 50x^2$ lb·ft for (5 ft $\leqslant x \leqslant$ 15 ft); $V = -100$ lb, $M = 2000 - 100x$ lb·ft for (15 ft $\leqslant x \leqslant$ 20 ft)
8.14 $V_{max} = 10$ kN, $M_{max} = -5$ kN·m (both at $x = 0$)

8.16 $V_{max} = 4000$ lb at (3 ft $\leqslant x \leqslant$ 5 ft) or -4000 lb at (11 ft $\leqslant x \leqslant$ 13 ft), $M_{max} = 10\,000$ lb·ft (at $x = 9$ ft)

8.18 $V_{max} = -17.5$ kN, $M_{max} = -12.5$ kN·m (both at $x = 4$ m)

8.20 $V_{max} = -2.17F$ (at $x = 3a$), $M_{max} = 1.06\,Fa$ (at $x = 1.949a$)

8.22 22.9%

8.24 Square cross section beam

8.26 179%

8.28 $\sigma_{t_{max}} = \sigma_{c_{max}} = 93.8$ MN/m^2 (at $x = 3$m)

8.30 $\sigma_{t_{max}} = 739$ psi (at $x = 11$ ft), $\sigma_{c_{max}} = 1680$ psi (at $x = 5$ ft)

8.32 4.4

8.34 (a) 305-mm × 89-mm (41.69 kg/m) (b) 305-mm × 102-mm (46.18 kg/m)

8.36 1690 lb

8.38 (a) 2700 psi (b) 2700 psi

8.40 $|\sigma| = \dfrac{14}{3}\dfrac{|M|}{bh^2}\left(\dfrac{2|y|}{h}\right)^{1/3}$

8.42 103.3 mm

8.44 261%

8.46 1.95 MN/m^2 (at *a-a*), 3.89 MN/m^2 (at *b-b*), 3.24 MN/m^2 (at *c-c*)

8.48 133 psi

8.50 14 kN

8.52 $\tau_{beam} = 89$ psi, $\tau_{glue} = 63$ psi (both at $x = 12$ ft)

8.54 127.5 kN

8.56 22.8 mm

8.58 0.41 in.

8.62 $v(x) = \dfrac{\gamma}{120EI}(-x^5 + 5L^4x - 4L^5)$;

$v(0) = \dfrac{-\gamma L^5}{30EI}$, $\theta(0) = \dfrac{\gamma L^4}{24EI}$

8.64 $v(x) = \dfrac{q}{48EI}[3Lx^3 - 2x^4 - L^3x]$, $(0 \leqslant x \leqslant L)$; $v(x) = \dfrac{q}{48EI}[3Lx^3 - 2x^4 - L^3x + 9L(x-L)^3]$, $\left(L \leqslant x \leqslant \dfrac{3L}{2}\right)$;

$\theta\left(\dfrac{3L}{2}\right) = \dfrac{-qL^3}{48EI}$; $v_{max} = \dfrac{-qL^4}{128EI}$ $\left(at\ x = \dfrac{3L}{2}\right)$

8.66 -1.27 in., $-1.39°$

8.68 162 mm

8.70 49.3 lb/ft

8.72 $\mp\dfrac{6EI\,\Delta}{L^2}$ (at $x = 0, L$)

8.74 544 lb/ft

8.76 $FL/2$

8.78 75%, 96%

8.80 15 710 lb·in.

8.82 62.3 MN/m^2, 703 N

8.84 4900 lb, -1.96 in.

8.86 1.44

8.88 1.67 kN

8.90 8.33 kips/ft

8.92 (a) 0% (b) 13.2%

8.94 9670 lb/ft

8.96 $q_{limit} = 16\,M_{fp}/L^2$

8.98 $q_{limit} = 27.4\,M_{fp}/L^2$

8.100 $q_{limit} = 4\,M_{fp}/L^2$

8.102 $C_{limit} = 25.34\,\sigma_{YS}$ lb·in.

CHAPTER 9

9.2 $\sigma_{t_{max}} = 6.51$ ksi (at B), $\sigma_{c_{max}} = 4.81$ ksi (at A)

9.4 $w \geqslant 49.7$ mm

9.6 10.1 kN

9.8 137 kips

9.10 Diamond-shaped region about the center with diagonals of length $a/3$ and $b/3$

9.12 2640 lb

9.14 $\sigma_x = 1.13$ MN/m^2, $\tau_{xy} = 0.30$ MN/m^2

9.16 At A: $\sigma_x = 2500$ psi, $\sigma_z = 5000$ psi, $\tau_{xz} = -698$ psi
At B: $\sigma_x = 3670$ psi, $\sigma_y = 5000$ psi, $\tau_{xy} = -230$ psi

9.18 $\sigma_n = 12\,540$ psi, $\sigma_t = 5460$ psi, $\tau_{nt} = -3540$ psi

9.20 $\sigma_n = -4380$ psi, $\tau_{nt} = -1710$ psi

9.22 $\sigma_n = 47.9$ MN/m^2, $\tau_{nt} = -4.42$ MN/m^2

9.24 $\sigma_t = -5.0$ ksi, $\tau_{nt} = 7.8$ ksi, and $\theta = 21°$ or $\sigma_t = -5.0$ ksi, $\tau_{nt} = -7.8$ ksi, and $\theta = -60°$

9.26 400 psi

9.28 $\sigma_n = -57.1$ psi, $\tau_{nt} = 833$ psi

9.30 $\sigma_1 = 4$ ksi, $\sigma_2 = -4$ ksi, $\tau_{max} = |(\tau_{nt})_{max}| = 4$ ksi

9.32 $\sigma_1 = 40$ MN/m^2, $\sigma_2 = -10$ MN/m^2, $\tau_{max} = |(\tau_{nt})_{max}| = 25$ MN/m^2

9.34 ± 693 psi

9.36 -40 MN/m$^2 \leqslant \sigma_x \leqslant 40$ MN/m^2, -17.3 MN/m$^2 \leqslant \tau_{nt} \leqslant 17.3$ MN/m^2

9.38 $\sigma_1 = 1.21$ MN/m^2, $\sigma_2 = -0.08$ MN/m^2, $|(\tau_{nt})_{max}| = 0.65$ MN/m^2

9.40 $\sigma_1 = 5180$ psi, $\sigma_2 = 2320$ psi, $|(\tau_{nt})_{max}| = 1430$ psi

9.42 7.6%

9.44 (a) 7.2 mm (b) 12.5 mm

9.46 yes

9.48 $(M/M_{YS})^2 + (T/T_{YS})^2 = 1$

9.50 5.8

9.52 30 MN/m^2; 33 MN/m^2

9.54 $\varepsilon_x = 480\mu$, $\varepsilon_y = -480\mu$, $\gamma_{xy} = -500\mu$

9.56 $\delta_x = -960$ μ in., $\delta_y = 180$ μ in., $\delta_z = -1380$ μ in.

9.58 25 ksi (T), 0.011 in.

9.60 480 kips (T)

9.62 -0.0174 in.; 96 kips (C)

9.64 $\delta = -\dfrac{(1-\nu^2)L\sigma}{E}$, $p = \dfrac{\nu\sigma t}{R}$

9.66 $\sigma_x = \sigma_y = \dfrac{-E\alpha\,\Delta T}{1-\nu}$, $\delta_z = \dfrac{(1+\nu)}{(1-\nu)}\alpha\,\Delta Th$

9.68 $\varepsilon_1 = -76\mu$, $\varepsilon_2 = -924\mu$, $|(\gamma_{nt})_{max}| = 848\mu$

9.70 $\varepsilon_1 = 366\mu$, $\varepsilon_2 = -766\mu$, $|(\gamma_{nt})_{max}| = 1130\mu$

9.72 $\sigma_x = 6860$ psi, $\sigma_y = 6040$ psi, $\tau_{xy} = -714$ psi

9.74 $a = b = 0.004$ mm

9.76 $\sigma_1 = -7.86$ MN/m^2, $\sigma_2 = -169.6$ MN/m^2

9.78 458μ

9.80 (a) -167μ (b) 500μ (c) -48μ

9.82 $\varepsilon_n = 346\mu$, $\varepsilon_t = -746$ μ, $\gamma_{nt} = -293\mu$

9.84 $\varepsilon_x = 294\mu$, $\varepsilon_y = -338\mu$, $\gamma_{xy} = 570\mu$

9.86 $\varepsilon_x = 294\mu$, $\varepsilon_y = 67\mu$, $\gamma_{xy} = -1220\mu$

9.88 $\varepsilon_1 = 278\mu$, $\varepsilon_2 = -56\mu$, $|(\gamma_{nt})_{max}| = 334\mu$

9.90 $\sigma_1 = 23\,430$ psi, $\sigma_2 = 3570$ psi, $|(\tau_{nt})_{max}| = 9930$ psi, $\tau_{max} = 11\,720$ psi

9.92 $F = 4500$ lb, $C = 2600$ lb·in.

9.94 $p = 121$ psi, $C = 51.5$ kip·in.

CHAPTER 10

10.2 $F_{cr} = 2k_T/L$

10.4 $F_{cr} = 2kr^2/(L+r)$

10.6 $m_2 = m_1/6$

10.8 139, 105.7 MN/m^2

10.10 6.0 ft

10.12 yes; 3.9

10.14 no; 33.8 mm

10.16 15.3 ft

10.18 76-mm × 76-mm (7.16 kg/m)

10.20 2-in. nominal diameter

10.22 yes

10.24 98 700 lb

10.26 29.7°C

10.28 356 lb

10.30 7.24 kN

10.32 no; brace at additional points

10.34 48 200 lb (approx.)

10.36 2.5 (approx.)

10.38 (a) 22 900 lb (approx.) (b) 27 400 lb (approx.)

10.40 13 740 lb (approx.); 0.884 in. (approx.)

10.42 20 200 psi

CHAPTER 11

11.2 842μ in./in.

11.4 3.5

11.6 455μ in./in.; no

11.8 yes; align gage in circumferential or longitudinal direction

11.10 no

11.12 250 psi

11.14 yes; output is 4 times that of a single gage

11.16 0.849μ in./in.

11.18 $\varepsilon = 2(maLt)/(EI)$

11.20 (a) gages 1 and 2 in adjacent arms, (b) gages 1 and 2 in opposite arms, dummy gages in other arms.

11.22 yes

11.24 59 880 ohms

11.26 $\varepsilon = (3v_{max}xt)/(2L^3)$; no
11.28 469 μ in./in.
11.30 66 MN/m^2; 2 gages, aligned parallel and perpendicular to crack.
11.32 1770 lb·in.
11.34 threshold strain = yield strain of material
11.36 4000 μ in./in.
11.38 500 psi
11.40 straight fringes parallel to x axis spaced 3.0 in. apart
11.42 14.2 (kN/m)/fringe
11.44 2560 psi
11.46 2.25

INDEX